The Most Successful Way of JEE Preparation

DPP

Daily Practice Problems

JEE Main & Advanced

Alkyl Halids to Amines

CHEMISTRY Vol-8

The Most Successful Way of JEE Preparation

DPP

Daily Practice Problems

JEE Main & Advanced

Alkyl Halids to Amines

CHEMISTRY Vol-8

Ranjeet Shahi

arihant
ARIHANT PRAKASHAN (SERIES), MEERUT

DPP
Daily Practice Problems

arihant

ARIHANT PRAKASHAN (SERIES), MEERUT

Administrative & Production Offices

Corporate Office 'Ramchhaya' 4577/15, Agarwal Road, Darya Ganj, New Delhi -110002
Tele: 011- 47630600, 43518550; Fax: 011- 23280316

Head Office Kalindi, TP Nagar, Meerut (UP) - 250002
Tel: 0121-2401479, 2512970, 4004199; Fax: 0121-2401648

Sales & Support Offices

Agra, Ahmedabad, Bengaluru, Bareilly, Chennai, Delhi, Guwahati, Hyderabad, Jaipur, Jhansi, Kolkata, Lucknow, Meerut, Nagpur & Pune

ISBN : 978-93-13193-47-0

Published by Arihant Publications (I) Ltd.

Production Team

Publishing Managers	Mahendra Singh Rawat & Keshav Mohan	*Inner Designer*	Mazher Chaudhary
Project Head	Sorabh Chaudhary	*Page Layouting*	Sunil Sharma, Hemant
Cover Designer	Shanu Mahad	*Proof Reading*	Salman

For further information about the products from Arihant, log on to www.arihantbooks.com or email to info@arihantbooks.com

/arihantpub /@arihantpub Arihant Publications /arihantpub

ABOUT DPP

Daily Practice Problems

Indian Institutes of Technology better known as IITs have always been the dream destination of each science group students who want to pursue their career as Engineers. These are the institutes which have got very high value not even in India but in the whole world. To get the admission in such institutes one need to face entrance Examinations; JEE Main & JEE Advanced, out of which JEE Main is considered as the preliminary exam & JEE Advanced is the final exam to get into the IITs. Approximately per year 12 lakh students enroll themselves for JEE Main and only the top 2.25 lakh in JEE Main merit can attempt for JEE Advanced, the fact clearly express the degree of toughness of JEE Main & JEE Advanced.

To have success in such an exam one needs to be very planned & systematic in its preparation. A very popular wise saying 'Rome was not built in a day' suggests that to achieve a big target like IITs we need to work continuously. For continuous work, one need to have proper planning that how much work he/she has to do on regular basis, i.e., on daily basis. Each student preparing for JEE has many books but unfortunately none of them gives the students a plan for daily studies, which is the most effective success key for JEE.

Keeping the daily practice needs of the students in mind we have come up with DPP Series (Daily Practice Problems) having approximately 26 books. The primary focus of this series is on achieving success through practice & proper planning. For that we have divided the whole syllabus of JEE Main & JEE Advanced; Physics, Chemistry & Mathematics, into different volumes (around 8 volumes of Mathematics & 9 volumes each of Physics & Chemistry), each volume has Daily Practice Problems Sheets having specific questions on various Topics of the individual chapter of the volume ensuring the complete Practice & Assessment of the topic.

DPP

Daily Practice Problems

In this way we have tried to give you an effective daily planning for JEE keeping in mind the daily hours that a student can devote to his practice, you are just required to practice one sheet of each subject on daily basis. This habit of daily practice will keep you away from exam stress and thoughts like; how much I have prepared/am I preparing properly or not/ do I need to study more on daily basis or what I am doing is ok, etc., which regularly tease your mind and can give you panic at any stage of your preparation.

Some special features that we have tried to incorporate in our Daily Practice Problems (DPP) series are discussed below

1. Micro Level coverage of all the topics of a Chapter/ Unit given in JEE Main & Advanced Syllabus.
2. Each DPP has various types of questions; Subjective Questions, Single/ Multiple Correct Options, Statements type, Comprehension based, Integer Value Answer type, Matching, etc., making you practice & ready for all the question formats of JEE.
3. New & innovative questions in each DPP challenging & improving your problem solving skills.
4. Along with Topical Coverage, Revisal DPPs for JEE Main & JEE Advanced with each Chapter, covering multidisciplinary questions proving comprehensive practice.
5. JEE Main & JEE Advanced Archive (collection of last 13 years' exams questions) with each chapter.
6. Complete Solutions for each DPP.

We hope that DPP series will prove to be very beneficial for the students preparing for JEE, by giving them daily practice in a systematic & planned manner. The suggestions/criticism from the students, teachers & parents about DPP series will be welcomed whole heartedly and we promise to update this series accordingly from time to time.

DPP
Daily Practice Problems

DAILY TASKS...

1 ALKYL HALIDES

2 ALIPHATIC ALCOHOLS, ETHERS AND EPOXIDES

DPP

Daily Practice Problems

3 ALIPHATIC ALDEHYDES AND KETONES

4 CARBOXYLIC ACIDS AND ACID DERIVATIVES

1. Alkyl Halides

DPP
DAILY PRACTICE PROBLEMS

DPP-1 Preparation of Alkyl Halides

Subjective Questions

Working Space

Direction *(Q. Nos. 1-5) These questions are subjective in nature, need to be solved completely on notebook.*

1 How an alkyl fluoride is conveniently prepared from other alkyl halides?

2 Three constitutional isomers of C_6H_{14} are shown below :

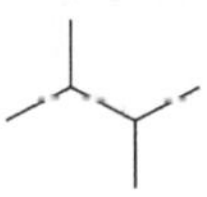
(I)

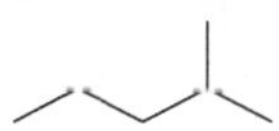
(II)

(III)

When treated with chlorine gas in light, isomer *A* gives a mixture of eight monochlorinated products. Under identical condition, isomer *B* gives a mixture of four monochlorinated products. Isomer *C* produces three different monochlorinated products in similar reaction condition. Identify *A*, *B* and *C*.

3 Discuss the mechanism of the following reaction.

$$CH_3CH_2OH + NaBr + H_2SO_4 \longrightarrow R\text{—}Br + NaHSO_4 + H_2O$$

4 Why direct free radical iodination reaction fail to produce iodoalkane? How this problem is overcome?

5 How iodoalkane is prepared from alkyl chloride by halogen exchange reaction? What is the limiting factor of this reaction?

Only One Option Correct Type

Direction *(Q. Nos. 6-12) This section contains 7 multiple choice questions. Each question has four choices (a), (b), (c) and (d), out of which ONLY ONE option is correct.*

6 When (S)-2-bromopentane is brominated, several 2,3-dibromopentanes are formed, which of the following is not formed?

a. 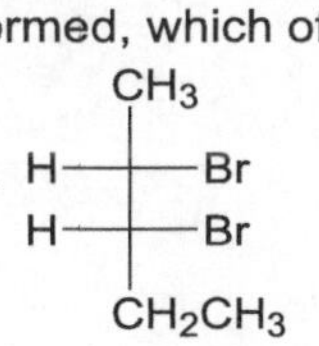

b. 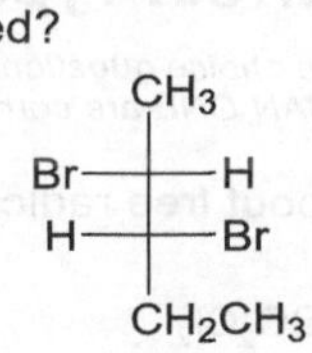

c. 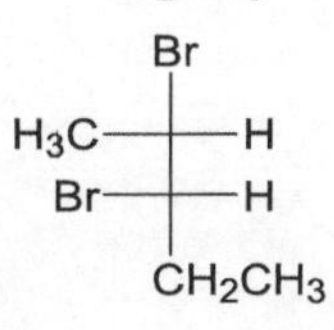

d. 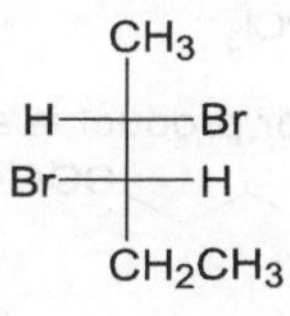

7 The yield of alkyl bromide obtained as a result of heating the dry silver salt of carboxylic acid with bromine in CCl_4 is

a. 1°> 3°> 2° bromides b. 1°> 2° > 3° bromides
c. 3°> 2° > 1° bromides d. 3°> 1°> 2° bromides

8 Which is incorrect about Hunsdiecker's reaction?

a. Only Cl_2 can give alkyl halide
b. I_2 will give ester when treated with *R*COOAg
c. The reaction proceeds through free radical
d. F_2 cannot give alkyl halide

9 The major product of the following reaction is

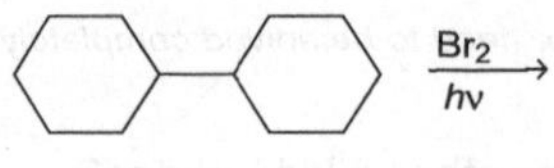

a. 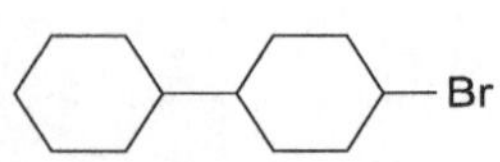

b. 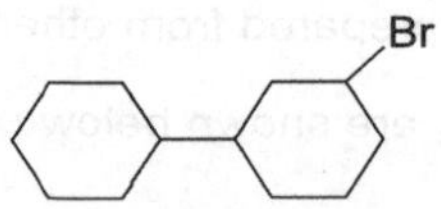

c. 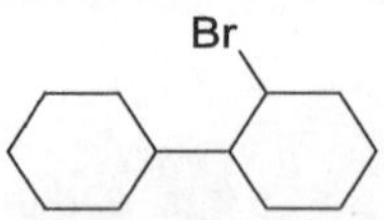

d. 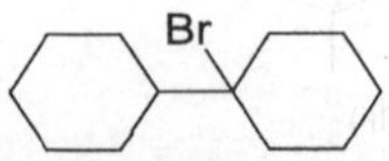

10 Racemic mixture is obtained due to the halogenation of

a. *n*-pentane b. *iso*pentane
c. *neo*pentane d. Both (a) and (b)

11 The reaction of $SOCl_2$ on alkanols to form alkyl chlorides gives good yields because

a. alkyl chlorides are immiscible with $SOCl_2$
b. the other products of the reaction are gaseous and escape out
c. alcohol and $SOCl_2$ are soluble in water
d. the reaction does not occur *via* intermediate formation of an alkyl chlorosulphite

12 Addition of bromine on propene in the presence of brine yields a mixture of

a. $CH_3CHClCH_2Br$ and $CH_3CHBrCH_2Cl$
b. $CH_3CHClCH_2Br$ and $CH_3CHBrCH_2Br$
c. $CH_3CHClCH_2Cl$ and $CH_3CHBrCH_2Br$
d. $CH_3CHClCH_2Cl$ and $CH_3CHBrCH_2Cl$

One or More than One Options Correct Type

Direction (Q. Nos. 13-17) *This section contains 5 multiple choice questions. Each question has four choices (a), (b), (c) and (d), out of which ONE or MORE THAN ONE are correct.*

13 Which of the following reagents can bring about free radical chlorination of propane?

a. $SOCl_2$ b. SO_2Cl_2
c. $Cl_2/h\nu$ d. PCl_3

14 In which of the following reaction(s), the major product is a primary alkyl bromide?

a. 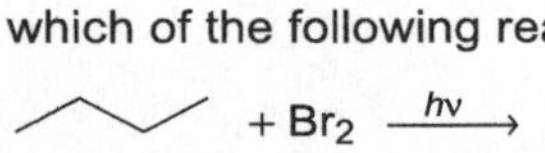

b. 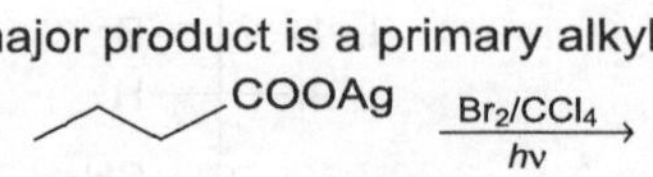

c. 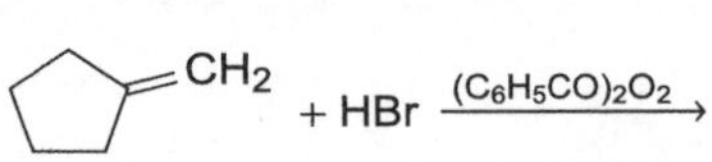

d. 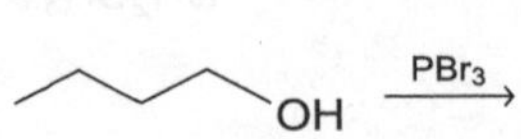

Working Space

15 Consider the following reaction,

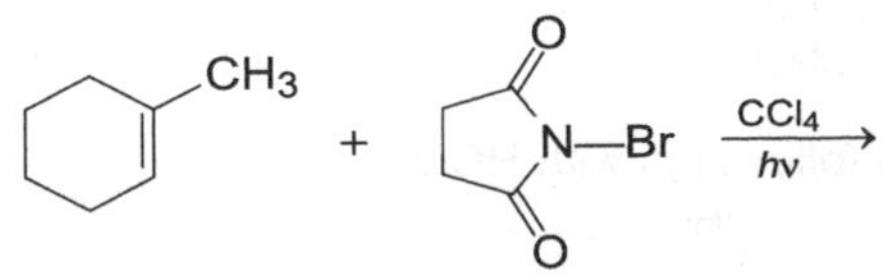

The expected product(s) is/are

a. 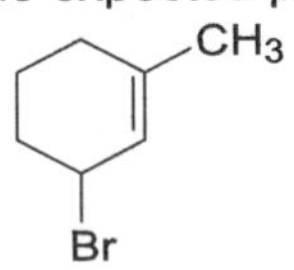 b. 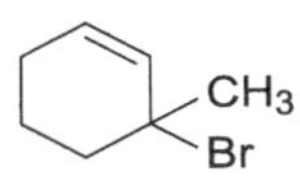 c. 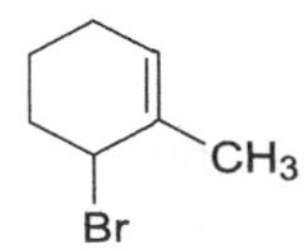 d. 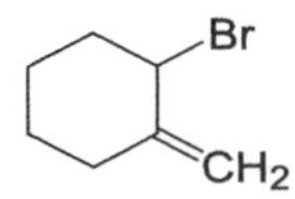

16 Consider the following reaction,

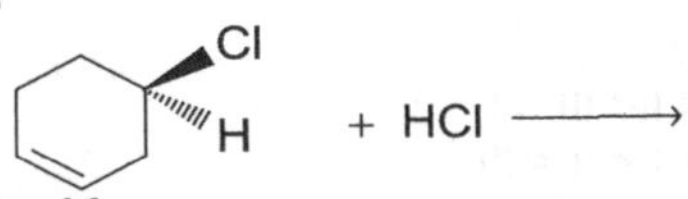

X

When a pure enantiomer of *X* is taken in the above reaction, correct completion regarding the reaction is/are

a. Four different dichlorocyclohexane are formed as significant products
b. a pair of enantiomers is formed
c. two pairs of diastereomers are formed
d. product mixture has zero specific rotation

17 Choose the correct statement(s) from the following regarding free radical chlorination and bromination reaction of alkane.

a. The first propagation step is exothermic in bromination reaction and endothermic in chlorination reaction
b. The first propagation step is endothermic in bromination reaction and exothermic in chlorination reaction
c. Chlorination reaction is more reactive process but very less selective, reverse is true for bromination
d. Chlorination is more selective but less reactive process and its reverse is true for bromination

Comprehension Type

Direction *(Q. Nos. 18-20) This section contains a paragraph, describing theory, experiments, data, etc. Three questions related to the paragraph have been given. Each question has only one correct answer among the four given options (a), (b), (c) and (d).*

Passage

An alcohol (R— OH) can be converted into alkyl chloride by the treatment with HCl. Reaction involves protonation of alcohol followed by the formation of carbocation intermediate. Carbocation intermediate in the final step undergo nucleophilic attack by Cl^- ion as :

$$R\text{—OH} + H^+ \xrightarrow{\text{Slow}} R\text{—}\overset{+}{O}H_2 \longrightarrow R^+ + H_2O$$

$$R^+ + Cl^- \longrightarrow R\text{—Cl}$$

18 Which of the following alcohol reacts most easily?

a. 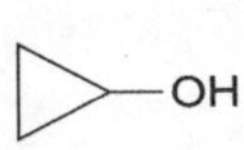 b. 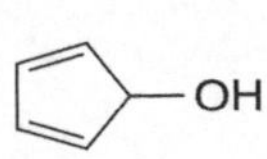 c. 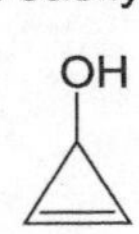 d. 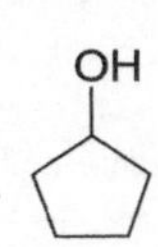

Working Space

19 Which of the following can catalyse the above reaction?

a. $AgNO_3$ b. $ZnCl_2$
c. CCl_4 d. H_3PO_4

20 What is the correct order of reactivity of the followings with HCl?

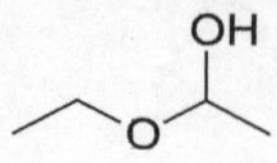

I.

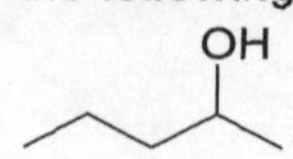

II.

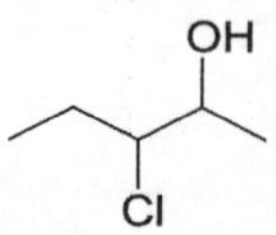

III.

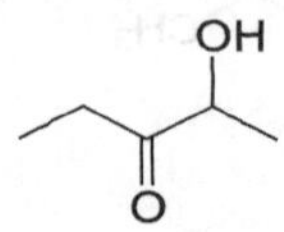

IV.

a. I < II < III < IV b. IV < II< III < I
c. III < II < I < IV d. IV < I < II < III

One Integer Value Correct Type

Direction *(Q. Nos. 21-24) This section contains 4 questions. When worked out will result in an integer from 0 to 9 (both inclusive).*

21 **If 2,4-dimethyl pentane is subjected to free radical chlorination reaction, how many different monochlorinated products would be formed?**

22 **On free radical chlorination reaction of butane, how many different, optically active, dichloroalkanes would be formed?**

23 **If 1, 3-butadiene is treated with excess of bromine in CCl_4, how many different tetrabromides would be formed?**

24 **Consider the following reaction,**

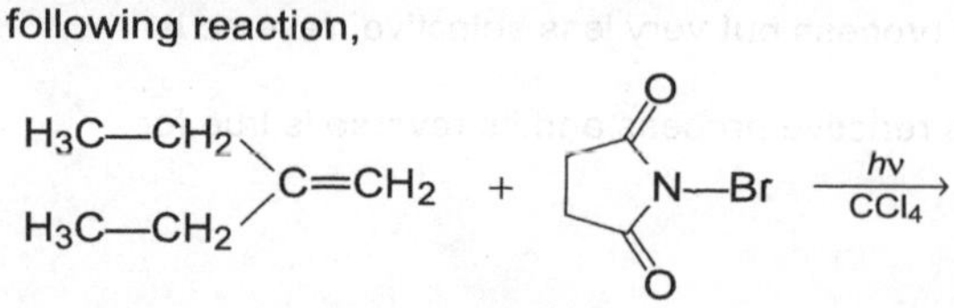

How many different monobromo derivatives would be produced?

DPP-2 Physical and General Properties of Alkyl Halides

Subjective Questions

Direction (Q. Nos. 1-5) *These questions are subjective in nature, need to be solved completely on notebook.*

1 Describe how cyclohexane can be deuterated?

2 CH_3Cl is gas at room temperature while, CCl_4 is liquid although former is polar while latter is non-polar. Give reason.

3 How primary, secondary and tertiary alkyl halides are conveniently reduced to corresponding alkanes?

4 When 2-chlorobutane is treated with Mg in presence of dry ether, besides Grignard reagent, some 2-butene is also produced, explain.

5 Which is more nucleophilic, an alkyl lithium (RLi) or a Grignard reagent (RMgX)?

Only One Option Correct Type

Direction (Q. Nos. 6-14) *This section contains 9 multiple choice questions. Each question has four choices (a), (b), (c) and (d), out of which ONLY ONE option is correct.*

6 What is the major product of the reaction?

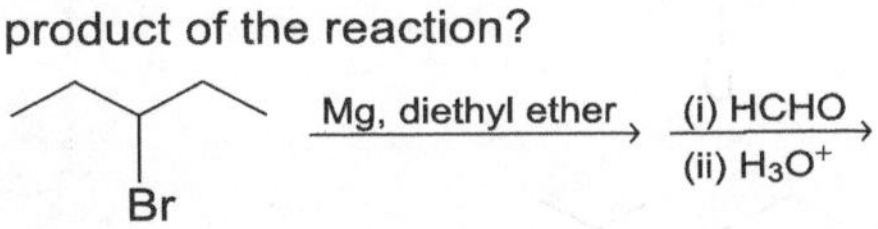

a. 2-ethyl-1-pentanol
b. 2-ethyl-1-butanol
c. 3-pentanol
d. 3-methyl1-1-pentanol

7 Which are the products of the following sequence of reaction?

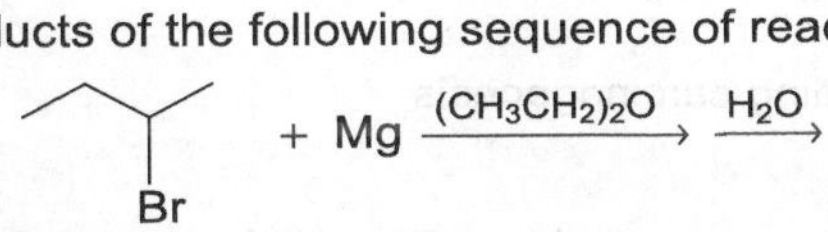

a. 2-butanol and Mg(OH)Br
b. Butanol and Mg(OH)Br
c. 2-butanol and MgHBr
d. Butane and Mg(OH) Br

Working Space

Working Space

8 What is the product of the reaction shown?

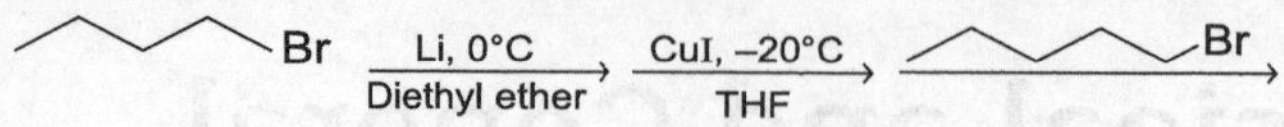

a. 4-nonene
b. nonane
c. 4-bromononane
d. 5-bromononane

9 What set of reagents can be used to make octane from butyl chloride?

a. Li and then CuI then more butyl chloride
b. Li then H_2O then CuI
c. Zn + Conc. HCl
d. Ni + H_2

10 Which of the following synthesis give 3-methyl-1-hexanol?

a. 2–bromopentane 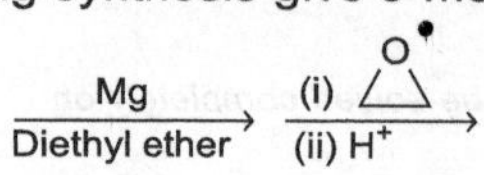

b. 2–bromohexane 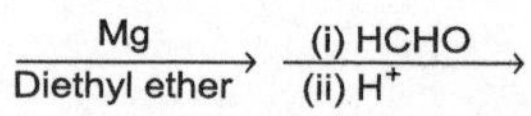

c. 3–bromopentane 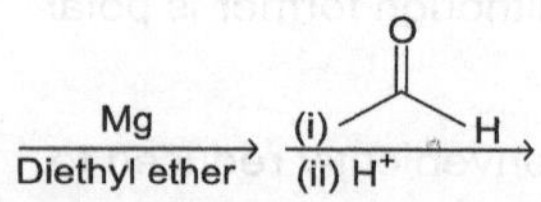

d. 1–bromopentane 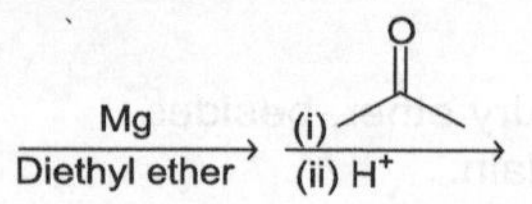

11 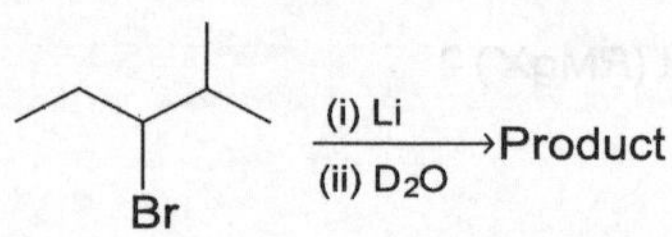

What is the product in the above reaction?

a. 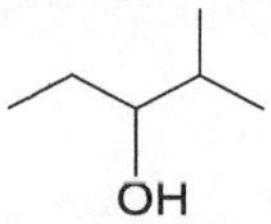

b. 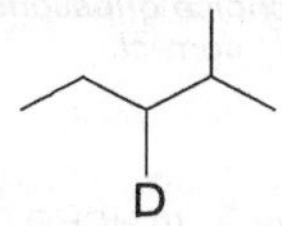

c. 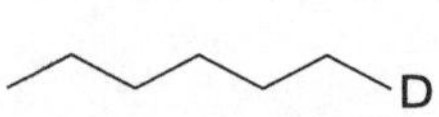

d. 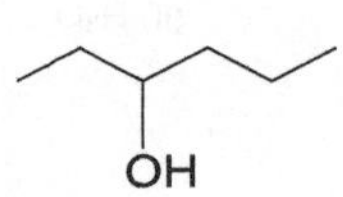

12 Chlorinated hydrocarbon

a. are commonly used as the solvents and degreasing agent in industries
b. generally have low solubility in water
c. commonly have a chronic toxic effect including carcinogenesis
d. All of the above

13 Which of the following alkyl halides cannot be converted into Grignard reagents?

a. CH_3I
b. C_6H_5Br
c. $HOCH_2CH_2CH_2Cl$
d. $CH_3OCH_2CH_2CH_2Cl$

14 Which one is liquid at room temperature?

a. CH_3Cl
b. C_2H_5Cl
c. CH_3Br
d. C_2H_5Br

Working Space

One or More than One Options Correct Type

Direction *(Q. Nos. 15-19) This section contains 5 multiple choice questions. Each question has four choices (a), (b), (c) and (d), out of which ONE or MORE THAN ONE are correct.*

15 1-chlorobutane can be reduced to butane by

a. $NaBH_4$ b. $LiAlH_4$
c. Zn/CH_3COOH d. $(C_6H_5)_3SnH$

16 Tertiary butyl chloride can be reduced to 2-methyl propane by

a. $NaBH_4$ b. $LiAlH_4$
c. Zn/CH_3COOH d. $(C_6H_5)_3SnH$

17 Which of the following float over water?

a. 1-chlorobutane b. 1-fluorobutane
c. 1-bromobutane d. 1-iodobutane

18 Which of the following halides are not suitable for Grignard's synthesis?

a. 1-bromobutene b. Chlorobenzene
c. 3-chloropropyne d. 2-chloroethanal

19 What is the expected product in the following reaction?

$$CH_3-CH=CH_2 + Br_2 + NaCl(aq) \longrightarrow$$

a. $H_3C-CH(Br)-CH_2Cl$ b. $H_3C-CH(Cl)-CH_2Br$
c. $H_3C-CH(Br)-CH_2Br$ d. $H_3C-CH(OH)-CH_2Cl$

Comprehension Type

Direction *(Q. Nos. 20-22) This section contains a paragraph, describing theory, experiments, data, etc. Three questions related to the paragraph have been given. Each question has only one correct answer among the four given options (a), (b), (c) and (d).*

Passage

In alkyl halides C—*X* bond is polar. Polarity of C—*X* bond is responsible for several characteristic reactions of alkyl halides *viz.*

$$R-X + H^- \text{ (from metal hydride)} \longrightarrow R-H + X^-$$

$$R-X + Mg \xrightarrow{(C_2H_5)_2O} RMgX \quad \text{(Grignard's reagent)}$$

$$2R-X + 2Na \xrightarrow{(C_2H_5)_2O} R-R + 2NaX \quad \text{(Wurtz's reaction)}$$

20 Which of the following will most easily be reduced by $LiAlH_4$?

a. 1-chlorobutane b. 2-chlorobutane
c. 2-chloro-2-methyl butane d. Chlorobenzene

21 Which forms Grignard's reagent with maximum difficulty?

a. 1-chlorobutane
b. 2-chlorobutane
c. 1-chloro-2,2-dimethyl butane
d. 2-chloro-2-methyl butane

22 Which of the following required more than one type of alkyl halides for its preparation by Wurtz's reaction?

a. Hexane b. Pentane
c. Cyclohexane d. Methyl cyclopentane

One Integer Value Correct Type

Working Space

Direction (Q. Nos. 23-25) This section contains 3 questions. When worked out will result in an integer from 0 to 9 (both inclusive).

23 Consider the following reactions, $CH_3Br + Mg \xrightarrow{Et_2O} X$

$$X \text{ (excess)} + HOC\text{—}CHO \xrightarrow[H_2O]{H^+} Y \text{ (diol)}$$

How many different diols are formed?

24 Dichloroalkanes with appropriate skeleton, on refluxing with Na in diethyl ether undergo cyclisation *via* Wurtz's reaction. How many different isomeric dibromoalkanes on refluxing with Na metal in diethyl ether can give the same 1,4-dimethyl cyclohexane?

25 How many of the following are denser than water?

i. CCl_4 ii. $CH_3CH_2CH_2CH_2Cl$
iii. Bromocyclohexane iv. Chlorocyclopentane
v. 1,3-difluorocyclopentane vi. CH_3I
vii. 1-fluorodecane viii. CH_2Br_2
ix. CH_3Cl x. $CHCl_3$

Statement Type

Direction (Q. Nos. 26 and 27) This section is based on Statement I and Statement II. Select the correct answer from the codes given below.

a. Both Statement I and Statement II are correct and Statement II is the correct explanation of Statement I
b. Both Statement I and Statement II are correct and Statement II is not the correct explanation of Statement I
c. Statement I is correct but Statement II is incorrect
d. Statement II is correct but Statement I is incorrect

26 **Statement I** CH_3F is less polar than CH_3Cl.

Statement II F has greater electronegativity than Cl.

27 **Statement I** A tertiary halide (R_3CX) is very less likely to form Grignard's reagent with Mg in THF.

Statement II Tertiary carbanion are highly unstable, very less likely to be formed.

DPP-3 S_N2 Reaction Basics

Subjective Questions

Working Space

Direction *(Q. Nos. 1-5) These questions are subjective in nature, need to be solved completely on notebook.*

1 Each of these compounds can be synthesised by an S_N2 reaction. Suggest a combination of haloalkane and nucleophile that will give each product.

i. CH_3OCH_3
ii. CH_3SH
iii. $CH_3CH_2CH_2PH_2$
iv. CH_3CH_2CN
v. $CH_3SCH_2C(CH_3)_3$
vi. $(CH_3)_3NH^+Cl^-$
vii. $C_6H_5COOCH_2C_6H_5$
viii. $CH_2{=}CHCH_2OCH(CH_3)_2$
ix. $CH_2{=}CHCH_2OCH_2CH{=}CH_2$

2 Complete the following nucleophilic substitution reactions. In each, show all electron pairs on both the nucleophile and the leaving group.

i. 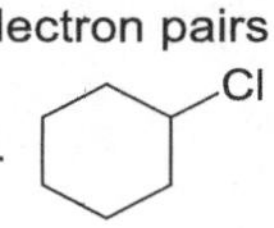 $+ CH_3COONa \xrightarrow{\text{Ethanol}}$

ii. 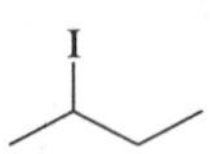 $+ CH_3CH_2SNa \xrightarrow{\text{Acetone}}$

iii. 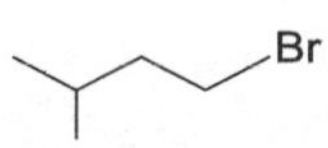 $+ H_3C{-}C{\equiv}C^-Na^+ \xrightarrow{\text{DMSO}}$

3 Indicate which compound in each of the following pairs would react faster?

i. CH_3CH_2Br or CH_3CH_2I with NaOH in acetone
ii. CH_3CH_2Br or CH_3CH_2I with NaOH in ethanol
iii. 1-bromo propane or 2-bromo propane with NaI in acetone
iv. *neo* pentyl iodide or 1-iodo pentane with CH_3OH
v. CH_3S^- or CH_3O^- with CH_3CH_2Br in DMF
vi. CH_3S^- or CH_3O^- with CH_3CH_2Br in ethanol
vii. $(CH_3)_3CO^-$ or $CH_3CH_2O^-$ with CH_3CH_2Cl in acetone
viii. CH_3NH_2 or CH_3OH with CH_3CH_2Br in methanol
ix. CH_3NH_2 or CH_3OH with CH_3CH_2Br in acetone
x. CH_3COO^- or $CH_3CH_2S^-$ with CH_3I in ethanol
xi. CH_3COO^- or $CH_3CH_2S^-$ with CH_3I in DMSO
xii. $HOCH_2CH_2CH_2Br$ or $HOCH_2CH_2CH_2CH_2Br$ with in presence of KOH

xiii. 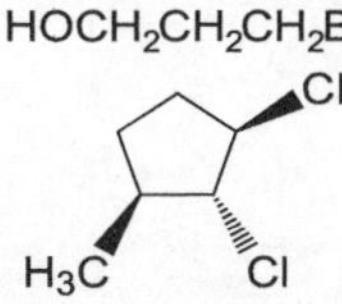 or 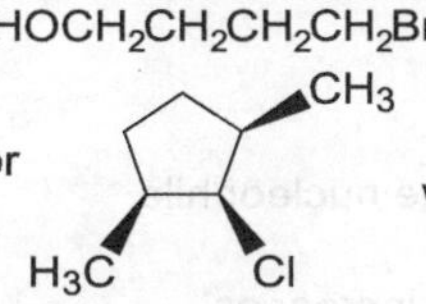 with CH_3S^- in acetone

xiv. 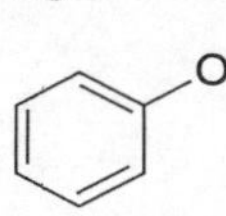 or 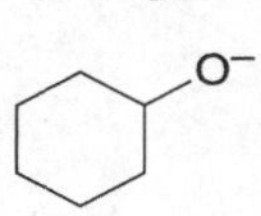 with $CH_3CH_2CH_2I$ in acetone

Working Space

4 In each of the following S_N2 reactions, you must make a choice between leaving groups or nucleophilic sites:

i. 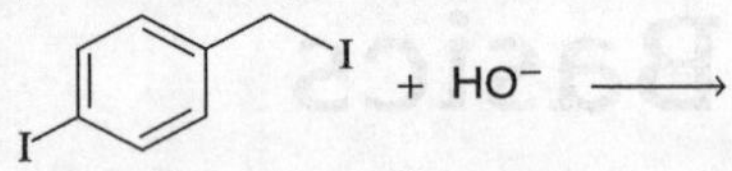

ii. 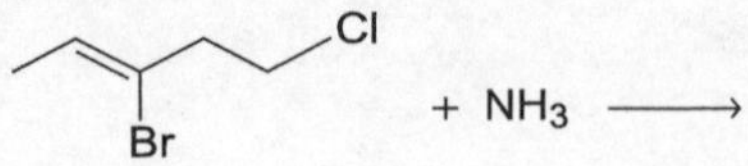

iii. 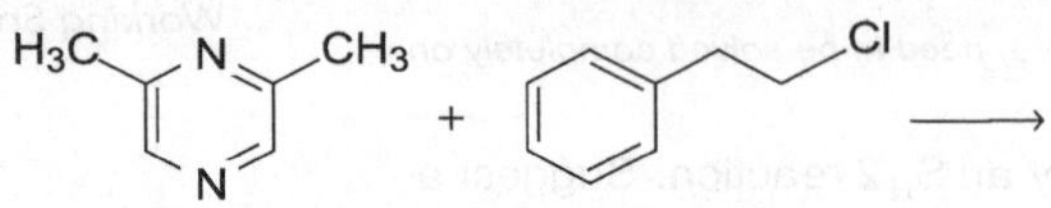

iv. 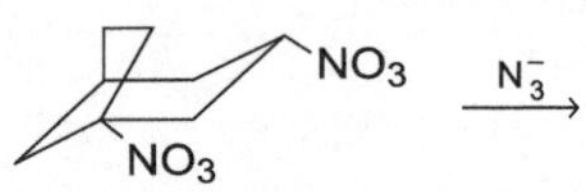

v. 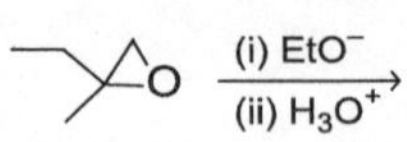

vi. 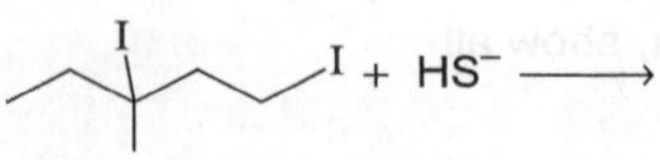

vii. 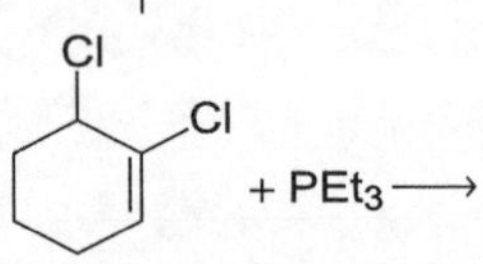

viii. 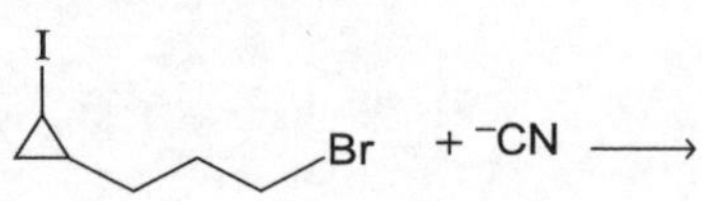

ix. 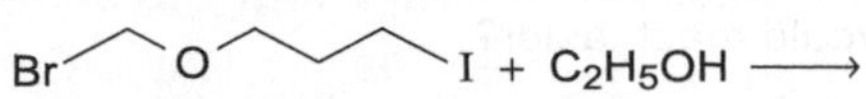

5 How does $R_4N^+Br^-$ acts as a catalyst in $R—Cl + NaCN \longrightarrow R—CN + NaCl$ reaction?

Only One Option Correct Type

Direction *(Q. Nos. 6-13) This section contains 8 multiple choice questions. Each question has four choices (a), (b), (c) and (d), out of which ONLY ONE is correct.*

6 Which of the S_N2 reaction is fastest?

a. $CH_3Br + HC{\equiv}C^- \longrightarrow CH_3C{\equiv}CH + Br^-$
b. $CH_3Br + HC{\equiv}CH \longrightarrow CH_3C{\equiv}CH + HBr$
c. $CH_3CH_2Br + HC{\equiv}C^- \longrightarrow CH_3CH_2C{\equiv}CH + Br^-$
d. $CH_3CH_2Br + HC{\equiv}CH \longrightarrow CH_3CH_2C{\equiv}CH + HBr$

7 Which statement is true about S_N2 mechanism?

a. The rate of reaction increases on increasing strength of the nucleophile
b. The reaction is faster in polar aprotic solvents
c. The rate of reaction increases as the leaving group ability increases
d. All of the above

8 A correct statement about transition state of S_N2 reaction is

a. the transition state proceeds and follow an unstable reaction intermediate
b. the transition state will always have net negative charge
c. existence of this state implies an exothermic reaction
d. the single transition state represents the point of maximum free energy of the reaction

9 What is the correct increasing order of reactivity of the followings in S_N2 reaction?

I. $CH_2{=}CH{-}CH_2{-}Br$ II. $CH_2{=}CH{-}I$
III. $CH_3CH_2CH_2{-}I$ IV. $CH_3OCH_2CH_2{-}I$

a. I < II < III < IV b. III < II < I < IV
c. II < III < IV < I d. II < I < III < IV

10 What is the correct increasing order of reactivity of the following in the S_N2 reaction?

I. $(CH_3)_2CCH_2Cl$ II. $CH_3CH_2CH(Cl)CH_3$
III. C_6H_5Cl IV. $p\text{-}O_2N{-}C_6H_4{-}CH_2Cl$

a. I < II < III < IV b. III < II < I < IV
c. II < III < IV < I d. III < I < II < IV

11 Consider the two lines shown in the diagram given below.

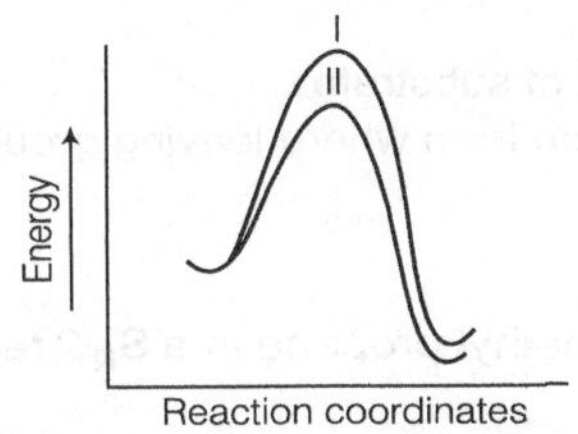

In a S_N2 reaction, these two lines compare the effect of the

a. substrate b. nucleophile
c. leaving group d. substrate or nucleophile

12 Consider the two lines shown in the diagram given below.

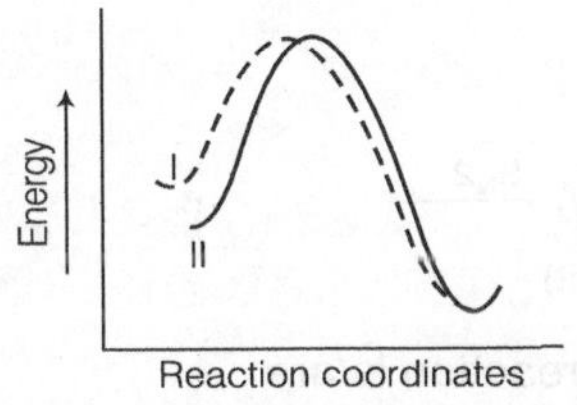

Which of the following apply appropriately to a S_N2 reaction?

a. Graph I could represent Cl^- and Graph II represents I^- leaving group
b. Graph I could represent HO^- and Graph II represents CH_3COO^- nucleophile
c. Graph I could represent H_2O and Graph II H_2S as nucleophile
d. Graph I could represent $(CH_3)_2NH$ and Graph II to $(CH_3)_2N^-$ nucleophiles

13 The correct statement concerning a S_N2 reaction is

a. the reaction mechanism involve atleast one reactive intermediate
b. transition state is pentavalent
c. product is formed after passing through several transition states
d. nucleophile attacks from front side on which leaving group is present

One or More than One Options Correct Type

Working Space

Direction (*Q. Nos. 14-19*) *This section contains 6 multiple choice questions. Each question has four choices (a), (b), (c) and (d), out of which ONE or MORE THAN ONE are correct.*

14 In a S_N2 reaction rate of reaction depends on

a. concentration of substrate
b. concentration of nucleophile
c. nature of leaving group
d. number of lone pairs on donor atom

15 Consider the following S_N2 reaction,

$$CH_3—Br + NaOH(aq) \longrightarrow CH_3—OH + NaBr$$

Which of the following could increase the reactivity (rate) of reaction?

a. Increasing NaOH concentration
b. Adding some NaI
c. Replacing CH_3Br by $^{14}CH_3—Br$
d. Using $NaO^{18}H$ in place of NaOH

16 A S_N2 reaction involves back side attack of nucleophile at the α-carbon of substrate because

a. both nucleophile and leaving group are electron rich, nucleophilic attack occur from most distant position
b. there is greater electron density on the back side of substrate
c. there is not enough physical space in the front side from where leaving group leaves the substrate
d. substrate possesses a hole in the backside

17 1-chlorobutane is more reactive than 2-chloro-2-methyl propane in a S_N2 reaction because

a. α-carbon is less crowded in 1-chlorobutane
b. α-carbon is less electropositive in 1-chlorobutane
c. electron donating inductive effect of three methyl group is greater in 2-chloro 2-methyl propane
d. α-carbon is more electropositive in 1-chlorobutane

18 Consider the reaction given below.

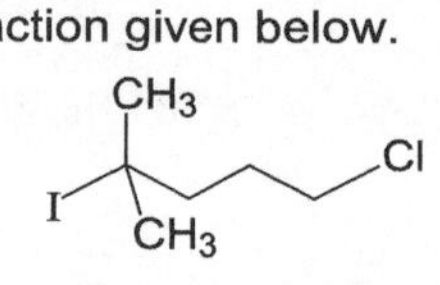

$+ NaCN(aq) \xrightarrow{S_N2}$

(1.0 equivalent)

The correct statement(s) applicable to the above reaction is/are

a. Cl^- is substituted predominantly
b. Cl^- is a better leaving group
c. substitution of I^- in the above reaction required greater activation energy than for Cl^-
d. addition of some NaI catalyse the substitution reaction

19 The correct statement regarding a S_N2 reaction is/are

a. reaction involving stronger bond formation to α-carbon occur at faster rate
b. reaction shows kinetic isotopic effect
c. substitution of I^- of $R—I$ by $C_6H_5SO_3^-$ lower the activation energy of reaction
d. presence of electron withdrawing group in substrate increases reactivity

Comprehension Type

Working Space

Direction *(Q. Nos. 20-22) This section contains a paragraph, describing theory, experiments, data, etc. Three questions related to the paragraph have been given. Each question has only one correct answer among the four given options (a), (b), (c) and (d).*

Passage

The general mechanism of a S_N2 reaction is as follows.

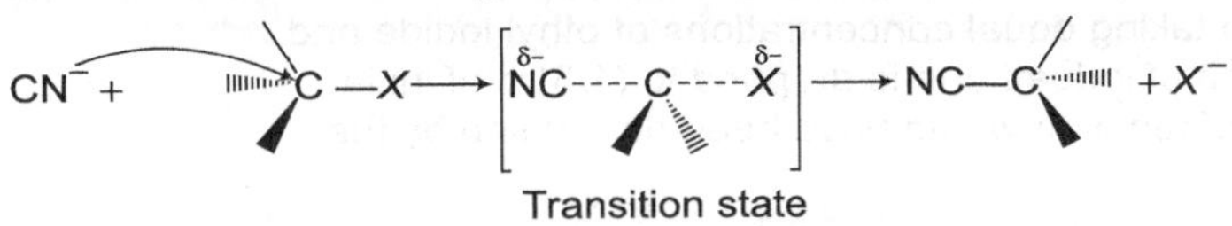

Factors that limit the rate of reaction are steric hindrance at α-cabron of substrate and strength of C—*X* bond. Any factor which stabilises transition state increases the rate of reaction.

20 Consider the following S_N2 reaction,

$$CH_3—I + HO^- \xrightarrow{k_1} CH_3—OH + I^- \quad ...(i)$$

$$CH_3—Cl + HO^- \xrightarrow{k_2} CH_3—OH + Cl^- \quad ...(ii)$$

Which of the following energy diagram is correctly labelled?

a. 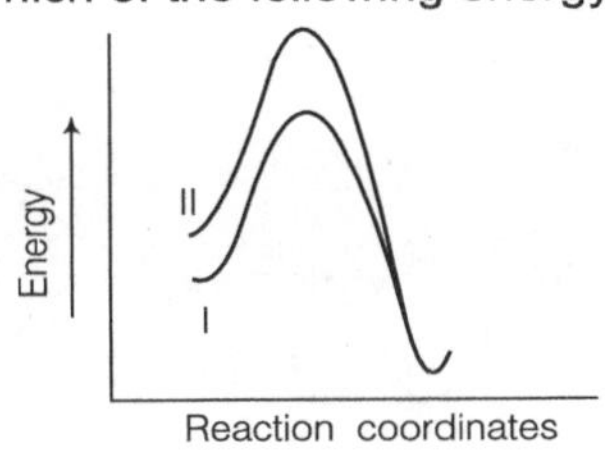

b. 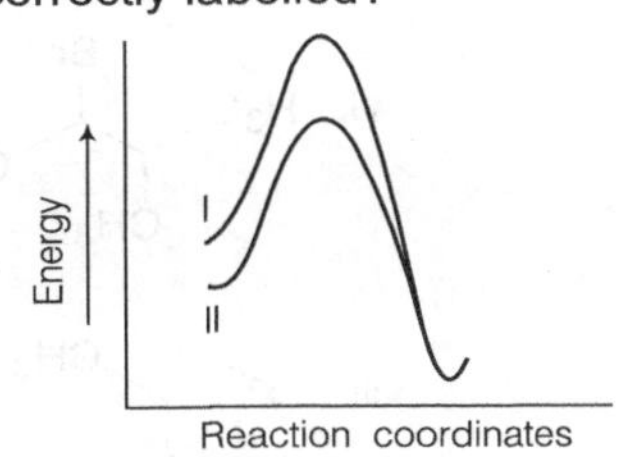

c. 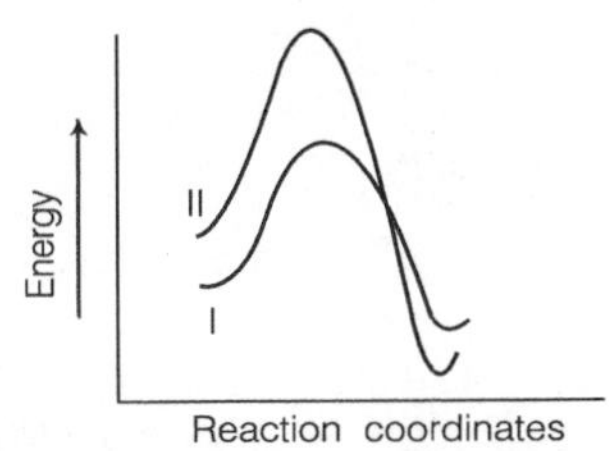

d. 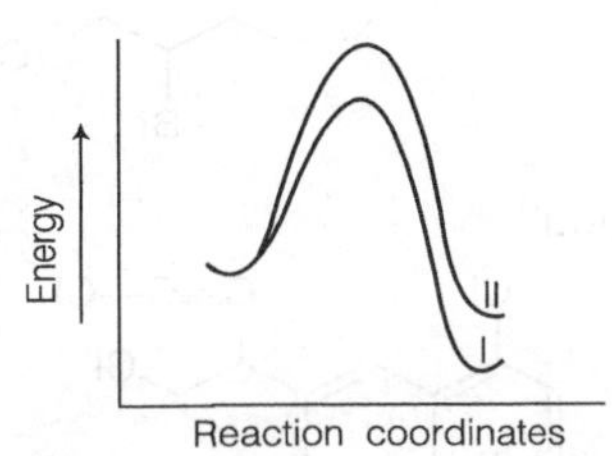

21 In the previous question, under identical reaction condition, i.e. temperature, concentration of substrate and nucleophile, the correct relationship between rate constant is

a. $k_I < k_{II}$ b. $k_I = k_{II}$ c. $k_I > k_{II}$ d. $k_I \approx k_{II}$

22 Which of the following is not a correct representation of S_N2 reaction?

a. Br— CH_2CH_2 — Br + NaS—CH_2CH_2—SNa ⟶ 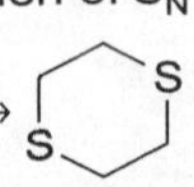

b. Br—$CH_2CH_2CH_2CH_2$—Br + Na_2S ⟶ 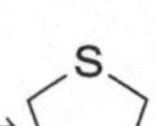

c. $CH_3CH_2OH + NaNH_2 \longrightarrow CH_3CH_2NH_2$

d. 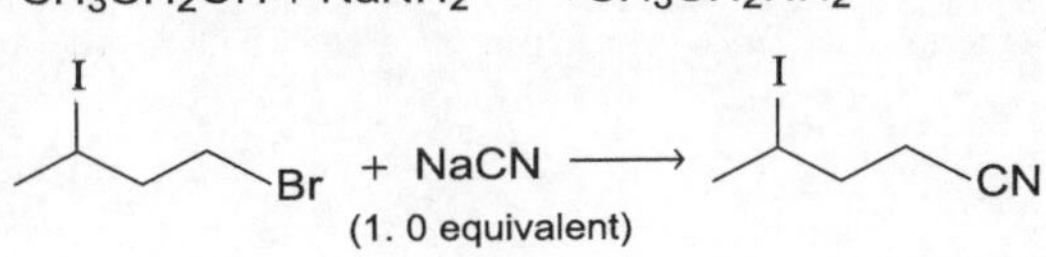

One Integer Value Correct Type

Working Space

Direction *(Q. Nos. 23-25) This section contains 3 questions. When worked out will result in an integer from 0 to 9 (both inclusive).*

23 Consider the following reaction,

$$CH_3CH_2I + NaOH \longrightarrow CH_3CH_2OH + NaI$$

The above reaction was started taking equal concentrations of ethyl iodide and NaOH. After 1.0 h, concentration of iodoethane is dropped to (1/3)rd of initial value. By what factor, the rate of reaction would have been decreased by the same time?

24 Under identical experimental condition, how many of the following substrate react at faster rate with aqueous NaOH than 2-bromobutane as substrate?

i. 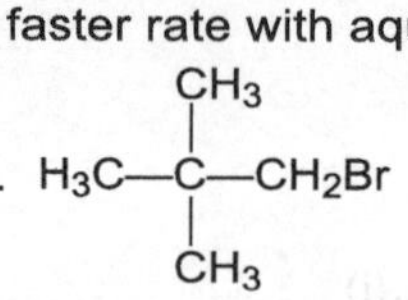

ii. $CH_3CH_2CH_2CH_2Cl$

iii. 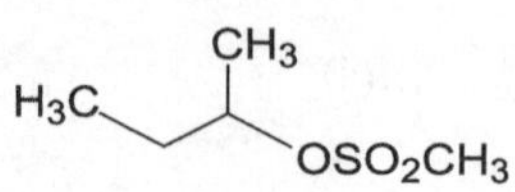

iv. 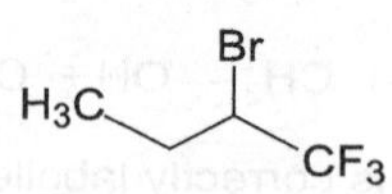

v. 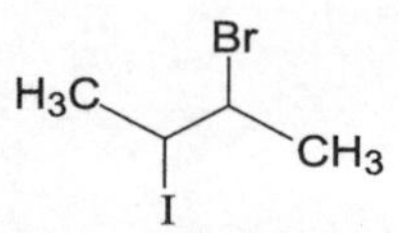

vi. 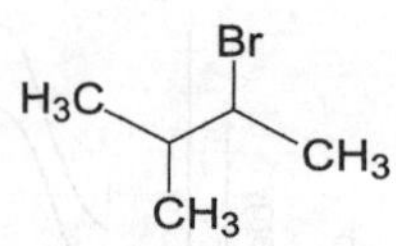

vii. 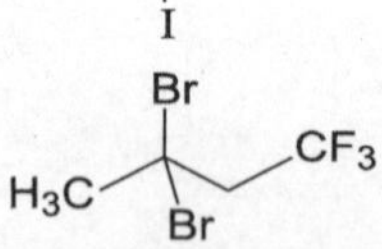

viii. 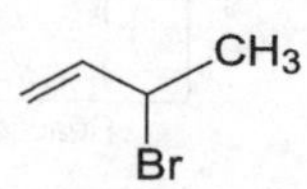

ix. 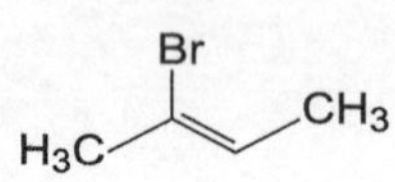

x. 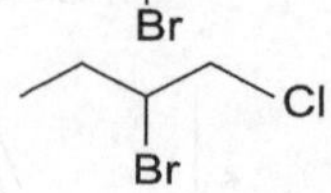

25 Consider the following compound,

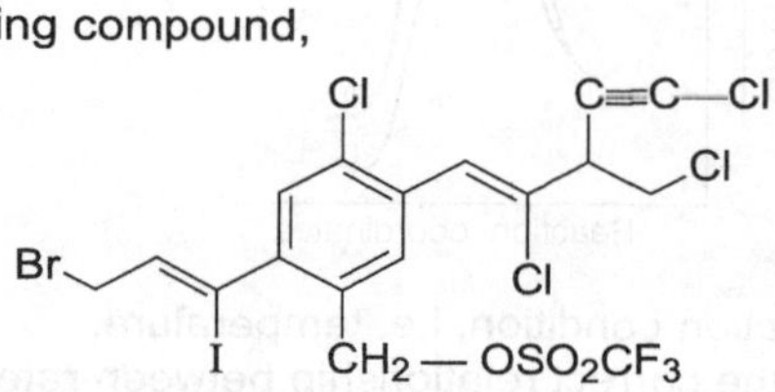

If the above compound is treated with excess of NaCN(*aq*), how many CN^- group would be incorporated by S_N2 reaction?

DPP-4 S_N2 Reaction Reactivity

Subjective Questions

Working Space

Direction (Q. Nos. 1-9) *These questions are subjective in nature, need to be solved completely on notebook.*

1 Rank the following molecules/ions according to their leaving group ability in a S_N2 reaction.

i. HO^-, H_2O, H_3O^+

ii. NH_2^-, NO_3^-, NO_2^-

iii. F^-, I^-, Cl^-, Br^-

iv. F_3CCOO^-, $F_3CCH_2COO^-$, $F_3CCH_2CH_2COO^-$

v. $CH_3CH_2O^-$, $CH_3CH_2^-$, CH_3COO^-

vi. 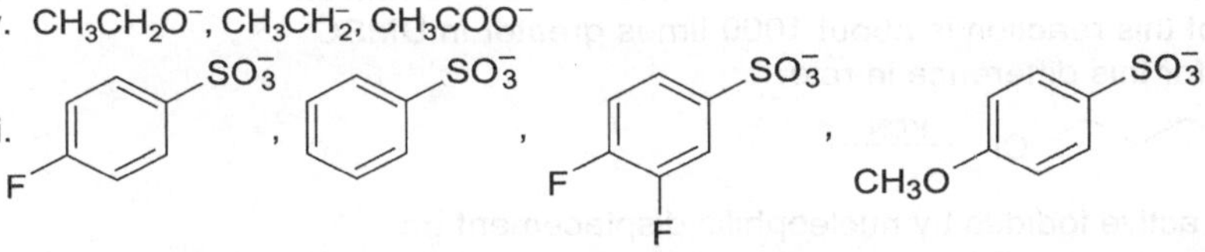

2 Complete the following S_N2 reactions. Pay attention to stereochemistry where ever applicable:

i. 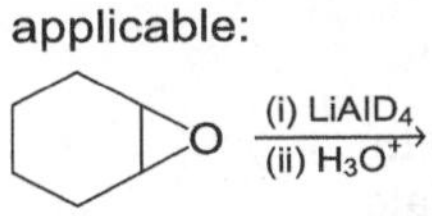

ii. 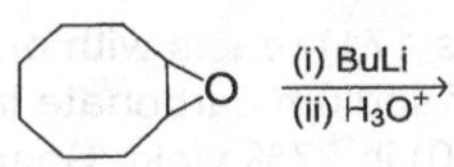

iii. 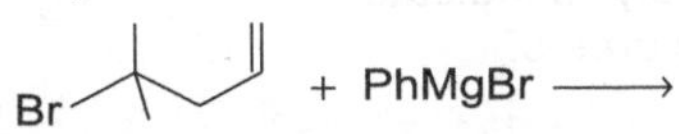

iv. 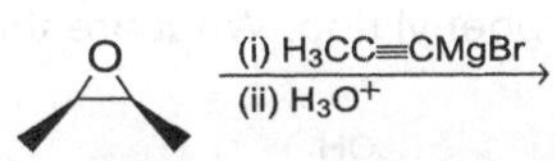

v. 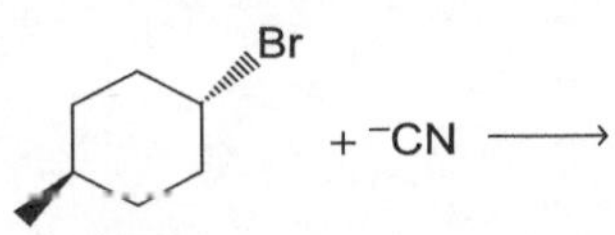

vi. 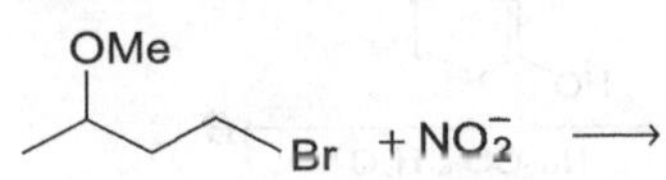

vii. 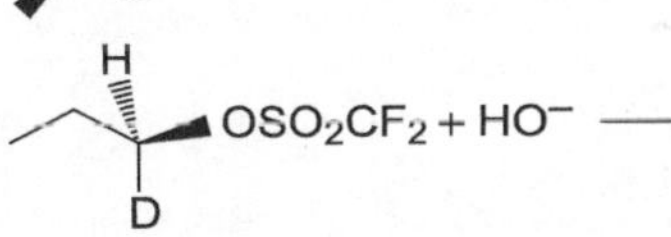

viii. 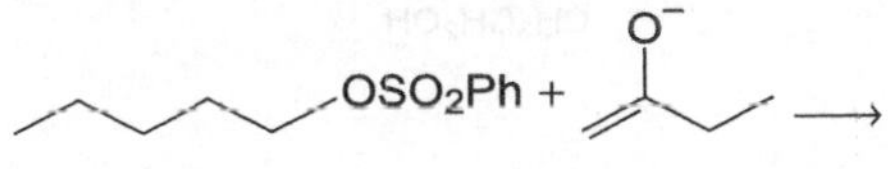

ix. 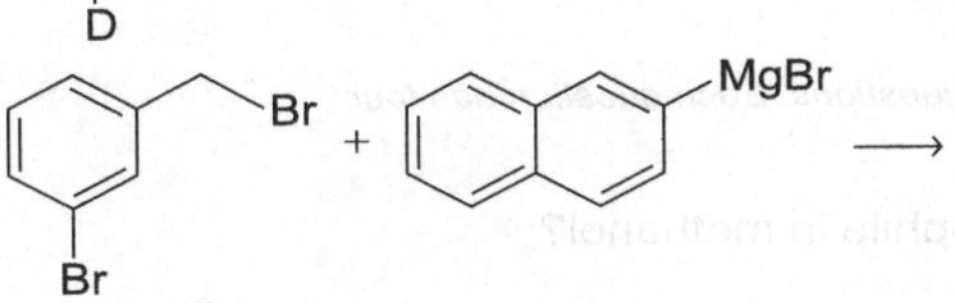

x. 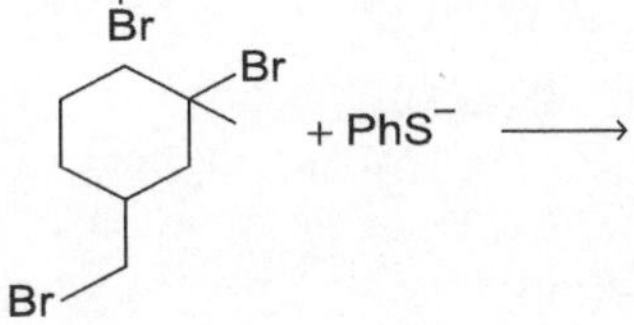

xi. 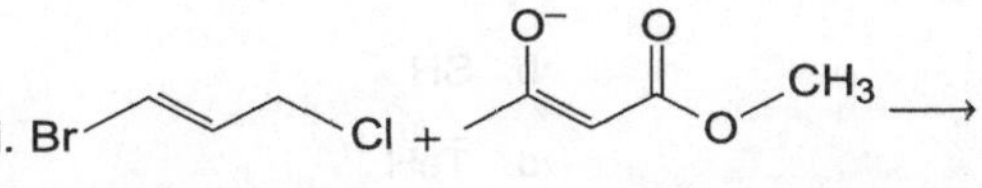

xii. 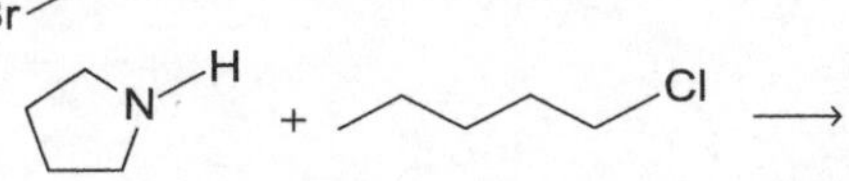

Working Space

3 Following 3-bromo-2-methoxybutane undergoes an S_N2 reaction with CH_3O^- to produce a product. What is the product? Is this product optically active?

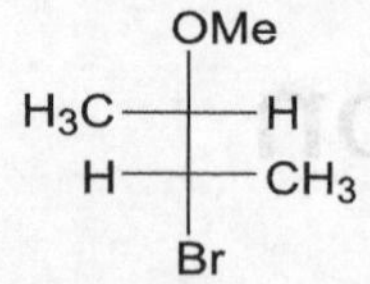

4 Acetate ion reacts 10 times faster than chloroacetate ion in an S_N2 reaction with methyl iodide in methanol. Explain this observation.

5 When treated with a solution of KBr, optically active 2-bromobutane racemises, but optically active 2-butanol does not racemise when treated with a solution of KOH. Explain the differences in these reactions.

6 From each pair, select the stronger nucleophile.

i. H_2O or OH^-

ii. CH_3COO^- or OH^-

iii. CH_3SH or CH_3S^-

iv. Cl^- or I^- (in DMSO)

v. Cl^- or I^- (in methanol)

7 Treatment of 1, 3-dichloropropane with potassium cyanide results in the formation of pentanedinitrile. The rate of this reaction is about 1000 times greater in DMSO than it is in ethanol. Account for this difference in rate.

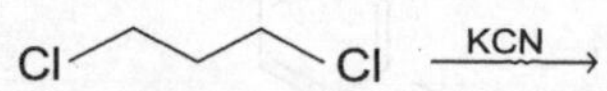

8 Attempts to prepare optically active iodides by nucleophilic displacement on optically active bromides using I^- normally produce racemic iodoalkanes. Why are the product of iodoalkanes racemic?

9 One mole of compound *A* (molar mass 171) reacts with one mole of 4-hydroxyphenol and one half mole of sodium carbonate in aqueous ethanol to produce compound *B* (molar mass 200) in 87% yield. Spectral analysis indicate that both *A* and *B* has at least one phenyl ring. What are the structures of *A* and *B* ?

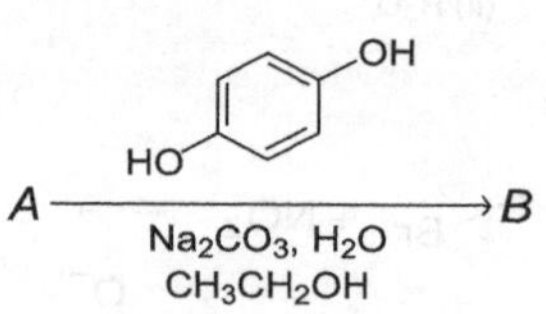

Only One Option Correct Type

Direction (*Q. Nos. 10-21*) *This section contains 12 multiple choice questions. Each question has four choices (a), (b), (c) and (d), out of which ONLY ONE is correct.*

10 Which of the following would be the strongest nucleophile in methanol?

a. $\bar{O}H$ b. $\bar{S}H$

c. $\bar{S}eH$ d. $\bar{T}eH$

Working Space

11 Predict the product in nucleophilic substitution reaction.

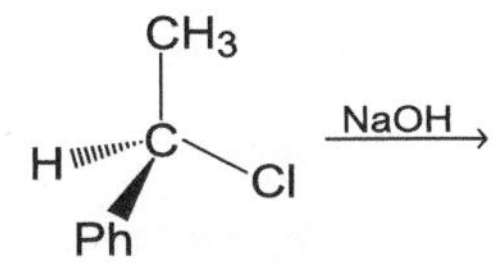

a. 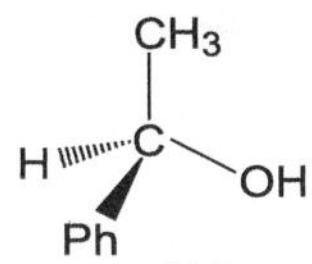

b. 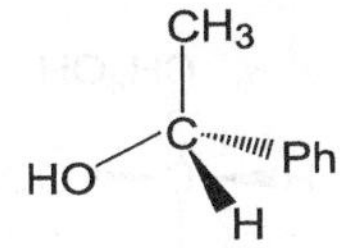

c. 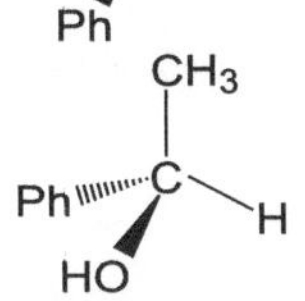

d. 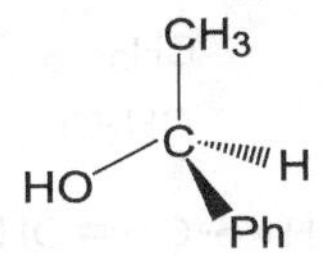

12 What is the product of the given reaction?

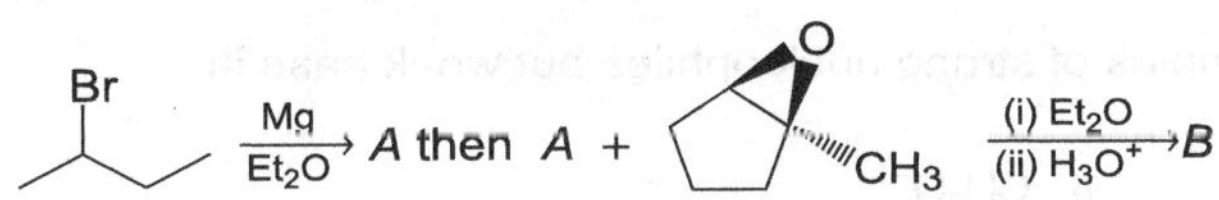

a. 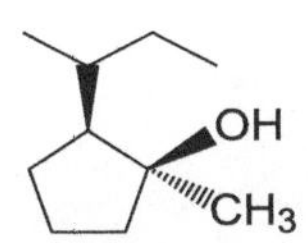

b. 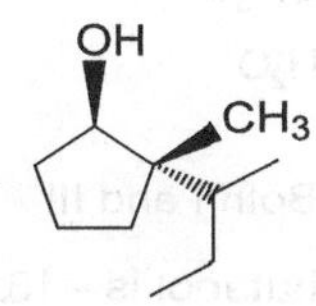

c. 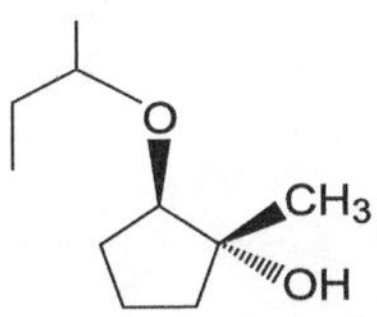

d. 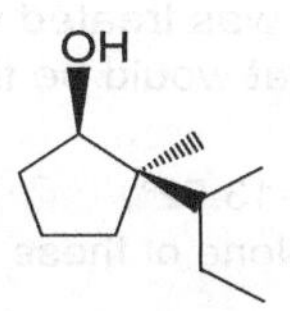

13 Which is the product of the given reaction?

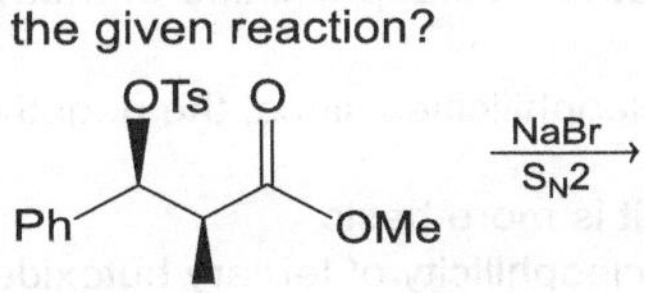

a. 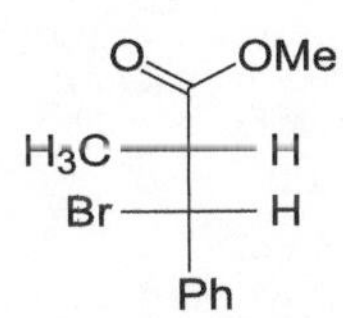

b. 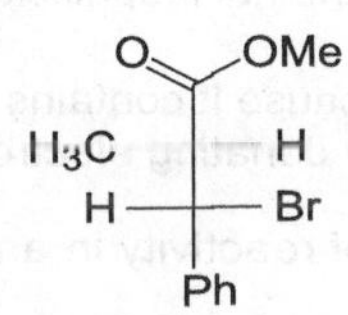

c. 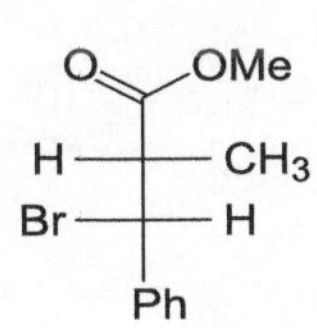

d. 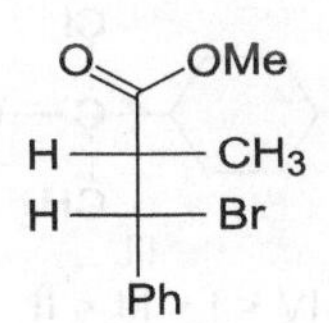

Working Space

14 What is the major product of the given reaction?

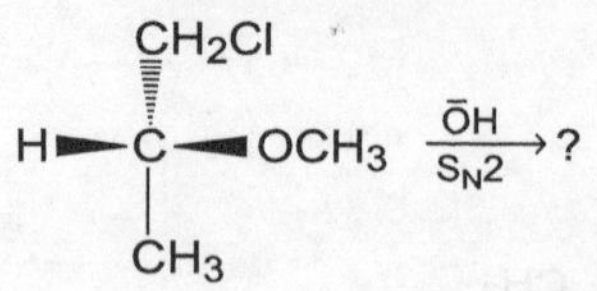

a. 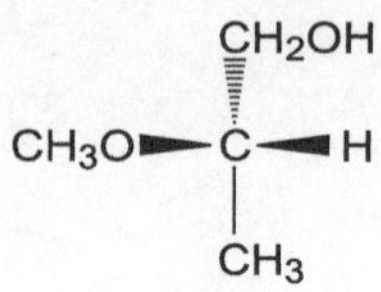

b. 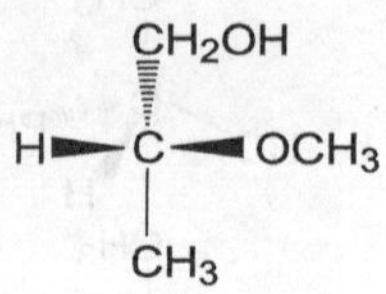

c. 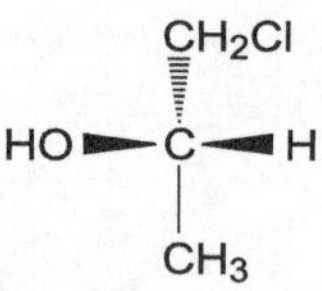

d. 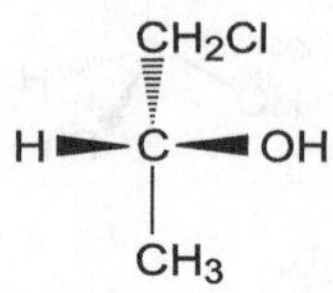

15 Which of the following are the examples of strong nucleophiles but weak base in protic solvents?

I. CH_3S^- II. CH_3O^-
III. I^- IV. H_2O
V. F^-

a. Both I and IV b. Both IV and V c. Both I and III d. Both III and IV

16 The specific rotation of optically pure (*R*)-2-butanol is −13.52° at 25°C. An optically pure sample of (*R*)-2-bromobutane was treated with aqueous NaOH in order to form 2-butanol *via* S_N2 reaction. What would be the specific rotation of the product assuming 100% yield?

a. 0° b. −13.52°
c. +13.52° d. None of these

17 Which of the following correctly describe the relative nucleophilicities of methoxide and tertiary butoxide ion?

a. These alkoxides have essentially the same nucleophilicities since, the negative charge in both is localised on an oxygen atom
b. Tertiary butoxide is more nucleophilic because it is more basic
c. Methoxide is more nucleophilic because the nucleophilicity of tertiary butoxide is diminished by steric effect.
d. Tertiary butoxide is more nucleophilic because it contains three methyl groups which increases the charge on oxygen by donating electron density

18 Arrange the following in increasing order of reactivity in an S_N2 reaction with KI in acetone solvent.

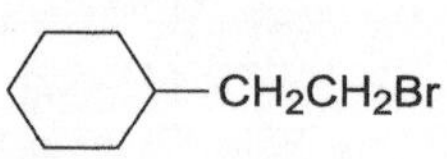

I.

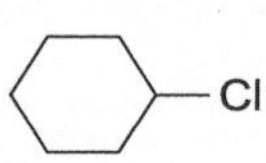

II.

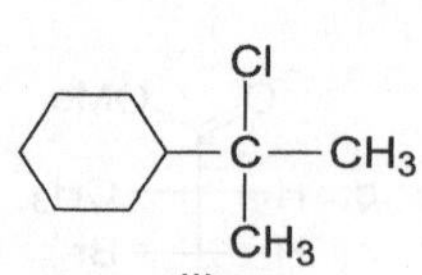

III.

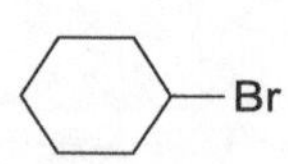

IV.

a. I < IV < III < II b. IV < I < III < II
c. II < III < I < IV d. III < II < I < IV

19 In the following set of nucleophiles, the strongest and the weakest nucleophile respectively are

I. CH_3S^- II. CH_3COO^- III. HO^- IV. $C_6H_5O^-$

a. Both I and II b. Both I and IV
c. Both III and II d. Both IV and II

20 The major product of the following reaction is:

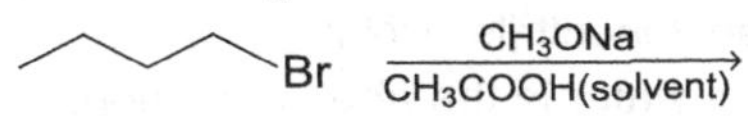

a. 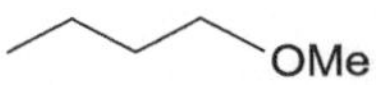

b. 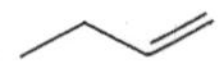

c. 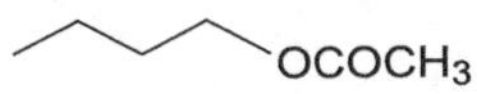

d. 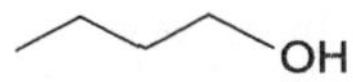

21 Which of the following on treatment with NaCN (*aq*) results in an achiral product?

a. 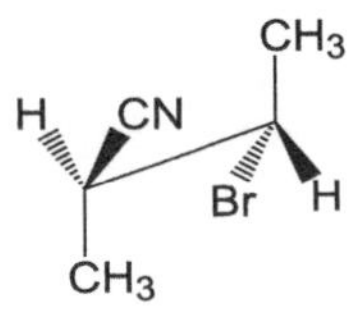

b. 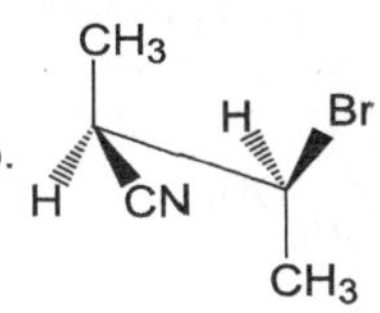

c. 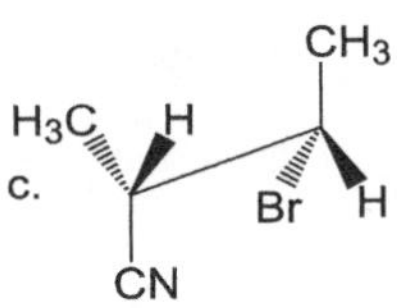

d. 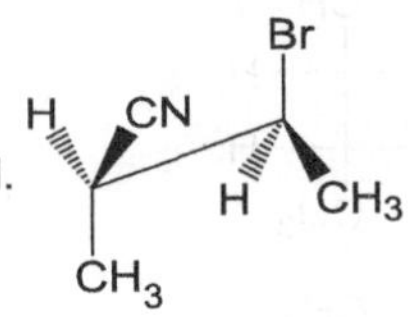

One or More than One Options Correct Type

Direction *(Q. Nos. 22-26) This section contains 5 multiple choice questions. Each question has four choices (a), (b), (c) and (d), out of which ONE or MORE THAN ONE are correct.*

22 In which of the following pair(s), the first one is stronger nucleophile than the second one?

a. CH_3O^-, $(CH_3)_3CO^-$

b. $(CH_3)_2N^-$, NH_2^-

c. $C_6H_5O^-$, $(CH_3)_2CHO^-$

d. HS^-, HO^-

23 In which of the following pair(s), the first one is better leaving group than the second one?

a. Br^-, I^-

b. I^-, $CF_3SO_3^-$

c. $CF_3SO_3^-$, $CH_3SO_3^-$

d. p-O_2N-C_6H_4-SO_3^-, $C_6H_5SO_3^-$

24 In which of the following reaction, inversion of configuration at chiral carbon takes place?

a. 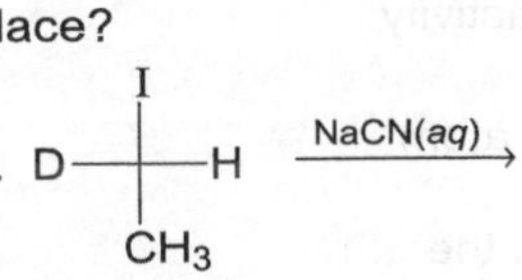

b. 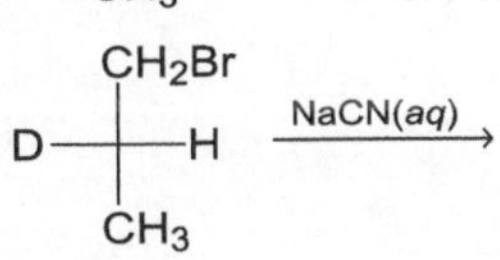

c. 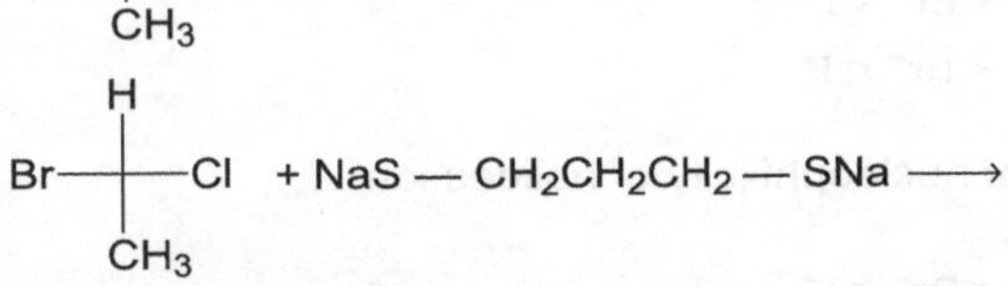

d. 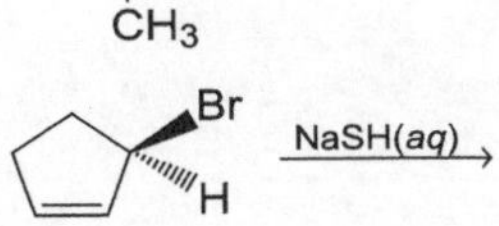

Working Space

25 Consider the following S_N2 reaction,

1-bromobutane + NaCN(*aq*) $\longrightarrow$ pentanenitrile + NaBr

Which of the following solvent(s) when replaces water in the above reaction, increases the rate?

a. Acetone

b. Dimethyl sulphoxide

c. N, N-dimethyl formamide

d. Tetrahydrofurane

26 In which of the following reaction(s), reactant and product are correctly matched?

a. 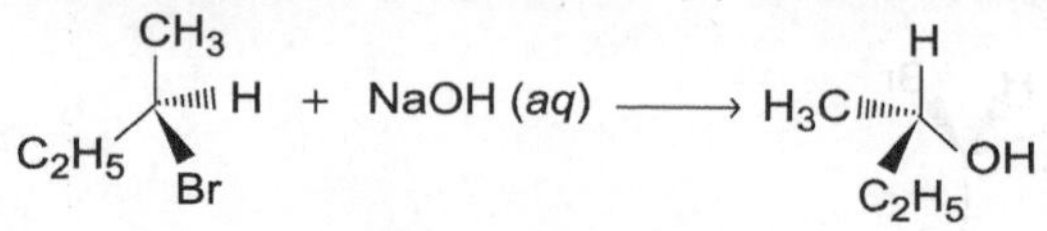

b. 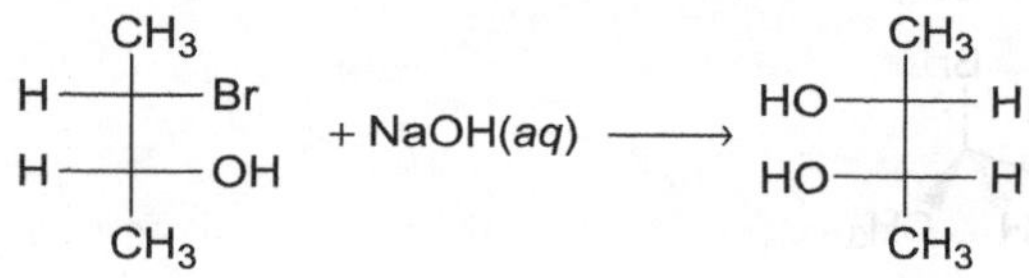

c. 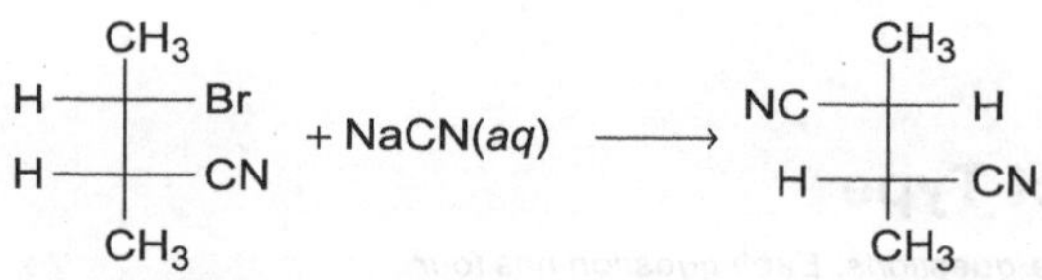

d. 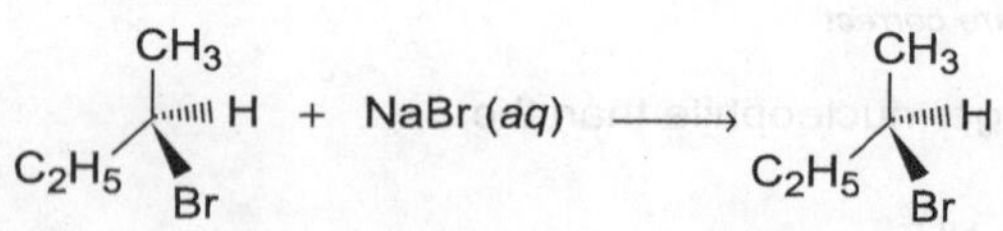

(+)-*dextro* (±)-racemic

Comprehension Type

Direction *(Q. Nos. 27-32) This section contains 2 paragraphs, each describing theory, experiments, data, etc. Six questions related to the paragraphs have been given. Each question has only one correct answer among the four given options (a), (b), (c) and (d).*

Passage I

In a S_N2 reaction, solvent plays a very important role. Hydroxylic solvents like water, alcohols, carboxylic acids etc., solvate nucleophile, decreases its reactivity in the reaction.

Size of donor atom of nucleophile also affect nucleophilicity. A larger donor atom is more easily polarised and less solvated, hence a stronger nucleophile. Therefore, in a given group of periodic table, nucleophilicity increases down the group. However solvation effect is more predominant than polarisability.

27 In water, the order of nucleophilicity of halides is

a. $I^- < Br^- < Cl^- < F^-$

b. $F^- < Cl^- < Br^- < I^-$

c. $I^- = Br^- = Cl^- = F^-$

d. $F^- = Cl^- < Br^- < I^-$

28 In dimethyl sulphoxide (DMSO) solvent, the order of nucleophilicity of halides ion is

a. $I^- < Br^- < Cl^- < F^-$

b. $F^- < Cl^- < Br^- < I^-$

c. $I^- = Br^- = Cl^- = F^-$

d. $F^- = Cl^- < Br^- < I^-$

29 Which of the following describe correctly the role of H_2O as solvent in a S_N2 reaction?

a. It stabilises nucleophile but not to transition state
b. It stabilises transition state but not nucleophile
c. It stabilises both transition state and nucleophile but to a greater extent to nucleophile
d. It stabilises both transition state and nucleophile but to a greater extent to transition state

Passage II

In a S_N2 reaction, back side attack takes place which leads to inversion of configuration at α-carbon which is known as Walden's inversion :

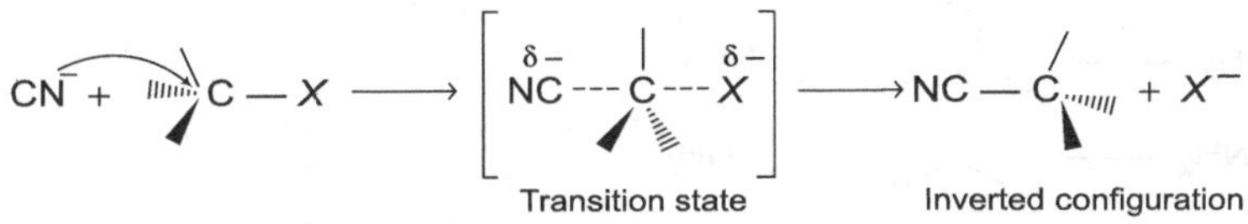

30 What is true regarding the following reaction?

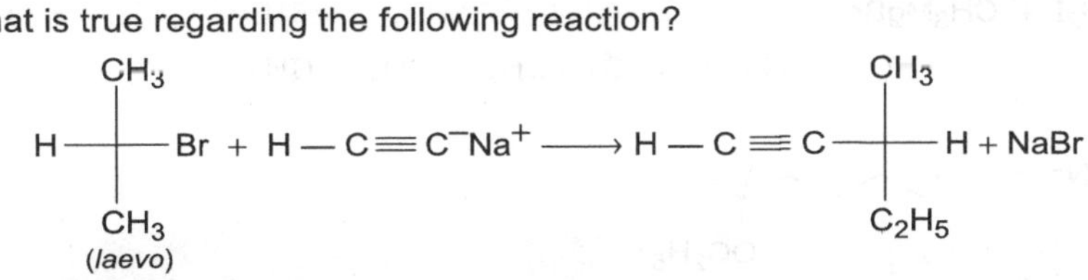

a. The product would be a *dextro* isomer
b. The product would be a *laevo* isomer
c. The product would be a racemic mixture
d. The product would be either *laevo* or *dextro* isomer

31 What can be correctly predicted regarding product of the following reaction?

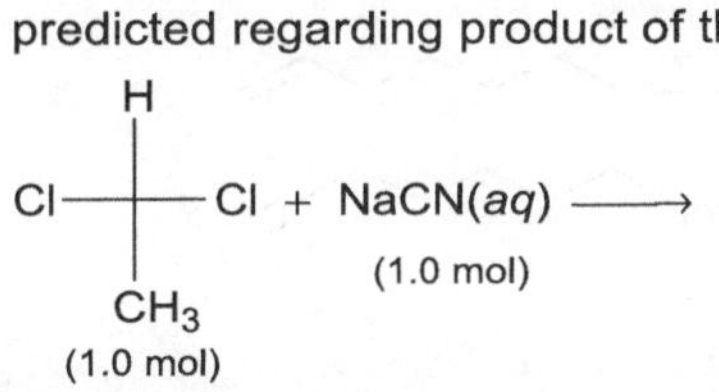

a. The product would be achiral
b. The product would be optically active
c. The product would be racemic mixture
d. The product would contain unequal amount of enantiomers

32 What can be predicted correctly about product of the following S_N2 reaction?

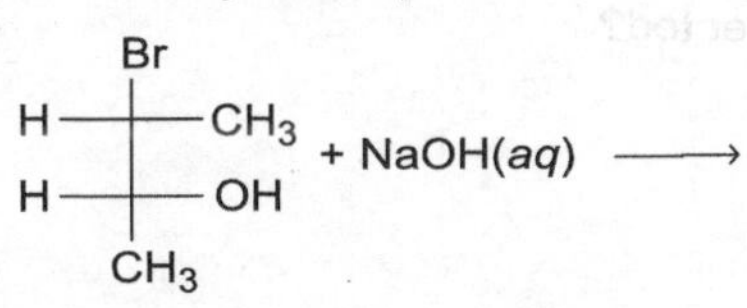

a. Product would be enantiomer of starting compound
b. Product would be a pure enantiomer
c. Product would be a racemic mixture
d. Product would be achiral

One Integer Value Correct Type

Working Space

Direction *(Q. Nos. 33 and 34) This section contains 2 questions. When worked out will result in an integer from 0 to 9 (both inclusive).*

33 How many of the following reactions fail to give major product by S_N2 reaction as indicated?

i. 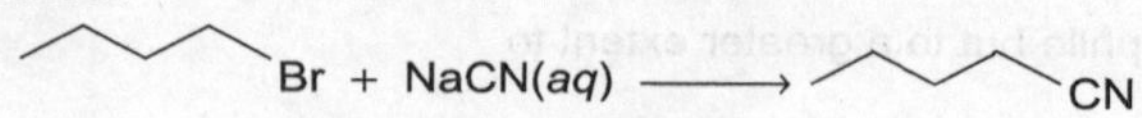

ii. 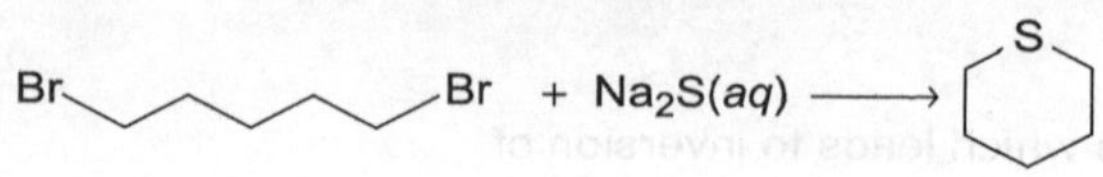

iii. 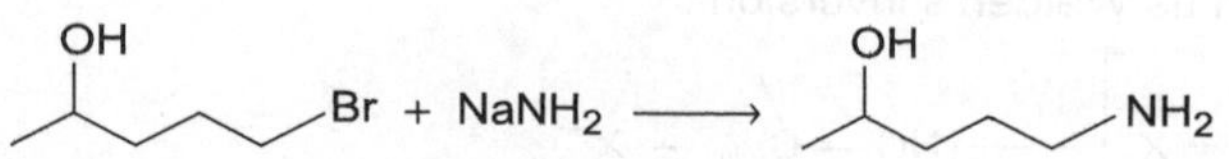

iv. 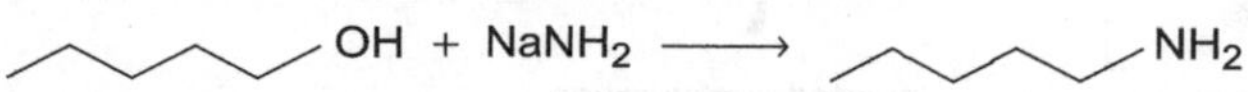

v. $H-C\equiv C-CH_2-CH_2I + CH_3MgBr \longrightarrow H-C\equiv C-CH_2-CH_2-CH_3$

vi. 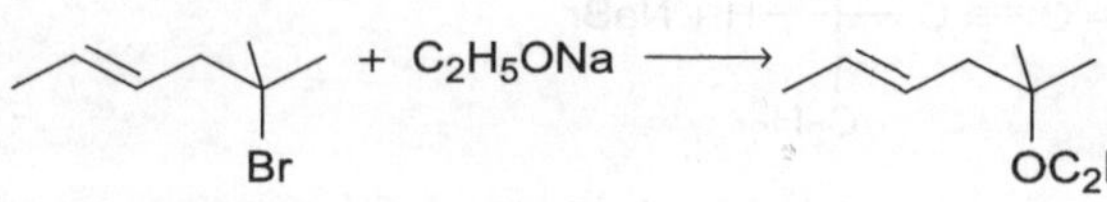

vii. 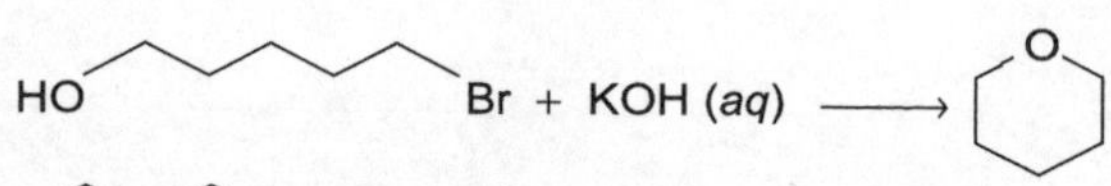

viii. 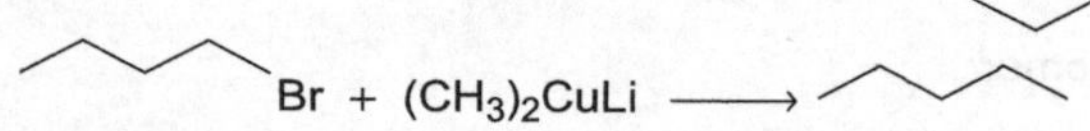

ix. 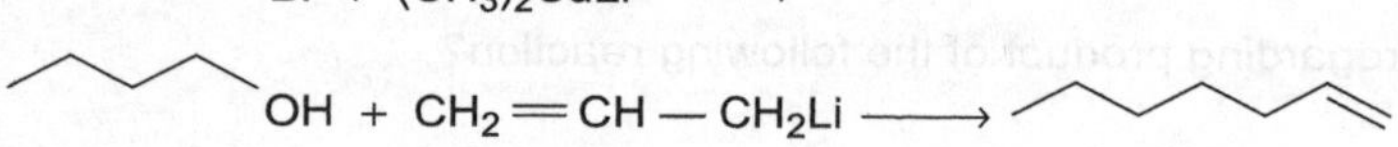

x. 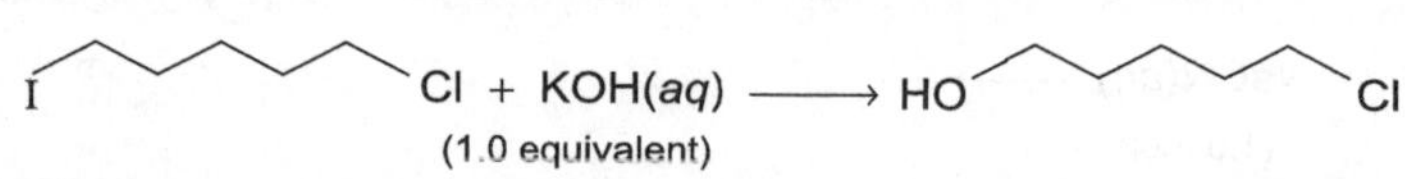

34 In the following reaction,

$$\underset{(1.0\ mol)}{H_3C-\overset{Cl}{\underset{Cl}{\overset{|}{\underset{|}{C}}}}-\overset{Cl}{\underset{Cl}{\overset{|}{\underset{|}{C}}}}-CH_3} + \underset{(2.0\ mol)}{NaCN(aq)} \longrightarrow$$

How many different dicyano products are expected?

DPP-5 Ambident Nucleophiles and NGP in S_N2 Reaction

Subjective Questions

Working Space

Direction *(Q. Nos. 1-6) These questions are subjective in nature, need to be solved completely on notebook.*

1 In each pair of compounds, which compound is more nucleophilic in a polar protic solvent?

i. CH_3O^- or CH_3S^- ii. $(CH_3)_3N$ or $(CH_3)_3B$

iii. CH_3Cl or CH_3OH iv. $(CH_3)_3N$ or $(CH_3)_3P$

v. ^-SH or Cl^- vi. CH_3OCH_3 or CH_3SCH_3

2 Mustard gas is about 14,000 times more reactive in nucleophilic substitutions than in chloroethane. Explain. $Cl—CH_2CH_2—S—CH_2—CH_2—Cl$ (Mustard gas)

3 Knowing what you do about the regioselectivity of S_N2 reactions, predict the product of hydrolysis of this compound

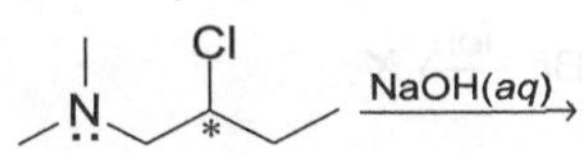

4 When an alkyl halide reacts with the thiocyanate ion, ^-SCN, reaction takes place at the sulphur atom. When an alkyl halide reacts with the cyanate ion, ^-OCN, reaction takes place at the nitrogen atom. Draw the Lewis structures for both the cyanate and thiocyanate ions and explain the difference in the reactivity of these ions.

$$R—X + {}^-SCN \longrightarrow RSCN + X^-$$

$$R—X + {}^-OCN \longrightarrow RNCO + X^-$$

5 The following nucleophilic substitution occurs with rearrangement. Suggest a mechanism for formation of the observed product. If the starting material has the *S* configuration, what is the configuration of the chiral center in the product.

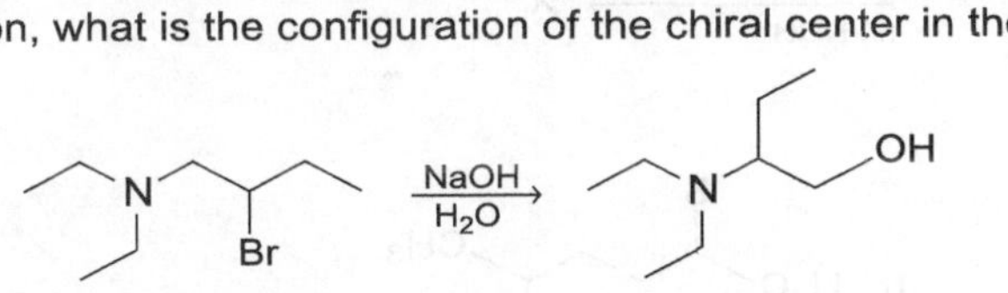

6 Complete the following reaction by indicating stereochemistry of product properly.

i. 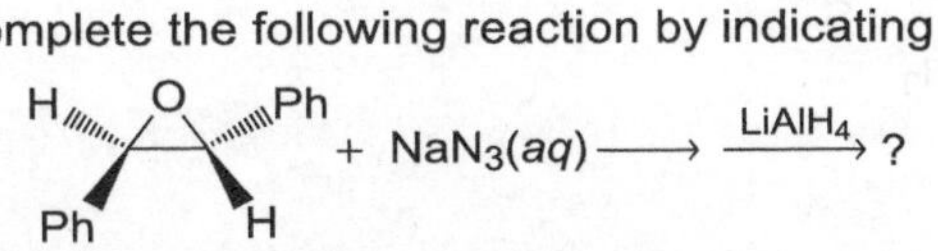

ii. 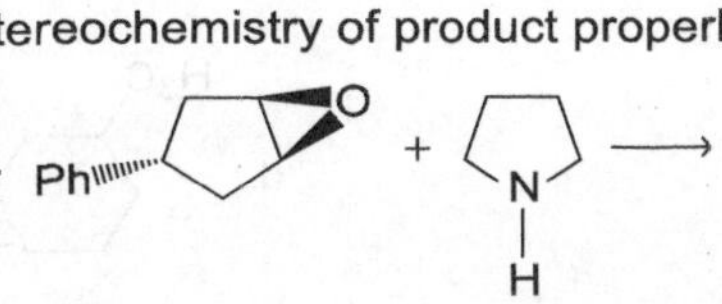

iii. 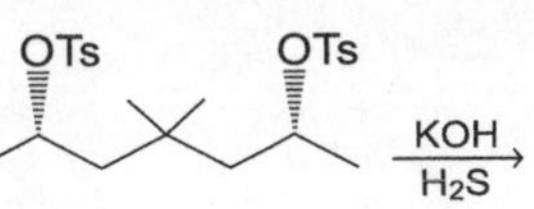

Only One Option Correct Type

Working Space

Direction *(Q. Nos. 7-14) This section contains 8 multiple choice questions. Each question has four choices (a), (b), (c) and (d), out of which ONLY ONE is correct.*

7 Cyanide anion has two atoms that have lone pair of electrons. Either could act as nucleophile in the reaction. Yet in the vast majority of the cases, carbon acts as nucleophile and forms a bond to the substrate, why?

a. Nitrogen is strongly solvated in hydroxylic solvents, it will be less reactive
b. Carbon is more reactive in polar aprotic solvents, normally used for these S_N2 reactions
c. Carbon atoms are more basic than nitrogen
d. The lone pair of electrons on C is in *sp* hybrid orbital

8 In the following two phase reaction, catalyst works by

$$C_6H_5CH_2Br + KCN \xrightarrow[H_2O/\text{heptane}]{C_6H_5CH_2\overset{+}{N}(CH_3)_3\overset{-}{Cl}} C_6H_5CH_2CN + KBr$$

a. transforming CN^- from the aqueous phase to organic containing $C_6H_5CH_2Br$
b. transforming $C_6H_5CH_2Br$ from organic phase to the aqueous phase containing CN^-
c. removing Br^- from the organic base to aqueous phase
d. removing K^+ from the organic phase to make CN^- more nucleophilic

9 Which of the following would react most rapidly with sodium ethoxide to produce an ether?

a. Chlorobenzene
b. 2-nitrotoluene
c. *p*-nitrochlorobenzene
d. *m*-(chloromethyl)-toluene

10 In the given reaction,

$$CH_3—CH_2—\ddot{\underset{\cdot\cdot}{S}}—\overset{*}{C}H_2—CH_2—Br \xrightarrow{HOH} X$$

X will be

a. $CH_3—CH_2—S—\overset{*}{C}H_2—CH_2—OH$
b. $CH_3—CH_2—S—CH_2—{}^{*}CH_2OH$
c. 1 : 1 mixture of (a) and (b)
d. 2 : 1 mixture of (a) and (b)

11 In the given reaction,

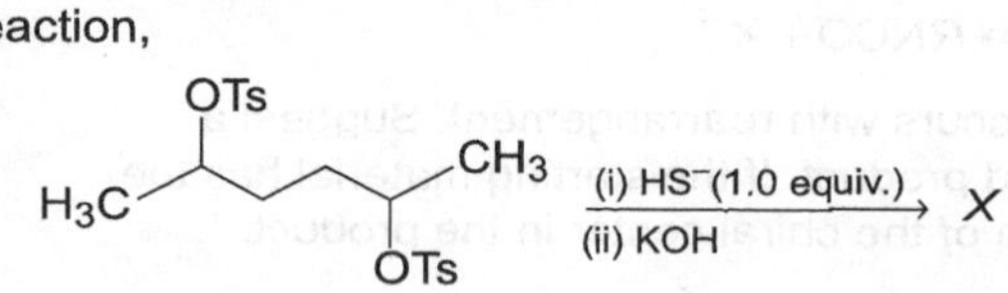

X will be

a. 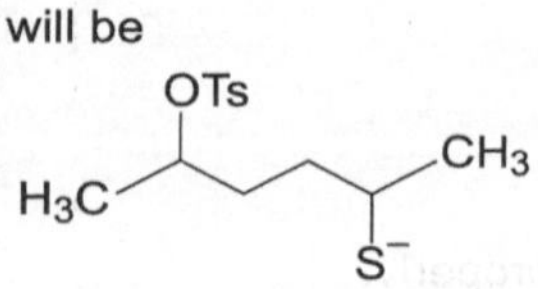

b. 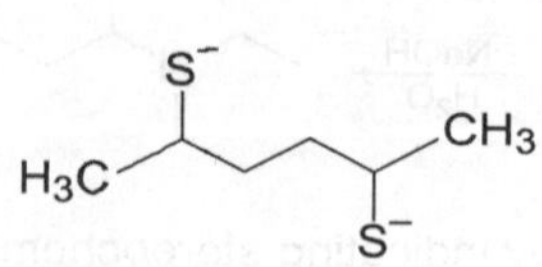

c. 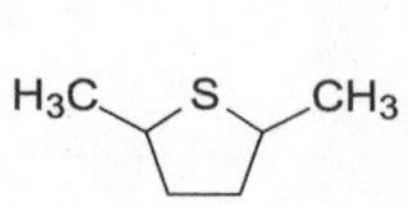

d. 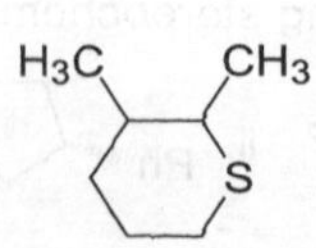

Working Space

12 Consider the following reaction,

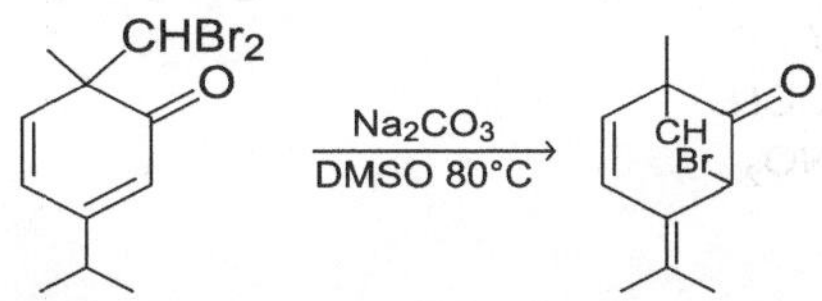

What can be correctly predictied regarding this reaction?

a. S_N1 reaction is occurring
b. S_N2 reaction is occurring
c. Neighbouring group participation and S_N2 reaction are occurring
d. Neighbouring group participation and $S_N1$reaction are occurring

13 Pick out the alkyl bromide which proceeds with retention of configuration in an S_N2 reaction with CH_3ONa (*aq*).

a. 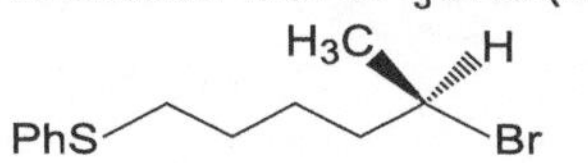
b. 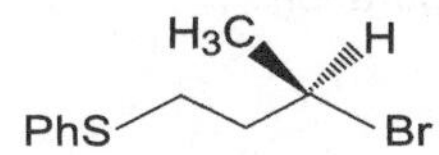
c. 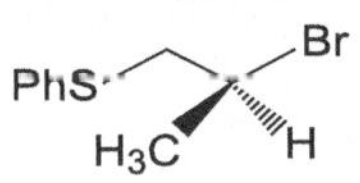
d. 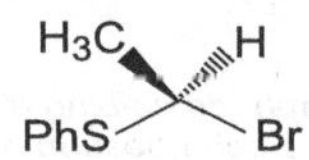

14 For the following reaction, pick out the best term which describe its mechanism

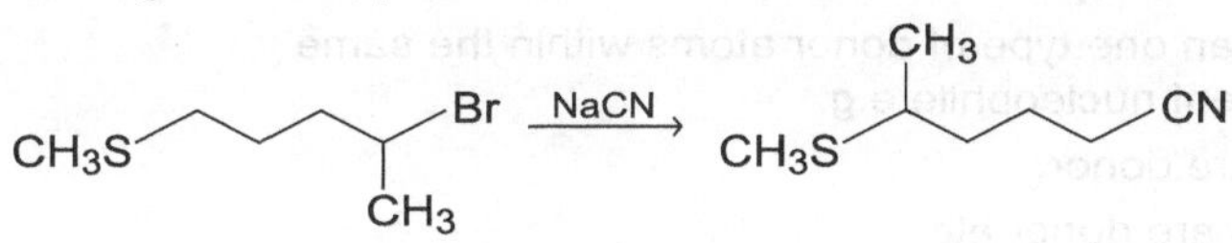

a. S_N1
b. S_N2
c. S_N2 with neighbouring group participation
d. addition followed by substitution

One or More than One Options Correct Type

Direction (Q. Nos. 15-19) This section contains 5 multiple choice questions. Each question has four choices (a), (b), (c) and (d), out of which ONE or MORE THAN ONE are correct.

15 Pick out the alkyl bromides which proceeds with retention of configuration in an S_N2 reaction with $CH_3ONa(aq)$

a. 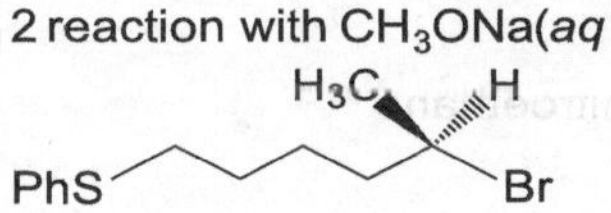
b. 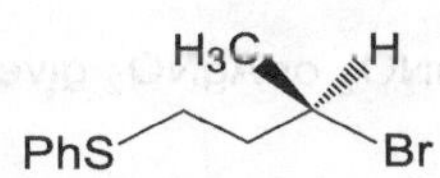
c. 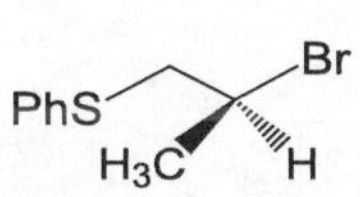
d. 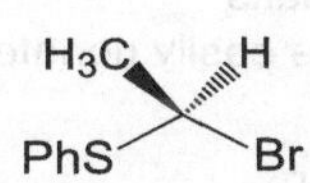

16 Which of the following is/are ambident nucleophile?

a. CN^-
b. OCN^-
c. SCN^-
d. NO_2^-

17 Which of the following gives the same substitution (S_N2) product with C_2H_5Br no matter whether sodium or silver salt is used?

a. CN^-
b. OCN^-
c. SCN^-
d. NO_2^-

Working Space

18 Which of the following gives the different substitution product with CH_3CH_2Br when their sodium and silver salts are used?

a. CN^-
b. OCN^-
c. SCN^-
d. NO_2^-

19 Consider the following substitution reaction,

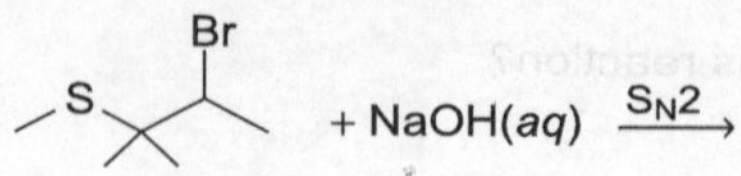

The correct statement(s) is/are

a. reaction occur at much faster rate than when substrate is 2-bromo-3,3-dimethyl pentane
b. reaction occur with inversion of configuration
c. reaction occur with retention of configuration
d. E2 product is preferred over S_N2

Comprehension Type

Direction (Q. Nos. 20-22) *This section contains a paragraph, describing theory, experiments, data, etc. Three questions related to the paragraph have been given. Each question has only one correct answer among the four given options (a), (b), (c) and (d).*

Passage

Nucleophiles which have more than one type of donor atoms within the same donor group are known as ambident nucleophile e.g.

CN^- : Both carbon and nitrogen are donor.

$\dot{O}CN^-$: Both oxygen and nitrogen are donor etc.

In case of free anionic and ambident nucleophiles, the better donor atom donate its lone pair of electron and forms bond with α-carbon after substitution reaction. In a given period, donor ability decreases from left to right. In a group, donor ability increases from top to bottom.

20 When CH_3CH_2Br is treated with aqueous NaCN, CH_3CH_2CN is formed while CH_3CH_2NC is formed when AgCN is used in the similar reaction. It is due to

a. negative charge on carbon atom in NaCN
b. covalent nature of Ag—C bond in AgCN
c. covalent nature of C—N bond in NaCN
d. ionic nature of Ag—C bond in AgCN

21 Treatment of CH_3CH_2Br with either $NaNO_2$ or $AgNO_2$ gives the same nitroethane as major product because

a. negative charge on nitrogen atom in both salts
b. nitrogen lone pair is free in both salts, more easily donated than oxygen lone pair
c. $NaNO_2$ gives free NO_2^- in solution
d. $AgNO_2$ has Ag covalently bonded to oxygen

22 When iodomethane is treated with sodium isocyanate (NaNCO), CH_3NCO is the major product while AgNCO gives CH_3CH_2OCN as major product because

a. negative charge on nitrogen in NaNCO and on oxygen in AgNCO
b. negative charge on oxygen atom in both salts
c. free NCO^- in case of NaNCO and covalent Ag—N bond in AgNCO
d. greater basicity of nitrogen in NaNCO and oxygen in AgNCO

Working Space

One Integer Value Correct Type

Direction *(Q. Nos. 23 and 24) This section contains 2 questions. When worked out will result in an integer from 0 to 9 (both inclusive).*

23 If *meso* form of the 2,4-dibromopentane is treated with excess of KCN(*aq*), how many S_N2 products with both bromide substituted, would result?

24 Consider the following S_N2 reactions.

I. 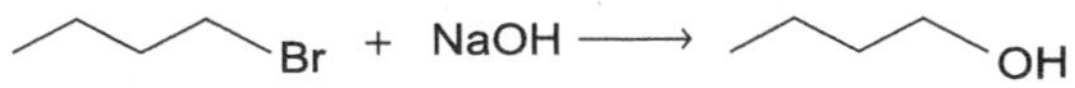

II. 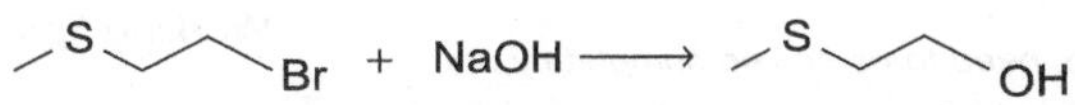

Only neighbouring effect increases reactivity of (II) by a factor of 5×10^4. If reaction (I) is carried out with initial concentration of C_4H_9Br = 0.05 M and NaOH 0.01 M and reaction (II) is carried out with initial concentration of both $CH_3SCH_2CH_2Br$ and NaOH = 0.1M each, rate of reaction (II) would be 10^x times greater than the rate of reaction (I) under identical experimental condition. What is the value of *x*?

Matching List Type

Direction *(Q. No. 25) Choices for the correct combination of elements from Column I and Column II are given as options (a), (b), (c) and (d), out of which one is correct.*

25 Match the Column I with Column II and mark the correct option from the codes given below.

	Column I		Column II
i.	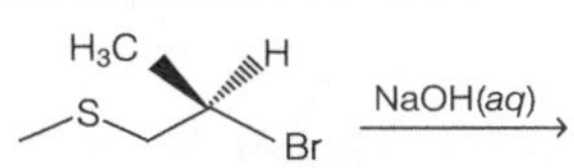	p.	Net inversion of configuration in S_N2 reaction.
ii.	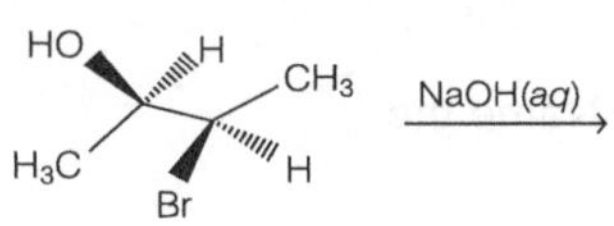	q.	Net retention of configuration in S_N2 reaction
iii.	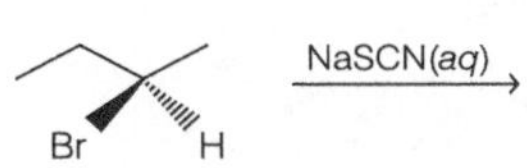	r.	Formation of achiral product
iv.	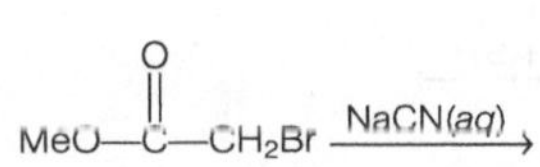	s.	Neighbouring group participation

Codes

	i	ii	iii	iv
a.	q,s	p,r	p	r,s
b.	q,s	s	p,s	q,s
c.	p	q,r	s	p
d.	p,r	q,s	p,r	q

DPP-6 S_N1 Reaction

Subjective Questions

Working Space

Direction *(Q. Nos. 1-5) These questions are subjective in nature, need to be solved completely on notebook.*

1 **Complete the following SN reactions. In each case, identify the mechanism of substitution (S_N1 or S_N2), nucleophile, leaving group, substrate carbon:**

i. $HO^- + CH_3Br \longrightarrow$ ii. $CH_3O^- + CH_3CH_2CH_2CH_2ONO_2 \longrightarrow$

iii. $CH_3CH_2COCH_2^- + PhCH_2I \longrightarrow$ iv. $2EtNH_2 + EtCl \longrightarrow$

v. N_3^- + 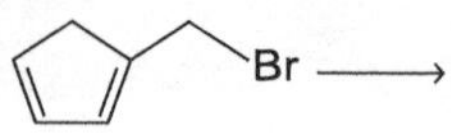

vi. $MeO{-}\overset{O}{\underset{O}{\overset{\|}{\underset{\|}{S}}}}{-}OMe + H_2O(excess) \longrightarrow$

vii. $MeOH + H_3COSO_2F \longrightarrow$

viii. $MeOH + H_3COSO_2CF_3 \longrightarrow$

ix. Br^- + 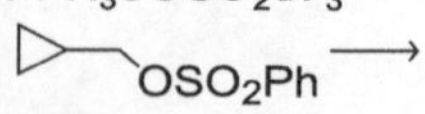

x. 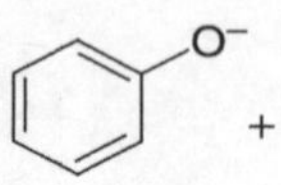

$+ CH_3CH_2Cl \longrightarrow$

2 **What product(s), would you expect from the following reactions? Name the mechanism (S_N1 or S_N2) that gives each product(s).**

i. $CH_3CH_2CH_2Br \xrightarrow[CH_3OH]{CH_3ONa}$

ii. 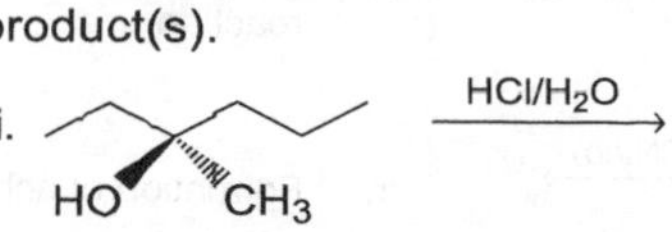

iii. $CH_3CH_2CH_2Br \xrightarrow[Excess]{NH_3}$

iv. 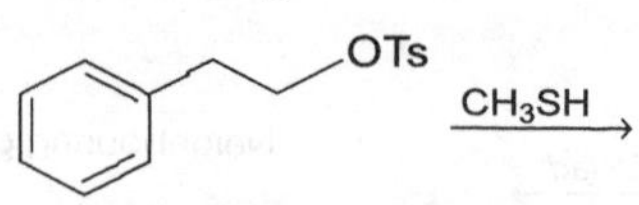

v. 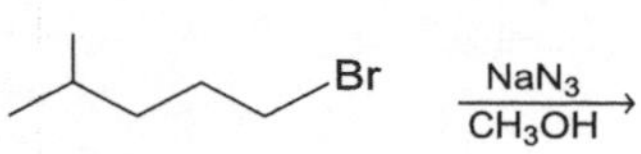

vi. 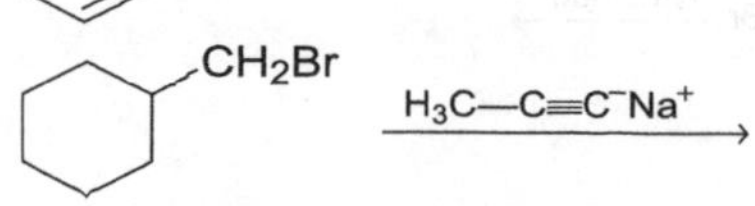

vii. 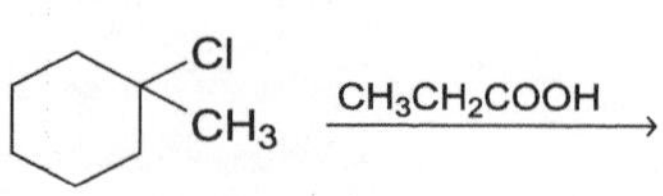

viii. 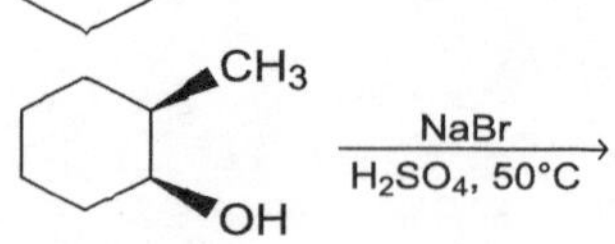

ix. 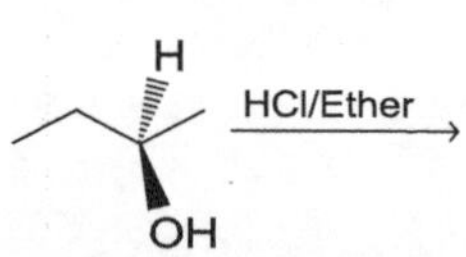

x. 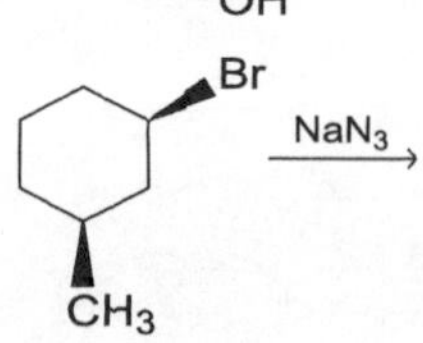

xi. 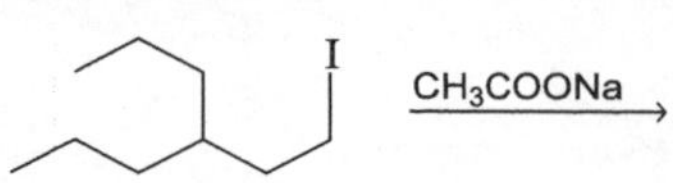

xii. 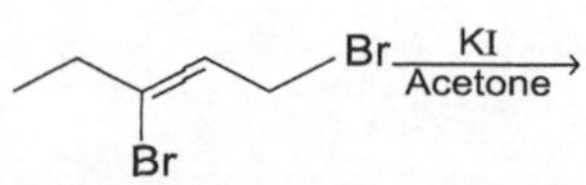

Working Space

3 The order of reactivity of alkyl bromides with sodium iodide in acetone is 1-bromobutane > 1-bromo-2-methylpropane >> 2-bromo-2-methylpropane. A group of students in an organic laboratory attempted to experimentally prove this order of reactivity using freshly washed, but not dry, test tubes. They could not confirm this order of reactivity until they carefully dried the test tubes. Why does this order of reactivity proceed only under dry conditions?

4 The rate of nucleophilic substitution is the same for the reaction of *tert*-butyl alcohol with either HCl or HBr. However, using an equimolar mixture of HCl and HBr, *tert*-butyl bromide is the major product and *tert*-butyl chloride is the minor product. Explain these results.

5 Account for the following relative rates of solvolysis under experimental conditions favouring S_N1 reaction.

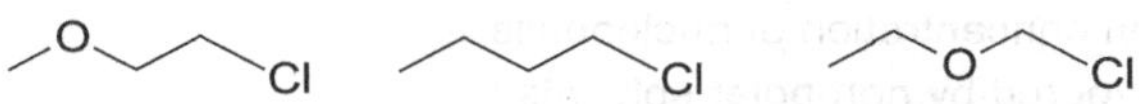

0.2   1.0   9

Relative rate of S_N1 reaction given.

Only One Option Correct Type

Direction (Q. Nos. 6-20) *This section contains 15 multiple choice questions. Each question has four choices (a), (b), (c) and (d), out of which ONLY ONE is correct.*

6 Which one of the two iodine will be more reactive in the S_N1 and S_N2 reactions?

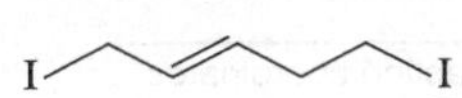

A   *B*

a. *A* will be faster in S_N1 reaction but slower in S_N2
b. *A* will be a faster both in S_N1 and S_N2 reactions
c. *A* and *B* will be equally reactive
d. *B* will be faster in both S_N1 and S_N2 reactions

7 In S_N1 reaction, racemisation occurs if the reaction occurs at a stereogenic centre, however, 50:50 mixture of enantiomers are rarely obtained, why?

a. Usually one enantiomer is more stable than other
b. Retention of configuration is always favoured in reaction
c. Hydroxylic solvent favour retention of configuration
d. There is steric hindrance to the approach of nucleophile from the front side as some of the leaving group are not departed completely which favour inversion of configuration

8 Pick out the most reactive alkyl halide for an S_N1 reaction.

a. 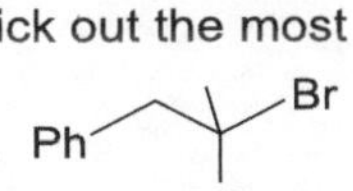
b. 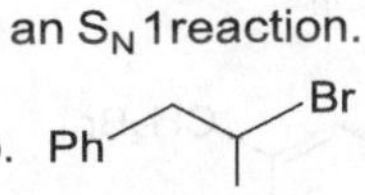

c. 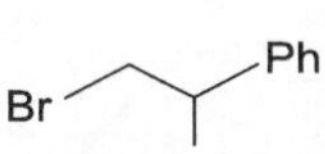
d. 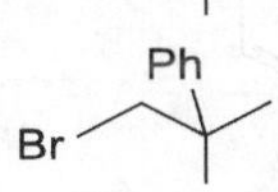

9 Pick out the compound which reacts fastest in the presence of $AgNO_3$.

a. $(CH_3)_3CCl$ b. $(CH_3)_2CHCH_2Cl$
c. $(CH_3)_2CHCl$ d. CH_3CH_2Cl

Working Space

10 From the following, pick out the explanation for why molecule *Y* hydrolyses faster than molecule *Z*.

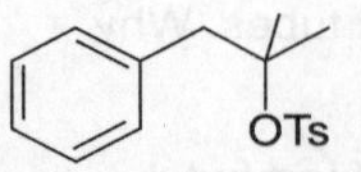

Y

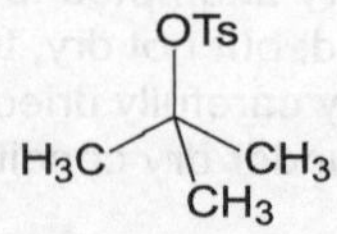

Z

a. Less steric hindrance b. Neighbouring group participation
c. Greater carbocation stability d. Better solvent

11 Which of the following statement(s) is/are true for S_N1 reaction?

I. The rate of S_N1 reaction depends on concentration of alkyl halide
II. The rate of S_N1 reaction depends on concentration of nucleophile
III. S_N1 reactions of alkyl halides are favoured by non-polar solvents

a. Only I b. Only II
c. Only III c. Both I and III

12 From the following, pick out the potential energy profile for a S_N1 reaction.

a.
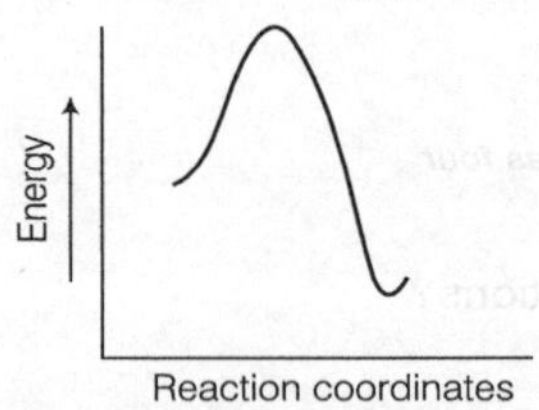

b.
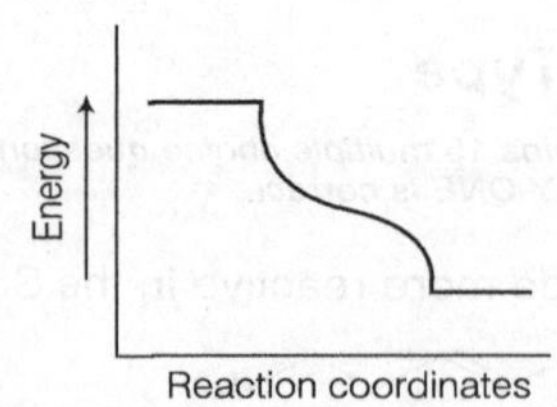

c.
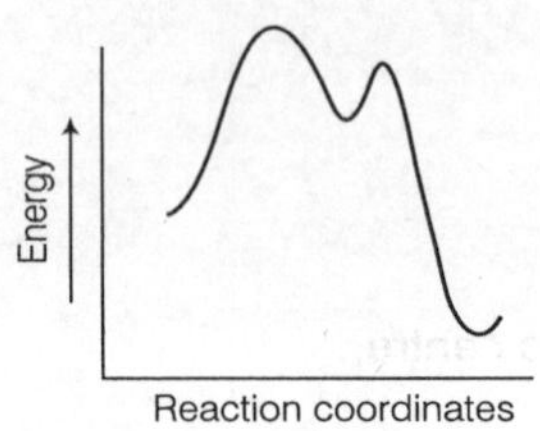

d.
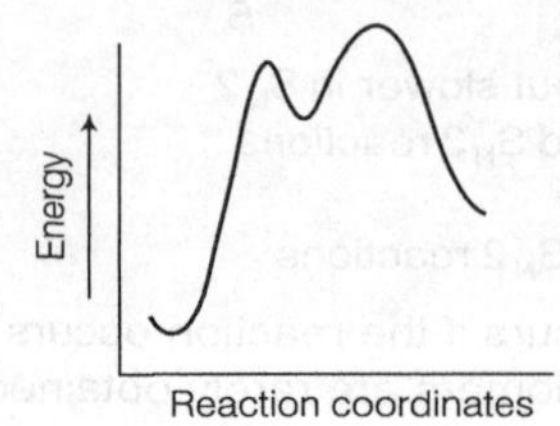

13 The most and least reactive electrophiles respectively in a S_N1 reaction are

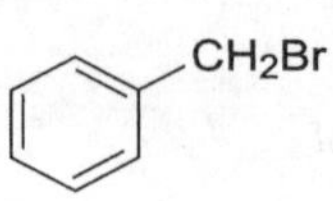

I.

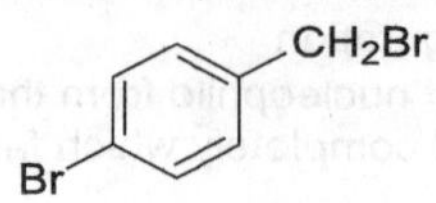

II.

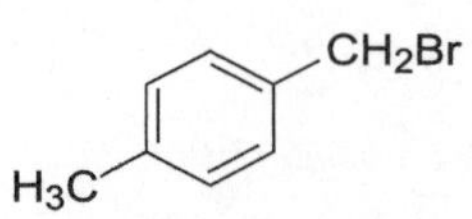

III.

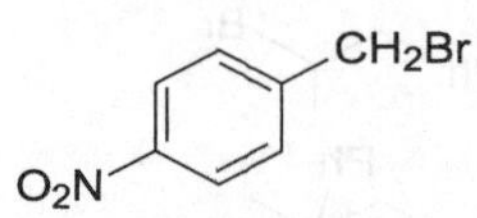

IV.

a. Both I and II b. Both II and III
c. Both I and III d. Both III and IV

14 Pick out the strongest substrate(s) for a S_N1 reaction.

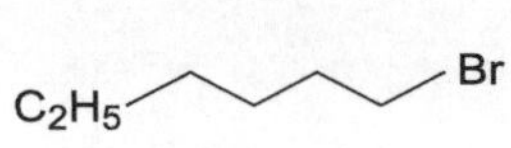

I.

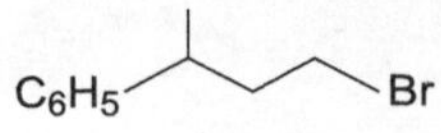

II.

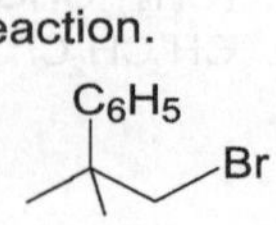

III.

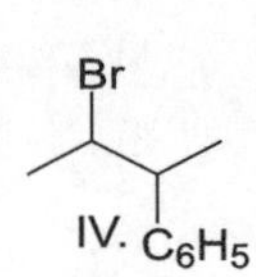

IV.

a. Only I b. Only II
c. Only III d. Only IV

Working Space

15 Pick out the following factor(s) which promote a S_N1 reaction :

I. Temperature
II. Concentration of nucleophile
III. Concentration of alkyl halide
IV. Aprotic solvents

a. Both I and II
b. Both II and III
c. Both III and IV
d. Both I and III

16 Which of the following alkyl halides is respectively most and least electrophilic in S_N1 reaction?

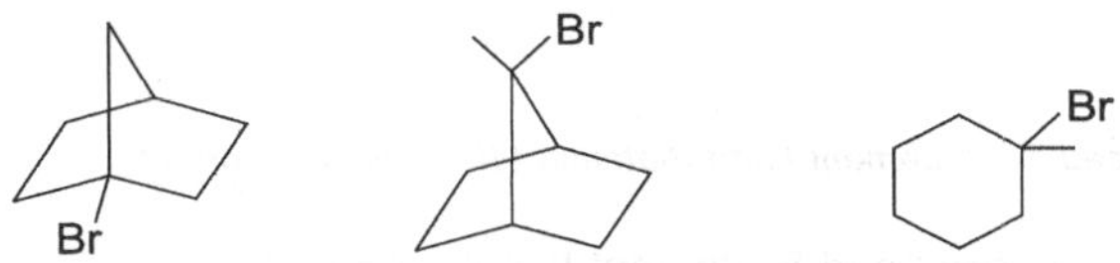

I. II. III.

a. Both I and III
b. Both II and III
c. Both III and I
d. Both I and II

17 Which of the following is true regarding a S_N1 reaction?

a. It would be faster at 25° C than at 50° C
b. It would be faster in ethanol than in pentane
c. Keeping the moles of reactants constant but doubling the quantity of solvent would decrease the rate by a factor of 4
d. Stereochemical inversion occurs exclusively

18 Rank the following molecules increasing in order of relative rate of S_N1 solvolysis with methanol and heat.

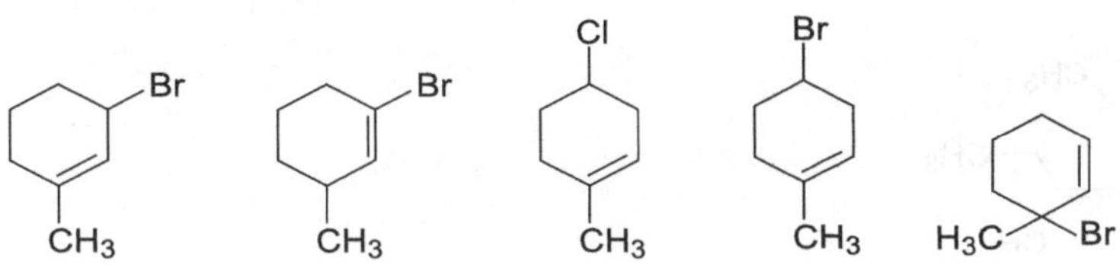

I. II. III. IV. V.

a. III < II < IV < V < I
b. II < III < IV < I < V
c. V < IV < III < II < I
d. II < III < IV < V < I

19 Which is the most likely product when the following iodide is heated with water?

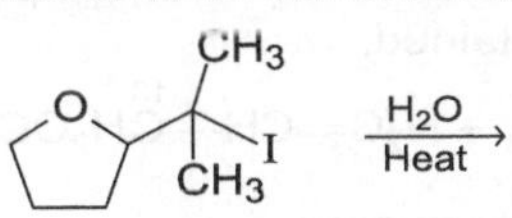

a. 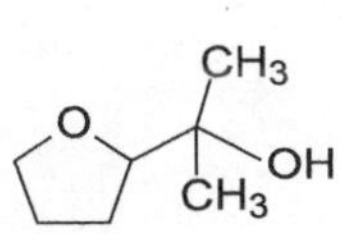

b. 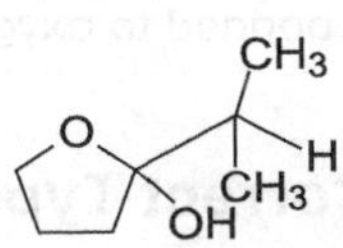

c. 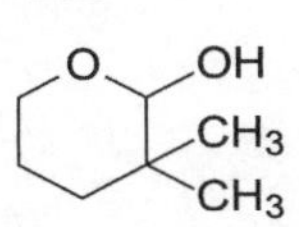

d. 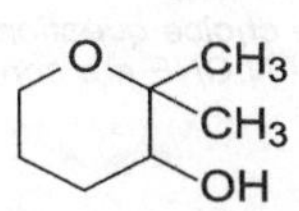

20 What is the correct order of reactivity of the followings in hydrolysis reaction at elevated temperature?

Working Space

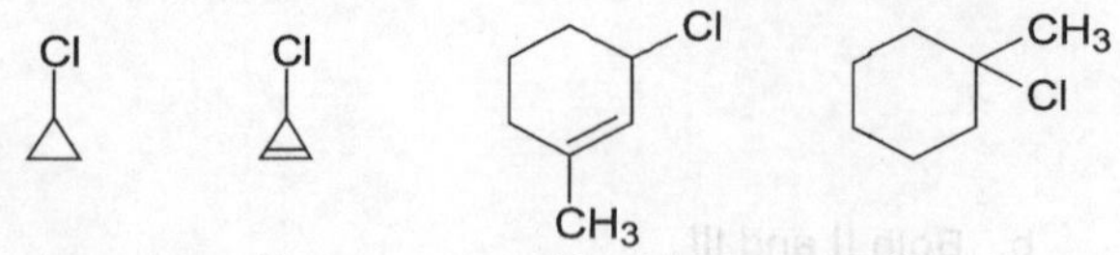

I. II. III. IV.

a. I < II < III < IV
b. IV < III < II < I
c. I < III < IV < II
d. I < IV < III < II

Statement Type

Direction (*Q. Nos. 21 and 22*) *This section is based on Statement I and Statement II. Select the correct answer from the codes given below.*

a. Both Statement I and Statement II are correct and Statement II is the correct explanation of Statement I
b. Both Statement I and Statement II are correct and Statement II is not the correct explanation of Statement I
c. Statement I is correct but Statement II is incorrect
d. Statement II is correct but Statement I is incorrect

21 Consider the following two bromides I and II, undergoing solvolysis reaction in boiling ethanol :

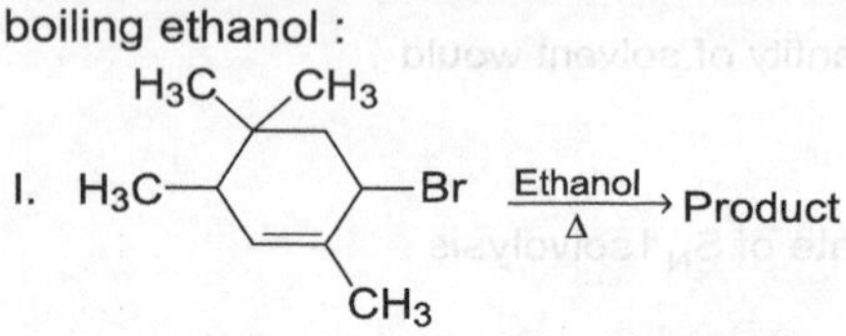

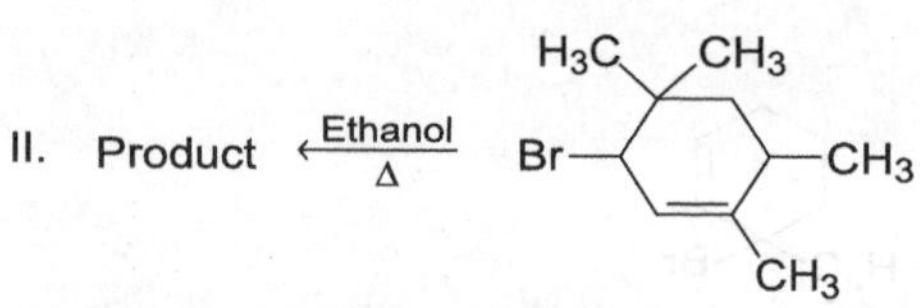

Statement I I is less reactive than II in the given solvolysis reaction.

Statement II Resonance stabilisation available with the intermediate formed from II is the important driving force.

22 **Statement I** When 3-bromo propene, which contain a labelled ^{13}C at C-1 position is refluxed with methanol, following products were obtained.

$$H_2\overset{13}{C}{=}CH{-}CH_2Br \xrightarrow[\text{Boil}]{CH_3OH} H_2\overset{13}{C}{=}CH{-}CH_2OCH_3 + H_2C{=}CH{-}\overset{13}{C}H_2OCH_3$$

Statement II Methanol has an acidic proton bonded to oxygen.

One or More than One Options Correct Type

Direction (*Q. Nos. 23-27*) *This section contains 5 multiple choice questions. Each question has four choices (a), (b), (c) and (d), out of which ONE or MORE THAN ONE are correct.*

23 What is/are true regarding a S_N1 reaction?

a. Rate of reaction is faster in H_2O than in C_2H_5OH
b. The reaction usually have more than one transition states
c. If α-carbon is chiral, partial racemisation and net inversion takes place
d. Reaction occurs at faster rate if $CH_3O^{18}H$ is used in place of CH_3OH

Working Space

24 When the reactants shown below undergo substitution, which of the products will form?

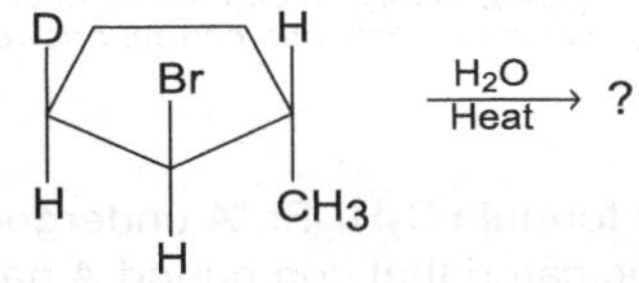

a. 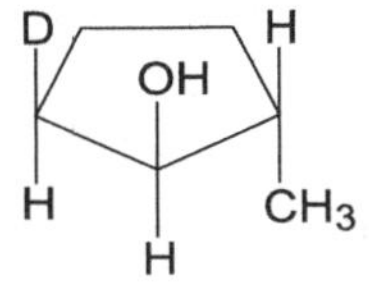
b. 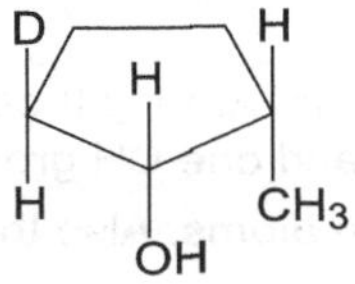

c. 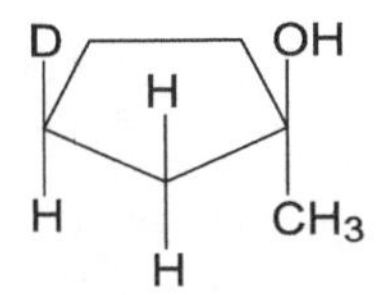
d. 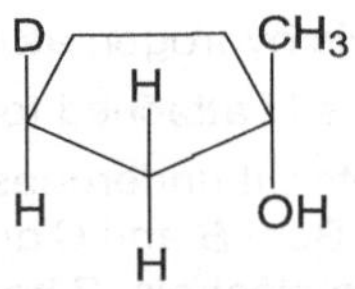

25 If a pure dextrorotatory enantiomer of the substrate of the following reaction is boiled with water, the correct statement(s) regarding $S_N1$product(s) is/are

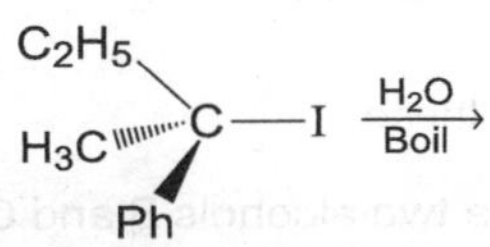

a. product would be pure laevorotatory
b. product would consists of both laevo and dextrorotatory enantiomers
c. product would be completely racemic mixture
d. product would be optically active with enantiomer of inverted configuration predominating

26 What is/are the expected solvolysis product(s) in the following reaction?

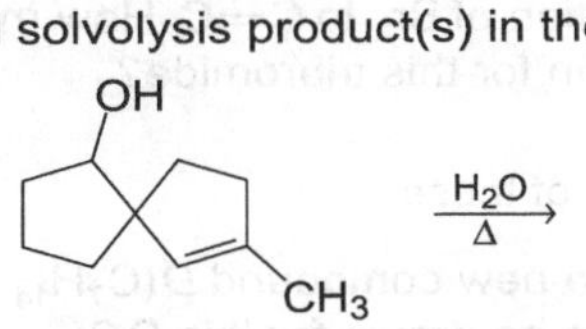

a. 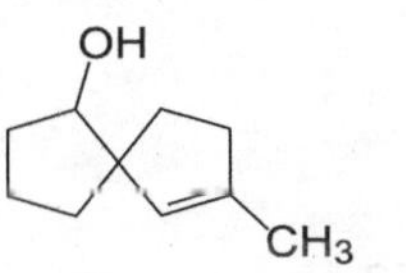
b. 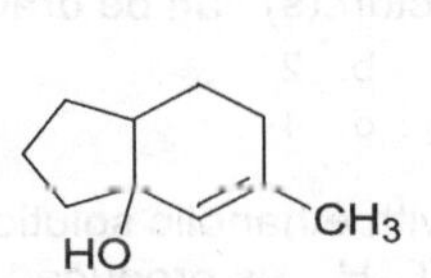

c. 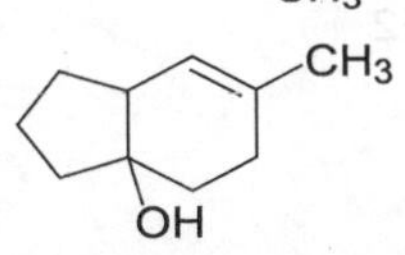
d. 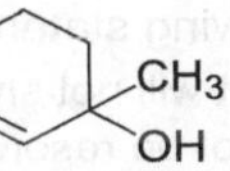

27 Which of the following can catalyse the following $S_N1$reaction?

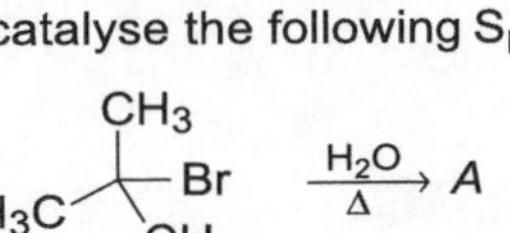

a. $ZnCl_2$
b. NaOH
c. NaCl
d. $AgNO_3$

Comprehension Type

Working Space

***Direction** (Q. Nos. 28-32) This section contains a paragraph, describing theory, experiments, data, etc. Five questions related to the paragraph have been given. Each question has only one correct answer among the four given options (a), (b), (c) and (d).*

Passage

A chemist isolated a compound *A* with molecular formula $C_7H_{13}Br$. *A* undergoes very fast S_N1 reaction. Spectroscopic evidence indicated that compound *A* has the following structural characteristics :

It contains five sp^3-hybridised carbon atoms. Among those five sp^3 carbon atoms, three are methyl groups, one CH_2 group and one CH group.

It also contains two sp^2-hybridised carbon atoms. Also there is only one hydrogen atom attached to sp^2 carbons.

The compound contains a total of six allylic hydrogen atoms.

The carbon atom that holds the Br has one H attached to it.

When compound *A* reacts with boiling water, it undergoes a S_N1 reaction and produces two principal products *B* and *C*. Both *B* and *C* are alcohols with their molecular formula $C_7H_{14}O$. Among the two alcohols, *B* has the —OH group attached to a sp^3 carbon atom that has no H -atoms bonded to it.

28 How many stereoisomers are possible for *A*?

a. 2
b. 3
c. 4
d. None of these

29 What can be said about the isomerism shown by the two alcohols *B* and *C*?

a. Both *B* and *C* show stereoisomerism and can be resolved into enantiomers
b. *B* and *C* are stereoisomers
c. Both *B* and *C* show stereoisomerism but only *C* can be resolved into enantiomers
d. Neither *B* nor *C* can be resolved into enantiomers

30 If the starting compound *A* is brominated in gas phase in the presence of a Lewis acid catalyst, a tribromide would result from addition of Br_2 to C=C. How many different structures of stereoisomers can be drawn for this tribromide?

a. 2
b. 4
c. 8
d. None of these

31 If the original compound *A* is treated with $LiAlH_4$ a new compound $D(C_7H_{14})$ would be produced. How many different structure(s) can be drawn for this *D*?

a. 1
b. 2
c. 3
d. 4

32 If the starting compound *A* is treated with ethanolic solution of KOH in boiling condition, a further new compound *E* (C_7H_{12}) is produced as major product by E2 elimination reaction. Which of the following statement will be consistent with E?

a. It will have no stereogenic centre i.e. it will not show stereoisomerism
b. It will show stereoisomerism but cannot be resolved into enantiomers
c. It will have more than one stereogenic centres and more than two, different structures of stereoisomers can be drawn for *E*
d. None of the above appropriately applies to *E*

Working Space

One Integer Value Correct Type

Direction *(Q. Nos. 33-35) This section contains 3 questions. When worked out will result in an integer from 0 to 9 (both inclusive).*

33 Consider the following solvolysis reaction.

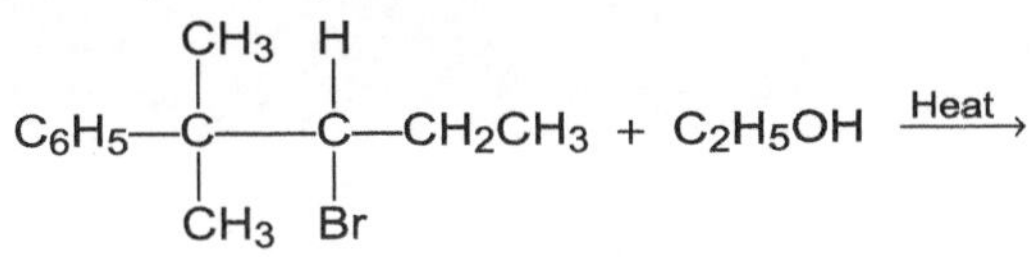

How many principal products would be formed by S_N1 reaction?

34 How many of the following undergo solvolysis reaction faster than benzyl chloride?

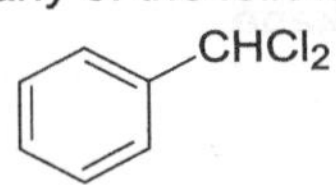
i.

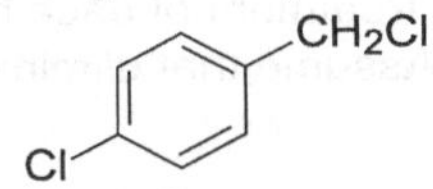
ii.

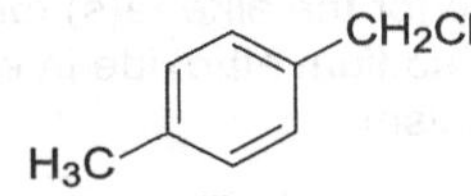
iii.

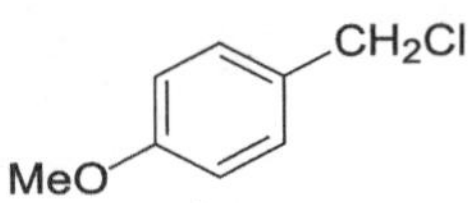
iv.

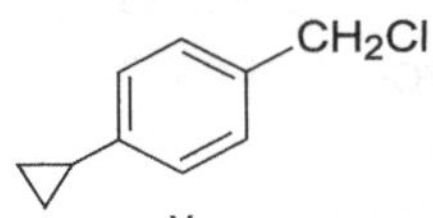
v.

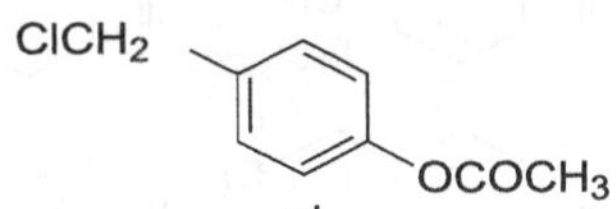
vi.

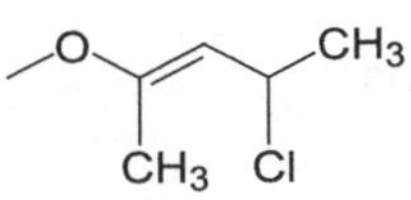
vii.

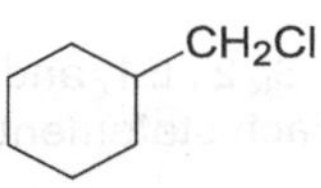
viii.

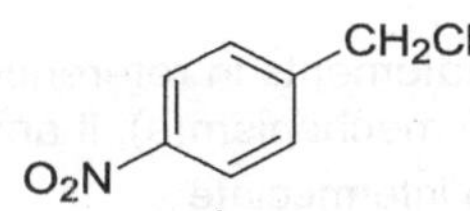
ix.

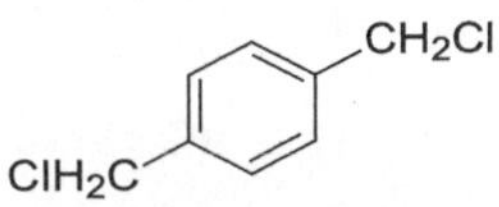
x.

35 Following compound when heated in ethanol, S_N1 reaction occur involving rearrangement of carbocation.

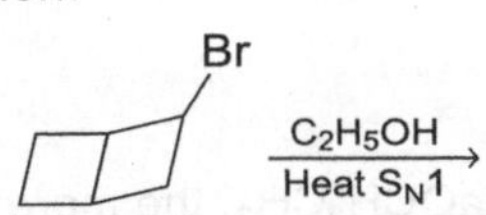

In the major product, how many carbon atoms are present in the single largest ring?

DPP-7 E2 Reaction Basics

Subjective Questions

Direction *(Q. Nos. 1-5) These questions are subjective in nature, need to be solved completely on notebook.*

Working Space

1 Draw structural formulae for the alkene(s) formed by treatment of each haloalkane or halocycloalkane with sodium ethoxide in ethanol. Assume that elimination occurs by an E2 mechanism.

i. 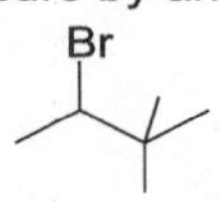
ii. 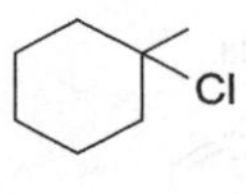
iii. 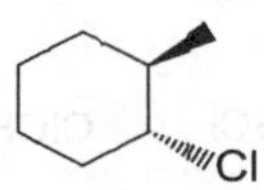

iv. 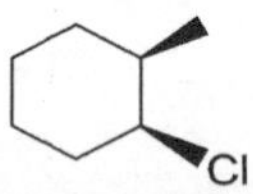
v. 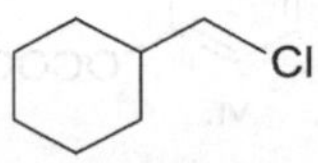
vi. 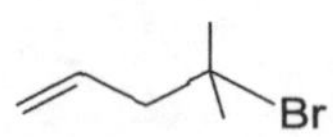

2 Consider the following statements in reference to S_N1, S_N2, E1, and E2 reactions of haloalkanes. To which mechanism(s), if any, does each statement apply?

i. Involves a carbocation intermediate
ii. Is first order in haloalkane and first order in nucleophile
iii. Involves inversion of configuration at the site of substitution
iv. Involves retention of configuration at the site of substitution
v. Substitution at a chiral center gives predominantly a racemic product
vi. Is first order in haloalkane and zero order in base
vii. Is first order in haloalkane and first order in base
viii. Is greatly accelerated in protic solvents of increasing polarity
ix. Rearrangements are common
x. Order of reactivity is 3° > 2° > 1°
xi. Order of reactivity is methyl > 1° > 2° > 3°

3 When 1-iodo-1-methylcyclohexane is treated with $NaOCH_2CH_3$, the more highly substituted alkene product predominates. When $KOC(CH_3)_3$ is used instead, the less highly substituted alkene product predominates. Offer an explanation.

4 Arrange the following haloalkanes in increasing order of ratio of E2 to S_N2 products observed, on reaction of each with sodium ethoxide in ethanol.

i. Bromoethane
ii. 1-bromo-2-methyl propane
iii. 2-chloro-2-methyl butane
iv. 1-bromo-3-methyl butane

5 The following ethers can, in principle, be synthesised by two different combinations of haloalkane or halocycloalkane and metal alkoxide. Show one combination that forms ether bond (1) and another that forms ether bond (2). Which combination gives the higher yield of ether and why?

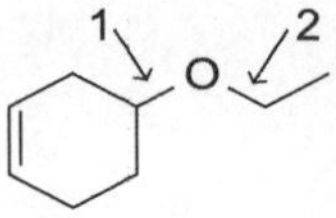
I.

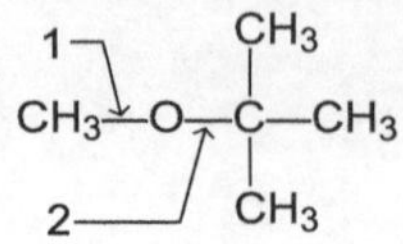

II.

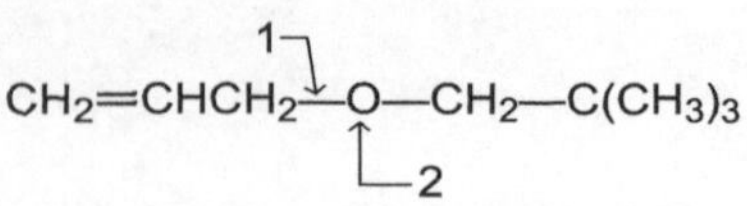

III.

Only One Option Correct Type

Working Space

Direction *(Q. Nos. 6-20) This section contains 15 multiple choice questions. Each question has four choices (a), (b), (c) and (d), out of which ONLY ONE is correct.*

6 Provide the structure of the major organic product which results in the following reaction.

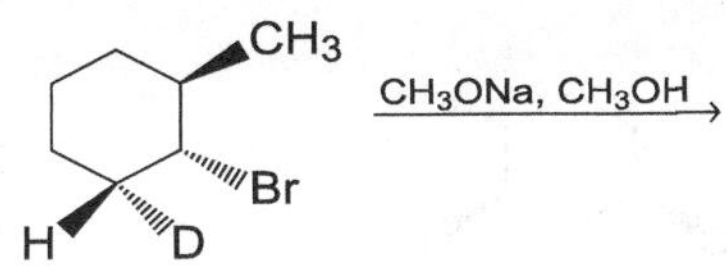

a. 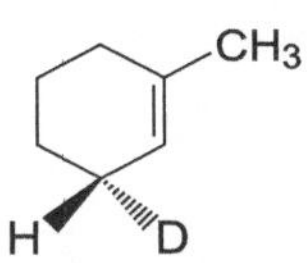

b. 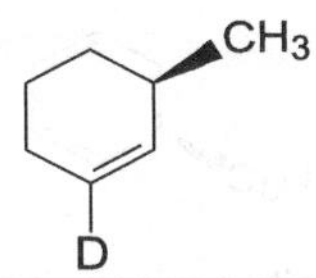

c. 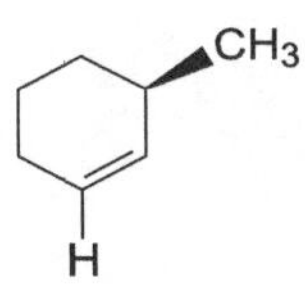

d. 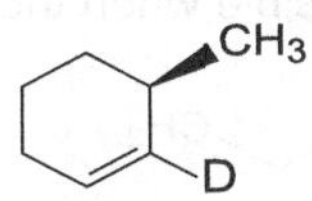

7 Provide the structure of the major organic product which results in the following reaction.

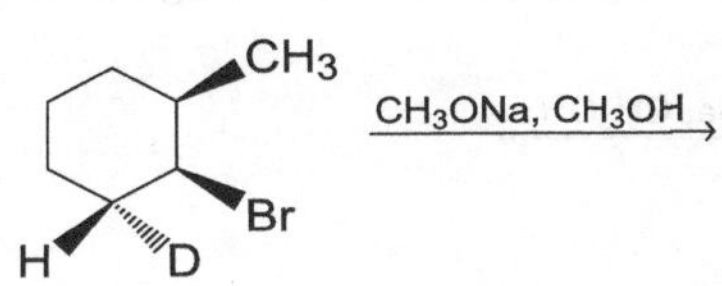

a. 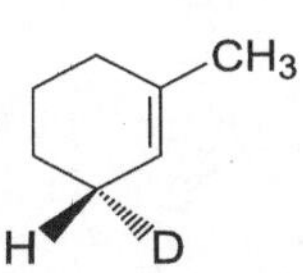

b. 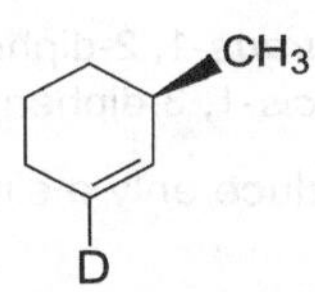

c. 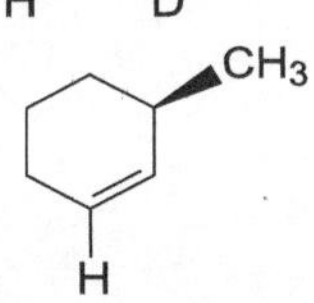

d. 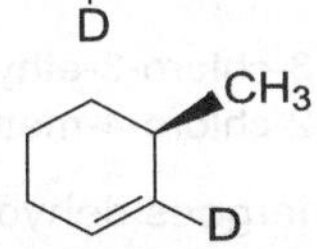

8 Provide the structure of the major organic product which results in the following reaction.

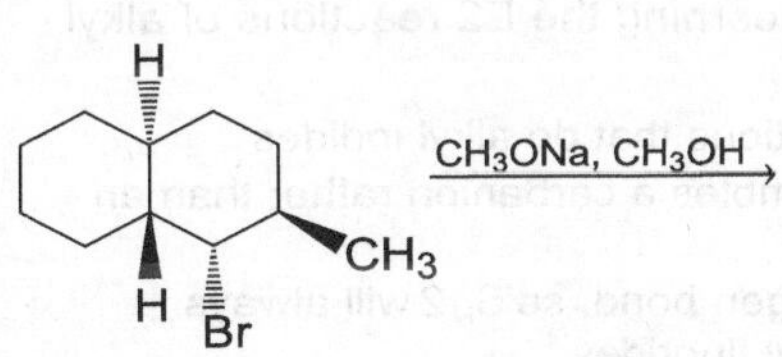

a. 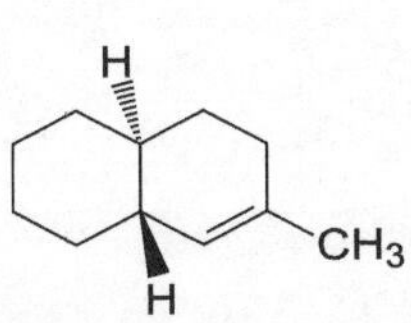

b. 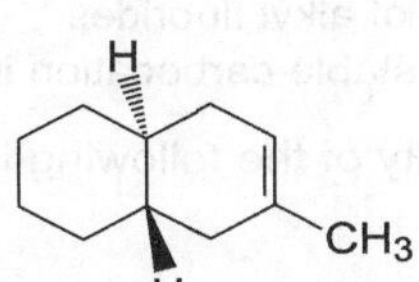

c. 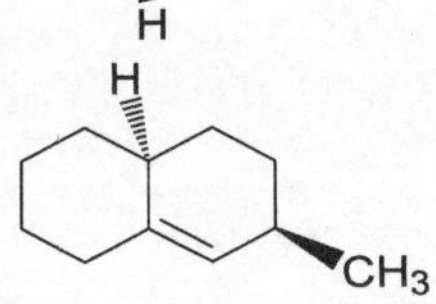

d. 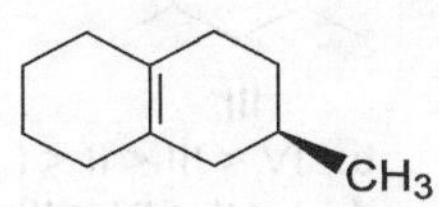

Working Space

9 Provide the structure of the major organic product which results in the following reaction.

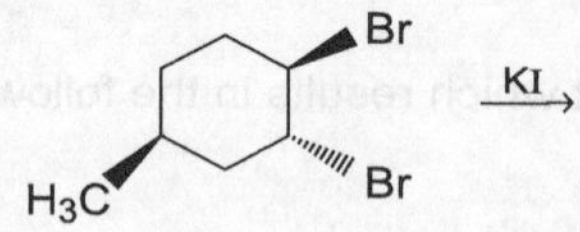

a. 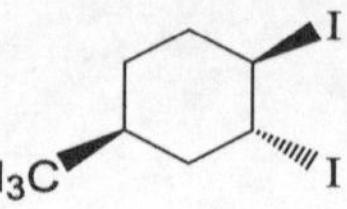

b. 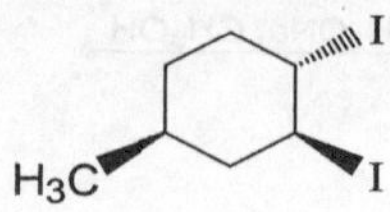

c. 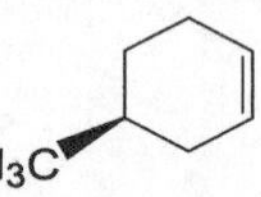

d. 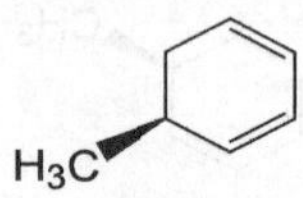

10 How many distinct alkene products are possible when the alkyl iodide given below undergoes E2 elimination?

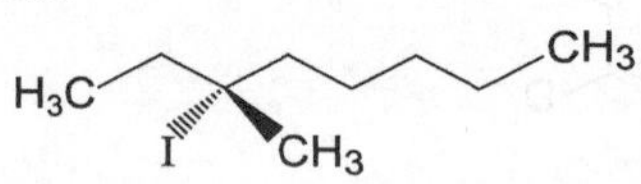

a. 1 b. 2 c. 3 d. 5

11 Provide the structure of the major organic product given in the following reaction.

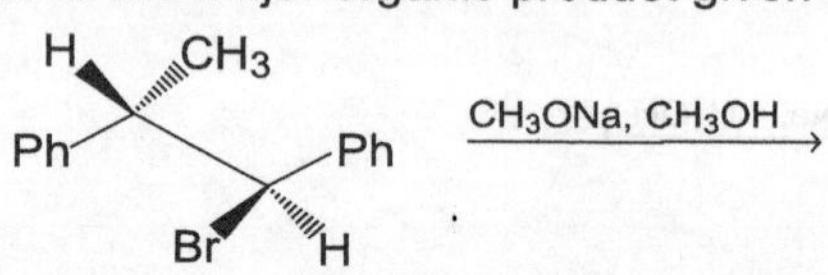

a. *cis*-1, 2-diphenyl propene
b. *trans*-1, 2-diphenyl propene
c. equal mixture of (a) and (b)
d. *cis*-1, 3-diphenyl propene

12 Which of the following alkyl halides can produce only a single alkene product when treated with sodium methoxide?

a. 2-chloro-2-methyl pentane
b. 3-chloro-3-ethyl pentane
c. 3-chloro-2-methyl pentane
d. 2-chloro-4-methyl pentane

13 Which of the alkyl chlorides listed below undergoes dehydrohalogenation in the presence of a strong base to give 2-pentene as the only alkene product?

a. 1-chloropentane
b. 2-chloropentane
c. 3-chloropentane
d. 1-chloro-2-methyl butane

14 Which of the following statements is true concerning the E2 reactions of alkyl fluorides?

a. Alkyl fluorides react more readily in E2 reactions that do alkyl iodides
b. The transition state of this E2 reaction resembles a carbanion rather than an alkene
c. The C—F bond is the weakest carbon-halogen bond, so S_N2 will always predominate over E2 in the reactions of alkyl fluorides
d. Alkyl fluorides react to form the most stable carbocation intermediate

15 What is the increasing order of reactivity of the following in an E2 reaction with ethanolic KOH solution?

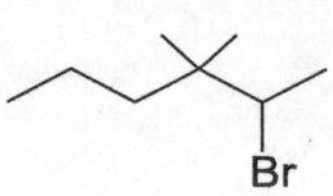

I.

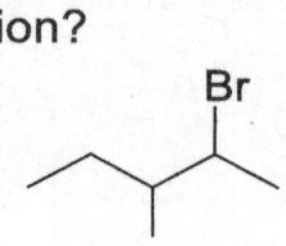

II.

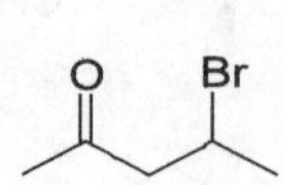

III.

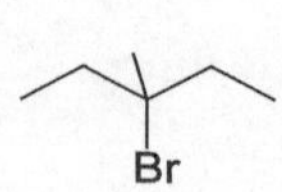

IV.

a. I < II < III < IV
b. IV < III < II < I
c. III < I < II < IV
d. I < II < IV < III

16 Consider the following reaction,

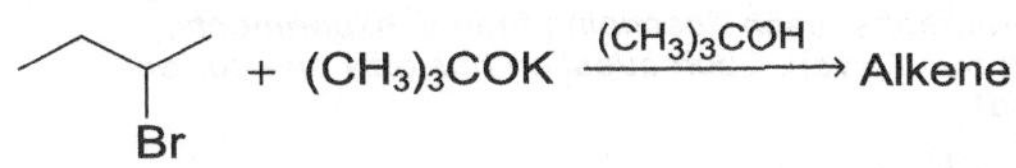

The correct statement regarding the above reaction is
a. major product is *trans*-2-butene
b. major product is 1-butene
c. activation energy diagram has only one transition state
d. reaction proceeds *via* carbocation intermediate

17 Consider the following reaction,

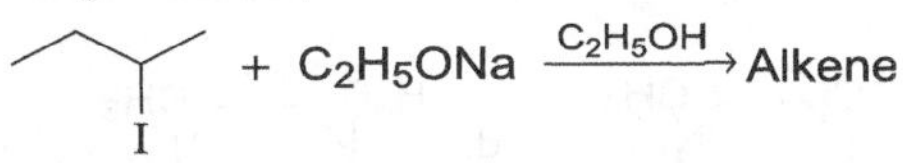

The correct statement concerning the above reaction is
a. rate of reaction is quadrupled on doubling concentration of both iodoalkane and base at constant temperature
b. replacing the normal I^{127} by radioactive I^{131} in the above reaction decreases rate
c. *trans*-2-butene is formed as the major product
d. a total of three alkene isomers are formed in principle

18 Consider the following reaction,

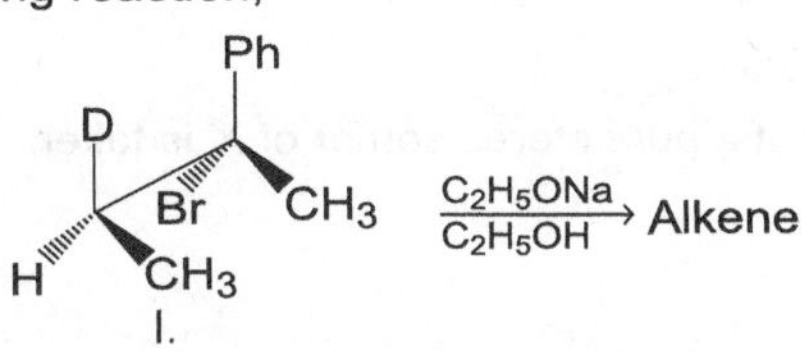

The correct statement concerning the above reaction is
a. DBr is the eliminated and *trans* isomer is formed as the major product
b. HBr is eliminated and *trans* isomer is formed as the major product
c. diastereomer of (I) gives *cis* alkene as the major product by elimination of HBr
d. diastereomer of (I) gives *cis* alkene as the major product by elimination of DBr

19 In the reaction the given below ,

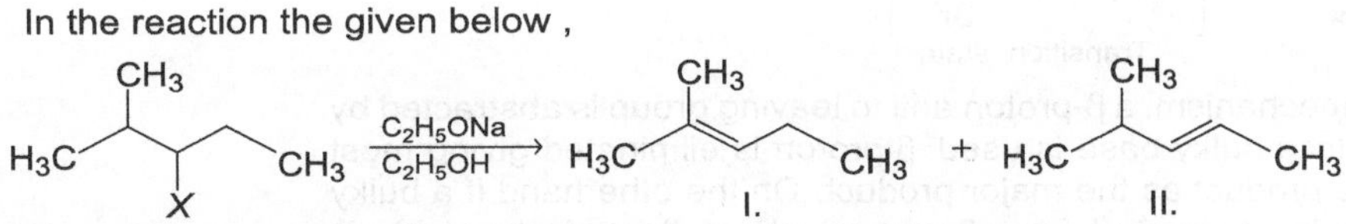

The correct statement(s) concerning the above reaction is/are
a. (I) is the major elimination product
b. (II) is formed at faster rate than (I)
c. (I) is formed at faster rate than (II)
d. increasing order of reactivity with different *X* is F < Cl < Br < I

20 Consider the following two reactions,

I. 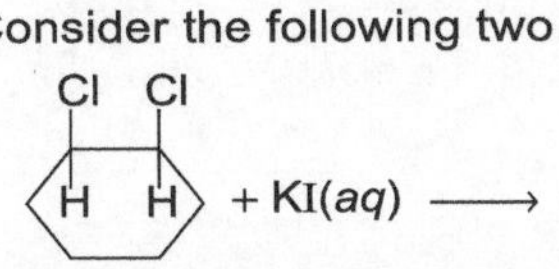

II. 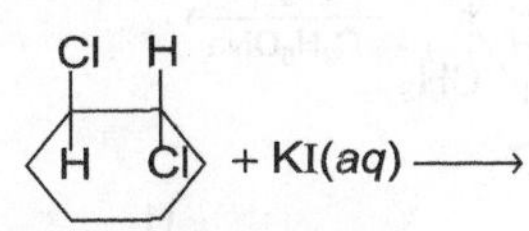

The correct statement concerning the above reaction is
a. cyclohexene is the major product in both reactions
b. cyclohexene is the major product in reaction (II) but not in (I)
c. in (I) S_N2 reaction occurs predominantly
d. S_N2 reaction is the major reaction in both cases

Comprehension Type

Working Space

***Direction** (Q. Nos. 21-26) This section contains 2 paragraphs, each describing theory, experiments, data, etc. Six questions related to the paragraphs have been given. Each question has only one correct answer among the four given options (a), (b), (c) and (d).*

Passage I

An alkyl halide *X* shows diastereomerism but cannot be resolved into enantiomers. *X* when treated with KOH solution gives alkene as well as alcohol in comparable amounts by E2 and S_N2 reactions respectively. Also, E2 reaction above gives alkene *Y* as single structural isomer which exists as stereoisomers. The S_N2 reaction above produced alcohol *Z* which show stereoisomerism but cannot be resolved into enantiomers.

21 The structure of *X* is

a. 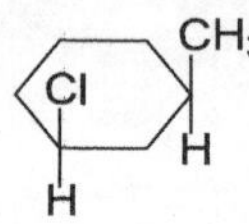
b. 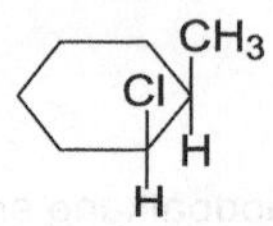
c. 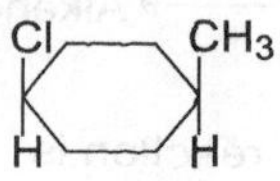
d. 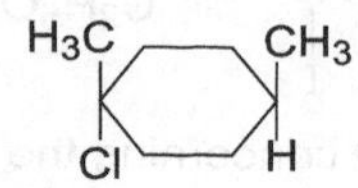

22 The correct statement concerning *Y* is

a. *Y* exists as a pair of geometrical isomers
b. *Y* consists of a pair of enantiomers in equal amount
c. *Y* possesses a plane of symmetry
d. *Y* consists of a pair of enantiomers but in unequal amount

23 The correct statement concerning alcohol (*Z*) is

a. It is formed as a pair of geometrical isomers if a pure stereoisomer of *X* is taken
b. It possesses a plane of symmetry
c. It possesses a centre of symmetry
d. It is formed as racemic mixture

Passage II

In an E2 reaction, following one step mechanism is involved.

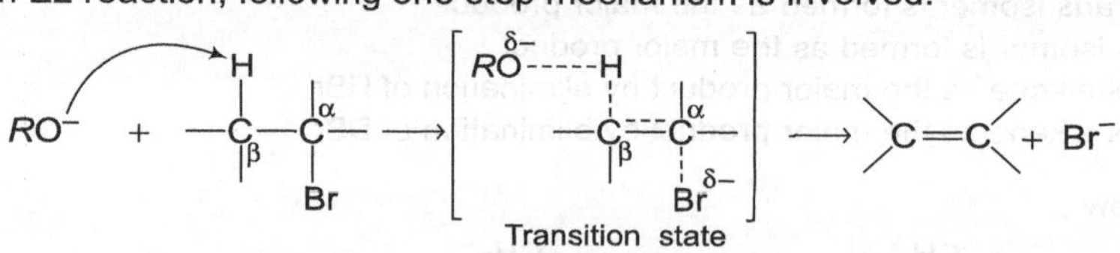

As shown in the above mechanism, a β-proton anti to leaving group is abstracted by the alkoxide base. If a less bulky base is used, β-proton is eliminated giving most substituted, most stable product as the major product. On the otherhand if a bulky base is used, β-proton is abstracted from β-carbon where there is least steric hindrance and this gives least substituted, although least stable, alkene as the major product.

24 What is the major product in the following elimination reaction?

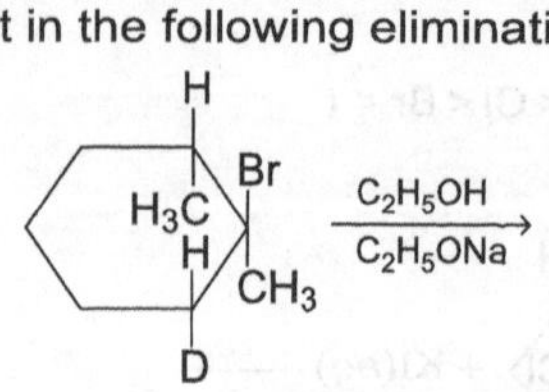

a. 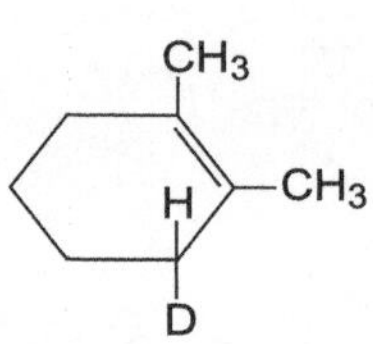
b. 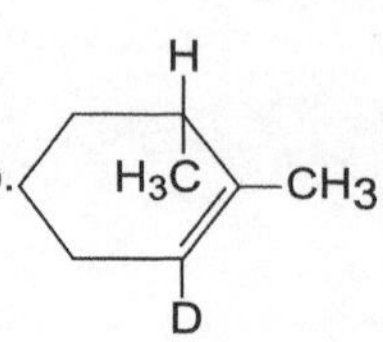
c. 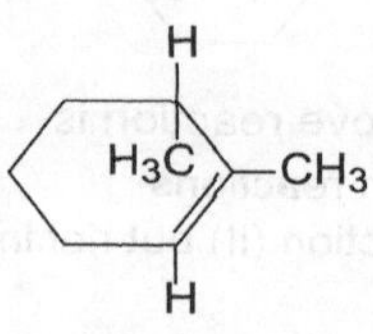
d. 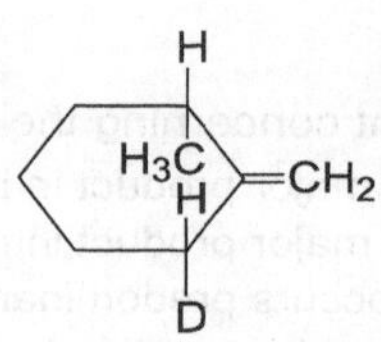

25 In the following reaction, how many elimination products would be formed in principle by E2 mechanism?

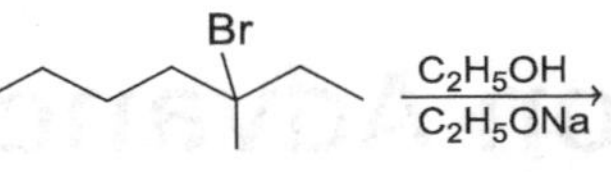

a. 2
b. 3
c. 4
d. 5

26 What is the major elimination product in the following reaction?

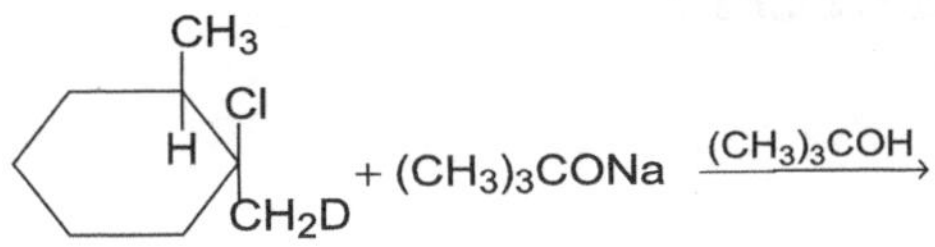

a. 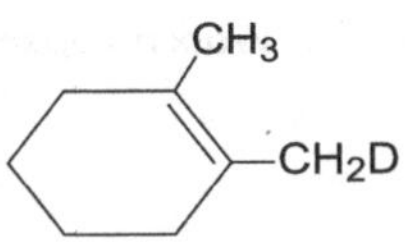

b. 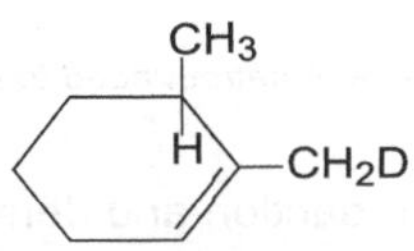

c. 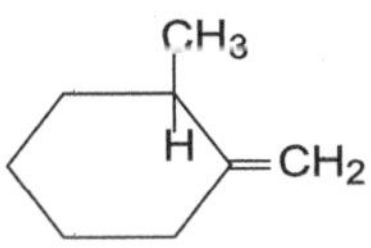

d. 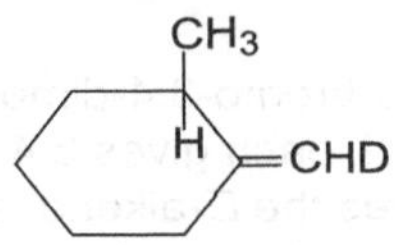

One Integer Value Correct Type

Direction (Q. Nos. 27-30) This section contains 4 questions. When worked out will result in an integer from 0 to 9 (both inclusive).

27 If *trans*-3-methylchlorocyclopentane is treated with KOH solution, both S_N2 and E2 occur simultaneously. How many total substitution and elimination products are formed in principle?

28 How many different alkenes would be formed in the following reaction?

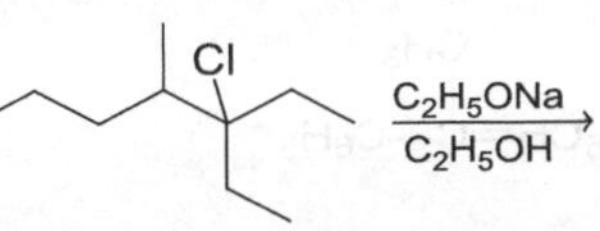

29 How many different isomers of an alkyl bromide, upon treatment with C_2H_5ONa in C_2H_5OH can give 3-methyl-2-hexene as one of the elimination product by E2 mechanism?

30 An alkene *X* has molecular formula $C_{11}H_{24}$. *X* on free radical chlorination gives three monochloro alkanes derivatives *P*, *Q* and *R* as positional isomers. Only *Q* is resolvable into enantiomers. *P* does not undergo E2 elimination reaction with ethanolic KOH solution while *Q* and *R* both give the same alkene with ethanolic KOH which is capable of showing geometrical isomerism. How many carbon atoms are present in the parent chain of *X*?

Working Space

DPP-8 E2 Reaction Advanced, E1 cb and Hofmann's Elimination

Subjective Questions

Working Space

Direction *(Q. Nos. 1-3) These questions are subjective in nature, need to be solved completely on notebook.*

1 Propose mechanism of E1cb elimination reaction and derive rate law. What factors favour E1cb mechanism?

2 Following are diastereomers *A* and *B* of 3-bromo-3,4-dimethylhexane. On treatment with sodium ethoxide in ethanol, each gives 3,4-dimethyl -3-hexene as the major product. One diastereomer gives the *E*-alkene, and the other gives the *Z*-alkene. Which diastereomer gives which alkene? Account for the stereoselectivity and stereospecificity of each β-elimination.

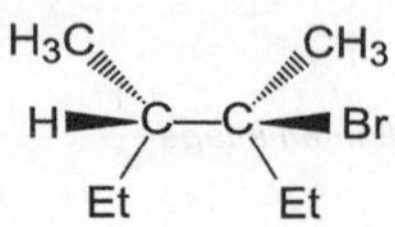

A

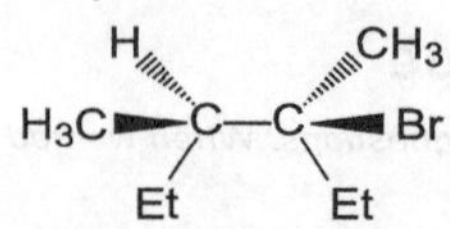

B

3 Treatment of the following stereoisomer of 1-bromo-1, 2-diphenylpropane with sodium ethoxide in ethanol gives a single stereoisomer of 1,2-diphenylpropene. Predict whether the product has the *cis*-configuration or the *trans* configuration.

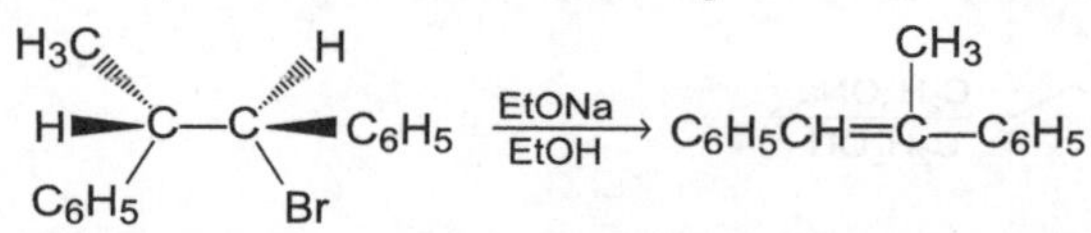

Only One Option Correct Type

Direction *(Q. Nos. 4-8) This section contains 5 multiple choice questions. Each question has four choices (a), (b), (c) and (d), out of which ONLY ONE option is correct.*

4 What is the major alkene product in the following reaction?

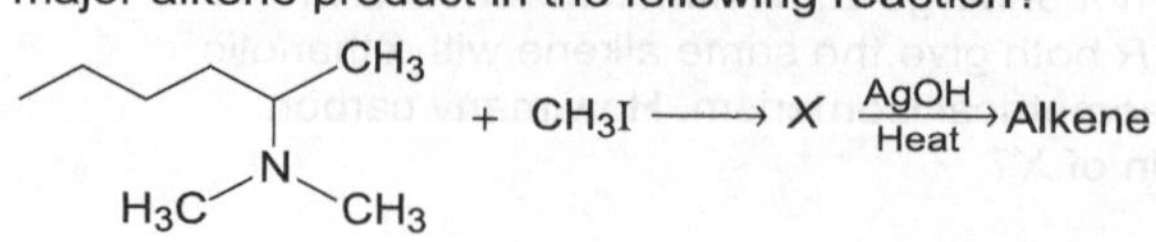

a. 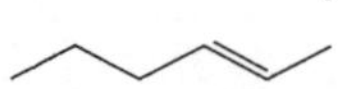

b. 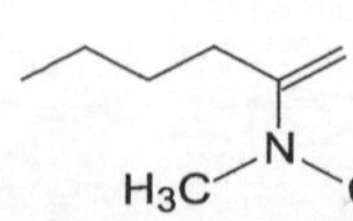

c. 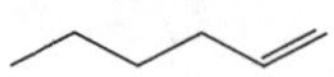

d. 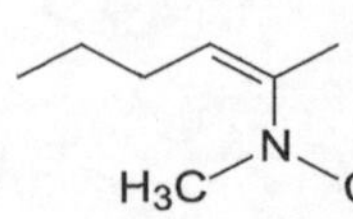

5 Which of the following gives 1-butene as the major product most easily on heating with AgOH?

a. 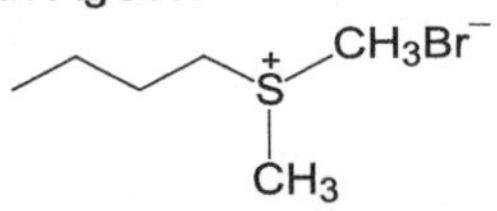

b. 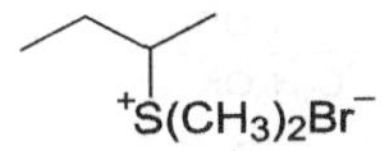

c. 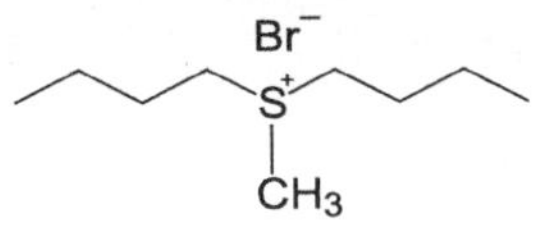

d. 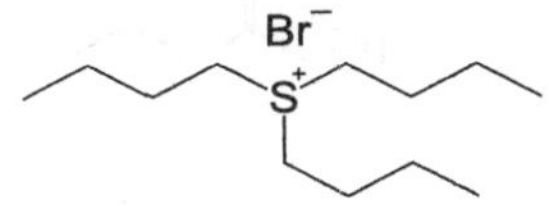

6 1-butene would be formed most easily in the following reaction when *X* is

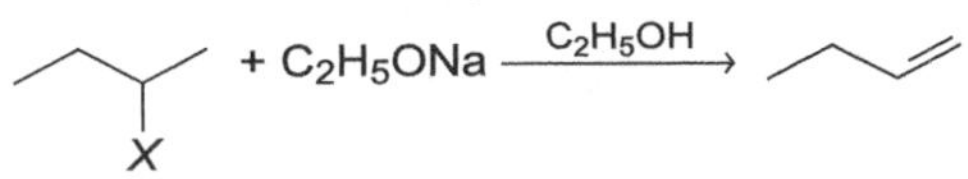

a. I^- b. Br^-

c. Cl^- d. F^-

7 Which of the following compound is most likely to follow E1cb mechanism when treated with C_2H_5ONa in ethanol?

a. 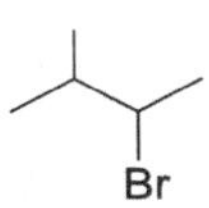

b. 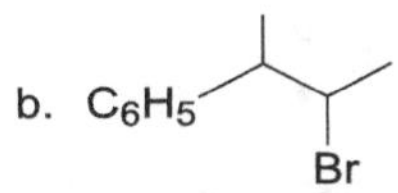

c. 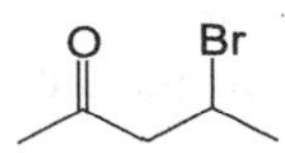

d. 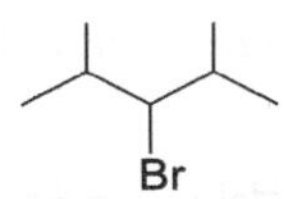

8 Consider the following reaction and the product formed.

$$A{-}CH_2{-}\underset{\displaystyle Br}{\underset{|}{CH}}{-}CH_3 \xrightarrow[C_2H_5ONa]{C_2H_5OD} A{-}\overset{\displaystyle D}{\overset{|}{C}}{=}CH{-}CH_3$$

One of the various product

The most likely mechanism of the above reaction is

a. E2 b. E_2C c. E1 d. E1cb

One or More than One Options Correct Type

Direction *(Q. Nos. 9-11) This section contains 3 multiple choice questions. Each question has four choices (a), (b), (c) and (d), out of which ONE or MORE THAN ONE are correct.*

9 Which of the following gives methanol as the major product when heated with AgOH?

a. $(CH_3)_4NI$ b. $CH_3CH_2{-}N(CH_3)_3I$

c. $(C_6H_5)_2SCH_3I$ d. $(CH_3)_3SI$

10 Which of the following gives *cis*-2,3-diphenyl, 1-2-butene as the major E2 product, when treated with ethanolic KOH solution?

a. 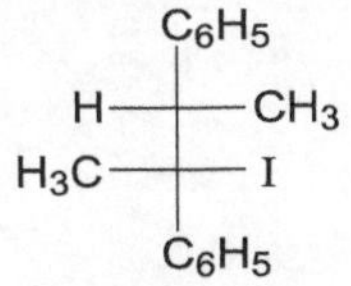

b. 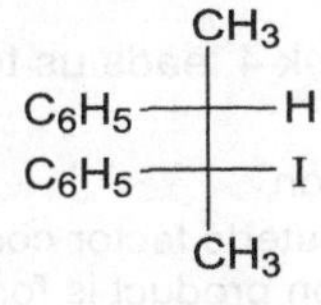

c. 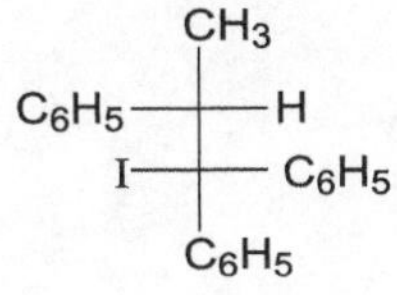

d. 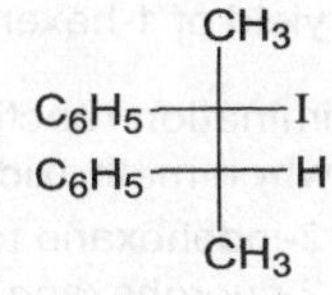

Working Space

Working Space

11 The following elimination reaction, if proceeds by E1cb mechanism, equilibrium mixture would consist of

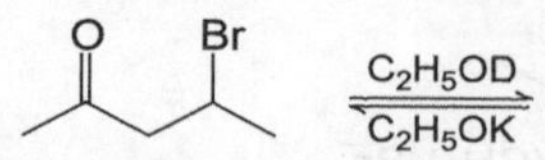

a. 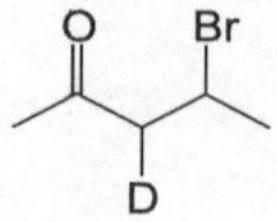

b. 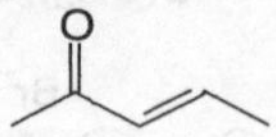

c. 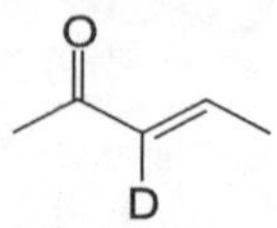

d. 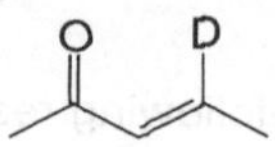

Comprehension Type

Direction (Q. Nos. 12-20) *This section contains 2 paragraphs, each describing theory, experiments, data, etc. Nine questions related to the paragraphs have been given. Each question has only one correct answer among the four given options (a), (b), (c) and (d).*

Passage I

The following E2 is carried out with different halogen substituent:

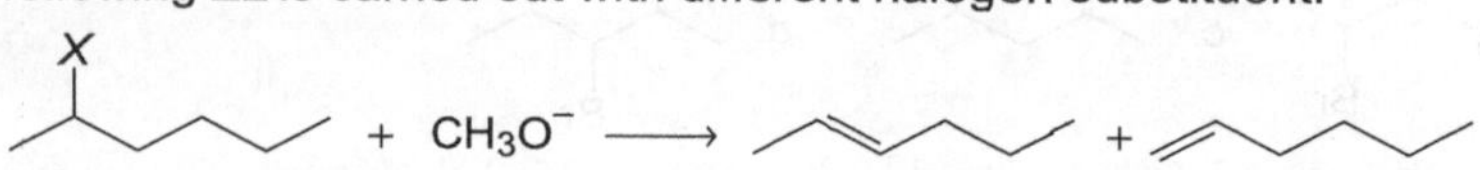

2-hexene 1-hexene

Four different reaction flasks, containing equimolar concentration of sodium methoxide in methanol solvent, were charged with equal moles of one of the four halide in each flask and allowed to react under identical experimental conditions as shown above. After a week, the percentage yield of two alkenes in each flask were measured and listed below :

Flask No.	Leaving group	Conjugate, acid pK_a	% yield of 2-hexene	% yield of 1-hexene
1.	I^-	−10	81	19
2.	Br^-	−9	72	28
3.	Cl^-	−7	67	33
4.	F^-	3.2	30	70

From the percentage yield of two products observed with different leaving group, it was concluded that both ease of leaving group and steric hindrance to the approach of base CH_3O^- to β-H affect the orientation of elimination reaction.

12 Based to the above observation, which flask will have the largest amount of 1-hexene?

a. 1 b. 2 c. 3 d. 4

13 Highest percentage yield of 1-hexene in flask 4 leads us to conclude that

a. 1-hexene is more stable than 2-hexene
b. Leaving group is not important in E2 reaction
c. Fluoride, being the poorest leaving group, steric factor control the orientation of elimination, no matter how much elimination product is formed.
d. None of the above correctly explain the % yield of 1-hexene in yield

14 During formation of products in the given elimination reaction, the highest energy transition state would have been produced when methoxide ion reacts with

a. 2-iodohexane to form 2-hexene b. 2-iodohexane to form 1-hexene
c. 2-fluorohexane to form 1-hexene d. 2-fluorohexane to form 2-hexene

Working Space

15 If we consider *X* to be iodine in the given elimination reaction, the number of different elimination products present in the flask-1 would be?

a. 2 b. 3
c. 4 d. 5

16 Transition state with maximum double bond character will be formed when methoxide reacts with

a. 2-iodohexane to form 2-hexene
b. 2-iodohexane to form 1-hexene
c. 2-fluorohexane to form 1-hexene
d. 2-fluorohexane to form 1-hexene

17 Transition state with least double bond character will be formed when methoxide reacts with

a. 2-iodohexane to form 2-hexene b. 2-iodohexane to form 1-hexene
c. 2-fluorohexane to form 2-hexene d. 2-fluorohexane to form 1-hexene

Passage II

When compound 1 is heated with C_2H_5ONa compound 2 and 3 are formed:

Reaction I

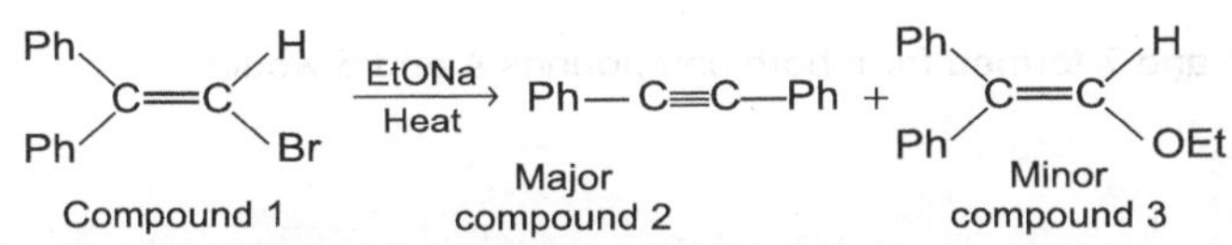

Two mechanisms were proposed for reaction I.

Mechanism A HBr is eliminated from compound 1 to form a symmetrical vinyl carbene intermediate *A*, which then rearranges to compound 2.

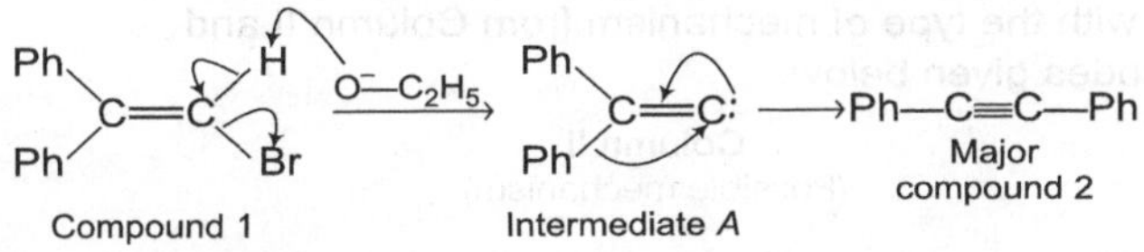

Mechanism B Ethoxide ion first abstract a proton to form a carbanion intermediate *B* which then rearranges with loss of bromide ion to form compound 2.

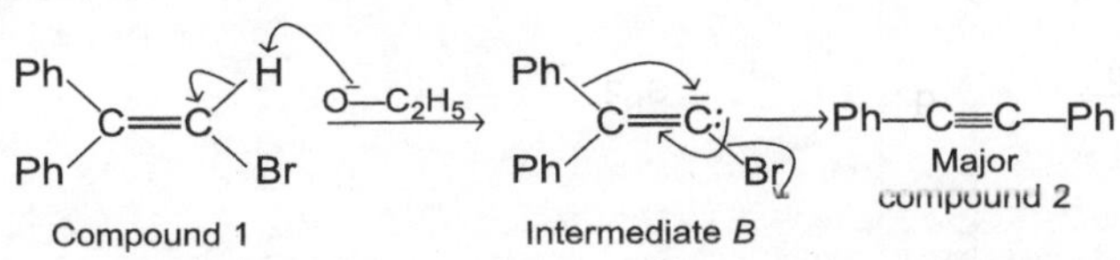

To distinguish between the two machanisms, an isotopic labeling experiment was designed. Two compounds (Compound 4 and 5) were labelled with C-14 and each was treated separately with sodium ethoxide under identical experimental condition where following results were obtained.

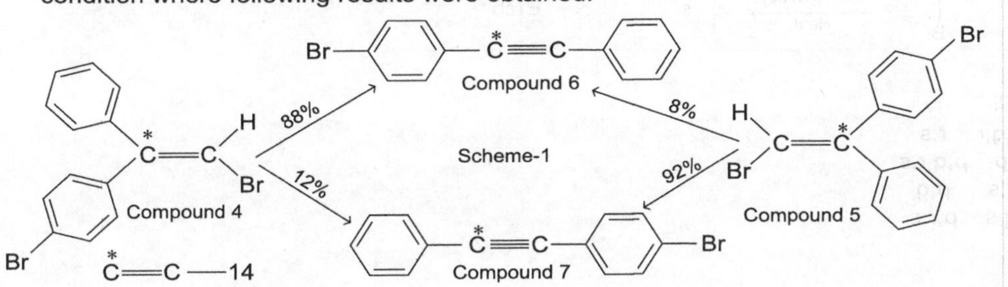

Working Space

18 Based on the results of scheme 1, the chemist most likely ruled out mechanism *A* because they assumed that intermediate *A* should have formed

a. compound 6 only
b. compound 7 only
c. equal amounts of 6 and 7
d. neither compound 6 nor 7

19 Compound 2 and 6 can be distinguished from each other by all of the following techniques except :

a. gas chromatography
b. mass spectrometry
c. dipole moment measurement
d. polarimetry

20 In reaction scheme 1, had the α-C to bromine be labelled with C-14

a. no change in observation regarding % of various products would have been observed
b. percentage of the compounds 6 and 7 formed from compound 4 would have been interchanged
c. percentage of the compounds 6 and 7 formed from compound 5 would have been interchanged
d. percentage of the compounds 6 and 7 formed from both compounds 4 and 5 would have been interchanged

Matching List Type

Direction *(Q. Nos. 21 and 22) Choices for the correct combination of elements from Column I and Column II are given as options (a), (b), (c) and (d), out of which one is correct.*

21 Match the reaction from Column I with the type of mechanism from Column II and mark the correct option from the codes given below.

	Column I (Reaction)		Column II (Possible mechanism)
i.	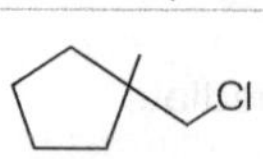 $\xrightarrow{NaOH(aq)}$	p.	S_N1
ii.	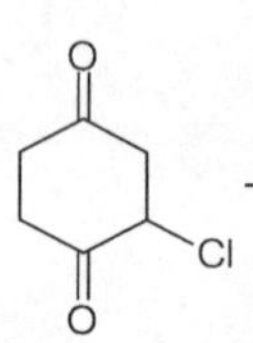 $\xrightarrow{NaOH(aq)}$	q.	S_N2
iii.	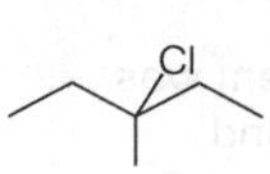 $\xrightarrow{H_2O/Heat}$	r.	E2
iv.	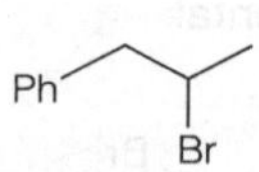 $\xrightarrow[Heat]{C_2H_5ONa}$	s.	E1cb

Codes

	i	ii	iii	iv
a.	p	p,q	p,q,r	r,s
b.	q	q,r,s	p	p,q,r,s
c.	s	p	r,s	p,q
d.	p,q	r,s	q,s	p,r,s

Working Space

22 Match the alkyl halides in the Column I with the properties of their products produced in E2 elimination reaction in Column II and mark the correct option from the codes given below.

	Column I (Alkyl halides)		Column II (Properties)
i.	$CH_3CH_2CHClCH_3$	p.	Fail in E2 reactions
ii.	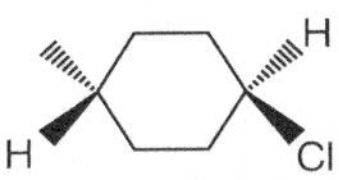	q.	Produces two alkenes in E2 reaction which are not stereomers
iii.	$(CH_3)_2CClCH_2CH_3$	r.	Produces pair of enantiomers in E2 reaction
iv.	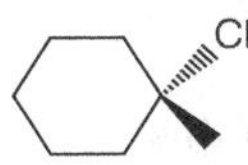	s.	Produces three alkenes in E2 reaction

Codes

	i	ii	iii	iv
a.	s	p,s	p	q,s
b.	p,q	r	s	p
c.	s	r	q	p
d.	q	r	p	r,s

One Integer Value Correct Type

Direction *(Q. Nos. 23 and 24) This section contains 2 questions. When worked out will result in an integer from 0 to 9 (both inclusive).*

23 In the reaction given below how many elimination products are formed in principle if reaction proceeds by E1 cb mechanism?

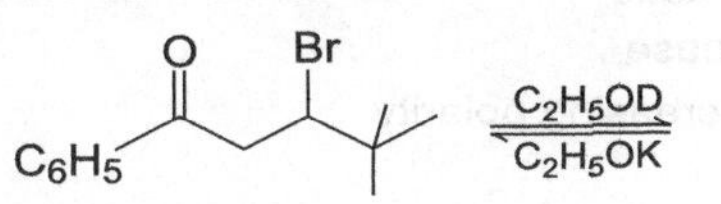

24 Consider the following reaction,

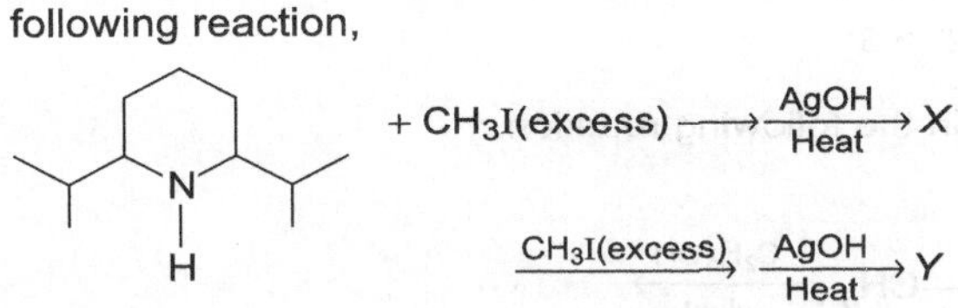

How many different stereoisomers of the major elimination product Y are possible?

DPP-9 E1 Reaction and Related Reactions

Subjective Questions

Working Space

Direction (*Q. Nos. 1-5*) ***These questions are subjective in nature, need to be solved completely on notebook.***

1 Propose a mechanism for the formation of these products in the solvolysis of this alkyl bromide.

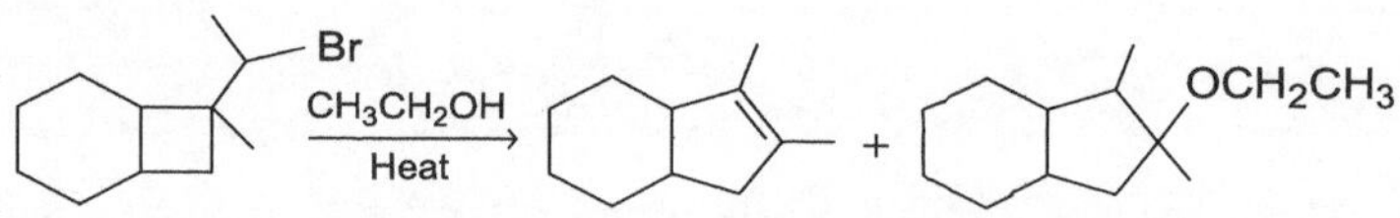

2 Consider the following statements in reference to S_N1, S_N2, E1, and E2 reactions of haloalkanes. To which mechanism(s), if any, does each statement apply?

i. Involves a carbocation intermediate
ii. Is first order in haloalkane and first order in nucleophile
iii. Involves inversion of configuration at the site of substitution
iv. Involves retention of configuration at the site of substitution
v. Substitution at a chiral center gives predominantly a racemic product
vi. Is first order in haloalkane and zero order in base
vii. Is first order in haloalkane and first order in base
viii. Is greatly accelerated in protic solvents of increasing polarity
ix. Rearrangements are common
x. Order of reactivity is 3° > 2° > 1°
xi. Order of reactivity is methyl > 1 ° > 2° > 3°

3 Predict the major elimination product in the following reaction.

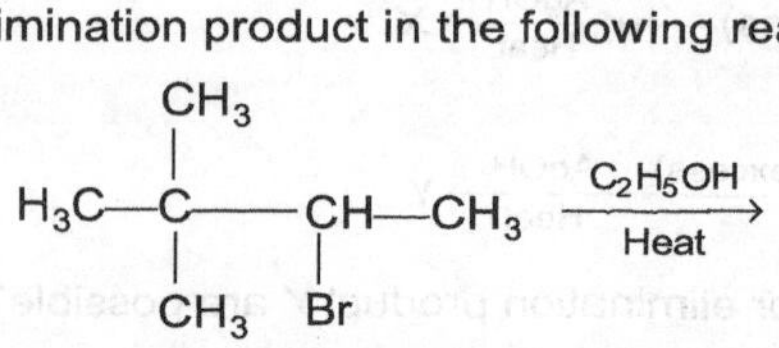

Propose mechanism and derive rate law of the above reaction. Also draw appropriate activation energy diagram.

4 How the following factors affect the reactivity in an E1 reaction?

i. Structure of substrate
ii. Solvents
iii. Lewis acid
iv. Strength of base

5 Consider the following two substrate and compare their reactivity in E1 reaction when heated in ethanol.

i. 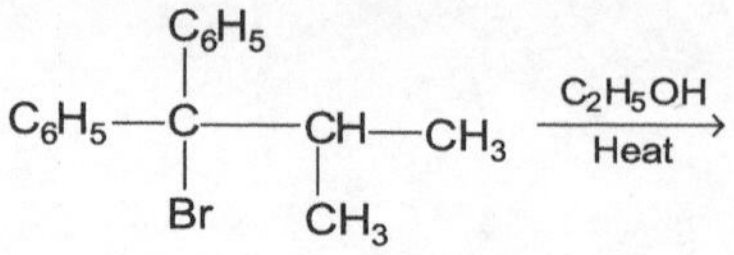

ii. 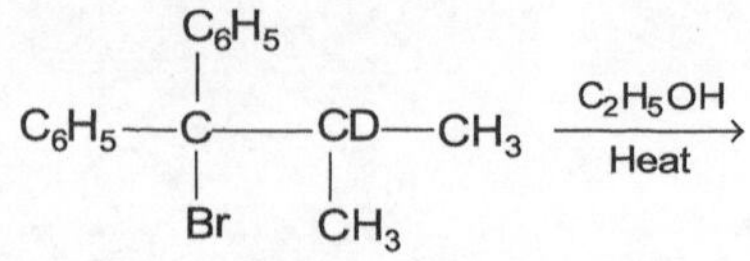

Working Space

Only One Option Correct Type

Direction *(Q. Nos. 6-13) This section contains 8 multiple choice questions. Each question has four choices (a), (b), (c) and (d), out of which ONLY ONE option is correct.*

6 The incorrect statement concerning E1 reaction is

a. polar protic solvent increases reactivity
b. lewis acid catalyses reaction
c. reaction proceeds *via* carbocation intermediates
d. reaction proceeds *via* a single transition state

7 The correct statement regarding a unimolecular elimination reaction is

a. Rearrangement of carbon skeleton may occur during elimination reaction
b. Reaction is stereospecific
c. If ^{14}C replaces α-carbon of substrate, reactivity remains unaltered
d. In the given condition of E1 reaction, no substitution reaction takes place

8 Consider the following reaction,

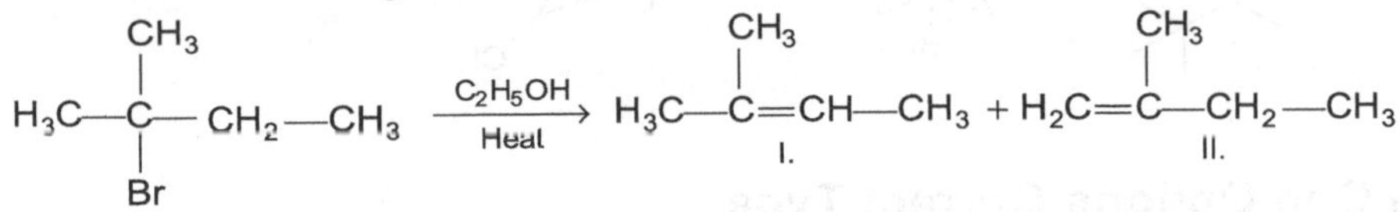

The correct statement concerning I and II is

a. I is the major product as it is formed at faster rate than II
b. II is the major product as it is formed at faster rate than I
c. I is major product as it is more stable than II
d. I and II are formed in comparable amounts

9 Consider the following elimination reaction,

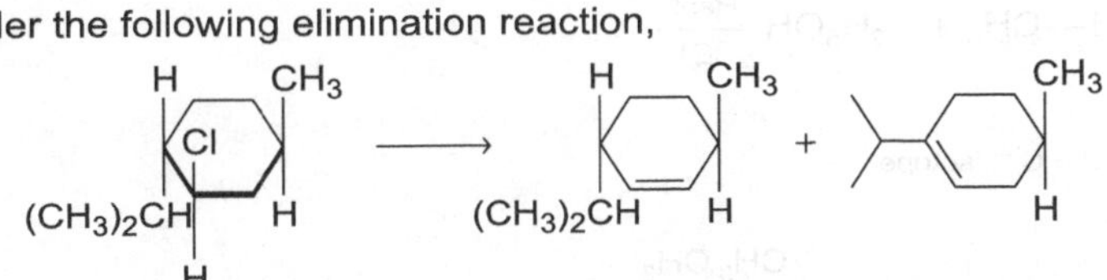

Menthyl chloride 2-menthene 3-menthene

When reaction is carried out with C_2H_5ONa in ethanol, 2-menthene is the major product while 3-menthene is the major product if reaction is substrate is heated in ethanol only. It is due to

a. with strong base, less substituted alkene is always the major product.
b. with strong base, reaction proceeds by E2 mechanism while with weak base, E1 reaction takes place.
c. methyl group gives steric hindrance to strong base in E2 reaction.
d. isopropyl group gives steric hindrance to strong base in E2 reaction.

10 The reaction of tertiary butyl chloride in water to yield tertiary butyl alcohol is not appreciably affected by dissolved NaF; in DMSO, however, NaF bring about rapid formation of isobutene. It is because of

a. in DMSO, E1 reaction predominates over S_N1
b. E2 reaction is preferred over S_N1 in DMSO
c. E2 reaction is preferred over S_N2 in DMSO
d. fluoride ion is not solvated by DMSO

11 When $(CH_3)_3C—Cl$ is heated in ethanol, both alkene and ether are formed simultaneously by E1 and S_N1 reaction. Which of the following will have adverse effect on both of the above reaction?

a. Presence of Lewis acid
b. Presence of some water in ethanol
c. Presence of some NaCl
d. Presence of some $CHCl_3$

Working Space

12 Consider the following reaction,

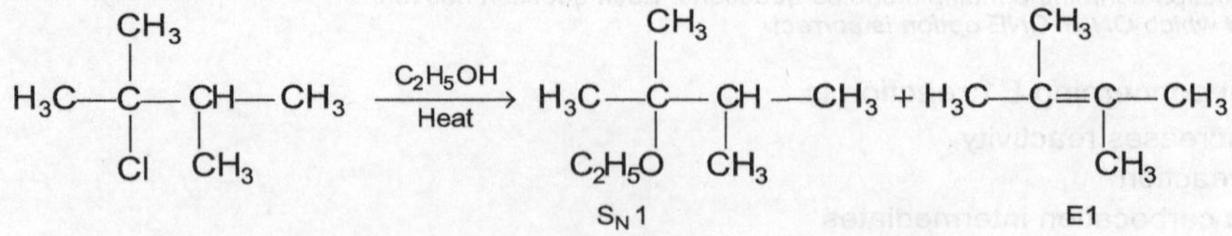

Increasing temperature increases the proportion of elimination (E1) product because

a. E1 product is enthalpy favoured
b. E1 product is entropy favoured
c. S_N1 product is entropy favoured
d. S_N1 product is enthalpy favoured

13 Which is the least reactive halide in both E1 and S_N1 reaction?

a. 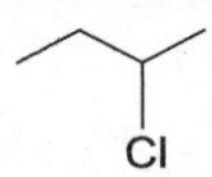
b. 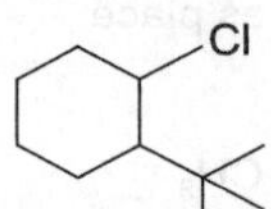
c. 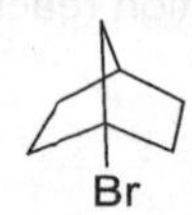
d. 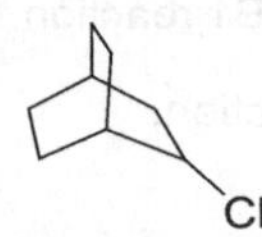

One or More than One Options Correct Type

***Direction** (Q. Nos. 14-16) This section contains 3 multiple choice questions. Each question has four choices (a), (b), (c) and (d), out of which ONE or MORE THAN ONE are correct.*

14 Consider the following reaction,

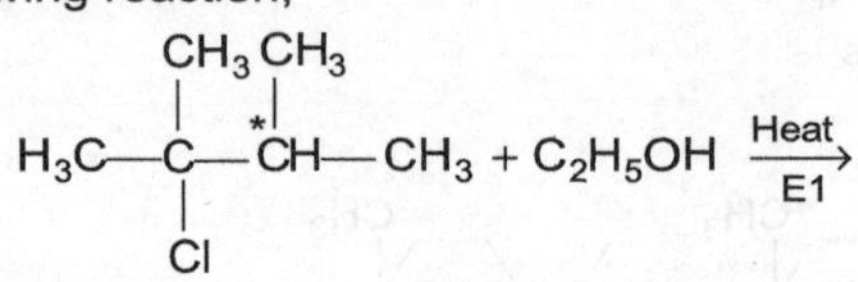

$^*C = C^{14}$ isotope

The possible alkene(s) is/are

a. 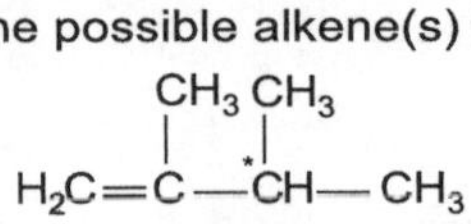
b. 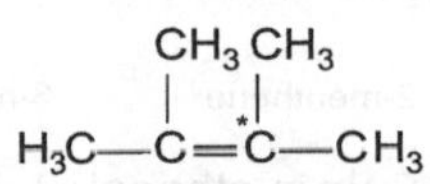
c. 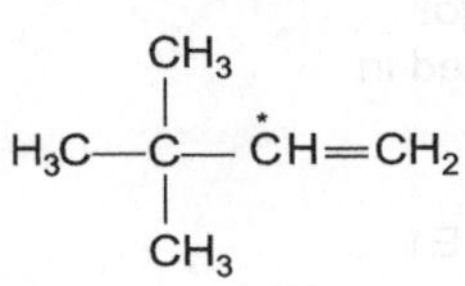
d. 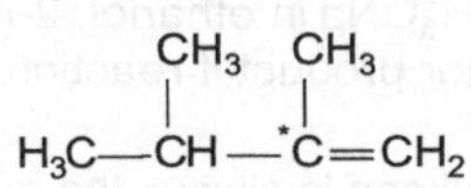

15 Consider the following elimination reaction,

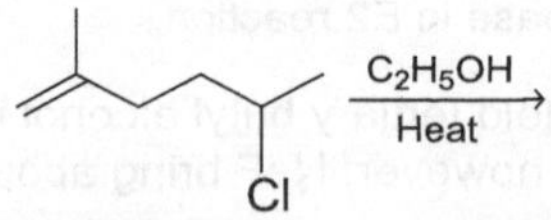

The possible elimination product(s) is/are

a. 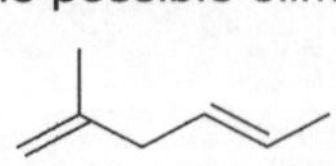
b. 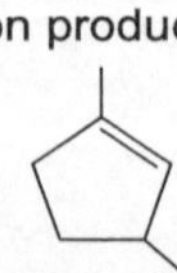
c. 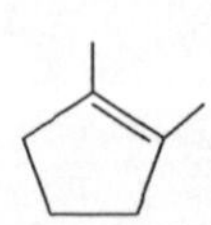
d. 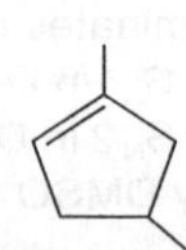

16 The correct statement regarding an E1 reaction is/are ?

a. Reaction rate is same for deuterated and un-deuterated halides
b. R—Cl and R—OTs reacts at same rate (—OTs = tosylate)
c. $R—I^{127}$ is more reactive than $R—I^{131}$
d. Stronger base increases the reactivity

Working Space

Comprehension Type

***Direction** (Q. Nos. 17-19) This section contains a paragraph, describing theory, experiments, data, etc. Three questions related to the paragraph have been given. Each question has only one correct answer among the four given options (a), (b), (c) and (d).*

Passage

Consider the following reaction. The tertiary butyl group can be *cis* or *trans* to tosylate group. Answer the following questions based on the knowledge of E2 and E1 reaction mechanism.

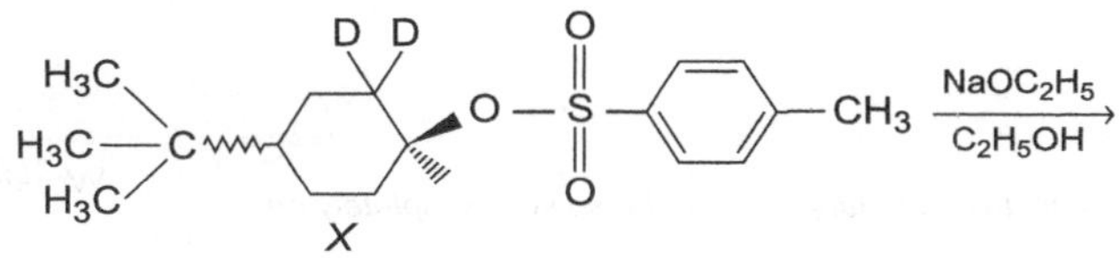

17 The correct statement concerning above elimination reaction is

a. when tertiary butyl group is *trans*, reaction occurs rapidly and deuterium is lost from β-carbon
b. when tertiary butyl group is *trans*, reaction occurs rapidly and hydrogen is lost from β-carbon
c. when tertiary butyl group is *cis*, reaction occurs rapidly and deuterium is lost from β-carbon
d. when tertiary butyl group is *cis*, reaction occurs rapidly and hydrogen is lost from β-carbon

18 When tertiary butyl group and tosylate group are *trans*, the correct statement is

a. reaction occurs slowly and hydrogen is lost from β-carbon
b. reaction occurs slowly and both hydrogen and deuterium are lost from β-carbon
c. reaction occurs rapidly and hydrogen is lost from β-carbon
d. reaction occurs rapidly and deuterium is lost from β-carbon

19 If the rate of reaction is independent of NaOEt concentration, it is concluded that

a. tertiary butyl group is *cis* to tosylate
b. tertiary butyl group is *trans* to tosylate
c. substitution reaction occur predominantly
d. NaOEt is in large excess

One Integer Value Correct Type

***Direction** (Q. Nos. 20 and 21) This section contains 2 questions. When worked out will result in an integer from 0 to 9 (both inclusive).*

20 Consider the following reaction,

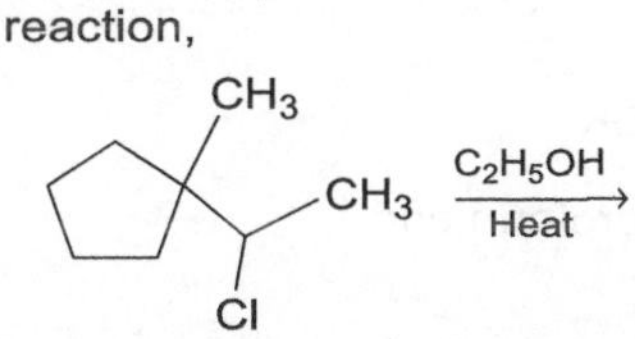

In principle, how many different alkenes are possible by the above elimination reaction?

21 E1 reaction always competes with S_N1 reaction. How many S_N1 products are formed in principle in the question number 20?

DPP-10 Dihalides and Polyhalides

Subjective Questions

Working Space

Direction *(Q. Nos. 1-4) These questions are subjective in nature, need to be solved completely on notebook.*

1 During free radical chlorination reaction of methane, what factor will favour formation of CCl_4?

2 If 2-butyne is first treated with Br_2 - H_2O followed by Cl_2 - CCl_4 in the presence of $AlCl_3$, dichlorodibromo ethane is formed. Discuss stereochemistry of the product.

3 How the following transformations can be carried out?

i. 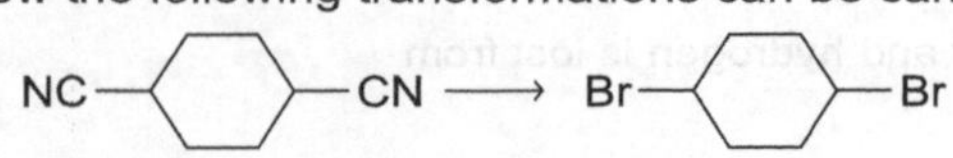

ii. 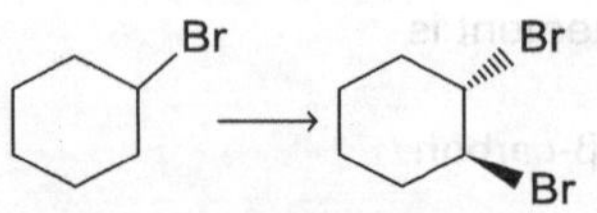

iii. $(CH_3)\ C{=}O$ to $CHCl_3$

4 How ethane is converted into C_2F_6? How only monofluoroethane can be obtained from ethane?

Only One Option Correct Type

Direction *(Q. Nos. 5-17) This section contains 13 multiple choice questions. Each question has four choices (a), (b), (c) and (d), out of which ONLY ONE option is correct.*

5 Consider the following reaction,

$$CCl_4 + KOH(aq) \longrightarrow$$

the major product in the above reaction is

a. K_2CO_3 b. CO_2
c. $C(OH)_4$ d. $HCOOK$

6 Nitro chloropicrin is prepared by the action of chloroform and

a. hot $KNO_3\ (aq)$
b. hot $NaNO_2(aq)$
c. hot Conc. HNO_3
d. hot dil. $NaNO_2(aq) + NaNO_2(aq)$

7 On heating glycerol with excess of HI gives

a. allyl iodide b. isopropyl iodide
c. propylene d. 1,2,3-tri iodopropane

8 A *geminal* dichloride is formed in all of the following reactions except

a. CH_3CHO and PCl_5 b. CH_3COCH_3 and PCl_5
c. $CH_2{=}CH_2$ and Cl_2 d. $CH_2{=}CHCl$ and HCl

Working Space

9 $A + KOBr \longrightarrow CHBr_3 + CH_3COOK$; Here, *A* is

a. isopropanol
b. methanol
c. ethanoic acid
d. *n*-butyl alcohol

10 The product obtained when bleaching powder is distilled with acetone

a. CCl_4
b. $CHCl_3$
c. CH_2Cl_2
d. CH_3Cl

11 What are the products in the following reaction?

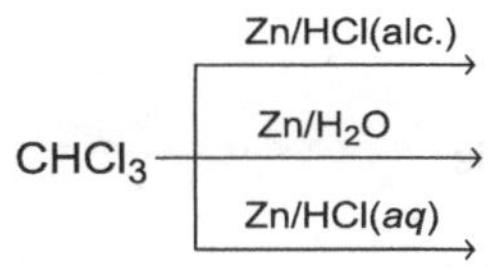

a. CH_2Cl_2, CH_3Cl, CH_4
b. CH_4, CH_3Cl, CH_2Cl_2
c. CH_3Cl, CH_2Cl_2, CH_4
d. CH_3Cl, CH_4, CH_2Cl_2

12 The end product in the following reaction is

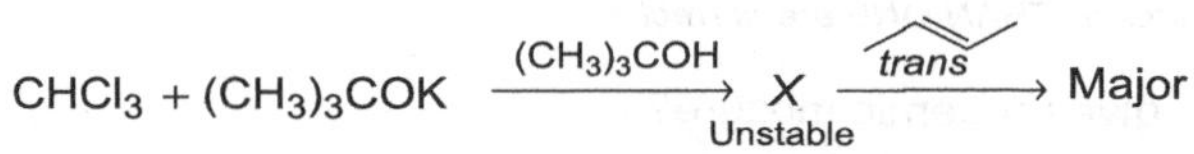

a. 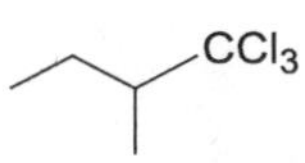

b. 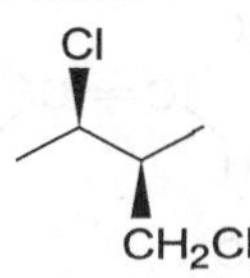

c. 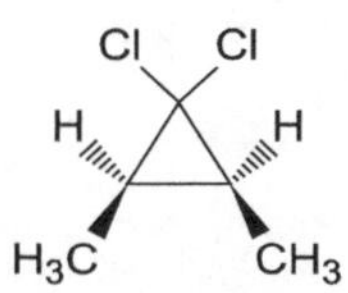

d. 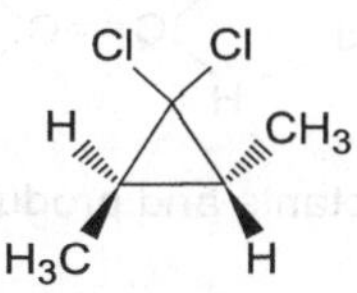

13 What is the product formed in the following reaction?

$$(CH_3)_3CCH_2Cl \xrightarrow{300°C} \text{Product}$$

a. 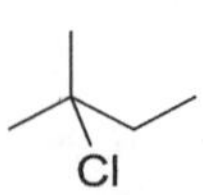

b. 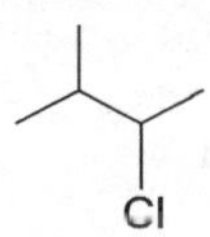

c. 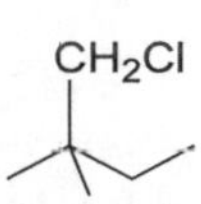

d. 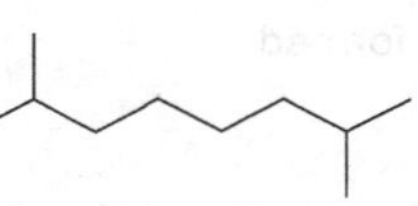

14 What is the major product in the following reaction?

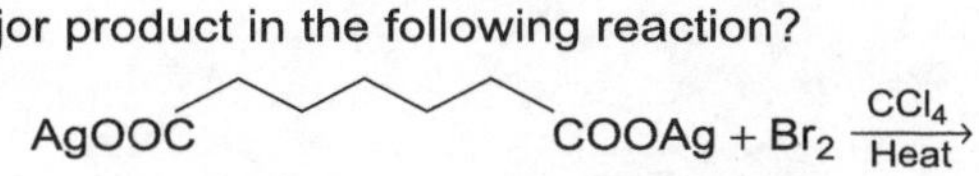

a. 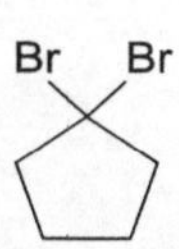

b. 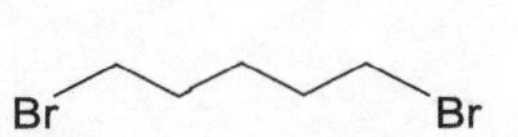

c. 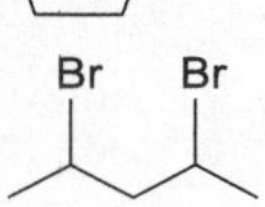

d. 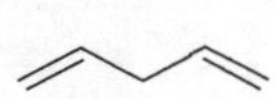

Working Space

15 What is the major product in the following reaction?

$$CH_3CCl_2CH_3 + KOH(aq) \longrightarrow$$

a. 2-propanol
b. 2-chloro-2-propanol
c. Propanone
d. Propanal

16 What is formed predominantly in the following reaction?

$$CS_2 + 3Cl_2 \xrightarrow{AlCl_3} ? + S_2Cl_2$$

a. $CSCl_2$
b. CCl_4
c. C_2Cl_6
d. C_2SCl_4

17 The major organic product in the following reaction is

$$CH_3CCl_2CH_3 + SbF_3 \longrightarrow$$

a. $CH_3CFClCH_3$
b. $CH_3CF_2CH_3$
c. $FCH_2CF_2CH_2F$
d. $F_2CHCCl_2CHF_2$

One or More than One Options Correct Type

Direction (Q. Nos. 18-20) *This section contains 3 multiple choice questions. Each question has four choices (a), (b), (c) and (d), out of which ONE or MORE THAN ONE are correct.*

18 Which upon reaction with Br_2 in CCl_4 gives racemic mixture?

a. 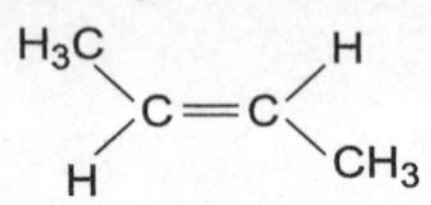

b. 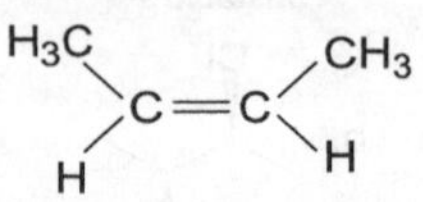

c. $H_3C-CH=CH_2$

d. 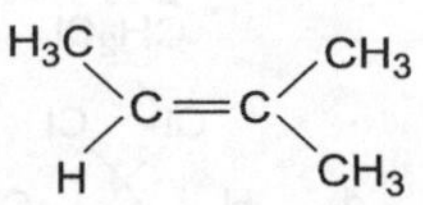

19 In which of the following reaction(s), reactants and products are correctly matched?

a. $2CH_2F—CH_2F + \text{heat} \longrightarrow 4HF + CH_2=CH—CH=CH_2$
b. $CH_2F—CH_2F + 2H_2O \longrightarrow HOCH_2CH_2OH + 2HF$
c. $CHF_3 + H_2O \longrightarrow HCOOH + 3HF$
d. $CCl_4 + 2HF \longrightarrow CF_2Cl_2 + 2HF$

20 Which is/are correct about $CHCl_3$?

a. With concentrated HNO_3, chloropicrin is formed
b. With CH_3COCH_3/KOH, chloretone is formed
c. It is an ionisable liquid at room temperature
d. When heated with Ag, acetylene is formed

SET - 1

Revisal Problems
for JEE Main

Only One Option Correct Type

Direction *(Q. Nos. 1-26) This section contains 26 multiple choice questions. Each question has four choices (a), (b), (c) and (d), out of which ONLY ONE is correct.*

Working Space

1 Which of the following statements about a S_N1 mechanism is true?

a. The reaction is fastest with 3° halides
b. The rate of reaction increases when the solvent is changed from acetone to ethanol
c. The identity of the leaving group does not affect the reaction rate
d. Both (a) and (b)

2 Place the following alkyl halides in increasing order of reactivity in a S_N1 reaction.

I. $CH_3CHClCH_3$ II. $CH_3CH_2CH_2Cl$
III. $(CH_3)_3CCl$ IV. CH_3Cl

a. II, IV, I, III   b. I, III, IV, II
c. II, I, IV, II   d. IV, II, I, III

3 Which of the following will form a white precipitate on treatment with $AgNO_3(aq)$?

a. 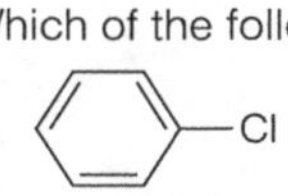
b. 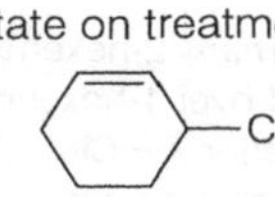
c. 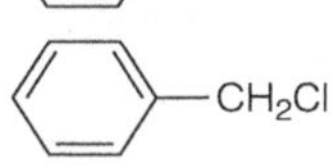
d. Both (b) and (c)

4 Which of the following is most reactive in S_N1 reaction?

a. 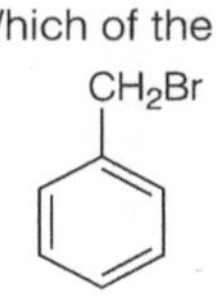
b. 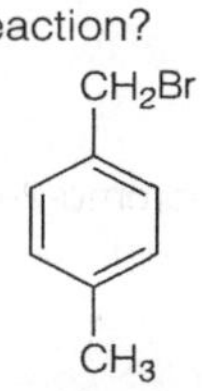
c. 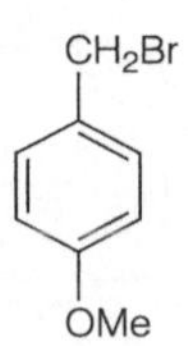
d. 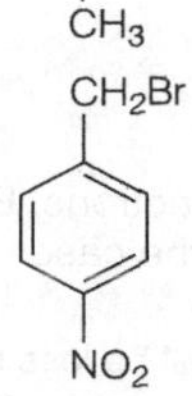

5 Which of the following is most reactive in S_N1 reaction?

a. 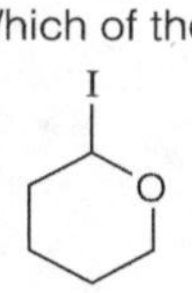
b. 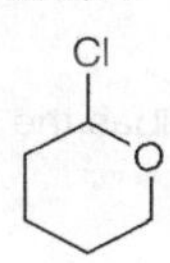
c. 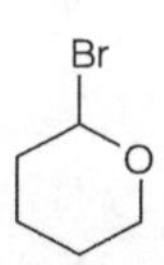
d. 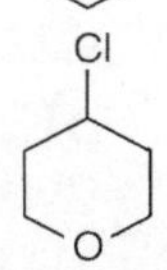

6 A solution of (+)-1-chloro-1-phenylethane in toluene racemises slowly in the presence of a small amount of $SbCl_5$ due to the formation of

a. carbanion   b. carbine
c. free radical   d. carbocation

Working Space

7 Which of the following compounds is most rapidly hydrolysed by S_N1 mechanism?

a. C_6H_5Cl
b. $H_2C{=}CH{-}CH_2{-}Cl$
c. $(C_6H_5)_3CCl$
d. $C_6H_5CH_2Cl$

8 Which is the correct order of reactivity towards β-elimination with a strong base?

a. 1-chlorobutane < 2-chlorobutane
b. 4-chloro-1-butene < 2-chlorobutane
c. 2-iodobutane < 2-bromobutane
d. chloroethane < penta deuterated chloroethane

9 Consider the elimination reaction given below :

$$CH_2{=}CH{-}CH_2{-}CHCl{-}CH(CH_3)_2 \xrightarrow{HO^-}$$

Major product is

a. 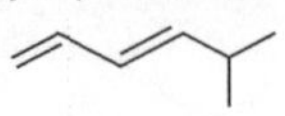
b. 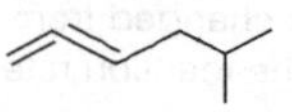
c. 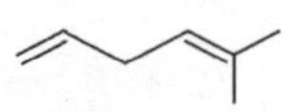
d. 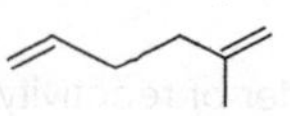

10 Choose the correct statement about the elimination reaction shown (ignore competing substitution).

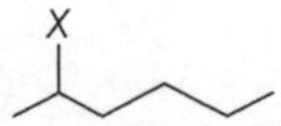

+ CH_3CH_2ONa ⟶

a. When X = Cl, 1-hexene will be favoured over *cis* and *trans* 2-hexene
b. When X = Br, *cis* and *trans*-2-hexene will be favoured over 1-hexene
c. When X = F, the product distribution will be same as for X = Cl
d. When X = Cl, ionisation takes place followed by rearrangements

11 Which of the following undergoes E1 reaction most readily?

a. 1-chloropentane
b. 2-chloropentane
c. 2-chloro-2-methyl butane
d. 2, 2-dimethyl-1-chloropropane

12 What is the major product of the E2 reaction of *trans*-1-bromo-2-methylcylcohexane with strong base?

a. methylcyclohexene
b. 2-methylcyclohexene
c. 3-methylcyclohexene
d. 4-methylcyclohexene

13 Cyanide anion has two atoms that have lone pair of electrons. Either could act as nucleophile in the reaction. Yet in the vast majority of the cases, carbon acts as nucleophile and forms a bond to the substrate, why?

a. Nitrogen is strongly solvated in hydroxylic solvents, it will be less reactive
b. Carbon is more reactive in polar aprotic solvents, normally used for these S_N2 reactions
c. Carbon atoms are more basic than nitrogen
d. The lone pair of electrons on C is in *sp* hybrid orbital

14 In an aprotic solvent, which relative ordering best describes the nucleophilicity of the halide ions?

a. iodide > bromide > chloride >> fluoride
b. iodide > bromide > chloride > fluoride
c. iodide < bromide < chloride < fluoride
d. iodide < bromide > chloride > fluoride

15 Which statement is true about S_N2 mechanism?

a. The rate of reaction increases on increasing strength of the nucleophile
b. The reaction is faster in polar aprotic solvents
c. The rate of reaction increases as the leaving group ability increases
d. All statements are true

Working Space

16 Predict the product in nucleophilic substitution reactions?

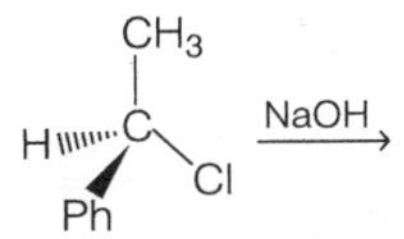

a. 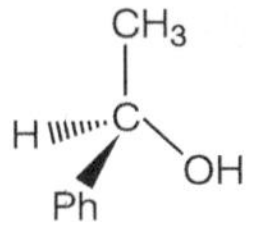

b. 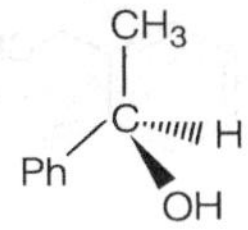

c. 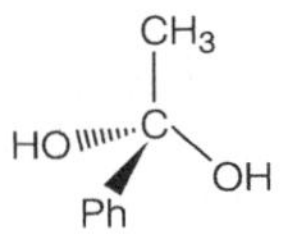

d. 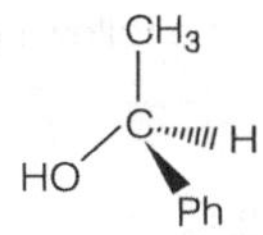

17 The most reactive in the following in reaction with NaCN is

a. 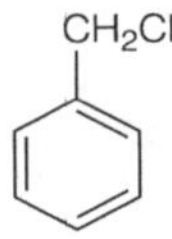

b. 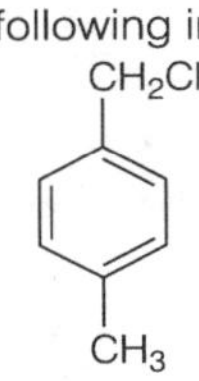

c. 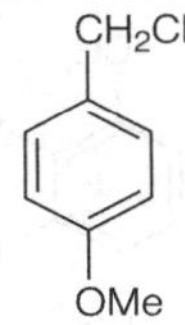

d. 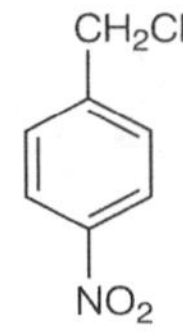

18 Which S_N2 reaction would take place most rapidly?

a. $H\bar{S} + CH_3Cl \xrightarrow[25°C]{OH_2} CH_3SH + \bar{Cl}$

b. $H\bar{O} + CH_3Cl \xrightarrow[25°C]{OH_2} CH_3OH + \bar{Cl}$

c. $CH_3OH + CH_3Cl \xrightarrow[25°C]{CH_3OH} CH_3OCH_3 + \bar{Cl}$

d. $CH_3COO^- + CH_3Cl \xrightarrow[25°C]{CH_3COOH} CH_3O\overset{O}{\overset{\|}{C}}\,CH_3 + \bar{Cl}$

19 The reactivity of 2-bromo-2-methyl butane (I), 1-bromo pentane (II) and 2-bromo pentane (III) towards S_N2 reaction is

a. I > II > III b. I > III > II
c. II > III > I d. II > I > III

20 Which of the following is correct increasing rate of S_N2 reaction?

I. 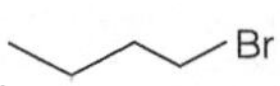 II. 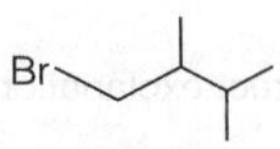 III. 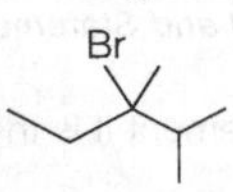 IV. 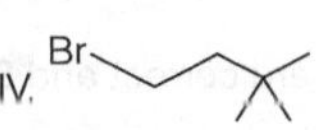

a. II < III < I < IV b. IV < III < II < I c. III < II < IV < I d. I < IV < II < III

21 Chloroethane is treated separately with aqueous NaSCN and NaOCN. Which of the following is true statement regarding the substitution product?

a. In both case, C—N bond is mainly formed
b. In no case, the major product involves formation of C—N bond
c. With NaSCN, mainly C—S bond is formed while with NaOCN, mainly C—N bond is formed
d. With NaSCN, mainly C—N bond is formed while with, mainly C—O bond is formed

22 Consider the reaction coordinate diagram for a S_N2 reaction, that is, for a reaction in which the charges are more distributed in the transition state than in the reactants. How does a change to a more polar solvent affect this reaction coordinate diagram?

a. Energy of reactants is raised, and transition state energy remains about the same
b. Energies of reagents and transition state are lowered by about the same amount
c. Energy of reactants are lowered slightly, and energy of transition state structure are lowered significantly more
d. Energy of reactants is lowered significantly, and energy of transition state structure is lowered significantly less

Working Space

23 The major product in the following reaction is

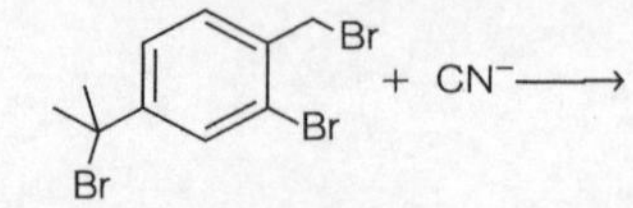

a. 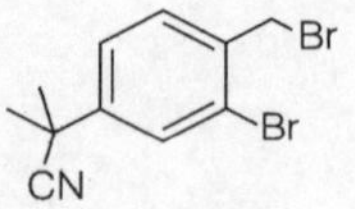 b. 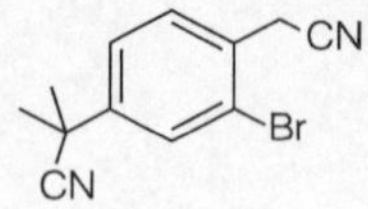 c. 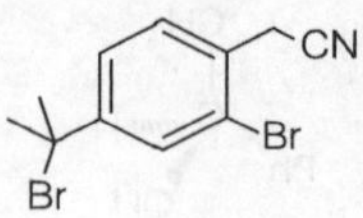 d. 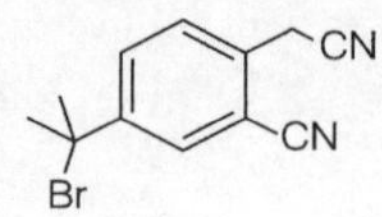

24 What will be the major monobromination product in the following reaction?

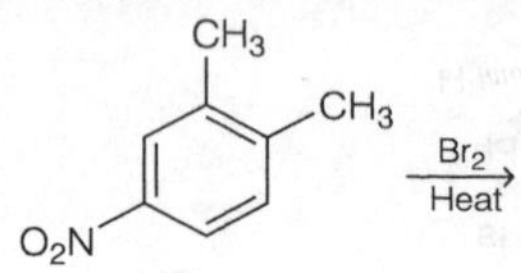

a. 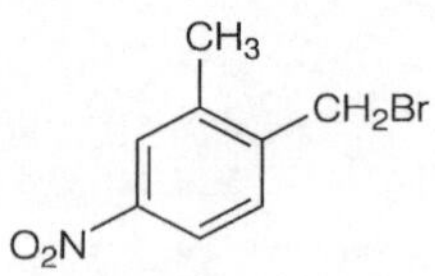

b. 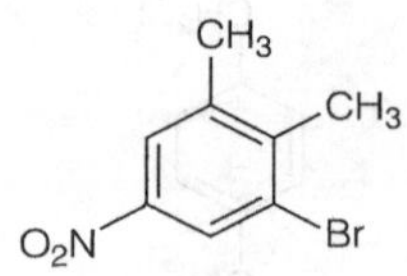

c. 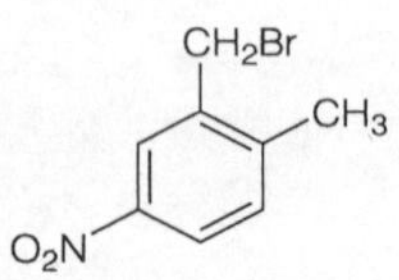

d. 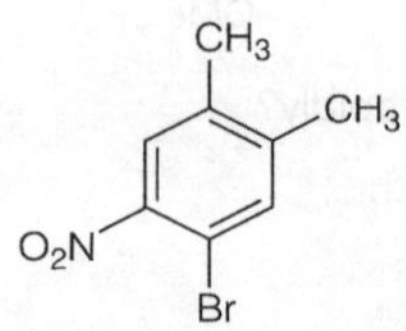

25 In the following groups :

OAc (I), —OMe(II), —OSO_2Me(III), —OSO_2CF_3(IV) the order of leaving group ability is

a. I > II > III > IV b. IV > III > I > II

c. III > II > I > IV d. II > III > IV > I

26 Grignard's reagent when exposed to moisture

a. gets oxidised b. gets hydrolysed

c. gets decomposed to give hydrocarbons d. remains unaffected

Statement Type

***Direction** (Q. Nos. 27-30) This section is based on Statement I and Statement II. Select the correct answer from the codes given below.*

a. Both Statement I and Statement II are correct and Statement II is the correct explanation of Statement I

b. Both Statement I and Statement II are correct but Statement II is not the correct explanation of Statement I

c. Statement I is correct but Statement II is incorrect

d. Statement II is correct but Statement I is incorrect

27 **Statement I** Direct fluorination of ethane usually fails to produce monofluorination product.

Statement II Fluorination of ethane produces hexafluoro ethane.

28 **Statement I** When CH_3CH_2I is treated with a saturated AgCN solution, CH_3CH_2NC is formed as the major product.

Statement II Cyanide is an ambident ligand.

29 **Statement I** 1-chloropropane when treated with NaI in acetone, 1-iodopropane is formed.

Statement II Iodide is stronger nucleophile than chloride ion.

30 **Statement I** Bimolecular elimination (E2) reaction is stereospecific.

Statement II Strong base approaches from *anti* position to leaving group at β-carbon.

SET - 2

Revisal Problems
for JEE Advanced

Only One Option Correct Type

Direction *(Q. Nos. 1-12) This section contains 12 multiple choice questions. Each question has four choices (a), (b), (c) and (d), out of which ONLY ONE is correct.*

Working Space

1 What is the major product of the reaction?

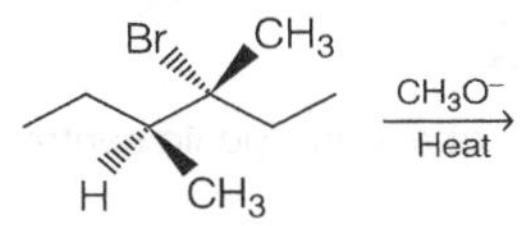

a. 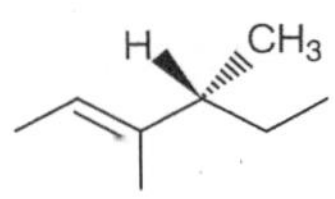

b. 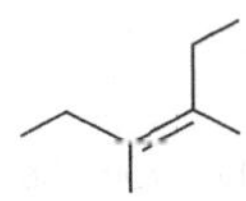

c. 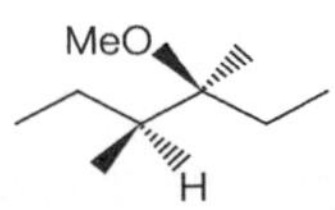

d. 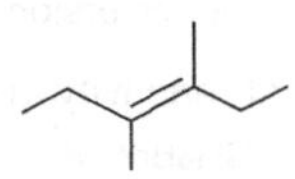

2 What is the major elimination product of the following reaction?

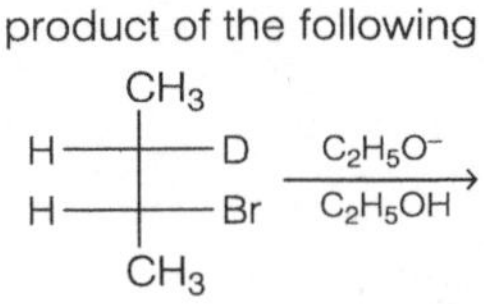

a. 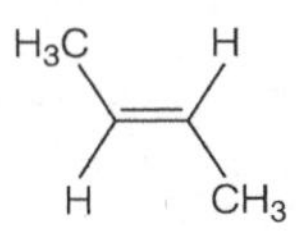

b. 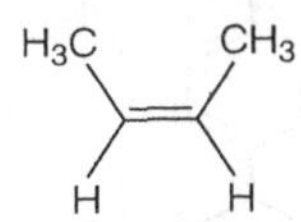

c. 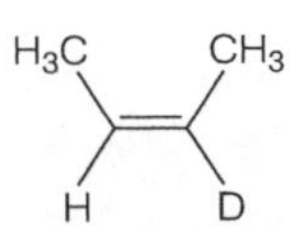

d. 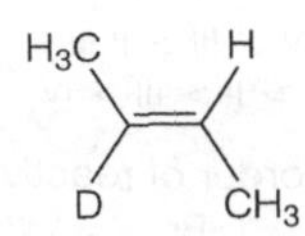

3 Which of the following is missing reagent in the following reaction sequence?

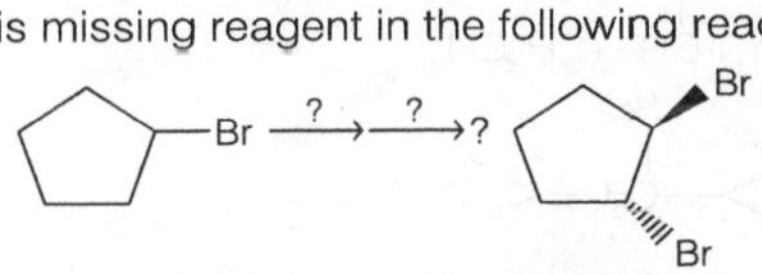

a. $(CH_3)_3COK$, Br_2

b. $(CH_3)_3COK$, HBr

c. $(CH_3)_3COK$, Br_2/H_2O

d. H_2SO_4, Br_2

4 Predict substitution product in the following reaction.

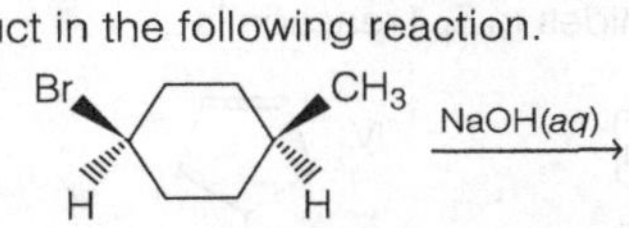

a. 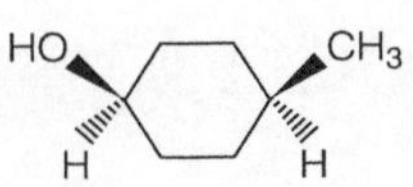

b. 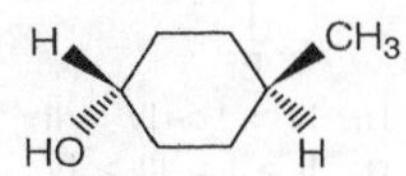

c. Both (a) and (b)

d. None of these

5 Propose product with appropriate stereochemistry in the following S_N2 reaction.

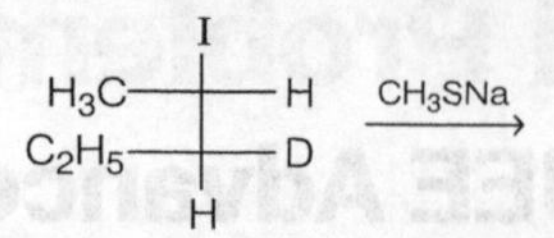

a. 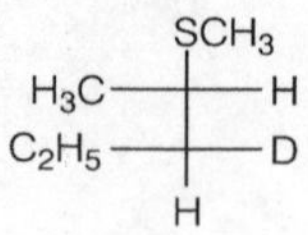

b. 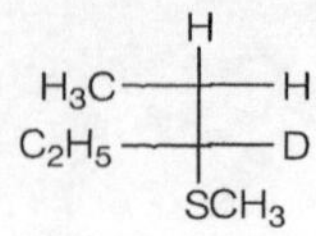

c. 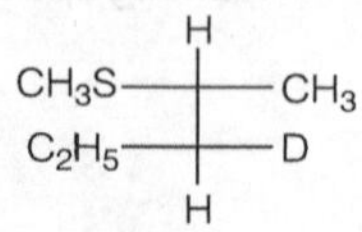

d. 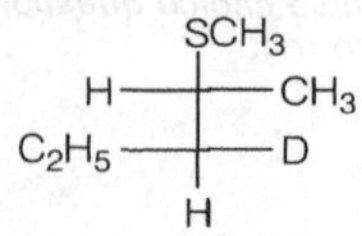

6 In S_N1 reaction, racemisation occurs if the reaction occurs at a stereogenic centre, however 50 : 50 mixture of enantiomers are rarely obtained, why?

a. Usually one enantiomer is more stable than other
b. Retention of configuration is always favoured
c. Hydroxylic solvent favour retention of configuration
d. There is steric hindrance to the approach of nucleophile from the front side as some of the leaving group are not departed completely which favour inversion of configuration

7 Arrange the following halides in decreasing order of reactivity in S_N1 reaction.

I. 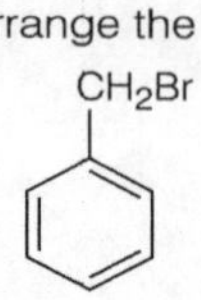

II. 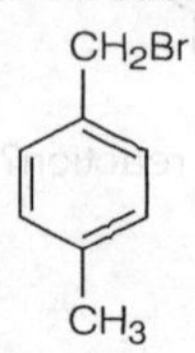

III. 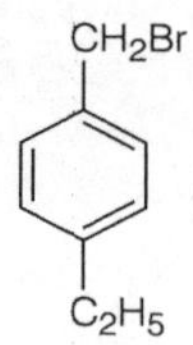

IV. 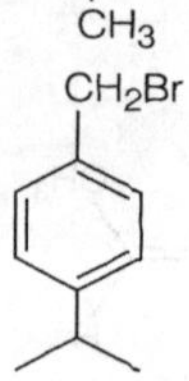

a. II > III > IV > I
b. IV > III > II > I
c. III > IV > II > I
d. I > II > III > IV

8 Arrange the following compounds in decreasing order of reactivity for hydrolysis reaction.

I. $PhCH_2Br$

II. Ph—CH(Br)—CH_2CH_3

III. 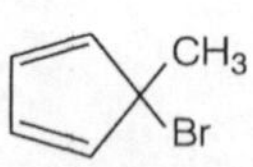

IV. 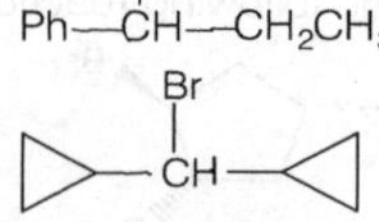

a. I > II > III > IV
b. IV > II > I > III
c. III > IV > II > I
d. IV > III > II > I

9 The increasing order of reactivity of the following bromides in S_N1 reaction is

I. 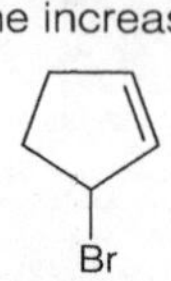

II. 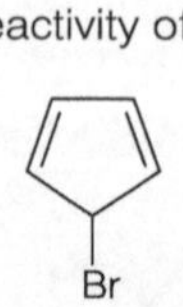

III. 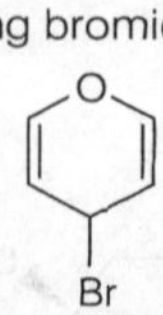

IV. 

a. III > I > II > IV
b. III > I > IV > II
c. II > III > I > IV
d. II > I > III > IV

Working Space

10 Which is the least reactive in solvolysis reaction?

a. 

b. 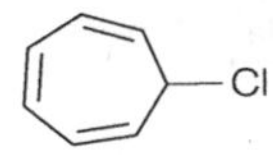

c. 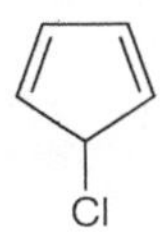

d. 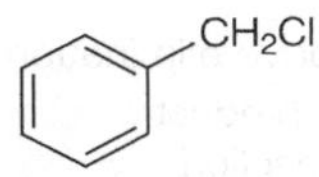

11 Among the following given bromides, the correct decreasing order of reactivity in S_N1 reaction is

I. 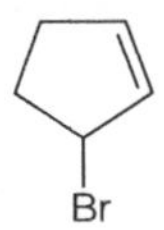

II. 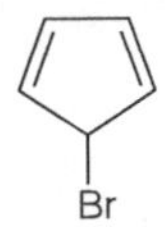

III. 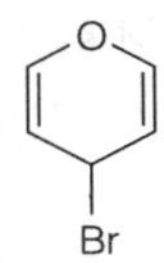

a. III > II > I
b. III > I > II
c. II > III > I
d. II > I > III

12 Arrange the following compounds in increasing order of their rate hydrolysis by S_N1 mechanism.

I. 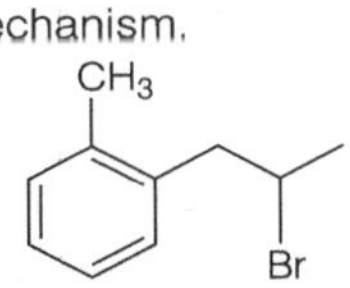

II. 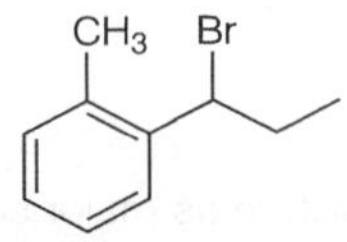

III. 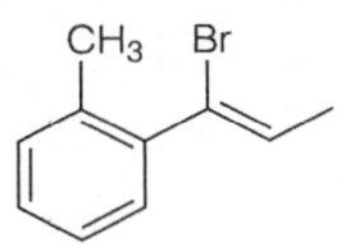

IV. 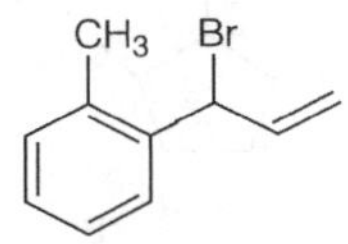

a. I < II < III < IV
b. III < I < II < IV
c. II < I < IV < III
d. III < II < I < IV

One or More than One Options Correct Type

***Direction** (Q. Nos. 13-22) This section contains 10 multiple choice questions. Each question has four choices (a), (b), (c) and (d), out of which ONE or MORE THAN ONE are correct.*

13 What are the expected products of the following reaction?

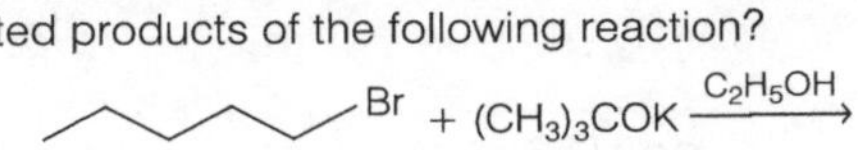

a. 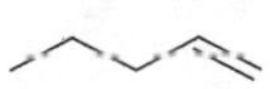

b. 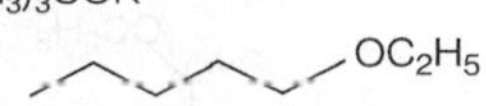

c. 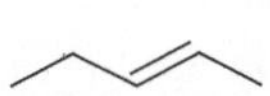

d. 

14 What is (are) true about E1cb reaction?

a. The overall order of reaction is 2
b. Electron withdrawing group adjacent to β-carbon favour this mechanism
c. The rate determining step is unimolecular
d. If deuterated solvent is used, deuterium incorporation in substrate may be observed

15 Which of the following is expected to give more than two products in an E2 elimination reaction?

a. 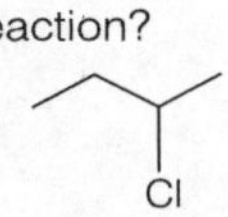

b. 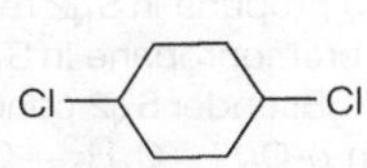

c. 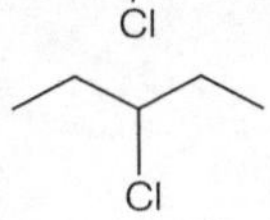

d. 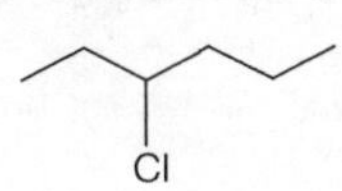

Working Space

Working Space

16 Consider the following elimination reaction,

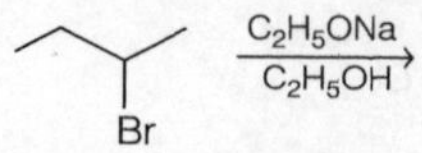

The statement that is (are) true regarding the above elimination reaction is

a. It produces a pair of diastereomers as important product
b. Deuteration at β-carbon slows down the rate of reaction
c. Its chloride counterpart will undergo slower elimination reaction
d. Replacing base by $(CH_3)_3CONa$ results in less stable product as the major product

17 Consider the following reactions:

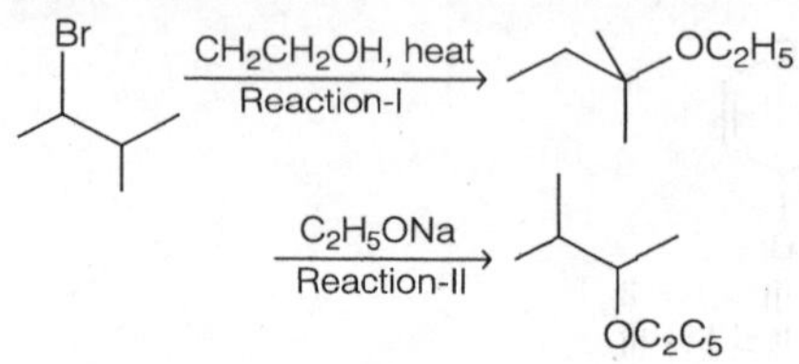

The correct statement regarding above substitution reaction is/are

a. Both reactions proceeded by S_N2 mechanism
b. Both reactions proceeded by S_N1 mechanism
c. Reaction-I proceeded by S_N1 mechanism
d. Reaction-II proceeded by S_N2 mechanism

18 Which of the following will result in racemic mixture as major solvolysis products?

a. 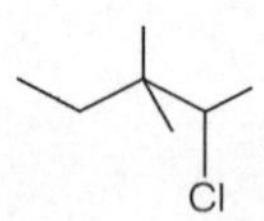

b. 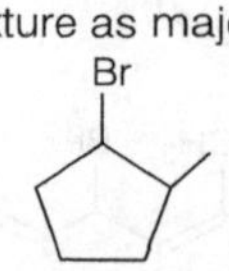

c. 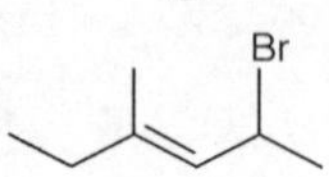

d. 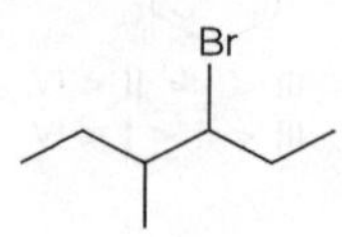

19 Consider the following solvolysis reaction,

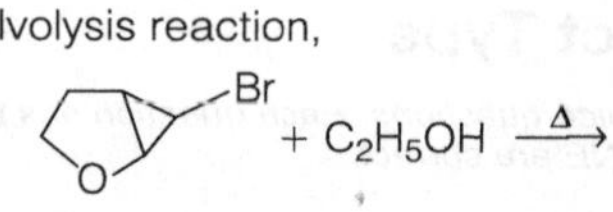

The expected product(s) is/are

a. 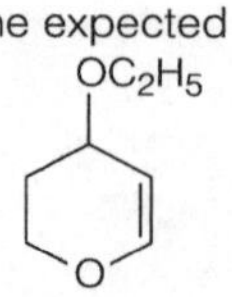

b. 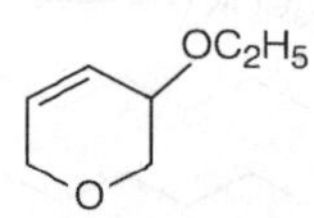

c. 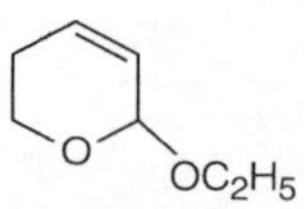

d. 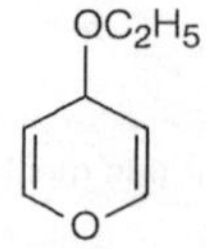

20 Select the correct statement.

a. 1-iodopropane reacts more rapidly than 2-iodo propane in S_N2 reaction
b. 1-chloropropane has greater reactivity than 1-bromopropane in S_N2 reaction
c. 1-bromopropane reacts faster than $(CH_3)_3CCH_2Br$ under S_N2 condition
d. p-$CH_3O—C_6H_4—Br$ has greater reactivity than p-$O_2N—C_6H_4—CH_2Br$ in S_N2 reaction

Working Space

21 Consider the following S_N2 reaction,

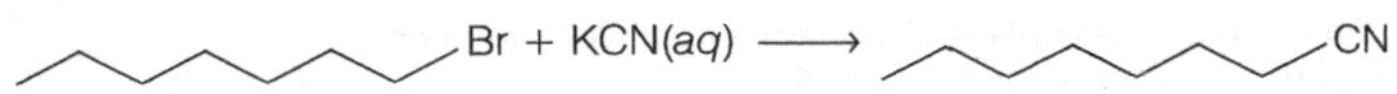

The correct statement is (are)

a. Addition of $(C_4H_9)_4NCl$ catalyse reaction.
b. Reaction occur smoothly even in the absence of any catalyst
c. Labeling α-^{14}C will increase the rate of reaction
d. Substituting KCN by AgCN (saturated) will yield isocyanide as the major product

22 What is (are) true about the following free radical chlorination reaction ?

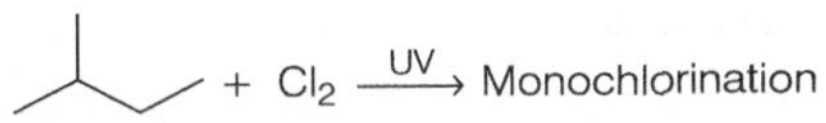

a. A total of six isomers are formed
b. Two pair of enantiomers are formed
c. Tertiary chloride is formed as the major product
d. The product mixture will rotate the plane polarised light

Comprehension Type

Direction (Q. Nos. 23-25) ***This section contains a paragraph, describing theory, experiments, data, etc. Three questions related to the paragraph have been given. Each question has only one correct answer among the four given options (a), (b), (c) and (d).***

Passage

Compounds *A*, *B* and *C* are alkyl bromides with the molecular formulate $C_5H_{11}Br$. Treatment of *A* or *B* with potassium *tert*-butoxide gives the same alkene *D*. If the base is changed to sodium ethoxide, *A* and *C* give the same alkene *E*. Treatment of *C* with potassium *tert*-butoxide gives yet another alkene *F*.

23 The correct statement regarding *A*, *B* and *C* is

a. Compound *B* reacts the faster with NaCN than *C* in DMSO
b. Compound *A* reacts the fastest with NaCN in DMSO
c. Compound *C* reacts the fastest with NaCN in DMSO
d. Compounds *B* and *C* have comparable reactivity with NaCN in DMSO

24 The correct statement concerning *A*, *B* and *C* is

a. All of them are chiral
b. Only *A* and *B* are chiral
c. Only *A* and *C* are chiral
d. Only *B* and *C* are chiral

25 Which forms most stable Grignard's reagent when treated with Mg in THF?

a. *A*
b. *B*
c. *C*
d. Both *B* and *C*

One Integer Value Correct Type

Direction (Q. Nos. 26-29) ***This section contains 4 questions. When worked out will result in an integer from 0 to 9 (both inclusive).***

26 Methyl cyclopentane on treatment with chlorine gas in sunlight undergo free radical chlorination reaction. How many of the above monochlorination products upon treatment with ethanolic KOH solution can give 1-methyl cyclopentane as one of the product?

27 In principle, how many different alkyl chlorides would be formed on treatment of 3-methyl-2-pentanol with concentrated HCl in the presence of $ZnCl_2$?

28 How many different isomers of alkyl bromide on treatment with ethanolic KOH solution can result in 3-methyl-3-hexene?

29 If $Br_2CH—CHBr_2$ is treated with excess of NaCN solution such that all bromides are substituted by nucleophiles, how many different substitution products would be formed in principle which are chiral?

Matching List Type

Working Space

***Direction** (Q. Nos. 30 and 31) Choices for the correct combination of elements from Column I and Column II are given as options (a), (b), (c) and (d), out of which one is correct.*

30 Match the reactions from Column I and their corresponding properties from Column II and mark the correct option from codes given below.

	Column I		Column II
i.	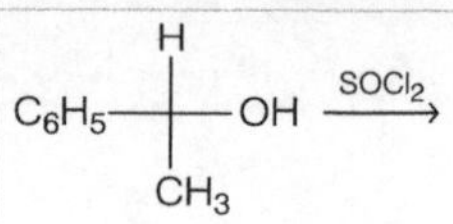	p.	Complete retention of configuration
ii.	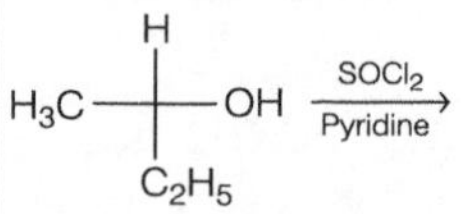	q.	Racemic mixture is formed
iii.	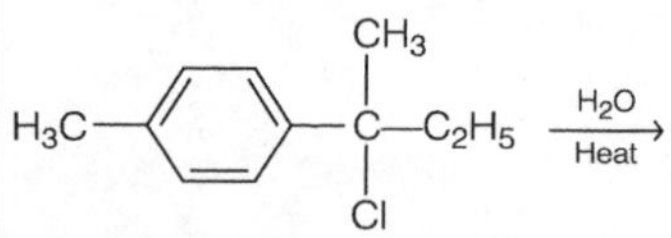	r.	S_N2
iv.	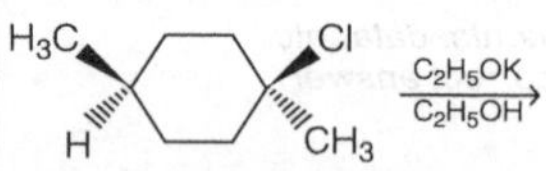	s.	S_N1

Codes

	i	ii	iii	iv
a.	p,s	r	q	q
b.	p	q	r	s
c.	s	q	p	r
d.	r	s	p	q

31 Match the reactions from Column I and their corresponding properties from Column II and mark the correct option from codes given below.

	Column I		Column II
i.	$RCH_2CH_2Br \xrightarrow[C_2H_5OD]{C_2H_5ONa} RCD = CH_2$	p.	E1
ii.	$H_3C-C(CH_3)(Cl)-CH_2CH_3 \xrightarrow[C_2H_5OH]{C_2H_5ONa} (CH_3)_2C = CHCH_3$	q.	Order = 2
iii.	$R-CH(F)-CH(Cl)-R \xrightarrow[C_2H_5OH]{C_2H_5ONa \text{ (one equ.)}} RCH = CClR$	r.	E2
iv.	$H_3C-C(CH_3)(CH_3)-CH(Cl)-CH_3 \xrightarrow[Heat]{C_2H_5OH} (H_3C)_2C = C(CH_3)_2$	s.	E1cb

Codes

	i	ii	iii	iv
a.	q,s	q,r	s	p
b.	q	p	s	r
c.	p	q	r	s
d.	s,q	r	p	q

JEE Main & AIEEE Archive

(Compilation of Last 13 Years' Questions)

1 The synthesis of alkyl fluorides is best accomplished by **(JEE Main 2015)**

a. Free radical fluorination b. Sandmeyer's reaction
c. Finkelstein reaction d. Swarts reaction

2 The major organic compound formed by the reaction of 1, 1, 1 –trichloroethane with silver powder is **(JEE Main 2014)**

a. acetylene b. ethene
c. 2 –butyne d. 2 –butene

3 Compound *A*, (C_8H_9Br) gives a white precipitate when warmed with alcoholic $AgNO_3$. Oxidation of *A* gives an acid *B*, $(C_8H_6O_4)$. *B* easily forms anhydride on heating. Identify the compound *A*. **(JEE Main 2013)**

a. 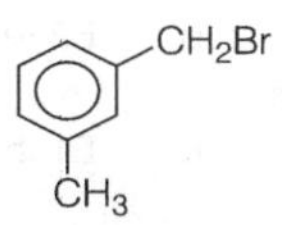
b. 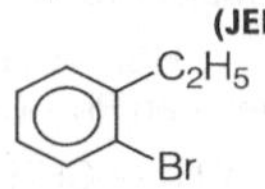
c. 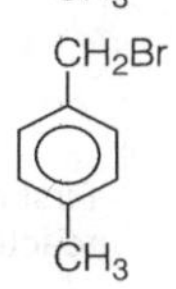
d. 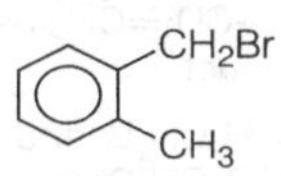

4 A solution of (–1)-chloro-1-phenylethane in toluene racemises slowly in the presence of a small amount of $SbCl_5$, due to the formation of **(JEE Main 2013)**

a. carbanion b. carbene
c. carbocation d. free radical

5 How many chiral compounds are possible on monochlorination of 2-methyl butane? **(AIEEE 2012)**

a. 8 b. 2 c. 4 d. 6

6 Iodoform can be prepared from all except **(AIEEE 2012)**

a. ethyl methyl ketone b. isopropyl alcohol
c. 3-methyl-2-butanone d. isobutyl alcohol

7 Consider the following bromides,

(*A*) 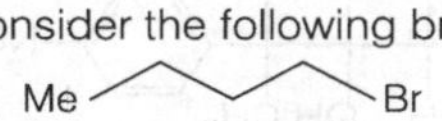

(*B*) 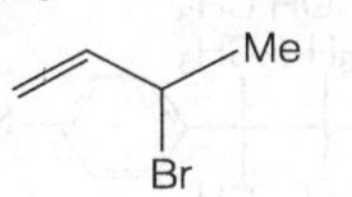

(*C*) 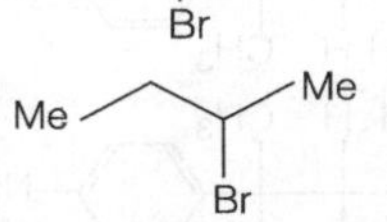

(AIEEE 2010)

The correct order of S_N1 reactivity is

a. (*B*) > (*C*) > (*A*) b. (*B*) > (*A*) > (*C*)
c. (*C*) > (*B*) > (*A*) d. (*A*) > (*B*) > (*C*)

8 The organic chloro compound, which shows complete stereochemical inversion during an S_N2 reaction is **(AIEEE 2008)**

a. $(C_2H_5)_2$ CHCl b. $(CH_3)_3CCl$
c. $(CH_3)_2CHCl$ d. CH_3Cl

9 Which of the following reactions will yield 2, 2-dibromopropane? **(AIEEE 2007)**

a. $CH_3—C\equiv CH + 2HBr \longrightarrow$
b. $CH_3CH=CHBr + HBr \longrightarrow$
c. $CH\equiv CH + 2HBr \longrightarrow$
d. $CH_3—CH=CH_2 + HBr \longrightarrow$

10 $CH_3Br + Nu^- \longrightarrow CH_3—Nu + Br^-$

The decreasing order of the rate of the above reaction with nucleophile (Nu^-) *A* and *D* is

$[Nu = (A)\ PhO^-, (B) AcO^-, (C) HO^-, (D) CH_3O^-]$ **(AIEEE 2006)**

a. $D > C > A > B$ b. $D > C > B > A$
c. $A > B > C > D$ d. $B > D > C > A$

11 Alkyl halides react with dialkyl copper reagents to give **(AIEEE 2005)**

a. alkenyl halides
b. alkanes
c. alkyl copper halides
d. alkenes

12 2-methyl butane on reacting with bromine in the presence of sunlight gives mainly

a. 1-bromo-3-methylbutane **(AIEEE 2005)**
b. 2-bromo-3-methylbutane
c. 2-bromo-2-methylbutane
d. 1-bromo-2-methylbutane

13 Elimination of bromine from 2-bromobutane results in the formation of

a. predominantly 2-butyne **(AIEEE 2005)**
b. predominantly 1-butene
c. predominantly 2-butene
d. equimolar mixture of 1 and 2-butene

14 Reaction of one molecule of HBr with one molecule of 1, 3-butadiene at 40°C gives predominantly **(AIEEE 2005)**

a. 1-bromo-2-butene under kinetically controlled conditions
b. 3-bromobutene under thermodynamically controlled conditions
c. 1-bromo-2-butene under thermodynamically controlled conditions
d. 3-bromobutene under kinetically controlled conditions

JEE Advanced & IIT JEE Archive

(Compilation of Last 10 Years' Questions)

1 Match the chemical conversion in Column I with the appropriate reagents in Column II and select the correct answer using the code given below the lists. **(2013 Adv., Matching Type)**

	Column I		Column II
i.	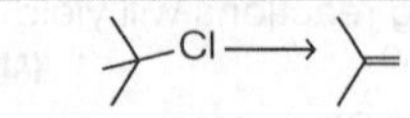	p.	(i) $Hg(OAc)_2$; (ii) $NaBH_4$
ii.	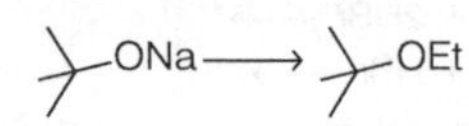	q.	NaOEt
iii.	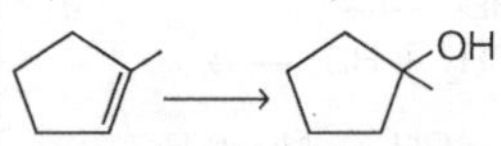	r.	Et-Br
iv.	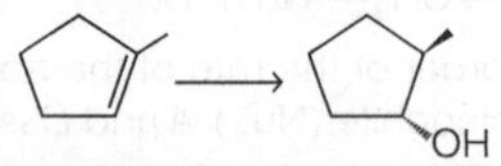	s.	(i) BH_3 ; (ii) H_2O_2 / NaOH

Codes

	i	ii	iii	iv
a.	q	r	p	s
b.	r	q	p	s
c.	q	r	s	p
d.	r	q	s	p

2 KI in acetone, undergoes S_N2 reaction with each *P*, *Q*, *R* and *S*. The rates of the reaction vary as **(2013 Adv., Only One Option Correct Type)**

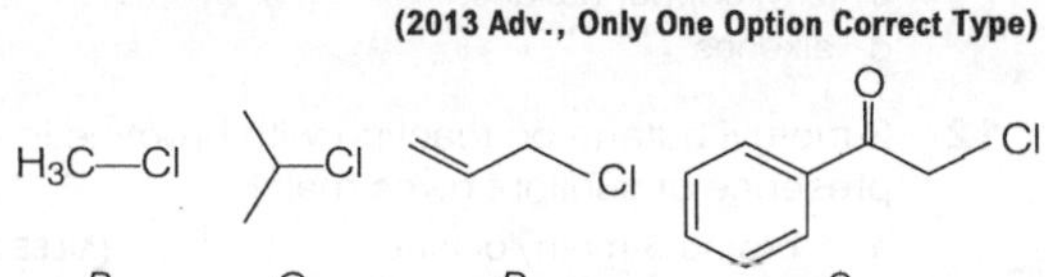

P *Q* *R* *S*

a. $P > Q > R > S$ b. $S > P > R > Q$
c. $P > R > Q > S$ d. $R > P > S > Q$

3 The total number of alkenes possible by dehydrobromination of 3-bromo-3-cyclopentylhexane using alcoholic KOH is **(2011, Integer Type)**

4 The major product of the following reaction is **(2008, Only One Option Correct Type)**

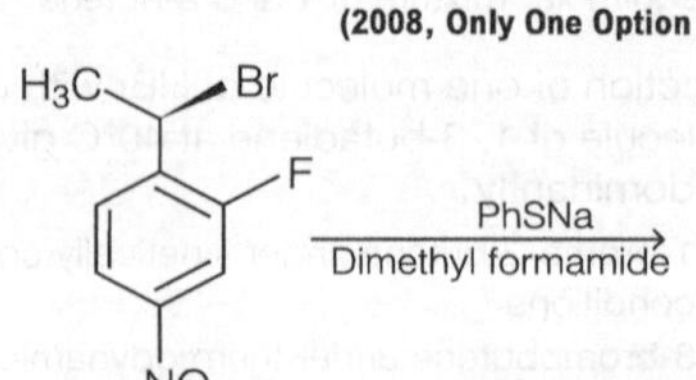

a. 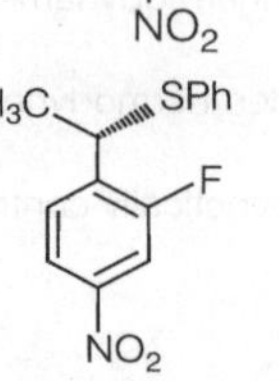 b. 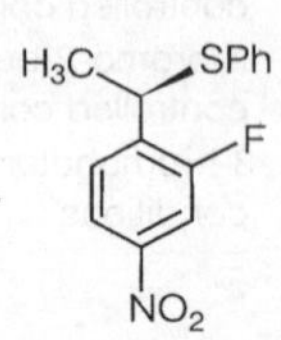

c. 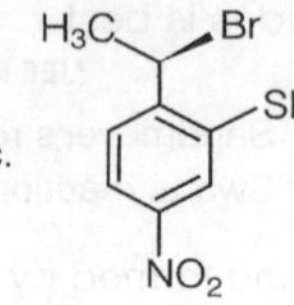 d. 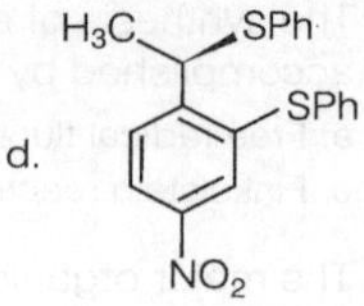

5 Match the Column I with Column II and mark the correct option from the codes given below. **(2006, Matching Type)**

	Column I		Column II
i.	CH_3—CHBr—CD_3 on treatment with alc. KOH gives CH_2=CH—CD_3 as the major product.	p.	E1 reaction
ii.	Ph — CHBr— CH_3 reacts faster than Ph—CHBr— CD_3	q.	E2 reaction
iii.	Ph—CH_2 —CH_2Br on treatment with C_2H_5OD / $C_2H_5O^-$ gives Ph—CD=CH_2 as the major product.	r.	E1cb reaction
iv.	$PhCH_2CH_2Br$ and $PhCD_2CH_2Br$ react with same rate.	s.	First order reaction

Codes

	i	ii	iii	iv
a.	q	q	r	p,s
b.	p,s	p	q	r,s
c.	q	q,r	p,q,r	q
d.	q	s	q,r	p,r,s

6 The following compound on hydrolysis in aqueous acetone will give **(2005, Only One Option Correct Type)**

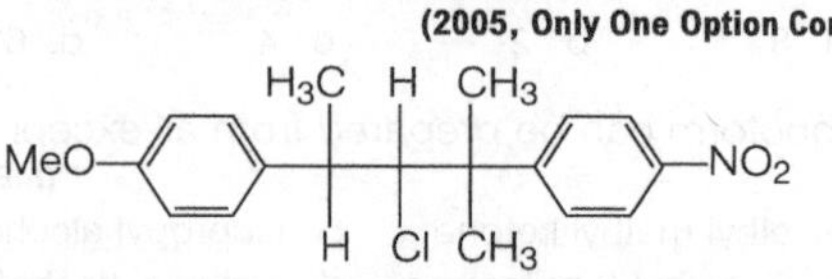

(*K*) 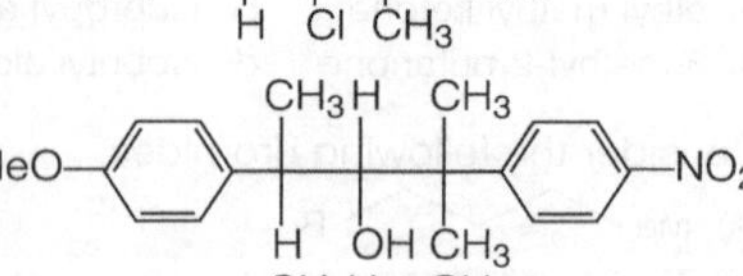

(*L*) 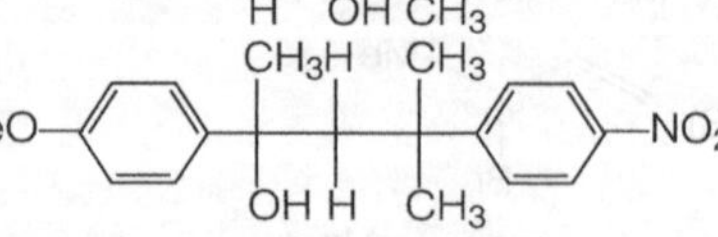

(*M*) 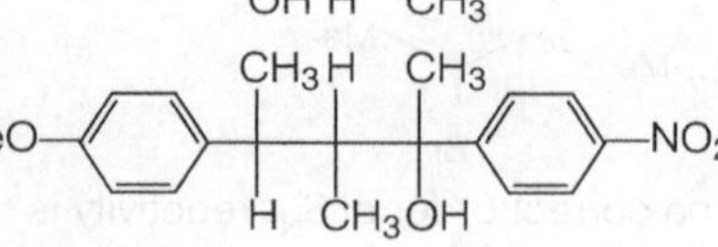

It mainly gives

a. *K* and *L* b. Only *K*
c. *L* and *M* d. Only *M*

2. Aliphatic Alcohols, Ethers and Epoxides

DPP
DAILY PRACTICE PROBLEMS

DPP-1 Preparation of Alcohols

Subjective Questions

Direction (Q. Nos. 1-5) *These questions are subjective in nature, need to be solved completely on notebook.*

1 $C_5H_{12}O$ has several alcohol isomers. Identify which isomers can be synthesised by $LiAlH_4$ reduction of

i. an aldehyde
ii. a ketone
iii. a carboxylic acid
iv. an ester $RCOOCH_3$

2 Bring about the following transformation in efficient manner.

i. $C_6H_6 \longrightarrow C_6H_5CH(OH)CH_3$
ii. $C_6H_5CH_2CH_3 \longrightarrow C_6H_5CH_2CH_2OH$
iii. $C_6H_5CH_2CH_3 \longrightarrow C_6H_5CH(OH)CH_3$
iv. $C_6H_5-\overset{\overset{\large O}{\|}}{C}-CH_3 \longrightarrow C_6H_5CH_2CH_2OH$

3 Propose the structure of *A* through *G*

$$A\text{(a bromide)} \xrightarrow[\text{(ii) } CH_3(CH_2)_3CHO \text{ (iii) } H_3O^+]{\text{(i) Mg/Ether}} B \xrightarrow[\text{Heat}]{H_2SO_4\text{ (Conc.)}} C \xrightarrow[\text{Zn-}H_2O]{O_3} \text{Butanal + Pentanal}$$

$$C \xrightarrow{Br_2} D \xrightarrow[H_2O]{NaNH_2} E$$

$$E \xrightarrow[\text{(ii) } H_2O_2/NaOH]{\text{(i) } B_2H_6} F$$

$$E \xrightarrow{NaNH_2} \xrightarrow[\text{(ii) } H_3O^+]{\text{(i) Butanone}} G$$

4 Synthesise the following compounds from any starting materials containing not more than six carbons.

i. 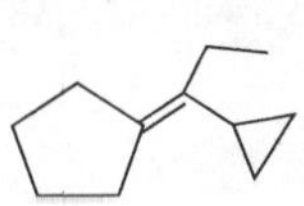

ii. 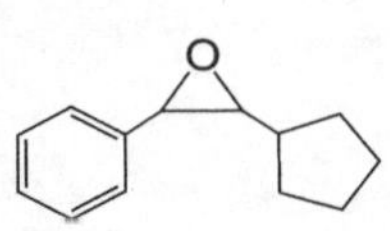

iii. 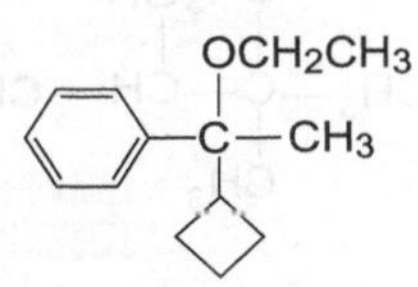

5 Bring about the following transformation.

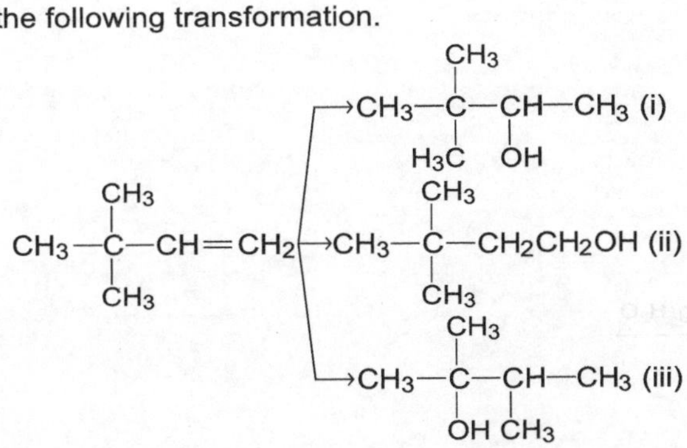

Working Space

Only One Option Correct Type

Working Space

Direction *(Q. Nos. 6-13) This section contains 8 multiple choice questions. Each question has four choices (a), (b), (c) and (d), out of which ONLY ONE is correct.*

6 Which of the following pairs of compounds can be used as starting material in the synthesis of 2-phenyl-2-pentanol?

a. $CH_3—(CH_2)_2—CH_2Br$ and C_6H_5COOH
b. $(CH_3)_2CHCH_2Br$ and $PhCOCH_3$
c. PhBr and $CH_3CH_2CH_2COCH_3$
d. PhBr and $(CH_3)_2CHCH_2COCH_3$

7 What is the product of the reaction of methyl cyclohexene with B_2H_6 in THF followed by the oxidation with alkaline H_2O_2?

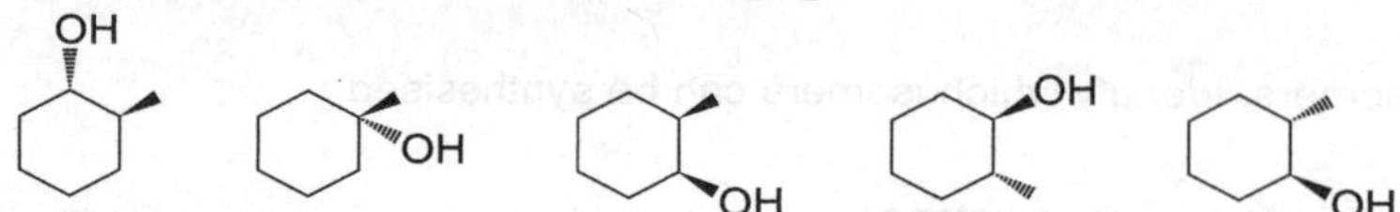

I. II. III. IV. V.

a. Both II and III
b. Only II
c. Both III and IV
d. Both I and IV

8 Choose the reagent and reactant that would produce 2-methyl-2-butanol as the major product.

I. 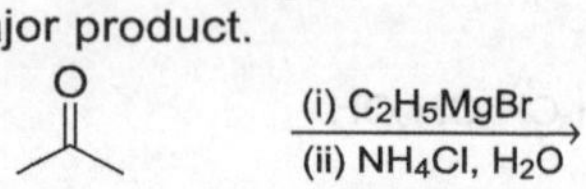

II. 

III. 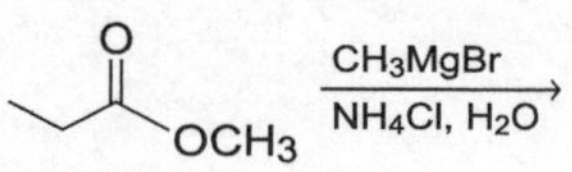

a. I, II and III
b. Both I and III
c. Only I
d. Only III

9 How the following compound can be synthesised ?

$$CH_3—C(CH_3)_2—CH(OH)—CH_3$$

a. $CH_3—C(CH_3)(OH)—C(CH_3)(OH)—CH_3 \xrightarrow{H_2SO_4}$

b. $(CH_3)_3C—CH{=}CH_2 \xrightarrow[H_2O]{H_2SO_4}$

c. $(CH_3)_3C—CH{=}CH_2 \xrightarrow[\text{(ii) } H_2O_2/HO^-]{\text{(i) } BH_3}$

d. $(CH_3)_3C—CH{=}CH_2 \xrightarrow[\text{(ii) } NaBH_4]{\text{(i) } (CH_3COO)_2Hg/H_2O}$

10 What is the correct structure for the major compound produced by the following reaction sequence?

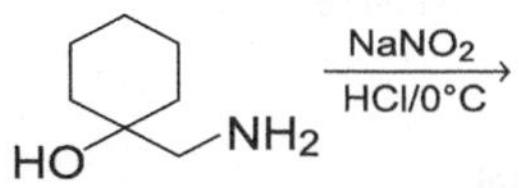

a. 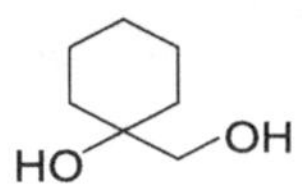

b. 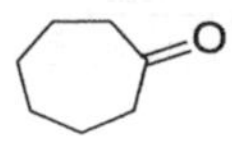

c. 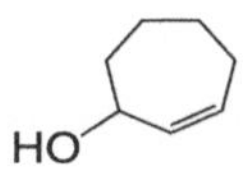

d. 

Working Space

11 In the reaction given below,

$$CH_3—CH=CH—CH_3 + CO + H_2 \xrightarrow[\text{Heat}]{[CoH(CO)_4]} \xrightarrow[H_2]{\text{Cu-Zn}}$$

The final major organic product is

a. $CH_3—CH=CH—CH_2CH_2OH$

b. $CH_3CH_2—\underset{\underset{CH_3}{|}}{CH}—CH_2OH$

c. $CH_3—CH_2—\overset{\overset{OH}{|}}{\underset{\underset{CH_3}{|}}{C}}—CH_3$

d. $CH_3—CH_2—\underset{\underset{CH_3}{|}}{CH}—CHO$

12 Primary alcohol can easily be prepared from primary alkyl halide *via* S_N2 reaction with aqueous NaOH. However, similar method does not work for the preparation of tertiary alcohol. Which reaction can be used for the efficient preparation of tertiary alcohol (tertiary butanol) from tertiary butyl bromide?

a. $(CH_3)_3CBr + KOH \xrightarrow{C_2H_5OH} \xrightarrow[\text{Heat}]{H_2SO_4}$

b. $(CH_3)_3CBr + CH_3COOAg(aq) \longrightarrow \xrightarrow{H_3O^+}$

c. $(CH_3)_3CBr \xrightarrow[(C_2H_5)_2O]{Mg} \xrightarrow[\text{(ii) } H_3O^+]{\text{(i) } O_2}$

d. $(CH_3)_3CBr \xrightarrow{\text{KOH/Ethanol}} \xrightarrow[\text{(ii) } H_2O_2/NaOH]{\text{(i) } B_2H_6}$

13 What is the major product in the following reaction?

$$C_6H_5—CH=CH—CH_3 \xrightarrow[H_2O]{\text{(i) }(CH_3COO)_2Hg} \xrightarrow[H_3O^+]{\text{(ii) } NaBH_4}$$

a. $C_6H_5—CH_2—\underset{\underset{OH}{|}}{CH}—CH_3$

b. $C_6H_5—\underset{\underset{OH}{|}}{CH}—CH_2CH_3$

c. $C_6H_5CH_2—CH=CH_2$

d. $C_6H_5—\overset{\overset{CH_3}{|}}{\underset{\underset{OH}{|}}{C}}—CH_3$

One or More than One Options Correct Type

Working Space

Direction (Q. Nos. 14-20) *This section contains 7 multiple choice questions. Each question has four choices (a), (b), (c) and (d), out of which ONE or MORE THAN ONE are correct.*

14 In the following reaction,

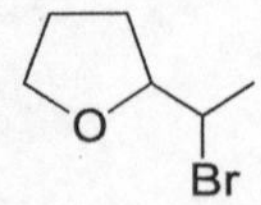 + $AgNO_3$ (*aq*) $\xrightarrow{\text{Heat}}$

The possible substitution product(s) is/are

a. 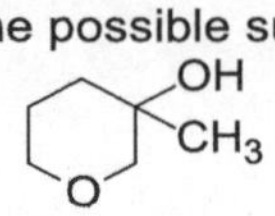

b. 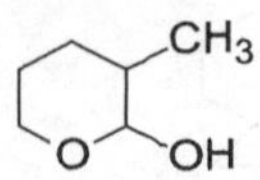

c. 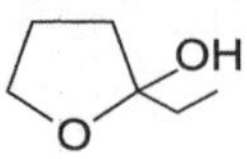

d. 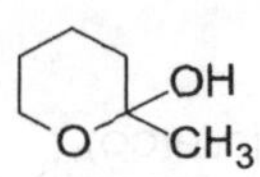

15 In the following reaction,

$$CH_3CH_2CH{=}CH_2 + CO + H_2 \xrightarrow{[CoH(CO)_4]} \xrightarrow[H_2/\Delta]{\text{Cu-Zn}}$$

Possible product(s) is/are

a. $CH_3CH_2{-}CH{=}CH{-}CH_2OH$

b. $CH_3CH_2CH_2CH_2CH_2OH$

c. $CH_3{-}CH_2{-}CH(CH_3){-}CH_2OH$

d. $CH_3{-}CH_2{-}C(CH_3)(OH){-}CH_3$

16 Which of the following Grignard's synthesis can result into 2-cyclopentyl-2-butanol?

a. 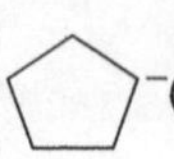 $^{-}(MgBr)^{+}$ + 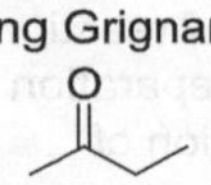 $\xrightarrow{H_3O^+}$

b. 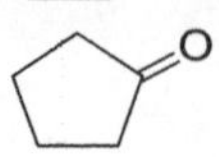 + $H_3C{-}CH_2{-}CH(CH_3)MgBr \xrightarrow{H_3O^+}$

c. 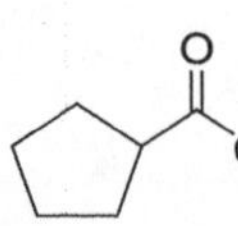 $CH_3 + CH_3CH_2MgBr \xrightarrow{H_3O^+}$

d. 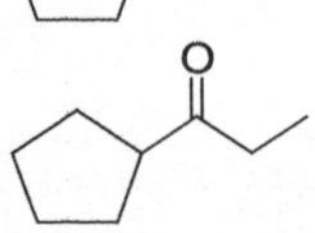 + $CH_3MgBr \xrightarrow{H_3O^+}$

17 Aliphatic alcohol when treated with dilute H_2SO_4, undergo isomerisation *via* reversible reaction. In the following reaction, which of the isomers are expected to be present at equilibrium?

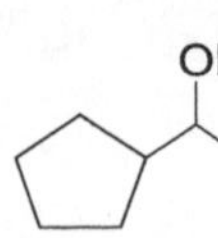 $\xrightarrow{\text{Dil. } H_2SO_4}$

a. 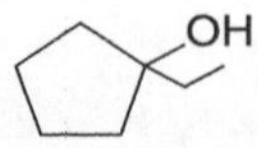

b. 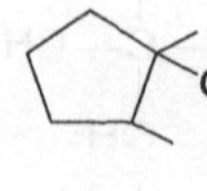

c. 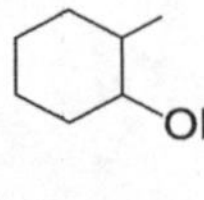

d. 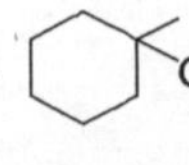

18 In the following reaction,

$$CH_3-\underset{C_6H_5}{\overset{CH_3}{C}}-CH{=}CH_2 + H_2SO_4\,(\text{dil.}) \xrightarrow{\text{Boil}}$$

The alcohol(s) formed in significant yield is/are

a. $CH_3-\underset{C_6H_5}{\overset{CH_3}{C}}-\overset{OH}{CH}-CH_3$

b. $C_6H_5-\underset{CH_3}{\overset{OH}{C}}-\overset{CH_3}{CH}-CH_3$

c. $CH_3-\underset{CH_3}{\overset{OH}{C}}-\overset{CH_3}{CH}-C_6H_5$

d. $CH_3-\underset{H}{\overset{OH}{C}}-\underset{CH_3}{\overset{CH_3}{C}}-C_6H_5$

19 In the reaction given below,

$$CH_3-CH_2-\overset{O}{\overset{\|}{C}}-CH_3 + LiAlD_4 \longrightarrow \xrightarrow{H_3O^+}$$

The correct statement regarding the outcome of the above reaction is/are

a. a pure enantiomer of alcohol is formed
b. racemic mixture of alcohol is formed
c. product alcohol has deuterium attached to oxygen
d. product alcohol has deuterium attached to α-carbon

20 Consider the following reaction,

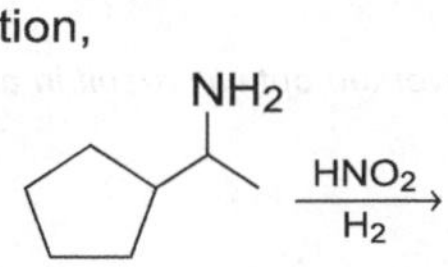

Possible product(s) is/are

a. 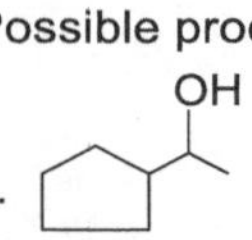

b. 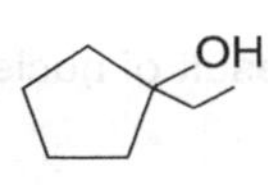

c. 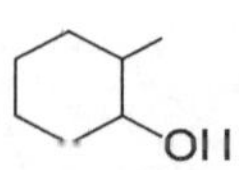

d. 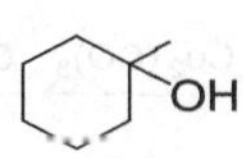

Comprehension Type

Direction *(Q. Nos. 21-23) This section contains a paragraph, describing theory, experiments, data, etc. Three questions related to the paragraph have been given. Each question has only one correct answer among the four given options (a), (b), (c) and (d).*

Passage

Consider the following sequence of reaction,

$$\text{Compound } A(C_7H_{11}Br) \xrightarrow[Et_2O]{Mg} \xrightarrow[H_3O^+]{CH_3-\overset{O}{\overset{\|}{C}}-CH_3} B$$

$$B \xrightarrow[\text{Heat}]{H_2SO_4\,(\text{Conc.})} C(C_{10}H_{16}) \xrightarrow[Pt/\Delta]{H_2} \text{2-methylpropylcyclohexane}$$

21 The structure of compound *B* is

a. 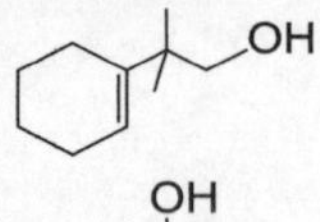

b. 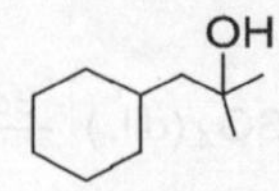

c. 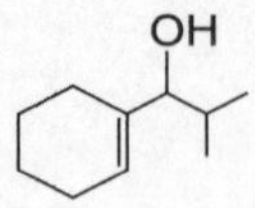

d. 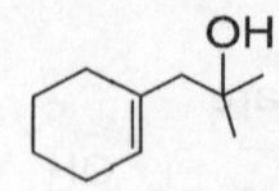

22 $A \xrightarrow{LiAlH_4} \xrightarrow[\text{(ii) } H_2O_2/NaOH]{\text{(i) } B_2H_6} X$

Hence, *X* is

a. 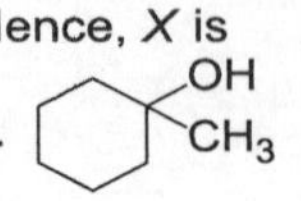

b. 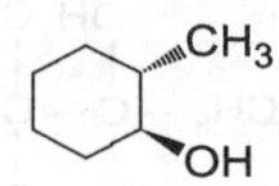

c. 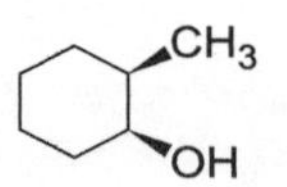

d. 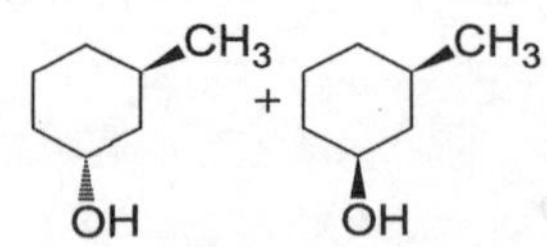

23 If *C* is treated with excess of $Br_2(l)$, how many different isomers of bromination product(s) result?

a. 4 b. 6

c. 7 d. 8

One Integer Value Correct Type

Direction *(Q. Nos. 24-27) This section contains 4 questions. When worked out will result in an integer from 0 to 9 (both inclusive).*

24 In the following reaction,

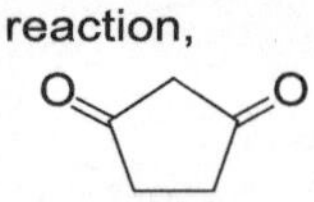 $+ \underset{\text{Excess}}{CH_3MgBr} \xrightarrow{(C_2H_5)_2O} \xrightarrow{H_3O^+}$

How many different diols are formed as a result of nucleophilic addition reaction?

25 In the reaction given below,

$$CH_3-CH_2-\underset{\underset{CH_3}{|}}{C}=CH-CH_3 + CO + H_2 \xrightarrow{Co_2(CO)_8} \xrightarrow[H_2]{Cu\text{-}Zn} \text{Alcohols}$$

How many different products are expected?

26 If 3, 3-dimethyl-2, 4-pentanedione is treated with a Grignard reagent consisting of mixture of CH_3MgBr and C_2H_5MgBr and finally hydrolysing product with dilute H_2SO_4 results in the formation of how many different diols?

27 If a pure enantiomer of 3-methyl-1-pentene is treated with boiling solution of dilute H_2SO_4, how many different alcohols are expected in principle?

DPP-2 Reactions of Alcohols-I

Subjective Questions

Working Space

Direction *(Q. Nos. 1-5) These questions are subjective in nature, need to be solved completely on notebook.*

1 Complete the following acid-base reactions.

i. $CH_3CH_2OH + H_3O^+ \longrightarrow$

ii. 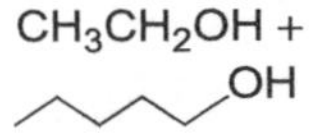 $+ HI \longrightarrow$

iii. 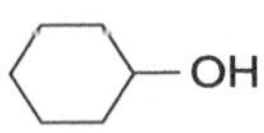 $+ BF_3 \longrightarrow$

iv. 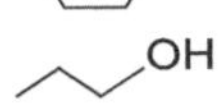 $+ CH_3MgBr \longrightarrow$

v. 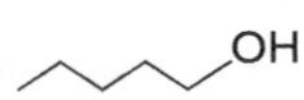 $+ NaH \longrightarrow$

vi. 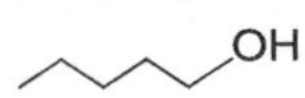 $+ CH_3—C \equiv CNa \longrightarrow$

2 Select the stronger acid from each of the following pairs.

i. H_2O, H_2CO_3

ii. CH_3OH, CH_3COOH

iii. CH_3CH_2OH, $CH_3C \equiv CH$

iv. CH_3CH_2OH, CH_3CH_2SH

v. 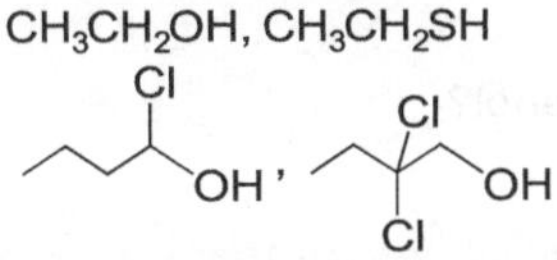

3 In each of the following, estimate the position of equilibrium.

i. $CH_3CH_2O^- + CH_3—C \equiv CH \rightleftharpoons CH_3CH_2OH + CH_3C \equiv C^-$

ii. $CH_3CH_2O^- + HCl \rightleftharpoons CH_3CH_2OH + Cl^-$

iii. $CH_3CH_2O^- + C_6H_5OH \rightleftharpoons CH_3CH_2OH + C_6H_5O^-$

iv. 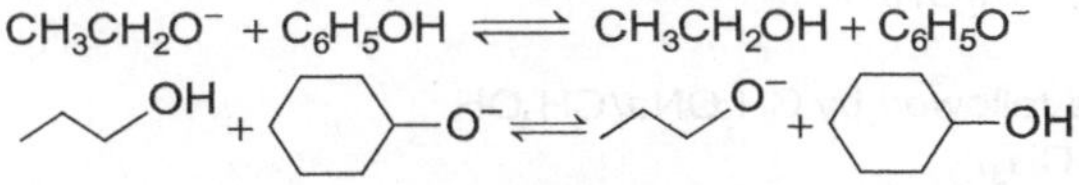

4 When ethanol is treated with D_2O in trace amount of DCl, entire ethanol is deuterated as C_2H_5OD, explain.

5 Consider the compound I and compound II given below,

$CH_3CH(OH)CH_3$ (I) and $F_3C—CH(OH)—CF_3$(II)

Explain the following observation.

i. Compound II has almost triple the molecular weight of compound I but compound II has lower boiling point than compound I.

ii. Dipole moment of compound II is 0.32 D while that of compound I is 1.66 D.

iii. Compound II is much stronger acid than compound I.

Only One Option Correct Type

Working Space

Direction (Q. Nos. 6-13) *This section contains 8 multiple choice questions. Each question has four choices (a), (b), (c) and (d), out of which ONLY ONE is correct.*

6 The correct order of acidic strength of the following alcohols is

I. CH_3OH II. C_2H_5OH III. $CH_3—CH(OH)—CH_3$ IV. $CH_3—C(CH_3)_2—OH$

a. I < II < III < IV b. IV < III < II < I c. III < IV < I < II d. III < II < IV < I

7 Which of the following does not give effervescence when added to ethanol ?

a. CH_3MgBr b. NaH c. $NaNH_2$ d. $NaHCO_3$

8 Which compound given below has the highest solubility in water ?

a. CH_3OH b. CH_3CH_2OH
c. $HO—CH_2CH_2—OH$ d. $HO—CH_2—CH(OH)—CH_2OH$

9 What is the correct increasing order of acidity of the following?

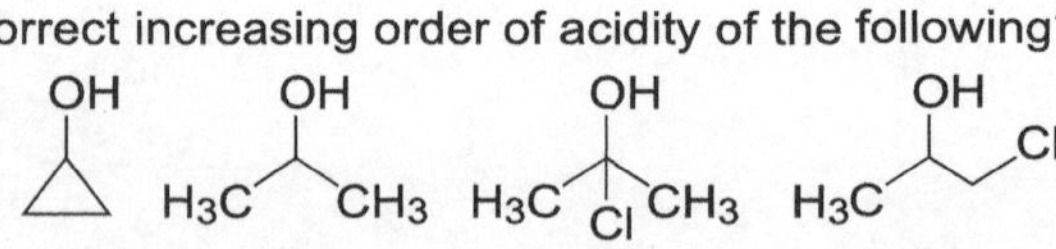

I. II. III. IV.

a. I < II < III < IV b. IV < III < II < I c. I < II < IV < III d. II < I < IV < III

10 What is the order of solubility of the following in water?

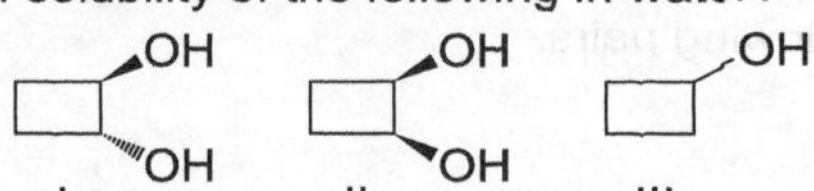

I. II. III.

a. I < II < III b. III < II < I c. III < I < II d. II < I < III

11 Which of the following can be used for the distinction of ethanol from phenol?

a. $NaHCO_3$ b. Na c. CH_3MgBr d. NaOH

12 Consider the following reaction,

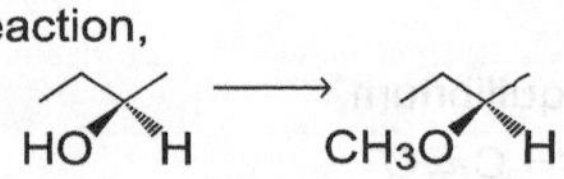

The above reaction can best be brought about by

a. Conc. H_2SO_4 /heat followed by the treatment with CH_3OH/H^+
b. Treatment with CH_3OH/H^+ only
c. Treatment with *p*-toluene sulphonic acid chloride followed by CH_3ONa/CH_3OH
d. Treatment with Na metal followed by addition of CH_3I

13 Consider the following reaction,

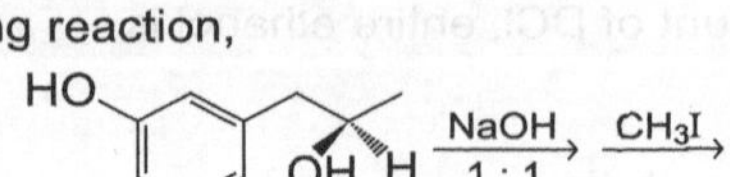

The major product is

a. 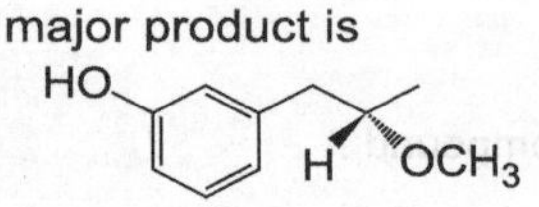

b. 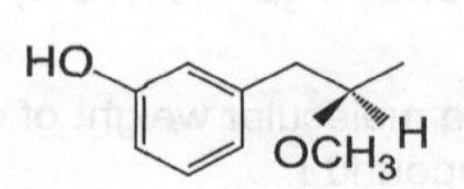

c. 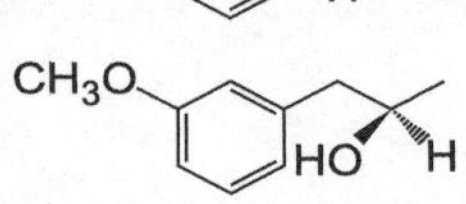

d. 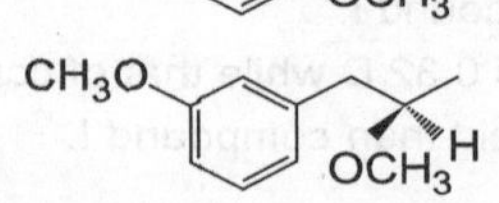

One or More than One Options Correct Type

Direction (Q. Nos. 14-17) *This section contains 4 multiple choice questions. Each question has four choices (a), (b), (c) and (d), out of which ONE or MORE THAN ONE are correct.*

14 Consider the compound shown below,

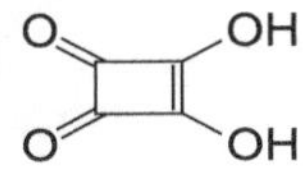

Select the correct statement(s).

a. It is more acidic than methanol
b. It is more acidic than acetic acid
c. It reacts very fast with Lucas reagent
d. It is diacidic

15 Methanol and ethanol can be distinguished by

a. heating with I_2/NaOH
b. treatment with Schiff's base
c. treatment with H_2CrO_4
d. treatment with acidic $KMnO_4$

16 Consider the following substituted ethanol G—CH_2CH_2OH

Which group(s) when present as G gives greater equilibrium of *gauche* conformer than its *anti* counterpart ?

a. —OH b. —F
c. —NH_2 d. —NO_2

17 Which of the statement given below concerning 3-methyl-2-butanol is/are correct ?

a. Treatment with cold, concentrated H_2SO_4 convert it into achiral 2-methyl-2-butanol
b. Treatment with NaOH solution convert it into 2-methyl-2-butanol
c. Treatment of its pure enantiomer with PCl_5 gives mainly a pure enantiomer of chloroalkane
d. It changes the orange colour of acidic $K_2Cr_2O_7$ solution to blue green

Comprehension Type

Direction (Q. Nos. 18-20) *This section contains a paragraph, describing theory, experiments, data, etc. Three questions related to the paragraph have been given. Each question has only one correct answer among the four given options (a), (b), (c) and (d).*

Passage

An organic compound X ($C_{10}H_{12}O_3$) is not soluble in water or $NaHCO_3$. A solution of Br_2 in CCl_4 is decolourised by X forming $C_{10}H_{12}O_3Br_2$. X on controlled ozonolysis followed by the treatment with $(CH_3)_2S$ gives Y ($C_8H_8O_3$) and $C_2H_4O_2$. Y can also be obtained by reaction between *ortho* methoxy phenol with $CHCl_3$ in KOH solution followed by acid hydrolysis.

18 What is the correct structure of X?

a. 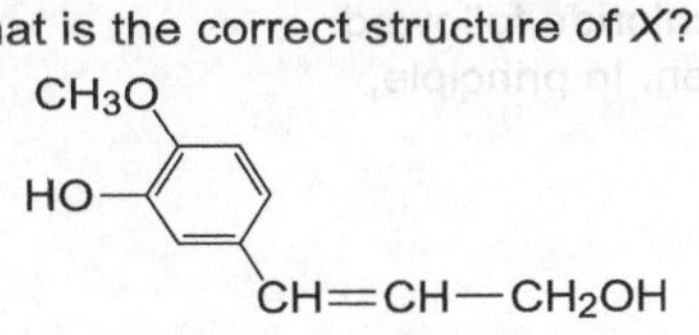

b. 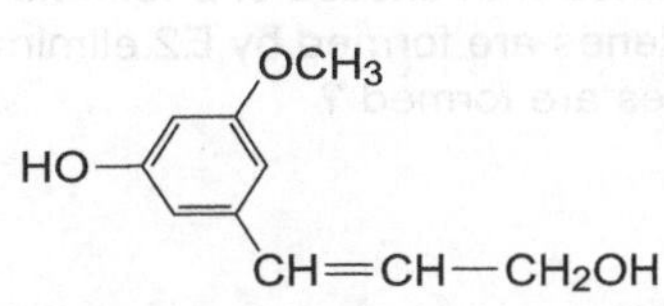

c. 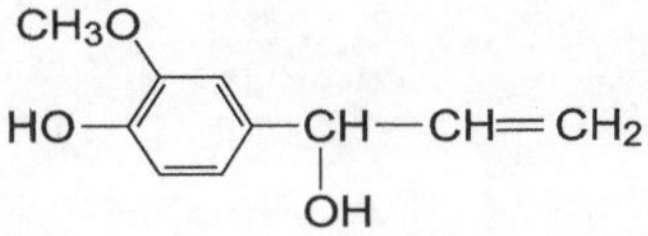

d. 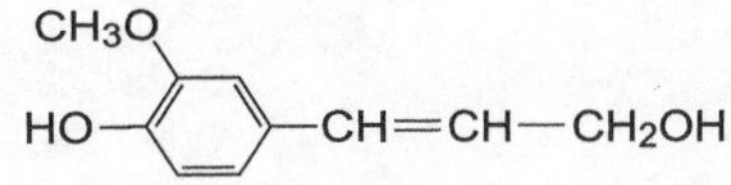

Working Space

19 If X is treated with cold HBr, the major product would be

a. 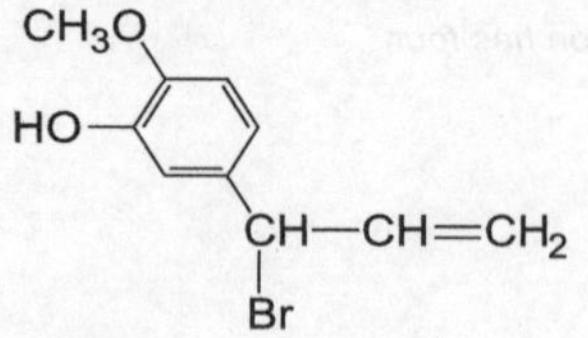

b. 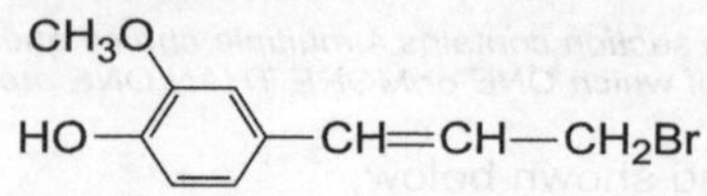

c. 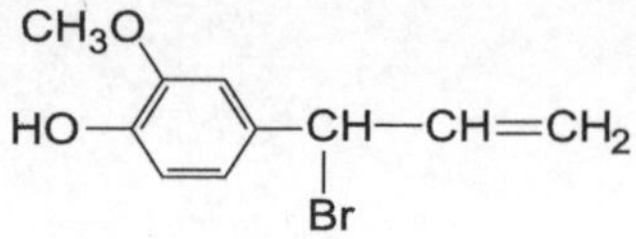

d. 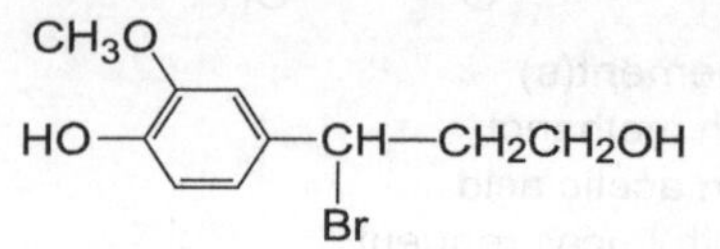

20 What would be the major product if X is treated with cold concentrated H_2SO_4?

a. 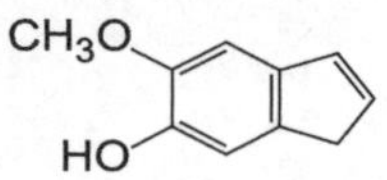

b. 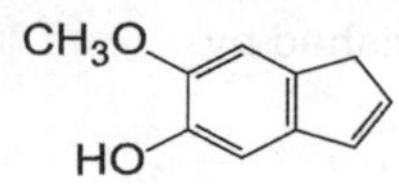

c. 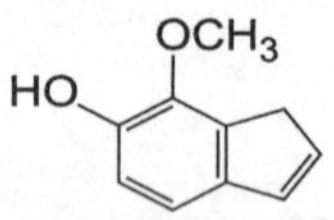

d. 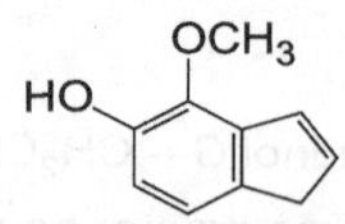

One Integer Value Correct Type

***Direction** (Q. Nos. 21-24) This section contains 4 questions. When worked out will result in an integer from 0 to 9 (both inclusive).*

21 How many reagents from the list given below, gives visible change when treated with 2-propanol?

i. $HCl/ZnCl_2$
ii. NaOH
iii. $K_2Cr_2O_7/H_2SO_4$
iv. MnO_2
v. $NaNH_2$
vi. C_2H_5MgBr
vii. NaH
viii. Cl_2/H_2O
ix. Cu/heat
x. HIO_3

22 If 4-methyl-2-pentene is refluxed with dilute H_2SO_4, hydration reaction takes place. In principle, how many different alcohols are formed ?

23 Consider the following reaction, how many different organic products are formed at the end of the reaction?

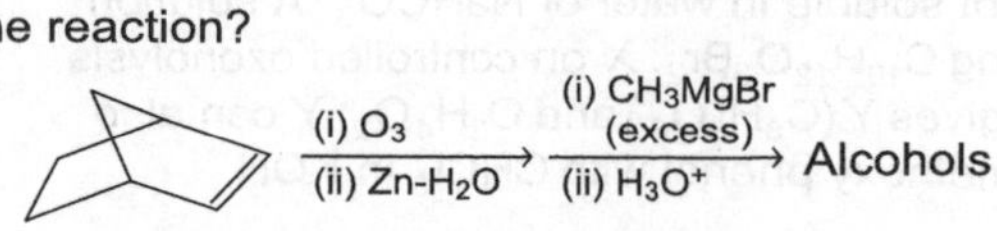

24 If pentane-2, 4-diol is treated with excess of p-toluene sulphonyl chloride followed by C_2H_5ONa/C_2H_5OH, dienes are formed by E2 elimination reaction. In principle, how many different dienes are formed ?

DPP-3 Reactions of Alcohols-II

Subjective Questions

Working Space

Direction (Q. Nos. 1-5) These questions are subjective in nature, need to be solved completely on notebook.

1 Two diastereomeric sets of enantiomer pairs *A*/*B* and *C*/*D* exist for 3-bromo-2-butanol:

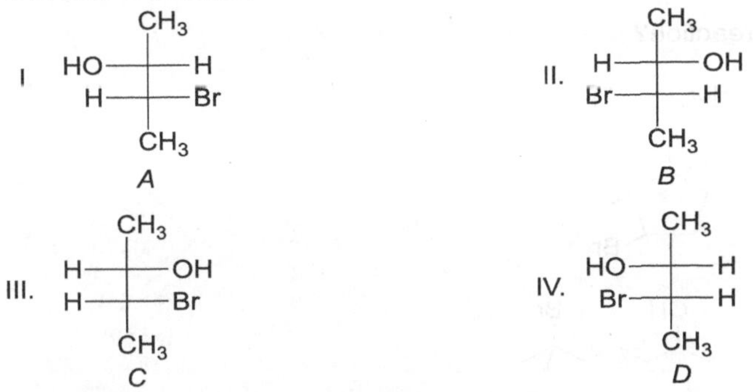

When *A* or *B* is treated with HBr, only racemic 2, 3-dibromobutane is formed, no *meso* isomer is formed. When *C* or *D* is treated with HBr, only *meso* 2, 3-dibromobutane is formed, explain.

2 Propose the mechanism of indicated products in the following reaction.

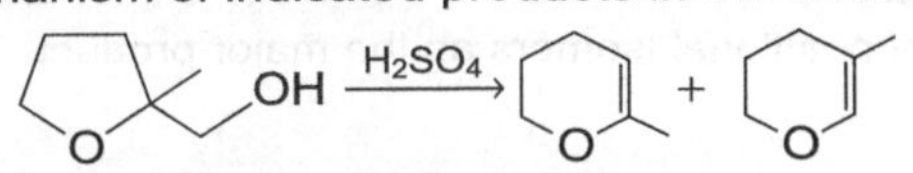

Which is the major product ?

3 Propose the structures of intermediates *A* through *E*:

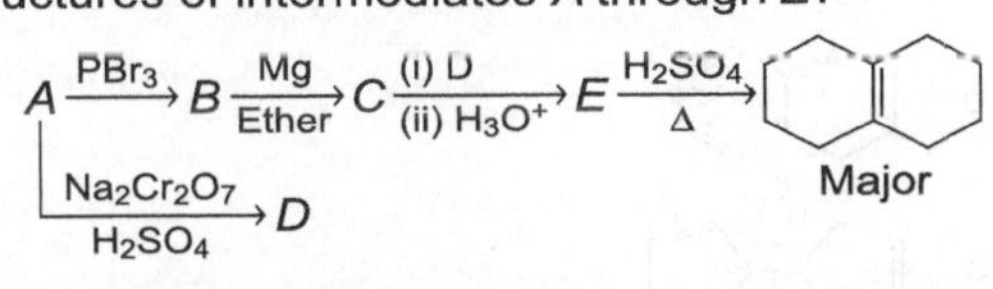

4 Compound *A* is an optically active alcohol. Treatment with chromic acid converts *A* into a ketone *B*. In a separate experiment, *A* is treated with PBr_3 converting *A* into *C*. Compound *C* is purified and then it is allowed to react with Mg in ether to give a Grignard reagent *D*. Compound *B* is added to the resulting solution of Grignard reagent. After hydrolysis, this solution is found to contain 3, 4-dimethyl hexane-3-ol. Propose the structure of *A* to *D*.

5 Discuss the stereochemistry and offer a mechanistic explanation for the product formed when a pure enantiomer of 2-butanol reacts with $SOCl_2$ (i) in the presence of pyridine (ii) in the absence of pyridine.

Only One Option Correct Type

Working Space

Direction *(Q. Nos. 6-13) This section contains 8 multiple choice questions. Each question has four choices (a), (b), (c) and (d), out of which ONLY ONE is correct.*

6 Arrange the following alcohols in the increasing order of reactivity with HBr.

I. benzyl alcohol
II. *p*-methyl benzyl alcohol
III. *p*-nitrobenzyl alcohol
IV. *p*-chlorobenzyl alcohol

a. I < II < III < IV
b. IV < III < II < I
c. II < I < IV < III
d. III < IV < I < II

7 Which is the most appropriate reagent for the following oxidation reaction?

$$CH_3—CH=CH—CH_2OH \longrightarrow CH_3—CH=CH—COOH$$

a. $KMnO_4/H_2SO_4$
b. $KMnO_4/NaOH$
c. $CrO_3/HCl/Pyridine$
d. $CrO_3/Acetone/H_2O$

8 What is the major product of the following reaction?

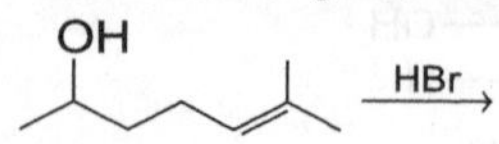

a. 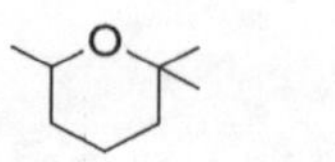
b. 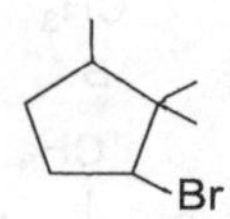
c. 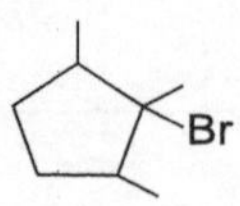
d. 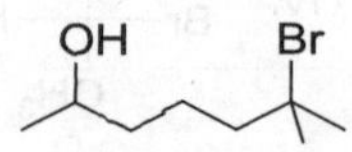

9 Which of the following is true statement regarding reaction of *cis* and *trans* 2-hexene with CH_3OH/H^+?

a. Both react at same rate
b. *Cis* isomer reacts faster than *trans* isomer
c. Both *cis* and *trans* isomers give mixture of positional isomers as the major product
d. Both (a) and (c)

10 The major product formed in the following reaction is

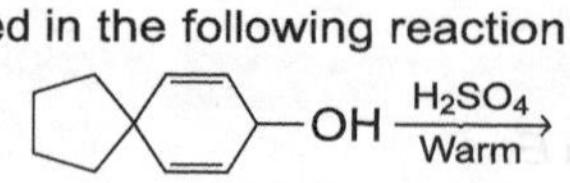

a. 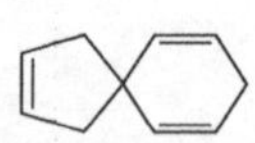
b. 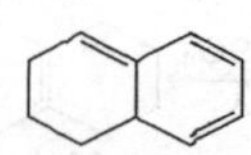
c. 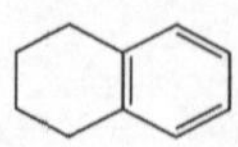
d. 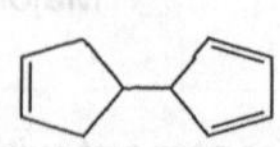

11 The relative rate of reaction of the following with H_2SO_4 is

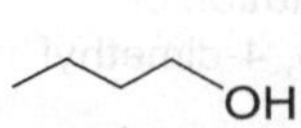
I.

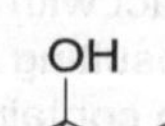
II.

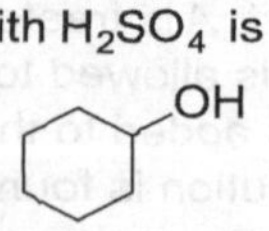
III.

a. I > II > III
b. I > III > II
c. II > I > III
d. II > III > I

Working Space

12 The major product in the following reaction is

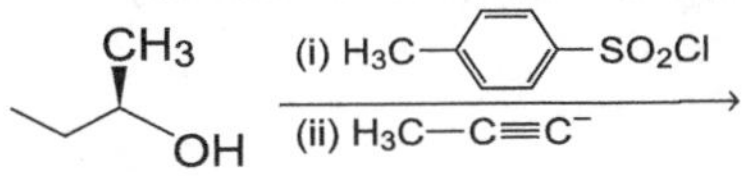

a. 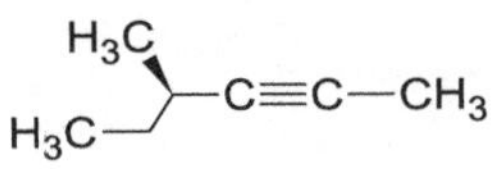
b. 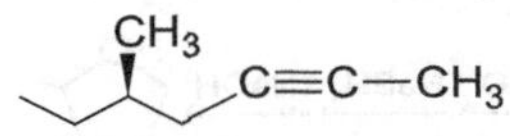

c. 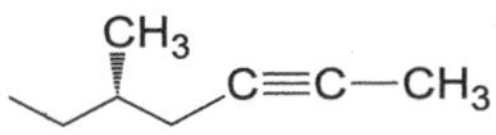
d. 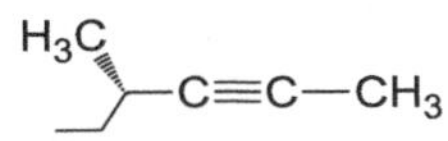

13 What is the major product of the following reaction?

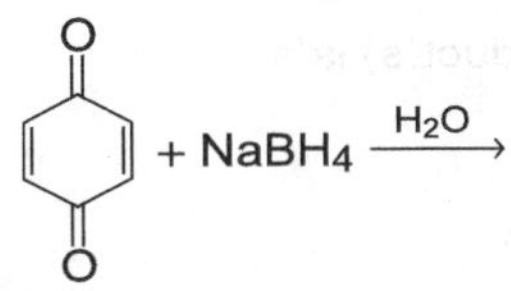

a. 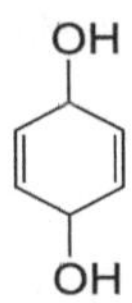
b. 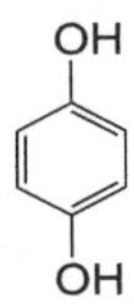
c. 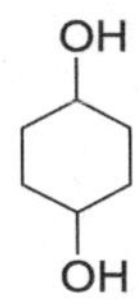
d. 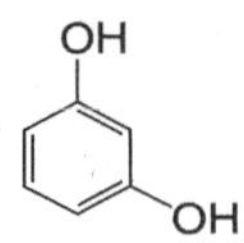

One or More than One Options Correct Type

Direction (Q. Nos. 14-19) *This section contains 6 multiple choice questions. Each question has four choices (a), (b), (c) and (d), out of which ONE or MORE THAN ONE are correct.*

14 What is/are the expected product(s) in the following reaction?

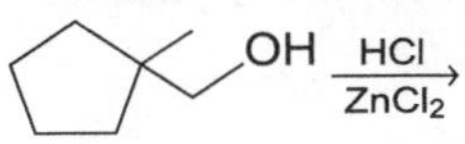

a. 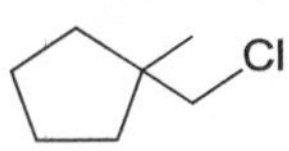
b. 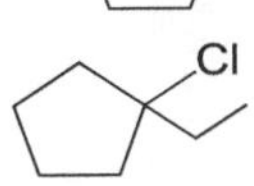
c. 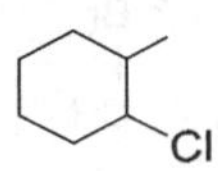
d. 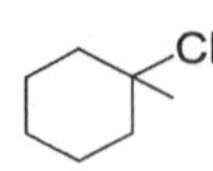

15 What is/are the possible dehydration product(s) in the following reaction?

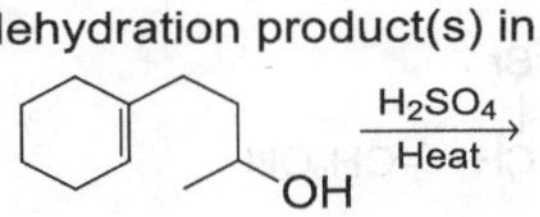

a. 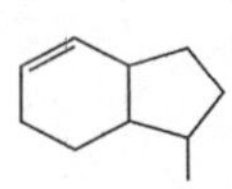
b. 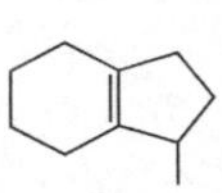
c. 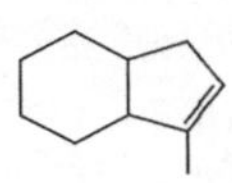
d. 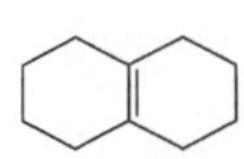

16 What is/are the possible dehydration product(s) in the following reaction?

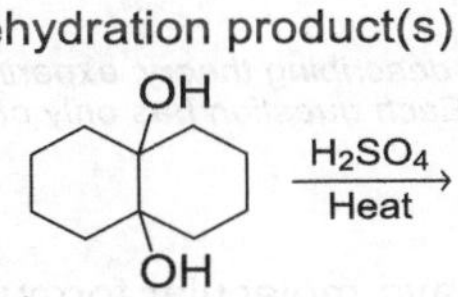

a. 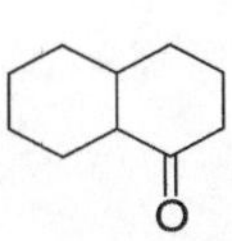
b. 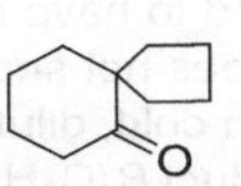

c. 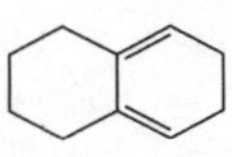
d. 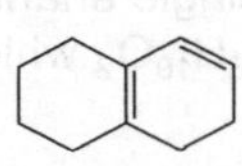

17 In which of the following reaction(s), reactant and product(s) are correctly matched ?

Working Space

a. 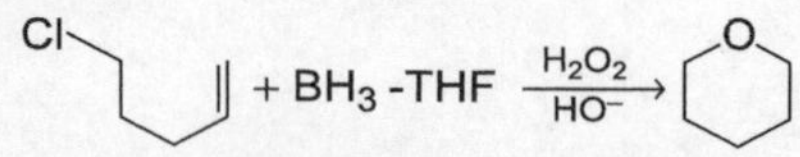

b. 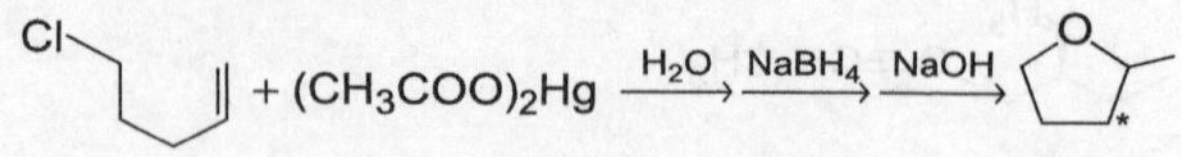

c. 

d. $C_6H_5Br + (CH_3)_3CONa \longrightarrow C_6H_5OC(CH_3)_3$

18 In the following dehydration reaction, expected product(s) is/are

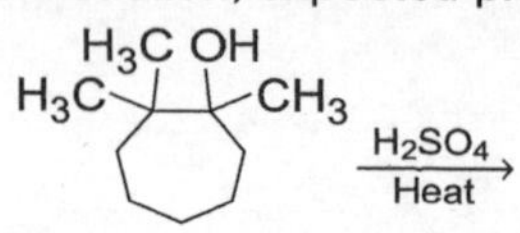

a. 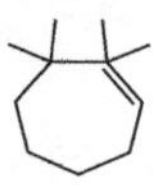

b. 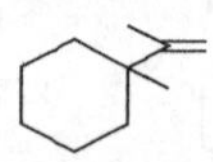

c. 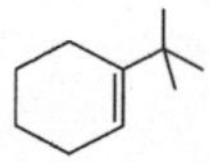

d. 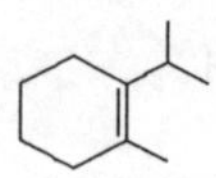

19 Upon treatment with bromine water, allyl bromide gives chiefly primary alcohol $BrCH_2CHBrCH_2OH$. What are the expected primary alcohols in the following reaction ?

$$\underset{\substack{| \\ Br^{82}}}{CH_2}-CH{=}CH_2 \xrightarrow[H_2O]{Br_2} \text{[Normal bromine is } Br^{80}]$$

a. $\underset{\substack{| \\ Br}}{CH_2}-\overset{\substack{Br \\ |}}{CH}-CH_2OH$

b. $\underset{\substack{| \\ ^{82}Br}}{CH_2}-\overset{\substack{^{82}Br \\ |}}{CH}-CH_2OH$

c. $\underset{\substack{| \\ ^{82}Br}}{CH_2}-\overset{\substack{Br \\ |}}{CH}-CH_2OH$

d. $\underset{\substack{| \\ Br}}{CH_2}-\overset{\substack{^{82}Br \\ |}}{CH}-CH_2OH$

Comprehension Type

Direction (Q. Nos. 20-24) *This section contains 2 paragraphs, each describing theory, experiments, data, etc. Five questions related to the paragraphs have been given. Each question has only one correct answer among the four given options (a), (b), (c) and (d).*

Passage I

An unknown organic compound *A* is determined to have molecular formula $C_6H_{12}O$ and passing it through chiral column does not separate it into enantiomers. *A* does not react with Br_2 nor with cold, dilute alkaline $KMnO_4$. Heating *A* with concentrated H_2SO_4 gives product *B* (C_6H_{10}) which can be separated into enantiomers. Ozonolysis of a single enantiomer of *B* produces *C*, an optically active keto-aldehyde of formula $C_6H_{10}O_2$ which gives positive iodoform test.

20 Structure of *B* that satisfy the above criteria is

a. 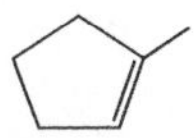 b. 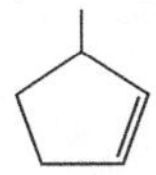 c. 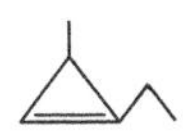 d. 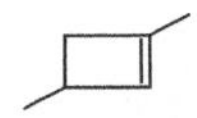

21 The correct statement regarding *C* is

a. Reduction of a pure enantiomer of *C* with $NaBH_4$ gives achiral product
b. Reduction of a pure enantiomer of *C* with $NaBH_4$ gives racemic mixture
c. Reduction of a pure enantiomer of *C* with $NaBH_4$ gives a pair of diastereomers
d. Treatment of a pure enantiomer of *C* with excess of CH_3MgBr followed by acid hydrolysis gives racemic mixture of diols

22 The correct statement regarding the outcome of the following reaction is

$$B \xrightarrow[\text{(ii) } H_2O_2/HO^-]{\text{(i) } B_2H_6} X$$

a. *X* is an enantiomer of *A*
b. *X* is a diastereomer of *A*
c. If a pure enantiomer of *B* is taken, *X* could consist of a pair of diastereomers
d. If a pure enantiomer of *B* is taken, *X* would be a racemic mixture

Passage II

An organic compound $P(C_{10}H_{18}O)$ when treated with $Br_2(l)$ gives $Q(C_{10}H_{18}OBr_4)$. *P* on vigorous oxidation gives the following compounds

$$CH_3-\overset{O}{\overset{\|}{C}}-CH_3,\ HO-\overset{O}{\overset{\|}{C}}-\overset{O}{\overset{\|}{C}}-OH,\ CH_3-\overset{O}{\overset{\|}{C}}-CH_2CH_2-COOH$$

Also, both *P* and the following compound *X* upon treatment with HBr gives the same compound $C_{10}H_{17}Br$.

$$CH_3-\overset{CH_3}{\overset{|}{C}}=CH-CH_2-CH_2-\underset{OH}{\underset{|}{\overset{CH_3}{\overset{|}{C}}}}-CH=CH_2$$

X

23 What is the structural formula of *P*?

a. $CH_3-\overset{CH_3}{\overset{|}{C}}=CH-CH=CH-\underset{OH}{\underset{|}{\overset{CH_3}{\overset{|}{C}}}}-CH_2CH_3$

b. $CH_3-\overset{CH_3}{\overset{|}{C}}=CH-CH=CH-CH_2-CH_2-\overset{OH}{\overset{|}{C}H}-CH_3$

c. $CH_3-\overset{CH_3}{\overset{|}{C}}=CH-CH=\overset{CH_3}{\overset{|}{C}}-CH_2CH_2CH_2OH$

d. $CH_3-\overset{CH_3}{\overset{|}{C}}=CH-CH_2CH_2-\underset{CH_3}{\underset{|}{C}}=CH-CH_2OH$

Working Space

24 P on treatment with dilute acid undergo isomerisation to a more stable isomer R. What is the most likely structure of R?

a. 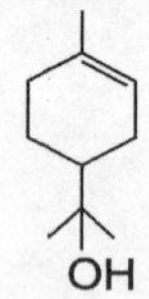

b. 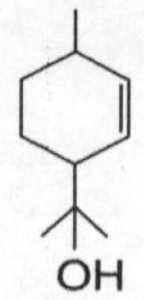

c. 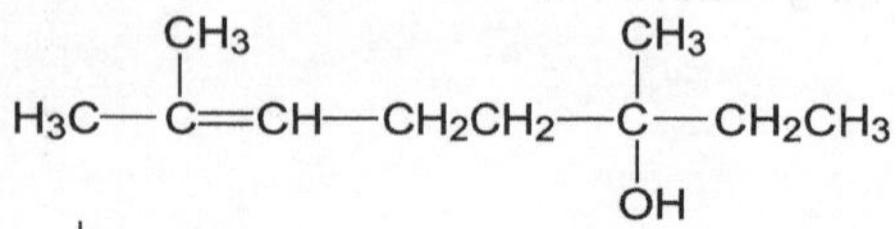

d. 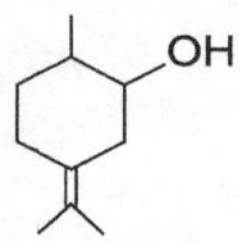

One Integer Value Correct Type

***Direction** (Q. Nos. 25-28) This section contains 4 questions. When worked out will result in an integer from 0 to 9 (both inclusive).*

25 In the reaction given below isotopic exchange occur

$$\text{Glycerol} \xrightarrow[D_2SO_4]{D_2O^{18}}$$

What is the maximum gain in molecular mass observed?

26 Consider the following dehydration reaction,

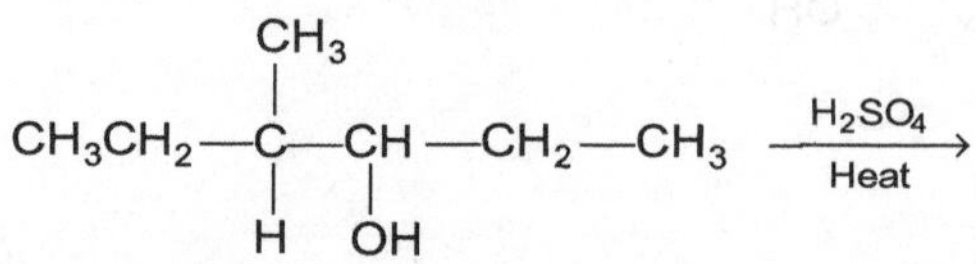

If a pure enantiomer of alcohol is taken, then how many different alkenes are formed?

27 In the following reaction how many different diols, are formed?

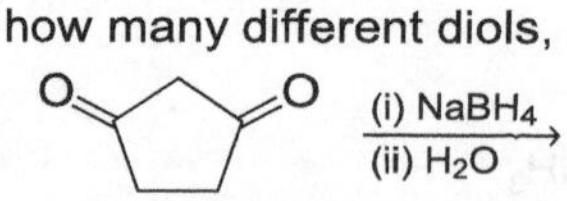

28 If all carbonyls isomers of molecular mass = 86 u are separately treated with CH_3MgBr followed by acid hydrolysis, how many of them will give racemic mixture?

DPP-4 Tests and Distinctions of Alcohols

Subjective Questions

Working Space

Direction *(Q. Nos. 1-5) These questions are subjective in nature, need to be solved completely on notebook.*

1 How are primary, secondary and tertiary alcohols distinguished using acidic solution of CrO_3 and any other reagents needed?

2 1.0 mole of each of the following are treated separately with HIO_4. What products are formed and how many moles of HIO_4 are consumed in each cases?

3 How 1-butanol, 2-butanol and tertiary butanol can be differentiated using Lucas reagent?

4 How 1-butanol, 2-butanol and tertiary butanol can be differentiated using Victor Meyer test?

5 Describe simple chemical tests that would serve to distinguish between

i. *n*-butyl alcohol and 1-octene
ii. 1-propanol and 1-bromopentane
iii. 1-butanol and 3-butene-1-ol
iv. 3-pentanol and 1-pentanol
v. α-phenyl ethyl alcohol and β-phenyl ethyl alcohol

Only One Option Correct Type

Direction *(Q. Nos. 6-13) This section contains 8 multiple choice questions. Each question has four choices (a), (b), (c) and (d), out of which ONLY ONE is correct.*

6 Which reagent can best distinguishes 3-pentanol and 3-methyl-3-pentanol?

a. I_2/NaOH
b. $KMnO_4$/NaOH
c. HIO_4
d. CrO_3/H_2SO_4

7 An organic compound $X(C_5H_{12}O)$ gives effervescence with NaH. X on treatment with acidic CrO_3 solution turns solution to blue-green forming $C_5H_{10}O_2$. Also X can be resolved into enantiomers. The correct statement regarding X is

a. It is 2-pentanol
b. It is 3-pentanol
c. With HBr, it gives achiral alkyl bromide
d. The oxidation product $C_5H_{10}O_2$ is achiral

8 An organic compound $C_4H_{10}O(X)$ on reaction with I_2/red-P gives C_4H_9I which on further reaction with $AgNO_2$ gives $C_4H_9NO_2(Y)$. Y on treatment with HNO_2 forms a blue solution which turns to red on making solution slightly alkaline. The possible identity of X is

a. 1-butanol
b. 2-methyl-1-pentanol
c. 2-butanol
d. Either (a) or (b)

Working Space

9 An organic compound $X(C_7H_{16}O)$ on treatment with concentrated HCl solution forms immediate turbidity even in the absence of $ZnCl_2$. Also, X is resolvable into enantiomers. The correct statement concerning X is

a. It is 3-methyl-2-hexanol
b. It turns orange CrO_3/H_2SO_4 solution green
c. It is 3-methyl-3-hexanol
d. It forms yellow precipitate with I_2/NaOH

10 Which reagent given below does not produce any visible change when added to ethylene glycol?

a. $NaNH_2$ b. CrO_3/H_2SO_4
c. H_2CO/H_2SO_4 d. NaH

11 Which of the following is least likely to form turbidity with HCl in the presence of $ZnCl_2$ at room temperature?

a. 
b. 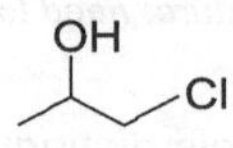
c. 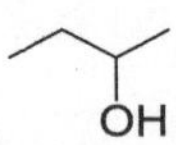
d. 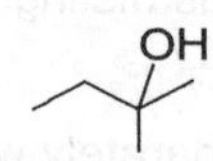

12 The correct statement regarding 3-ethyl-3-hexanol is

a. It changes colour of CrO_3/H_2SO_4
b. It is oxidised on heating with copper metal, producing ketone
c. It gives yellow precipitate with NaOH/I_2
d. It changes colour of cerric nitrate $[Ce(NO_3)_4]$ from yellow to red

13 An alcohol has molecular formula $C_6H_{12}O$ X and it gives immediate turbidity with cold, concentrated HCl even in the absence of $ZnCl_2$. X can also be obtained by treatment of an ether with excess of CH_3MgBr followed by acid hydrolysis. Hence, the correct statement regarding X is

a. It is 3-methyl-3-pentanol
b. It is 2-methyl-3-pentanol
c. It is 2-methyl-2-pentanol
d. Either (b) or (c)

One or More than One Options Correct Type

Direction *(Q. Nos. 14-17) This section contains 4 multiple choice questions. Each question has four choices (a), (b), (c) and (d), out of which ONE or MORE THAN ONE are correct.*

14 Which reagent(s) can be used to differentiate between 2-pentanol and 1-pentanol?

a. I_2/NaOH b. HCl/$ZnCl_2$
c. $CrO_3 - H_2SO_4$ d. $NaNH_2$

15 Alcohols given below that behaves like 1°-aliphatic alcohol in Lucas test is/are

a. 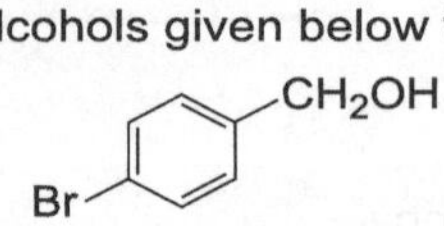

b. 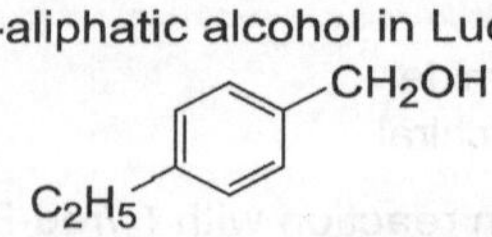

c. 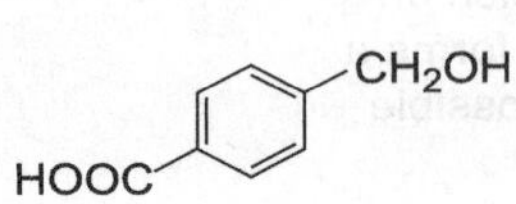

d. 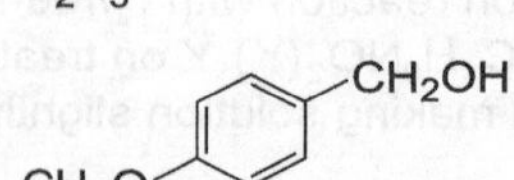

Working Space

16 Consider the following reaction sequence,

$$R\text{—}OH \xrightarrow{I_2/\text{P-red}} R\text{—}I \xrightarrow{AgNO_2} R\text{—}NO_2 \xrightarrow{HNO_2} \text{Blue}$$

Hence, R—OH could be

a. C_6H_5—CH_2OH

b. CH_3CH_2OH

c. 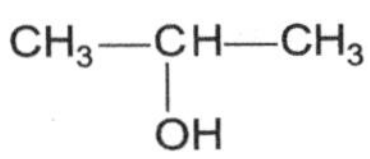

d. 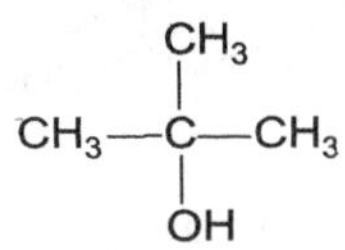

17 Consider the following reaction,

$$X(C_6H_{12}O) \xrightarrow{Na} \text{Gas evolved}$$

$$\underset{\text{Chiral}}{X} \xrightarrow{H_2/Ni} \underset{\text{Achiral}}{Y(C_6H_{14}O)}$$

The correct statement(s) concerning X and Y is/are

a. Both form immediate turbidity with HCl in the presence of $ZnCl_2$

b. Both change colour of CrO_3 - H_2SO_4

c. X gives yellow solid with NaOH/I_2

d. X decolourises Br_2 - CCl_4 solution forming $C_6H_{12}OBr_2$

Comprehension Type

Direction (Q. Nos. 18-20) *This section contains a paragraph, describing theory, experiments, data, etc. Three questions related to the paragraph have been given. Each question has only one correct answer among the four given options (a), (b), (c) and (d).*

Passage

An organic compound $X(C_9H_{12}O)$ gives the following reactions :

i. Na – Slow gas bubble formation
ii. Acetic anhydride – Pleasent smelling liquid
iii. CrO_3 - H_2SO_4 – Blue-green solution
iv. Hot $KMnO_4$ – Benzoic acid
v. Br_2 - CCl_4 – No decolouration
vi. I_2 + NaOH – Yellow solid is formed
vii. X rotates the plane polarised light

18 The structure of X is

a. 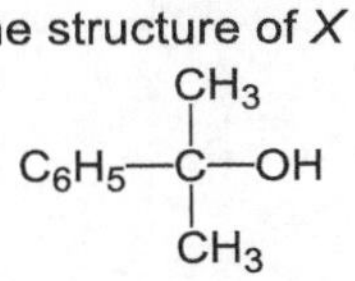

b. 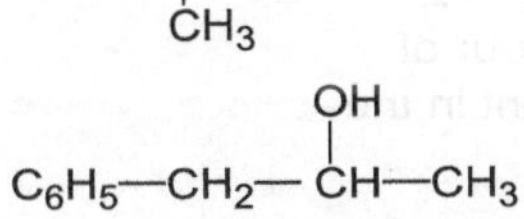

c. C_6H_5— $CH_2CH_2CH_2OH$

d. 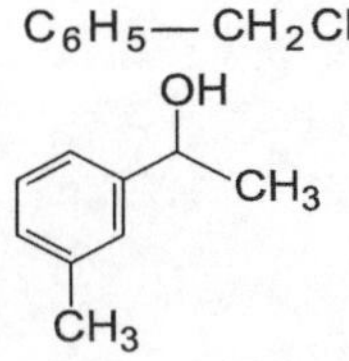

Working Space

19 If X is treated with HCl in the presence of $ZnCl_2$, the major product would be

a. 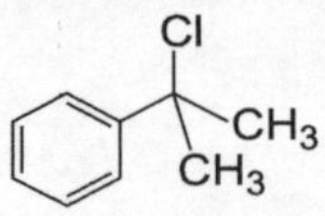

b. 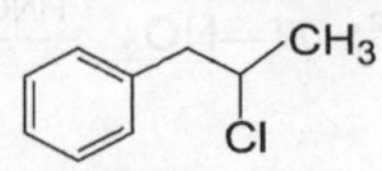

c. 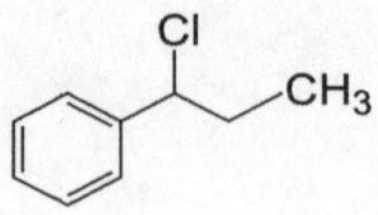

d. 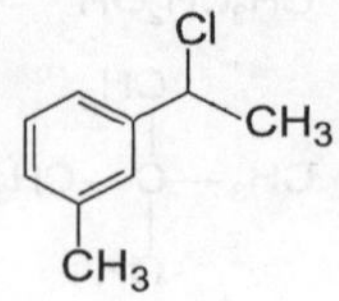

20 An isomer of X, Y was also found to be optically active. It showed the same reactions as X except (vi). Hence, Y is

a. 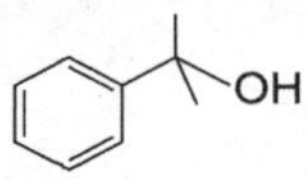

b. 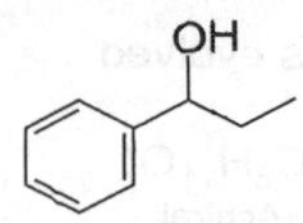

c. 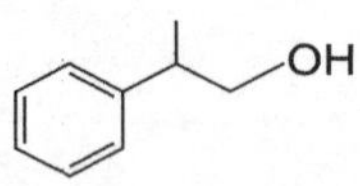

d. 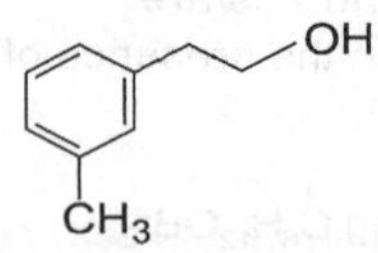

One Integer Value Correct Type

Direction *(Q. Nos. 21-24) This section contains 4 questions. When worked out will result in an integer from 0 to 9 (both inclusive).*

21 An organic compound $X(C_5H_{12}O)$ is chiral. Also

$$X \xrightarrow{I_2/\text{Red-P}} C_5H_{11}I \xrightarrow{AgNO_2} \xrightarrow{HNO_2} \text{Blue} \xrightarrow{NaOH} \text{Red}$$

How many primary hydrogens are present in X?

22 An alcohol $X(C_7H_{16}O)$ gives immediate turbidity with $HCl/ZnCl_2$ and can be resolved into enantiomers. Also X has a tertiary hydrogen. How many CH_3— groups are present in X?

23 An alcohol $X(C_4H_{10}O_3)$ is chiral and absorbs two moles of HIO_4 per mole of X. How many stereoisomers exist for X?

24 An organic compound $X(C_7H_{16}O)$ gives effervescence with Na. X has both enantiomers and diastereomers. X gives yellow precipitate with alkaline iodine solution. With CrO_3 -H_2SO_4, X is converted into $C_7H_{14}O(Y)$ which has enantiomers but not diastereomers. X on refluxing with H_2SO_4 isomerises to Z which neither gives yellow precipitate with alkaline iodine nor changes colour of CrO_3 -H_2SO_4. What is the minimum number of carbons that can be present in the parent chain of X?

DPP-5 Ethers

Subjective Questions

Working Space

Direction (Q. Nos. 1-6) *These questions are subjective in nature, need to be solved completely on notebook.*

1 1-propanol on heating with concentrated H_2SO_4 at 140°C gives *n*-propyl ether as the major product while on similar treatment 2-propanol gives mainly propene, explain.

2 Predict the major organic product of the following reaction.

i. 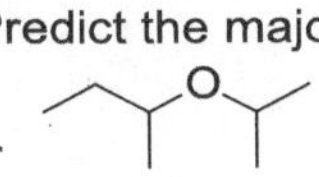 + HBr (excess) $\xrightarrow{\text{Heat}}$

ii. 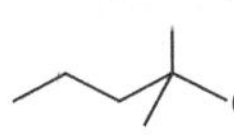 OC_2H_5 + HBr (excess) $\xrightarrow{\text{Heat}}$

iii. $C_6H_5—O—C_2H_5 \xrightarrow[\text{Heat}]{\text{HI}}$

iv. 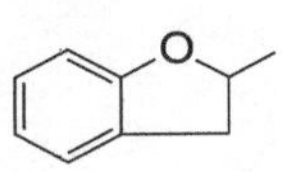 + HBr (conc.) $\xrightarrow{\text{Heat}}$

3 Propyl ether stored for long time, on subjecting to distillation undergoes explosion, explain.

4 Show how would you convert 1-hexene into

i. 2-methoxy hexane?
ii. 1-methoxy hexane?
iii. 1-methoxy hexan-2-ol?
iv. 2-methoxyhexan-1-ol?
v. 1-phenyl hexane-2-ol?
vi. 2-methoxy-1-phenyl hexane?

5 Synthesise the following ethers from the indicated starting materials.

i. cyclopentyl-*n*-propyl ether from cyclopentanol and 1-propanol
ii. 2-ethoxy octane from octene
iii. 1-ethoxy-1-methylcyclohexane from 2-methylcyclohexanol
iv. 1-methoxy decane from1-decene

6 There are two different ways of making 2-ethoxy octane from octan-2-ol using Williamson's ether synthesis. When pure (–)octane-2-ol of specific rotation – 8.24° is treated with sodium metal and then ethyl iodide, the product is 2-ethoxy octane with specific rotation of – 15.6°. When pure (–)2-octanol is treated with tosyl chloride then with sodium ethoxide, the product is again 2-ethoxyoctane. Discuss the stereochemistry of product formed by tosylate route.

Only One Option Correct Type

Working Space

Direction *(Q. Nos. 7-14) This section contains 8 multiple choice questions. Each question has four choices (a), (b), (c) and (d), out of which ONLY ONE is correct.*

7 Identify the final major product of the reaction sequence.

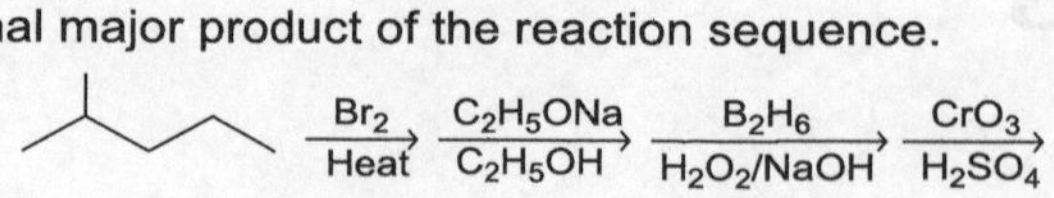

a. 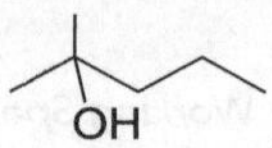   b. 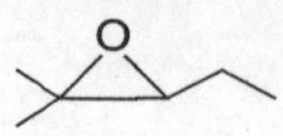

c. 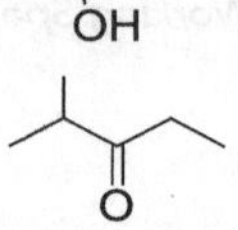   d. 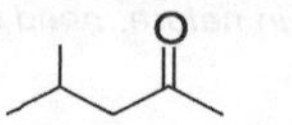

8 Which is the best reaction for preparation of *t*-butyl ethyl ether?

a. $CH_3-C(CH_3)_2-I + CH_3CH_2ONa \longrightarrow$

b. $CH_3-C(CH_3)_2-ONa + CH_3CH_2Br \longrightarrow$

c. $CH_3-C(CH_3)_2-ONa + CH_3CH_2OH \longrightarrow$

d. $CH_3-C(CH_3)_2-OH + C_2H_5OH \xrightarrow[140^\circ C]{H_2SO_4}$

9 The major organic product formed in the following reaction is

$$CH_3-C(CH_3)_2-CH=CH_2 + C_2H_5OH \xrightarrow[\Delta]{\text{Conc } H_2SO_4}$$

a. $CH_3-C(CH_3)_2-CH_2CH_2-OC_2H_5$

b. $CH_3-C(CH_3)=C(CH_3)-CH_3$

c. $CH_3-C(CH_3)_2-CH(OC_2H_5)-CH_3$

d. $CH_3-C(CH_3)(OC_2H_5)-CH(CH_3)-CH_3$

Working Space

10 Consider the following chain of reactions :

The major product formed in the final step is

a. 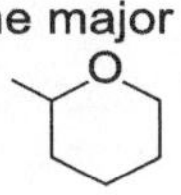

b. 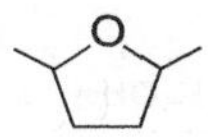

c. 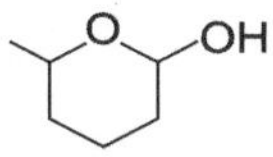

d. 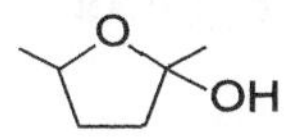

11 The best set of reagents for the following transformation is

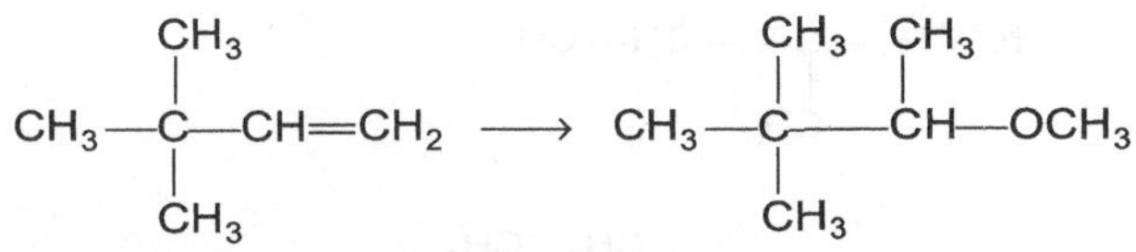

a. CH_3OH/H_2SO_4
b. (I) $(CH_3COO)_2Hg$-H_2O, (II) $NaBH_4$, (III) CH_3OH/H_2SO_4
c. $(CH_3COO)_2Hg$-CH_3OH followed by $NaBH_4$
d. (I) H_2SO_4/heat (II) Na (III) CH_3I

12 The major organic product formed in the following reaction is

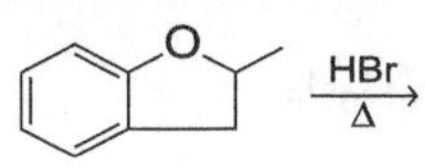

a. 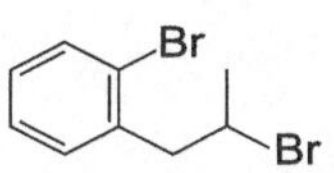

b. 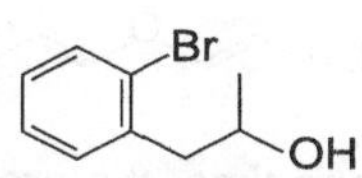

c. 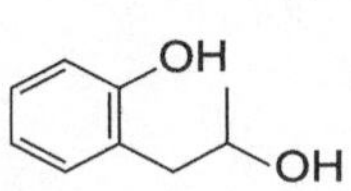

d. 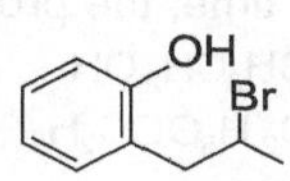

13 Inorganic salts are usually insoluble in organic solvents e.g. $KMnO_4$ is insoluble in benzene. However presence of one of the following compound in benzene makes $KMnO_4$ soluble and gives a purple coloured solution. Which is that compound?

a. $CH_3CH_2OCH_2CH_3$

b. 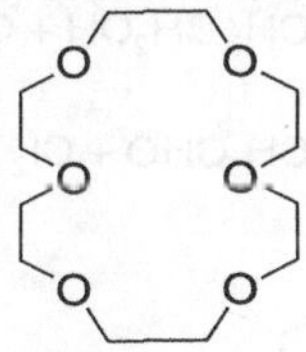

c. CH_3CH_2OH

d. H_2O

14 Consider the following roadmap reaction :

$$X\ (\text{Hydrocarbon}) \xrightarrow[h\nu]{Br_2} \xrightarrow[(CH_3)_3COH]{(CH_3)_3COK} Y\ (\text{major}) \xrightarrow[H_2O_2]{HBr} Z$$

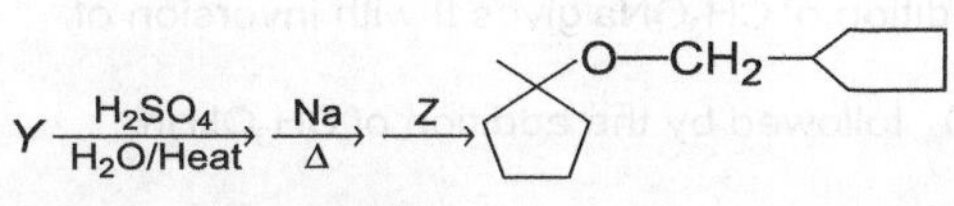

The most probable structure of *X* is

a. 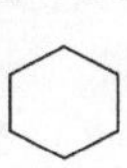

b. 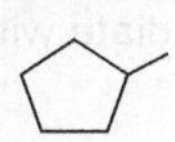

c. 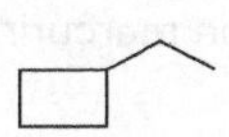

d. 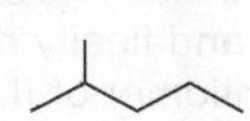

One or More than One Options Correct Type

Working Space

Direction (Q. Nos. 15-20) *This section contains 6 multiple choice questions. Each question has four choices (a), (b), (c) and (d), out of which ONE or MORE THAN ONE are correct.*

15 Consider the following reaction,

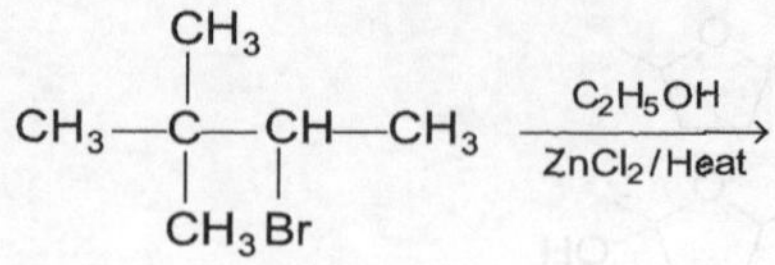

The expected substitution product(s) is/are

a. 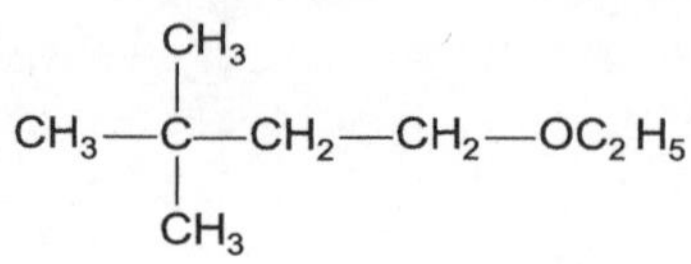

b. 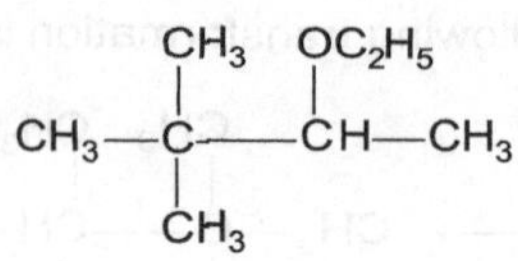

c. 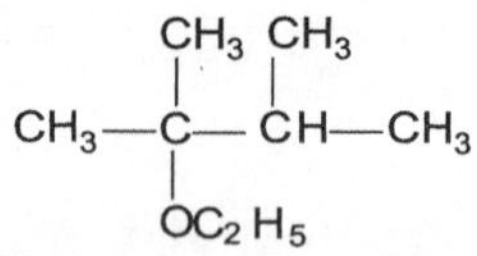

d. 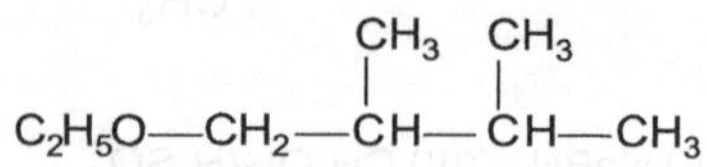

16 Which of the following ethers can be obtained in good yield by intermolecular dehydration of alcohols?

a. 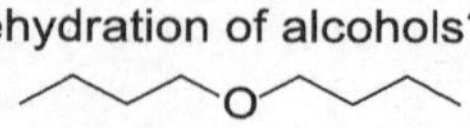

b. $C_6H_5—O—C_6H_5$

c. 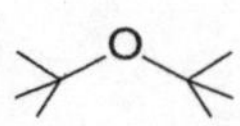

d. 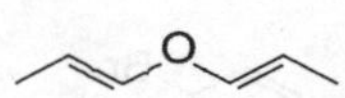

17 If diethyl ether is left in contact of air for long time, the product formed is/are

a. CH_3COOH

b. CH_3CH_2OH

c. $CH_3CH_2O—\underset{\displaystyle OOH}{\underset{|}{C}H}—CH_3$

d. $C_2H_5OOC_2H_5$

18 In which of the following reaction, the major product is an acyclic ether?

a. $C_6H_5OH + (CH_3)_2SO_4 \xrightarrow{\Delta}$

b. $CH_3CH_2OH + CH_2N_2 \xrightarrow{\Delta}$

c. $C_2H_5OH + \text{conc. } H_2SO_4 \xrightarrow{140°C}$

d. $CH_3CHO + Cl_2 \xrightarrow{CH_3COOH} \xrightarrow{NaBH_4} \xrightarrow{KOH}$

19 Consider the following reaction,

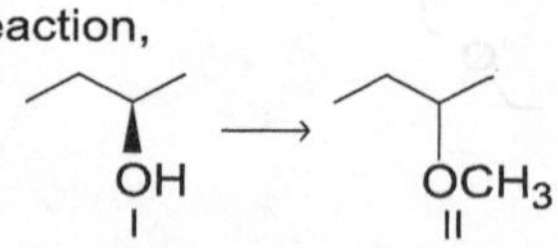

The correct statement(s) concerning the above transformation is/are

a. If I is treated with Na followed by addition of CH_3I gives II with the retention of configuration

b. If I is treated with TsCl followed by the addition of CH_3ONa gives II with inversion of configuration

c. If I is first heated with concentratred H_2SO_4 followed by the addition of CH_3OH in dil. H_2SO_4 gives racemic mixture of II

d. If I is heated with concentrated H_2SO_4 followed by the treatment with $(CH_3COO)_2Hg—CH_3OH$ and finally reducing the mercurinium intermediate with $NaBH_4$ gives a pure enantiomer of II

Working Space

20 Consider the following reaction,

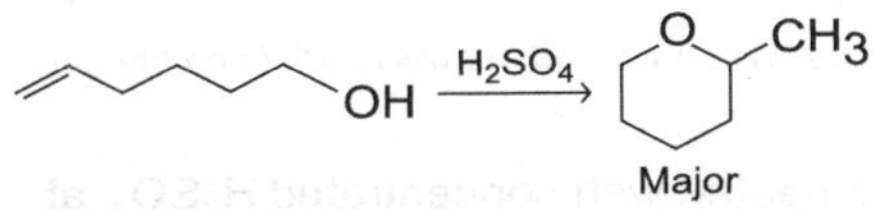

The correct statement regarding the above reaction is/are

a. In first step, olefinic bond is protonated

b. In first step, hydroxy group is protonated which later forms carbocation intermediate

c. Product would be a racemic mixture

d. Besides the shown product, some five membered cyclic ether is also formed

Comprehension Type

Direction (Q. Nos. 21-23) *This section contains a paragraph, describing theory, experiments, data, etc. Three questions related to the paragraph have been given. Each question has only one correct answer among the four given options (a), (b), (c) and (d).*

Passage

An organic compound *X* has molecular formula $C_{11}H_{16}O$ and it can be resolved into enantiomers. *X* does not evolve any gas with Na metal. *X* when treated with concentrated HBr gives C_6H_5OH and $Y(C_5H_{11}Br)$. *Y* is chiral and has the same configuration as that of compound *X*.

21 Which reaction gives *X* as the major product?

a. $C_6H_5ONa + CH_3—\underset{|}{\underset{Br}{CH}}—CH_2CH_2CH_3 \longrightarrow$

b. $C_6H_5ONa + CH_3—\underset{Br}{\underset{|}{CH}}—\underset{CH_3}{\underset{|}{CH}}—CH_3 \longrightarrow$

c. $C_6H_5ONa + BrCH_2—\underset{CH_3}{\underset{|}{CH}}—CH_2CH_3 \longrightarrow$

d. $C_6H_5ONa + CH_3—\overset{CH_3}{\overset{|}{\underset{Br}{\underset{|}{C}}}}—CH_2CH_3 \longrightarrow$

22 If *Y* is treated with C_2H_5ONa/C_2H_5OH, then how many different E2 products would be formed?

a. 1 b. 2 c. 3 d. 4

23 If *X* is treated with H_2SO_4 it undergoes a rearrangement to give *Z* as the major organic product. What is *Z*?

a. $C_6H_5—O—\overset{CH_3}{\overset{|}{\underset{CH_3}{\underset{|}{C}}}}—C_6H_5$

b. $C_6H_5—\overset{CH_3}{\overset{|}{\underset{CH_3}{\underset{|}{C}}}}—\overset{OH}{\overset{|}{CH}}—CH_3$

c. $C_6H_5—\overset{CH_3}{\overset{|}{\underset{CH_3}{\underset{|}{C}}}}—CH_2—CH_2OH$

d. 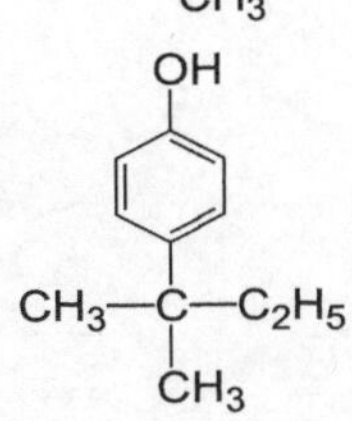

One Integer Value Correct Type

Working Space

Direction (Q. Nos. 24-28) This section contains 5 questions. When worked out will result in an integer from 0 to 9 (both inclusive).

24 If a mixture of a 1-propanol and 2-butanol is heated with concentrated H_2SO_4 at 140° C, how many different ethers would be formed?

25 How many lone pairs are present in 12-crown-4?

26 When 2-ethyl-3-methyl-1-pentene is treated with CH_3OH in H_2SO_4, how many different methoxy ethers would be formed in significant amount?

27 How many of the following reactions give the cyclic ether as the major organic product?

i. 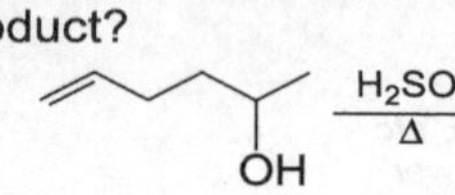

ii. 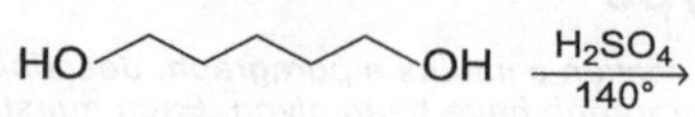

iii. $CH_3-\overset{O}{\overset{\|}{C}}-CH_3 + CH_2{=}C{=}O \xrightarrow{\Delta}$

iv. $CH_3-CH{=}CH_2 + Ag_2O \xrightarrow{\Delta}$

v. 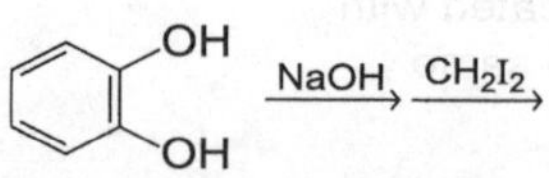

vi. 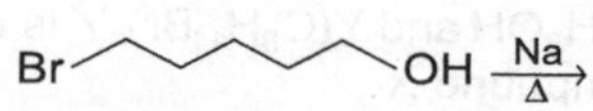

vii. 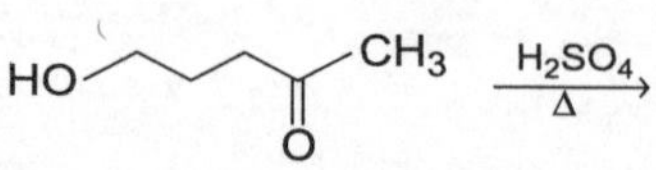

viii. $CH_3-\overset{O}{\overset{\|}{C}}-CH_3 + \underset{OH}{CH_2}-\underset{OH}{CH_2} \xrightarrow[\Delta]{H_2SO_4}$

ix. $HO-CH_2-CH_2-OH \xrightarrow[\Delta]{CH_2N_2}$

x. 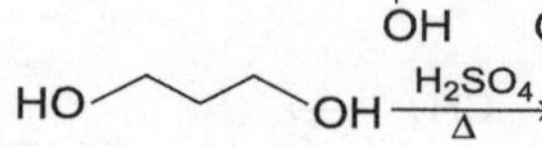

28 How many ether isomers exist for $C_5H_{12}O$?

DPP-6 Epoxides

Subjective Questions

Working Space

Direction (Q. Nos. 1-5) *These questions are subjective in nature, need to be solved completely on notebook.*

1 Explain the product formed in the following rearrangement :

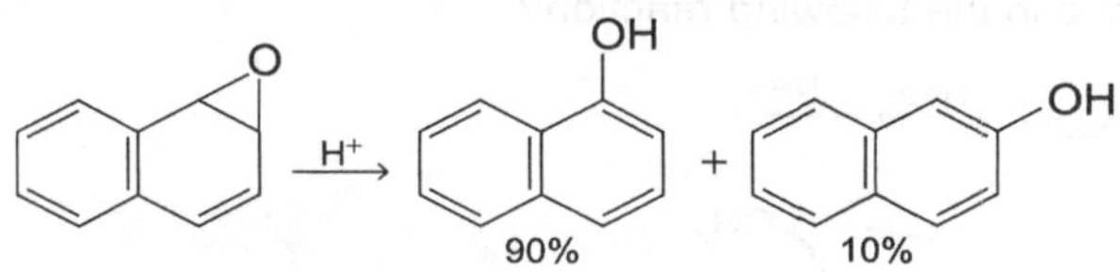

2 Predict the major product in the following reactions.

i. 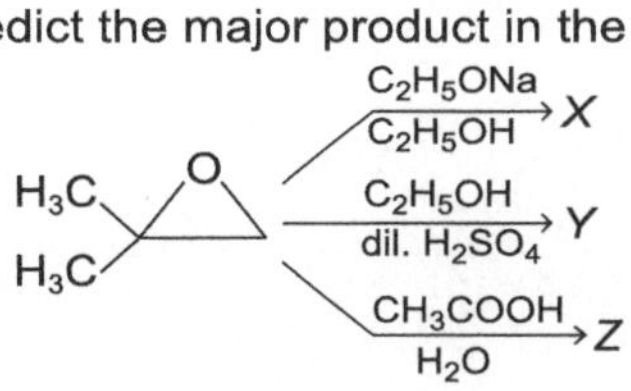

ii. 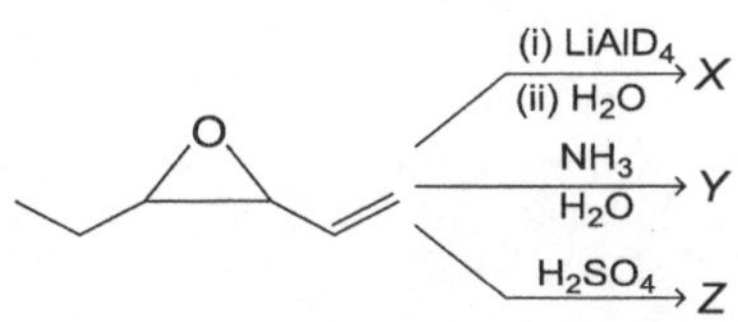

3 Identify products *X*, *Y* and *Z* in the following chain reactions.

$$C_6H_5-\overset{\overset{\large O}{\|}}{C}-CH_3 \xrightarrow[CH_3COOH]{Cl_2} X \xrightarrow[\text{(ii) } H_2O]{\text{(i) } NaBH_4} Y \xrightarrow{CaO} Z$$

4 Using 1-butanol and ethanol as only source of carbon, synthesise racemic mixture of 2-ethoxy-1-butanol.

5 Predict the products, mentioning stereochemistry where applicable, in the following reactions.

i. 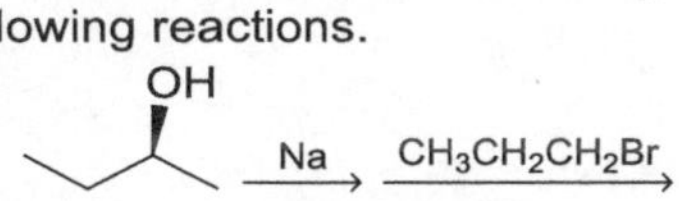

ii. 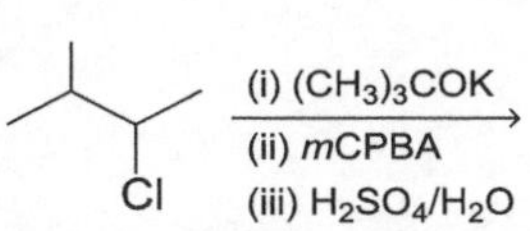

iii. 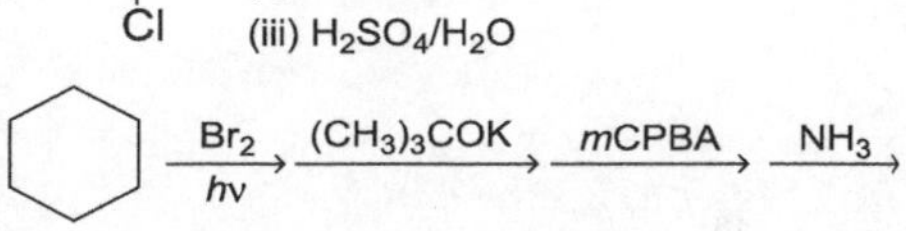

Only One Option Correct Type

Working Space

Direction *(Q. Nos. 6-13) This section contains 8 multiple choice questions. Each question has four choices (a), (b), (c) and (d), out of which ONLY ONE is correct.*

6 The most appropriate reagents that can bring about the following transformation is

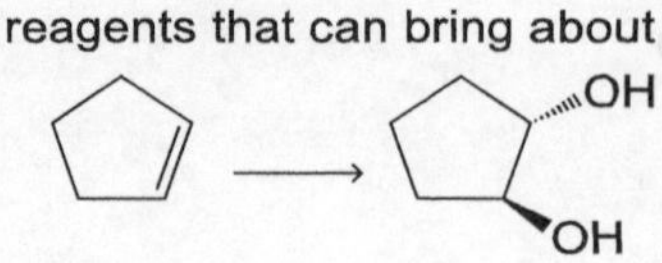

a. $KMnO_4/NaOH(aq)$
b. $H_2SO_4(aq)$/heat
c. Br_2 - H_2O then NaOH
d. CF_3CO_3H then $H_2SO_4(H_2O)$

7 Which major organic product is formed in the following reaction?

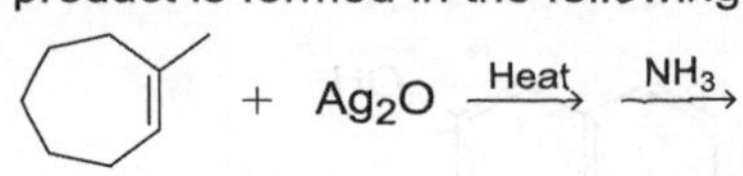

a. 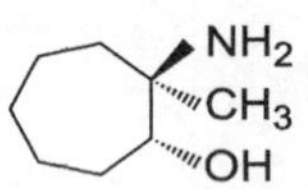
b. 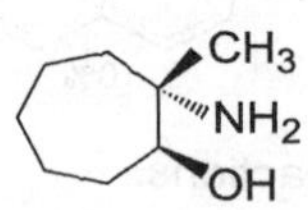

c. 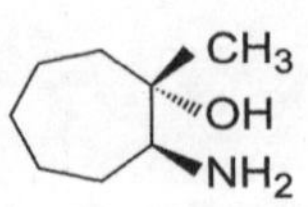
d. 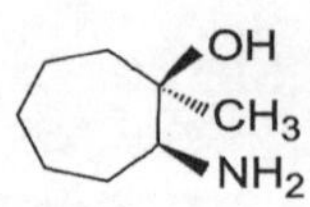

8 The major organic product formed in the reaction given below is

$$CH_3-\overset{O}{\overset{\|}{C}}-CH_3 \xrightarrow[\text{Heat}]{CH_2N_2} \xrightarrow[H_2SO_4]{C_2H_5OH}$$

a. $CH_3-C(OH)(CH_3)-CH_2-OC_2H_5$
b. $CH_3-C(OC_2H_5)(CH_3)-CH_2OH$
c. $CH_3-C(OH)(CH_3)-CH_2OH$
d. $CH_3-C(OC_2H_5)(CH_3)-CH_2-OC_2H_5$

9 In the following hydrolysis reaction, the major final product is

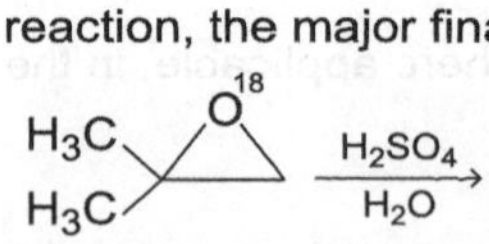

a. $CH_3-C(OH)(CH_3)-CH_2-{}^{18}OH$
b. $CH_3-C({}^{18}OH)(CH_3)-CH_2OH$
c. $CH_3-C(OH)(CH_3)-CH_2OH$
d. $CH_3-C({}^{18}OH)(CH_3)-CH_2-{}^{18}OH$

Working Space

10 In the reaction given below,

$$cis\text{-}3\text{-hexene} + CF_3CO_3H \longrightarrow X \xrightarrow[H_2O]{H_2SO_4} Y$$

The correct deduction concerning *X* and *Y* is

a. Both *X* and *Y* are *meso* isomers
b. *X* is *meso* isomer while *Y* is racemic mixture
c. *X* is racemic mixture while *Y* is *meso*
d. Both *X* and *Y* are racemic mixture

11 An alkene on treatment with *meta* chloro perbenzoic acid undergoes epoxidation reaction to give oxirane. What is the correct order of increasing reactivity of the following in the above mentioned epoxidation reaction?

I. $CH_2{=}CH_2$ II. $CH_3{-}CH{=}CH_2$

III. $CH_3{-}CH{=}C(CH_3){-}CH_3$ IV. $CF_3{-}CH{=}CH_2$

a. I < II < III < IV b. IV < III < II < I
c. IV < I < II < III d. III < II < IV < I

12 Which of the following compound gives more than one oxirane upon epoxidation with Ag_2O/heat?

a. $CH_3{-}C(CH_3){=}CH_2$

b. 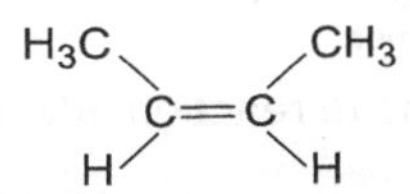

c. 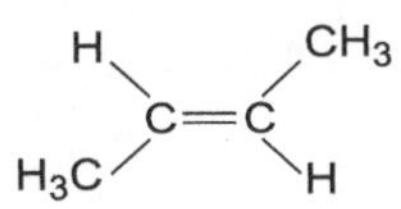

d. 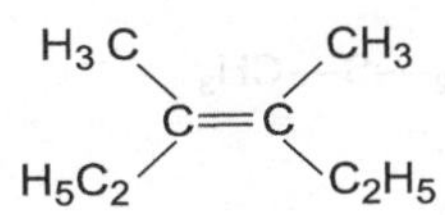

13 The reaction given below that gives oxirane as the major organic product is

a. $CH_3CH_2OH + CH_2N_2 \xrightarrow{Heat}$
b. $CH_2{=}CH_2 \xrightarrow[H_2O]{Br_2} \xrightarrow{CaO}$
c. $CH_3{-}\overset{O}{\overset{\|}{C}}{-}CH_3 + Ag_2O \xrightarrow{Heat}$
d. $CH_3{-}CH{=}CH_2 + CH_2{=}S(O)(CH_3)_2 \xrightarrow{Heat}$

One or More than One Options Correct Type

Direction (Q. Nos. 14-19) *This section contains 6 multiple choice questions. Each question has four choices (a), (b), (c) and (d), out of which ONE or MORE THAN ONE are correct.*

14 Consider the following reaction sequence,

$$CH_2{=}CH{-}CH_3 \xrightarrow[Heat]{Cl_2} I \xrightarrow[H_2O]{NaOH} II(C_3H_6O)$$

$$II \xrightarrow[H_2O]{Cl_2} III \xrightarrow{Ca(OH)_2} IV\ (C_3H_6O_2) \xrightarrow[HCl]{H_2O} HO{-}CH_2{-}\underset{}{CH(OH)}{-}CH_2OH$$

Glycerine

The intermediate products (I – IV) in the above reaction is/are

a. $CH_2(Cl){-}CH(Cl){-}CH_3$

b. 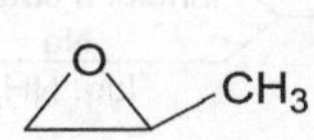

c. $CH_2{=}CH{-}CH_2OH$

d. 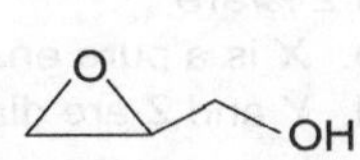

Working Space

15 The reagent that can bring about the following transformation in good yield is/are

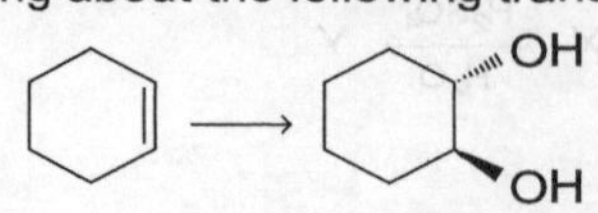

a. $O_3/Zn - H_2O$
b. $KMnO_4/H_2O/NaOH$
c. CF_3CO_3H/H_2O (boil)
d. $Cl_2 - H_2O$ then CaO then H_2SO_4(dil.)

16 In the reaction below,

$$CH_3-C\equiv C-CH_3 \xrightarrow[\text{Liq. } NH_3]{Na} X \xrightarrow[\text{Heat}]{Ag_2O} Y \xrightarrow[H_2O]{NaOH} Z$$

The correct statement concerning the above reaction is/are
a. Y is a racemic mixture while Z is achiral
b. Both Y and Z are racemic mixture
c. A diastereomer of Y gives racemic Z
d. X show diastereomerism but not enantiomerism

17 Consider the reaction given below,

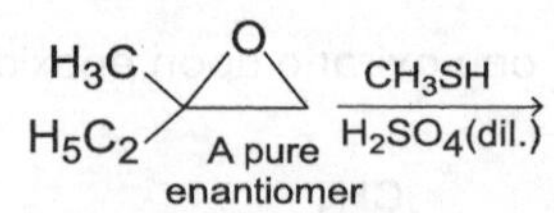

The correct statement regarding the above reaction is/are

a. The major product is $C_2H_5-C(OH)(CH_3)-CH_2-S-CH_3$

b. The major product is $C_2H_5-C(SCH_3)(CH_3)-CH_2OH$

c. The product would be a single enantiomer
d. The product would consist of a racemic mixture

18 Consider the following reaction sequence,

$$HOCH_2-C\equiv C-CH_2CH_2OH \xrightarrow[\text{Lindlar's catalyst}]{H_2/Pd} I \xrightarrow[\text{Heat}]{H_2SO_4} II \xrightarrow{CH_3CO_3H} III$$

I, II and III could be

a. 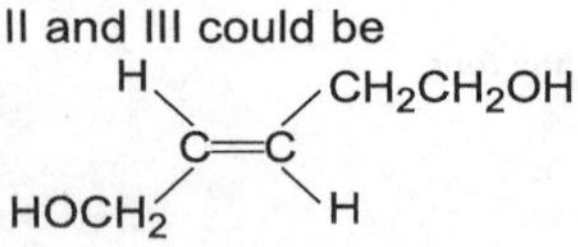

b. 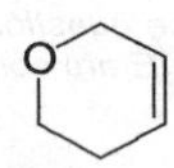

c. 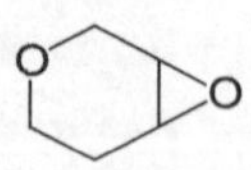

d. 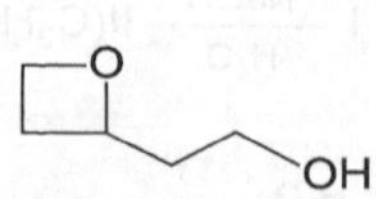

19 Consider the following reaction sequence,

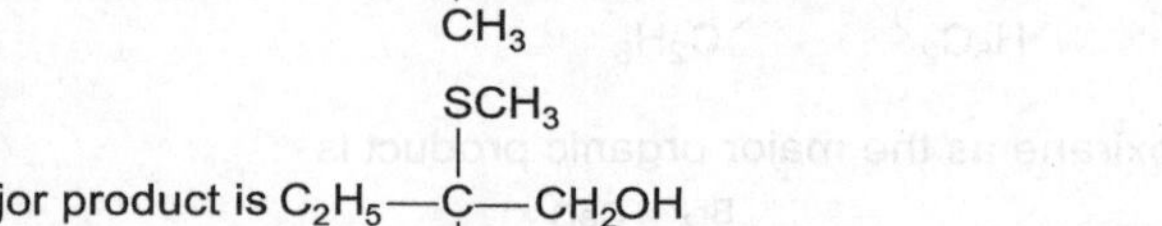

The correct statement regarding X, Y and Z is/are
a. X is a racemic mixture
b. X is a pure enantiomer
c. Y and Z are enantiomers
d. Y and Z are diastereomers

Comprehension Type

Direction *(Q. Nos. 20-22) This section contains a paragraph, describing theory, experiments, data, etc. Three questions related to the paragraph have been given. Each question has only one correct answer among the four given options (a), (b), (c) and (d).*

Passage

Oxirane is a reactive ether, undergo nucleophilic attack at α-carbon in the presence of both acidic and basic medium. The generalised mechanism in the two medium are :

I. Acidic medium

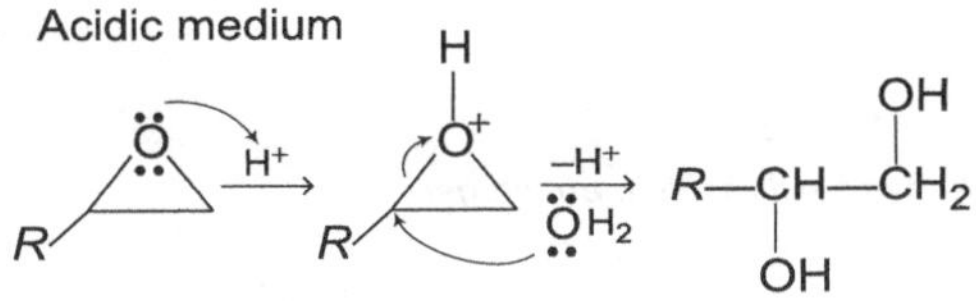

II. Basic medium

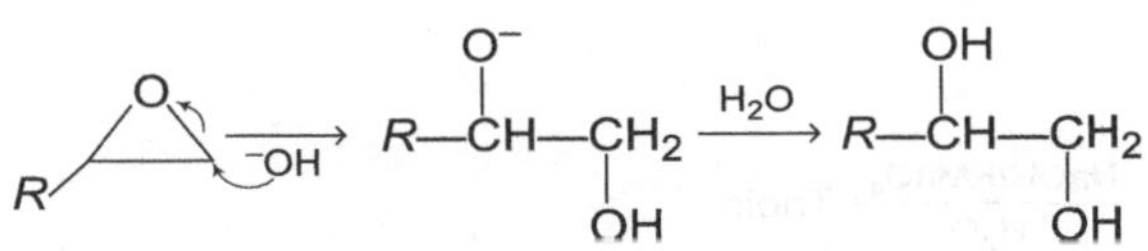

As shown above, in acidic medium, nucleophilic attack occur at more substituted α-carbon while in basic medium, nucleophilic attack occur preferably at less substituted α-carbon.

20 In the reaction given below, the final major organic product Y is

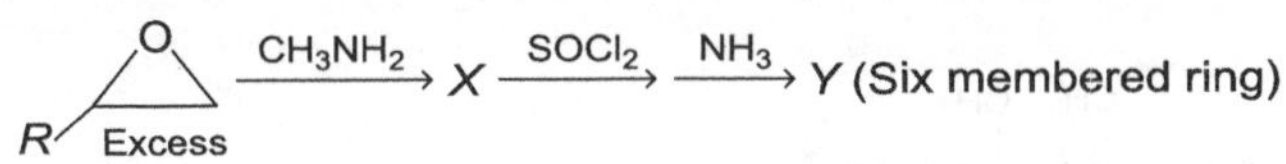

a. 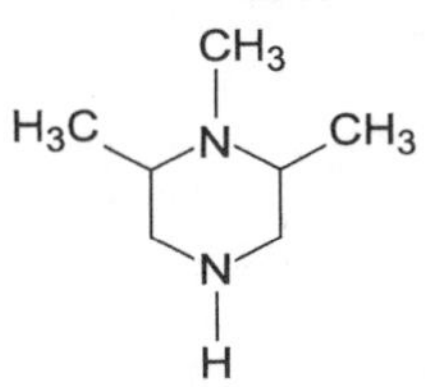

b. 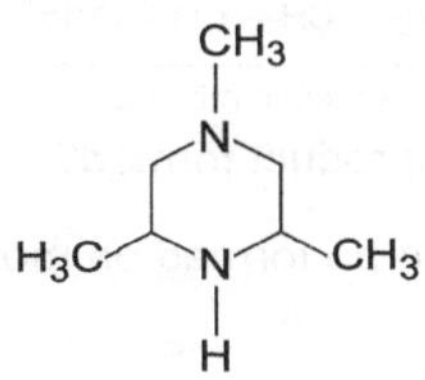

c. 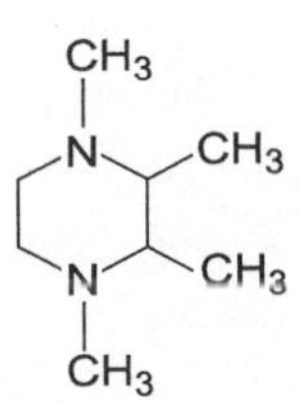

d. 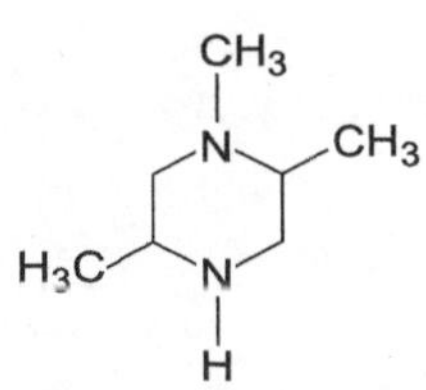

21 If X is heated with concentrated H_2SO_4 a cyclic compound is formed which is

a. 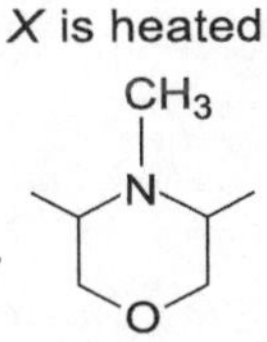

b. 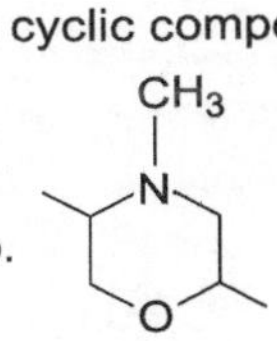

c. 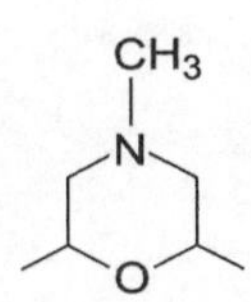

d. 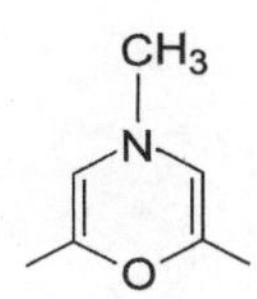

Working Space

22 What would be formed if 2-phenyloxirane is treated with CH_3MgBr followed by acid hydrolysis?

a. $C_6H_5—CH(OH)—CH_2CH_3$

b. $C_6H_5—CH(CH_3)—CH_2OH$

c. $C_6H_5—CH_2CH_2CH_2—OH$

d. $C_6H_5—CH_2—CH(OH)—CH_3$

One Integer Value Correct Type

***Direction** (Q. Nos. 23-27) This section contains 5 questions. When worked out will result in an integer from 0 to 9 (both inclusive).*

23 How many different ether isomers are possible for C_3H_6O?

24 Consider the reaction given below,

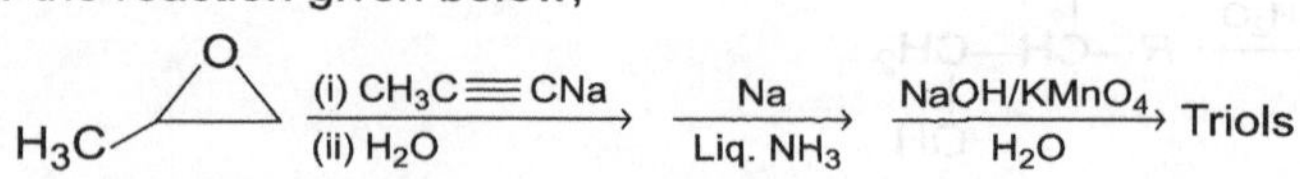

How many different triols isomers are formed?

25 If propene undergoes oxidation to methyl oxirane, what is the net change in oxidation number of carbon atoms?

26 In the following reaction,

$$CH_3—CHO \xrightarrow[\Delta]{CH_2N_2} \xrightarrow[\text{Racemic mixture}]{C_2H_5—CH(CH_3)—CH(CH_3)—ONa} \xrightarrow{H_2O}$$

How many different stereoisomers of product formed?

27 How many different arene oxide would be formed on mono epoxidation of phenanthrene?

DPP-7 Diols and Polyols

Only One Option Correct Type

Direction (Q. Nos. 1-8) This section contains 8 multiple choice questions. Each question has four choices (a), (b), (c) and (d), out of which ONLY ONE is correct.

Working Space

1 Consider the following reaction,

$$CH_2{=}CH_2 + Cl_2\text{-}H_2O \longrightarrow ClCH_2{-}CH_2OH \xrightarrow{X} HOCH_2CH_2OH$$

The most appropriate reagent X for the above reaction is

a. $NaOH(aq)$ b. $NaHCO_3(aq)$ c. $NH_3(aq)$ d. H_2SO_4/H_2O

2 Which is the most suitable condition for the following transformation?

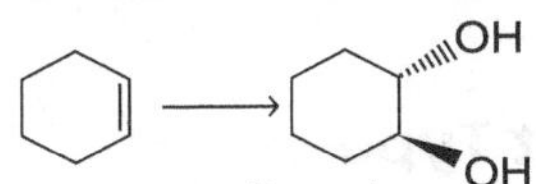

Racemic

a. $O_3/Zn\text{-}H_2O$ then $NaBH_4$
b. CF_3CO_3H then $NaBH_4$
c. CF_3CO_3H then $NaOH(aq)$
d. $KMnO_4$ – $NaOH(aq)$

3 What is the increasing order of solubility of the following in water?

CH_3OH (I), $HOCH_2CH_2OH$ (II), $CH_3{-}CH(OH){-}CH_2OH$ (III)

$HOCH_2{-}\underset{\large OH}{\underset{|}{CH}}{-}CH_2OH$ (IV)

a. I < II < III < IV
b. I < III < II < IV
c. IV < III < II < I
d. IV < II < III < I

4 How many different diols product are formed in the following reaction?

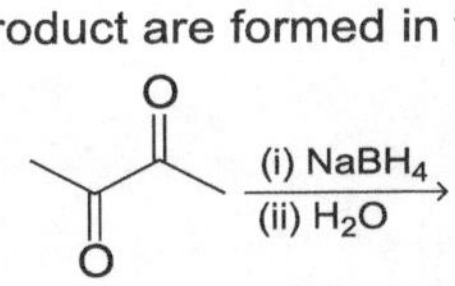

a. 1
b. 2
c. 3
d. 4

5 Which does not form dihalide when treated with ethylene glycol?

a. PCl_3
b. $SOCl_2$
c. PBr_3
d. PI_3

6 What is the major organic product of the following reaction?

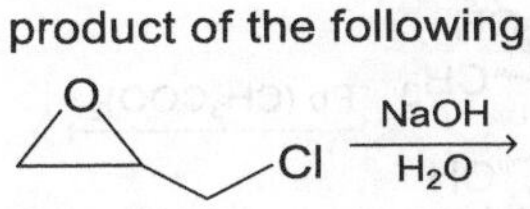

a. 

b. 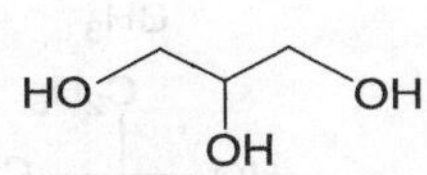

c. 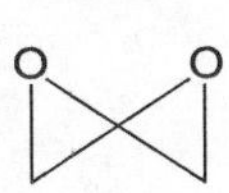

d. 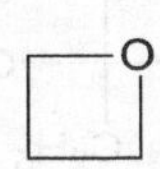

Working Space

7 What is formed as the major product in the reaction given below?

$$CH_2{=}CH{-}\overset{\overset{\Large O}{\|}}{C}{-}CH_3 \xrightarrow[H^+]{HOCH_2CH_2OH} \xrightarrow[Pt]{H_2} \xrightarrow[H_2O]{H_2SO_4}$$

a. $CH_2{=}CH{-}\underset{}{\overset{OH}{\overset{|}{C}}}H{-}CH_3$ b. $CH_3CH_2{-}\overset{OH}{\overset{|}{C}}H{-}CH_3$

c. $CH_3{-}CH_2{-}\overset{O}{\overset{\|}{C}}{-}CH_3$ d. 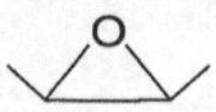

8 The major organic product formed below is

$$HOCH_2CH_2OH \xrightarrow{Conc.\ H_2SO_4}$$

a. $CH_2{=}CH{-}\overset{OH}{\overset{|}{C}}H{-}CH3$ b. $CH_3CH_2{-}\overset{OH}{\overset{|}{C}}H{-}CH_3$

c. $CH_3{-}CH_2{-}\overset{O}{\overset{\|}{C}}{-}CH_3$ d. 

One or More than One Options Correct Type

***Direction** (Q. Nos. 9-14) This section contains 6 multiple choice questions. Each question has four choices (a), (b), (c) and (d), out of which ONE or MORE THAN ONE are correct.*

9 Which of the following reaction(s) gives diols?

a. $CH_3{-}CH{=}CH_2 + KMnO_4 \xrightarrow[H_2O]{NaOH}$

b. $CH_3{-}CH{=}CH_2 + Cl_2\text{-}H_2O \xrightarrow{NaOH}$

c. $CH_3{-}CH{=}CH_2 \xrightarrow[H_2SO_4]{CF_3CO_3H}$

d. $CH_3{-}CH{=}CH_2 + Br_2(l) \longrightarrow \xrightarrow[H_2O]{Na_2CO_3}$

10 Which reaction(s) gives dioxane?

a. $HOCH_2CH_2OH$ + conc. $H_2SO_4 \xrightarrow{Heat}$

b.  + dil. $H_2SO_4 \xrightarrow{Heat}$

c. $Cl{-}CH_2CH_2{-}OH \xrightarrow{NaOH}$

d. 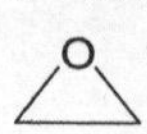 + NaOH (*aq*) $\xrightarrow{Heat}$

11 Which reaction(s) given below gives dicarbonyl?

a. 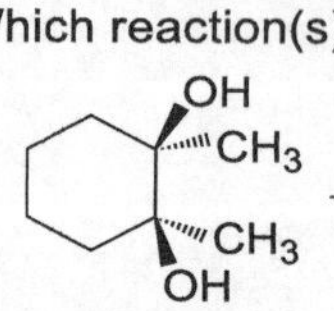 $\xrightarrow{HIO_4}$ b. 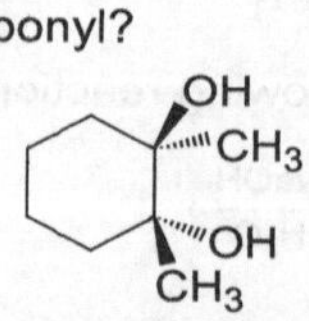 $\xrightarrow{Pb(CH_3COO)_2}$

c. 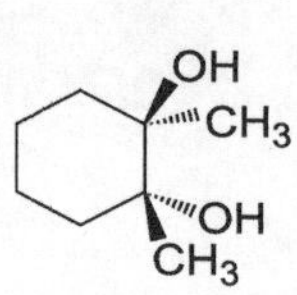 $\xrightarrow{HIO_4}$ d. 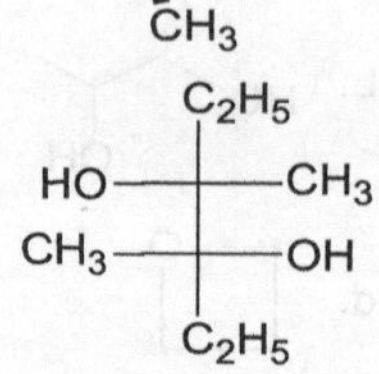 $+ HIO_4 \longrightarrow$

Working Space

12 In the following rearrangement, possible product(s) is/are

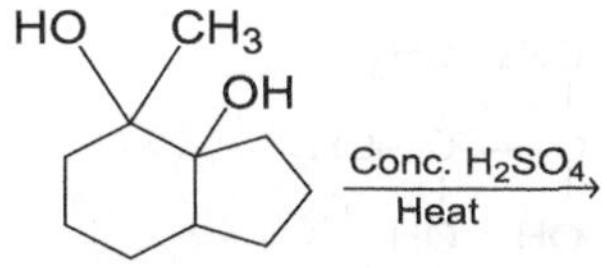

a. 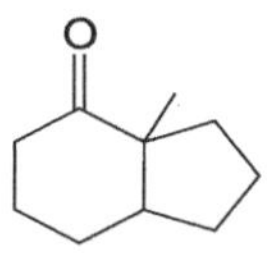

b. 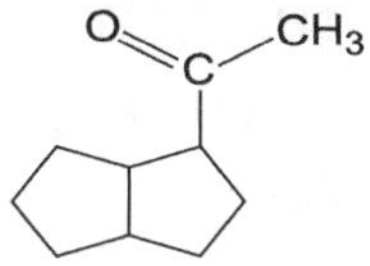

c. 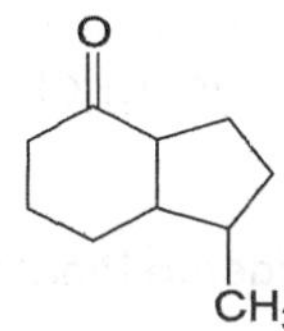

d. 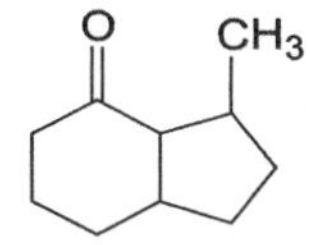

13 Which reaction(s) given below gives glycerol?

a. $CH_3—CH{=}CH_2 \xrightarrow[450°C]{Cl_2} \xrightarrow[H_2O/heat]{Na_2CO_3} \xrightarrow{HOCl} \xrightarrow{NaOH}$

b. 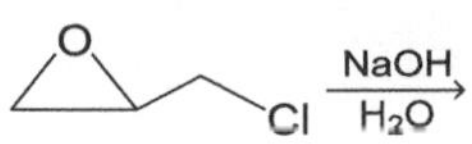

c. $CH_2{=}CH—CHO \xrightarrow[Ni/\Delta]{H_2} \xrightarrow[H_2O]{H_2O_2}$

d. $CH_2{=}CH—CHO \xrightarrow[\text{(ii) } H_2O]{\text{(i) } NaBH_4} \xrightarrow{H_2O_2}$

14 In the following rearrangement, reaction possible product(s) is/are

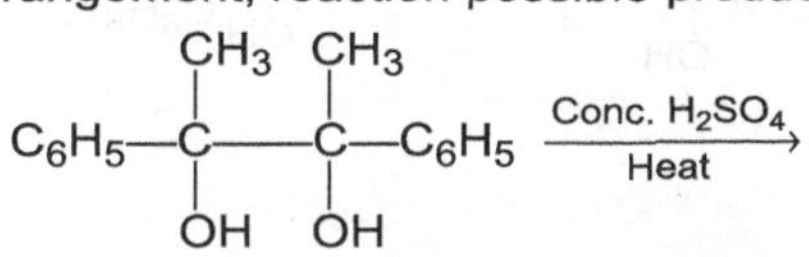

a. 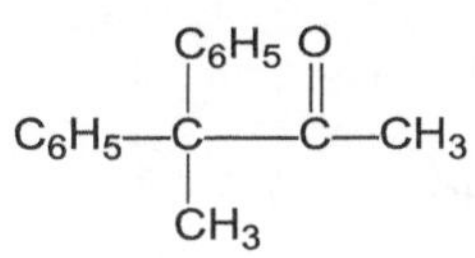

b. 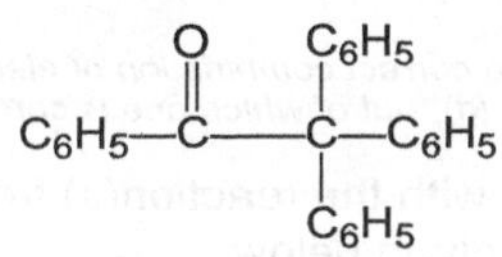

c. 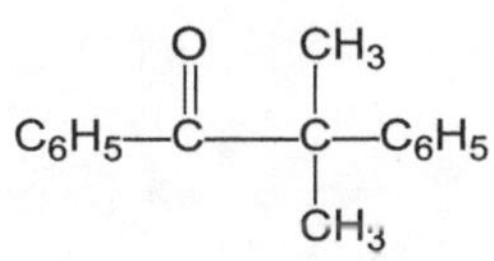

d. 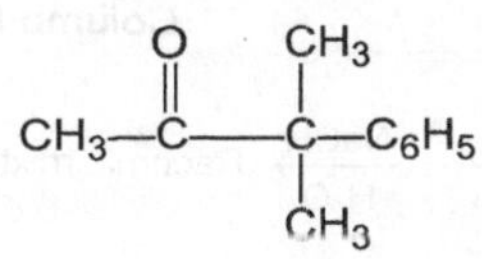

Comprehension Type

***Direction** (Q. Nos. 15-17) This section contains a paragraph, describing theory, experiments, data, etc. Three questions related to the paragraph have been given. Each question has only one correct answer among the four given options (a), (b), (c) and (d).*

Passage

Glycol when heated with concentrated H_2SO_4, undergo a variety of reactions as

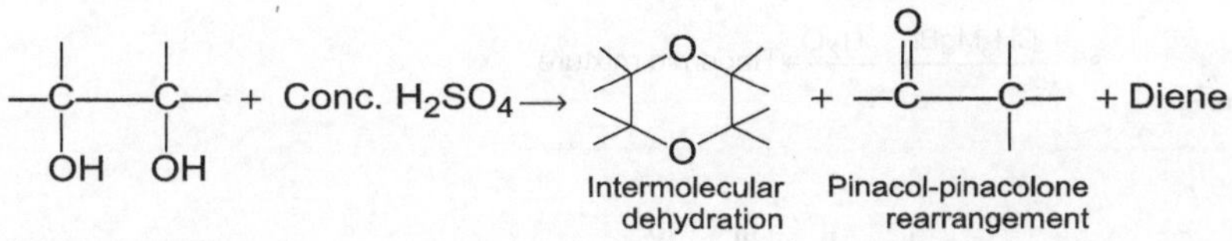

Nature of predominant product depends upon the nature of other groups present at the two α-carbons.

15 Which is most likely to undergo intermolecular dehydration to give dioxane or substituted dioxane?

a. 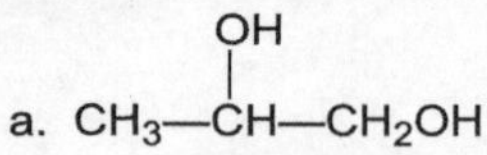

b. 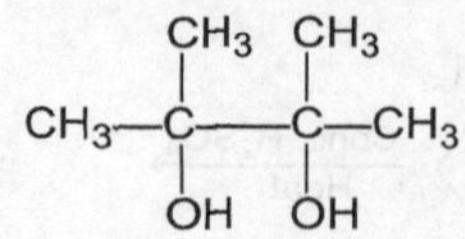

c. 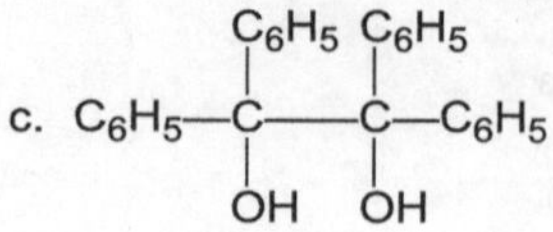

d. 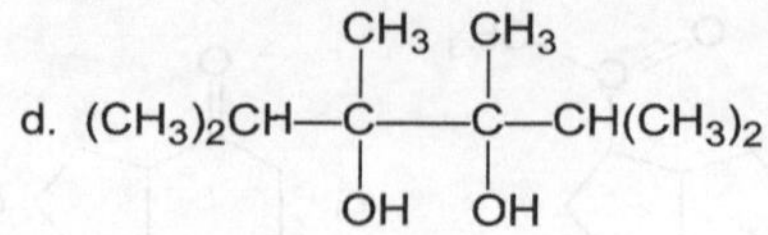

16 Which of the following is most likely to undergo Pinacol-pinacolone rearrangement?

a. 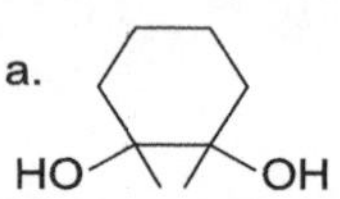

b. 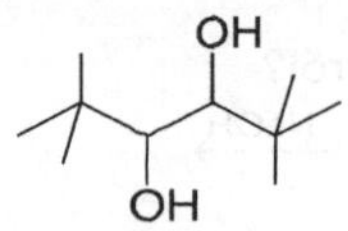

c. 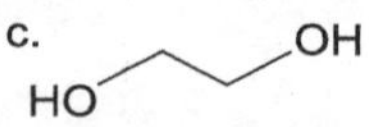

d. 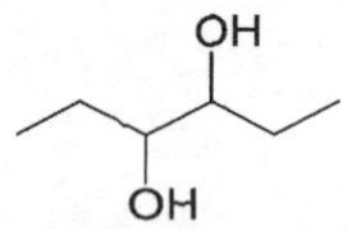

17 Which diol given below is most likely to form diene in the above reaction?

a. 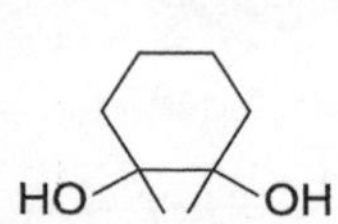

b. 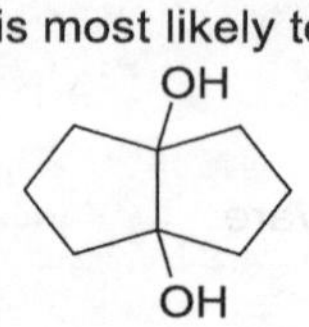

c. 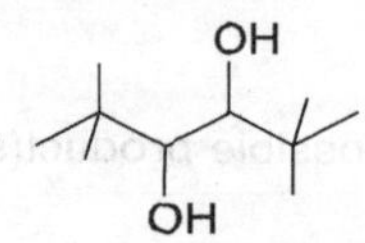

d. 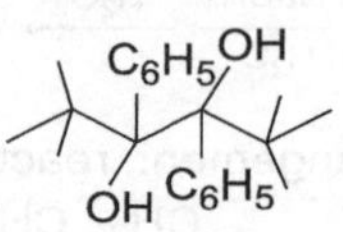

Working Space

Matching List Type

Direction *(Q. Nos. 18 and 19) Choices for the correct combination of elements from Column I and Column II are given as options (a), (b), (c) and (d), out of which one is correct.*

18 Match the reactant from Column I with the reaction(s) from Column II and mark the correct option from the codes given below.

	Column I		Column II
i.	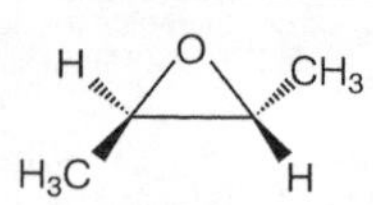	p.	$\xrightarrow[H_2O]{NaOH}$ Racemic mixture
ii.	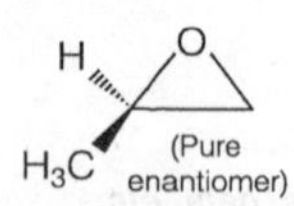	q.	$\xrightarrow[H_2O]{NaOH}$ Pure, single enantiomer.
iii.	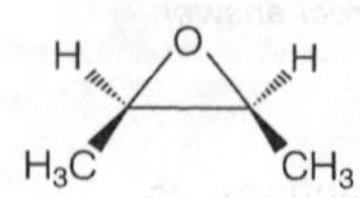	r.	$\xrightarrow[H_2O]{NaOH}$ *Meso* isomer
iv.	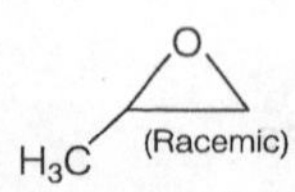	s.	$\xrightarrow{CH_3MgBr}\xrightarrow{H_2O}$ Racemic mixture

Codes

	i	ii	iii	iv		i	ii	iii	iv
a.	r,s	q	p,s	p,s	b.	p	q	r	s
c.	q	r	s	p	d.	q,p	s	p	r

19 Match the compounds in Column I with their reaction and product(s) in Column II and mark the correct option from the codes given below.

Working Space

	Column I		Column II
i.	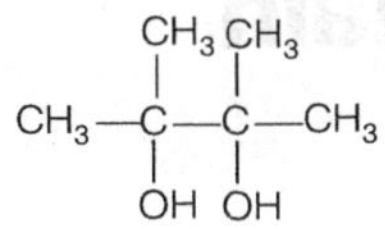	p.	$HIO_4 \longrightarrow$ Single carbonyl compound.
ii.	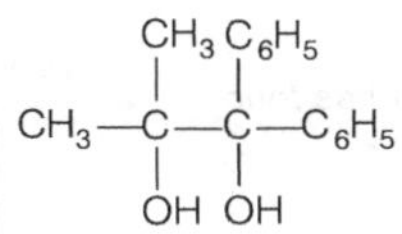	q.	$HIO_4 \longrightarrow$ More than one carbonyl compound.
iii.	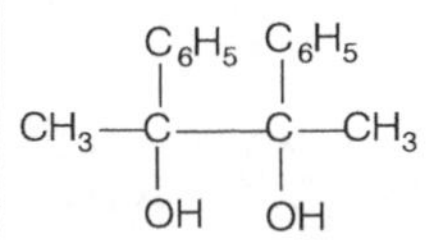	r.	H_2SO_4 heat $\longrightarrow$ Single pinacolone.
iv.	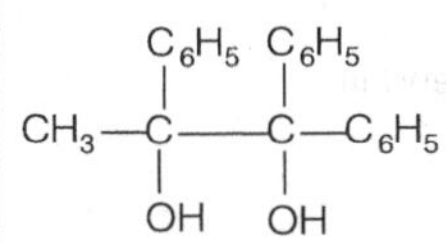	s.	H_2SO_4 heat $\longrightarrow$ More than one pinacolones.
		t.	PCl_3 (excess), then NH_3 (excess) heat $\rightarrow$ Single cyclic amine with nitrogen as a part of ring.

Codes

	i	ii	iii	iv
a.	p,r,t	q,s,t	p,s	q,s
b.	p,r	s	p	q
c.	p	q,s	s	s,t
d.	p,q	r	r,s	t

SET - 1

Revisal Problems for JEE Main

Only One Option Correct Type

Direction *(Q. Nos. 1-26) This section contains 26 multiple choice questions. Each question has four choices (a), (b), (c) and (d), out of which ONLY ONE is correct.*

Working Space

1 What is the major product of the reaction?

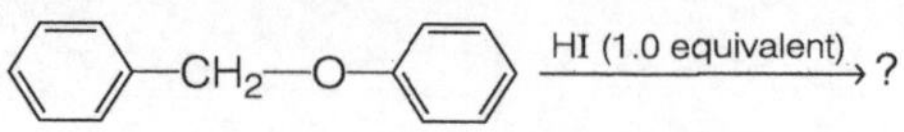

I. 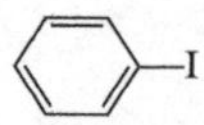

II. 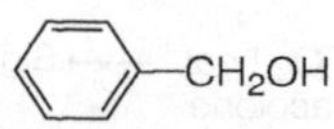

III. 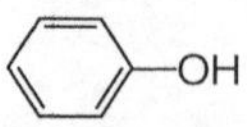

IV. 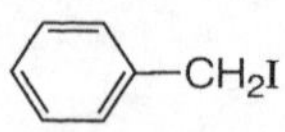

a. Both I and II b. Both III and IV c. Both I and III d. Both II and III

2 What is the major product of the reaction?

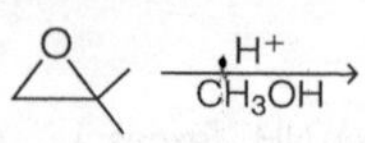

a. 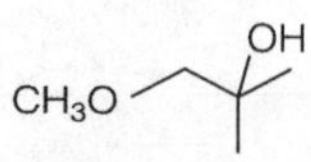

b. 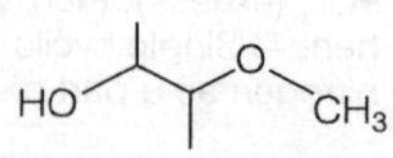

c. 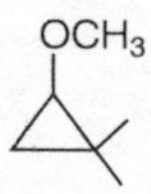

d. 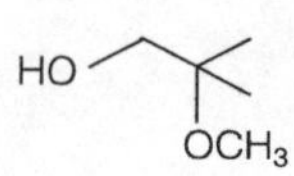

3 What is the product of the reaction?

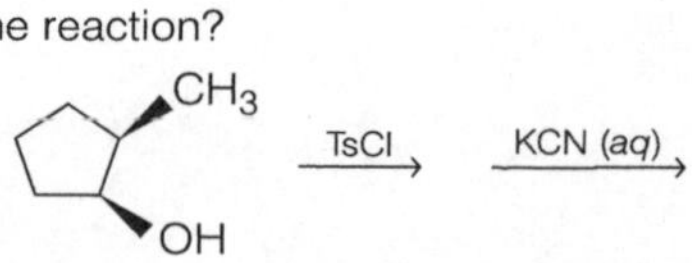

a. 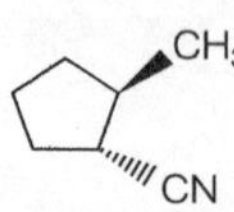

b. 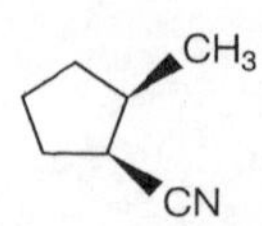

c. 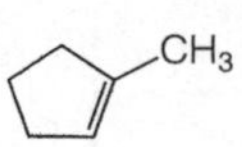

d. 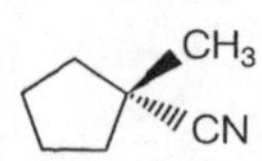

4 What is/are the possible product of the reaction?

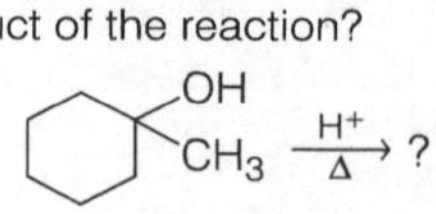

I. 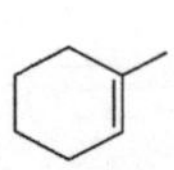

II. 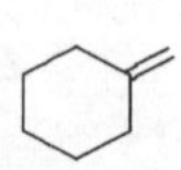

III. 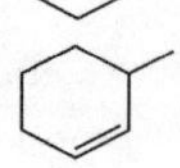

a. Only I b. Only II
c. Only III d. Both I and II

Working Space

5 What is the way to synthesise?

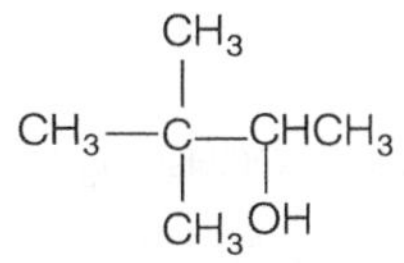

a. $(CH_3)_2\underset{OH}{C}-\underset{OH}{C}-(CH_3)_2 \xrightarrow{H_2SO_4}$ b. $(CH_3)_3CCH{=}CH_2 \xrightarrow{H^+}$

c. $(CH_3)_3CCH{=}CH_2 \xrightarrow[\text{(ii) } H_2O_2/OH]{\text{(i) } BH_3}$ d. $(CH_3)_3CCH{=}CH_2 \xrightarrow[\text{(ii) } NaBH_4]{\text{(i) } Hg(OAc)_2/H_2O/THF}$

6 Which reaction will best perform the synthetic transformation?

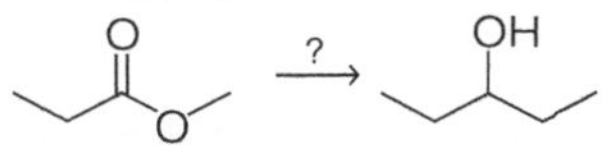

a. (i) CH_3CH_2MgBr, diethylether, (ii) HCl, H_2O
b. (i) CH_3CH_2CHO, $LiN[CH(CH_3)_2]_2$, –78°C (ii) $HClH_2O$, Δ (iii) $NaBH_4$ (iv) HCl, H_2O
c. (i) $NaOCH_3$, CH_3OH, (ii) HCl, H_2O, Δ
d. (i) $(CH_3)_2CHCH_2]_2$ AlH, –78°C (ii) H_2O (iii) CH_3CH_2MgBr, diethylether, (iv) HCl, H_2O

7 Which organic compound, will undergo the synthetic transformation?

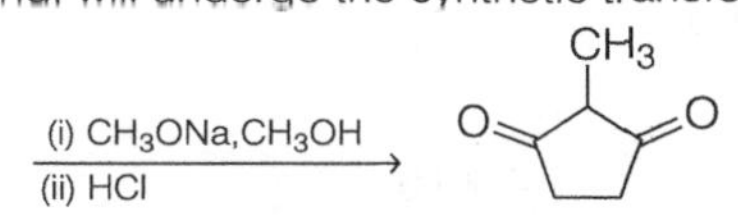

a. 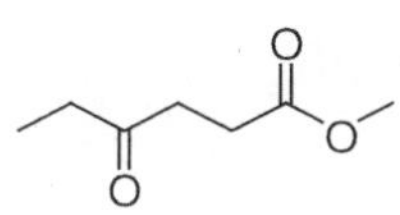 b. 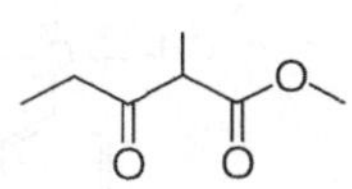

c. 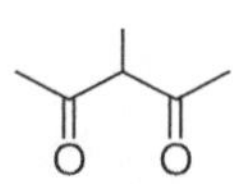 d. 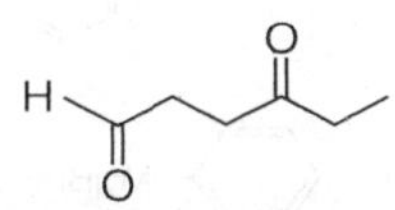

8 Which would be the best reagent to convert compound *A* to *B*?

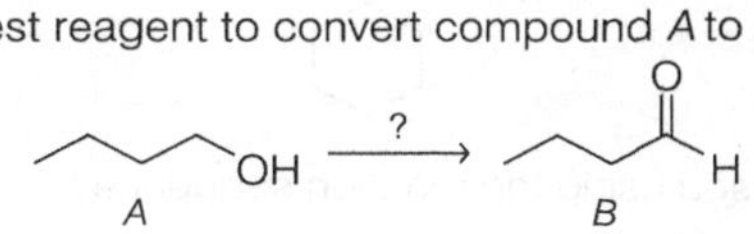

a. 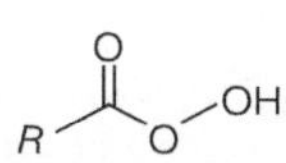 c. $Na_2Cr_2O_7$, H_2SO_4, H_2O

b. (i) Ag_2O, NH_3, (ii) H_3O^+ d. Pyridinium chlorochromate (PCC)

9 Which of the following diol(s) will not undergo a periodate (HIO_4) cleavage?

I. 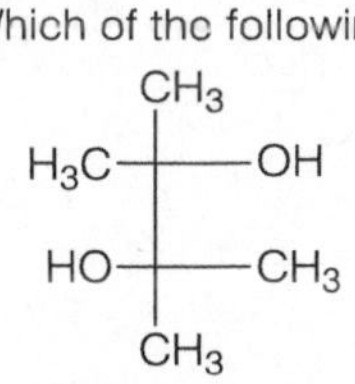 II. 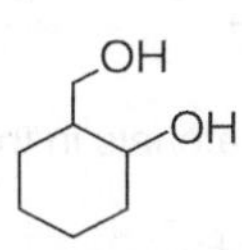 III. 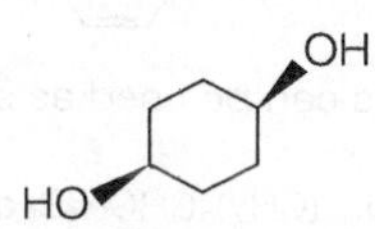

a. Only III b. Only II c. I, II and III d. Both II and III

10 What is the major product of the reaction?

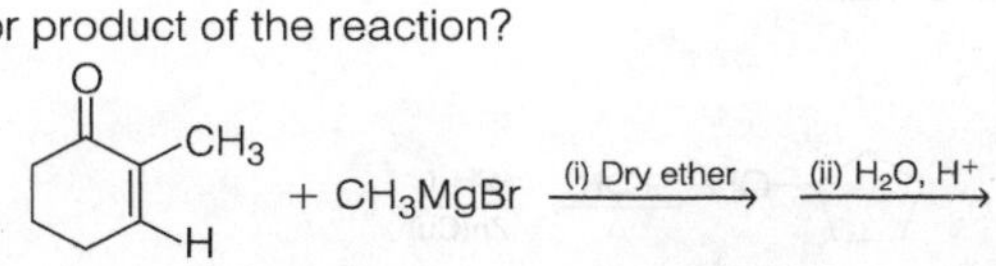

a. 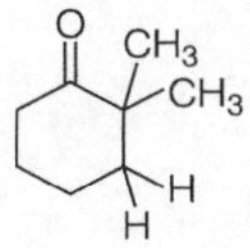 b. 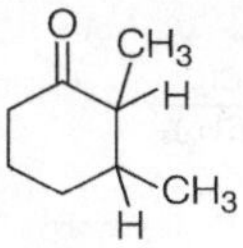 c. 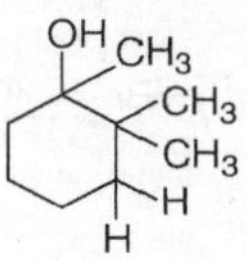 d. 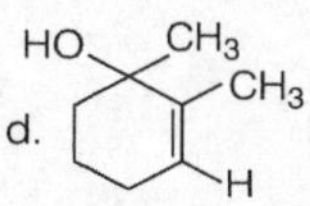

Working Space

11 Which of the following is not a reactive intermediate in the mechanism of the following reaction producing either product ?

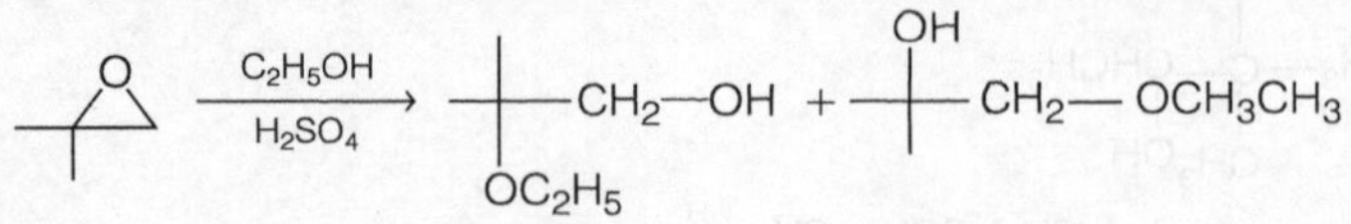

a. 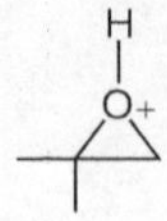

b. 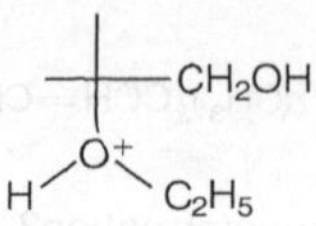

c. 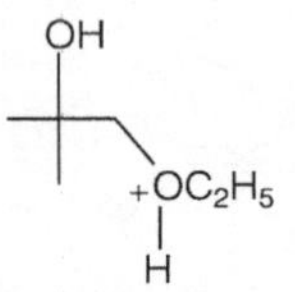

d. 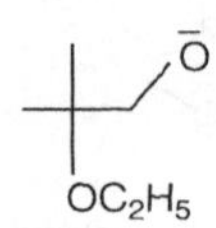

12 Predict the major product of the reaction.

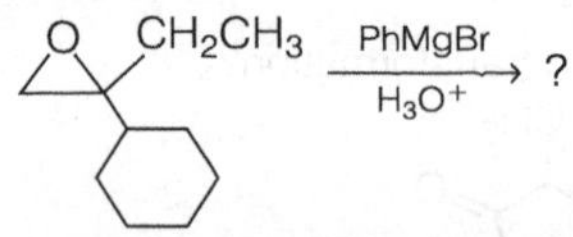

a. 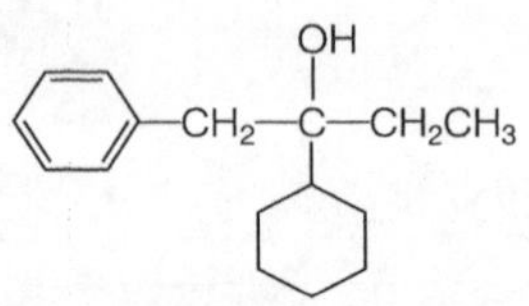

b. 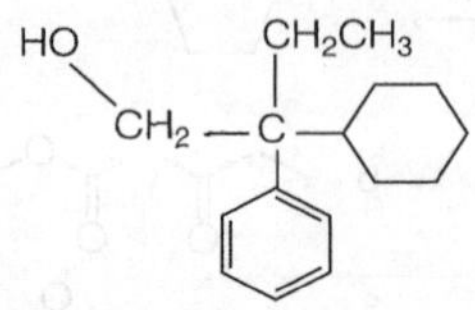

c. 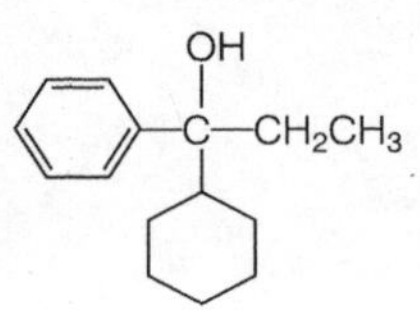

d. 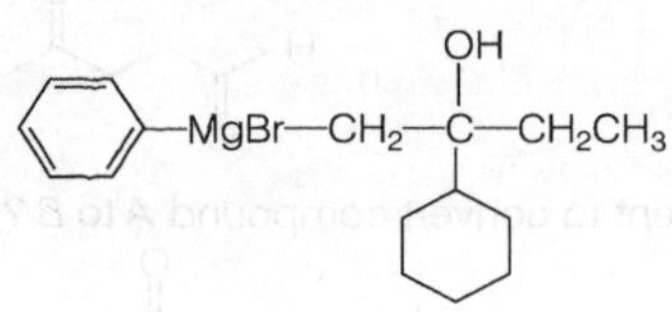

13 Which of the following alcohols cannot be synthesised using the reaction sequence?

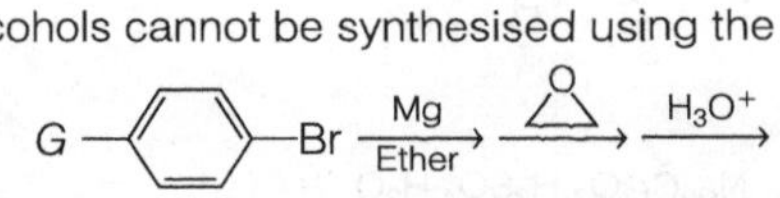

a. 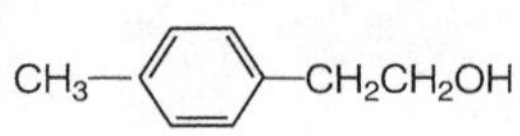

b. 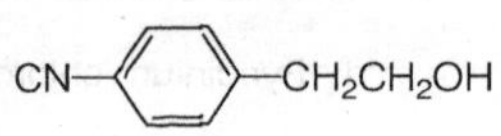

c. 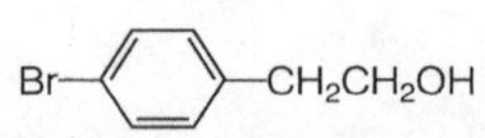

d. 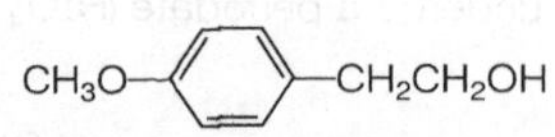

14 Which of the following pairs of the compounds can be used as starting materials in the synthesis of 2-phenyl-2-hexanol?

a. $CH_3(CH_2)_3Br$ and $PhCOCH_3$

b. $(CH_3)_2CHCH_2Br$ and $PhCOCH_3$

c. PhBr and $CH_3CH_2CH_2COCH_3$

d. PhBr and $(CH_3)_2CHCH_2COCH_3$

15 Which of the following sequence can be used to make

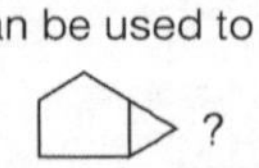

a. 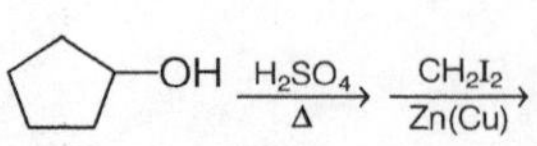

b. 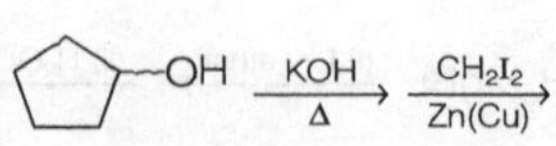

c. 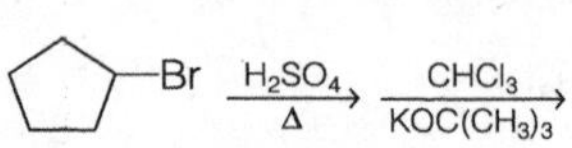

d. 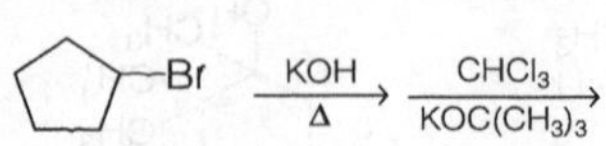

Working Space

16 Which of the following reaction will most efficiently synthesise compound X (3-methyl-3-methoxy hexane) from 2-butanone, 1-chloropropane and methyl iodide?

a. 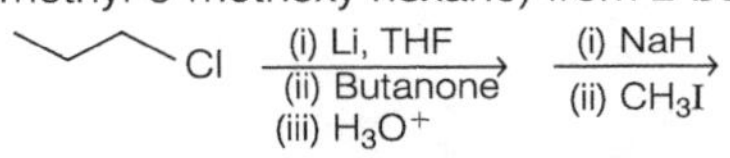

b. 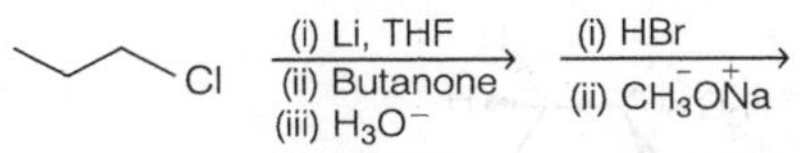

c. 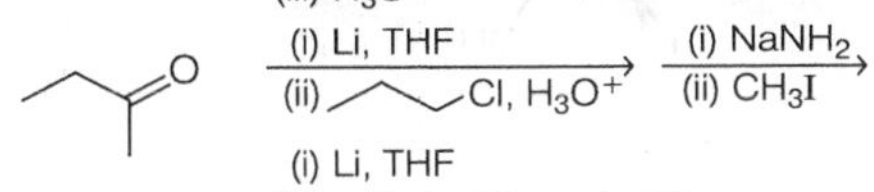

d. 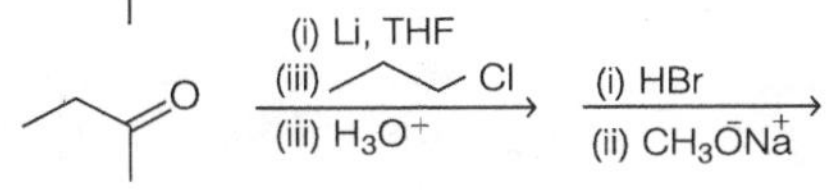

17 When 4-pentene-1-ol is treated with aqueous bromide, a cyclic bromosubstituted ether is formed rather than expected bromohydrane. Select the explanation that the best account for the result.

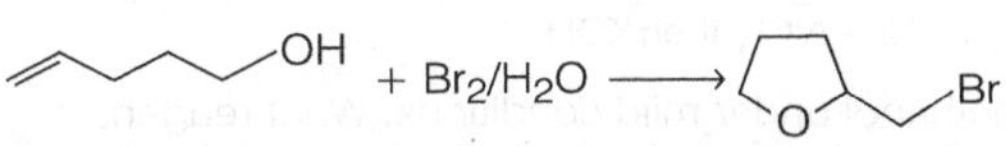

a. The ether is the result of a hydride shift in the initially formed carbocation
b. The ether is formed in the intramolecular nucleophilic attack on the initially formed bromonium ion
c. The reaction is initiated by the protonation of the π-system by the OH^- group
d. The initially formed bromonium ion undergoes an alkyl shift to form new C—O bond

18 Which of the following will react with periodic acid to produce an aldehyde product ?

a. 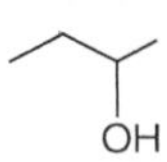

b. 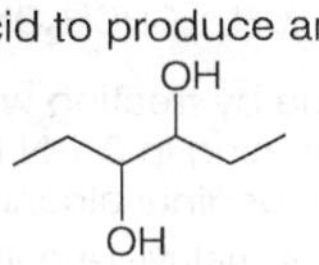

c. 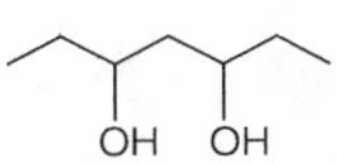

d. 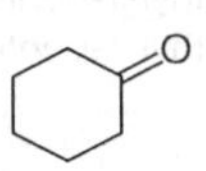

19 Which reaction condition would be best to perform the following transformation?

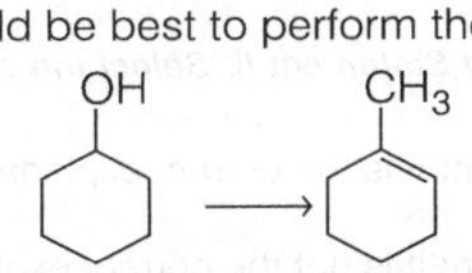

a. (i) PCC (ii) MeMgBr (iii) $POCl_3$

b. (i) 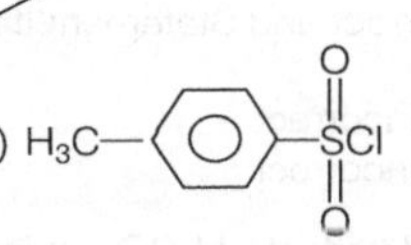

(ii) MeMgBr

c. (i) HBr (ii) Mg (iii) MeI

d. (i) NaOH (ii) MeI

20 What will be the major product of the reaction ?

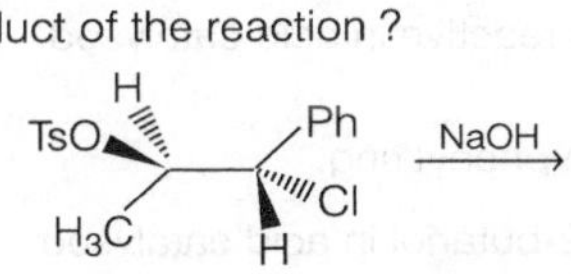

a. 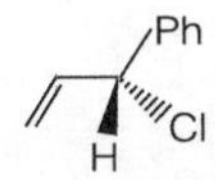

b. 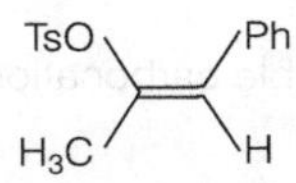

c. 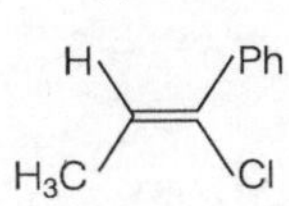

d. 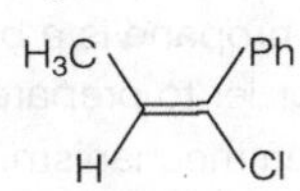

Working Space

21 What is the major product of the reaction?

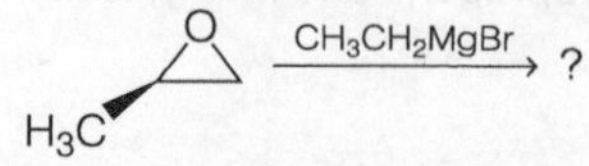

a. 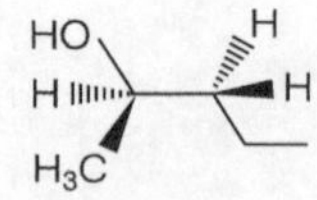 b. 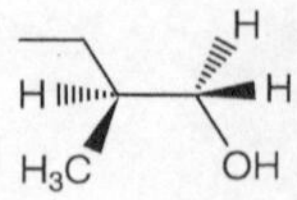 c. 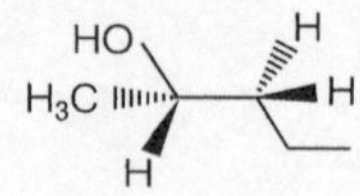 d. 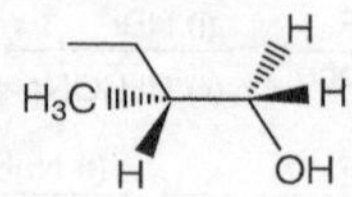

22 Consider the molecule, 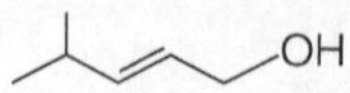

Which reagent will not give a positive test with this compound ?

a. Cold conc. H_2SO_4
b. Br_2/CCl_4
c. NaOH(*aq*)
d. Dilute $KMnO_4/H_2O$

23 Which reagent would provide the product shown ? Propene $\longrightarrow$ 2-methyl oxirane

a. Br_2-H_2O then KOH
b. CH_2N_2 /Heat
c. CrO_3/pyridine, Cold
d. Cl_2 - $AlCl_3$ then KOH

24 Chloroethane, C_2H_5Cl, does not react with methanol under mild conditions. What reagent could be added to the reaction mixture to increase the rate of substitution?

a. HCl (Conc.)
b. NaOH
c. NH_4OH
d. $AgNO_3$

25 Which reaction conditions would be best for the synthesis of isobutyl *sec*-butyl ether $CH_3CH_2CH(CH_3)$-O-$CH_2CH(CH_3)_2$?

a. $(CH_3)_2CHCH_2OH + H_2SO_4$ + Heat
b. $CH_3CH_2CH(CH_3)OH + H_2SO_4$ + Heat
c. $CH_3CH_2CH(CH_3)ONa + (CH_3)_2CHCH_2Br$
d. $(CH_3)_2CHCH_2ONa + CH_3CH_2CH(CH_3)Br$

26 A $C_6H_{14}O$ chiral alcohol is converted to a bromide by treating with PBr_3. Reaction of this bromide, first with Mg in ether, followed by quenching in 0.1 N HCl produces an achiral C_6H_{14} hydrocarbon. Which of the following is the original alcohol?

a. 2-ethyl-1-butanol
b. 4-methyl-1-pentanol
c. 3-methyl-3-pentanol
d. 3-methyl-1-pentanol

Statement Type

Direction (Q. Nos. 27-30) ***This section is based on Statement I and Statement II. Select the correct answer from the codes given below.***

a. Both Statement I and Statement II are correct and Statement II is the correct explanation of Statement I
b. Both Statement I and Statement II are correct and Statement II is not the correct explanation of Statement I
c. Statement I is correct but Statement II is incorrect
d. Statement II is correct but Statement I is incorrect

27 **Statement I** When 1-propanol is refluxed with dil. H_2SO_4, it isomerises to propanol.

Statement II 2-propanol is more stable than 1-propanol.

28 **Statement I** Diphenyl ether (Ph—O—Ph) is very less reactive in acid catalysed hydrolysis to phenol.

Statement II Oxygen is in resonance on both side with phenyl ring.

29 **Statement I** 3-methyl-2-butanol is more reactive than 2-butanol in acid catalysed dehydration to alkene.

Statement II 3-methyl-2-butanol forms more stable carbocation than 2-butanol during dehydration reaction.

30 **Statement I** Reaction of C_2H_5ONa with 2-chloro propane is a better method than the reaction of $(CH_3)_2CHONa$ with chloro ethane in order to prepare ethyl-isopropyl ether.

Statement II Here ether is formed by S_N2 reaction mechanism.

Revisal Problems
for JEE Advanced

Only One Option Correct Type

Working Space

Direction *(Q. Nos. 1-12) This section contains 12 multiple choice questions. Each question has four choices (a), (b), (c) and (d), out of which ONLY ONE is correct.*

1 Consider the following reaction,

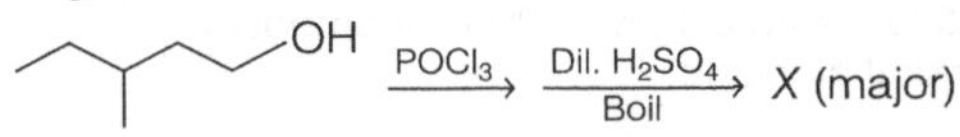

The correct statement regarding *X* is

a. It would change the orange colour of acidic dichromate solution
b. It will give yellow ppt with alkaline iodine solution
c. Its saturated aqueous solution will give immediate turbidity with conc HCl
d. It gives a major alkene with $POCl_3$ which show stereoisomerism

2 Reaction of (+)-3-methylcyclopentene with (i) Cl_2, H_2O and (ii) NaH will provide

a. a mixture of diastereomers
b. a racemic mixture
c. a single stereomer
d. positional isomers

3 From which alcohol and which alkyl halide could *A* be best prepared through Williamson ether synthesis?

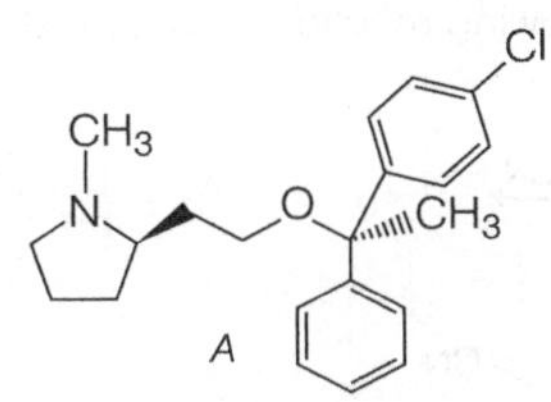

a. 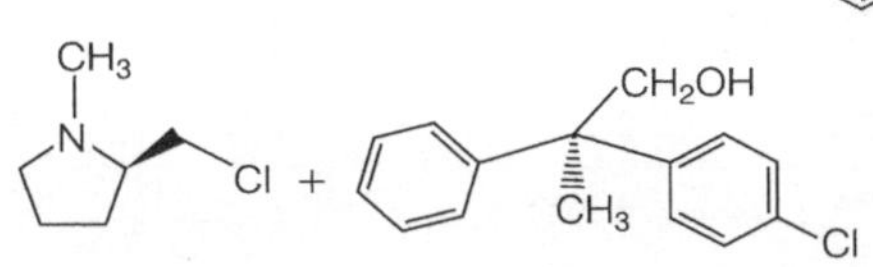

b. 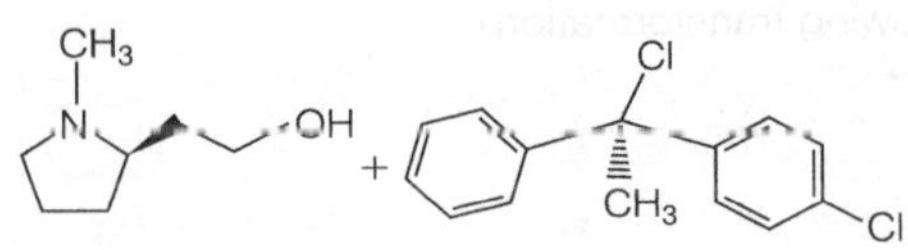

c. 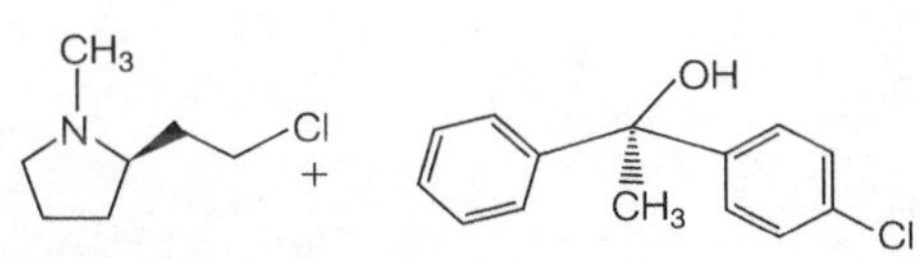

d. Both (a) and (b)

4 A synthesis of 2,5-dimethyl-2-hexanol from 2-methylpropene requires the formation of two four carbon intermediates, *X* and *Y*. These intermediates combine to give the desired product after the usual hydrolysis work-up. Select the appropriate methods of preparing *X* and *Y* from 2-methylpropene.

a. *X* add HBr, then react with Mg in ether *Y* add water, acid-catalysis
b. *X* add HBr (peroxides), then react with Mg in ether *Y* react with $C_6H_5CO_3H$ in CH_2Cl_2
c. *X* add HOBr *Y* add B_2H_6 in ether, then NaOH
d. *X* add HOBr *Y* add HBr (peroxides), then react with Mg in ether

Working Space

5 What is the major product of the following reaction?

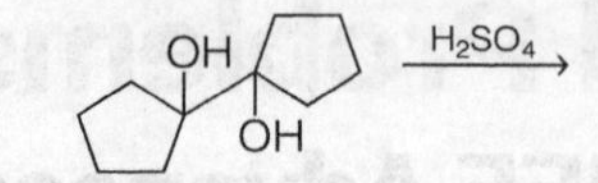

a. 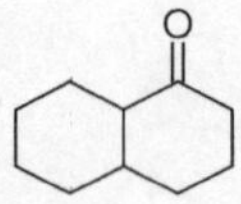

b. 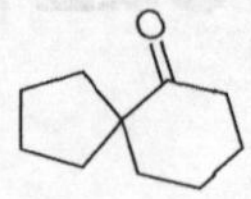

c. 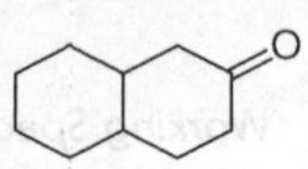

d. 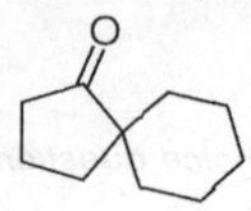

6 A chiral $C_5H_{10}O$ alcohol is reduced by catalytic hydrogenation to an achiral $C_5H_{12}O$ alcohol. The original alcohol is oxidised by activated MnO_2 to an achiral carbonyl compound (C_5H_8O). Which of the following might be the chiral alcohol?

a. 1-penten-3-ol
b. 4-penten-2-ol
c. 3-methyl-2-buten-1-ol
d. 2-methyl-2-buten-1-ol

7 The Lucas test is used to distinguish small (7 or fewer carbons) 1°, 2° and 3° alcohols. The alcohol to be tested is added to a solution of anhydrous $ZnCl_2$ in conc. HCl at room temperature. Which of the following statements is not correct ?

a. 1° alcohols dissolve, but do not react
b. 3° alcohols react quickly to give an insoluble alkyl chloride
c. 3° alcohols rapidly dehydrate, and the gaseous alkene bubbles out of the test solution
d. 2° alcohols dissolve and react slowly to give an insoluble alkyl chloride

8 If the following ^{13}C labelled compound undergo the following reaction, which product would be obtained ?

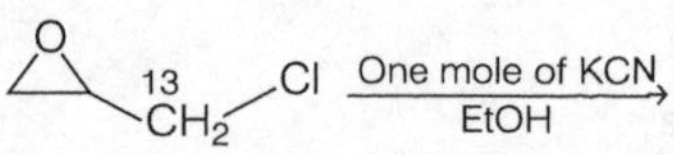

a. 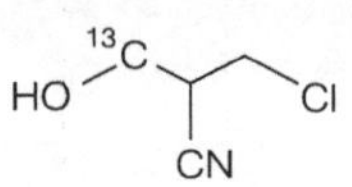

b. 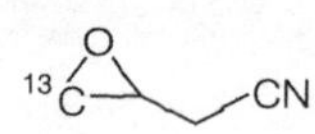

c. 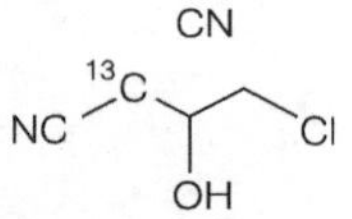

d. 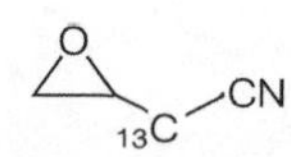

9 Which sets of reagents shown below would accomplish the following transformation?

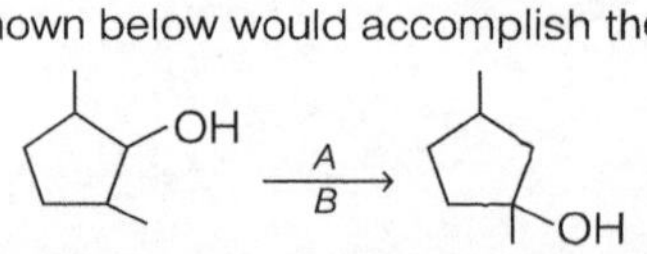

	Column I		Column II
i.	BH_3, H_2O_2/NaOH	p.	H_2SO_4
ii.	H_2/Pd/C	q.	BH_3/H_2O_2/NaOH
iii.	H_3PO_4	r.	$Hg(OAc)_2$; $NaBH_4$
iv.	EtOK	s.	BH_3/H_2O_2/NaOH

a. (i) → q
b. (ii) → s
c. (iii) → r
d. (iv) → p

Working Space

10 A $C_5H_{12}O$ compound is optically active, and is oxidised by PCC in CH_2Cl_2 to an optically active $C_5H_{10}O$ product, which is racemised in acid or base. Which of the following best fits these facts ?

a. 2-pentanol b. 2-methoxybutane
c. 2-methyl-1-butanol d. 3-methyl-1-butanol

11 What is the major product of the reaction?

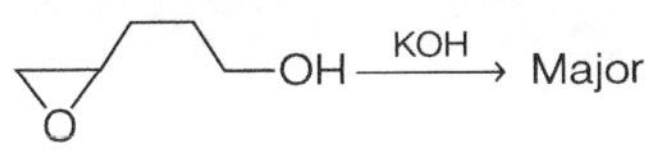

a. 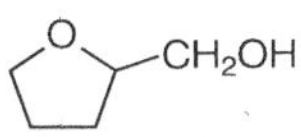 b. 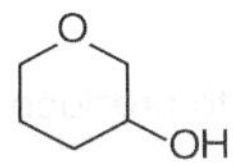

c. 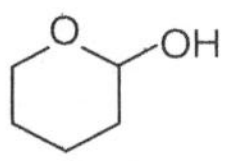 d. 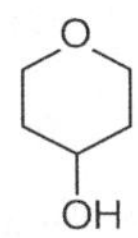

12 When treated with acid, the strained ether (I) is protonated on oxygen, and ring opens to give a resonance stabilised carbocation. Which diene listed below when treated with acid will give the same carbocation ?

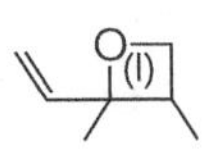

a. 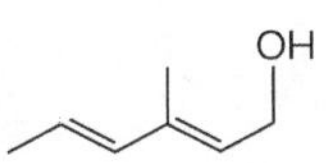 b. 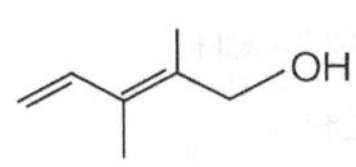

c. 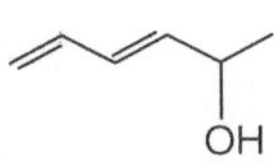 d. 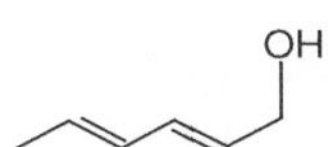

One or More than One Options Correct Type

Direction (Q. Nos. 13-18) *This section contains 6 multiple choice questions. Each question has four choices (a), (b), (c) and (d), out of which ONE or MORE THAN ONE are correct.*

13 The reaction producing 2-ethyl-2-methyl oxirane is/are

a. 2-methyl-1-butene + Ag_2O + heat $\longrightarrow$ b. Butanone + CH_2N_2 + heat $\longrightarrow$
c. 2-methyl-1-butene + Br_2-H_2O $\xrightarrow{KOH}$ X d. 2-methyl-2-butene $\xrightarrow{CF_3CO_3H}$

14 Compounds which on reduction can produce alcohols is/are

a. CH_3CH_2COCl b. CH_3CONH_2
c. methyl oxirane d. acetic anhydride

15 Lucas test is used for distinguishing primary, secondary.and tertiary alcohols as:

$$R\text{OH} + \text{Conc. HCl} \xrightarrow{ZnCl_2} R\text{Cl (white turbidity)} + H_2O$$

The correct statement regarding the above test is/are

a. ROH behaves as a base
b. Greater the value of pK_a of alcohol, greater the reactivity with HCl and thus sooner the formation of white turbidity
c. Alcohols that react faster with Na metal, will give turbidity at fastest rate
d. Alcohols that do not change the orange colour of acidic dichromate solution produce immediate turbidity in this test

16 3-methyl-3-hexanol can be prepared by the reaction of

a. CH_3MgBr and 3-hexanone followed by hydrolysis
b. C_2H_5MgBr and 2-pentanone followed by hydrolysis
c. propyl-MgBr + 2-butanone followed by hydrolysis
d. C_4H_9MgBr and propanone followed by hydrolysis

17 The correct statement regarding the product(s) of the following reaction is/are

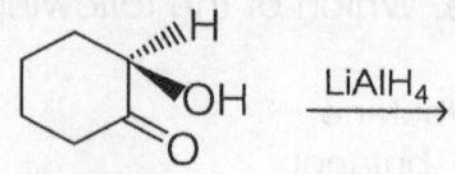

A pure enantiomer

a. a pair of enantiomers in equal amount
b. a pair of diastereomers would be produced
c. the product mixture will have non-zero optical rotation
d. one of the product alcohol would be in the *meso* form

18 Which of the following Grignard synthesis would fail to produce the desired alcohols?

a. $H—C{\equiv}C—CH_2—\underset{\underset{O}{\|}}{C}—CH_3 + CH_3MgBr \xrightarrow{Ether} \xrightarrow[Boil]{H_3O^+} H—C{\equiv}C—CH_2—\overset{\overset{CH_3}{|}}{\underset{\underset{OH}{|}}{C}}—CH_3$

b. $ClCH_2CH_2CHO + CH_3MgBr \xrightarrow{Ether} \xrightarrow[Boil]{H_3O^+} ClCH_2CH_2\underset{\underset{OH}{|}}{CH}—CH_3$

c. $CH_3O—CH_2—\underset{\underset{O}{\|}}{C}—CH_3 + CH_3MgBr \xrightarrow{Ether} \xrightarrow[Boil]{H_3O^+} CH_3O—CH_2—\overset{\overset{CH_3}{|}}{\underset{\underset{OH}{|}}{C}}—CH_3$

d. $C_6H_5CHO + CH_3MgBr \xrightarrow{Ether} \xrightarrow[Boil]{H_3O^+} C_6H_5\underset{\underset{OH}{|}}{CH}—CH_3$

Comprehension Type

***Direction** (Q. Nos. 19-24) This section contains 2 paragraphs, each describing theory, experiments, data, etc. Six questions related to the paragraphs have been given. Each question has only one correct answer among the four given options (a), (b), (c) and (d).*

Passage I

A and *B* are two structural isomers with their molecular formula $C_7H_{14}O$ and both gives-off effervescence on heating with Na-metal. However, neither *A* nor *B* decolourise brown colour of Br_2-H_2O. Treating either *A* or *B* with hot conc. sulphuric acid solution results in the formation of the same compound *C* (C_7H_{12}). *C* on treatment with O_3 followed by Zn-H_2O gives *D* ($C_7H_{12}O_2$). *D* gives *E* ($C_7H_{16}O_2$) on treatment with $NaBH_4$. *E* has only one methyl group and forms yellow ppt with I_2/KOH. Also, *A* gives delayed turbidity on treatment with conc. HCl/$ZnCl_2$ while *B* gives immediate turbidity on similar treatment.

19 Which of the following statement regarding *A* and *B* is correct?
a. Both *A* and *B* are capable of showing optical activity
b. Only *A* can show enantiomerism, not *B*
c. Only *B* can show enantiomerism not *A*
d. Neither *A* nor *B* show enantiomerism

20 Which of the following statement regarding *A* and *B* is incorrect?
a. Only *A* will change the colour of dichromate solution not *B*
b. *C* on reaction with $(CH_3COO)_2Hg/H_2O$ followed by $NaBH_4$ gives *B*
c. *C* on treatment with B_2H_6/H_2O_2 followed by alkaline work-up gives *A*
d. Treating either *A* or *B* with $SOCl_2$ followed by the reduction with $LiAlH_4$ results in the formation of same product

21 Which of the following statement regarding acid catalysed isomerisation of *A* and *B* is true?
a. *A* on treatment with dil. H_2SO_4 isomerises to *B* and at equilibrium, *A* predominate
b. *A* on treatment with dil. H_2SO_4 isomerises to *B* and at equilibrium, *B* predominate
c. Isomerisation of *A* to *B* or *B* to *A* in the presence of acid catalyst involve the same carbocation in the last final step
d. Both (b) and (c) are correct

Working Space

Working Space

Passage II

An organic compound $A(C_9H_{10}O)$ is chiral and does not evolve any gas on treatment with Na-metal but on hydrolysis with dil. H_2SO_4 gives $B(C_9H_{12}O_2)$ which on further treatment with alkaline solution of iodine does not give an yellow precipitate. Also A on treatment with excess of conc. HBr gives $C(C_9H_{11}OBr)$ as the major product. C on further treatment with C_2H_5ONa/C_2H_5OH followed by acidification of product gives D (an isomer of A). D on ozonolysis followed by work-up with dimethyl sulphide gives $E(C_8H_8O_2)$ as one of the product.

22 The most likely structure of starting compound A is

a. 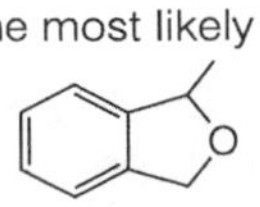 b. 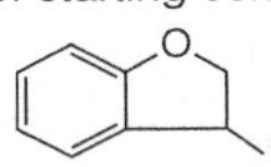 c. 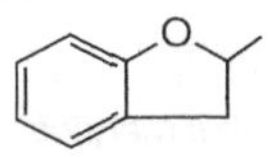 d. 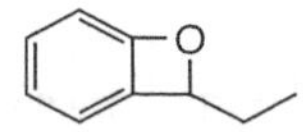

23 Which of the following statement regarding B is correct?

a. B is an optically inactive substance
b. B is an optically active substance with same configuration as that of A
c. B is an optically active substance with opposite configuration to that of A
d. A pure enantiomer of A will produce, on hydrolysis reaction, a racemic mixture of B

24 The statement that is true regarding A to D is

a. If oxygen of A is labelled by ^{18}O, D will retain ^{18}O
b. If A is hydrolysed with H_2O^{18}/H^+, B will have ^{18}O on phenyl ring
c. A can show both enantiomerism and diastereomerism
d. B can show both enantiomerism as well as diastereomerism

Matching List Type

Direction (Q. Nos. 25-27) ***Choices for the correct combination of elements from Column I and Column II are given as options (a), (b), (c) and (d), out of which one is correct.***

25 Match the Column I with Column II and mark the correct option from the codes given below.

	Column I		Column II
i.	1-butanol	p.	Treatment with H^+/H_2O gives racemic mixture.
ii.	2-butanol	q.	Changes the colour of acidic $K_2Cr_2O_7$.
iii.	(+)-3-methyl-3-hexanol	r.	Gives turbid solution with $ZnCl_2$/Conc. HCl at room temperature.
iv.	(−)-2-ethyl oxirane	s.	With $LiAlH_4$, gives another compound from column II.

Codes

	i	ii	iii	iv		i	ii	iii	iv
a.	p,q	p,q,r	p,r	s	b.	q	r	p	q
c.	p,s	r	p	q	d.	p,q	r	s	p

26 Match the Column I with Column II and mark the correct option from the codes given below.

	Column I		Column II
i.	Pentanal	p.	Treatment with H_2/Ni gives achiral alcohol.
ii.	3-pentanone	q.	Treatment with butyl magnesium bromide followed by hydrolysis gives achiral alcohol.
iii.	Pent-1-en-3-ol	r.	Treatment with methyl magnesium bromide followed by hydrolysis gives chiral alcohols.
iv.	2-pentanone	s.	Treatment with ethyl magnesium bromide followed by hydrolysis gives chiral alcohols.

Codes

	i	ii	iii	iv		i	ii	iii	iv
a.	p,q,rs	p,q	p,r,s	s	b.	p,q	r	s	q
c.	p,r	q	p,s	q	d.	q,r	p	q	r

Working Space

27 Match the Column I with Column II and mark the correct option from the codes given below.

	Column I		Column II
i.	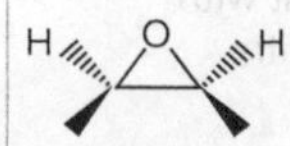	p.	Acid hydrolysis gives racemic mixture.
ii.	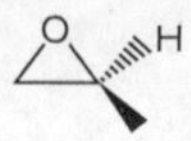	q.	Acid hydrolysis gives *meso* isomer.
iii.	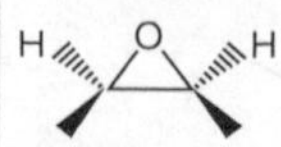	r.	Treatment with methyl magnesium bromide followed by hydrolysis gives racemic mixture.
iv.		s.	Treatment with HCl gives chiral halohydrin.

Codes

	i	ii	iii	iv
a.	p,s	s	p,r,s	q,s
b.	q	r	s	p
c.	p	q	r	s
d.	p,q	r	s	p

One Integer Value Correct Type

Direction *(Q. Nos. 28-30)* ***This section contains 3 questions. When worked out will result in an integer from 0 to 9 (both inclusive).***

28 An ether, on treatment with sulphuric acid and sodium iodide, yields a single iodoalkane that contains 74.7% iodine by weight. How many carbon atoms are present in a molecule of this ether ? [Molar mass of I = 127]

29 An organic compound $A(C_{10}H_{18}O_8)$ on treatment with excess of CH_3COCl gives a fully acetylated product whose molar mass is found to be 518 g/mol. How many hydroxyl functional groups are present in A?

30 How many different triols would be formed when the following compound is treated with excess of CH_3MgBr followed by hydrolysis ?

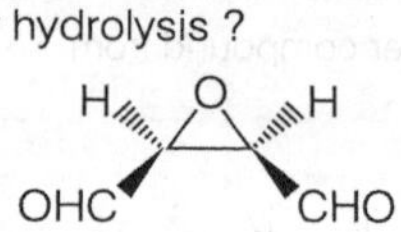

JEE Main & AIEEE Archive

(Compilation of Last 13 Years' Questions)

1 The most suitable reagent for the conversion of $R—CH_2—OH \longrightarrow R—CHO$ is **(JEE Main 2014)**

a. $KMnO_4$ b. $K_2Cr_2O_7$
c. CrO_3 d. PCC (Pyridinium chlorochromate)

2 An unknown alcohol is treated with the 'Lucas reagent' to determine whether the alcohol is primary, secondary or tertiary. Which alcohol reacts fastest and by what mechanism? **(JEE Main 2013)**

a. Secondary alcohol by S_N1 b. Tertiary alcohol by S_N1
c. Secondary alcohol by S_N2 d. Tertiary alcohol by S_N2

3 Consider the following reaction, $C_2H_5OH + H_2SO_4 \longrightarrow$ Product

Among the following, which one cannot be formed as a product under any conditions? **(AIEEE 2011)**

a. Ethyl hydrogen sulphate b. Ethylene
c. Acetylene d. Diethyl ether

4 From amongst the following alcohols the one that would react fastest with conc. HCl and anhydrous $ZnCl_2$ is **(AIEEE 2010)**

a. 2-butanol b. 2-methylpropan-2-ol
c. 2-methylpropanol d. 1-butanol

5 The main product of the following reaction is

$$C_6H_5CH_2CH(OH)CH(CH_3)_2 \xrightarrow{\text{Conc. } H_2SO_4} ?$$

(AIEEE 2010)

a. 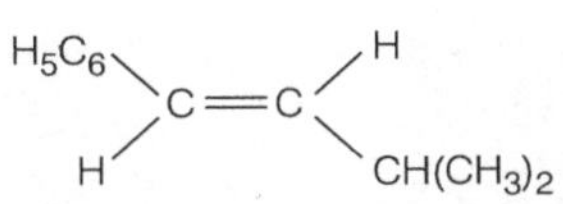

b. 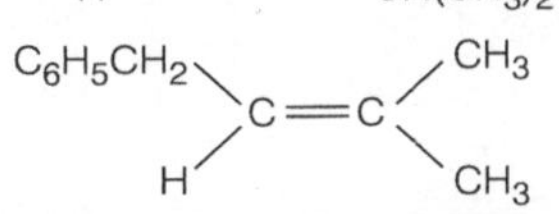

c. 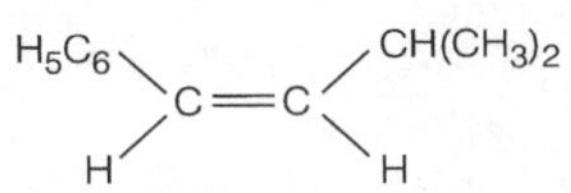

d. 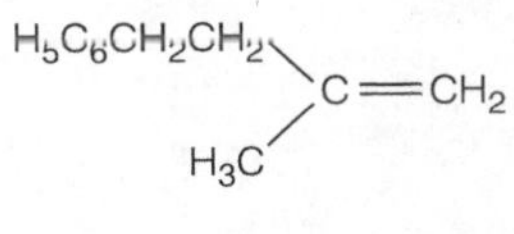

6 In the following sequence of reactions,

$$CH_3CH_2OH \xrightarrow{P + I_2} A \xrightarrow[\text{ether}]{Mg} B \xrightarrow{HCHO} C \xrightarrow{H_2O} D$$

the compound D is **(AIEEE 2007)**

a. butanal b. *n*-butyl alcohol
c. *n*-propyl alcohol d. propanal

7 Among the following the one that gives positive iodoform test upon reaction with I_2 and NaOH is **(AIEEE 2006)**

a. $CH_3CH_2CH(OH)CH_2CH_3$ b. $C_6H_5CH_2CH_2OH$

c. 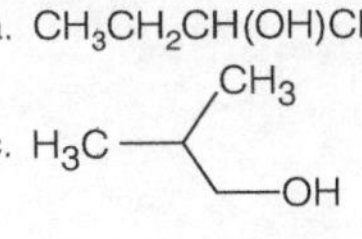 d. $PhCHOHCH_3$

8 Acid catalysed hydration of alkenes except ethene leads to the formation of **(AIEEE 2005)**

a. mixture of secondary and tertiary alcohols
b. mixture of primary and secondary alcohols
c. secondary or tertiary alcohol
d. primary alcohol

9 Among the following compounds which can be dehydrated very easily? **(AIEEE 2004)**

a. $CH_3CH_2CH_2CH_2CH_2OH$

b. $CH_3CH_2CH_2\overset{OH}{\overset{|}{C}}HCH_3$

c. $CH_3CH_2\overset{CH_3}{\overset{|}{\underset{OH}{\underset{|}{C}}}}CH_2CH_3$

d. $CH_3CH_2\underset{CH_3}{\underset{|}{C}}HCH_2CH_2OH$

10 An ether is more volatile than an alcohol having the same molecular formula. This is due to **(AIEEE 2003)**

a. dipolar character of ethers
b. alcohols having resonance structures
c. intermolecular hydrogen bonding in ethers
d. intermolecular hydrogen bonding in alcohols

11 During dehydration of alcohols to alkenes by heating with concentrated H_2SO_4 the initiation step is **(AIEEE 2003)**

a. protonation of alcohol molecule
b. formation of carbocation
c. elimination of water
d. formation of an ester

12 Maximum dehydration takes place in that of **(AIEEE 2002)**

a. 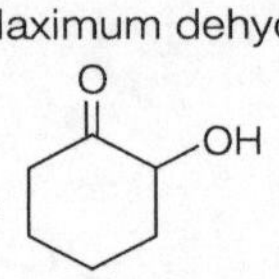

b. 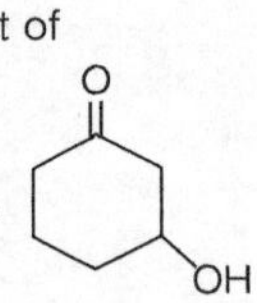

c. 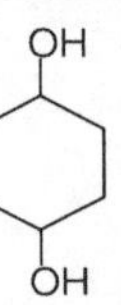

d. 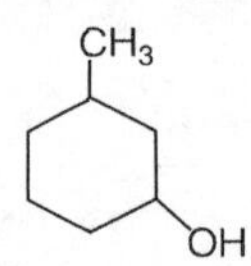

JEE Advanced & IIT JEE Archive

(Compilation of Last 10 Years' Questions)

Passage for Q. Nos. (1 and 2)

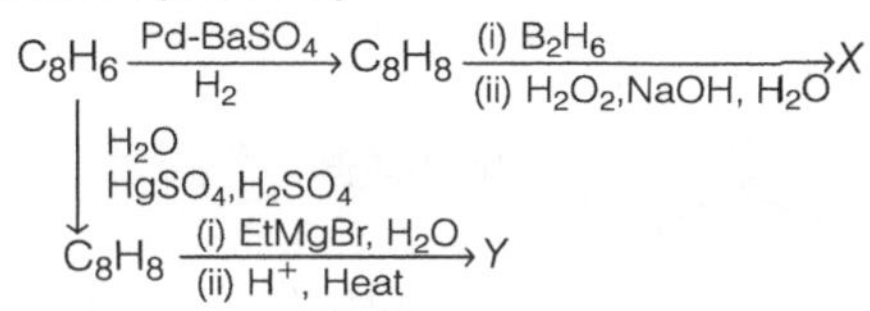

(2015 Adv., Comprehension Type)

1 Compound X is

a. 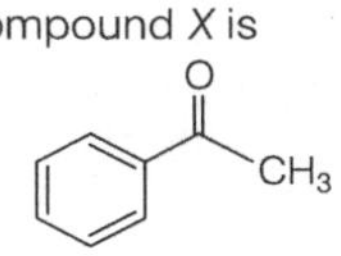

b. 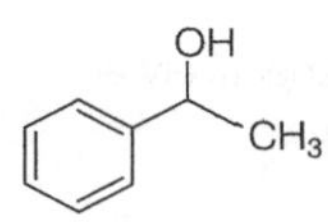

c. 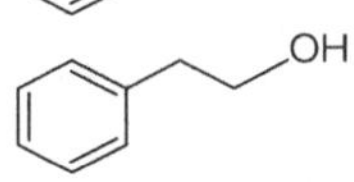

d. 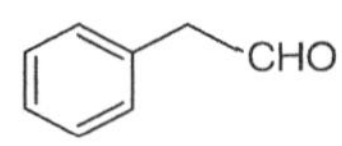

2 The major compound Y is

a. 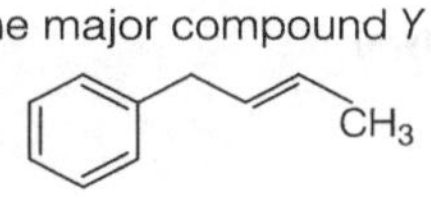

b. 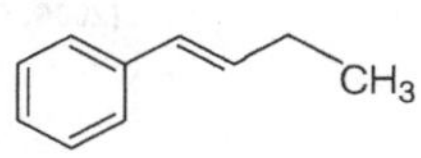

c. 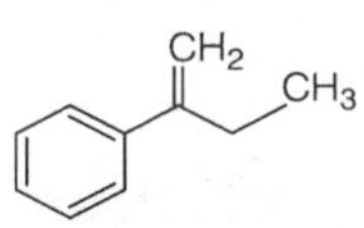

d. 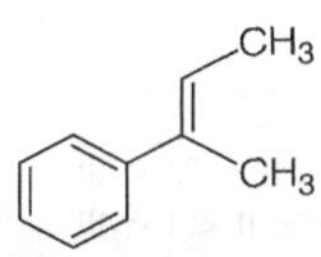

3 The acidic hydrolysis of ether X shown below is fastest when

(2014 Adv., Only One Option Correct Type)

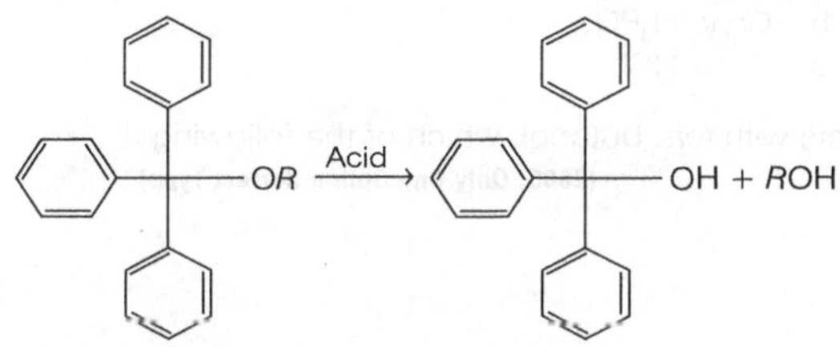

X

a. one phenyl group is replaced by a methyl group
b. one phenyl group is replaced by a *para*-methoxyphenyl group
c. two phenyl groups are replaced by two *para*-methoxyphenyl groups
d. no structural change is made to X

4 The major product of the following reaction is

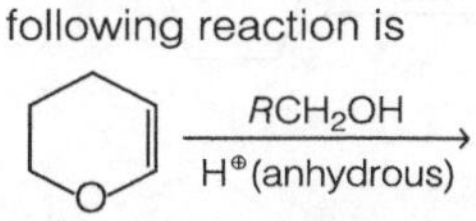

(2011, Only One Option Correct Type)

a. a hemiacetal
b. an acetal
c. an ether
d. an ester

Passage for Q. Nos. (5 and 6)

A tertiary alcohol H upon acid catalysed dehydration gives a product I. Ozonolysis of I leads to compounds J and K. Compound J upon reaction with KOH gives benzyl alcohol and a compound L, whereas K on reaction with KOH gives only M.

(2008, Comprehension Type)

$M =$ 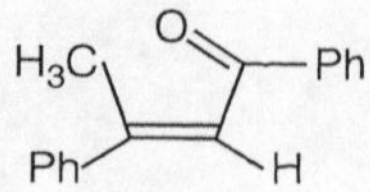

5 Compound H is formed by the reaction of

a. 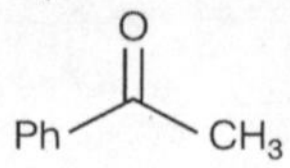 ; PhMgBr

b. 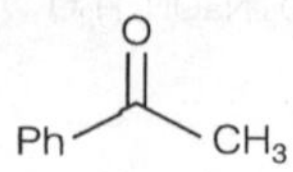 ; $PhCH_2MgBr$

c. 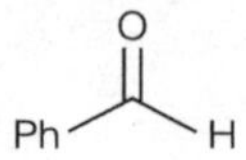 ; $PhCH_2MgBr$

d. 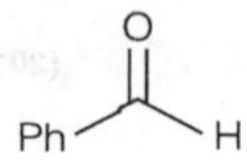 ; 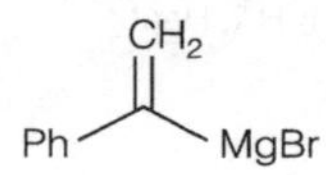

6 The structures of compounds J, K and L respectively are

a. $PhCOCH_3$, $PhCH_2COCH_3$ and $PhCH_2COO^-K^+$

b. PhCHO, $PhCH_2CHO$ and $PhCOO^-K^+$

c. $PhCOCH_3$, $PhCH_2CHO$ and $CH_3COO^-K^+$

d. PhCHO, $PhCOCH_3$ and $PhCOO^-K^+$

7 The increasing order of boiling points of the following mentioned alcohols is

(2006, Only One Option Correct Type)

I. 1, 2-dihydroxy benzene

II. 1, 3-dihydroxy benzene

III. 1, 4-dihydroxy benzene

IV. Hydroxy benzene

a. I < II < III < IV

b. I < II < IV < III

c. IV < I < II < III

d. IV < II < I < III

8 The best method to prepare cyclohexene from cyclohexanol is by using

(2005, Only One Option Correct Type)

a. conc. HCl + $ZnCl_2$

b. Conc. H_3PO_4

c. HBr

d. Conc. HCl

9 When phenyl magnesium bromide reacts with *tert*. butanol, which of the following is formed?

(2005, Only One Option Correct Type)

a. *Tert*. butyl methyl ether

b. Benzene

c. *Tert*. butyl benzene

d. Phenol

10 Consider the given reaction,

(2005, Subjective Type)

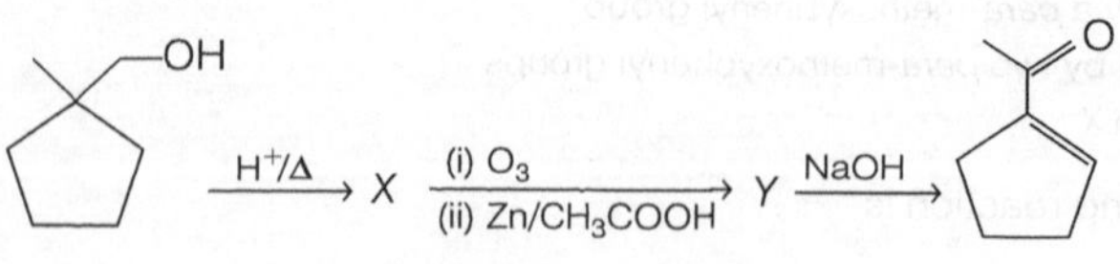

Identify X and Y.

JEE Archive

3. Aliphatic Aldehydes and Ketones

DPP
DAILY PRACTICE PROBLEMS

DPP-1 Preparation of Aldehydes and Ketones

Subjective Questions

Working Space

Direction *(Q. Nos. 1-5) These questions are subjective in nature, need to be solved completely on notebook.*

1 How an aldehyde is prepared by the controlled oxidation of primary alcohol?

2 How a carboxylic acid chloride be efficiently reduced Into corresponding aldehyde?

3 How an acid chloride or anhydride be efficiently reduced into aldehyde using derivatives of $LiAlH_4$?

4 How an ester or a nitrile be converted into corresponding aldehyde?

5 How an aldehyde is made industrially by

i. oxo process?

ii. carbonylation of alkyl borane?

Only One Option Correct Type

Direction *(Q. Nos. 6-17) This section contains 12 multiple choice questions. Each question has four choices (a), (b), (c) and (d), out of which ONLY ONE is correct.*

6 An optically active organic compound has molecular formula $C_5H_{12}O(X)$. X on oxidation with CrO_3/H_2SO_4 gives an achiral $C_5H_{10}O$. Hence, X could be

a. 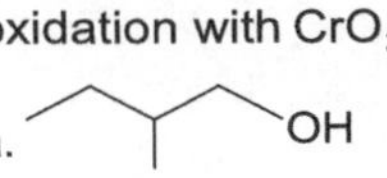

b. 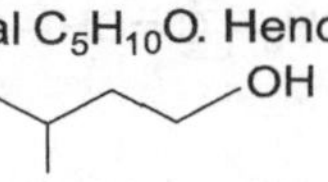

c. 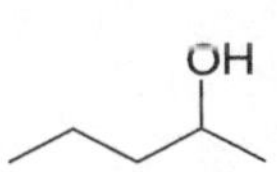

d. 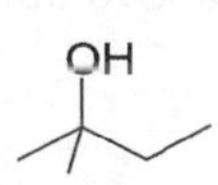

7 Which of the following reaction will not produce an aldehyde?

a. $CH_3CH_2CN \xrightarrow[HCl]{SnCl_2}$

b. $CH_3CH_2OH \xrightarrow{(CF_3CO)_2O} \xrightarrow{DMSO}$

c. 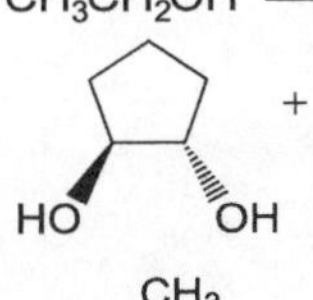 $+ HIO_4 \longrightarrow$

d. 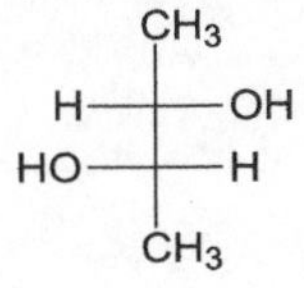 $+ Pb(CH_3COO)_4 \xrightarrow{H_2O}$

8 Which reagent below cannot reduce an acid chloride to an aldehyde?
a. $Pd/BaSO_4$
b. $Li[\{(CH_3)_3CO\}_3AlH]$ at $-80°C$
c. Na/C_2H_5OH
d. $[HAl\{CH_2CH(CH_3)_2\}_2]$ at $-80°C$

9 The incorrect statement regarding oxo process for synthesis of an aldehyde is
a. Mixture of CO and H_2 is allowed to react with an alkene
b. $Co_2(CO)_8$ may be used as a catalyst
c. $[Co(CO)_4H]$ may act as a catalyst
d. This process can also be used in the same manner for the synthesis of ketone

10 All of the following reaction gives atleast one ketone as a significant organic product except
a. 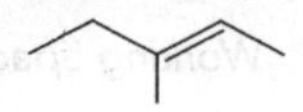 $+ O_3 \xrightarrow{Zn\text{-}H_2O}$
b. 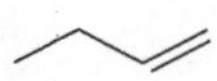 $\xrightarrow[CO]{(C_2H_5)_2BH} \xrightarrow[HO^-]{H_2O_2}$
c. $CH_3CH_2CN\,(excess) + CH_3CH_2MgBr \xrightarrow{H_3O^+}$
d. 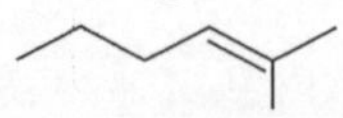 $\xrightarrow[H_2O_2\text{-}NaOH]{B_2H_6} \xrightarrow[H_2SO_4]{CrO_3}$

11 All of the following results in the formation of an aldehyde except
a. 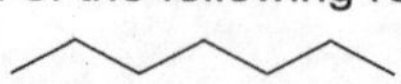 $\overset{+}{N}H_3Cl^- + (CH_2)_6N_4 \xrightarrow[H_2O(V)]{CH_3COOH}$
b. $(CH_3CH_2CH_2CO_2)_2Ca \xrightarrow{Heat}$
c.  $C{\equiv}C{-}H + B_2H_6 \xrightarrow[H_2O_2]{NaOH}$
d. $CH_3CH_2CH_2COOH + HCOOH \xrightarrow[300°C]{MnO}$

12 Consider the following reaction,

$$CH_3{-}\underset{|}{\overset{CH_3}{C}}{=}CH{-}\underset{OH}{CH}{-}CH_3 \xrightarrow{?} CH_3{-}\overset{CH_3}{C}{=}CH{-}\overset{O}{\overset{\|}{C}}{-}CH_3$$

All of the following reagents can bring about the above transformation except
a. MnO_2
b. $[(CH_3)_3CO]_3Al$/Acetone
c. CrO_3/Pyridine/HCl
d. $KMnO_4/OH^-$

13 A hydrocarbon $X(C_7H_{12})$ on ozonolysis followed by the treatment with $(CH_3)_2S$ gives $C_7H_{12}O_2$ which gives positive Tollen's test as well as positive iodoform test. The compound below satisfying the criteria of X is

a. 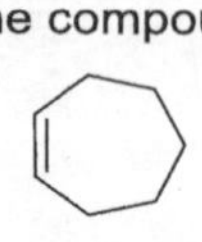
b. 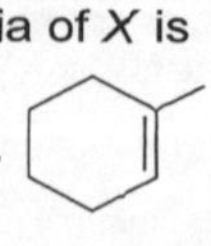
c. 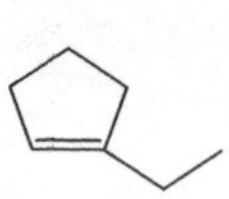
d. 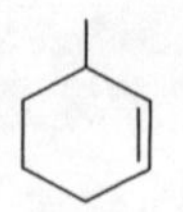

Working Space

14 A hydrocarbon X has molecular formula C_5H_{10} · X on treatment with B_2H_6 in H_2O_2/NaOH gives an optically active $C_5H_{12}O$ which on treatment with CrO_3/HCl / pyridine gives $C_5H_{10}O$ which is still chiral. Which of the following can be a product of reductive ozonolysis of X?

a. $CH_3-\overset{\overset{O}{\|}}{C}-CH_3$

b. $CH_3-CH_2-\overset{\overset{O}{\|}}{C}-CH_3$

c. $CH_3-\underset{}{\overset{\overset{CH_3}{|}}{C}H}-CHO$

d. 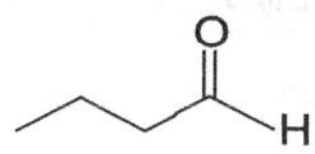

15 What is the final major product of the following reaction

$$CH_3-CH_2-CH_2-C\equiv C-H \xrightarrow{NaNH_2} \xrightarrow{CH_3CH_2CH_2-Br} \xrightarrow[H_2SO_4]{HgSO_4} ?$$

a. 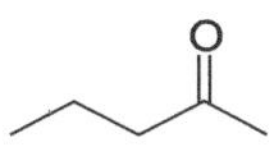

b. 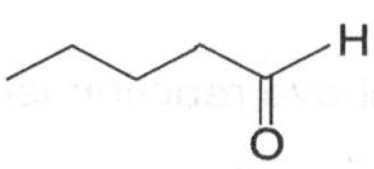

c. 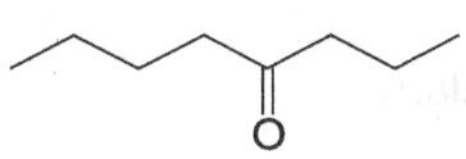

d. 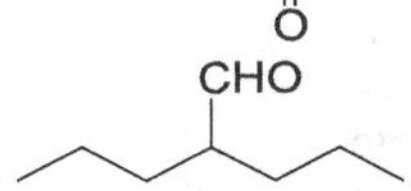

16 Consider the following reaction,

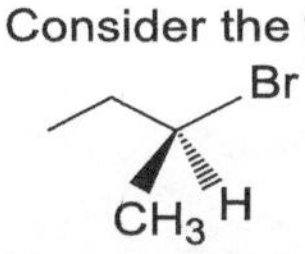 $\xrightarrow{NaCN(aq)} \xrightarrow[HCl]{SnCl_2}$ X (major organic product)

(A pure enantiomer)

The incorrect statement regarding X is

a. It is an aldehyde
b. Product has inverted configuration at the α-carbon
c. X is a racemic mixture
d. Product can be either laevorotatory or dextrorotatory isomer

17 Consider the following reaction sequence,

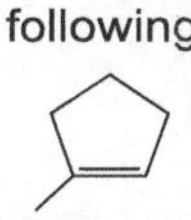 $\xrightarrow[HO^-]{KMnO_4(dil.)} \xrightarrow[H_3O^+]{NaIO_4}$ Final major product X.

The correct statement regarding X is

a. It has an aldehyde functional group
b. It has a ketone functional group
c. It has both aldehyde and ketone functional groups
d. It has a ketone and an acid functional group

One or More than One Options Correct Type

Direction (Q. Nos. 18-22) *This section contains 5 multiple choice questions. Each question has four choices (a), (b), (c) and (d), out of which ONE or MORE THAN ONE are correct.*

18 In which of the following reactions, an aldehyde is formed as major product?

a. $CH_3-C\equiv CH + (CH_3-\overset{\overset{CH_3}{|}}{C}H-\underset{\underset{CH_3}{|}}{C}H)_2BH \xrightarrow[NaOH]{H_2O_2}$

b. $CH_3-CH_2-C\equiv CH + KMnO_4 \xrightarrow[Cold]{H_2O}$

c. $CH_3-CH_2CH_2CH_2-OH + Cu \xrightarrow{\Delta}$

d. $CH_3-CH_2CH_2CH_2-COOH + HCOOH \xrightarrow[Heat]{MnO}$

Working Space

19 In which of the following reactions, ketone is formed as the major organic product?

a. CH_3CH_2CN (excess) + $CH_3MgBr \xrightarrow{Et_2O} \xrightarrow{H_3O^+}$

b. $CH_3CH_2—C≡C—CH_3 + KMnO_4 \xrightarrow[\text{Cold}]{\text{NaOH (Dil.)}}$

c. $CH_3CH_2CH_2—COCl$ (excess) + $CH_3MgBr \xrightarrow{Et_2O}$

d. $H—COOC_2H_5$ (excess) + $CH_3CH_2MgBr \xrightarrow{Et_2O}$

20 Consider the following reaction,

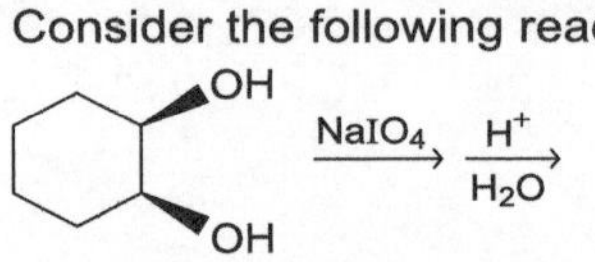

The correct statement(s) regarding the above reaction is/are

a. It involves cleavage of sigma C—C bond

b. Product is a dialdehyde

c. Reaction proceeds *via* cyclic intermediate

d. Optically active diastereomers of the above mentioned reactant behaves similarly in the same reaction

21 Consider the reaction mentioned below,

$$CH_3—CH_2—CH_2—CH{=}CH_2 + CO + H_2 \xrightarrow{Co_2(CO)_8}$$

The expected organic product(s) is/are

a. $CH_3CH_2CH_2—\underset{\displaystyle CHO}{\underset{|}{CH}}—CH_2—CHO$

b. 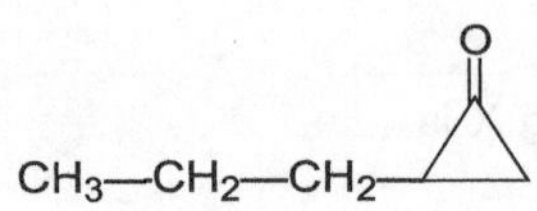

c. $CH_3CH_2CH_2CH_2CH_2CHO$

d. $CH_3—CH_2CH_2—\overset{\displaystyle CHO}{\overset{|}{CH}}—CH_3$

22 Consider the following reaction,

$CH_3CH_2CH_2COCl \xrightarrow{?} CH_3CH_2CH_2CHO$

Reagent(s) that can bring about the above reaction successfully is/are

a. $Pd/BaSO_4$

b. $[(CH_3)_2CHO]_3Al$

c. $Li[\{(CH_3)_3CO\}_3AlH]$ at – 80°C

d. CuO/Heat

Comprehension Type

***Direction** (Q. Nos. 23-25) This section contains a paragraph, describing theory, experiments, data, etc. Three questions related to the paragraph have been given. Each question has only one correct answer among the four given options (a), (b), (c) and (d).*

Passage

A hydrocarbon *A* ($C_{10}H_{18}$) is capable of showing both enantiomerism as well as diastereomerism. Treatment of *A* either with $HgSO_4/H_2SO_4$ or B_2H_6/H_2O_2-NaOH results in the same carbonyl compound *B*. Also,

$A \xrightarrow[C_2H_5OH\text{ (dry)}]{Na} B \xrightarrow[Zn\text{-}H_2O]{O_3} C\ (C_5H_{10}O)$ *C* can also be obtained as one of the product in the following reaction.

$$1\text{-butene} + CO + H_2 \xrightarrow{Co_2(CO)_8} C + \text{Other product}$$

23 What is the most likely structure of *B*?

a. 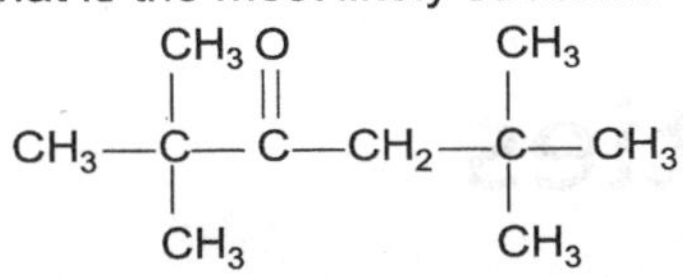

b. 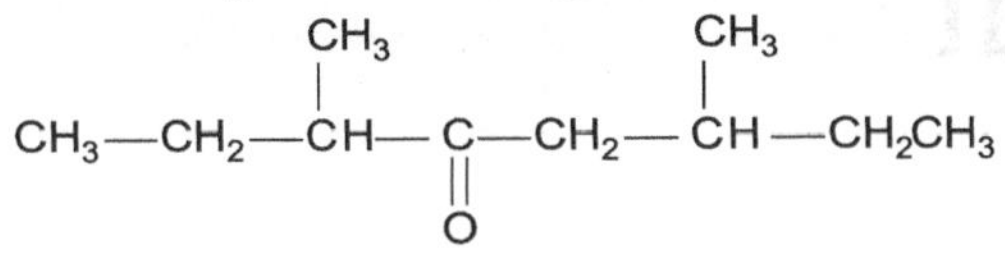

c. 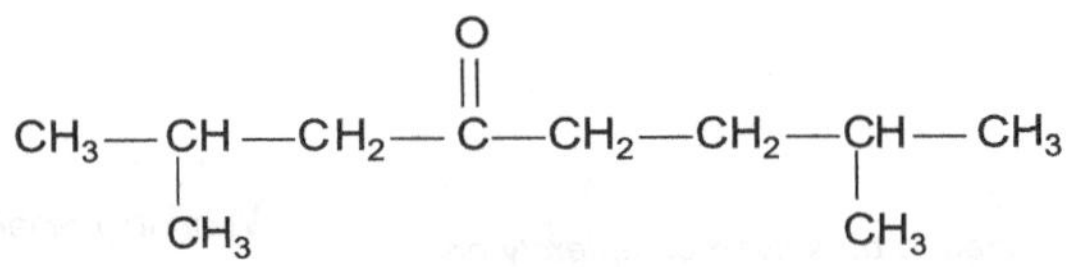

d. 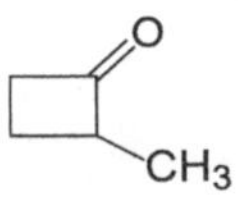

24 What is the structure of compound *C*?

a. 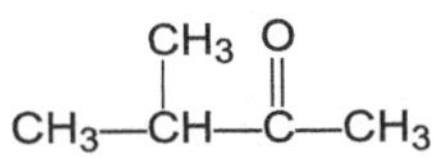

b. 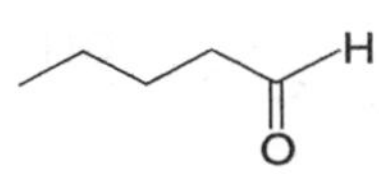

c. 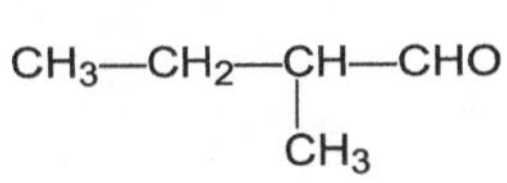

d. 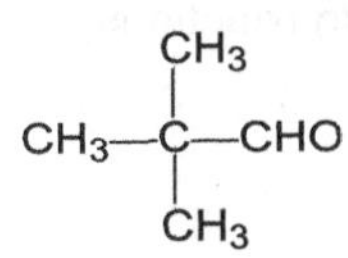

25 Consider the reaction given below,

$$C + CH_3MgBr \xrightarrow{Et_2O} \xrightarrow[H_2O]{H^+} \text{Alcohols}$$

How many different alcohols are expected?

a. 1 b. 2 c. 3 d. 4

One Integer Value Correct Type

Direction *(Q. Nos. 26-30) This section contains 5 questions. When worked out will result in an integer from 0 to 9 (both inclusive).*

26 How many different aldehyde isomers exist for $C_5H_{10}O$?

27 How many different ketones isomer exist for $C_6H_{12}O$?

28 How many different alcohol isomers with molecular formula $C_5H_{12}O$ can be oxidised to ketones using $K_2Cr_2O_7$-H_2SO_4?

29 If all the ketone isomers of $C_6H_{10}O$ are reduced independently with $NaBH_4$, how many of them will produce racemic mixture of alcohols?

30 Consider the following reaction,

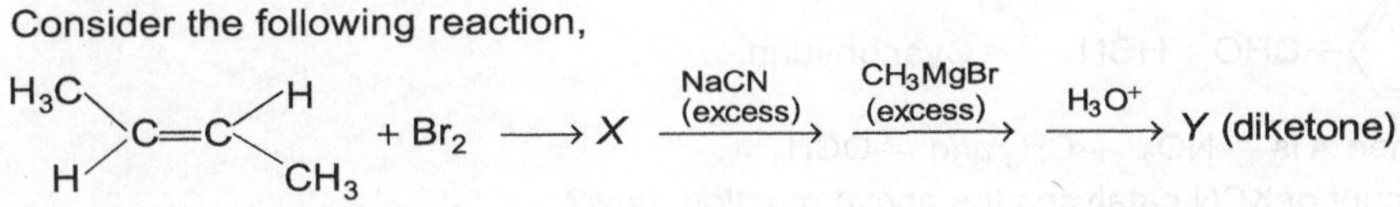

$$Y \xrightarrow{NaBH_4} \xrightarrow{H_3O^+} \text{diol}$$

How many different diols are expected at the end of the above reaction?

Working Space

DPP-2 Physical Properties and General Reactions

Subjective Questions

Working Space

Direction *(Q. Nos. 1-6) These questions are subjective in nature, need to be solved completely on notebook.*

1 Compare and comment on the boiling points of

i. 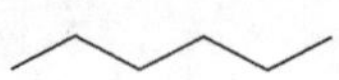

ii. 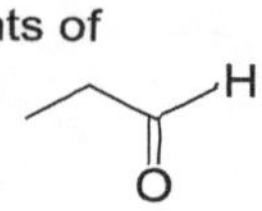

iii. $CH_3-\overset{O}{\overset{\|}{C}}-CH_3$

iv. 

2 Provide suitable reagent(s) for the following reactions.

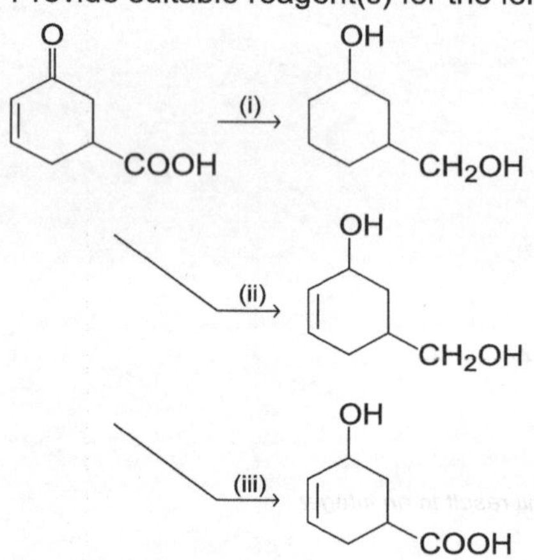

3 How a mixture containing propanal and ethanol can be separated using $NaHSO_3$ solution?

4 What are trioxane and paraformaldehyde and what is their usefulness in the synthesis of formaldehyde?

5 How formaldehyde is stored at room temperature?

6 Consider the following reaction,

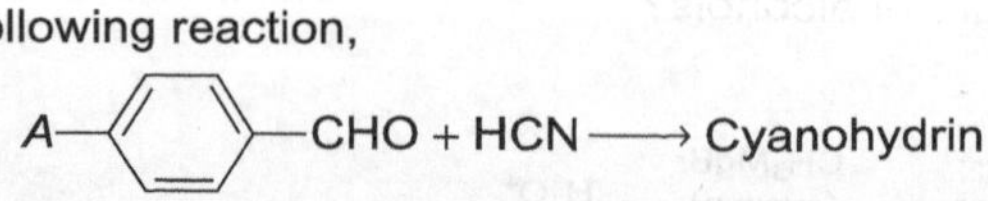

i. Compare reactivity when *A* is $-NO_2$, $-CH_3$ and $-OCH_3$.

ii. Presence of trace amount of KCN catalyses the above reaction, how?

Only One Option Correct Type

Working Space

Direction (Q. Nos. 7-15) *This section contains 9 multiple choice questions. Each question has four choices (a), (b), (c) and (d), out of which ONLY ONE is correct.*

7 Which of the following will give a racemic mixture on reduction with $NaBH_4$ followed by acid work-up?

a. 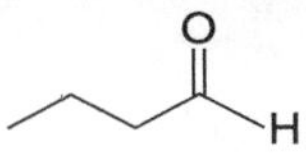

b. 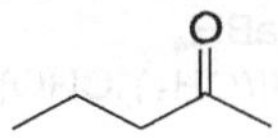

c. 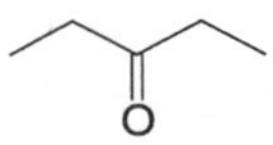

d. 

8 What would be the major product in the following reaction?

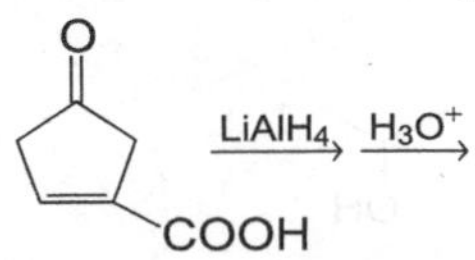

a. 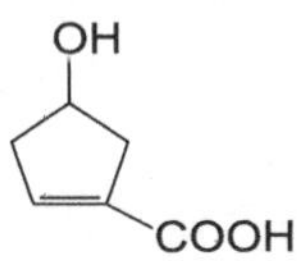

b. 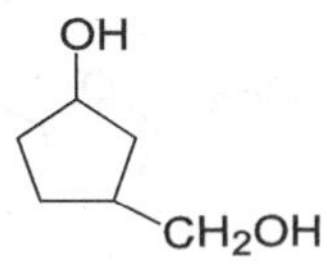

c. 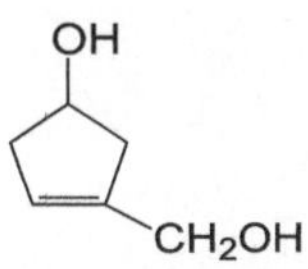

d. 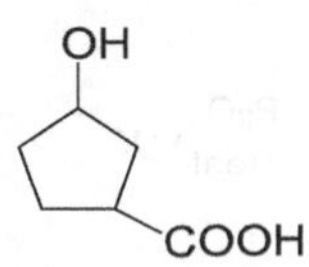

9 Which of the following on reaction with excess of $NaHSO_3$ in aqueous solution will give mixture of salts which can be separated into two fractions by fractional crystallisation?

a. 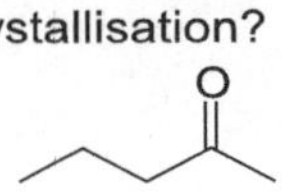

b. 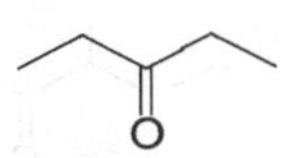

c. 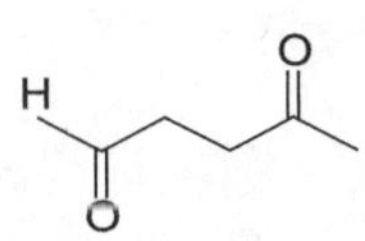

d. 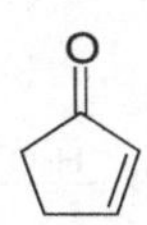

10 Which is the most suitable reagent for the following transformation?

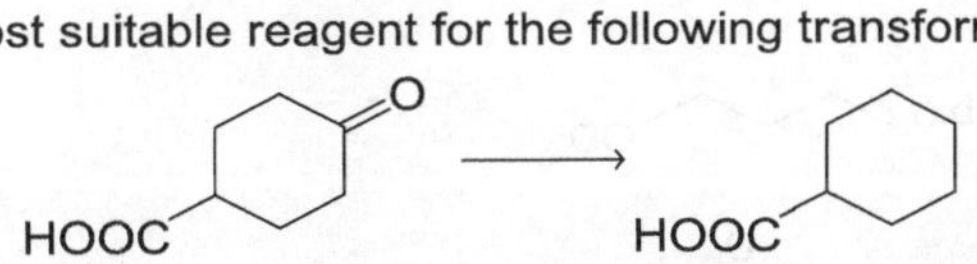

a. Zn(Hg)-HCl
b. N_2H_4/NaOH/Heat
c. $LiAlH_4$
d. $NaBH_4$

11 Which is the most suitable reagent for the following transformation?

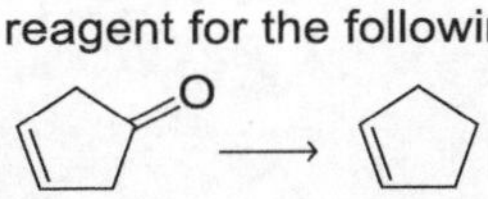

a. Zn(Hg)-HCl
b. N_2H_4/NaOH/Heat
c. $LiAlH_4$
d. $NaBH_4$

Working Space

12 The reagent which can best bring about the following transformation is

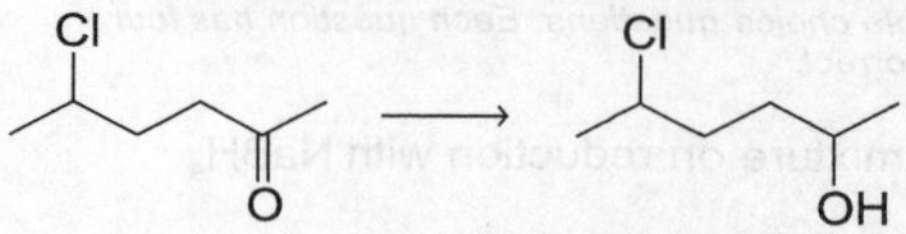

a. $LiAlH_4$
b. $NaBH_4$
c. Zn(Hg)-HCl
d. $Al[(CH_3)_2CHO]_3$

13 The most probable product of the following reaction is

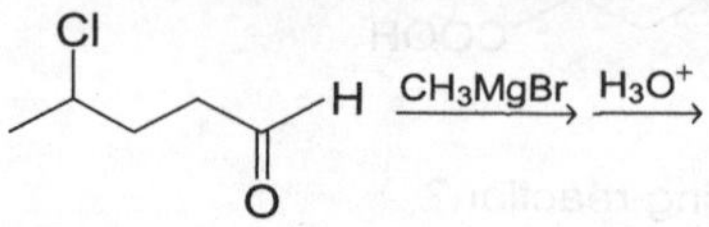

a. 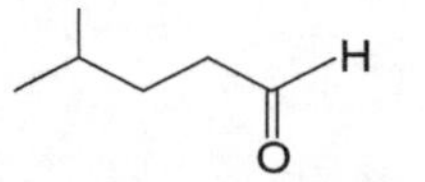

b. 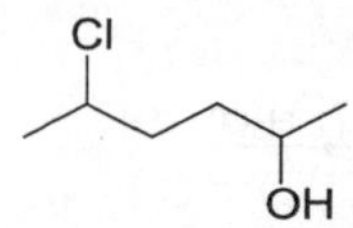

c. 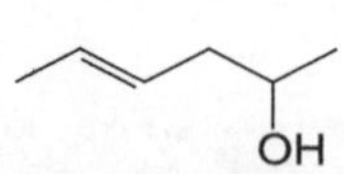

d. 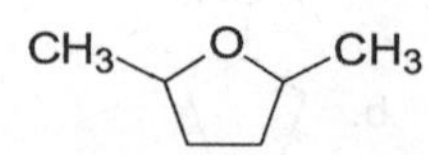

14 Consider the following reaction,

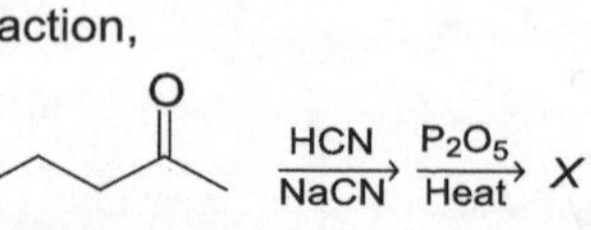

The most likely organic product *X* is

a. 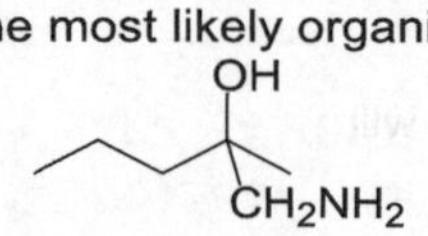

b. 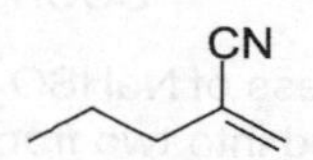

c. 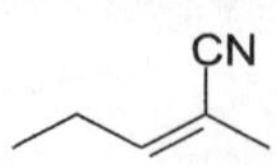

d. 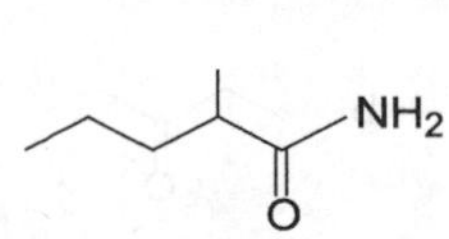

15 Consider the following reaction,

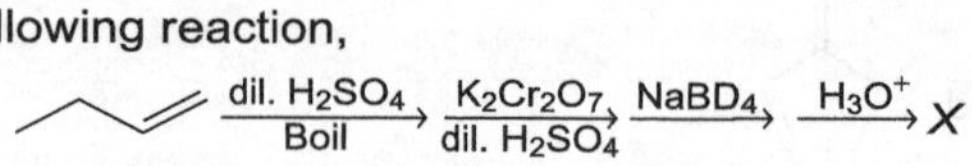

The most likely organic product *X* is

a. 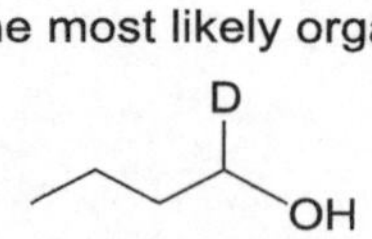

b. 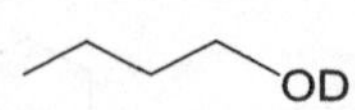

c. 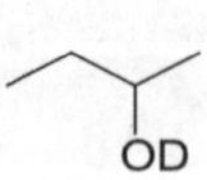

d. 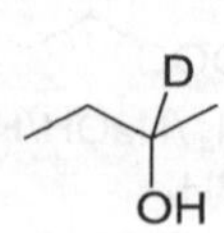

One or More than One Options Correct Type

Working Space

Direction *(Q. Nos. 16-19) This section contains 4 multiple choice questions. Each question has four choices (a), (b), (c) and (d), out of which ONE or MORE THAN ONE are correct.*

16 Consider the following reaction,

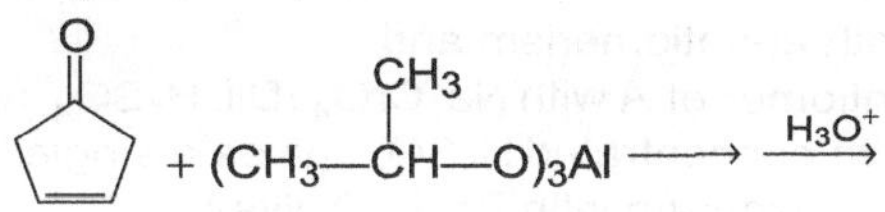

The organic product(s) formed above is/are

a. 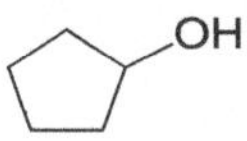

b. 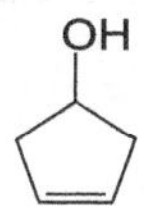

c. $CH_3-\overset{\overset{O}{||}}{C}-CH_3$

d. $CH_3-\underset{\underset{OH}{|}}{CH}-CH_3$

17 Concidor tho roaction givon bolow,

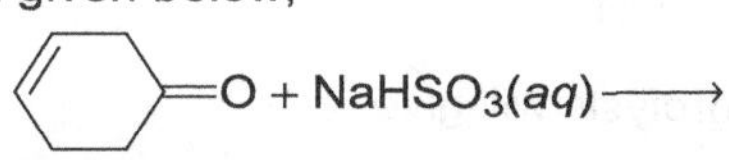

The correct observation regarding the above reaction is/are

a. Precipitate formation takes place
b. Racemic mixture of salts are formed
c. Salts on acid hydrolysis gives back the original reactant
d. If $NaHSO_3$ is in limited amount, no precipitation occur

18 Consider the reaction given below,

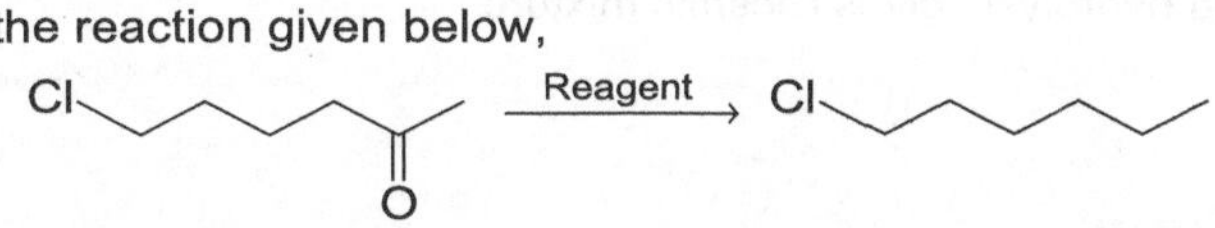

The correct deduction(s) regarding the above reagent is/are

a. Zn(Hg)-HCl can bring about the above reaction
b. If N_2H_4/NaOH/Heat is used, chloride may also be substituted by the hydroxide group
c. $NaBH_4$ followed by acid work-up can efficiently bring about the above reaction
d. Ni/H_2 at high temperature and pressure can be used in the above reaction

19 Consider the following reaction,

$$-\overset{\overset{O}{||}}{C}- + LiAlH_4 \longrightarrow \xrightarrow{H_3O^+}$$

The correct deduction(s) regarding mechanism of the above reaction is/are

a. Reaction involves nucleophilic addition at carbonyl carbon
b. Ketones are more reactive than aldehyde
c. If carbonyl carbon turns chiral, racemic mixture of products is formed provided no other chiral carbon is present
d. If the intermediate aluminium salt is hydrolysed with DCl-D_2O, deuterated alcohol (*R*—OD) is formed

Working Space

Comprehension Type

***Direction** (Q. Nos. 20-22) This section contains a paragraph, describing theory, experiments, data, etc. Three questions related to the paragraph have been given. Each question has only one correct answer among the four given options (a), (b), (c) and (d).*

Passage

An organic compound $A(C_7H_{16}O)$ shows both enantiomerism and diastereomerism. Treatment of a pure enantiomer of A with Na_2CrO_4 /Dil. H_2SO_4 gives B $(C_7H_{14}O)$. Also A on dehydration with concentrated H_2SO_4 gives a single alkene C (C_7H_{14}). Ozonolysis of C followed by work-up with Zn-H_2O gives D $(C_5H_{10}O)$ as one of the product which gives racemic mixture on reduction with $NaBH_4$.

20 The structure of A satisfying the above criteria is

a. 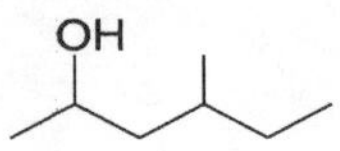

b. 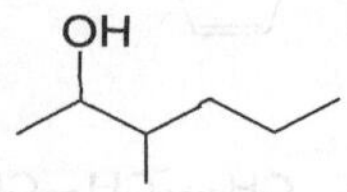

c. 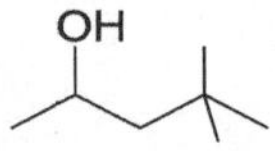

d. 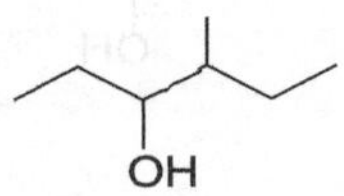

21 If B is reduced with $LiAlH_4$ followed by acid hydrolysis will give

a. a pure enantiomer
b. a racemic mixture
c. a pair of diastereomers
d. an achiral alcohol

22 The correct statement regarding the compound D is

a. With CH_3MgBr followed by acid hydrolysis gives racemic mixture
b. With $CH_3CH_2CH_2MgBr$ followed by acid hydrolysis gives racemic mixture
c. With CH_3CH_2MgBr followed by acid hydrolysis gives racemic mixture
d. Both (b) and (c) are correct

One Integer Value Correct Type

***Direction** (Q. Nos. 23-28) This section contains 6 questions. When worked out will result in an integer from 0 to 9 (both inclusive).*

23 If all the aldehyde isomers of $C_5H_{10}O$ is independently treated with HCN/NaCN solution, how many of them will give racemic mixture of cyanohydrin?

24 If one molecule of trioxane is heated, how many molecules of formaldehyde would be formed?

25 A mixture containing all isomeric cyclohexanedione is treated with excess of $NaHSO_3$ solution. How many different disulphite salts are formed?

26 Consider the following reaction,

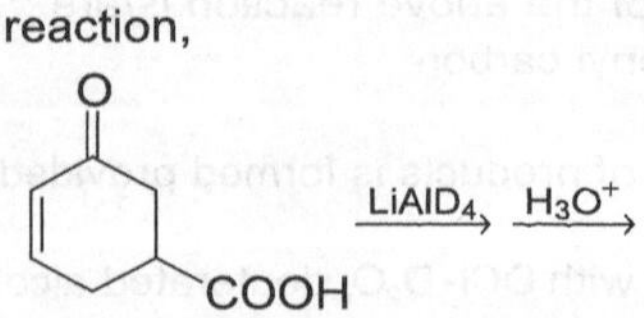

How many deuterium would be incorporated in the final product?

27 If glyoxal glycol is treated with a mixture of CH_3MgBr and C_2H_5MgBr in diethyl ether followed by acid hydrolysis, how many different diols would be formed, which are simultaneously optically active?

Working Space

28 Consider the reaction given below,

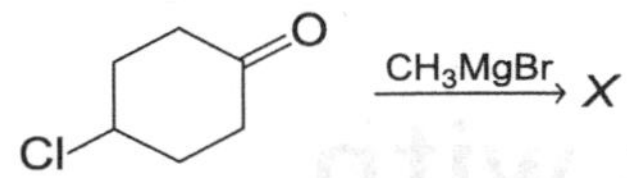

X is formed in which a new ring is formed. How many atoms are present in this new ring?

Matching List Type

Direction (Q. Nos. 29 and 30) *Choices for the correct combination of elements from Column I and Column II are given as options (a), (b), (c) and (d), out of which one is correct.*

29 Match the reactants from Column I with the reagents and expected outcomes from Column II. Mark the correct option form the codes given below.

	Column I		Column II
i.	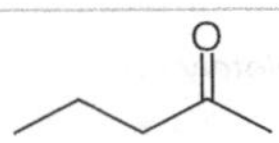	p.	$LiAlH_4$–racemic mixture of products.
ii.	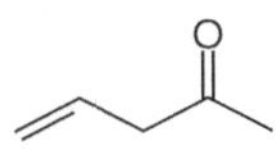	q.	$[(CH_3)_2CHO]_3Al$–racemic mixture of products.
iii.	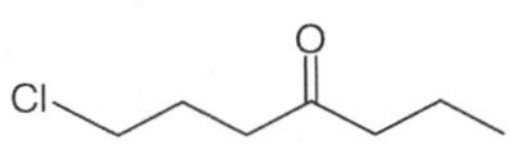	r.	Zn(Hg)-HCl–suitable for selective reduction of carbonyl group.
iv.	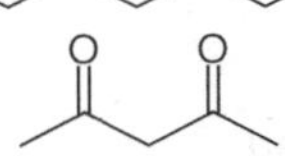	s.	N_2H_4/NaOH/Heat- suitable for selective reduction of carbonyl group.

Codes

	i	ii	iii	iv
a.	p,q,r,s	p,q,s	q,r	p,q,r,s
b.	p,r	q,r	s	r,r
c.	q,r	r,s	p,s	s
d.	p,r	q	r,s	s

30 Match the reactions from Column I with the properties of products from Column II. Mark the correct option form the codes given below.

	Column I		Column II
i.	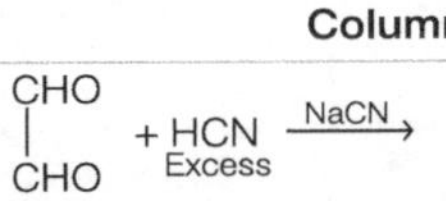	p.	Racemic mixture
ii.	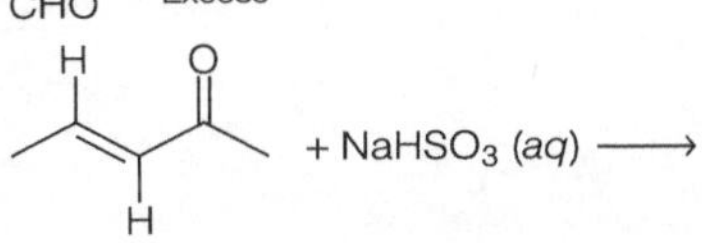	q.	Pair of diastereomers
iii.	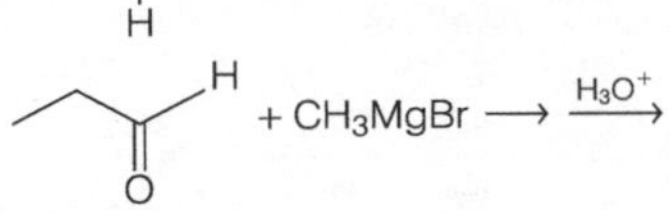	r.	*Meso* isomer
iv.	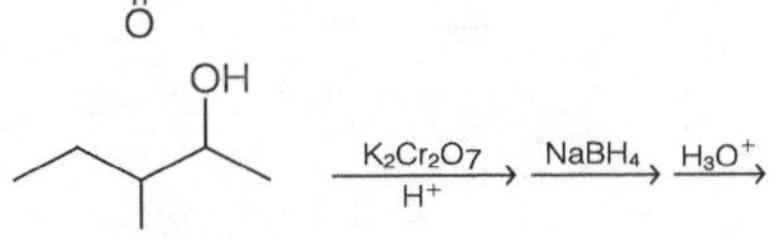 (+) 3-methyl-2-pentanol	s.	Product mixture can be separated into two fractions by chromatography

Codes

	i	ii	iii	iv
a.	p,q,r,s	q,s	p	q,s
b.	q,s	p,s	p,q	r,q
c.	p	q	r	s
d.	p,q,r	r,s	r	p,q

DPP-3 Reactions with Ammonia, Amines, Hydrazine and Its Derivatives

Subjective Questions

Direction *(Q. Nos. 1-5) These questions are subjective in nature, need to be solved completely on notebook.*

Working Space

1 Complete the following reactions.

i. $CH_2O + NH_3 \longrightarrow$

ii. $CH_3CHO + NH_3 \longrightarrow$

iii. $CH_3CH_2COCH_3 + NH_3 \longrightarrow$

iv. $CH_3CH_2CHO + CH_3NH_2 \longrightarrow$

2 When acetone is treated with CH_3NH_2 in acidic medium, imine is formed. However, the reaction is fastest when pH is 4.5, why?

3 Complete the following reactions.

i. $CH_3CH_2CHO + NH_2OH \longrightarrow$

ii. $CH_3CH_2CHO + N_2H_4 \longrightarrow$

iii. $CH_3CH_2CHO + H_2N—\overset{O}{\overset{\|}{C}}—NH—NH_2 \longrightarrow$

iv. 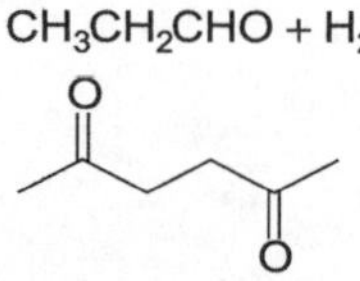 $+ N_2H_4 \longrightarrow$

v. 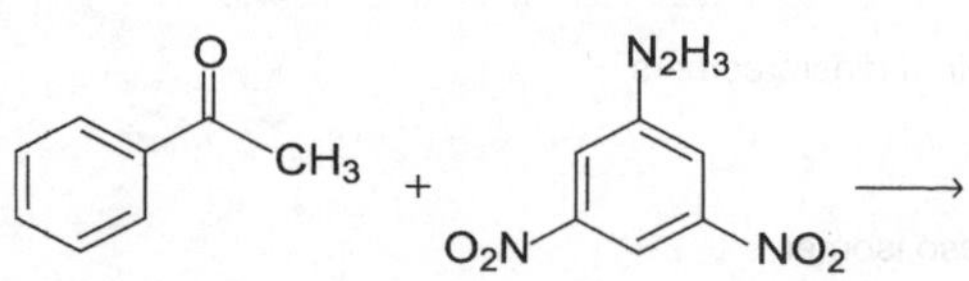

4 Complete the following reactions.

i. 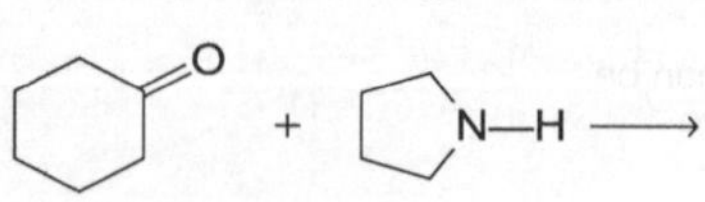

ii. $CH_3CH_2CHO + SeO_2 \longrightarrow$

iii. $CH_3CH_2CHO + Br_2 \longrightarrow$

iv. $CH_3CH_2COCH_3 + Cl_2 \xrightarrow{NaOH}$

5 Passing chlorine gas over ethanol at 60°C gives a trichloro derivative X. What is this X? What happens when X reacts with (i) hot, concentrated KOH, (ii) concentrated HNO_3 and (iii) $(C_2H_5O)_3Al$?

Only One Option Correct Type

Direction (Q. Nos. 6-15) *This section contains 10 multiple choice questions. Each question has four choices (a), (b), (c) and (d), out of which ONLY ONE is correct.*

6 Arrange the following in the increasing order of reactivity with NH_3.

I. CH_2O II. CH_3CHO

III. $CH_3—CO—CH_3$ IV. $CH_3—CH(CH_3)—C(=O)—CH(CH_3)—CH_3$

a. I < II < III < IV b. IV < III < II < I
c. III < IV < I < II d. II < I < IV < III

7 The major organic product formed in the following reaction is

$$\begin{matrix} CH_2—CHO \\ | \\ CH_2—CHO \end{matrix} + NH_3 \longrightarrow$$

a. $\begin{matrix} CH_2—CH{=}NH \\ | \\ CH_2—CH{=}NH \end{matrix}$ b. 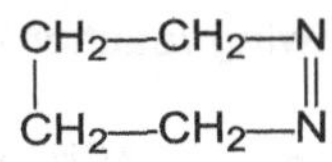

c. 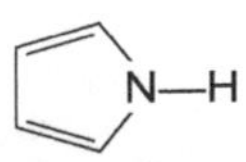 d. 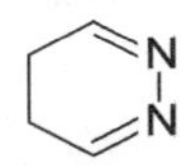

8 Which carbonyl compound below gives the reddish orange precipitate with 2, 4-dinitrophenyl hydrazine?

a. CH_3CH_2CHO b. $CH_3—CO—CH_3$

c. 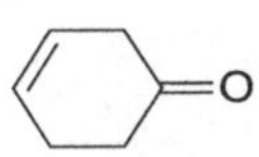 d. 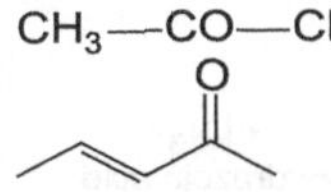

9 Which carbonyl compound below forms a single semicarbazone when allowed to react with semicarbazide?

a. 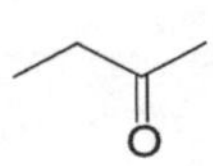 b. 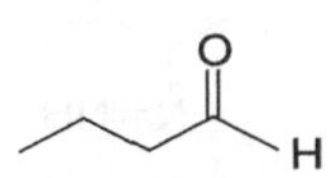

c. 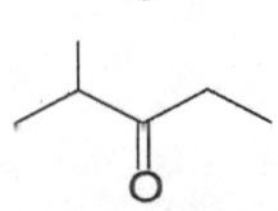 d. 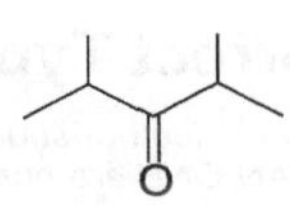

10 A hydrocarbon *X* (C_7H_{14}) on ozonolysis followed by work-up with Zn-H_2O gives $C_6H_{12}O$ as one of the product which on subsequent treatment with KOH/I_2 gives yellow precipitate and salt of a chiral acid. Hence, *X* could be

a. 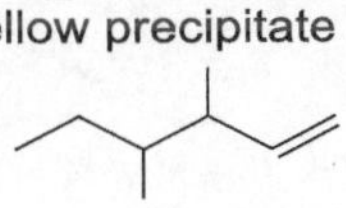 b. 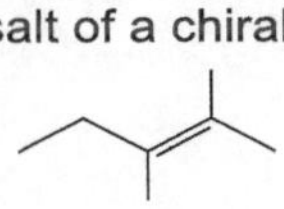 c. 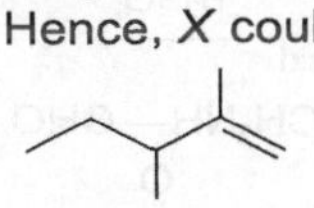 d. 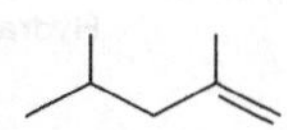

11 Which gives more than one oximes on treatment with hydroxylamine?

a. 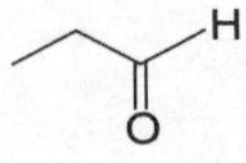 b. $H—C(=O)—H$

c. $C_6H_5—CO—C_6H_5$ d. 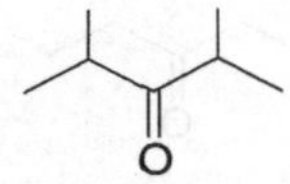

Working Space

12 Which forms an unsaturated amine on treatment with cyclopentanone in slightly acidic medium?

a. CH_3NH_2 b. $(CH_3)_3N$

c. NH_3 d. 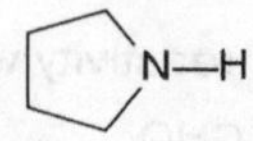

13 Consider the following reaction,

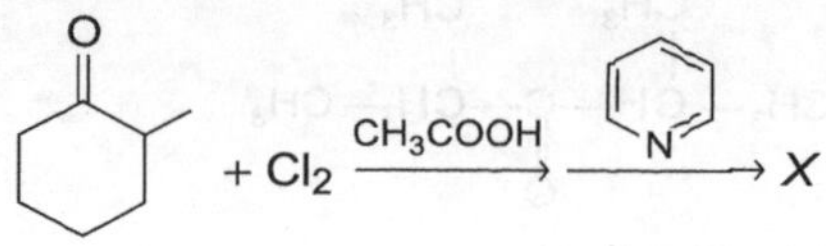

X (major) in the above reaction is

a. 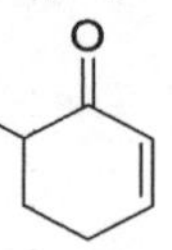 b. 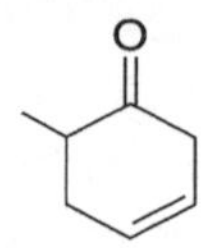 c. 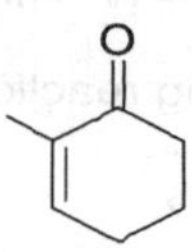 d. 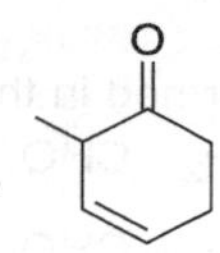

14 What is the final major product Y in the following reaction?

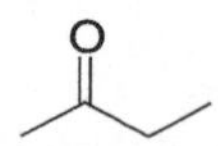 $+ PCl_5 \longrightarrow POCl_3 + X \xrightarrow[\text{Heat}]{\text{KOH}} Y$

a. 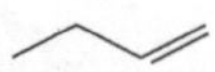 b. 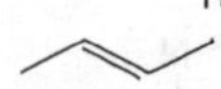

c. 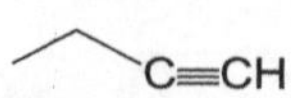 d. $CH_3—C≡C—CH_3$

15 What is the final major product in the following reaction?

$$CH_3CH_2—\overset{O}{\overset{\|}{C}}—CH_2CH_3 + \underset{\text{Hydrazoic acid}}{HN_3} \longrightarrow$$

a. 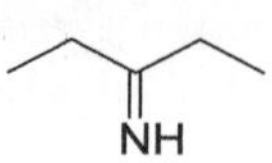 b. 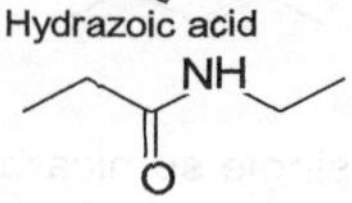

c. 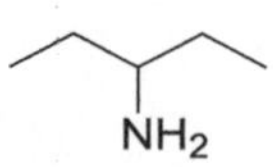 d. 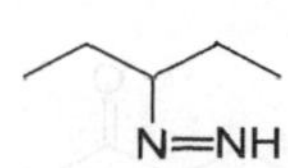

One or More than One Options Correct Type

***Direction** (Q. Nos. 16-19) This section contains 4 multiple choice questions. Each question has four choices (a), (b), (c) and (d), out of which ONE or MORE THAN ONE are correct.*

16 What is/are expected product(s) in the following reaction?

$$CH_3CHO + \underset{\text{Hydrazoic acid}}{HN_3} \xrightarrow{H_2SO_4}$$

a. CH_3CN b. $CH_3NH—CHO$

c. N_2 d. $CH_3—\overset{O}{\overset{\|}{C}}—NH_2$

17 Which of the following form enamine on heating with a secondary amine in weakly acidic medium?

a. $CH_3—\overset{CH_3}{\underset{CH_3}{C}}—CHO$ b. 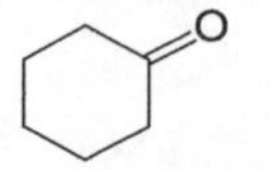 c. 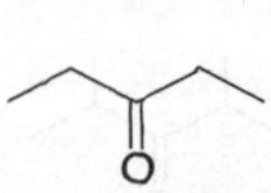 d. C_6H_5CHO

Working Space

18 The compound(s) which form a pair of diastereomers with hydroxylamine is/are

a. 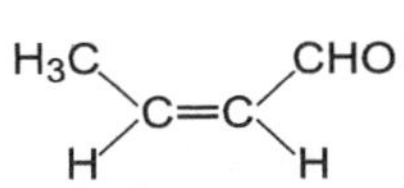

b. 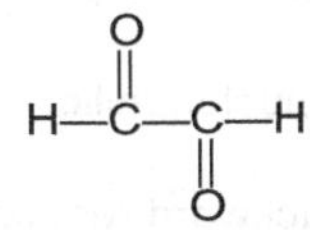

c. 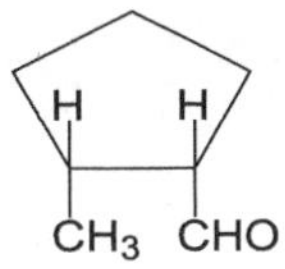

d. 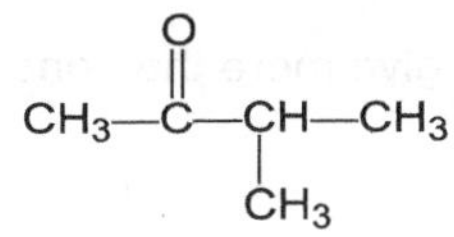

19 The compound(s) below that gives yellow precipitate with KOH/I_2 is/are

a. CH_3CH_2OH

b. $CH_3-\overset{O}{\overset{\|}{C}}-NH_2$

c. $CH_3CH_2-\overset{OH}{\overset{|}{C}H}-CH_3$

d. $Cl-CH_2-CHO$

Comprehension Type

***Direction** (Q. Nos. 20-22) This section contains a paragraph, describing theory, experiments, data, etc. Three questions related to the paragraph have been given. Each question has only one correct answer among the four given options (a), (b), (c) and (d).*

Passage

Aldehydes and ketones react with ammonia and primary amines to form imines in slightly acidic condition. A typical reaction mechanism with a primary amine is as follows

Step I 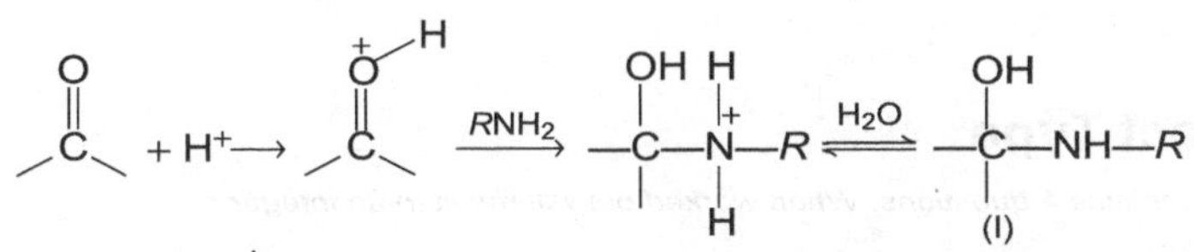

Step II 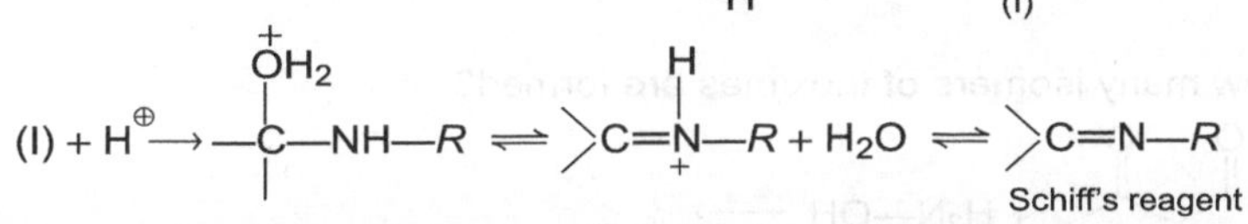

The proper pH control is crucial to imine formation otherwise reaction does not proceed at appreciable rate.

However, if a secondary amine is reacted, no hydrogen is left on nitrogen on intermediate (I) for the deprotonation in step II. Hence, the reaction proceeds as

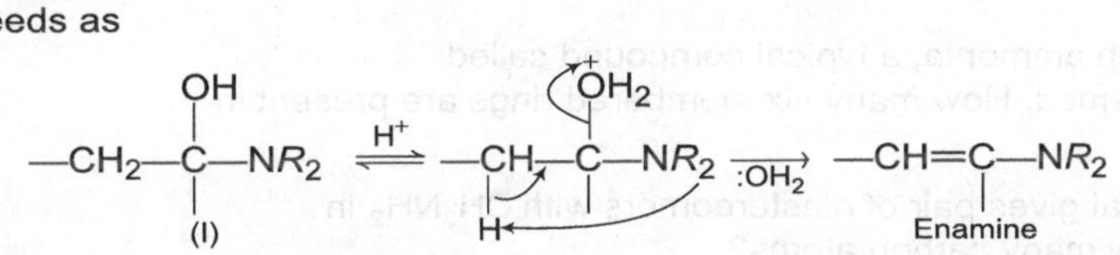

Enamine is a suitable intermediate route for α-alkylation of a carbonyl compound.

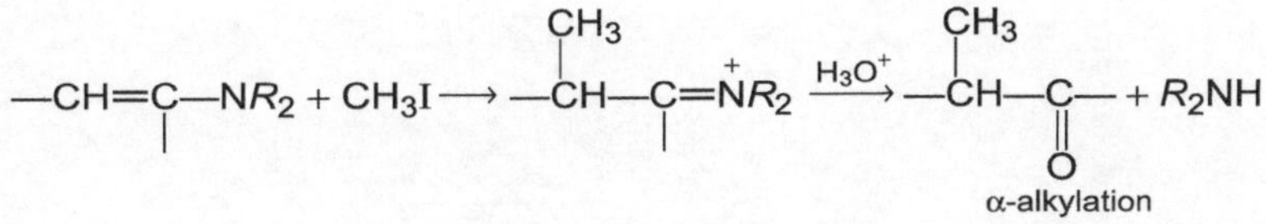

Working Space

20 What happens if imine formation is carried out at very low pH?

a. Reaction occur at very fast rate
b. Amine gets protonated and becomes non-nucleophilic
c. Imine formed gets hydrolysed very easily
d. Imine formed gets protonated favouring backward reaction

21 Which of the following is expected to give more than one imine when treated with CH_3NH_2?

a. CH_3CHO
b. $CH_3-\overset{O}{\overset{\|}{C}}-CH_3$
c. H_2CO
d. $(C_6H_5)_2CO$

22 What is the product X in the following reaction?

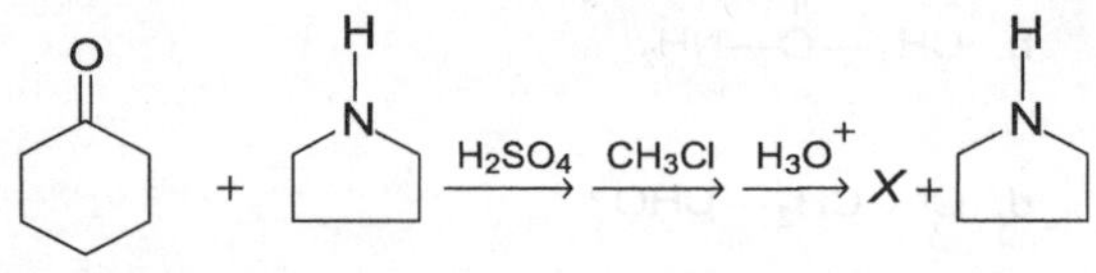

a. 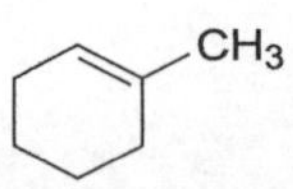
b. 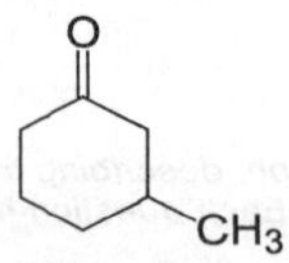
c. 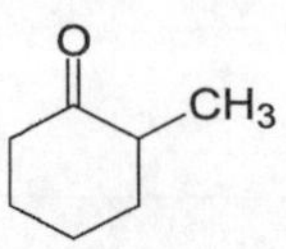
d. 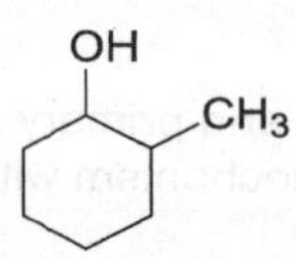

One Integer Value Correct Type

Direction *(Q. Nos. 23-26) This section contains 4 questions. When worked out will result in an integer from 0 to 9 (both inclusive).*

23 In the following reaction, how many isomers of trioximes are formed?

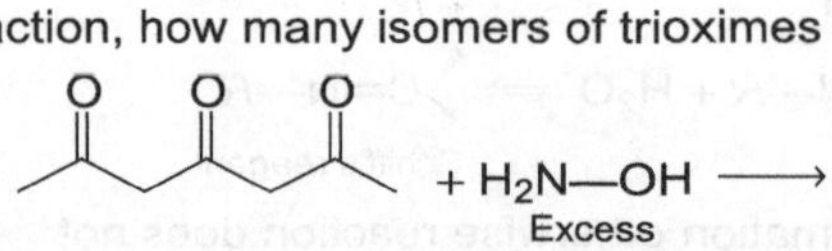

24 Consider the isomeric aldehydes with molar mass 100, if all the isomers (only structural) are treated independently with NH_2OH, how many of them would give more than two stereomeric oximes?

25 When formaldehyde reacts with ammonia, a typical compound called hexamethylene tetramine is formed. How many six membered rings are present in this compound?

26 The smallest acyclic ketone that gives pair of diastereomers with CH_3NH_2 in slightly acidic medium has how many carbon atoms?

DPP-4 Keto-Enol Tautomerism and Related Reactions

Subjective Questions

Working Space

Direction (Q. Nos. 1-6) *These questions are subjective in nature, need to be solved completely on notebook.*

1 Percentage of enol content of ethyl acetoacetate in different solvents are as follows :

Solvent	% Enol
H_2O	0.4
Acetone	7.3
Ethyl acetate	7.7
Methanol	6.9
Hexane	46.7

Explain the above observations.

2 Compound $(p\text{-}O_2N—C_6H_4)_2\text{-}CH—CHO$ is less stable than its enol tautomer, explain.

3 Butane-2,3-dione almost exclusively exist in keto form whereas cyclopentane-1, 2-dione exist exclusively in its enol form, explain.

4 Boiling point of nitromethane is 101° C but boiling point of its functional isomer methyl nitrite is only −12°C, give proper reason for this difference.

5 Acetone in dilute aqueous solution is almost 100% unhydrated. When acetone is dissolved in water enriched with H_2O^{18}, recovered acetone contains almost entirely O^{18}, why?

6 Explain the following observation.

i. The ketene, tropone 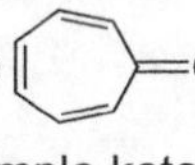 is exceptionally basic, forming much more stable salt with HCl than simple ketone.

ii. Semicarbazide (1 mole) is added to a mixture of cyclohexanone and benzaldehyde (1 mole each). If the product is isolated immediately, it consists almost entirely of semicarbazone of cyclohexanone while if the product is isolated after several hours, it consists almost entirely of semicarbazone of benzaldehyde.

Only One Option Correct Type

Working Space

Direction *(Q. Nos. 7-20) This section contains 14 multiple choice questions. Each question has four choices (a), (b), (c) and (d), out of which ONLY ONE is correct.*

7 In hexane-2,4-dione, how many different mono-enols are possible?

a. 2　　b. 3
c. 4　　d. 7

8 What is the correct order of equilibrium enol content of the following compounds?

I. CH_3COCH_3　　II. $CH_3COCH_2COOC_2H_5$
III. $CH_3COCH_2COCH_3$　　IV. CH_3COCH_2COH

a. I > II > III > IV　　b. I > IV > III > II
c. IV > II > III > I　　d. III > IV > II > I

9 In which of the following compounds, enol form exist?

a. $C_6H_5COCH_3$　　b. C_6H_5CHO

c. 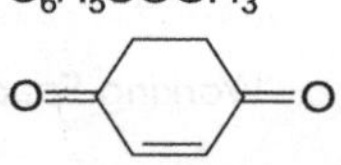　　d. Both (a) and (c)

10 Among the following compounds, one that will not show keto-enol tautomerism is

a. 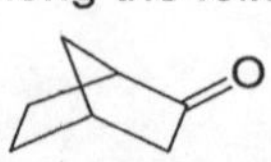　　b. 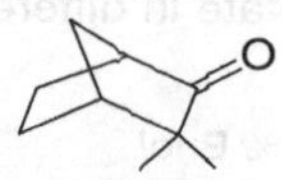

c. 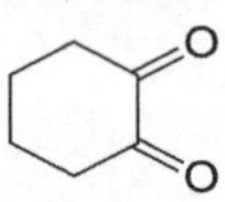　　d. 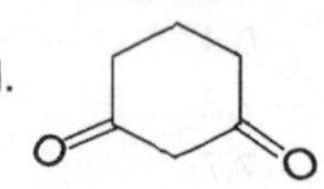

11 Arrange the following in the increasing order of hydrate content at equilibrium in aqueous solution

I. H_2CO　　II. $CH_3-\overset{O}{\overset{\|}{C}}-CH_3$

III. C_6H_5CHO　　IV. O_2N-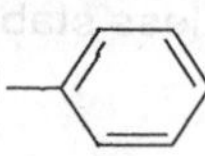$-CHO$

a. I < II < III < IV　　b. IV < III < II < I
c. III < IV < II< I　　d. III < IV < I < II

12 For which of the following equilibrium, $K_C < 1$?

a. $CF_3COCF_3 + H_2O \rightleftharpoons CF_3-C(OH)_2-CF_3$

b. $Cl_3C-CO-CCl_3 + H_2O \rightleftharpoons Cl_3C-C(OH)_2-CCl_3$

c. $CH_3-CO-CH_3 + H_2O \rightleftharpoons CH_3-C(OH)_2-CH_3$

d. $F-\overset{O}{\overset{\|}{C}}-H + H_2O \rightleftharpoons F-C(OH)_2-H$

Working Space

13 Arrange the following in the increasing order of stability of their most stable enol.

I. 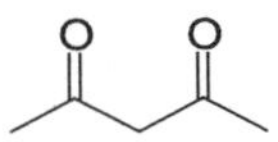

II. 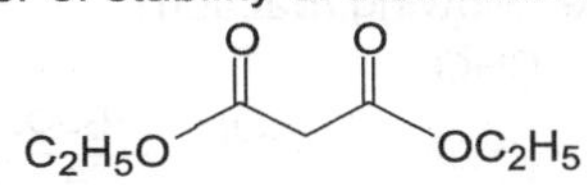

III. 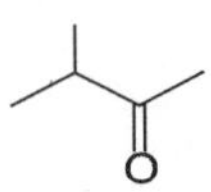

IV. 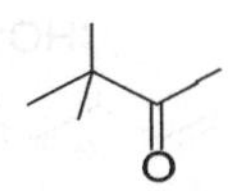

a. I < II < III < IV
b. IV < III < II < I
c. II < I < IV < III
d. III < IV < II < I

14 Which of the following observation does not establish the existence of keto-enol tautomerism in acetone?

a. When treated with DCl in D_2O, hexadeuterated acetone is recovered
b. When treated with $D_2O/NaOD$, hexadeuterated acetone is recovered
c. When treated with dil. H_2SO_4 in H_2O^{18}, $(CH_3)_2C{=}O^{18}$ is obtained
d. All of the above establishes keto-enol tautomerism

15 Hemiacetals are usually unstable whereas acetals are stable. Which has the most stable hemiacetal?

a. 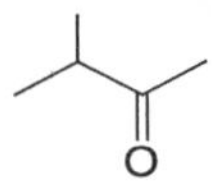

b. $C_6H_5{-}CHO$

c. 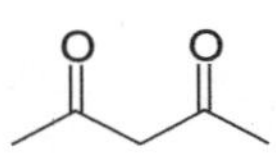

d. 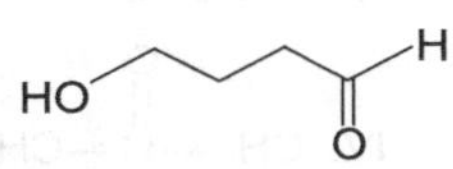

16 Which forms most stable acetal in the given condition?

a. $CH_3{-}\overset{O}{\overset{\|}{C}}{-}CH_3 + C_2H_5OH \text{ (excess)} \xrightarrow{H^+}$

b. 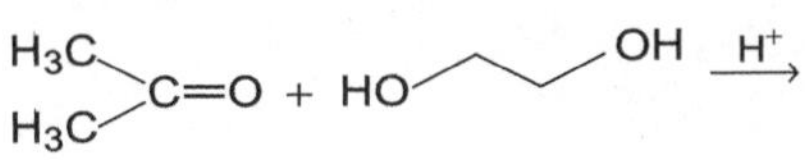

c. 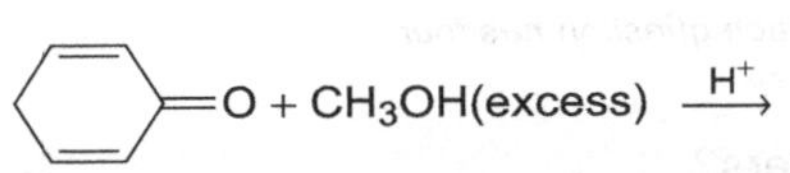

d. $C_6H_5{-}\overset{O}{\overset{\|}{C}}{-}C_6H_5 + CH_3OH\text{(excess)} \xrightarrow{H^+}$

17 Consider the following reaction,

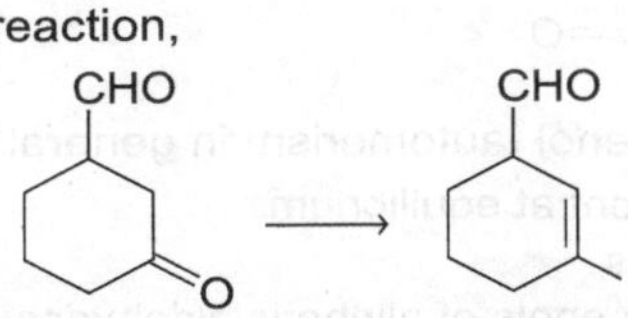

The appropriate sequence of reagent that can best bring about the above conversion is

a. (i) CH_3MgBr (1.0 equiv.) (ii) Conc. H_2SO_4/Heat
b. (i) CH_3Li (1.0 equiv.) (ii) Conc. H_2SO_4/Heat
c. (i) C_2H_5OH (2.0 equiv.); H^+ (ii) CH_3MgBr (iii) Conc. H_2SO_4/Heat
d. (i) $HOCH_2CH_2OH$; H^+ (ii) CH_3MgBr (iii) Conc. H_2SO_4/Heat (iv) H_3O^+/H_2O

Working Space

18 What is the major product in the following reaction?

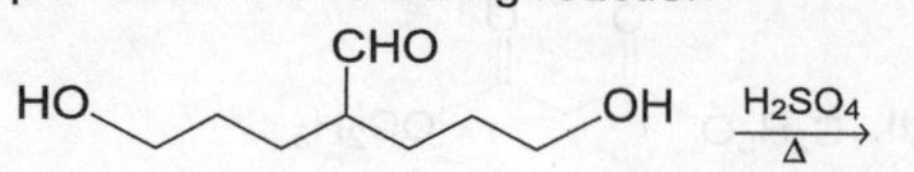

a. 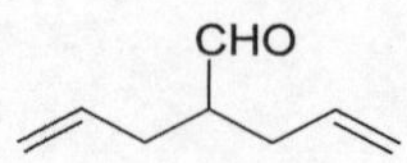

b. 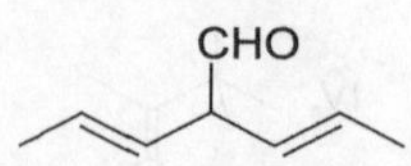

c. 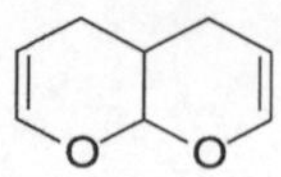

d. 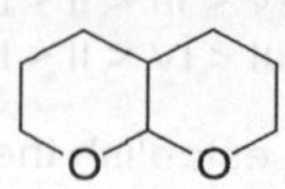

19 Which of the following has greater enol content than keto counter part?

a. $C_6H_5-CH_2CHO$

b. $CH_3-\overset{O}{\overset{\|}{C}}-CH_2-\overset{O}{\overset{\|}{C}}-CH_3$

c. 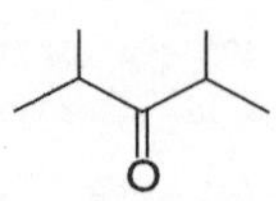

d. 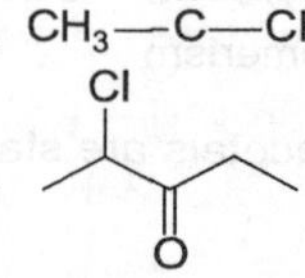

20 Arrange the following in the increasing order of acidic strength

I. H_2CO

II. CH_3CHO

III. $C_6H_5CH_2CHO$

IV. $CH_3-\overset{O}{\overset{\|}{C}}-CH_2-\overset{O}{\overset{\|}{C}}-CH_3$

a. I < II < III < IV

b. IV < III < II < I

c. III < IV < II < I

d. III < IV < I < II

One or More than One Options Correct Type

Direction (Q. Nos. 21-25) *This section contains 5 multiple choice questions. Each question has four choices (a), (b), (c) and (d), out of which ONE or MORE THAN ONE are correct.*

21 Which of the following has (have) more than one enol tautomers?

a. $CH_3-\overset{O}{\overset{\|}{C}}-CH_3$

b. $CH_3-\overset{CH_3}{\overset{|}{CH}}-\overset{O}{\overset{\|}{C}}-CH_3$

c. CH_3CH_2CHO

d. 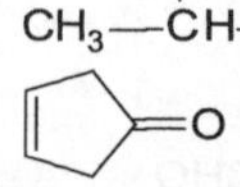

22 Which can be deduced correctly regarding keto-enol tautomerism in general?

a. Increasing temperature increases the enol content at equilibrium

b. Mono-enols are usually more stable than dienols

c. Enols of ketones are generally more stable than enols of aliphatic aldehydes

d. Keto-enol tautomerism is catalysed by both acidic and basic catalyst

23 The correct statement(s) regarding hydrates of aldehyde and ketone is/are

a. Usually hydrates have lower thermodynamic stability than anhydrous form

b. Hydrate content of acetone is greater in water than in hexane

c. CH_3CHO when treated with H_2O^{18} in acidic medium, gets converted into CH_3CHO^{18}

d. C_6H_5CHO has greater hydrate content than *p*-nitrobenzaldehyde

Working Space

24 For which of the following equilibrium, K_C is greater than 1?

a. 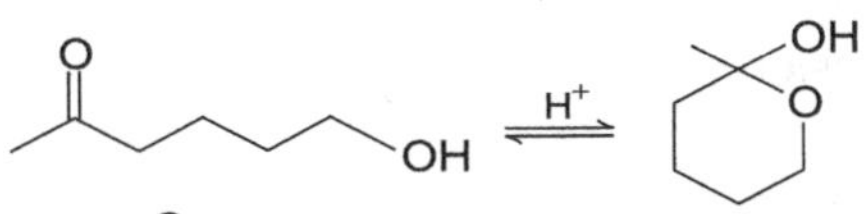

b. 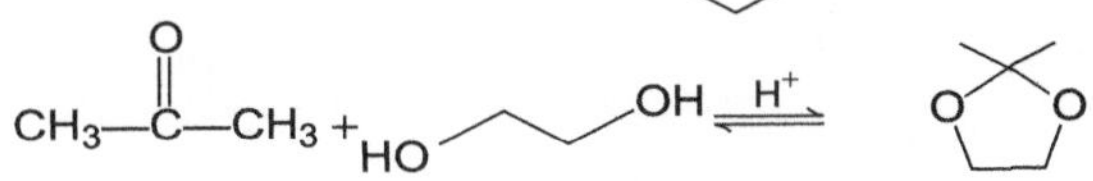

c. 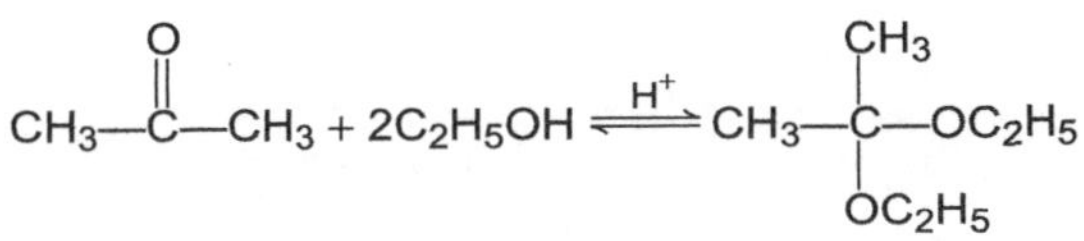

d. $Cl_3C—CHO + 2CH_3OH \xrightleftharpoons{H^+} Cl_3C—CH(OCH_3)_2$

25 Consider the reaction below,

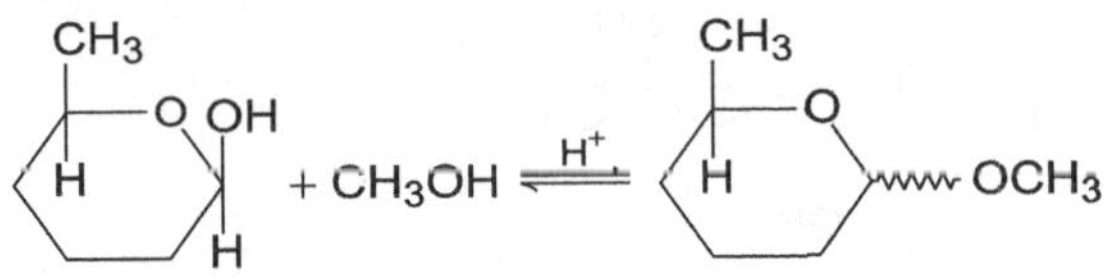

The correct statement(s) regarding the above process is/are

a. Equilibrium favours product
b. Pair of enantiomers of acetals are formed
c. Pair of diastereomers of acetals are formed
d. Equilibrium favours reactant

Comprehension Type

***Direction** (Q. Nos. 26-28) This section contains a paragraph, describing theory, experiments, data, etc. Three questions related to the paragraph have been given. Each question has only one correct answer among the four given options (a), (b), (c) and (d).*

Passage

In general, ketones are less reactive than aldehydes in nucleophilic addition reaction, for both steric and electronic reasons.
Hence, if a keto-aldehyde is treated with nucleophilic reagent, reaction occur first at aldehyde group.

26 The major final product in the following reaction is

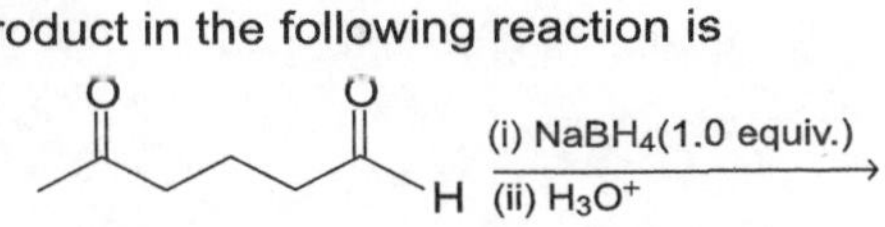

a. 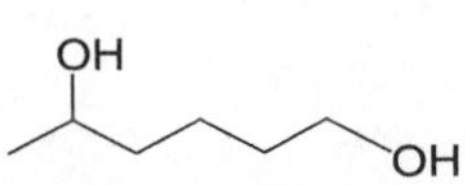

b. 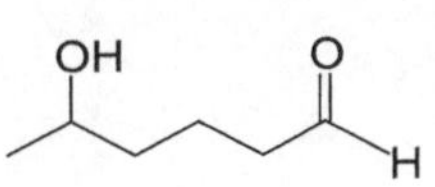

c. 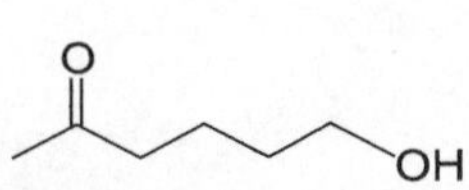

d. Mixture of (b) and (c)

Working Space

27 How the following transformation can be best brought about?

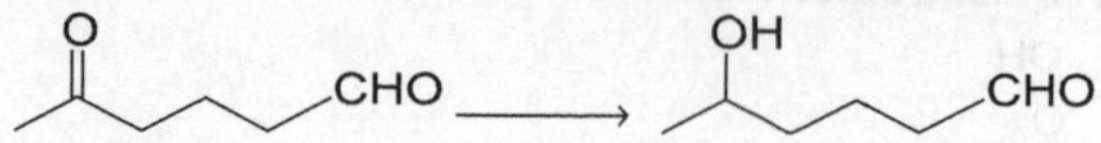

a. $\{(CH_3)_2CHO\}_3Al; H_3O^+$

b. $\underset{OH}{CH_2}$—$\underset{OH}{CH_2}$; H^+ then $NaBH_4$ followed by H_3O^+

c. H_2/Ni; high p and T

d. N_2H_4 /NaOH/Heat

28 Consider the following reaction,

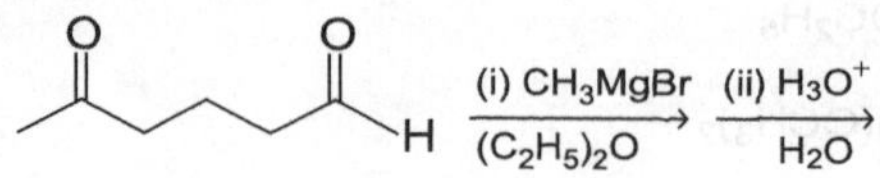

Major product is

a. 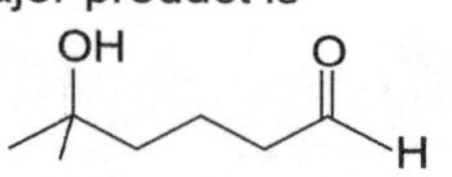

b. 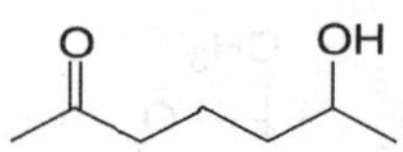

c. 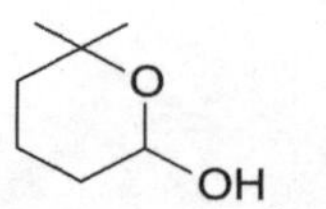

d. 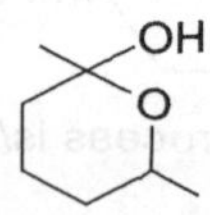

One Integer Value Correct Type

Direction *(Q. Nos. 29-32) This section contains 4 questions. When worked out will result in an integer from 0 to 9 (both inclusive).*

29 How many different enols exist for 4-methyl-3- hexanone?

30 If all the mono and dienols are considered, how many enols are possible for 2,4-pentanedione?

31 If butanone is treated with D_2O^{18} /DCl, isotopic exchange occur. What maximum gain in molecular mass is possible in the present case?

32 When acetaldehyde is treated with catalytic amount of H_2SO_4, a stable association product X is formed. How many lone pair of electrons are present in X?

DPP-5 Aldol Condensation

Subjective Questions

Working Space

Direction (Q. Nos. 1-5) *These questions are subjective in nature, need to be solved completely on notebook.*

1 Predict the product of the following aldol condensation.

i. 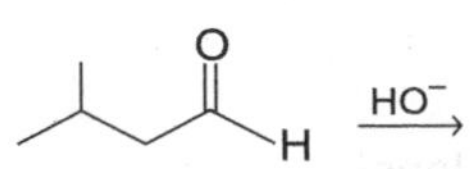

ii. 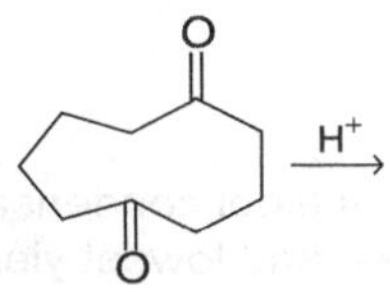

iii. $2Ph—CHO + CH_3—\overset{O}{\overset{\|}{C}}—CH_3 \xrightarrow{OH^-}$

iv. $Ph—\overset{O}{\overset{\|}{C}}—CH_3 + (CH_3)_3C—CHO \xrightarrow{OH^-}$

v. 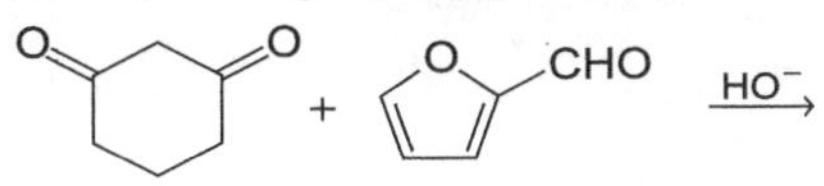

vi. 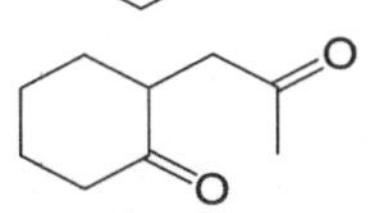

2 When acetaldehyde at fairly high concentration was allowed to undergo base catalysed aldol condensation in heavy water (D_2O), the product was found to contain almost no deuterium bound to carbon. This finding has been taken as one piece of evidence that slow step in this aldol condensation is the formation of carbanion. How would you justify this conclusion?

3 Complete the following reaction(s).

i 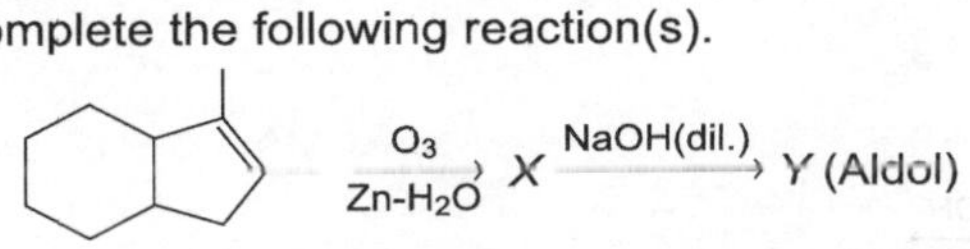

ii. 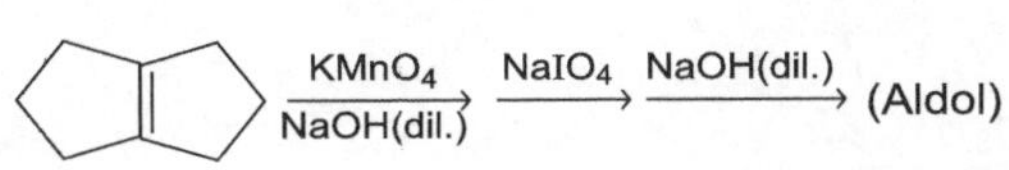

iii. 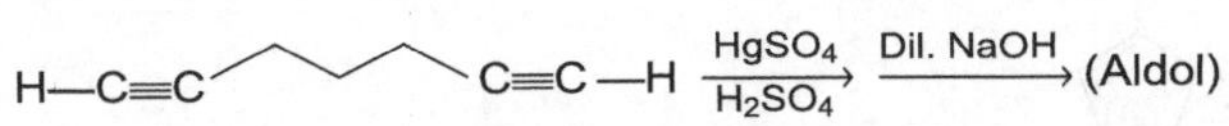

iv. 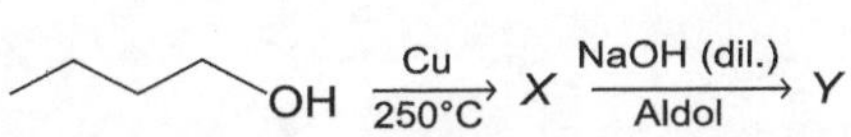

v. 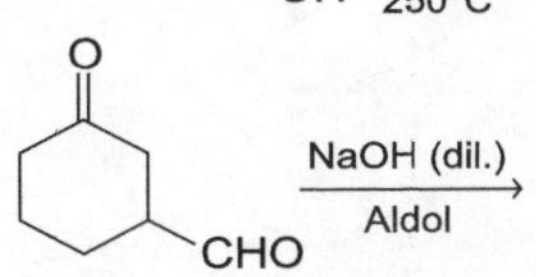

Working Space

4 Predict the product in the following aldol type reactions.

i. 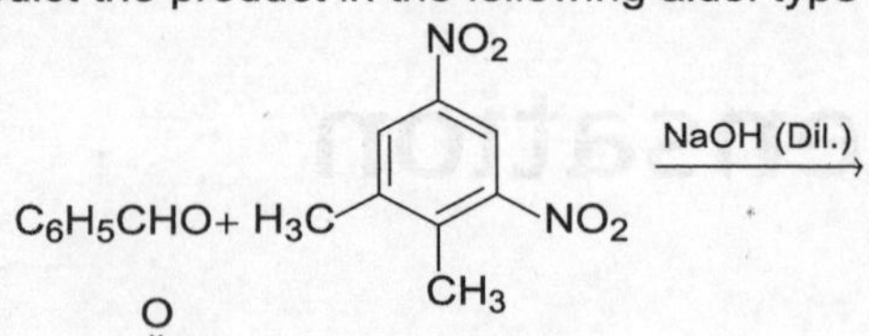

ii. 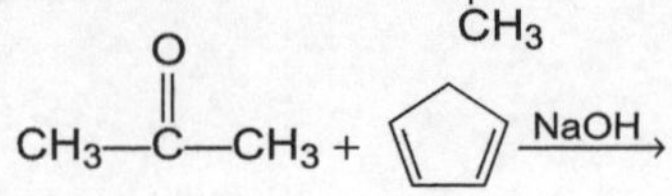

iii. $CH_3CHO + CHCl_3 \xrightarrow{NaOH}$

iv. $CH_3CHO + CH_3-C\equiv CH \xrightarrow{NaNH_2}$

v. $CH_3CHO + CH_3NO_2 \xrightarrow[Heat]{NaOH}$

5 Predict the products in the following cross aldol condensation reactions and label the product which is formed in the highest and lowest yield.

i. $CH_3CHO + CH_2O \xrightarrow{Dil.\ NaOH}$

ii. $CH_3-\overset{O}{\overset{\|}{C}}-CH_2-CH_3 \xrightarrow{Dil.\ NaOH}$

iii. $CH_3-\underset{}{\overset{CH_3}{\overset{|}{C}H}}-CHO + CH_3-CHO \xrightarrow{Dil.\ NaOH}$

iv. 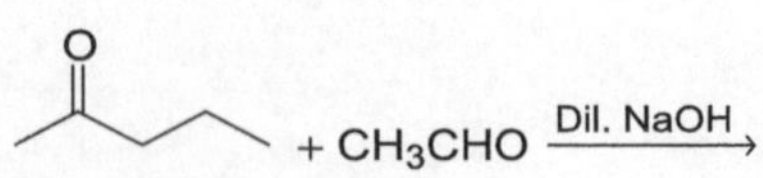

v. 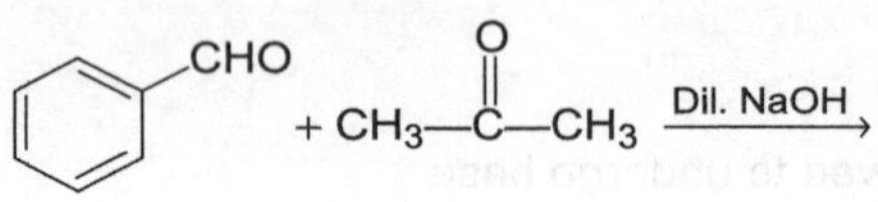

Only One Option Correct Type

Direction (Q. Nos. 6-15) *This section contains 10 multiple choice questions. Each question has four choices (a), (b), (c) and (d), out of which ONLY ONE is correct.*

6 Which is the major product in the following reaction?

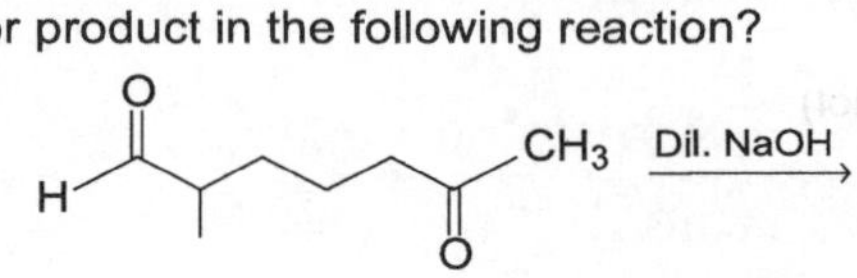

a. 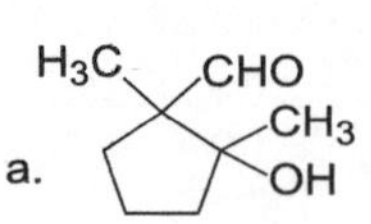

b. 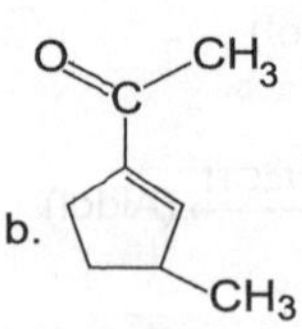

c. 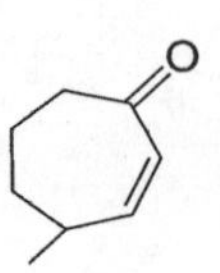

d. 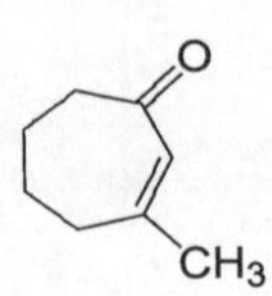

Working Space

7 Consider the following reaction,

$$\text{A dicarbonyl} \xrightarrow[\text{Aldol}]{\text{NaOH (Dil.)}} \xrightarrow{(Ph_3P)_3RhCl}$$ 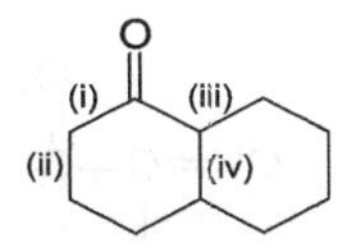

Which of the labelled C—C bond formation is not possible in the above reaction?

a. Only i b. Only ii
c. Only iii d. Only iv

8 In a mixed aldol condensation with ethanal as one aldehyde, which other from the following is expected to give maximum yield of cross condensation product?

a. $CH_3—\overset{\overset{O}{\|}}{C}—CH_3$

b. 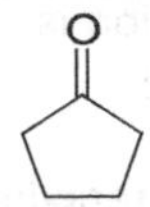

c. 

d. 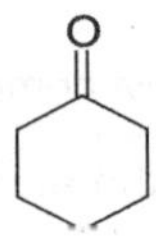

9 In the following reaction,

$$CH_2O + CH_3—NO_2 \xrightarrow{\text{Dil. NaOH}} \xrightarrow[\text{Heat}]{\text{Conc.} H_2SO_4}$$

The major organic product is

a. $H—\overset{\overset{O}{\|}}{C}—CH_2—NO_2$

b. $HO—CH_2—O—\underset{\underset{O^-}{|}}{\overset{+}{N}}=CH_2$

c. $CH_2=CH—NO_2$

d. 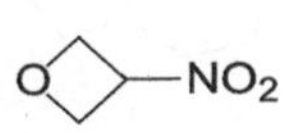

10 Consider the following aldol condensation reaction,

$$CH_3—CHO \xrightarrow{H_2SO_4} CH_3—CH=CH—CHO$$

The nucleophile is

a. $\bar{C}H_2—CHO$

b. $CH_2=CH—\bar{O}$

c. $CH_2=CH—OH$

d. $CH_3—\overset{\overset{\overset{+}{O}—H}{\|}}{C}—H$

11 In the following reaction,

$$\underset{\text{Excess}}{CH_2O} + CH_3CHO \xrightarrow{\text{NaOH (Dil.)}}$$

The major final product is

a. $HO—CH_2—CH_2—CHO$

b. $HO—CH_2—\overset{\overset{CH_2OH}{|}}{CH}—CHO$

c. $HO—CH_2—\underset{\underset{CH_2OH}{|}}{\overset{\overset{CH_2OH}{|}}{C}}—CHO$

d. $C(CH_2OH)_4$

Working Space

12 Identify the starting reagent needed to make the following compound by mixed aldol condensation.

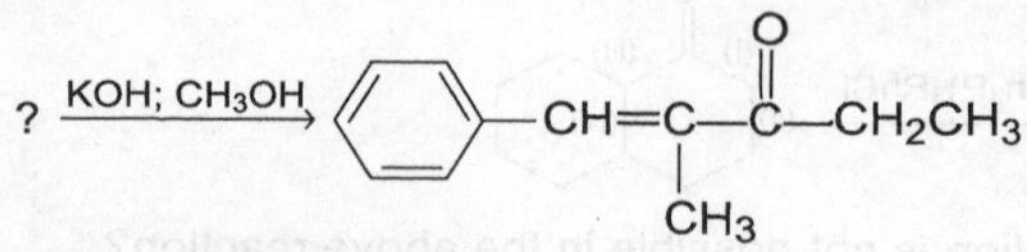

a. Benzaldehyde and 3-pentanone
b. Benzaldehyde and 2-pentanone
c. Acetophenone and 2-butanone
d. Acetophenone and butanal

13 The incorrect statement regarding aldol condensation is
a. It accomplishes the formation of a new C—C bond
b. The enolate is favoured at equilibrium
c. The key step in the mechanism of base catalysed reaction is attack of α-carbon of an enolate ion on the carbonyl carbon
d. Dehydration of aldol gives α, β-unsaturated carbonyl compound

14 Which of the following could result as a product in the aldol condensation reaction?
a. 4-methyl-3-penten-2-one
b. 4-methyl-4-penten-2-one
c. 4-methyl-5-hexen-2-one
d. 3-methyl-4-hexen-2-one

15 What is the major final product of the following sequence of reaction?

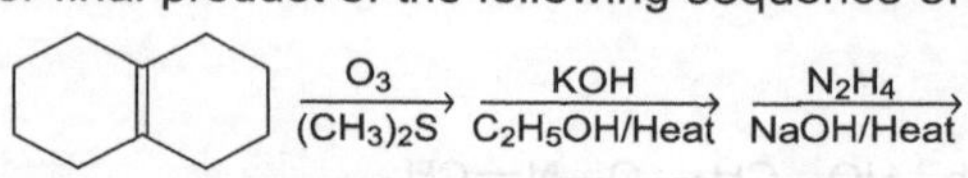

a. 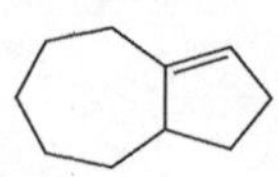
b. 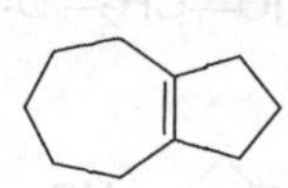
c. 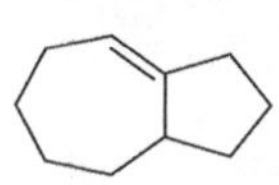
d. 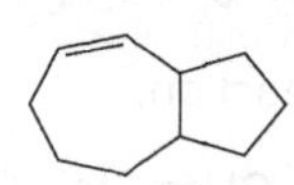

One or More than One Options Correct Type

***Direction** (Q. Nos. 16-20) This section contains 5 multiple choice questions. Each question has four choices (a), (b), (c) and (d), out of which ONE or MORE THAN ONE are correct.*

16 Successful mixed aldol condensation will be favoured by
a. treatment of electrophilic carbonyl component with stoichiometric strong base prior to the addition of the pre-nucleophilic carbonyl component
b. treatment of mixture of electrophilic carbonyl component and pre-nucleophilic component with stoichiometric strong base
c. continuous removal of aldol product from reaction mixture
d. treatment of pre-nucleophilic carbonyl component with stoichiometric strong base prior to the addition of electrophilic carbonyl component

17 Which of the following is/are true regarding aldol condensation?
a. Both acid and base can act as catalyst
b. A new carbon-carbon bond is always formed
c. CH_3CHO and D_3C—CHO react at same rate if all other conditions are similar
d. Propanal and propanone react at same rate if all other conditions are similar

Working Space

18 Which of the following indicated carbon-carbon bond can be formed *via* intramolecular aldol condensation of a dicarbonyl compound?

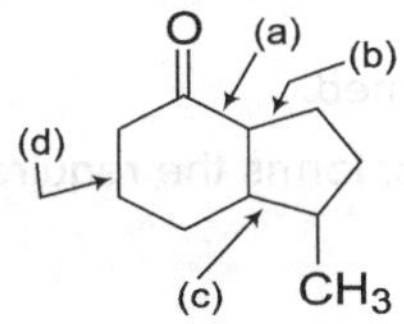

19 Which compound(s) below can react *via* an intramolecular aldol condensation to give a six membered ring?

a. 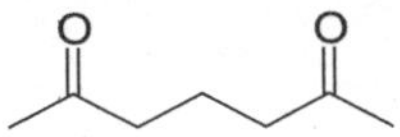

b. 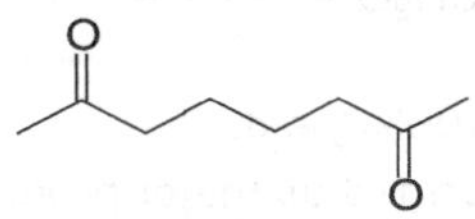

c. 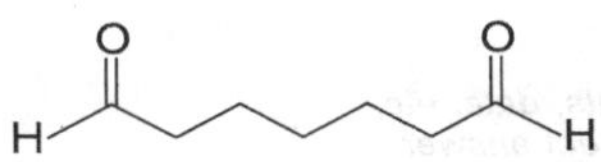

d. 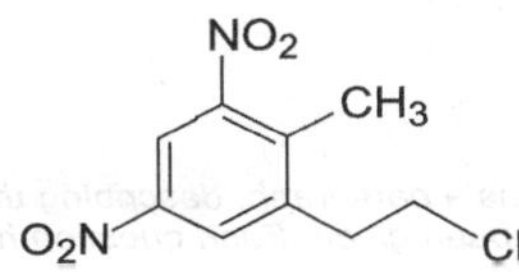

20 What is(are) true about the following aldol reaction?

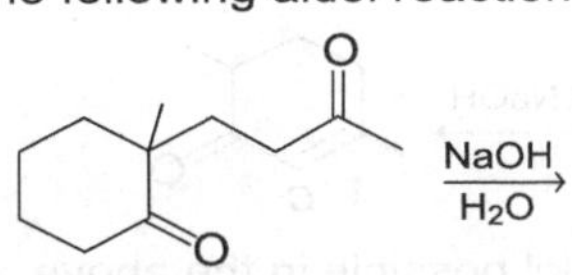

a. The major aldol is 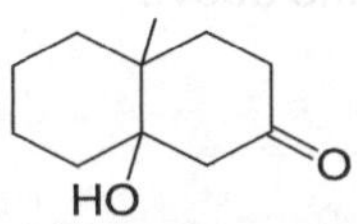

b. If a pure enantiomer of starting compound is taken, the racemic mixture of major aldol would be formed

c. Excluding stereoisomers, a total of three different aldol would be formed

d. Major aldol on dehydration would give 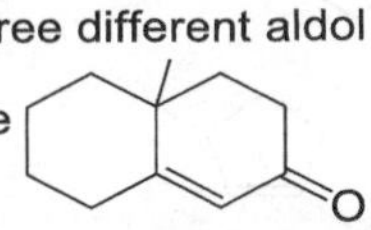

Statement Type

Direction *(Q. Nos. 21-24) This section is based on Statement I and Statement II. Select the correct answer from the codes given below.*

a. Both Statement I and Statement II are correct and Statement II is the correct explanation of Statement I

b. Both Statement I and Statement II are correct but Statement II is not the correct explanation of Statement I

c. Statement I is correct but Statement II is incorrect

d. Statement II is correct but Statement I is incorrect

21 **Statement I** In aldol condensation, enolates are nucleophilic and reversibly attack the carbonyl carbon of aldehyde or ketone in the aldol condensation.

Statement II An aldol on refluxing with dilute base gives back the carbonyl compound.

22 **Statement I** When a mixture of ethanal and propanal is treated with aqueous Na_2CO_3, four aldol (excluding stereoisomers) are formed.

Statement II In mixed aldol condensation, two self and two cross condensation products are always formed.

Working Space

23 **Statement I** When $CH_3-CH=CH-CHO$ is treated with dilute base.

$$CH_3-CH=CH-\underset{\displaystyle |}{\overset{\displaystyle OH}{CH}}-CH_2-CH=CH-CHO \text{ is formed.}$$

Statement II In the given compound, γ-H is most acidic, forms the required enolate.

24 **Statement I** In the reaction below,

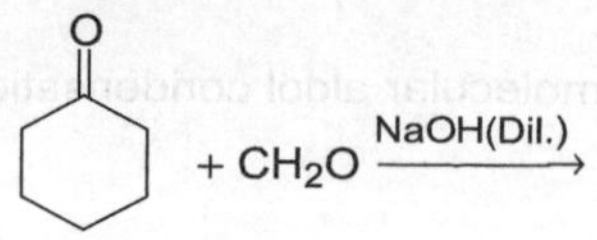

A single aldol product is formed in 100% yield.

Statement II Cross aldol product is formed as major product.

Comprehension Type

Direction *(Q. Nos. 25-27) This section contains a paragraph, describing theory, experiments, data, etc. Three questions related to the paragraph have been given. Each question has only one correct answer among the four given options (a), (b), (c) and (d).*

Passage

$$A(C_8H_{14}) \xrightarrow[(CH_3)_2S]{O_3} B(C_8H_{14}O_2) \xrightarrow{\text{Dil. NaOH}}$$ 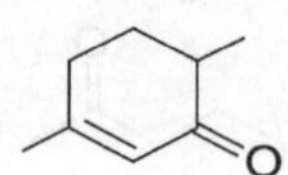

C

A is optically active and *C* is one of the several aldol possible in the above reaction.

25 The structure of *A* satisfying above criteria is

a. 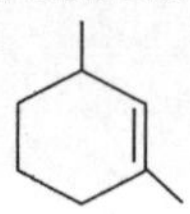

b. 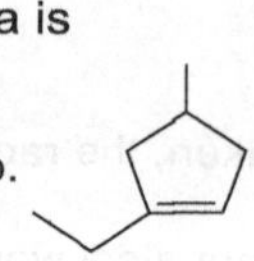

c. 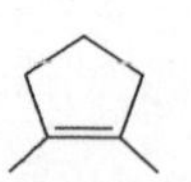

d. 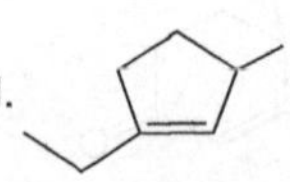

26 Besides *C*, the other six membered cyclic aldol formed in the above reaction is

a. 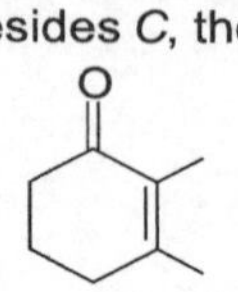

b. 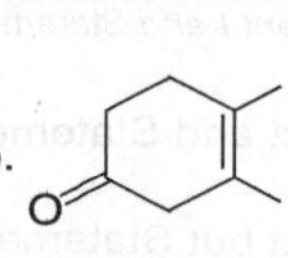

c. 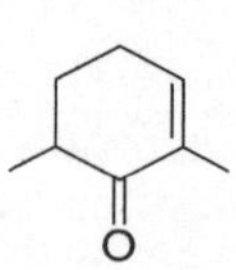

d. 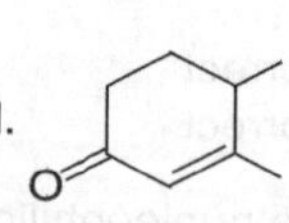

27 The product *B* is stereomeric. If a mixture containing all stereoisomers of *B* is treated with excess of $LiAlH_4$ followed by the acidification will give how many different isomeric diols?

a. 2 b. 4
c. 6 d. 8

One Integer Value Correct Type

Direction (Q. Nos. 28-32) *This section contains 5 questions. When worked out will result in an integer from 0 to 9 (both inclusive).*

28 Consider the following aldol condensation reaction,

$$\text{Butanone} + \text{NaOH (Dil.)} \longrightarrow \text{Aldols}$$

How many different isomers (including stereoisomers) are formed above?

29 Consider the following cross-aldol condensation reaction,

$$2C_6H_5\text{—CHO} + CH_3\text{—}\overset{\overset{\displaystyle O}{\|}}{C}\text{—}CH_3 \xrightarrow{\text{NaOH (Dil.)}} C_{17}H_{14}O \xrightarrow[H_3O^+]{[(CH_3)_2CHO]_3Al} C_{17}H_{16}O(X)$$

How many different isomeric X are formed?

30 Consider the following modified aldol condensation,

$$CH_3CHO\ (\text{excess}) + CH_3NO_2 \xrightarrow{\text{NaOH}}$$

How many ethanal, at the most, will react with one molecule of nitromethane?

31 Consider the following sequence of reaction,

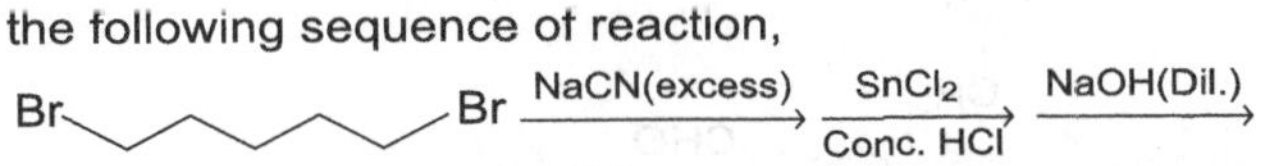

$$\underset{(X)}{\text{Aldol}} \xrightarrow{-H_2O} \alpha,\beta\text{-unsaturated carbonyl}$$

How many different aldol isomers of X are formed?

32 Consider the following sequence of reaction,

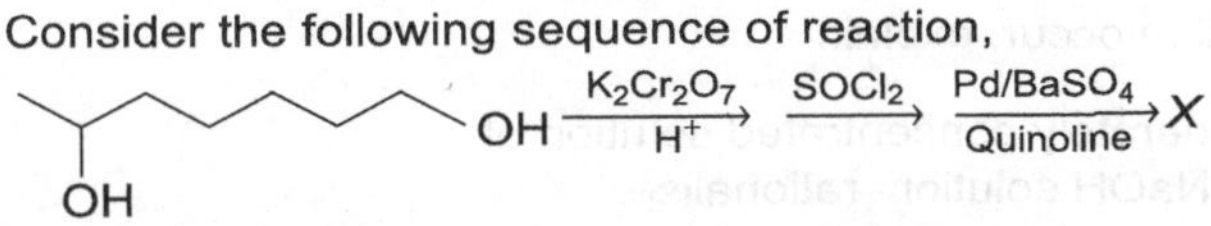

$$X \xrightarrow{\text{NaOH(Dil.)}} \underset{(X)}{\text{Aldol}} \xrightarrow{-H_2O} \alpha,\beta\text{-unsaturated carbonyl}$$

If all undehydrated aldols (X) formed above by intramolecular aldol condensation are considered, how many pairs of enantiomers are formed?

Working Space

DPP-6 Cannizaro Reaction

Subjective Questions

Working Space

Direction *(Q. Nos. 1-5) These questions are subjective in nature, need to be solved completely on notebook.*

1 When excess of methanal is reacted with CH_3CHO in NaOH solution, both $(HOCH_2)_3C—CHO$ and $C(CH_2OH)_4$ are formed while $C(CH_2OH)_4$ is not formed if reaction is carried out in aqueous K_2CO_3 solution, why?

2 Arrange the following in the increasing order of reactivity in Cannizaro reaction.

i. 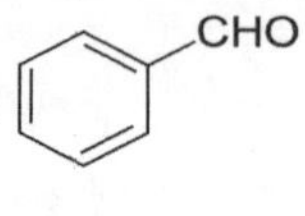

ii. 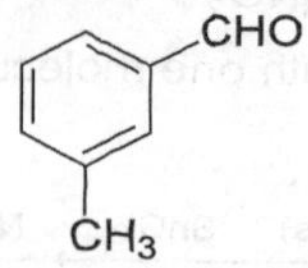

iii. 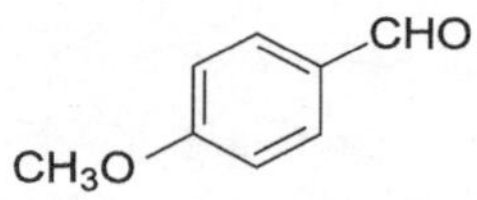

iv. 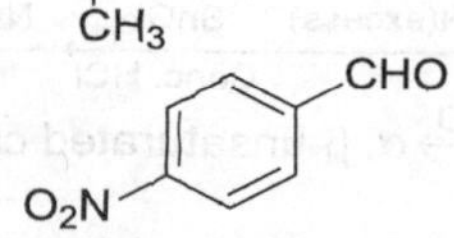

3 When methanal (CH_2O) is treated with NaOH (Conc.) in D_2O solvent, Cannizaro reaction takes place but no C—D bond formation occur, explain.

4 Cannizaro reaction is third order overall in moderately concentrated solution of NaOH but fourth order in highly concentrated NaOH solution–rationalise.

5 How a specific aldol can be prepared in 100% yield in mixed aldol condensation reaction? Synthesise $CH_3—\underset{CH_3}{\overset{OH}{\overset{|}{\underset{|}{C}}}}—CH_2—CHO$ as only aldol product using ethanal, acetone and any other reagents.

Only One Option Correct Type

Direction *(Q. Nos. 6-15) This section contains 10 multiple choice questions. Each question has four choices (a), (b), (c) and (d), out of which ONLY ONE is correct.*

6 The major organic product in the following reaction is

$$C_6H_5—\overset{O}{\overset{\|}{C}}—CHO \xrightarrow{\text{NaOH (Conc.)}} \xrightarrow{H_3O^+}$$

a. $C_6H_5—\overset{O}{\overset{\|}{C}}—COOH$

b. $C_6H_5—\overset{OH}{\overset{|}{CH}}—CHO$

c. $C_6H_5—\overset{OH}{\overset{|}{CH}}—COOH$

d. $C_6H_5—\overset{OH}{\overset{|}{CH}}—CH_2OH$

Working Space

7 Consider the following reaction,

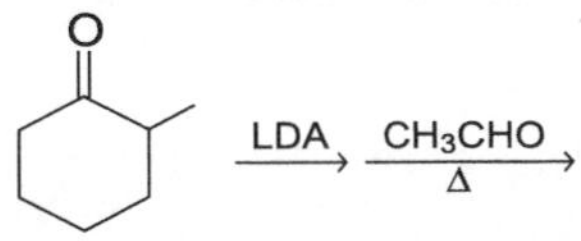

The major product is

a. 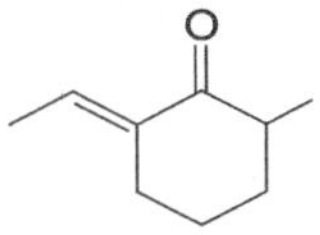

b. 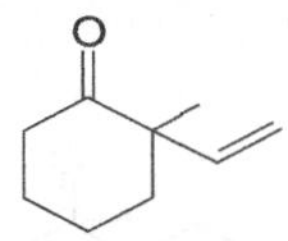

c. 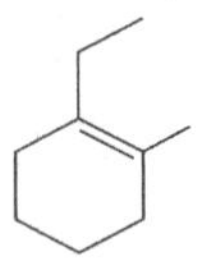

d. 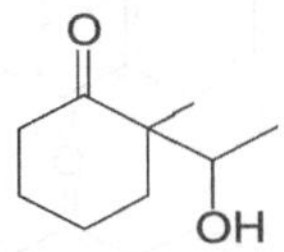

8 Which is the best hydride (H^-) donor in the key step of Cannizaro reaction?

a. 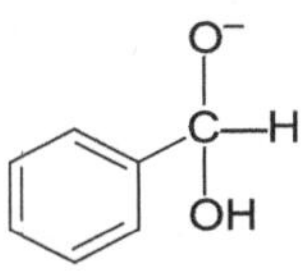

b. 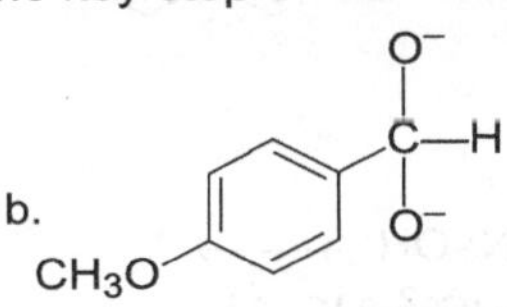

c. 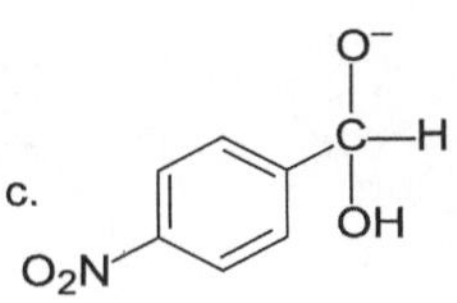

d. 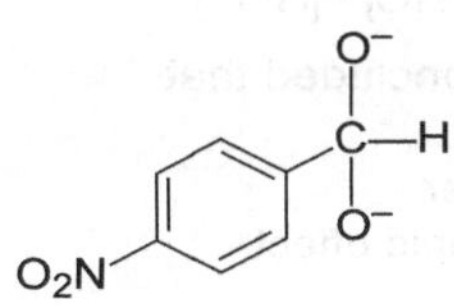

9 Consider the following reaction,

$$CH_3CHO + (C_2H_5O)_3Al \xrightarrow{H_3O^+}$$

The major organic product is

a. $CH_3CH{=}CH{-}CHO$

b. $CH_3CH_2OH + CH_3COOH$

c. CH_3CH_2OH

d. $CH_3COOC_2H_5$

10 What is the major organic product in the following reaction?

$$CH_2O\text{ (excess)} + CH_3CH_2CHO \xrightarrow{\text{Conc. NaOH}} \xrightarrow{H_3O^+}$$

a. $CH_3{-}C(CH_2OH)_3$

b. CH_3OH

c. HCOOH

d. Both (a) and (c)

11 Predict the major organic product in the following reaction.

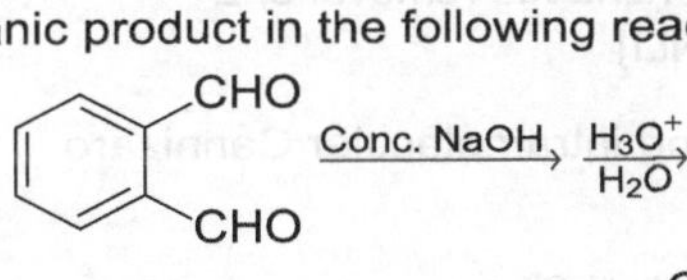

a. 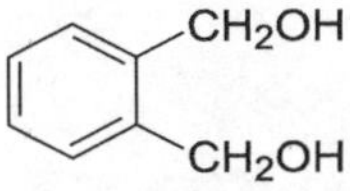

b. 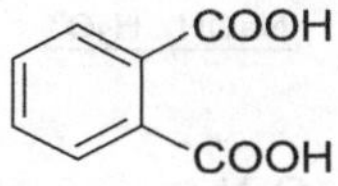

c. 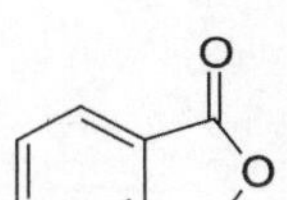

d. 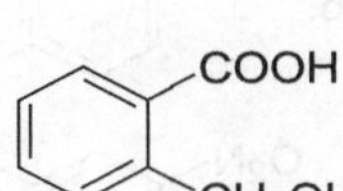

Working Space

12 What is the major organic product formed in the following reaction?

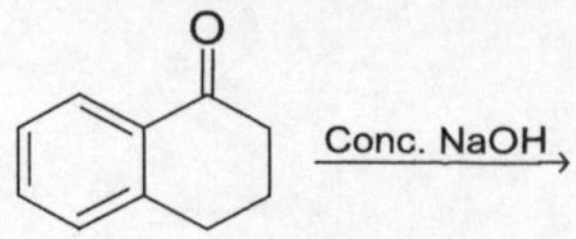

a. 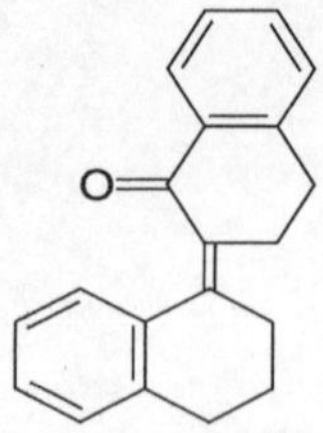

b. 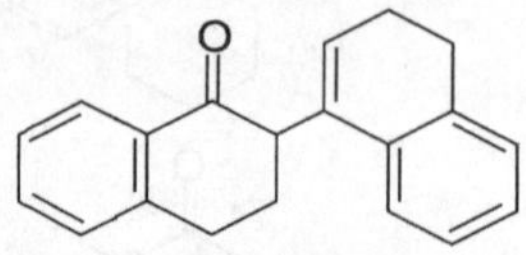

c. 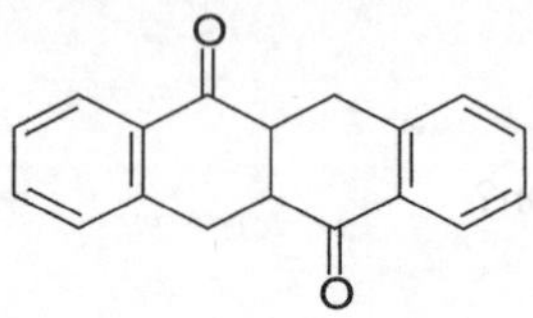

d. 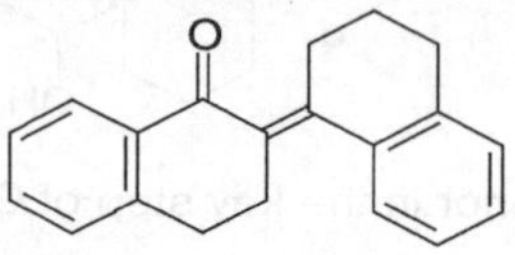

13 For a Cannizaro reaction,

$$R\text{—CHO} + \text{NaOH} \longrightarrow$$

Rate law is derived as : Rate = $k\,[RCHO]^2\,[HO^-]^2$

From the above rate law, it can be concluded that

a. reaction is pseudo 2nd order
b. reaction involves hydride ion transfer
c. reaction does not show kinetic isotopic effect
d. hydride donor is a dianion

14 Consider the following reaction,

$$\underset{X}{CH_3\text{—}CHO} + \underset{Y}{CH_3\text{—}\overset{O}{\overset{\|}{C}}\text{—}CH_3} \longrightarrow \underset{Z}{CH_3\text{—}\overset{OH}{\overset{|}{\underset{CH_3}{\underset{|}{C}}}}\text{—}CH_2\text{—}CHO}$$

How the product Z can be prepared selectively using X and Y and other reagents?

a. Treating Y first with LDA followed by subsequent addition of X
b. Treating X first with LDA followed by subsequent addition of Y
c. Treating mixture of X and Y with LDA
d. Treating mixture of X and Y with NaOH and continuous removal of Z
[LDA = Lithium diisopropyl amide $[(CH_3)_2CH_2NLi]$]

15 What is the major product formed in the following intramolecular Cannizaro reaction?

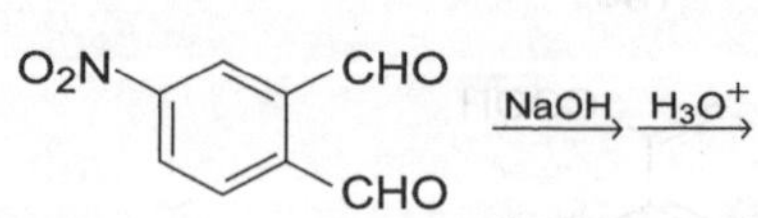

a. 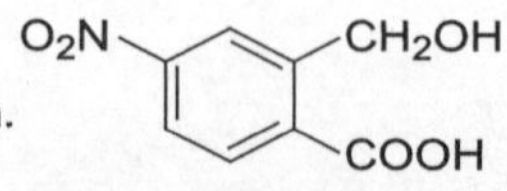

b. 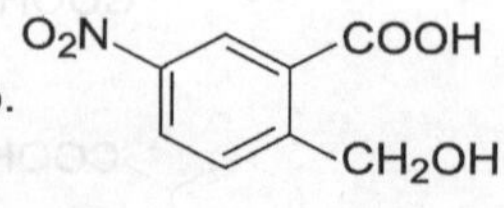

c. 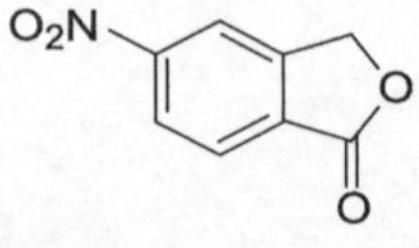

d. 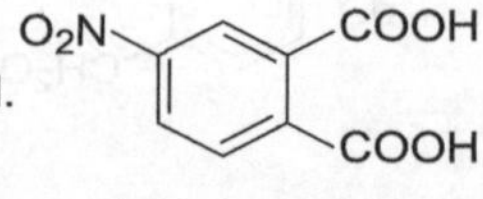

Working Space

One or More than One Options Correct Type

Direction *(Q. Nos. 16-20) This section contains 5 multiple choice questions. Each question has four choices (a), (b), (c) and (d), out of which ONE or MORE THAN ONE are correct.*

16 Consider the cross Cannizaro reaction given below

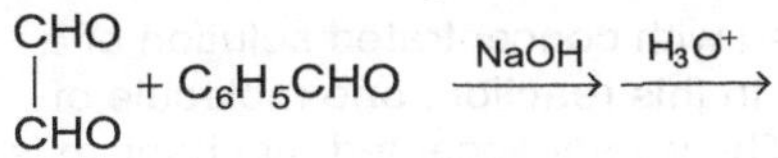

The expected product(s) is/are

a. 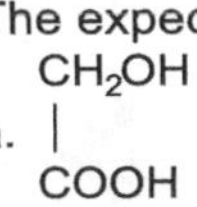
b. 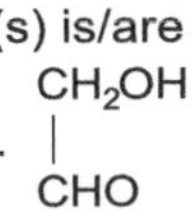
c. 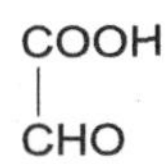
d. $C_6H_5CH_2OH$

17 In the Cannizaro reaction mentioned below

The possible product(s) is/are

a. CH_3OH
b. CH_3OD
c. $HCOOD$
d. H_2CDOH

18 What is/are the expected organic product(s) in the following reaction?

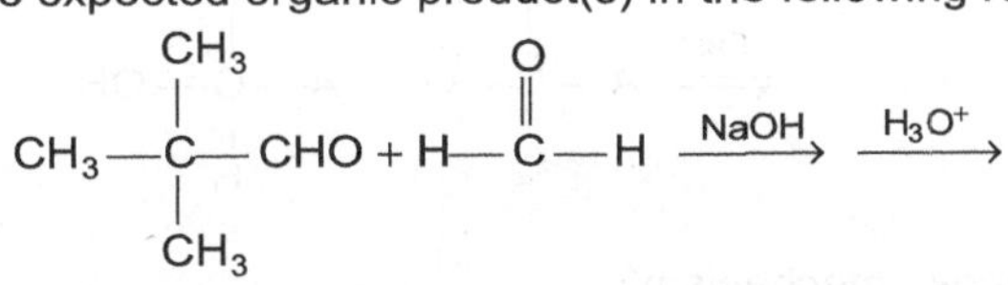

a. 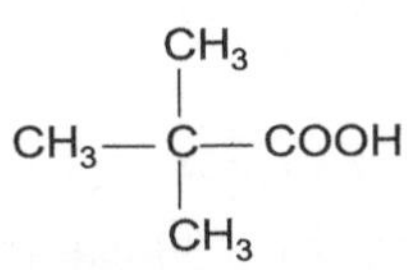
b. 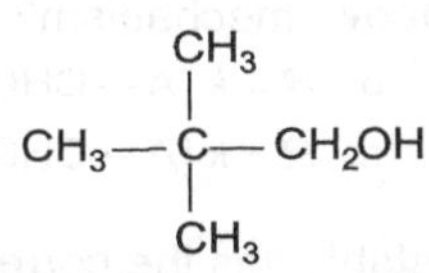
c. CH_3OH
d. $HCOOH$

19 Consider the following reaction,

$$CH_3CHO + CH_3CH_2CHO \xrightarrow[H_2O]{(CH_3CH_2O)_3Al}$$

The expected organic product(s) is/are

a. $CH_3COOCH_2CH_2CH_3$
b. $CH_3CH_2COOCH_2CH_3$
c. $CH_3COOCH_2CH_3$
d. $CH_3CH_2COOCH_2CH_2CH_3$

20 Consider the following reaction sequence,

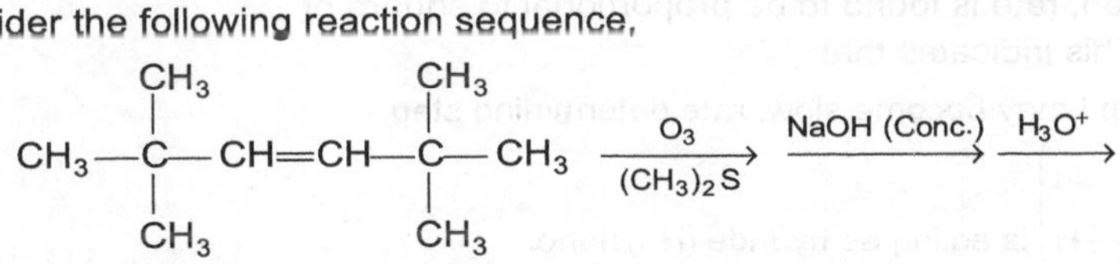

The expected organic product(s) is/are

a. 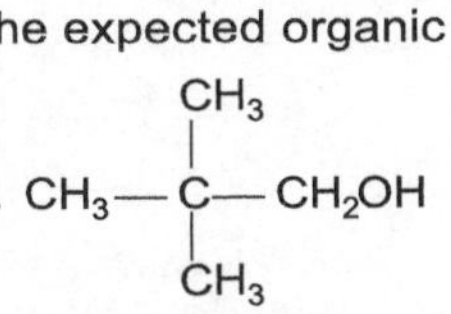
b. 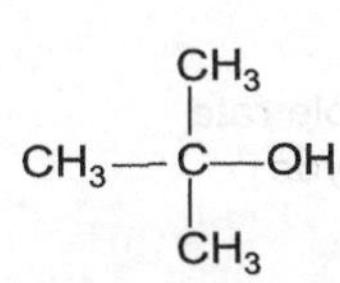
c. 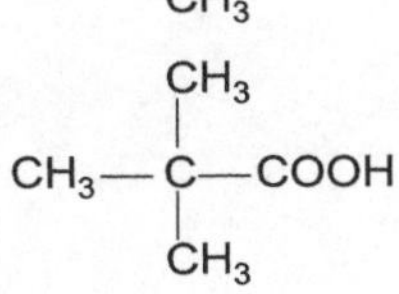
d. 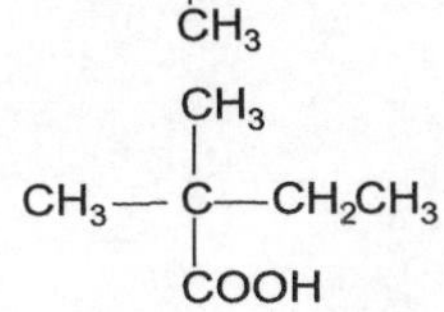

Comprehension Type

Working Space

Direction *(Q. Nos. 21-23) This section contains a paragraph, describing theory, experiments, data, etc. Three questions related to the paragraph have been given. Each question has only one correct answer among the four given options (a), (b), (c) and (d).*

Passage

When aldehydes lacking α-hydrogen is treated with concentrated solution of a strong base, Cannizaro reaction takes place. In this reaction, one molecule of aldehyde is oxidised while other is reduced. The widely accepted mechanism is

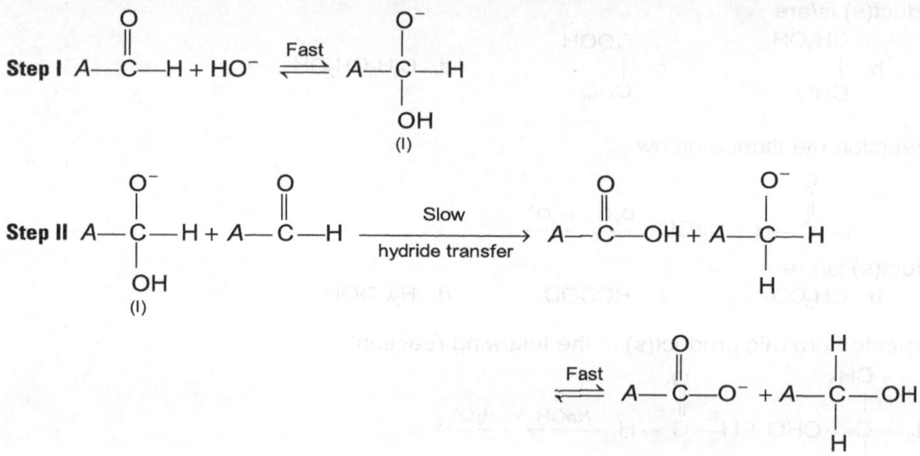

21 What is the rate law derived from the above mechanism?

a. $R = k\,[A\text{—CHO}]\,[HO^-]^2$
b. $R = k\,[A\text{—CHO}]^2\,[HO^-]^2$
c. $R = k\,[A\text{—CHO}]^2\,[HO^-]$
d. $R = k\,[A\text{—CHO}]^2$

22 Which of the following observed fact establishes the correctness of the above mechanism?

a. If A—COD is used, reaction occur at same rate provided all other conditions are same
b. In mixed Cannizaro reaction, if one aldehyde is H_2CO, it is always oxidised
c. If reaction is carried out in D_2O in the presence of NaOD, some C—D bond formation occur
d. If reaction is carried out in D_2O in the presence of NaOD, no C—D bond formation occur

23 In certain experimental condition, rate is found to be proportional to square of concentration of base (OH^-). This indicates that

a. with change in condition, Step I may become slow, rate determining step
b. dianion intermediate 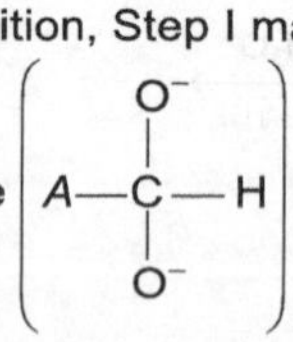 is acting as hydride (H^-) donor
c. Both Step I and Step II occur at comparable rate
d. mechanism changes with nature of aldehyde

Matching List Type

Direction (Q. Nos. 24 and 25) *Choices for the correct combination of elements from Column I and Column II are given as options (a), (b), (c) and (d), out of which one is correct.*

24 Consider the reactions of Column I and match with the products of Column II. Mark the correct option from the codes given below.

	Column I		Column II
i.	$F-\overset{\overset{O}{\|\|}}{C}-H + H-\overset{\overset{O}{\|\|}}{C}-H \xrightarrow[\text{(ii) } H_3O^+]{\text{(i) Conc. NaOH}}$ Cross products	p.	HCOOH
ii.	$C_6H_5CHO + CH_2O \xrightarrow[\text{(ii) } H_3O^+]{\text{(i) Conc. NaOH}}$ Cross products	q.	C_6H_5COOH
iii.	C_6H_5CHO + 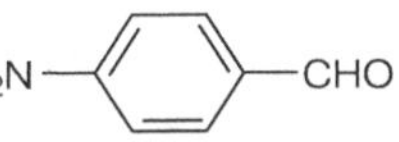 $\xrightarrow[\text{(ii) } H_3O^+]{\text{(i) Conc. NaOH}}$ Cross products	r.	$C_6H_5CH_2OH$
iv.	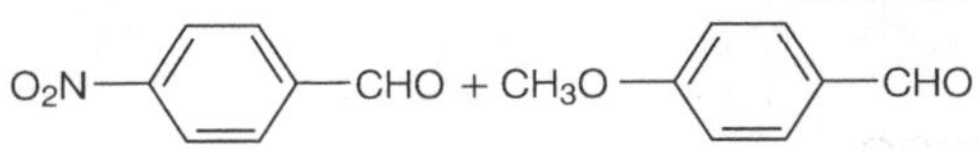 $\xrightarrow[\text{(ii) } H_3O^+]{\text{(i) Conc. NaOH}}$ Cross products	s.	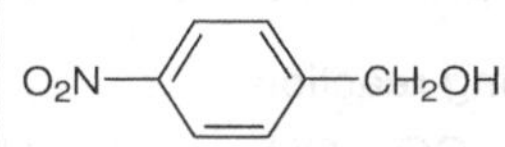
		t.	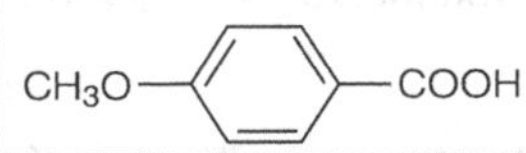

Codes

	i	ii	iii	iv
a.	p	p,r	q,s	s,t
b.	p,r	q	r	s
c.	q	s	p	r
d.	t	p	q	r

25 Match the reactions of Column I with the type of reactions from Column II. Mark the correct option from the codes given below.

	Column I		Column II
i.	$CH_3CHO + CH_2O \text{ (excess)} \xrightarrow[\text{(ii) } H_3O^+]{\text{(i) Conc. NaOH}}$	p.	Aldol condensation
ii.	$F-CHO + \text{Conc. NaOH} \xrightarrow{H_3O^+}$	q.	Cannizaro reaction
iii.	$(CH_3)_2CH-CHO \xrightarrow[\text{(ii) } H_3O^+]{\text{(i) Conc. NaOH}}$	r.	Claisen reaction
iv.	$CH_3CHO + C_6H_5CHO \xrightarrow[\text{(ii) } H_3O^+]{\text{(i) Conc. NaOH}}$	s.	Tischenko reaction

Codes

	i	ii	iii	iv
a.	q	s	p	r
b.	p,q	q	p,q	p,q,r
c.	p	q	r	s
d.	p	q	r	s

Working Space

One Integer Value Correct Type

Working Space

Direction (Q. Nos. 26-28) *This section contains 3 questions. When worked out will result in an integer from 0 to 9 (both inclusive).*

26 Consider the following Cannizaro reaction,

$$H_2CO + D{-}\overset{\overset{\large O}{\|}}{C}{-}D \xrightarrow[\text{(ii) } H_3O^+]{\text{(i) Conc. NaOH}}$$

How many different products are formed?

27 Consider the following Cannizaro reaction,

$$C_6H_5{-}\overset{\overset{\large O}{\|}}{C}{-}H + C_6H_5{-}\overset{\overset{\large O}{\|}}{C}{-}D \xrightarrow[\text{(ii) } H_3O^+]{\text{(i) Conc. NaOH}}$$

How many different products are expected in this reaction?

28 Consider the following reaction,

$$H_2CO + \underset{\text{Conc.}}{OH^-} \longrightarrow CH_3OH + HCOO^-$$

If CH_3OH is formed exclusively by hydride transfer from a dianion intermediate $\left(H{-}\overset{\overset{\large O^-}{|}}{\underset{\underset{\large O^-}{|}}{C}}{-}H\right)$, what is the overall order of reaction?

DPP-7 Beckmann Rearrangement

Subjective Questions

Working Space

Direction *(Q. Nos. 1-3) These questions are subjective in nature, need to be solved completely on notebook.*

1 Complete the following reaction,

$$CH_3CH_2CHO + NH_2OH \longrightarrow P + Q \xrightarrow{\text{Conc. } H_2SO_4} R + S$$

2 Write the major product of the following reactions.

i. $C_6H_5-\overset{\overset{\large O}{||}}{C}-CH_3 + CF_3CO_3H \longrightarrow$

ii. 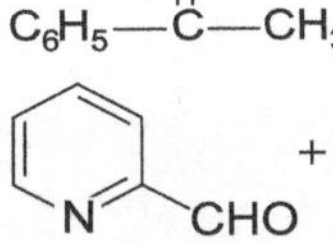 $+ NH_2OH \longrightarrow P + Q \xrightarrow{\text{Conc.}H_2SO_4} R + S$

iii. $CH_2O + CH_2N_2 \xrightarrow{\Delta}$

3 Complete the following reactions.

i. 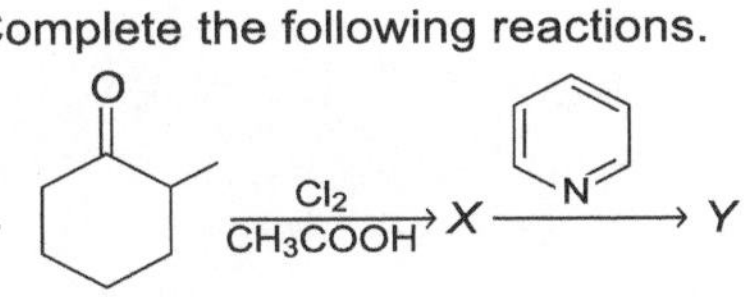

ii. $CH_3CHO \xrightarrow[CH_3COOH]{Cl_2} X \xrightarrow{NaCN} Y \xrightarrow[\text{Conc. HCl}]{SnCl_2} Z$

iii. 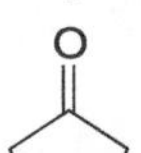 $+ CF_3CO_3H \longrightarrow$

Only One Option Correct Type

Direction *(Q. Nos. 4-10) This section contains 7 multiple choice questions. Each question has four choices (a), (b), (c) and (d), out of which ONLY ONE is correct.*

4 Which gives more than one amides on treatment with NH_2OH followed by PCl_5?

a. 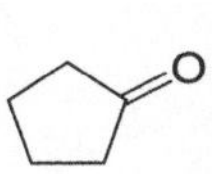

b. H_2CO

c. 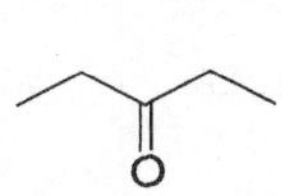

d. 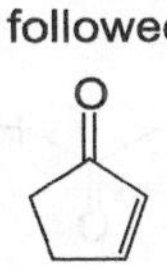

5 Which oximes on treatment with concentrated H_2SO_4 undergo rearrangement to give single amide?

a. 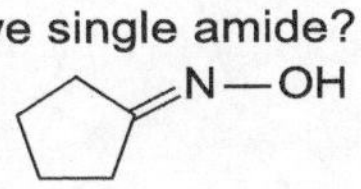

b. 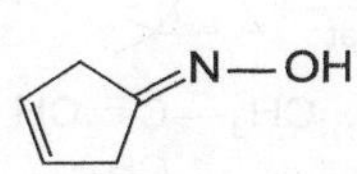

c. $CH_3-CH=N-OH$

d. 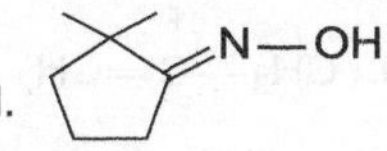

Working Space

6 What would be the major organic product in the following oxidation reaction?

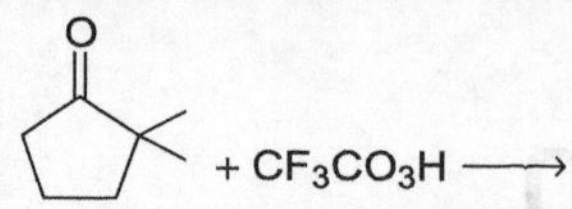

a. 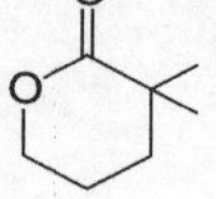

b. 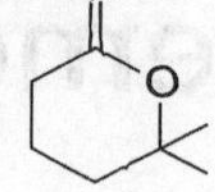

c. 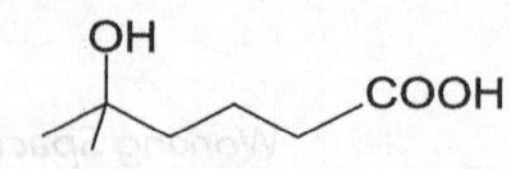

d. 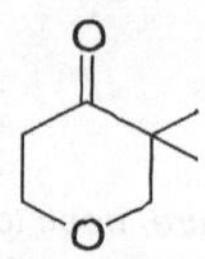

7 What is the final major product in the following reaction?

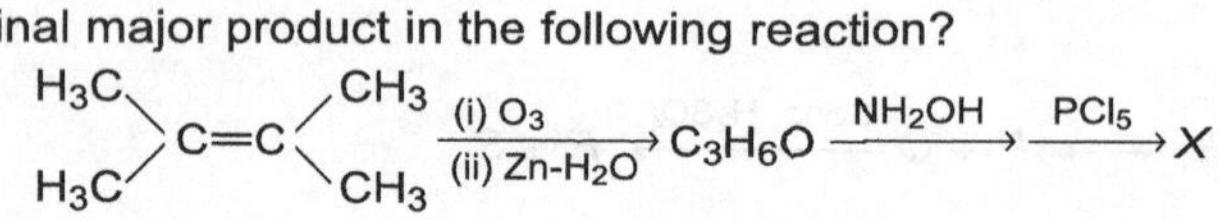

a. $CH_3-\overset{O}{\overset{\|}{C}}-NHCH_3$

b. $CH_3-\overset{O}{\overset{\|}{C}}-NH_2$

c. $CH_3-\underset{OH}{\underset{|}{\overset{CH_3}{\overset{|}{C}}}}-NH_2$

d. 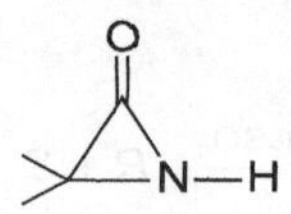

8 What is the major organic product in the following reaction?

$$C_6H_5COOH + CH_3Li\ (\text{Excess}) \longrightarrow \xrightarrow{H_3\overset{+}{O}} \xrightarrow{CF_3CO_3H}$$

a. $C_6H_5-\overset{O}{\overset{\|}{C}}-OCH_3$

b. $C_6H_5-\underset{CH_3}{\underset{|}{\overset{OH}{\overset{|}{C}}}}-CH_3$

c. $CH_3-\overset{O}{\overset{\|}{C}}-OC_6H_5$

d. $CH_3-\underset{C_6H_5}{\underset{|}{\overset{OH}{\overset{|}{C}}}}-C_6H_5$

9 What is *X* in the following reaction?

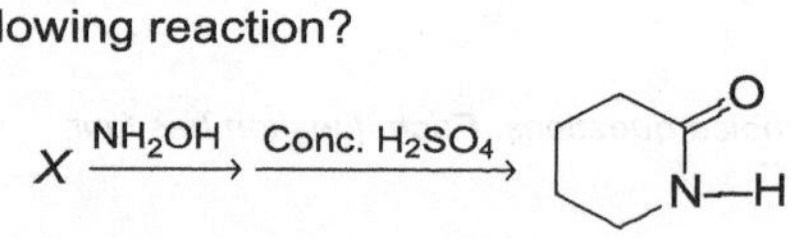

a. 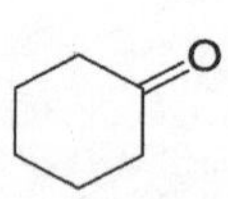

b. 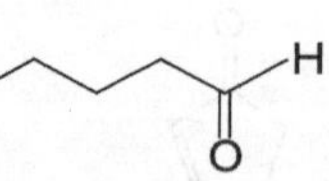

c. 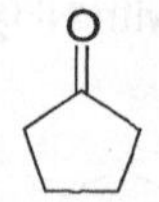

d. 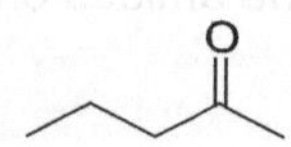

10 What is *X* in the following reaction?

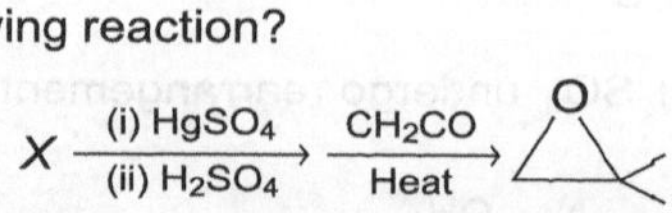

a. $CH_3-CH=CH_2$

b. $CH_3-C\equiv CH$

c. 

d. $CH_3-\overset{CH_3}{\overset{|}{C}}=CH_2$

Working Space

One or More than One Options Correct Type

Direction (Q. Nos. 11-14) This section contains 4 multiple choice questions. Each question has four choices (a), (b), (c) and (d), out of which ONE or MORE THAN ONE are correct.

11 In the reaction given below,

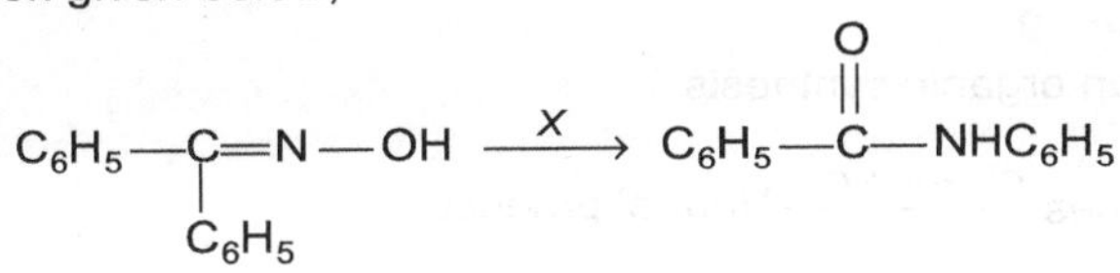

Reagent X could be

a. PCl_5 b. Conc. H_2SO_4 c. $SOCl_2$ d. HCl

12 Which reaction given below gives 2,2-dimethyl oxirane?

a. 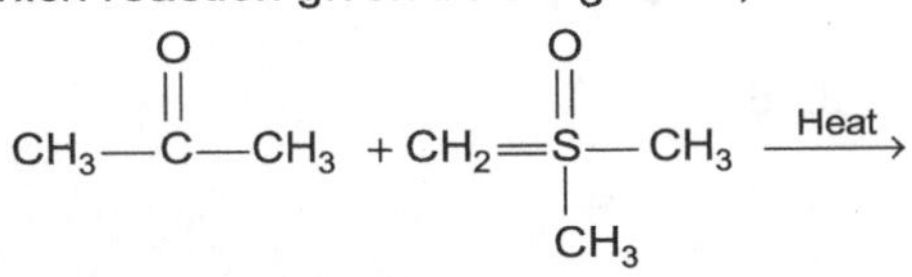

b. 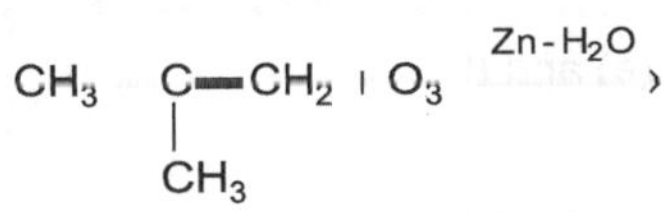

c. 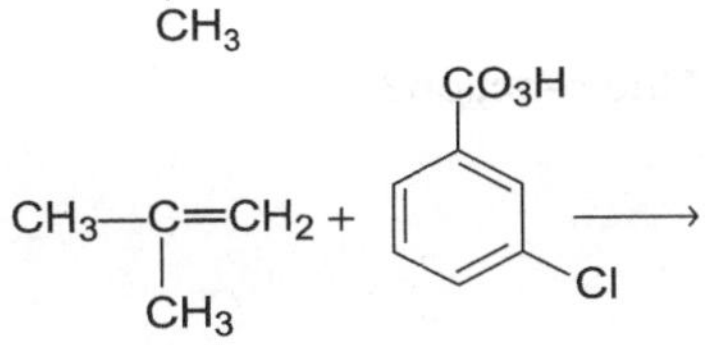

d. 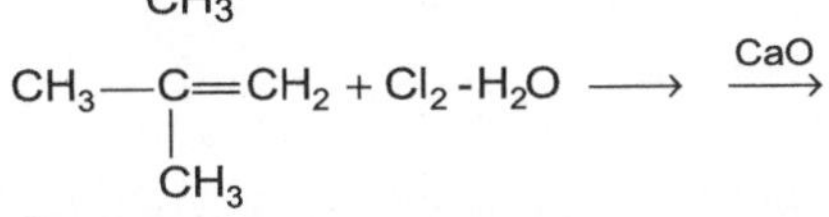

13 In the reaction given below,

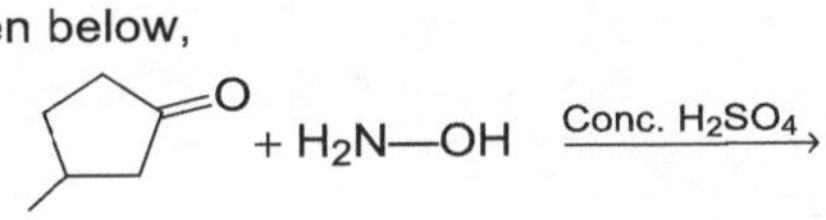

expected product(s) is/are

a. 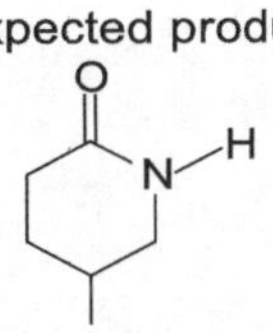

b. 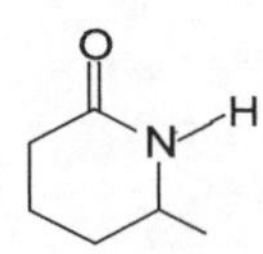

c. 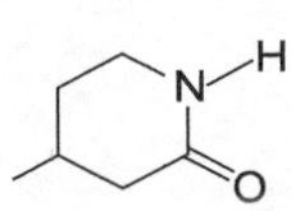

d. 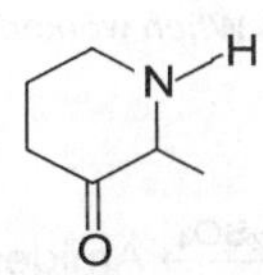

14 The correct statement(s) regarding the following reaction is/are

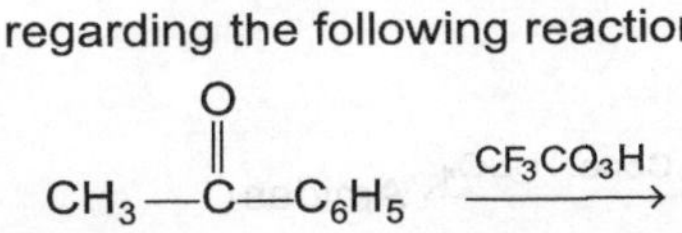

a. Product is a benzoic acid ester
b. Product is an acetic acid ester
c. Ketone is oxidised
d. Reaction involves migration of phenyl moiety onto oxygen atom

Comprehension Type

Working Space

Direction *(Q. Nos. 15-17) This section contains a paragraph, describing theory, experiments, data, etc. Three questions related to the paragraph have been given. Each question has only one correct answer among the four given options (a), (b), (c) and (d).*

Passage

Consider the following sequence of an organic synthesis

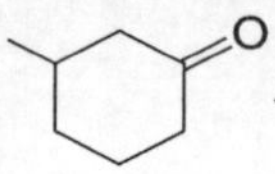

$+ NH_2OH \longrightarrow$ Oximes $\xrightarrow{\text{Conc.}H_2SO_4}$ Neutral products

Neutral products $\xrightarrow{H_3O^+/\text{Heat}}$ Polymers

15 If the starting compound has all its stereoisomers present, how many different oximes would be produced?

a. 2 b. 3 c. 4 d. 6

16 Which is not a part of neutral products?

a. 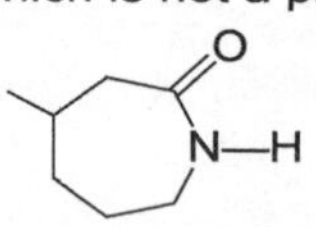 b. 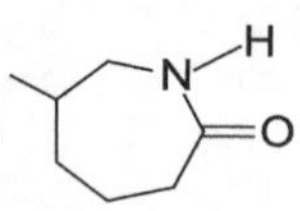 c. 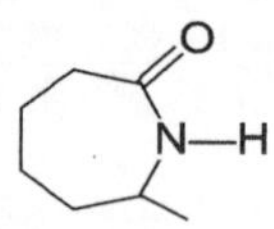 d. Both (a) and (b)

17 What is the possible repeating unit of polymer formed at the end of the reaction?

a. $-NH-\overset{O}{\overset{\|}{C}}-CH_2-\underset{CH_3}{\underset{|}{CH}}-CH_2CH_2CH_2-$

b. $-NH-\overset{O}{\overset{\|}{C}}-CH_2CH_2CH_2-\underset{CH_3}{\underset{|}{CH}}-CH_2-$

c. $-NH-\overset{O}{\overset{\|}{C}}-CH_2-CH_2-\underset{CH_3}{\underset{|}{CH}}-CH_2-CH_2-$

d. Both (a) and (b)

One Integer Value Correct Type

Direction *(Q. Nos. 18-20) This section contains 3 questions. When worked out will result in an integer from 0 to 9 (both inclusive).*

18 In the reaction 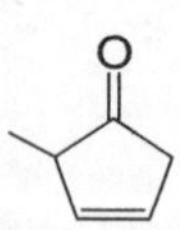 $+ NH_2OH \longrightarrow \xrightarrow{\text{Conc. } H_2SO_4}$ Amides, how many different amides are expected?

19 In the reaction 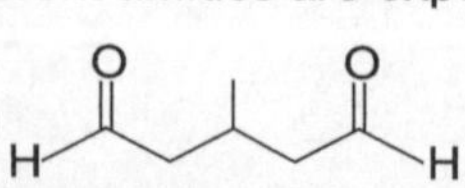 $\xrightarrow{NH_2OH} \xrightarrow{\text{Conc. } H_2SO_4}$ Amides, how many different amides are expected?

20 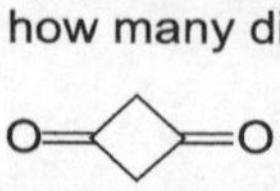 $+ NH_2OH \longrightarrow X \xrightarrow{\text{Conc. } H_2SO_4} Y \xrightarrow[H_2O]{H_3O^+} HOOC-CH_2-NH_2$ (Only product)

How many lone pair of electrons are present in *Y*?

DPP-8 Unsaturated Aldehydes and Ketones

Subjective Questions

Working Space

Direction *(Q. Nos. 1-5) These questions are subjective in nature, need to be solved completely on notebook.*

1 Propose an efficient synthesis of glyceraldehyde ($HOCH_2-CH(OH)-CHO$) from acrolein ($CH_2=CH-CHO$).

2 Identify products in the following reaction,

$$CH_3-\underset{\underset{CH_3}{|}}{C}=CH-\overset{\overset{O}{||}}{C}-CH_3 \xrightarrow[HCN]{NaCN} X \xrightarrow[\text{Conc. HCl}]{SnCl_2} Y \xrightarrow{\text{Dil. NaOH}} Z$$

3 Draw the structures of all products formed in solution containing C_6H_5CHO, $C_6H_5COCH_3$ and dilute NaOH. Identify the major product.

4 Explain, why the following exchange occur? Why does exchange not occur in other methyl group?

$$H-\overset{\overset{O}{||}}{C}-\overset{\overset{CH_3}{|}}{C}=CH-CH_3 \xrightarrow[NaOD]{D_2O} H-\overset{\overset{O}{||}}{C}-\overset{\overset{CH_3}{|}}{C}=CH-CD_3$$

5 What product is formed in the following reaction?

$$\underset{\text{(Methyl vinyl ketone)}}{CH_3-\overset{\overset{O}{||}}{C}-CH=CH_2} + Br_2 \longrightarrow$$

Compare reactivity of methyl vinyl ketone with 1-butene in the same reaction.

Only One Option Correct Type

Direction *(Q. Nos. 6-10) This section contains 5 multiple choice questions. Each question has four choices (a), (b), (c) and (d), out of which ONLY ONE is correct.*

6 Predict the major product in the given reaction, $CH_3-\overset{\overset{CH_3}{|}}{C}=CH-CHO \xrightarrow[\text{(dilute)}]{NaOH}$

a. $CH_3-\overset{\overset{CH_3}{|}}{C}=CH-COOH$

b. $CH_3-\overset{\overset{CH_3}{|}}{C}=CH-CH_2OH$

c. $CH_3-\overset{\overset{CH_3}{|}}{C}=CH-CH=CH-\underset{\underset{CH_3}{|}}{C}=CH-CHO$

d. Both (a) and (b)

Working Space

7 The major organic product in the following reaction is

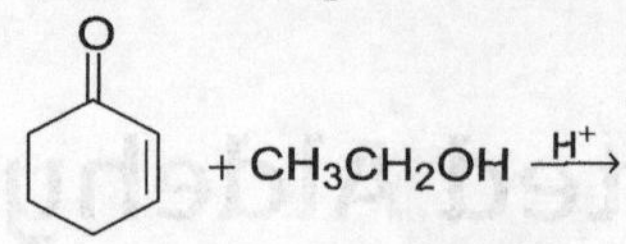

a. 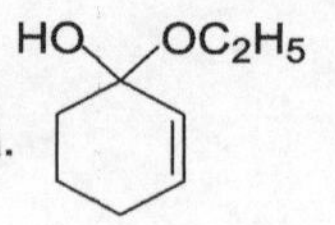

b. 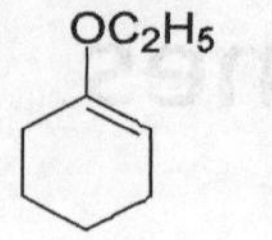

c. 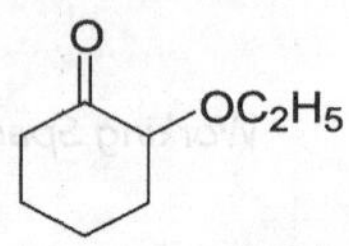

d. 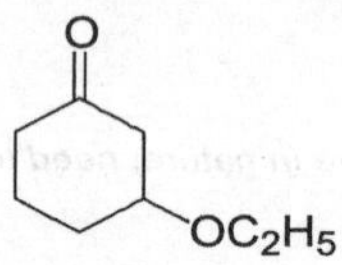

8 What is the major product in the following reaction?

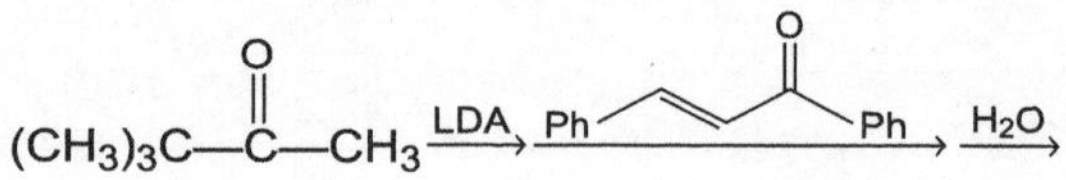

a. $(CH_3)_3C-C(=O)-CH_2-CH(Ph)-CH_2-C(=O)-Ph$

b. $(CH_3)_3C-C(=O)-CH_2-C(OH)(Ph)-CH=CH-Ph$

c. $(CH_3)_3C-C(=O)-CH=C(Ph)-CH=CH-Ph$

d. $(CH_3)_3C-C(OH)(CH_3)-CH_2-C(=O)-C(CH_3)_3$

9 The major organic product in the following reaction is

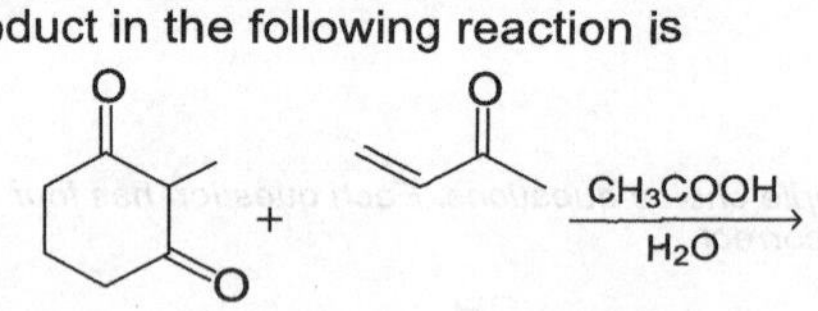

a. 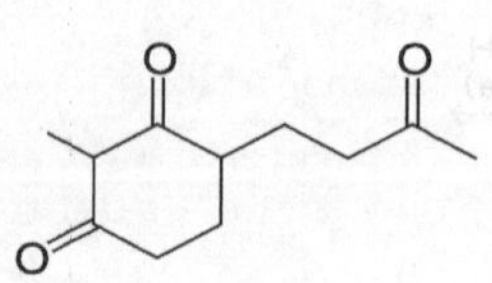

b. 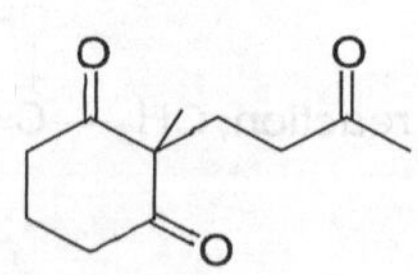

c. 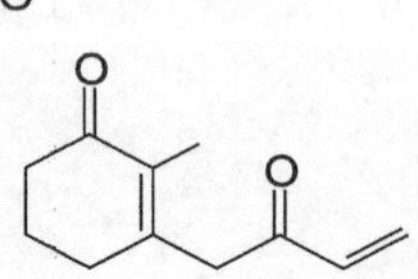

d. 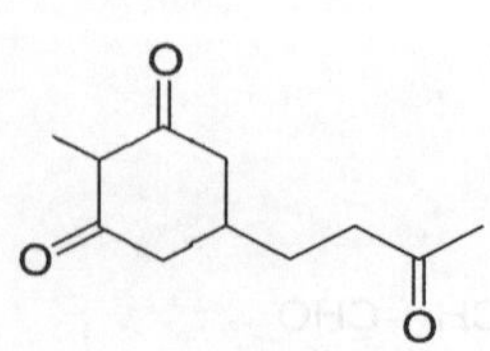

Working Space

10 In the reaction given below,

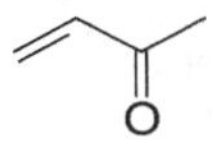 $+ (CH_3)_2CHMgBr \xrightarrow{(C_2H_5)_2O} \xrightarrow[\Delta]{Conc.\ H_2SO_4}$

The final major product is

a. 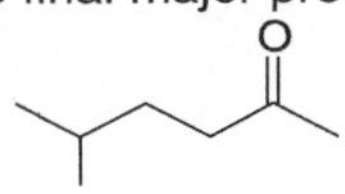
b. 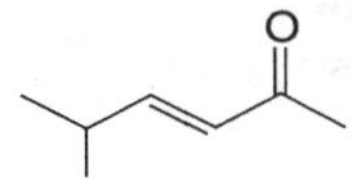
c. 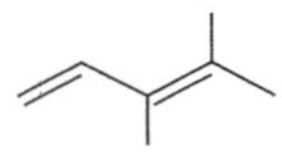
d. 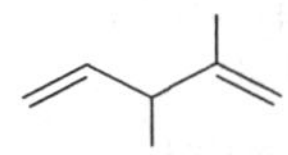

One or More than One Options Correct Type

Direction *(Q. Nos. 11-14) This section contains 4 multiple choice questions. Each question has four choices (a), (b), (c) and (d), out of which ONE or MORE THAN ONE are correct.*

11 In the reaction given below,

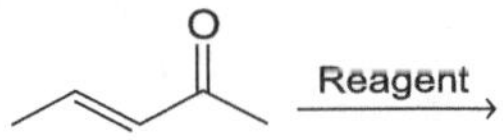

which reagent(s) below would undergo addition at olefinic bond?

a. HCl
b. NH_3
c. CH_3MgBr
d. NaH

12 In the reaction given below,

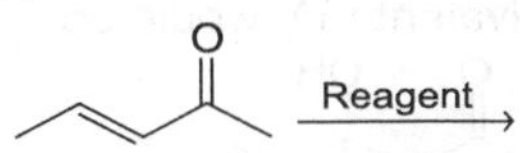

which reagent(s) will bring about nucleophilic addition at carbonyl carbon?

a. NH_3
b. HBr
c. CH_3MgBr
d. CH_3Li

13 What is/are true about nucleophilic addition reaction at α, β-unsaturated carbonyl compound of type 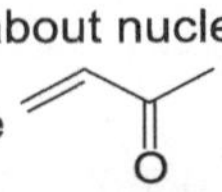 ?

a. $LiAlH_4$ brings about addition at carbonyl carbon
b. $(CH_3)_2CuLi$ brings about addition at conjugate (olefinic) position
c. CH_3Li brings about conjugate (olefinic) addition
d. CH_3NH_2 brings about conjugate (olefinic) addition

14 Consider the following reaction,

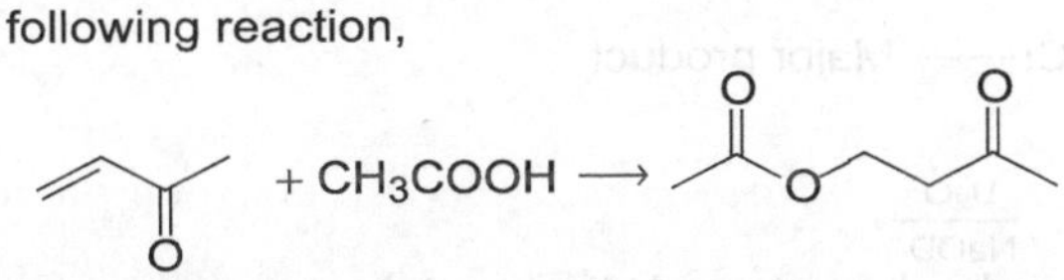

The correct statement(s) regarding mechanism of the above reaction and the reaction outcome is/are

a. The indicated product is major one
b. The reaction proceeds by nucleophilic attack of acid at protonated carbonyl at conjugate position
c. Some aldol product is also formed
d. Some cyclic ester is also formed

Working Space

Comprehension Type

Direction *(Q. Nos. 15-17) This section contains a paragraph, describing theory, experiments, data, etc. Three questions related to the paragraph have been given. Each question has only one correct answer among the four given options (a), (b), (c) and (d).*

Passage

Consider the following multistep synthesis :

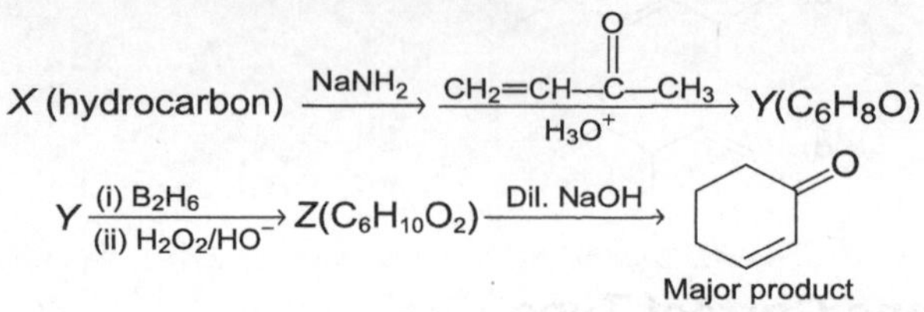

15 The hydrocarbon X is

a. C_2H_4 b. C_2H_2

c. $CH_3—C\equiv CH$ d. $CH_2=CH—CH=CH_2$

16 The structure of compound Y is

a. 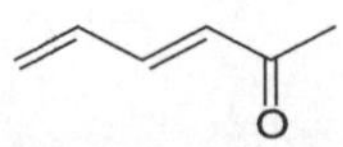

b. $H—C\equiv C—CH_2CH_2—\overset{O}{\overset{\|}{C}}—CH_3$

c. $CH_3—C\equiv C—CH_2CH_2—CHO$

d. $CH_3—C\equiv C—CH_2—\overset{O}{\overset{\|}{C}}—CH_3$

17 If X is replaced by pentane-2,4-dione, product equivalent to Y would be

a. 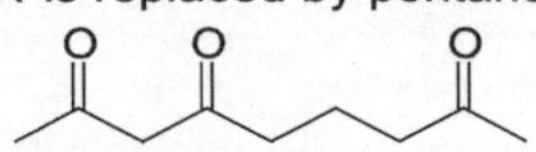

b. 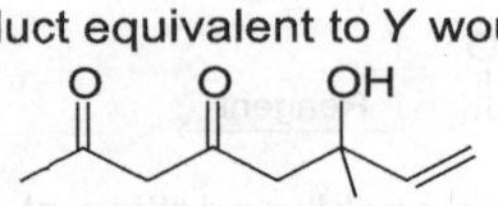

c. 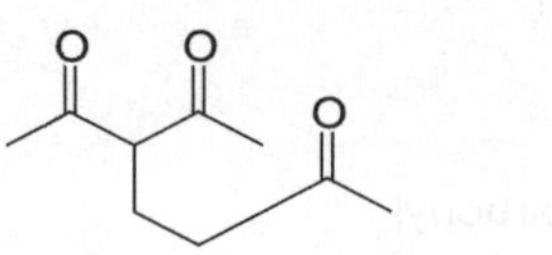

d. 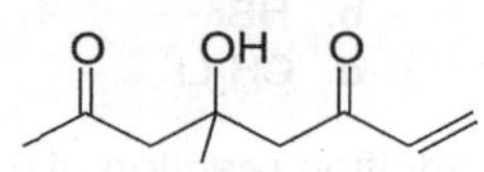

One Integer Value Correct Type

Direction *(Q. Nos. 18-20) This section contains 3 questions. When worked out will result in an integer from 0 to 9 (both inclusive).*

18 In the reaction given below, how many different stereoisomers of major product are possible?

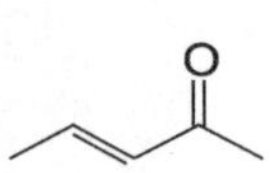 + ICl ⟶ Major product

19 In the reaction given as, 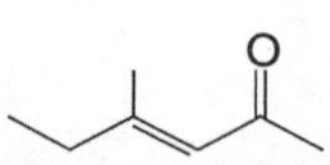 $\xrightarrow[\text{NaOD}]{D_2O}$ some of the hydrogen undergo exchange by deuterium. At the most, how many deuterium atoms incorporation are possible per molecule of ketone?

20 Consider the following sequence of reaction,

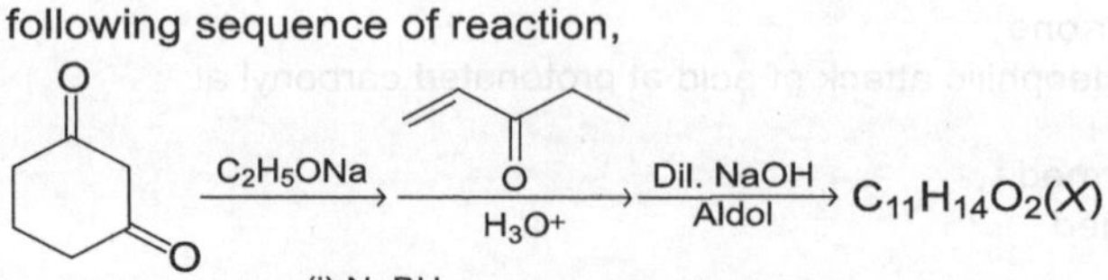

$X \xrightarrow[\text{(ii) } H_3O^+]{\text{(i) } NaBH_4}$ Diols

How many different stereoisomers of diol are formed?

DPP-9 Reformatsky, Witting Benzil-Benzilic Acid Rearrangements, etc.

Only One Option Correct Type

Direction (Q. Nos. 1-8) *This section contains 8 multiple choice questions. Each question has four choices (a), (b), (c) and (d), out of which ONLY ONE is correct.*

Working Space

1 What is the major product of the following reaction?

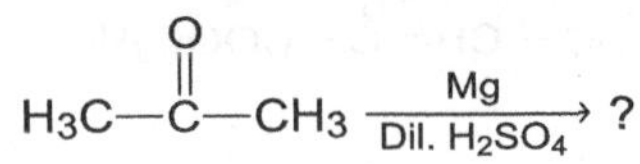

a. 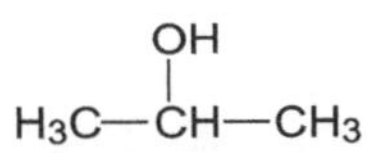

b. $H_3C-CH=CH_2$

c. 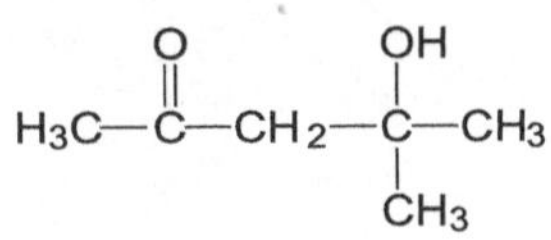

d. 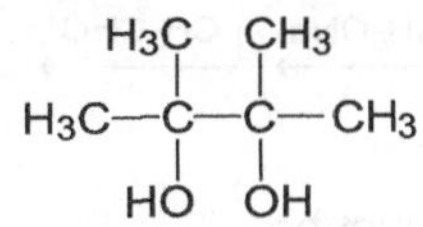

2 Predict the major product of the following reaction

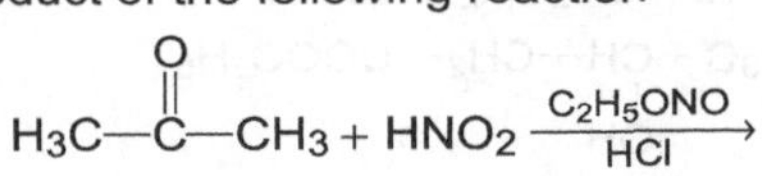

a. 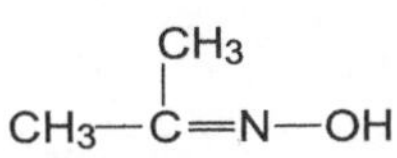

b. 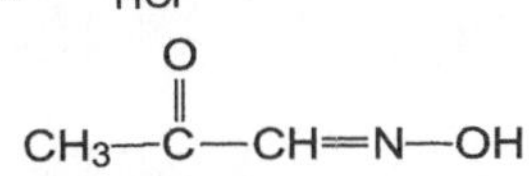

c. 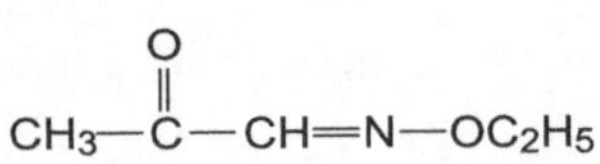

d. 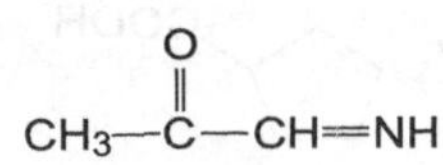

3 What is formed as predominant product in the following reaction?

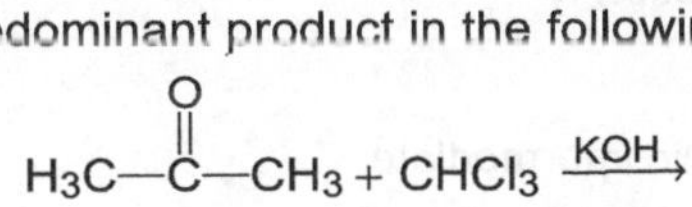

a. 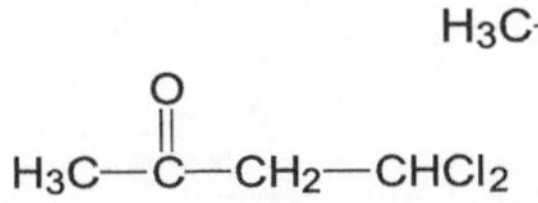

b. 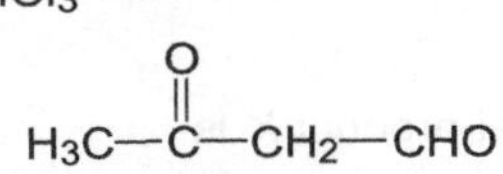

c. 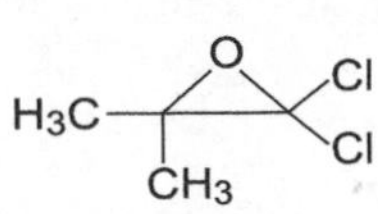

d. 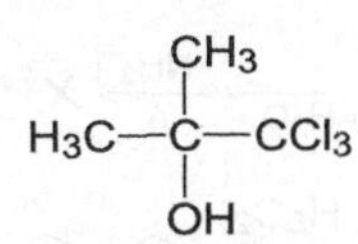

Working Space

4 Consider the following sequence of reaction,

$$BrCH_2COOEt \xrightarrow{Zn} \xrightarrow[H_3O^+]{Acetone} \xrightarrow[Heat]{H_2SO_4} X$$

If X is the major organic product, it must be

a. $H_3C-C(CH_3)=CH-COOH$ b. $H_3C-C(=O)-CH_2-CH_2-COOH$

c. $H_3C-C(OH)(CH_3)-CH_2COOH$ d. $H_3C-C(CH_3)=CH-COOEt$

5 Consider the given reaction,

$$CH_3CH_2CHO + CH_2(COOC_2H_5)_2 \xrightarrow[1:1/\Delta]{C_2H_5ONa} \xrightarrow[\Delta]{H_3O^+}$$

The major organic product is

a. $CH_3H_2C-CH=C(COOC_2H_5)-COOC_2H_5$ b. $CH_3H_2C-CH=C(COOH)-COOC_2H_5$

c. $H_3C-CH_2-CH=CH-COOEt$ d. $CH_3H_2C-CH=CH-COOH$

6 In the given reaction,

$$CH_2(Cl)-COOC_2H_5 \xrightarrow{C_2H_5ONa} \xrightarrow{CH_3CHO} X$$

if X is a condensation product then it must be

a. $OHC-CH_2CH_2-COOC_2H_5$ b. $OHC-CH_2CH_2-COOH$

c. 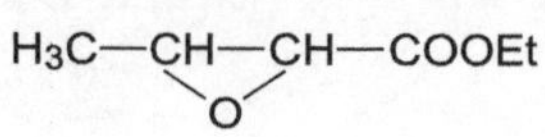 d. $H_3C-CH(OH)-CH_2-COOC_2H_5$

7 Consider the following reaction,

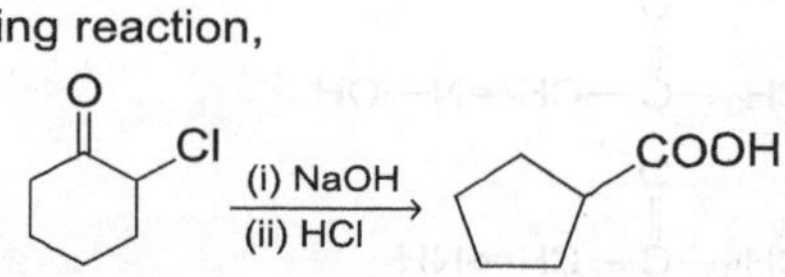

The correct statement regarding the above reaction is

a. It is a condensation reaction
b. It is a simple oxidation reaction
c. Reaction involves a rearrangement *via* a bicyclic intermediate
d. All of the above

8 In the reaction given below, the final product X is

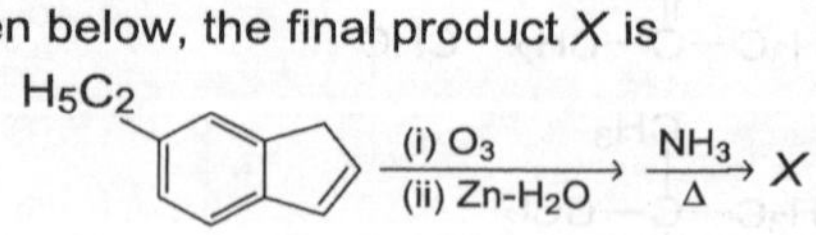

a. 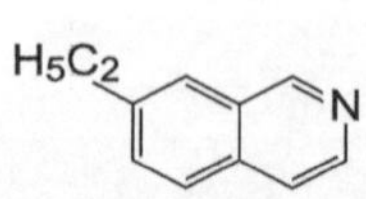 b. 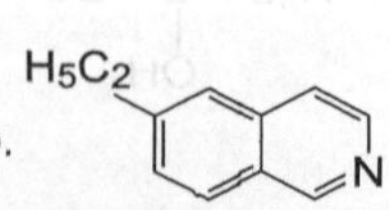

c. 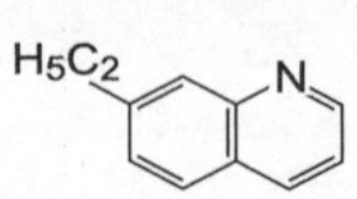 d. 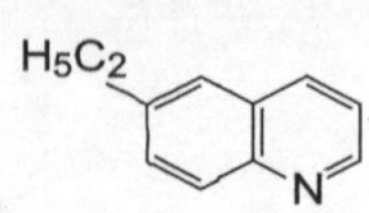

One or More than One Options Correct Type

Working Space

Direction *(Q. Nos. 9-14) This section contains 6 multiple choice questions. Each question has four choices (a), (b), (c) and (d), out of which ONE or MORE THAN ONE are correct.*

9 Consider the following Reformatsky reaction

$$H_3C-CH(Cl)-COOC_2H_5 \xrightarrow[DMSO]{Zn} \xrightarrow[(ii)\ H_3O^+]{(i)\ CH_3CHO}$$

The correct statements concerning the above reaction is/are

a. The major condensation product is $H_3C-CH(OH)-CH(CH_3)-COOC_2H_5$

b. The first step involves the formation the of an organometallic intermediate

c. Second step of reaction is equivalent to Grignard's synthesis

d. Reaction proceeds in the same manner if Mg is used in place of Zn

10 Consider the following chain reaction

$$Cl\ \ CH(CH_3)\ \ COOC_2H_5 \xrightarrow[C_2H_5OH]{C_2H_5ONa} \xrightarrow{Acetone} X \xrightarrow[\Delta]{H_3O^+} Y$$

The correct statements concerning the above reaction is/are

a. *X* is an α, β-epoxy ester

b. *Y* is an aldehyde

c. *Y* is a ketone

d. In the last step, an unstable enol is formed which tautomerises into stable keto form

11 Consider the following sequence of reaction,

$$H_5C_6-CH(OH)-C(=O)-C_6H_5 \xrightarrow{HNO_3} X \xrightarrow[(ii)\ HCl]{(i)\ KOH} Y$$

The correct statement(s) regarding the above reaction is/are

a. *X* is benzil $\left(C_6H_5-C(=O)-C(=O)-C_6H_5 \right)$

b. *Y* is $C_6H_5-C(C_6H_5)(OH)-COOH$

c. If *X* is treated with 1, 2-diamino benzene, following compound is formed

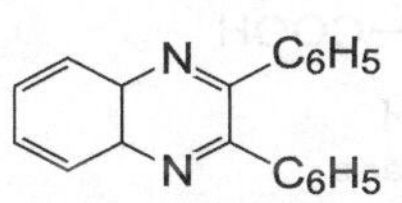

d. *Y* is $C_6H_5-CH(OH)-C_6H_5$

12 Consider the reaction sequence given below,

$$H_3C-CH_2Cl \xrightarrow[(ii)\ C_4H_9Li]{(i)\ Ph_3P} X \xrightarrow[\Delta]{CH_3CHO} Y$$

The correct statements regarding the above reaction is/are

a. *X* is $CH_3-CH{=}PPh_3$

b. *Y* is $CH_3-CH{=}CH-CH_3$

c. *Y* is $CH_3-CH_2-CH{=}CH_2$

d. *Y* is a mixture of diastereomers

Working Space

13 Consider the reaction sequence given below,

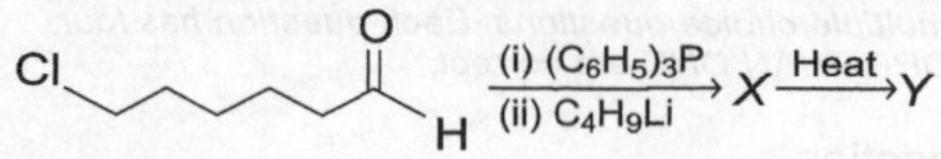

The correct deductions is/are

a. X is an unstable intermediate

b. Y is 1-methyl cyclopentene 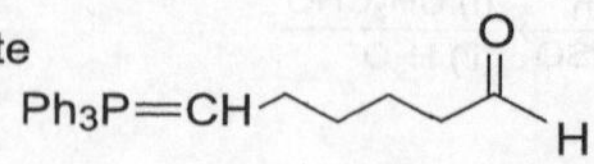

c. Y is cyclohexene

d. Y on treatment with cold, dilute and alkaline $KMnO_4$ gives racemic mixture of diols

14 Consider the following transformation,

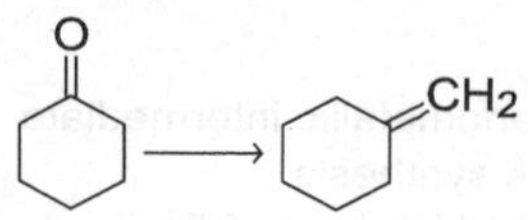

Which of the following sequence of reaction can bring about the above transformation in good yield?

a. 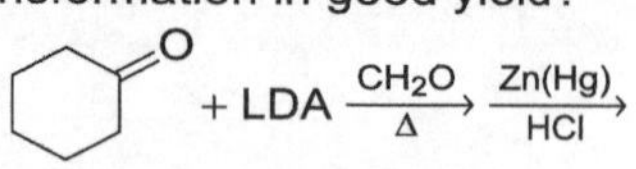

b. 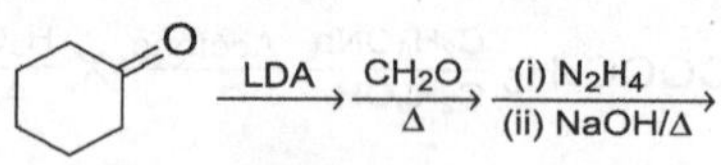

c. 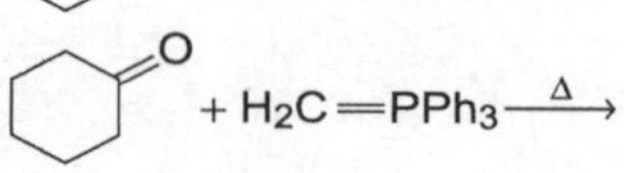

d. 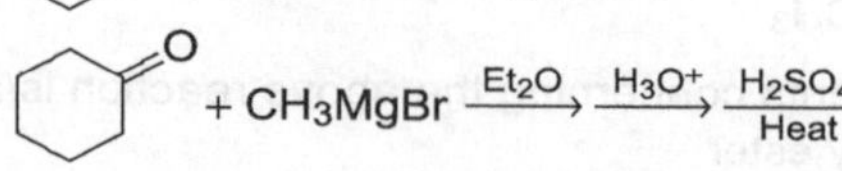

Comprehension Type

Direction *(Q. Nos. 15-17) This sections contains a paragraph, describing theory, experiments, data, etc. Three questions related to the paragraph have been given. Each question has only one correct answer among the four given options (a), (b), (c) and (d).*

Passage

Benzaldehyde when treated with KCN(aq) undergoes a condensation reaction, known as benzoin condensation forming benzoin A. Benzoin on oxidation with HNO_3 gives a diketone known as benzil B. Benzil on heating with ethanolic KOH undergo benzil-benzilic acid rearrangement as

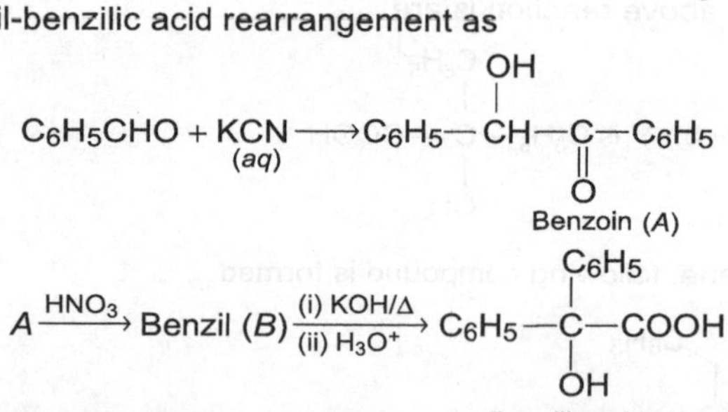

The widely accepted mechanism of last step rearrangement is

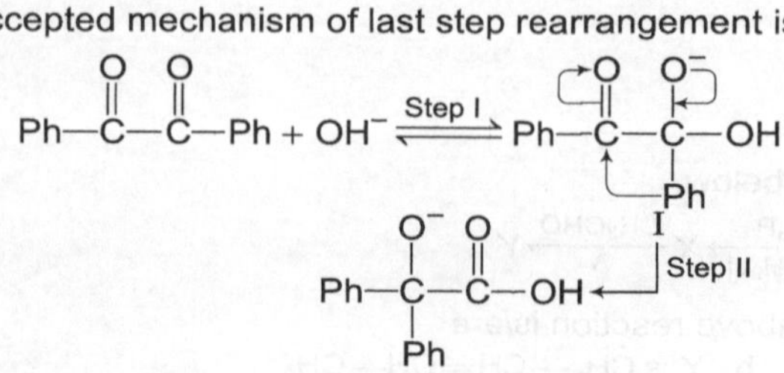

15 When benzil is treated with dilute $Na\overset{18}{O}H$; $H_2\overset{18}{O}$, both of its oxygens are replaced by O^{18} isotope. It proves that

a. attack of OH^- occur is first, fast reversible step
b. first step is the rate determining step
c. rearrangement shows kinetic isotopic effect
d. Both (b) and (c) are correct

16 In the following reaction ($^*C$ is C^{14} isotope)

The expected organic products is/are

a. 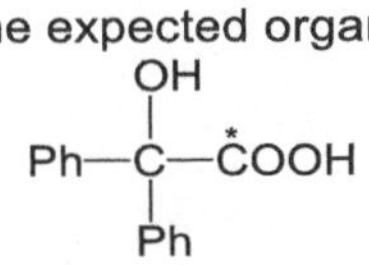

b. 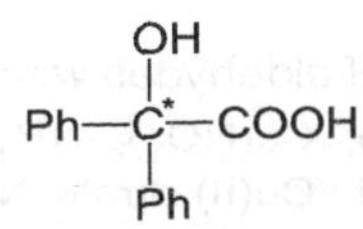

c. Both (a) and (b)
d. Either (a) or (b)

17 In the reaction given below,

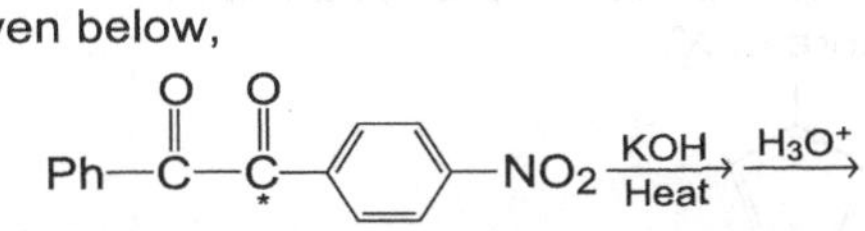

The major rearrangement product is

a. 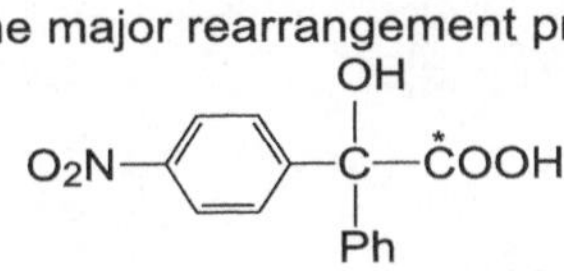

b. 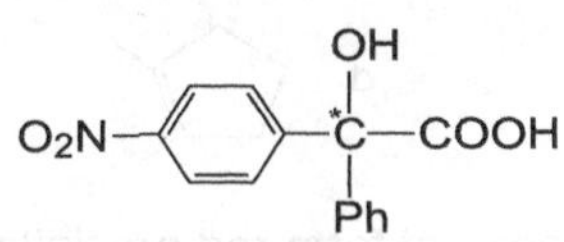

c. 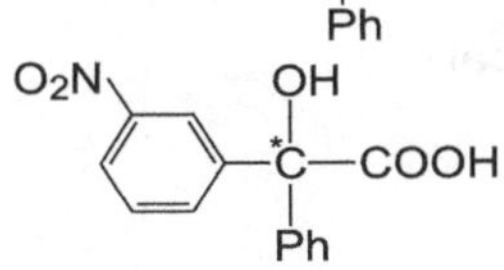

d. 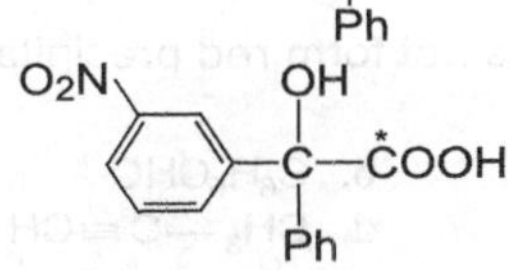

One Integer Value Correct Type

***Direction** (Q. Nos. 18-20) This section contains 3 questions. When worked out will result in an integer from 0 to 9 (both inclusive).*

18 Consider the following two step synthesis:

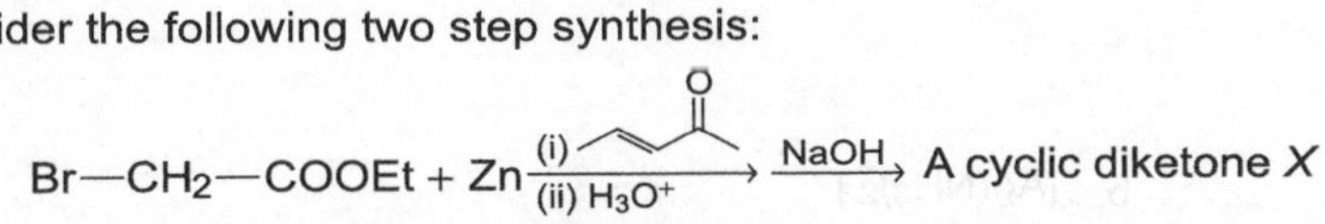

If X is finally treated with excess of $NaBH_4$ followed by acid work-up, how many different isomers of diols would be formed?

19 Consider the reaction given below,

$$Cl—CH_2—COOC_2H_5 \xrightarrow{C_2H_5ONa} \xrightarrow{CH_3CHO} \alpha, \beta\text{-epoxy ester}$$

In the above reaction, how many different isomers of esters are formed?

20 Consider the reaction given below,

$$CH_3CH_2Cl \xrightarrow[\text{(ii) } C_4H_9Li]{\text{(i) } (C_6H_5)_3P} X$$

$$X\,(\text{excess}) + OHC—CHO \xrightarrow{\Delta} Y + PO(C_6H_5)_3$$

How many different isomeric hydrocarbons of Y are formed?

DPP-10 Tests and Distinctions

Only One Option Correct Type

Working Space

Direction *(Q. Nos. 1-8) This section contains 8 multiple choice questions. Each question has four choices (a), (b), (c) and (d), out of which ONLY ONE is correct.*

1 Tollen's reagent used for the distinction of aldehydes with ketones is

a. $Pb(NO_3)_2$-$NH_3(aq)$
b. $Cu(NO_3)_2$-$NH_3(aq)$
c. $AgNO_3$-$NH_3(aq)$
d. Cu(II) citrate-$NH_3(aq)$

2 A hydrocarbon *X* on treatment with O_3 followed by the reduction of ozonide with Zn-H_2O gives *Y*. *Y* gives both Tollen's test as well as yellow precipitate with NaOH/I_2 solution. Which is a possible structure of *X*?

a. 
b. 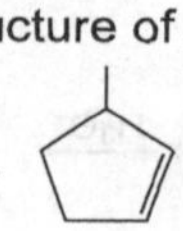
c. 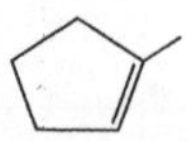
d. 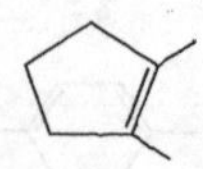

3 Which compound given below does not form red precipitate with ammoniacal solution of Cu(II) tartarate?

a. H_2CO
b. C_6H_5CHO
c. CH_3CHO
d. $CH_3—C≡CH$

4 Which reagent can differentiate between benzaldehyde and acetophenone?

a. $[Ag(NH_3)_2^+]$
b. NaOH/I_2
c. $[Cu(NH_3)_4]^{2+}$
d. Both (a) and (b)

5 Which reagent can be used to separate a mixture of ethanol and butanone into components?

a. 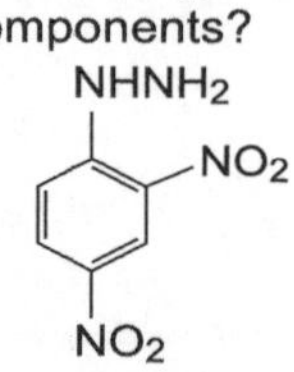
b. $[Ag(NH_3)_2^+]$
c. $[Cu(NH_3)_4]^{2+}$
d. NaOH/I_2

6 Reagent that can differentiate 2-propanol from acetone is

a. NaOH/I_2
b. $[Ag(NH_3)_2^+]$
c. $Ce(NO_3)_4$
d. $[Cu(NH_3)_4]^{2+}$

7 Which reagent given below can differentiate propanal from propanone?

a. Schiff's reagent
b. $NaHSO_3$
c. 2, 4-dinitrophenyl hydrazine
d. NaCN/HCN

Working Space

8 Which of the following would produce an orange coloured precipitate with 2, 4-dinitrophenyl hydrazine?

a. 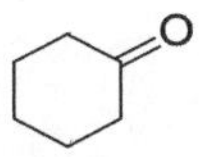

b. 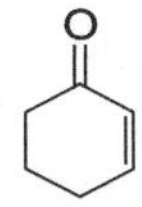

c. CH_3CHO

d. $H_3C-\overset{O}{\overset{||}{C}}-CH_3$

One or More than One Options Correct Type

Direction *(Q. Nos. 9-14) This section contains 6 multiple choice questions. Each question has four choices (a), (b), (c) and (d), out of which ONE or MORE THAN ONE are correct.*

9 Which of the following form(s) yellow precipitate with $NaOH/I_2$ solution?

a. CH_3CHO

b. $C_6H_5-\overset{O}{\overset{||}{C}}-CH_3$

c. $CH_3-\underset{}{\overset{OH}{\overset{|}{CH}}}-CH_3$

d. CH_3CH_2OH

10 Which of the following compounds reduce(s) ammoniacal silver nitrate solution?

a. CH_3CH_2CHO

b. $CH_3-\overset{OH}{\overset{|}{CH}}-\underset{O}{\underset{||}{C}}-CH_3$

c. CH_3CH_2OH

d. C_6H_5CHO

11 Which of the following fail(s) to reduce Fehling's solution?

a. H_2CO

b. C_6H_5CHO

c. $CH_3-\underset{OH}{\underset{|}{CH}}-\overset{O}{\overset{||}{C}}-CH_3$

d. $H-\overset{O}{\overset{||}{C}}-NH_2$

12 Which of the following reagents can be used for the separation of a mixture containing CH_3CHO and CH_3CN?

a. $NaHSO_3$

b. Schiff's reagent

c. $NaOH/I_2$

d. 2, 4-dinitrophenyl hydrazine

13 A hydrocarbon has molecular formula C_6H_{10}. Ozonolysis of this compound followed by reduction with $Zn\text{-}H_2O$ gives $C_6H_{10}O_2$ which reduces Tollen's reagent as well as forms yellow precipitate with alkaline iodine solution. The possible structure(s) of this hydrocarbon is/are

a. 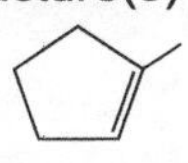

b. 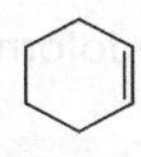

c. 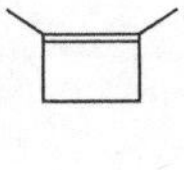

d. 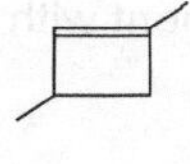

Working Space

14 Which compound(s) given below, upon acid hydrolysis gives a compound that forms a yellow precipitate with KOH/I_2 solution?

a. 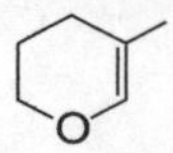

b. 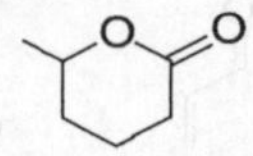

c. 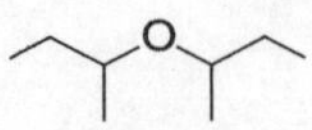

d. 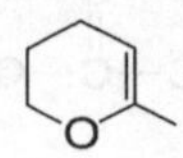

Comprehension Type

Direction (Q. Nos. 15-17) *This section contains a paragraph, describing theory, experiments, data, etc. Three questions related to the paragraph have been given. Each question has only one correct answer among the four given options (a), (b), (c) and (d).*

Passage

A neutral organic compound $A(C_{10}H_{20}O_2)$ neither reduces Tollen's reagent nor forms precipitate with 2, 4-dinitrophenyl hydrazine, but can be resolved into enantiomers. *A* on acid hydrolysis forms two compounds *B* and *C*, both are enantiomeric. *C* neither reduces Fehling's solution nor forms iodoform with alkaline iodine solution. *C* on oxidation with CrO_3/HCl/pyridine forms *D* which is still resolvable into enantiomers. *D* on further treatment with aqueous $(C_2H_5O)_3Al$ solution gives back *A*.

15 How many different stereoisomers exist for *A*?

a. 2
b. 4
c. 5
d. 6

16 What is true regarding *D*?

a. It forms a yellow precipitate with KOH/I_2
b. It forms an orange precipitate with 2, 4-dinitrophenyl hydrazine
c. On oxidation with $K_2Cr_2O_7/H_2SO_4$, it forms *B*
d. On treatment with $NaHSO_3$, it forms salt which cannot be separated by fractional crystallisation

17 The correct statement regarding the following reaction is

$$D + Cl_2 \xrightarrow{H_2SO_4} \alpha\text{-chlorination}$$

a. Product is achiral
b. Product is a racemic mixture
c. Product is a pair of diastereomer
d. No α-halogenation occur with *D*

One Integer Value Correct Type

Direction (Q. Nos. 18-20) *This section contains 3 questions. When worked out will result in an integer from 0 to 9 (both inclusive).*

18 In reaction of $C_6H_5COCH_3$ with KOH-I_2 to form iodoform, how many moles of KOH are consumed per mole of ketone?

19 How many different isomers of C_5H_8O, on treatment with 2, 4-dinitrophenyl hydrazine gives orange precipitate?

Working Space

20 Consider the following reaction,

$$CH_3—COOC_2H_5 + CH_3COCH_3 \xrightarrow[C_2H_5OH]{C_2H_5ONa} C_5H_8O_2(X)$$

$$X \xrightarrow[\text{Excess}]{NH_2OH} \xrightarrow{NiCl_2} Y \text{ (Red precipitate)}$$

The complex forming red precipitate has several ring structures. How many monocyclic rings are present in the complex?

Matching List Type

Direction (Q. No. 21) *Choices for the correct combination of elements from Column I and Column II are given as options (a), (b), (c) and (d), out of which one is correct.*

21 Match the qualitative tests listed in Column I with compounds from Column II that gives positive response to these tests and mark the correct option from the codes given below.

	Column I		Column II
i.	Yellow procipitate with 2, 4-dinitrophenyl hydrazine	p.	C_6H_5CHO
ii.	Orange precipitate with 2, 4-dinitrophenyl hydrazine	q.	Cyclohexanone
iii.	Red precipitate with ammoniacal Cu(II) tartarate	r.	3-hydroxy butanone
iv.	White precipitate with $NaHSO_3$	s.	Ethanal (CH_3CHO)

Codes

	i	ii	iii	iv
a.	q	p,s	p,s	r,s
b.	q,r,s	p	r,s	p,q,r,s
c.	s	q	r	p,q,r
d.	r,s	p,q	q,s	s

SET - 1

Revisal Problems for JEE Main

Only One Option Correct Type

Direction *(Q. Nos. 1-30) This section contains 30 multiple choice questions. Each question has four choices (a), (b), (c) and (d), out of which ONLY ONE is correct.*

Working Space

1 What is the IUPAC name of the following compound?

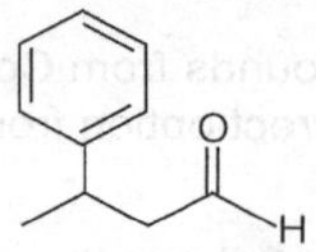

a. 3-methyl-3-phenyl propanol b. 3-phenyl butanal
c. 3-phenyl-1-butanone d. 3-phenyl butanoic acid

2 What is the major organic product obtained from the following reaction?

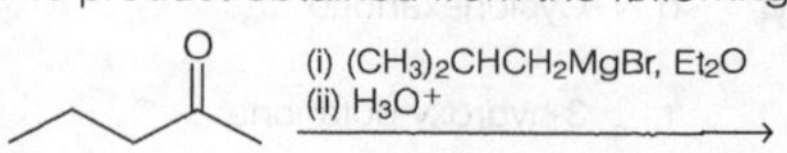

a. 2,3-dimethyl-3-heptanol b. 2,4-dimethyl-4-heptanol
c. 3,5-dimethyl-1-4-heptanol d. 3,5-dimethyl-3-heptanol

3 What is the major organic product obtained from the following reaction?

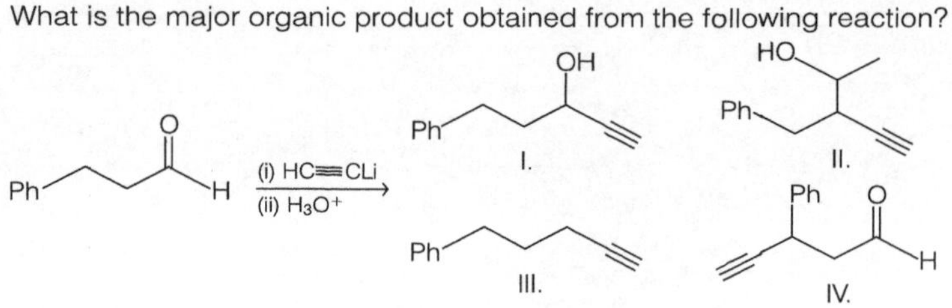

a. Only I b. Only II c. Only III d. Only IV

4 What is the major organic product obtained from the following reaction?

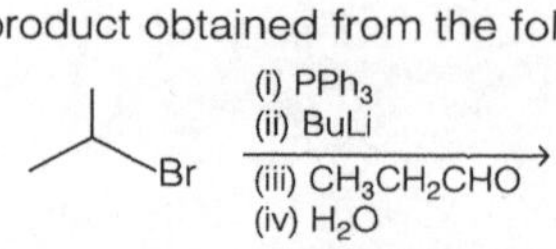

a. (*E*)-3-methyl-2-pentene b. (*Z*)-3-methyl-2-pentene
c. 2-methyl-2-pentene d. 4-methyl-1-pentene

5 Which ketone-diol undergoes cyclisation to form the following acetal?

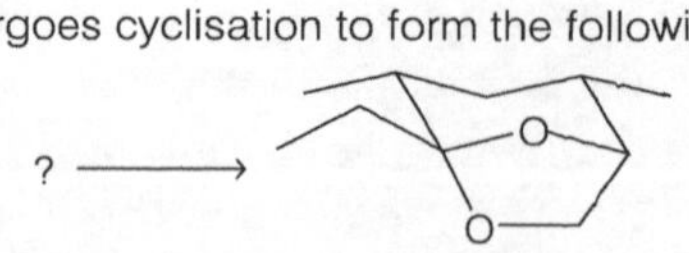

I. 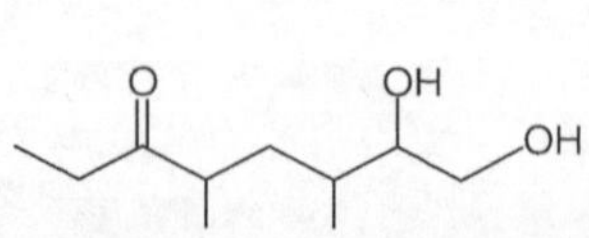

II. 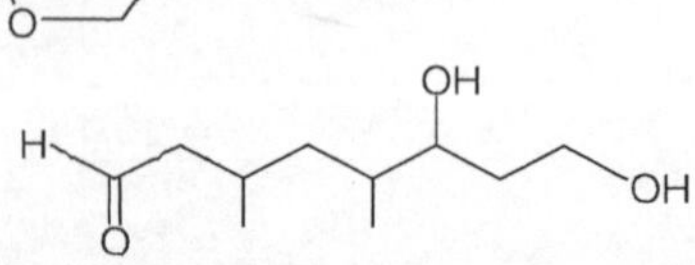

III. 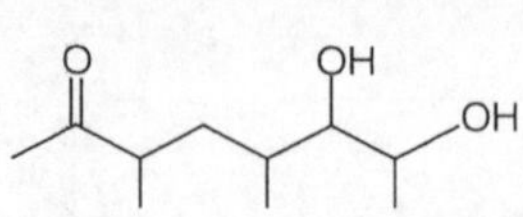

IV. 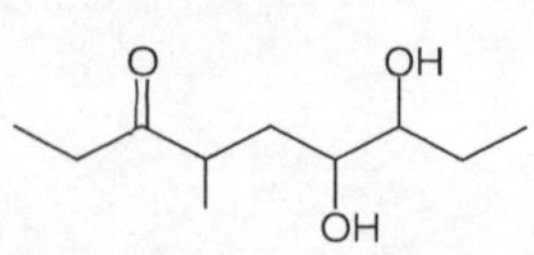

a. Only I b. Only II c. Only III d. Only IV

Working Space

6 What is the major organic product obtained from the following reaction?

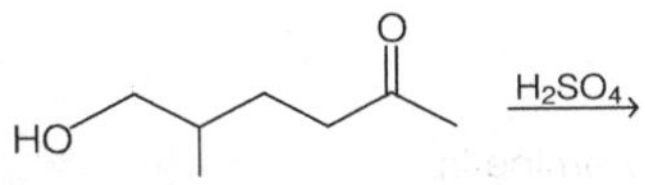

I. 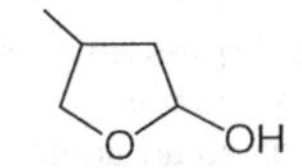 II. 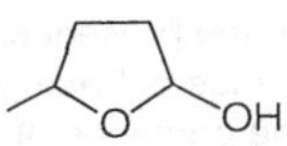

III. 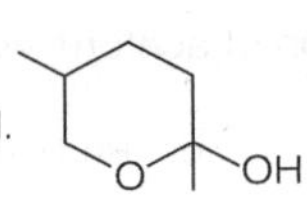 IV. 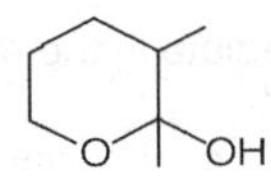

a. Only I b. Only II c. Only III d. Only IV

7 What is the major organic product obtained from the following reaction?

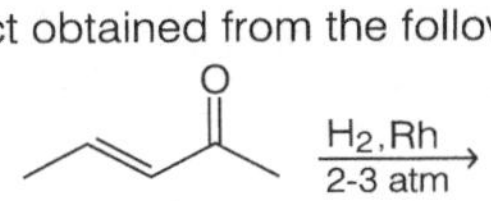

a. 2-pentanone
b. (*E*)-3-penten-2-ol
c. 2-pentanol
d. 4-hydroxy-2-pentanone

8 What is the major organic product obtained from the following sequence of reaction?

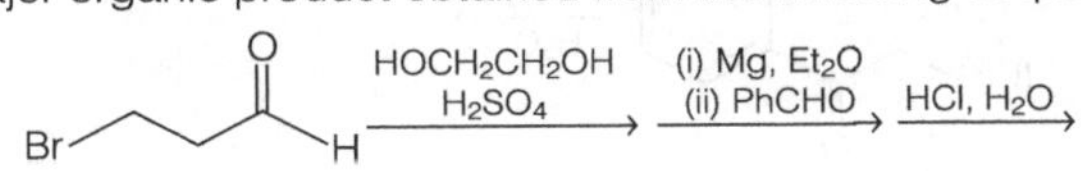

I. 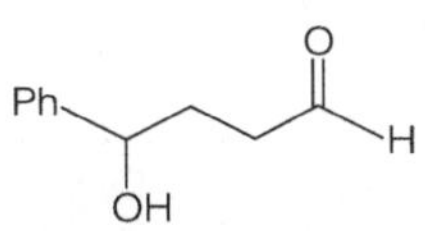 II. 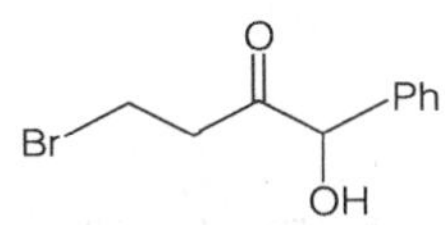

III. 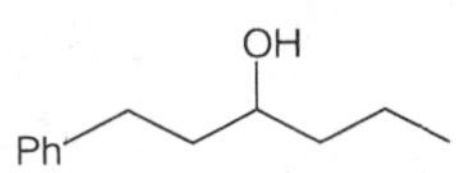 IV. 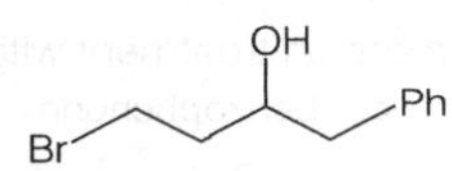

a. Only I b. Only II c. Only III d. Only IV

9 What is the major organic product obtained from the following sequence of reaction?

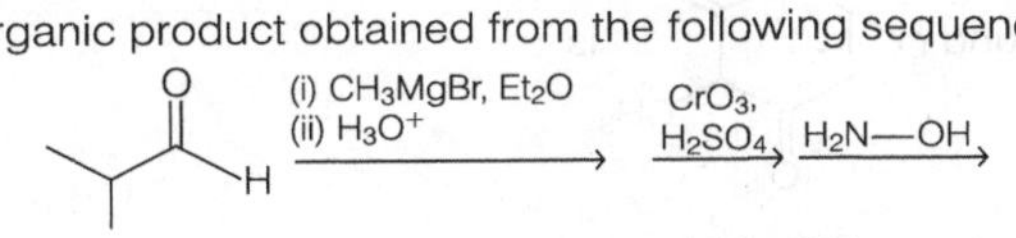

I. 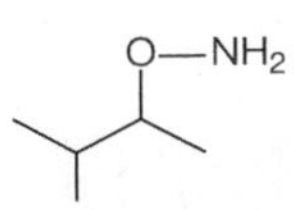 II. 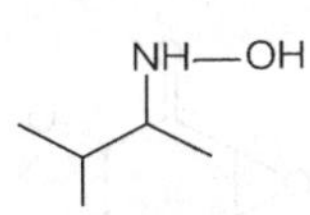

III. 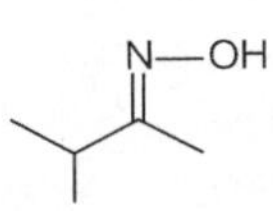 IV. 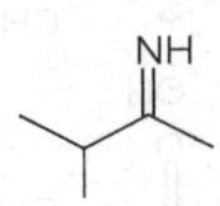

a. Only I b. Only II c. Only III d. Only IV

10 Which of the following has the highest K_{eq} for enolisation?

I. 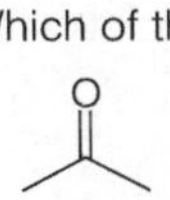 II. 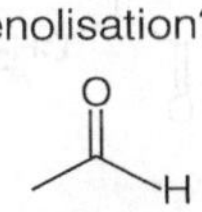

III. 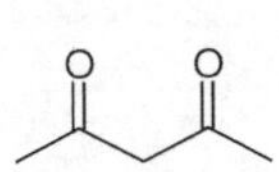 IV. 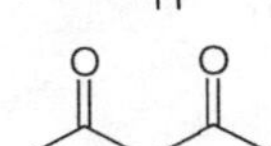

a. Only I b. Only II c. Only III d. Only IV

11 Which of the following is the best description of the mechanism of the reaction between a ketone and an amine to form an imine?

a. Concerted bimolecular substitution
b. Nucleophilic addition followed by elimination
c. Elimination followed by nucleophilic addition
d. Electrophilic addition followed by rearrangement

Working Space

12 Which of the following statement is not true?

a. Reaction of benzophenone with methanol in the presence of acid gives an acetal
b. Reaction of butanal with ethanol in the presence of base gives an acetal
c. Reaction of benzaldehyde with ethanol in the presence of acid gives an acetal
d. Reaction of acetone with methanol in the presence of base gives a hemiacetal

13 Which of the following is a reactive intermediate in the reaction of acetone with bromine in acetic acid to form 1-bromo-2-propanone?

a. Carbanion b. Enol c. Enolate d. Radical

14 Which of the following reactions can be used to prepare the compound given below?

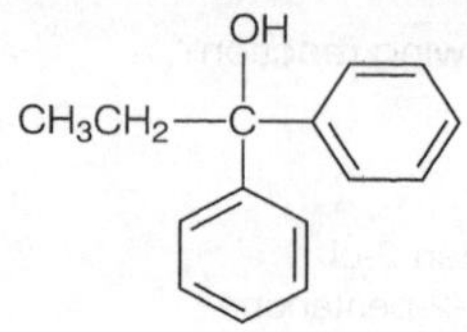

I. 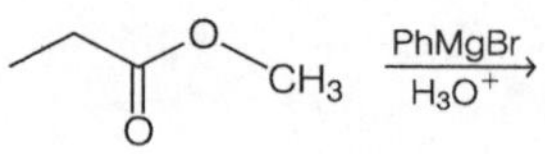

II. 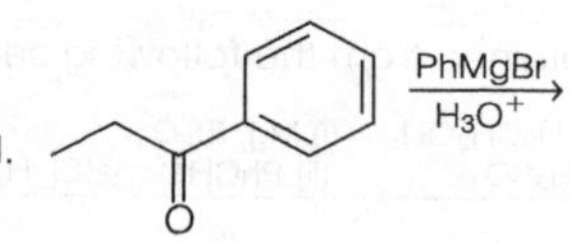

III. $C_6H_5-\overset{O}{\overset{\|}{C}}-C_6H_5 \xrightarrow[H_3O^+]{CH_3CH_2MgBr}$

a. Only I
b. Both I and II
c. Both II and III
d. I, II and III

15 Compound which does not give stereoisomers on treatment with hydroxylamine is

a. benzaldehyde b. acetophenone c. benzophenone d. 2-butanone

16 The most stable enol form of the compound 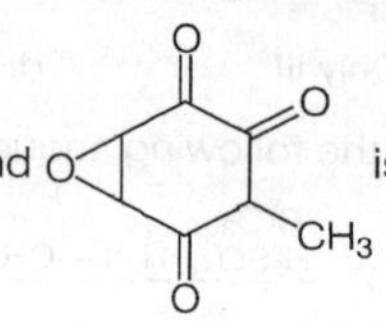 is

a. 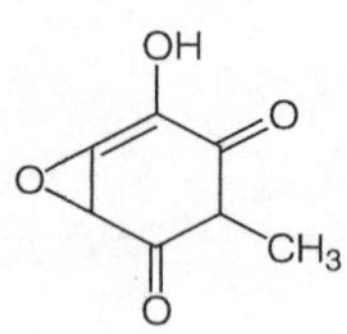

b. 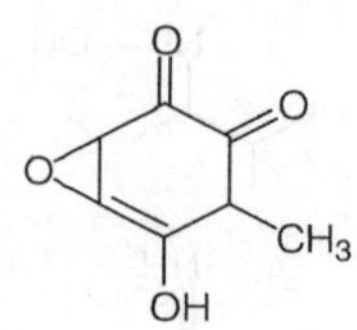

c. 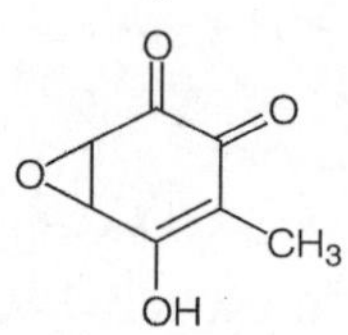

d. 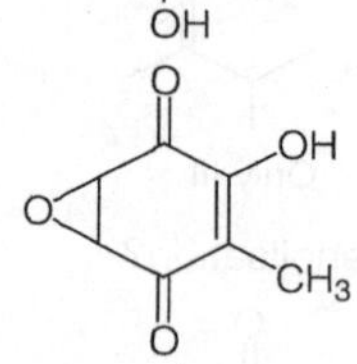

17 What is the major product in the following reaction?

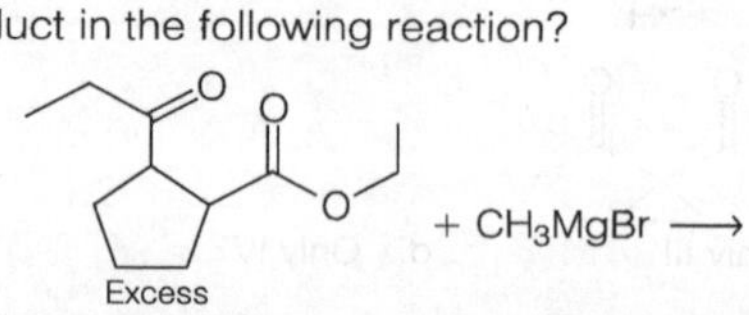

a. 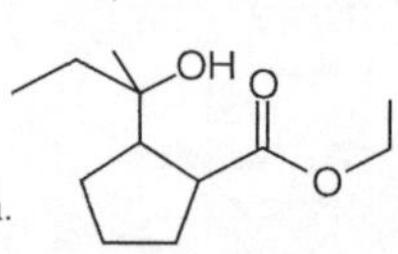

b. 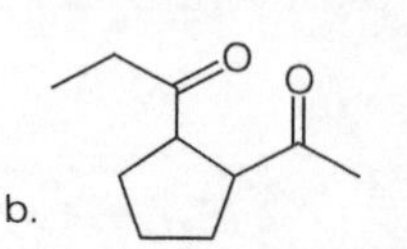

c. 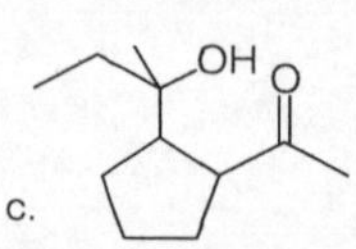

d. 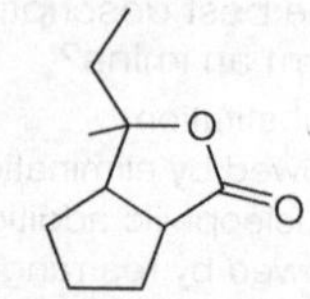

Working Space

18 Give the product of the following reaction.

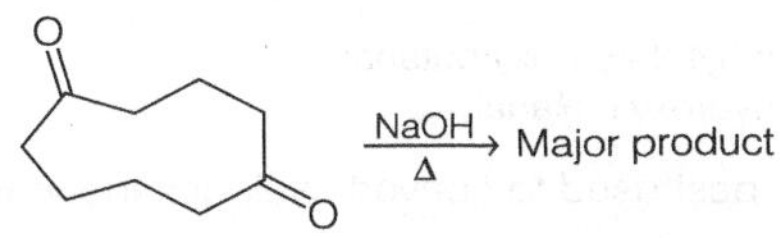

a. 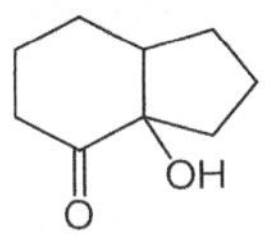

b. 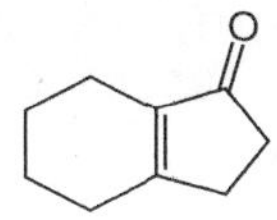

c. 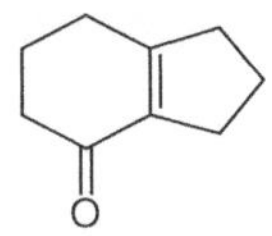

d. 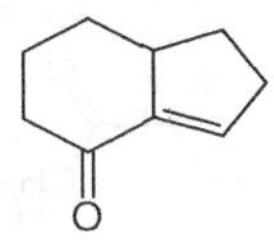

19 Which of the following reagent could be used to differentiate the following compounds by a visible change? 2-butanone and propanoic acid

a. $[Ag(NH_3)_2^+]/OH^-$ b. $LiAlH_4$
c. $KMnO_4/H^+/Cold$ d. $NaHCO_3$

20 An organic compound *X* produces *Y* in the first step of oxidation which on further oxidation gave an acid. *Y* on treatment with ammoniacal silver nitrate solution gave a black precipitate. *X* is likely to be a

a. 1° alcohol b. 3° alcohol
c. Ethanal d. 2-propanone

21 The major product of the following reaction is

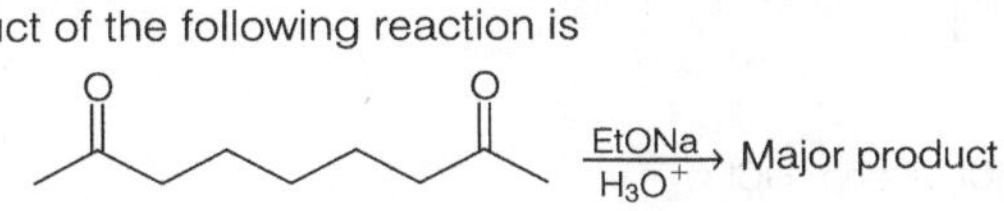

a. 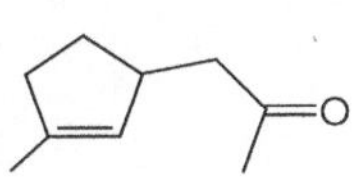

b. 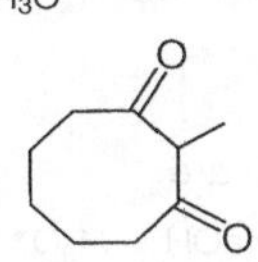

c. 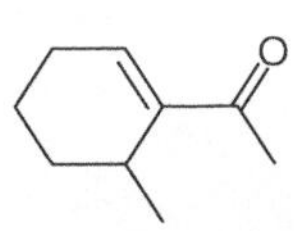

d. 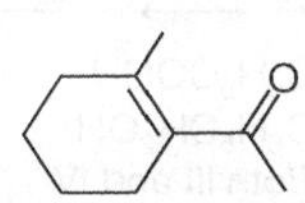

22 Which represents the best method for converting a carboxylic acid to an aldehyde?

a. Convert the acid to an acid chloride; and react the acid chloride with water
b. Reduce the acid with $LiAlH_4$
c. Convert the acid to *R*COCl; and reduce the *R*COCl with lithium aluminium tri(*t*-butoxy) hydride
d. Convert the acid to *R*COCl; react the *R*COCl with a *R*MgX; and reduce the product with $LiAlH_4$

23 You have two $C_6H_{10}O$ ketones, I and II. Both are optically active, but I is racemised by treatment with base and II is not. Wolff-Kishner reduction of both ketones give the same achiral hydrocarbon, formula C_6H_{12}. What reasonable structures may be assigned to I and II?

a. I is 3-methyl-4-penten-2-one; II is 4-methyl-1-penten-3-one
b. I is 2-methyl cyclopentanone; II is 3-methyl cyclopentanone
c. I is 3-methyl cyclopentanone; II is 2-methyl cyclopentanone
d. I is 2-ethyl cyclobutanone; II is 3-ethyl cyclobutanone

24 A $C_5H_{12}O$ compound is optically active, and is oxidised by PCC in CH_2Cl_2 to an optically active $C_5H_{10}O$ product, which is racemised in acid or base. Which of the following best fits these facts?

a. 2-pentanol b. 2-methoxy butane
c. 2-methyl-1-butanol d. 3-methyl-1-butanol

Working Space

25 Which of the following compounds would not be a possible product from the mixed aldol reaction of acetaldehyde and butanal?

a. 3-hydroxy butanal
b. 2-ethyl-3-hydroxy butanal
c. 3-ethyl-2-hydroxy hexanal
d. 3-hydroxy hexanal

26 Which reaction or sequence of reactions would be best used to convert cyclohexanone to *cis*-1,2-cyclohexanediol?

a. PCC in CH_2Cl_2 and base
b. (i) $NaBH_4$; (ii) H_3PO_4 and heat; (iii) $C_6H_5CO_3H$
c. (i) $NaBH_4$; (ii) H_3PO_4 and heat; (iii) OsO_4 in pyridine
d. (i) $NaBH_4$; (ii) OsO_4 in pyridine

27 Arrange the following in the increasing order of reactivity towards aldol condensation reaction

I. 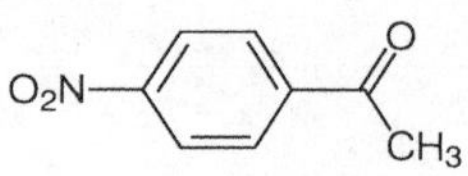

II. 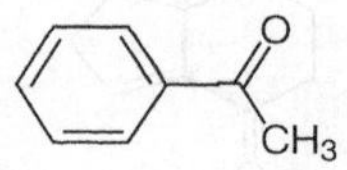

III. 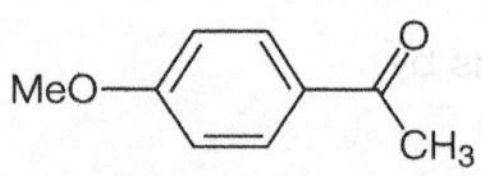

IV. 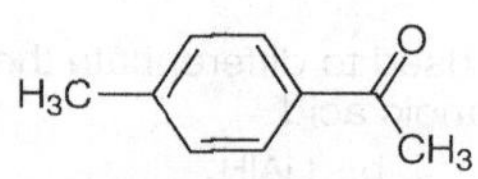

a. I < II < III < IV
b. IV < III < II < I
c. III < II < IV < I
d. III < IV < II < I

28 Provide the most suitable set of reagents required for the following transformation.

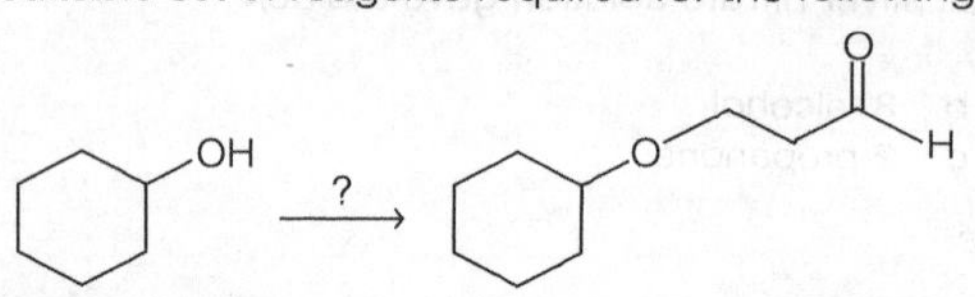

a. (i) $HOCH_2CH_2Br$; (ii) NaCN
b. (i) Na, ether; (ii) $ClCH_2CH_2I$; (iii) NaCN; (iv) $SnCl_2$/Conc. HCl
c. (i) Br_2, UV light; (ii) $HOCH_2CH_2CN$
d. (i) TsCl, pyridine; (ii) $BrCH_2CH_2CN$

29 The major organic products in the following reaction are

$$H_2CO + C_6H_5CHO \xrightarrow{\text{Conc. NaOH}} \xrightarrow{H_3O^+}$$

I. HCOOH
II. C_6H_5COOH
III. CH_3OH
IV. $C_6H_5CH_2OH$

a. Both I and II
b. Both III and IV
c. Both I and IV
d. Both II and III

30 The major organic product formed in the following reaction is

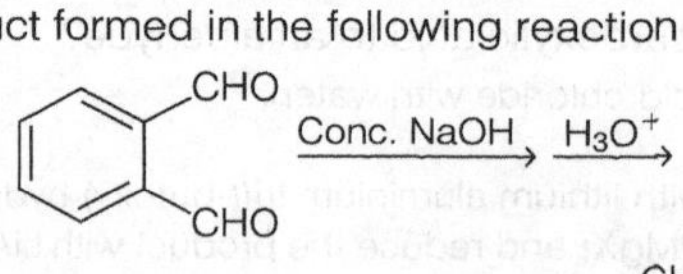

a. 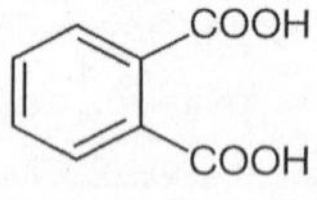

b. 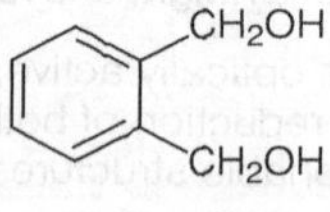

c. 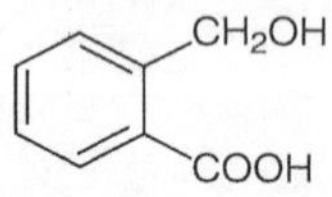

d. 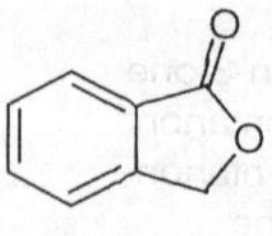

Revisal Problems for JEE Advanced

Only One Option Correct Type

Direction (Q. Nos. 1-8) *This section contains 8 multiple choice questions. Each question has four choices (a), (b), (c) and (d), out of which ONLY ONE is correct.*

Working Space

1 The best reagent that can bring about the following conversion is

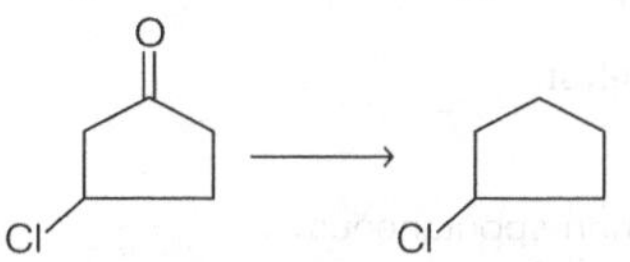

a. N_2H_4/OH^- b. $Zn(Hg)/C \cdot HCl$
c. $LiAlH_4$ d. HI/Red P

2 Which carbon-carbon bond could not be formed by an aldol condensation reaction?

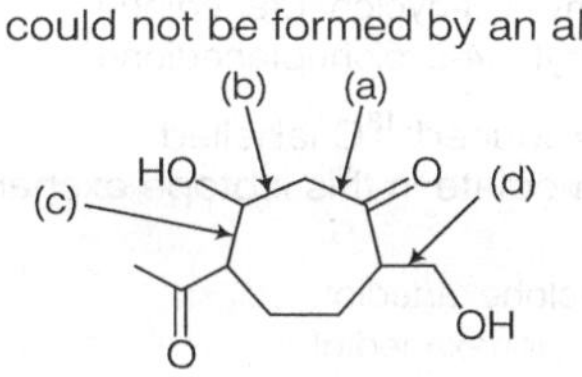

3 Arrange the following in the increasing order of hydrate content when dissolved in water.

I. 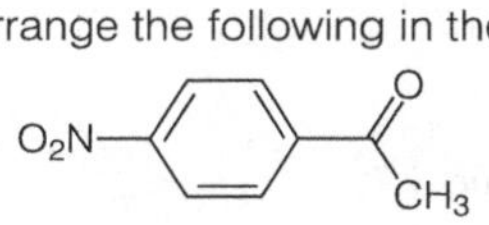

II. 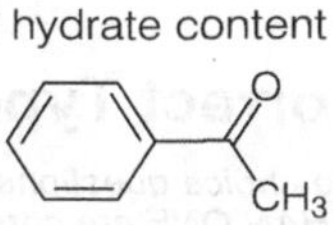

III. 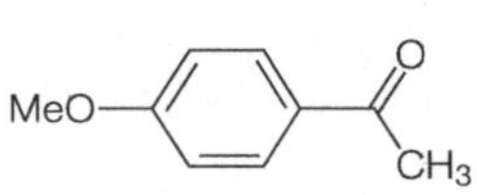

IV. 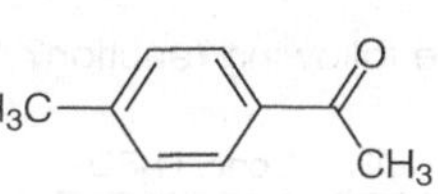

a. I < II < III < IV
b. IV < III < II < I
c. III < II < IV < I
d. III < IV < II < I

4 What is the major product in the following sequence of reaction?

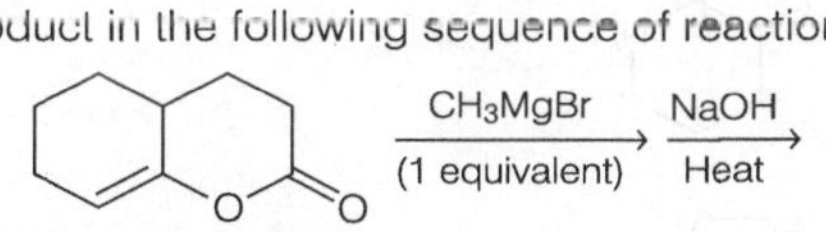

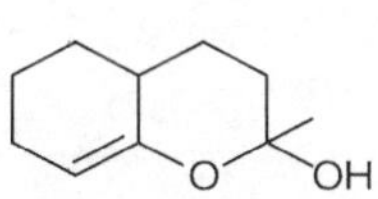

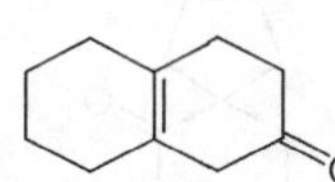

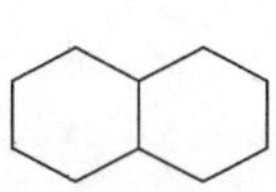

Working Space

5 An organic compound A ($C_6H_{12}O$) neither decolourise bromine water nor changes the colour of acidic dichromate solution. A on heating with H_2SO_4 produces an alkene which on oxidative ozonolysis gives B ($C_6H_{10}O_3$) which gives a yellow precipitate with $NaOH/I_2$. The most probable structure of A is

a. 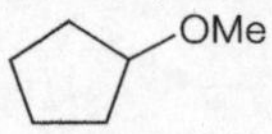
b. 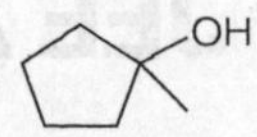

c. 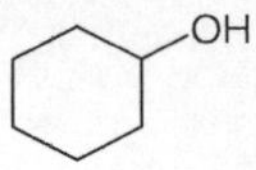
d. 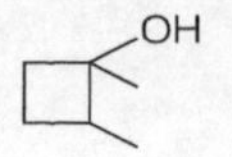

6 Which of the following is not true regarding the reaction given below?

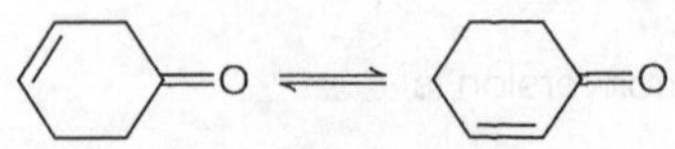

a. Reaction can be activated in the presence of acid catalyst
b. Reaction can be activated by alkali
c. Product is more stable than reactant
d. Reactant is more stable than product and reaction is non-spontaneous

7 Which of the following compounds exchanges the largest number of hydrogens from deuterium when treated with KOD in D_2O?

a. 3-methyl- 1,2-cycloheptanedione
b. 2-methyl-1,3-cycloheptanedione
c. 5-methyl-1,3-cycloheptanedione
d. 6-methyl-1,4-cycloheptanedione

8 Treatment of cyclohexanone with an excess of $H_2^{18}O$ produced ^{18}O labelled cyclohexanone. Which of the following is a likely intermediate in this isotope exchange? (the isotope is not named)

a. 1-cyclohexen-1-ol
b. 1,1-cyclohexanediol
c. 2-cyclohexen-1-one
d. 1, 2-cyclohexanediol

One or More than One Options Correct Type

***Direction** (Q. Nos. 9-14) This section contains 6 multiple choice questions. Each question has four choices (a), (b), (c) and (d), out of which ONE or MORE THAN ONE are correct.*

9 Which is/are the expected products in the following reaction?

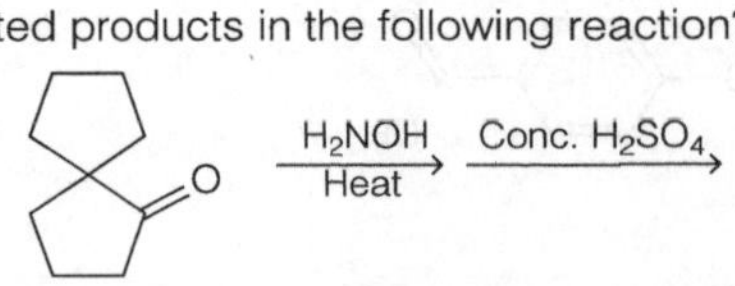

a. 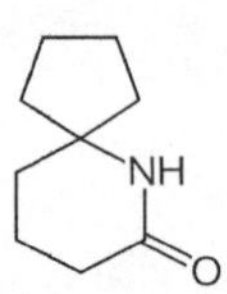
b. 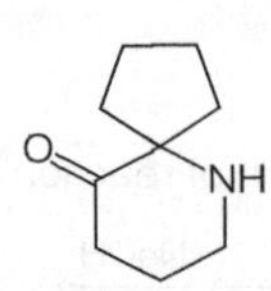

c. 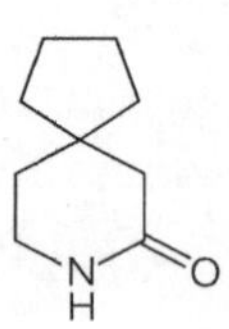
d. 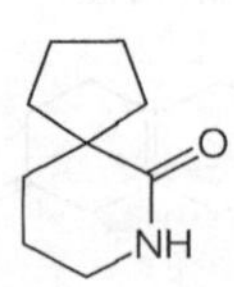

10 Treatment of A ($C_5H_{12}O$) with concentrated sulphuric acid results in the formation of three alkenes in differing yields. Also, A forms a yellow precipitate on treatment with $NaOH/I_2$. A turns colour of acidified dichromate solution to blue green, forming a new organic compound B which also forms yellow precipitate on treatment with $NaOH/I_2$. The most likely name of A is

a. 3-methyl-2-butanol
b. 3-methyl butanol
c. 2-pentanol
d. Either (b) or (c)

Working Space

11 The carbonyl compound(s) that will undergo racemisation on treatment with aqueous KOH is(are)

a. 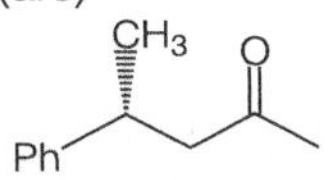

b. 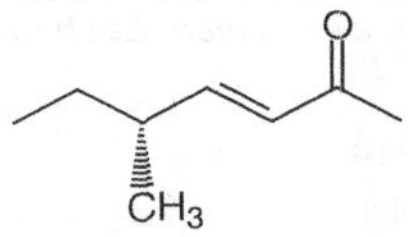

c. 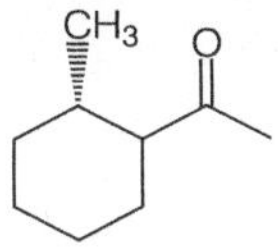

d. 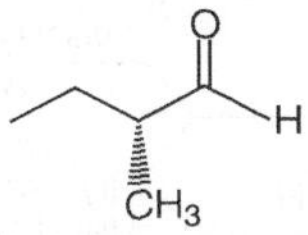

12 Which of the following reagents can bring about the transformation shown below?

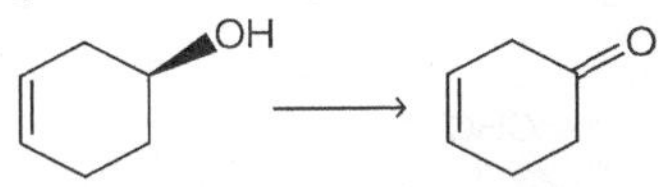

a. CrO_3/HCl/Pyridine
b. CrO_3/H_2SO_4/Acetone
c. $[(CH_3)_3CO]_3Al$
d. $KMnO_4/H_2SO_4$

13 Which of the following reactions has/have equilibrium constant (K_C) greater than one?

a. $CH_3CHO + KOH \rightleftharpoons CH_2{=}CH{-}OK + H_2O$

b. 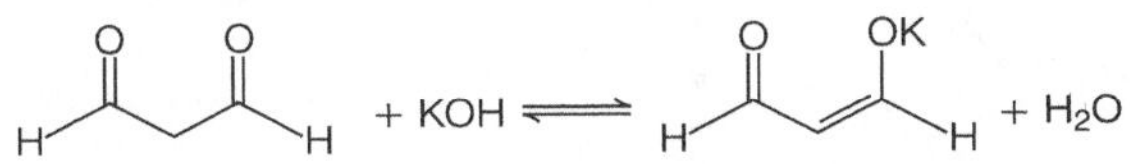

c. 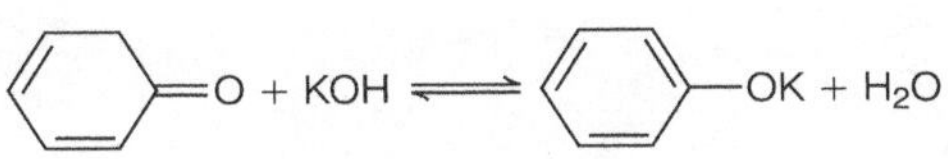

d. 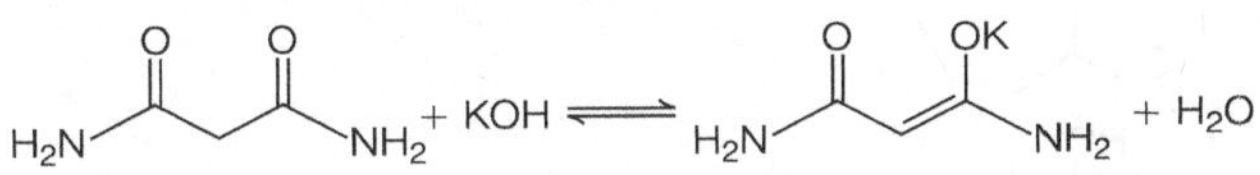

14 Which of the following, on treatment with $NaHSO_3(aq)$, will produce mixture of salts which can be separated by fractional crystallisation?

a. 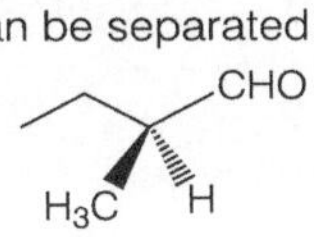

b. CHO–CHO

c. 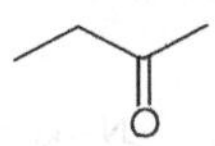

d. 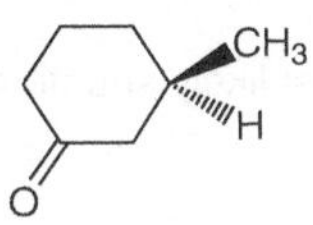

Statement Type

Direction *(Q. Nos. 15 and 16)* ***This section is based on Statement I and Statement II. Select the correct answer from the codes given below.***

a. Both Statement I and Statement II are correct and Statement II is the correct explanation of Statement I
b. Both Statement I and Statement II are correct but Statement II is not the correct explanation of Statement I
c. Statement I is correct but Statement II is incorrect
d. Statement II is correct but Statement I is incorrect

15 **Statement I** When benzaldehyde (C_6H_5CHO) is treated with concentrated NaOD solution in D_2O, Cannizaro reaction occur but no C—D (bonds with deuterium) is formed.
Statement II Cannizaro reaction involves hydride transfer mechanism.

16 **Statement I** Consider the reaction given below,

$$CH_3{-}\overset{O}{\overset{\|}{C}}{-}CH_3 + H_2N{-}\overset{O}{\overset{\|}{C}}{-}NH\cdot HNCH_3 \longrightarrow CH_3{-}\underset{CH_3}{\underset{|}{C}}{=}N{-}\overset{O}{\overset{\|}{C}}{-}NH\cdot HNCH_3$$

Statement II Aldehydes and ketones react with semicarbazide to form semicarbazone.

Comprehension Type

Direction (Q. Nos. 17-22) *This section contains 2 paragraphs, each describing theory, experiments, data, etc. Six questions related to the paragraphs have been given. Each question has only one correct answer among the four given options (a), (b), (c) and (d).*

Passage I

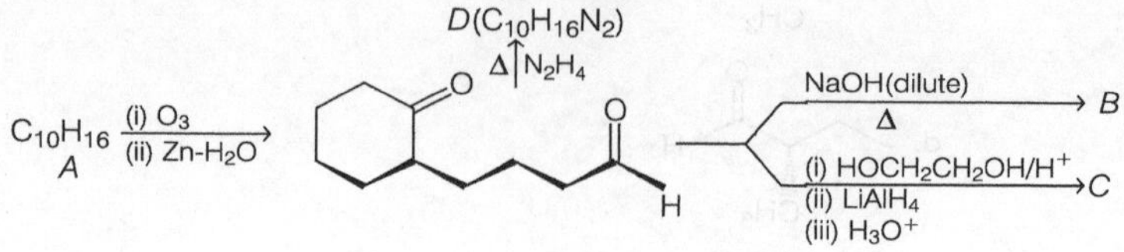

17 Which of the following cannot be *B*?

a. 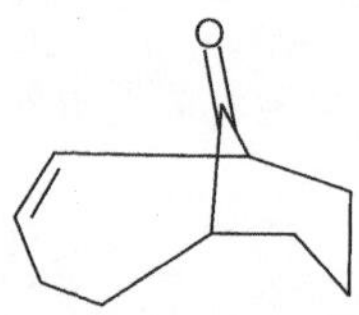

b. 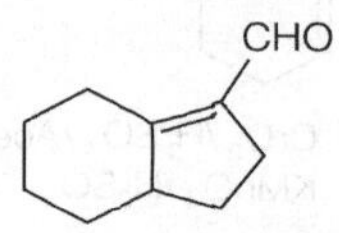

c. 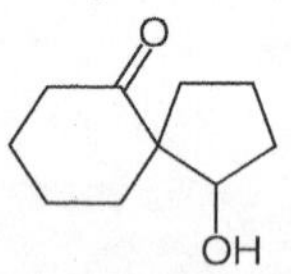

d. 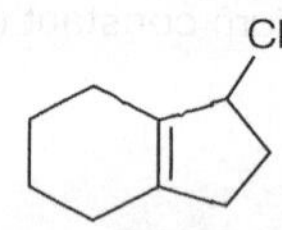

18 Which of the following is most likely structure of *C*?

a. 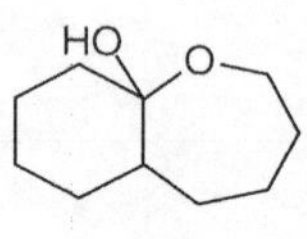

b. 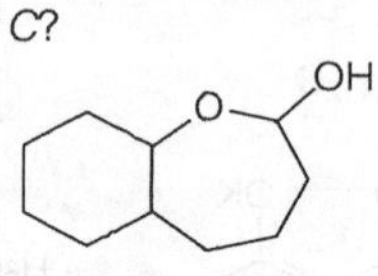

c. 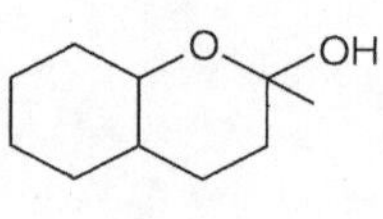

d. 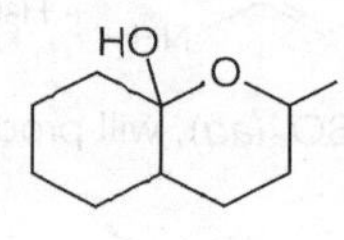

19 Which of the following most likely structure of *D*?

a. 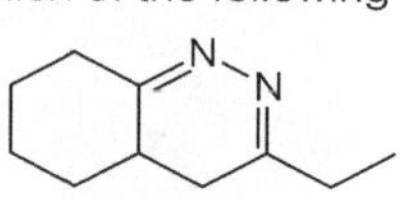

b. 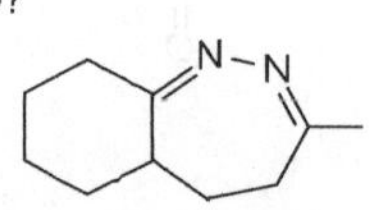

c. 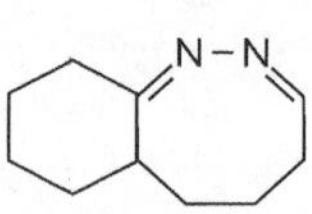

d. 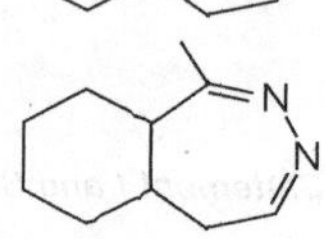

Passage II

A chiral organic compound *A* ($C_9H_{12}O_2$) does not give positive Tollen's test or iodoform test. However, *A* on treatment with 1.0 equivalent of CH_3MgBr followed by acid hydrolysis gives *B* ($C_{10}H_{16}O_2$) which gives positive iodoform test. *B* on refluxing with aqueous Na_2CO_3 solution gives the following compound as major product.

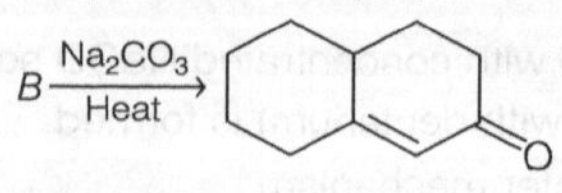

20 What is the most probable structure of *A*?

a. 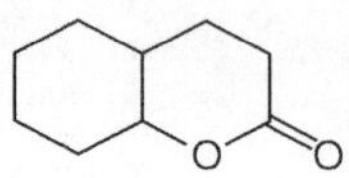

b. 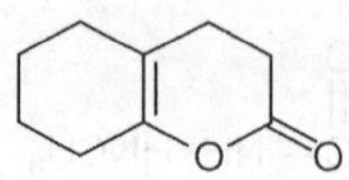

c. 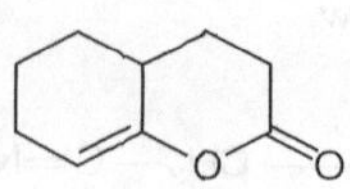

d. 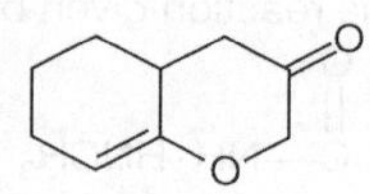

Working Space

Working Space

21 *B* on treatment with hydrazine (N_2H_4) gives a cyclic hydrazone. What is the structure of hydrazone?

a. 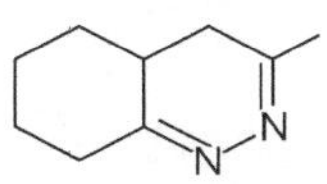

b. 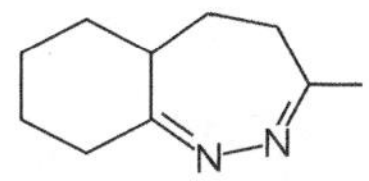

c. 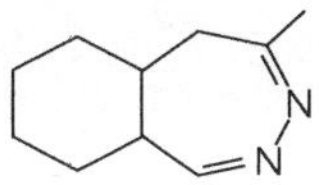

d. 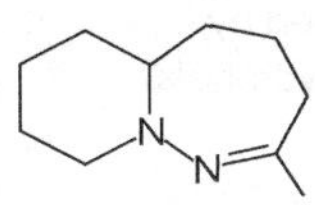

22 If compound *B* is treated with excess of CH_3MgBr followed by acid hydrolysis, how many different alcohols would be obtained?

a. 2　　b. 3　　c. 4　　d. 8

Matching List Type

Direction (Q. Nos. 23-25) ***Choices for the correct combination of elements from Column I and Column II are given as options (a), (b), (c) and (d), out of which one is correct.***

23 Match the Column I with Column II and mark the correct option from the codes given below.

	Compounds		Reagents		Products
i.	CH_2O	p.	H_2NOH/Conc. H_2SO_4	A.	A single amide
ii.	CH_3CHO			B.	Mixture of amide
iii.	$(CH_3)_2CHCHO$	q.	$(CH_3)_2NH$	C.	A single enamine
iv.	$(C_2H_5)_2CO$			D.	Isomeric enamine

Codes

	i	ii	iii	iv
a.	p,A	p,B	q,D	q,C
b.	q,C	p,B	q,C	p,A
c.	p,A	p,B	q,C	q,D
d.	q,C	p,B	q,D	p,A

24 Match the Column I with Column II and mark the correct option from the codes given below.

	Column I		Column II
i.	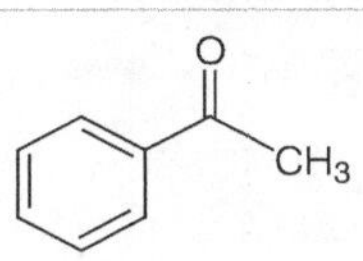	p.	Gives just one aldol only.
ii.	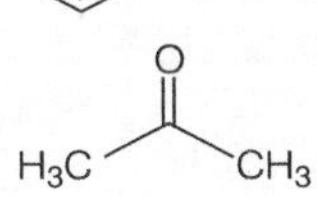	q.	Gives yellow precipitate with I_2/NaOH.
iii.	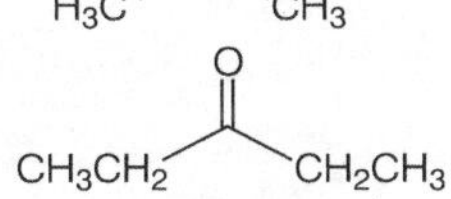	r.	Produces isomeric oximes with $HONH_2$
		s.	Gains more than 4 u in molar mass on treatment with NaOD/D_2O.
iv.	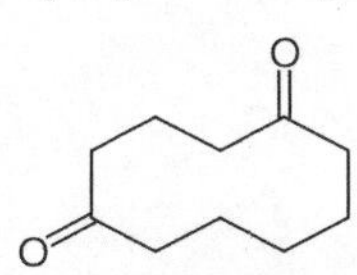	t.	Gives more than one aldol.

Codes

	i	ii	iii	iv		i	ii	iii	iv
a.	p	q	r	s,t	b.	q,r,t	p,q,s	t	r,s,t
c.	t,r,q	p,q,s	p	q	d.	q,s,t	p,r	r,s,q	s,r,t

25 Match the following reactions of Column I with products from the Column II.

Working Space

	Column I		Column II
i.	$D_2CO + NaOH\ (conc.) \longrightarrow$	p.	CD_3OH
ii.	$D_2CO + LiAlD_4 / H_2O \longrightarrow$	q.	$DCOOH$
iii.	$D_2CO + H_2CO + NaOH\ (conc.) \longrightarrow$ Followed by acid hydrolysis	r.	DCH_2OH
iv.	$D{-}\overset{O}{\overset{\Vert}{C}}{-}OC_2H_5 \xrightarrow{Li\ [\{(CH_3)_3CO\}_3\ AlD]} \xrightarrow{NaBH_4\ /H_2O}$	s.	D_2CHOH

Codes

	i	ii	iii	iv
a.	p,s	q	r,s	q,s
b.	p,q	r	s	q,s
c.	p,r	s	q,s	q,r,s
d.	p,q	p	q,r,s	s

One Integer Value Correct Type

Direction *(Q. Nos. 26-30) This section contains 5 questions. When worked out will result in an integer from 0 to 9 (both inclusive).*

26 If propanal (CH_3CH_2CHO) is treated with DCl in D_2O, keto-enol equilibria are established. At equilibrium, how many different enols would be existing?

27 Butanone + Dil. NaOH $\longrightarrow$ Aldol ($C_8H_{16}O_2$)

How many different aldol would be formed?

28 $H_2C{=}O + D_2C{=}O$ + Conc. NaOH $\longrightarrow$ Cannizaro reaction

How many different alcohols would be formed in the above reaction?

29 In the reaction given below, how many different oximes would be formed?

$$CH_3CH{=}CH{-}\overset{O}{\overset{\Vert}{C}}{-}CH{=}CHCH_3 + H_2NOH \longrightarrow \text{Oximes}$$

30 Consider the following reactions,

$$C_6H_{10}O \xrightarrow{H_2NOH} \xrightarrow{Conc.\ H_2SO_4} \text{Amides} \xrightarrow[(ii)\ H_3O^+]{(i)\ NaBH_4} \xrightarrow[Heat]{Conc.\ H_2SO_4} \xrightarrow{O_3/Zn\text{-}H_2O} \text{5-oxohexanal.}$$

How many different amides are formed in the above reactions?

JEE Main & AIEEE Archive

(Compilation of Last 13 Years' Questions)

1 In the given transformation, which of the following is the most appropriate reagent?

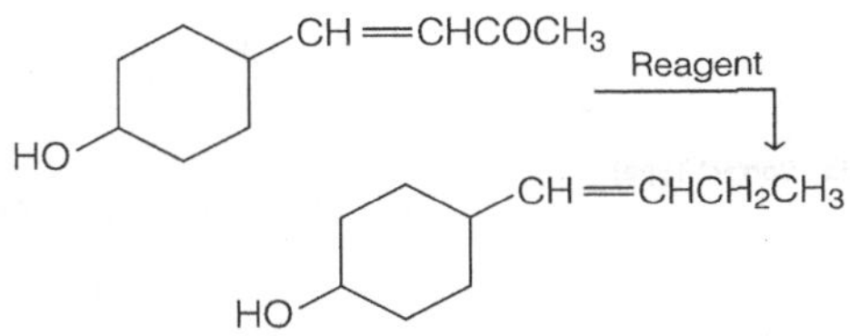

(AIEEE 2012)

a. $NH_2NH_2, \overset{\ominus}{O}H$ b. Zn-Hg /HCl

c. Na, Liq. NH_3 d. $NaBH_4$

2 Trichloroacetaldehyde was subjected to Cannizaro reaction by using NaOH. The mixture of the products contain sodium trichloroacetate ion and another compound. The other compound is **(AIEEE 2011)**

a. 2, 2, 2-trichloroethanol

b. trichloromethanol

c. 2, 2, 2-trichloropropanol

d. chloroform

3 Silver mirror test is given by which one of the following compounds? **(AIEEE 2011)**

a. Acetaldehyde b. Acetone

c. Formaldehyde d. Benzophenone

4 Which of the following on heating with aqueous KOH, produces acetaldehyde? **(AIEEE 2009)**

a. CH_3COCl b. CH_3CH_2Cl

c. CH_2ClCH_2Cl d. CH_3CHCl_2

5 The increasing order of the rate of HCN addition to compounds A-D is **(AIEEE 2006)**

A. HCHO B. CH_3COCH_3

C. $PhCOCH_3$ D. PhCOPh

a. A < B < C < D b. D < B < C < A

c. D < C < B < A d. C < D < B < A

6 The best reagent to convert pent-3-en-2-ol into pent-3-en-2-one is **(AIEEE 2005)**

a. pyridinium chlorochromate

b. chromic anhydride in glacial acetic acid

c. acidic dichromate

d. acidic permanganate

7 On vigorous oxidation by permanganate solution, $(CH_3)_2C═CHCH_2CHO$ gives **(AIEEE 2002)**

a. $(CH_3)_2CO$ and $OHCCH_2CHO$

b. $(CH_3)_2\underset{OH}{\underset{|}{C}}—\underset{OH}{\underset{|}{C}}HCH_2CHO$

c. $(CH_3)_2CO$ and $OHCCH_2COOH$

d. $(CH_3)_2CO$ and $CH_2(COOH)_2$

JEE Advanced & IIT JEE Archive

(Compilation of Last 10 Years' Questions)

1 The major product of the following reaction is

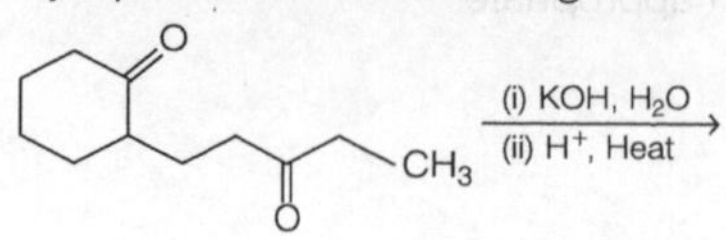

(2014 Adv., Only One Option Correct Type)

a. 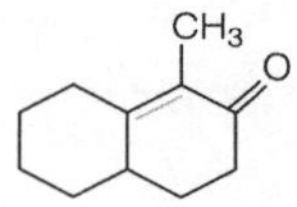

b. 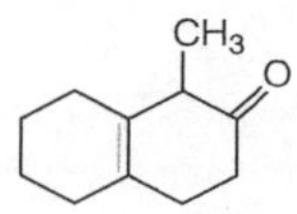

c. 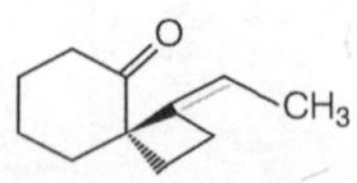

d. 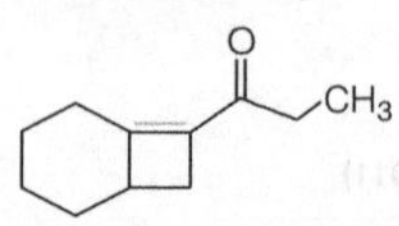

2 The major product in the following reaction is

(2014 Adv., Only One Option Correct Type)

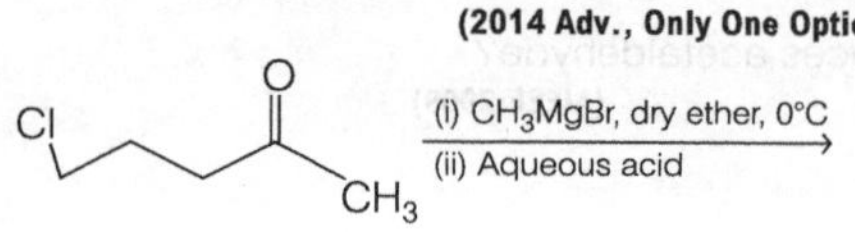

a. 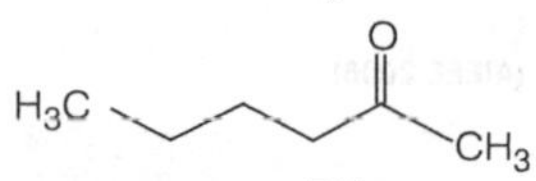

b. 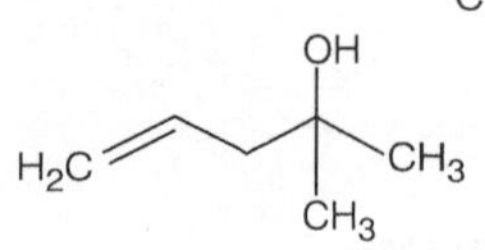

c. 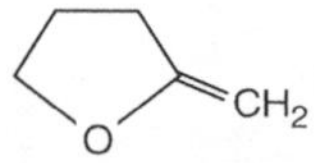

d. 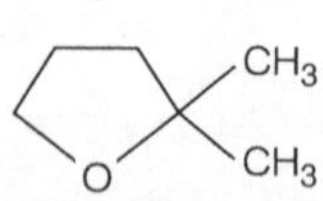

3 Consider all possible isomeric ketones including stereoisomers of MW = 100. All these isomers are independently reacted with $NaBH_4$. The total number of ketones that give a racemic product(s) is/are

Note *Stereoisomers are also reacted separately.*

(2014 Adv., Integer Type)

4 The major product H in the given reaction sequence is **(2012, Only One Option Correct Type)**

$$CH_3-CH_2-CO-CH_3 \xrightarrow{^-CN} G \xrightarrow{95\%\ H_2SO_4} H$$

a. $CH_3-CH=\underset{\underset{CH_3}{|}}{C}-COOH$

b. $CH_3-CH=\underset{\underset{CH_3}{|}}{C}-CN$

c. $CH_3-CH_2-\overset{\overset{OH}{|}}{\underset{\underset{CH_3}{|}}{C}}-COOH$

d. $CH_3-CH=\underset{\underset{CH_3}{|}}{C}-CO-NH_2$

5 The number of aldol reaction(s) that occurs in the given transformation is

(2012, Only One Option Correct Type)

$CH_3CHO + 4HCHO \xrightarrow{\text{Conc. } aq.\ NaOH}$ 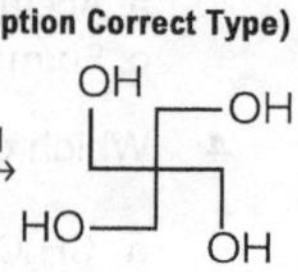

a. 1 b. 2

c. 3 d. 4

6 In the scheme given below, the total number of intramolecular aldol condensation products formed from 'Y' is

(2010, Integer Type)

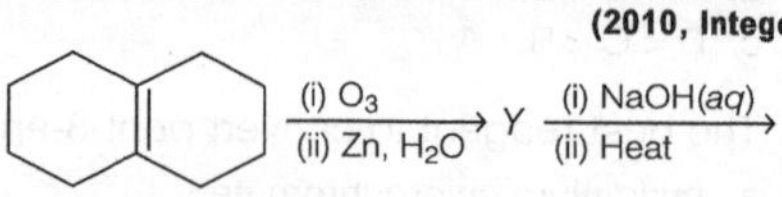

Passage for Q. Nos. (7-9)

Two aliphatic aldehydes P and Q react in the presence of aqueous K_2CO_3 to give compound R, which upon treatment with HCN provides compound S. On acidification and heating, S gives the product shown below :

(2010, Comprehension Type)

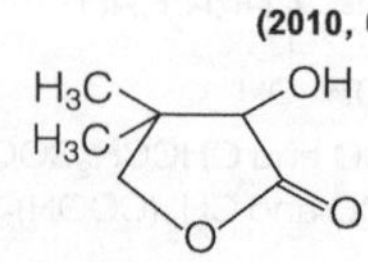

JEE Archive

7 The compounds *P* and *Q* respectively are

a. 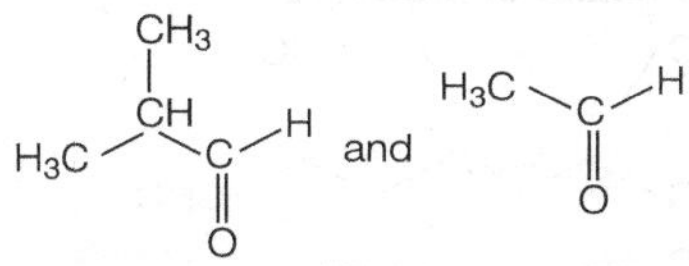

b. 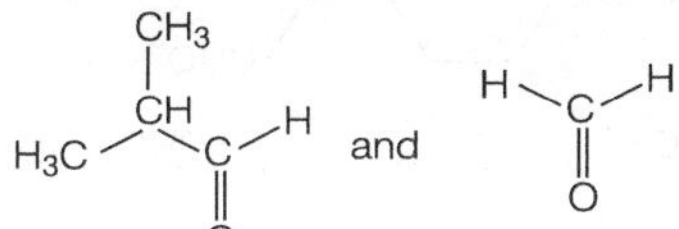

c. 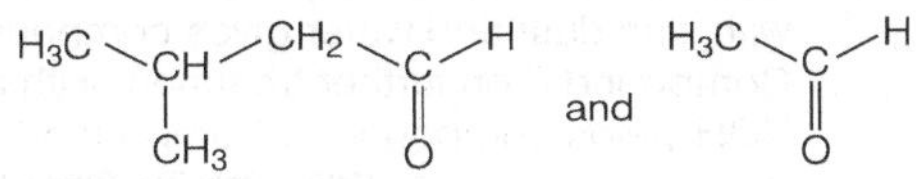

d. 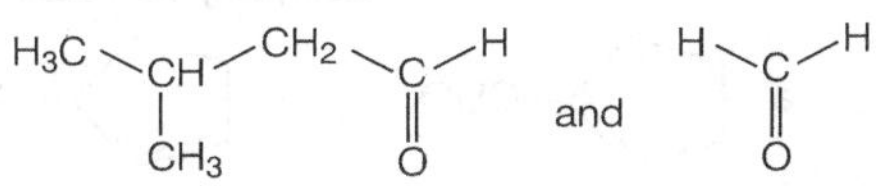

8 The compound *R* is

a. 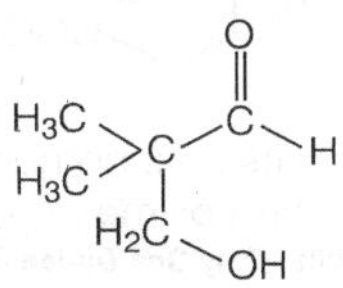
b. 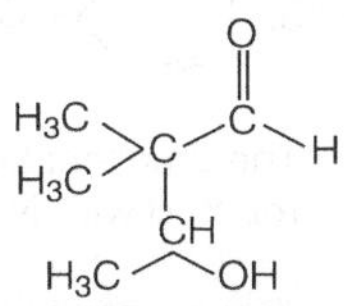

c. 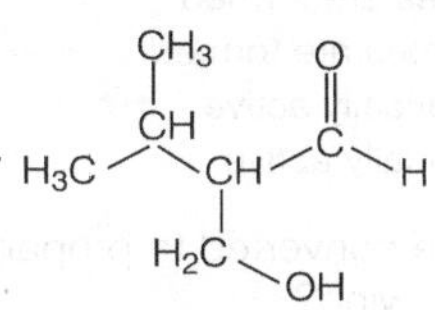
d. 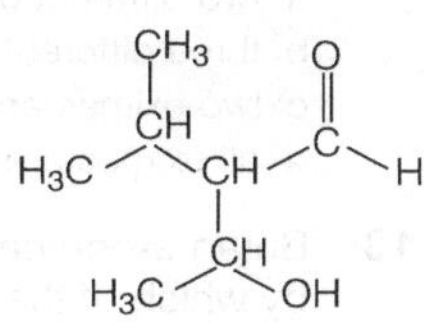

9 The compound *S* is

a. 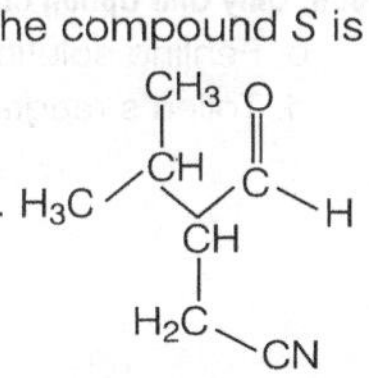
b. 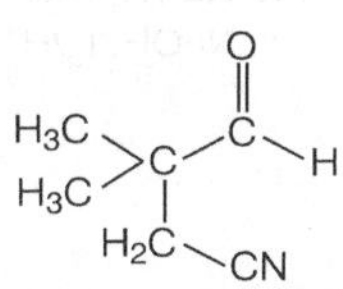

c. 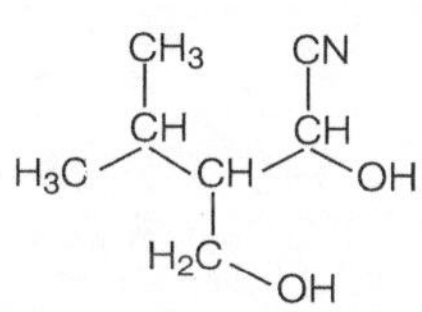
d. 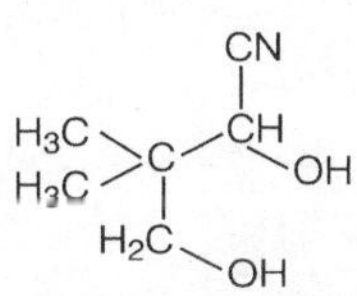

Passage for Q. Nos. (10-12)

A carbonyl compound *P*, which gives positive iodoform test, undergoes reaction with MeMgBr followed by dehydration to give an olefin *Q*. Ozonolysis of *Q* leads to a dicarbonyl compound *R*, which undergoes intramolecular aldo reaction to give predominantly *S*.

(2009, Comprehension Type)

$$P \xrightarrow[\text{2. } H^+,\ H_2O \quad \text{3. } H_2SO_4\ /\ \text{Heat}]{\text{1. MeMgBr}} Q \xrightarrow{O_3/Zn\text{-}H_2O} R \xrightarrow[\text{Heat}]{OH^-} S$$

10 The structure of the carbonyl compound *P*, is

a. 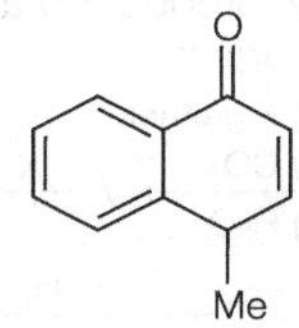
b. 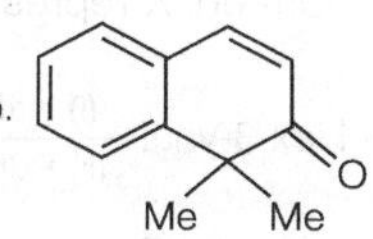

c. 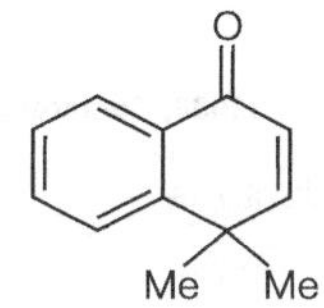
d. 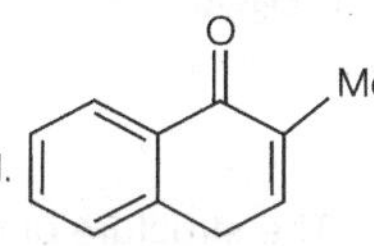

11 The structure of the product *S*, is

a. 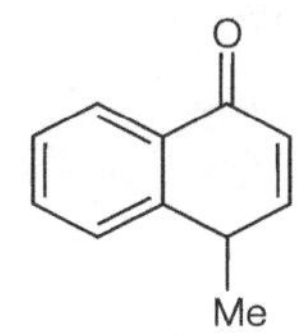
b. 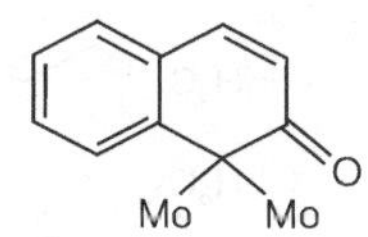

c. 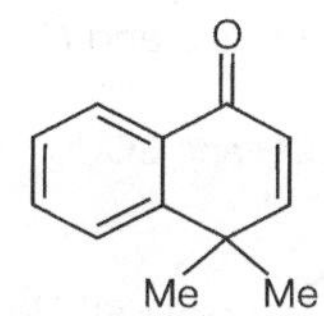
d. 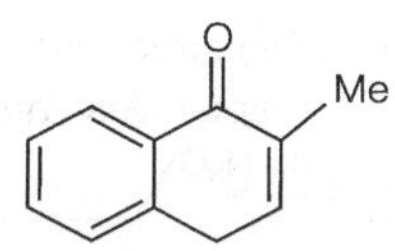

12 The structures of the products *Q* and *R*, respectively, are

a. 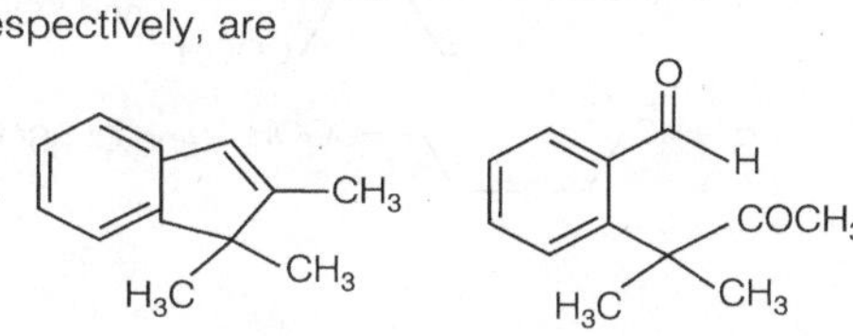

b. 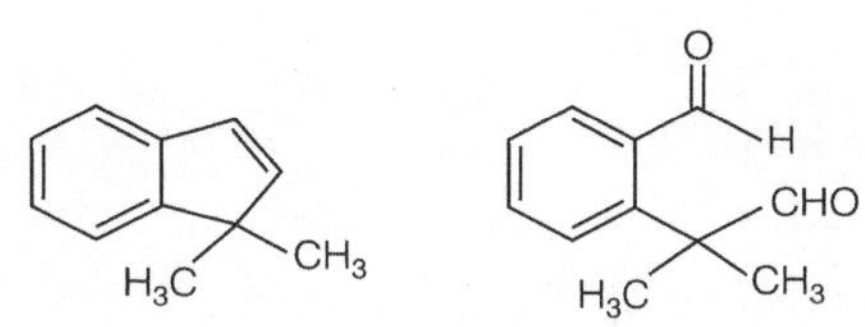

c. 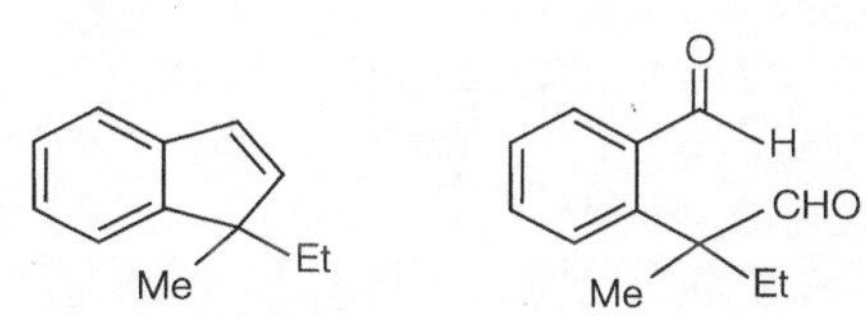

d. 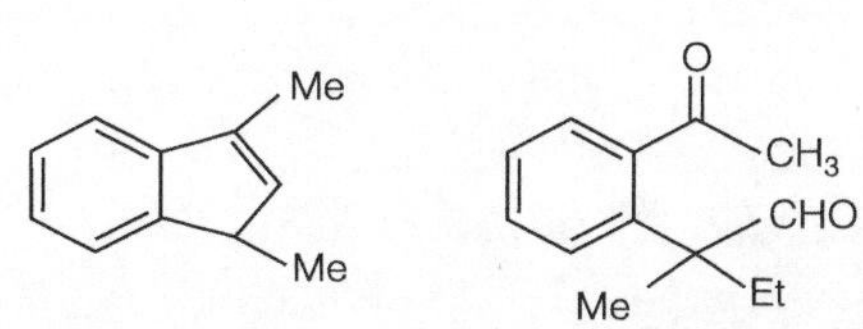

JEE Archive

Passage for Q. Nos. (13-15)

In the following sequence, product *I*, *J* and *L* are formed. *K* represents a reagent.

$$\text{Hex-3-ynal} \xrightarrow[\text{(ii) PBr}_3]{\text{(i) NaBH}_4} I \xrightarrow[\text{(iii) H}_3\text{O}^+]{\text{(i) Mg/ether, (ii) CO}_2} J \xrightarrow{K}$$

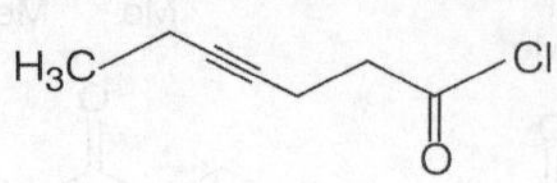 $\xrightarrow[\text{Pd/BaSO}_4\text{, Quinoline}]{H_2}$ *L*

(2008, Comprehension Type)

13 The structure of the product *I* is

a. 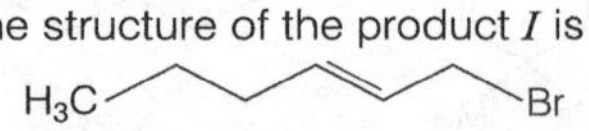

b. 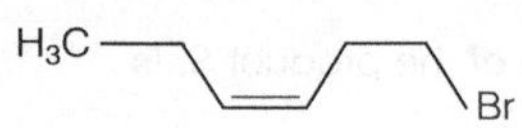

c. 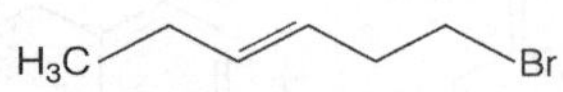

d. 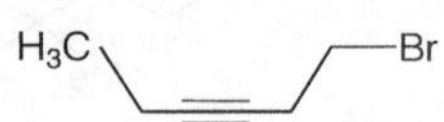

14 The structures of compounds *J* and *K*, respectively, are

a. 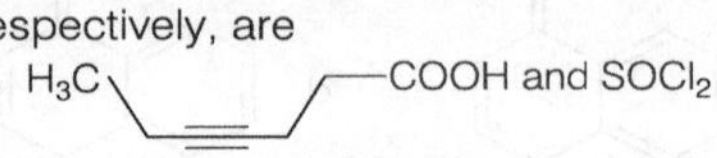 and $SOCl_2$

b. 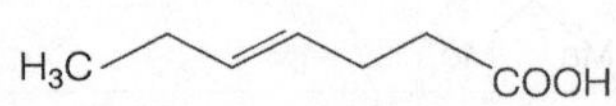 and SO_2Cl_2

c. 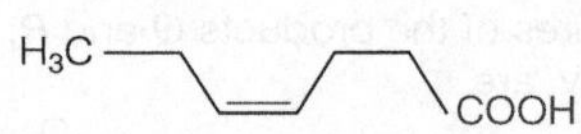 and $SOCl_2$

d.  and CH_3SO_2Cl

15 The structure of product *L* is

a. 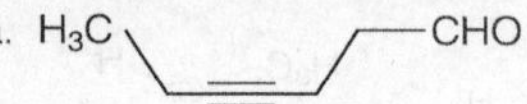

b. 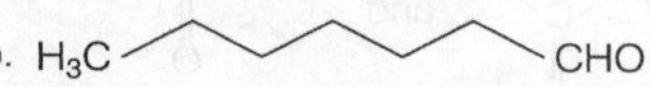

c. 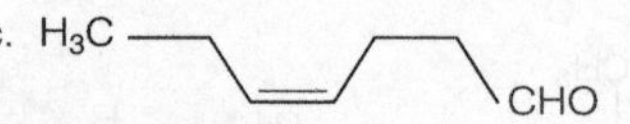

d. 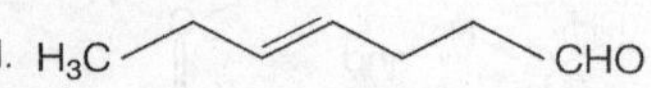

16 Cyclohexene on ozonolysis followed by reaction with zinc dust and water gives compound *E*. Compound *E* on further treatment with aqueous KOH yields compound *F*. Compound *F* is

(2007, Only One Option Correct Type)

a. 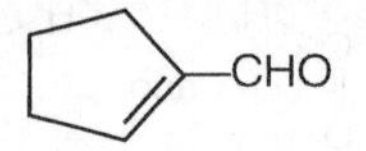 b. 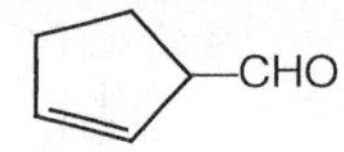

c. 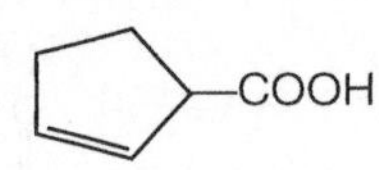 d. 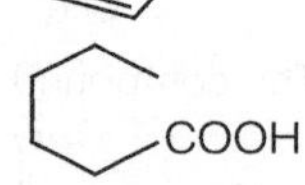

17 The smallest ketone and its next homologue are reacted with NH_2OH to form oxime

(2006, Only One Option Correct Type)

a. two different oximes are formed

b. three different oximes are formed

c. two oximes are optically active

d. all oximes are optically active

18 Butan-2-one can be converted to propanoic acid by which of the following?

(2006, Only One Option Correct Type)

a. NaOH, NaI/H^+ b. Fehling solution

c. NaOH, I_2/H^+ d. Tollen's reagent

4. Carboxylic Acids and Acid Derivatives

DPP-1 Preparation and General Properties

Subjective Questions

Working Space

Direction (Q. Nos. 1-5) *These questions are subjective in nature, need to be solved completely on notebook.*

1 Starting from butanedioic acid, synthesise the following.

i. *Meso*-tartaric acid

ii. Racemic tartaric acid

2 Starting from ethanoic acid, synthesise

i. propanoic acid

ii. butanoic acid

iii. propandioic acid

iv. butandioic acid

3 Convert the following.

i. Butanoic to propanoic acid

ii. C_6H_6 to *meta*-phthalic acid

4 How a mixture containing butanol, phenol and propanoic acid can be separated efficiently?

5 Write the detailed mechanism of both acid and base catalysed hydrolysis of propanenitrile. How the hydrolysis can be controlled in both cases to obtain intermediate propanamide as end product?

Only One Option Correct Type

Direction (Q. Nos. 6-13) *This section contains 8 multiple choice questions. Each question has four choices (a), (b), (c) and (d), out of which ONLY ONE is correct.*

6 Which is the most suitable reagent for the reaction?

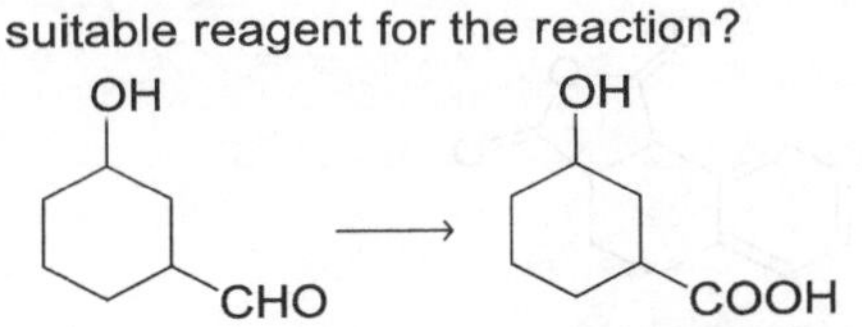

a. $KMnO_4$-H_2SO_4

b. $KMnO_4$-NaOH

c. $AgNO_3$-$NH_3(aq)$

d. CrO_3-H_2SO_4

7 What is the correct about the following reaction?

$$CH_3—CH_2—\underset{\displaystyle Br}{\underset{|}{CH}}—CH_3 \xrightarrow[(C_2H_5)_2O]{Mg} \xrightarrow[(ii)\ H_3O^+]{(i)\ CO_2}$$

a. The major product is achiral 2-methyl butanoic acid

b. The major product is racemic mixture of 2-methyl butanoic acid

c. If a pure enantiomer of starting compound is taken, the major product would be a pure enantiomer of 2-methyl butanoic acid

d. The major product would be *trans*-2-butene

Working Space

8 Predict the major organic product in the following reaction,

$$Ba\overset{*}{C}O_3 \xrightarrow{HCl} X(gas) \xrightarrow[(C_2H_5)_2O]{CH_3CH_2CH_2MgBr} \xrightarrow{H_3O^{\oplus}}$$

a. 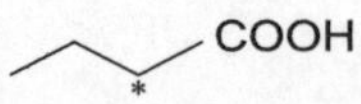
b. 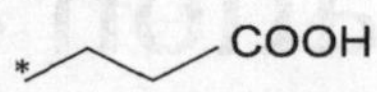
c. 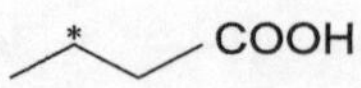
d. 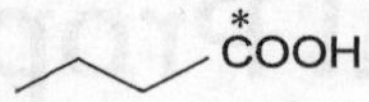

Note *Star represent carbon labelled C*[14]

9 Which reaction given below gives hexanedioic acid in good yield?

a.  $\xrightarrow[H_2SO_4/\Delta]{KMnO_4}$

b.  $\xrightarrow{O_3} \xrightarrow[H_2O]{Zn}$

c.  $\xrightarrow[\Delta]{Cl_2} \xrightarrow[C_2H_5OH]{C_2H_5ONa} \xrightarrow[H_2SO_4]{KMnO_4}$

d. $H_2N—CH_2CH_2CH_2CH_2CH_2CH_2—NH_2 \xrightarrow[H_2SO_4]{KMnO_4}$

10 Which reaction sequence given below gives 5-methyl hexanoic acid?

a. 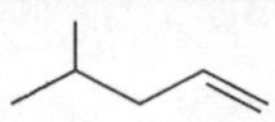 $\xrightarrow{HBr} \xrightarrow[Et_2O]{Mg} \xrightarrow[(ii)\ H_3O^{\oplus}]{(i)\ CO_2}$

b. 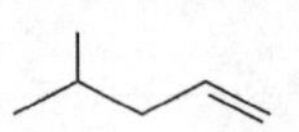 $\xrightarrow[H_2O_2]{HBr} \xrightarrow[Et_2O]{Mg} \xrightarrow[(ii)\ H_3O^{\oplus}]{(i)\ CO_2}$

c. 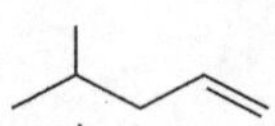 $\xrightarrow{HBr} \xrightarrow{NaCN} \xrightarrow{H_3O^{\oplus}}$

d. 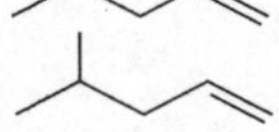 $\xrightarrow[H_2O/\Delta]{H_2SO_4} \xrightarrow[H_2SO_4]{CrO_3}$

11 What is the major organic product in the reaction given below?

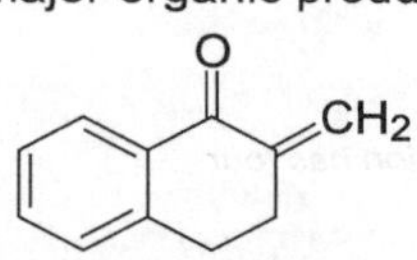 $\xrightarrow[(ii)\ H_2CrO_4]{(i)\ BH_3/H_2O_2/NaOH} \xrightarrow[Heat]{CH_2{=}PPh_3}$

a. 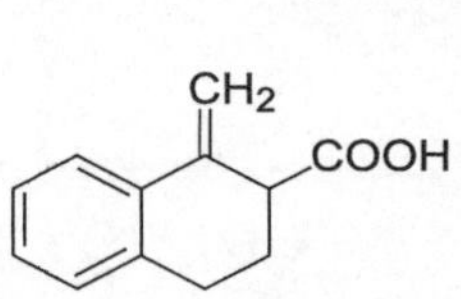
b. 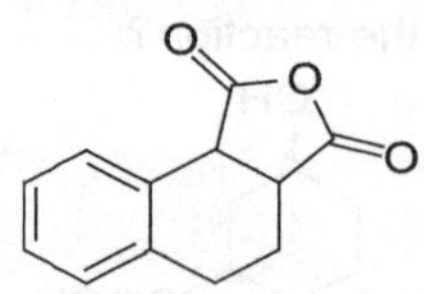
c. 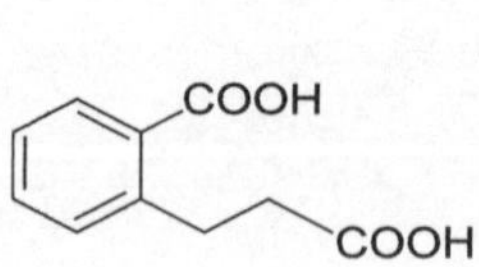
d. 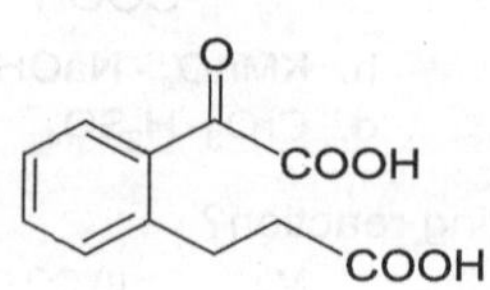

12 A hydrocarbon X has molar mass of 106. X on vigorous oxidation with hot, alkaline $KMnO_4$ followed by acid hydrolysis gives an organic compound whose neutralisation equivalent is 83. Also X on treatment with $Br_2/FeCl_3$ in principal produces three monobromination products. Hence, X could be

a. 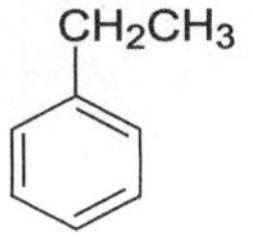

b. 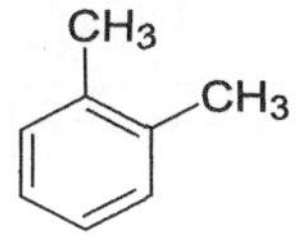

c. 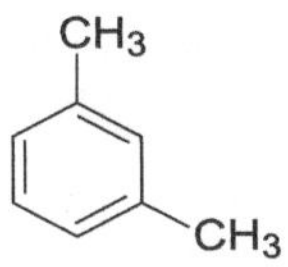

d. 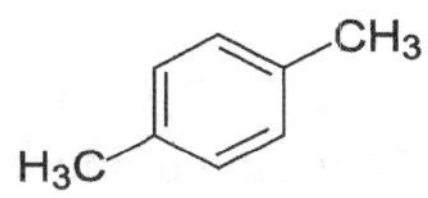

13 In the following synthesis, the major organic product formed is

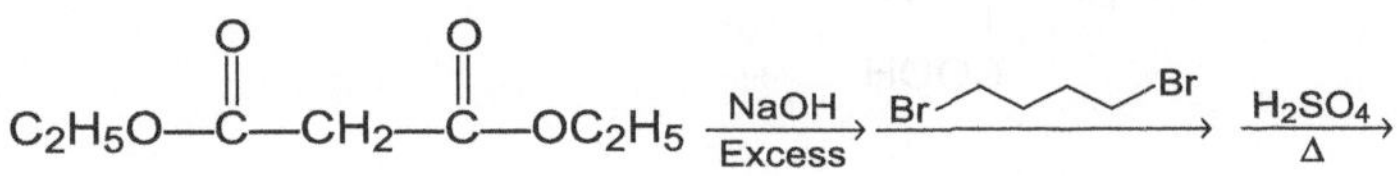

a. $HOOC—(CH_2)_4—COOH$

b. 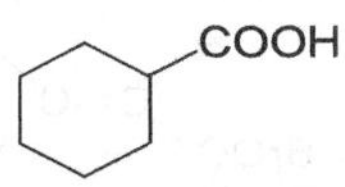

c. 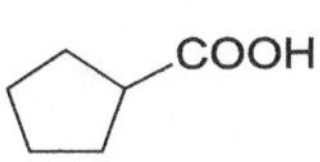

d. 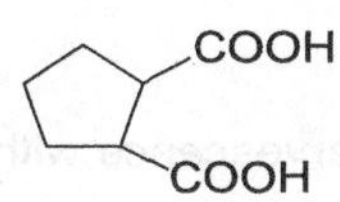

One or More than One Options Correct Type

Direction (Q. Nos. 14-19) *This section contains 6 multiple choice questions. Each question has four choices (a), (b), (c) and (d), out of which ONE or MORE THAN ONE are correct.*

14 Which is/are the correct set of reagents that can bring about the following transformation?

1-butene $\longrightarrow$ Pentanoic acid

a. $\xrightarrow[H_2O_2]{HBr} \xrightarrow{NaCN} \xrightarrow[H_2O]{H_2SO_4}$

b. $\xrightarrow[NaOH]{BH_3/H_2O_2} \xrightarrow[H_2SO_4]{CrO_3}$

c. $\xrightarrow[NaOH—H_2O]{BH_3/H_2O_2} \xrightarrow{SOCl_2} \xrightarrow{NaCN} \xrightarrow{H_2SO_4}$

d. $\xrightarrow[H_2O_2]{HBr} \xrightarrow[Et_2O]{Mg} \xrightarrow[H_2O]{CO_2}$

15 Which reagent(s) given below can be used to separate a mixture of butanol and butanoic acid from its ethereal solution?

a. $NaNH_2(aq)$

b. $NH_3(aq)$

c. NaOH

d. $NaHCO_3$

16 Which reaction(s) given below gives phthalic acid as the major organic product?

a. 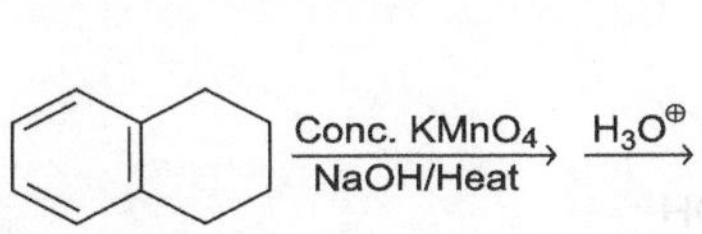

b. 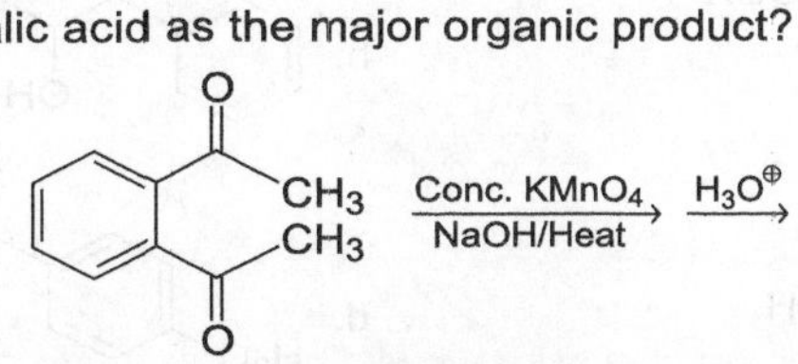

c. 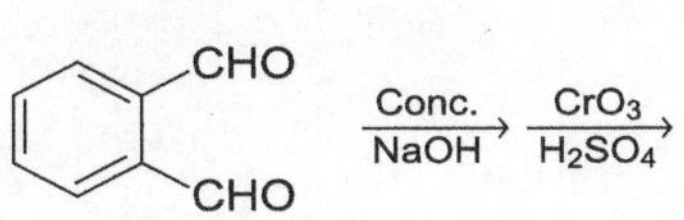

d. 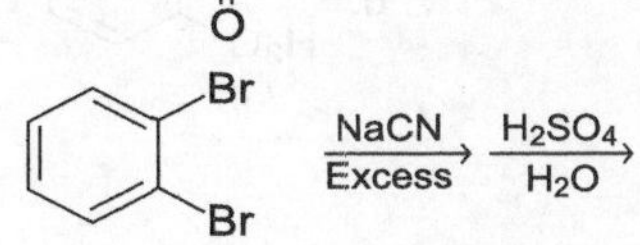

Working Space

Working Space

17 Which reaction sequence given below can convert pentanoic acid into butanoic acid?

a. $\xrightarrow[Br_2/CCl_4/\Delta]{Ag_2O} \xrightarrow[H_2O]{NaOH} \xrightarrow[H_2SO_4]{CrO_3}$

b. $\xrightarrow[(ii)\ H_2O]{(i)\ LiAlH_4} \xrightarrow[\Delta]{Al_2O_3} \xrightarrow[H_2SO_4/Heat]{KMnO_4}$

c. $\xrightarrow[(ii)\ H_2O]{(i)\ LiAlH_4} \xrightarrow{SOCl_2} \xrightarrow[C_2H_5OH]{C_2H_5ONa} \xrightarrow[H_2O_2]{O_3}$

d. $\xrightarrow[CH_3COOH]{Br_2} \xrightarrow{NaCN} \xrightarrow[H_2O/Heat]{H_2SO_4}$

18 If *cis*-2-butenedioic acid is treated with $Br_2(l)$, the product formed is/are

a. 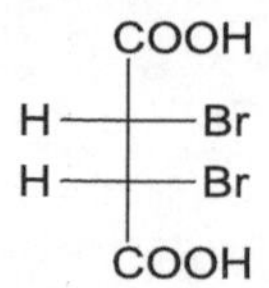

b. 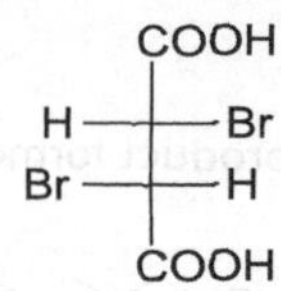

c. 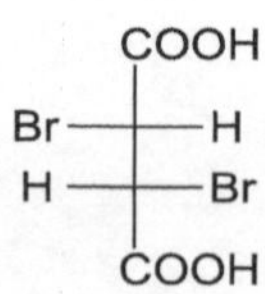

d. 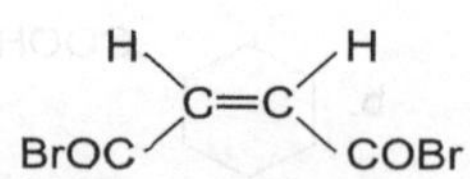

19 Which compound(s) given below gives effervescence with $NaHCO_3$?

a. C_6H_5OH

b. C_6H_5COOH

c. 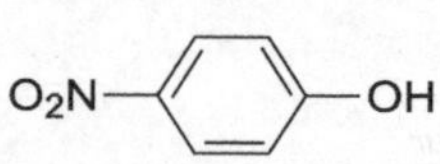

d. $C_6H_5SO_3H$

Comprehension Type

***Direction** (Q. Nos. 20-22) This section contains a paragraph, describing theory, experiments, data, etc. Three questions related to the paragraph have been given. Each question has only one correct answer among the four given options (a), (b), (c) and (d).*

Passage

An organic compound $A(C_9H_{10}O_3)$ is optically active. A changes orange colour of CrO_3-H_2SO_4 solution to blue-green. *A* on vigorous oxidation with hot, concentrated, alkaline $KMnO_4$ gives benzoic acid. Also *A* on treatment with HBr gives $B(C_9H_9O_2Br)$ with same configuration as that of *A*.

20 What is the structure of *A*?

a. 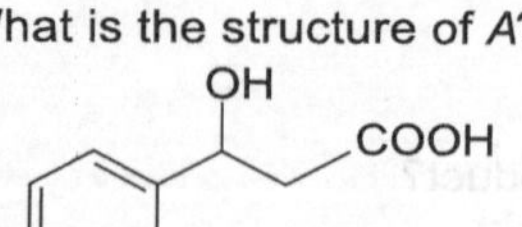

b. 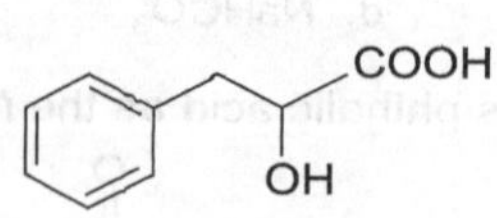

c. 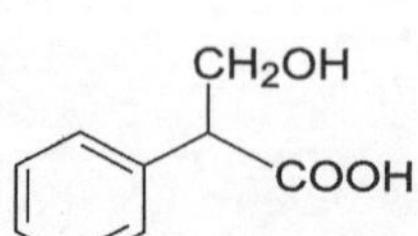

d. 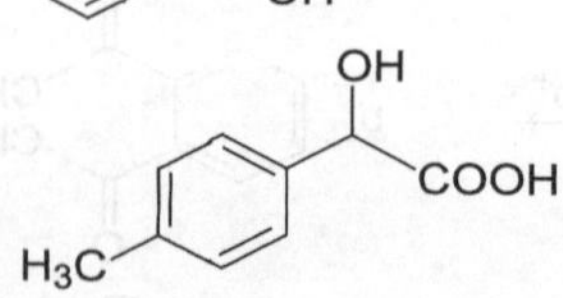

Working Space

21 Consider the following reaction,

$$B \xrightarrow[\text{(ii) } H_3O^+]{\text{(i) NaOH}} \xrightarrow[\text{Heat}]{H_3PO_4} \xrightarrow[H_2]{[(C_6H_5)_3P]_3RhCl} C$$

The structure of C is

a. 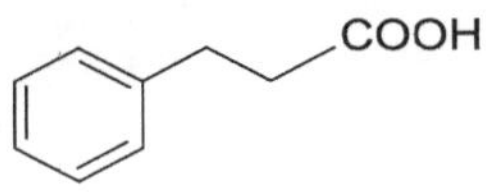

b. 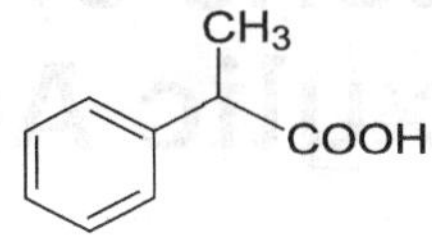

c. 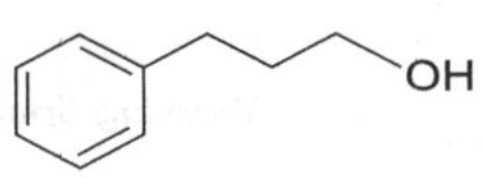

d. 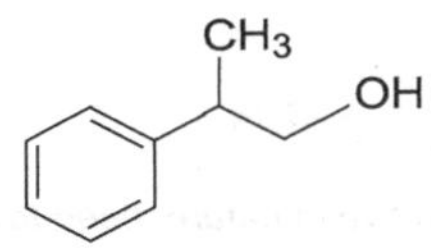

22 Which of the following sequence of reaction gives C as the major product?

a. $Ph—CH_2CH_2Cl \xrightarrow[Et_2O]{Mg} \xrightarrow[\text{(ii) } H_3O^+]{\text{(i) } CO_2}$

b. $Ph—\underset{\large Cl}{\underset{|}{C}H}—CH_3 \xrightarrow[\text{(ii) } CO_2]{\text{(i) Mg/}Et_2O} \xrightarrow{H_3O^+}$

c. $Ph—CH_2CH_2—CH{=}CH_2 \xrightarrow[H_2O_2]{O_3}$

d. $Ph—\underset{\large CH_3}{\underset{|}{C}H}—CH{=}CH_2 \xrightarrow[\text{NaOH/Heat}]{\text{Conc. } KMnO_4}$

One Integer Value Correct Type

Direction *(Q. Nos. 23-26) This section contains 4 questions. When worked out will result in an integer from 0 to 9 (both inclusive).*

23 Consider the following reaction,

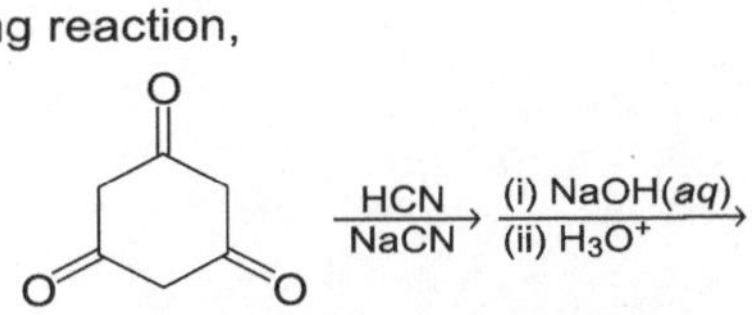

How many different tricarboxylic acids are formed?

24 How many different isomeric methyl cyclohexene, upon treatment with $KMnO_4/H_2SO_4$/Heat gives dicarboxylic acids?

25 A hydrocarbon X on treatment with hot, concentrated and alkaline $KMnO_4$, followed by acid work-up gives an acid whose neutralisation equivalent is 83, with same number of carbon atoms as in starting compound X. Also X has a phenyl ring. How many isomers of X exist, all containing a phenyl group?

26 An unsaturated, acyclic, monobasic carboxylic acid X has neutralisation equivalent of 156. X on treatment with $KMnO_4$/Heat/H_2SO_4 gives butanoic acid and a dibasic acid Y. How many carbon atoms in Y is /are sp^3-hybridised?

DPP-2 Reactions of Carboxylic Acids

Subjective Questions

Working Space

Direction (Q. Nos. 1-6) *These questions are subjective in nature, need to be solved completely on notebook.*

1 Complete the following reactions:

i. $C_6H_5—CH{=}CH—COOH \xrightarrow{SOCl_2} \xrightarrow{AlCl_3}$

ii. 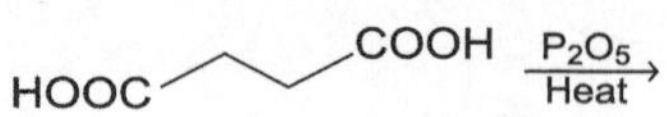

iii. $C_6H_5COOH + C_6H_5CH_2\overset{18}{O}H \xrightarrow{H_2SO_4}$

iv. 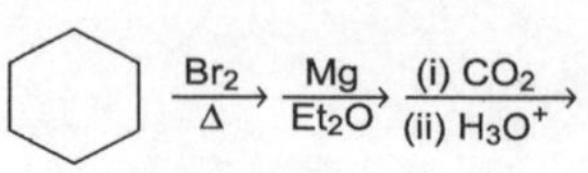

2 Prepare the following starting from benzene, toluene and alcohols of four carbons or fewer, using any necessary inorganic reagents.

i. Ethyl-α-methyl butyrate
ii. 3, 5-dinitro benzoic acid
iii. α-hydroxy propanoic acid
iv. 2-pentenoic acid
v. *p*-chloro benzoic acid

3 Arrange the monosodium salts of the following carboxylic acids in the increasing order of their basic strength.

i. Butanoic acid (I), 2-bromobutanoic acid (II), 3-bromobutanoic acid (III) 3-chlorobutanoic acid (IV).
ii. Benzoic acid (I), *p*-chlorobenzoic acid (II), 2,4-dichlorobenzoic acid (III) and 2,4,6-trichloro benzoic acid (IV).
iii. Benzoic acid (I), *p*-nitrobenzoic acid (II), *p*-toluic acid (III) and *m*-nitrobenzoic acid (IV).
iv. α-chlorophenyl acetic acid (I) *p*-chlorophenylacetic acid (II), phenyl acetic acid (III) and α-phenylpropanoic acid (IV).
v. *p*-nitrobenzoic acid (I), *p*-nitrophenylacetic acid (II) β-(*p*-nitrophenyl) propanoic acid (III) *p*-toluene sulphonic acid (IV).

4 Arrange the compounds of each set in the increasing order of reactivity in the indicated reaction.

i. Esterification by benzoic acid : 2-butanol (I), methanol (II) neopentyl alcohol (III), 1-propanol (IV).
ii. Esterification by ethanol : benzoic acid (I), 2, 6-dimethyl benzoic acid (II) *ortho* toluic acid (III), *p*-nitrobenzoic acid (IV).
iii. Reaction with CH_3NH_2 : acetic acid (I), 2-methyl propanoic acid (II), *p*-nitrobenzoic acid (III), *m*-nitrobenzoic acid (IV).

Working Space

5 Draw the stereochemical structures of major organic product formed in the following reactions.

i. Racemic β-bromobutyric acid + $Br_2/P \longrightarrow$

ii. Fumaric acid + $HCO_2OH \longrightarrow C_4H_6O_6$

iii. 1, 4-cyclohexadiene + $CHBr_3/t\text{-}BuOK \longrightarrow P(C_7H_8Br_2)$

$$P \xrightarrow[H_2SO_4]{KMnO_4} Q(C_7H_8Br_2O_4) \xrightarrow{NaBH_4} R(C_7H_{10}O_4)$$

6 Deduce the structures of *A* to *D* in the following synthetic scheme.

Acetylene + $CH_3MgBr \longrightarrow A + CH_4$

$$A + CO_2 \longrightarrow B \xrightarrow{H_3O^{\oplus}} C(C_3H_2O_2) \xrightarrow[H_2O]{H_2SO_4/HgSO_4} D(C_3H_4O_3)$$

$$D \xrightarrow{KMnO_4} CH_2(COOH)_2$$

Only One Option Correct Type

Direction (Q. Nos. 7-14) This section contains 8 multiple choice questions. Each question has four choices (a), (b), (c) and (d), out of which ONLY ONE is correct.

7 Which reaction sequence is best synthesis of adipoyl dichloride (hexanedioyl chloride) from cyclohexane?

a. $\xrightarrow[\Delta]{Cl_2} \xrightarrow[C_2H_5OH]{C_2H_5ONa} \xrightarrow[Zn\text{-}H_2O]{O_3} \xrightarrow[H_2O]{Cl_2}$

b. $\xrightarrow[\Delta]{Cl_2} \xrightarrow[C_2H_5OH]{C_2H_5ONa} \xrightarrow[H_2SO_4]{KMnO_4} \xrightarrow{SOCl_2}$

c. $\xrightarrow[\Delta]{Br_2} \xrightarrow[H_2O]{KOH} \xrightarrow[H_2SO_4]{K_2Cr_2O_7} \xrightarrow{PCl_5}$

d. $\xrightarrow[H_2SO_4]{KMnO_4} \xrightarrow{PCl_5}$

8 When heated carboxylic acid salts in which there is a good leaving group on β-carbon undergo, decarboxylation elimination to give an alkene. What would be formed as the major organic product in the reaction?

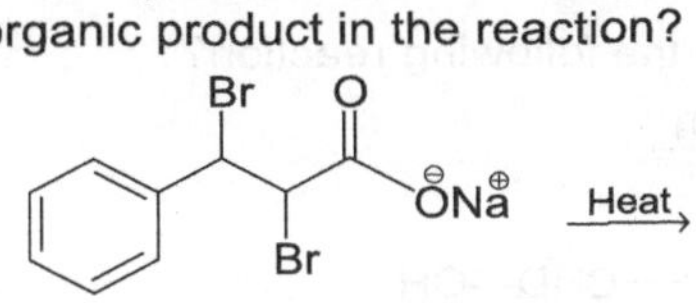

a. 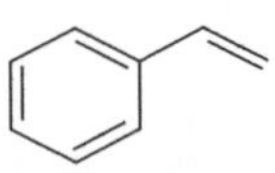

b. 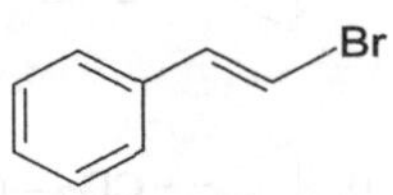

c. 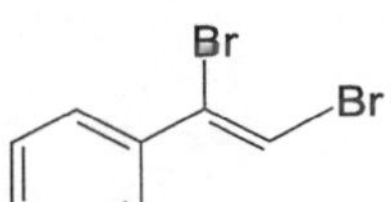

d. 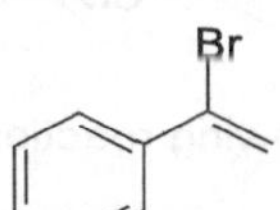

9 Which is the reaction sequence given below bring about best conversion of acetylene into 2-butyne-1, 4-diol?

a. $C_2H_2 \xrightarrow{Mg} \xrightarrow[(ii)\ H_3O^+]{(i)\ CO_2} \xrightarrow[(ii)\ H_3O^+]{(i)\ LiAlH_4}$

b. $C_2H_2 \xrightarrow[Excess]{CH_3MgBr} \xrightarrow[(ii)\ H_3O^+]{(i)\ CO_2} \xrightarrow[(ii)\ H_3O^+]{(i)\ LiAlH_4}$

c. $C_2H_2 + Br_2(aq) \longrightarrow \xrightarrow{NaCN} \xrightarrow[H_2O]{H_2SO_4} \xrightarrow[(ii)\ H_3O^+]{(i)\ LiAlH_4}$

d. $C_2H_2 \xrightarrow[Excess]{CH_3MgBr} \xrightarrow[(ii)\ H_3O^+]{(i)\ CO_2} \xrightarrow[(ii)\ H_3O^+]{(i)\ NaBH_4}$

Working Space

10 Ascorbic acid has acidic strength equivalent to that of any carboxylic acid. What is the correct structure of monosodium salt of ascorbic acid?

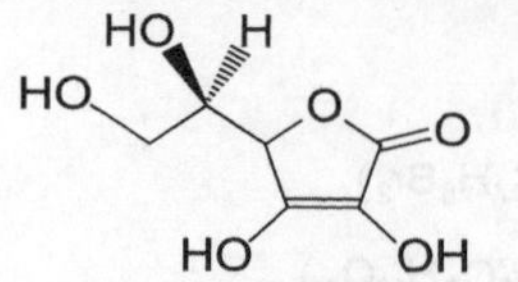

Ascorbic acid

a. 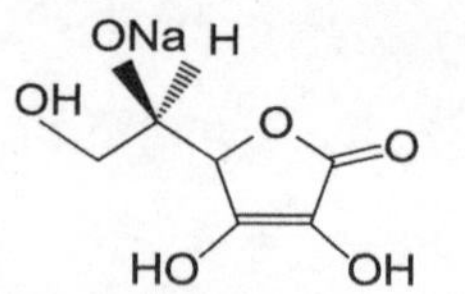

b. 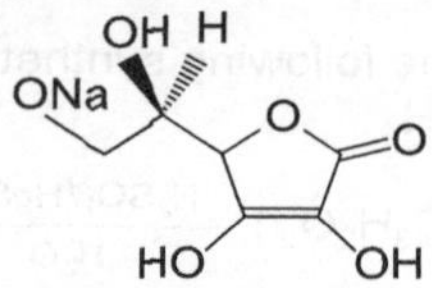

c. 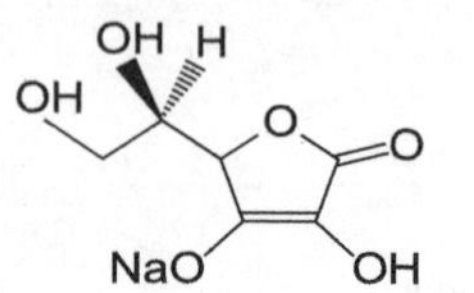

d. 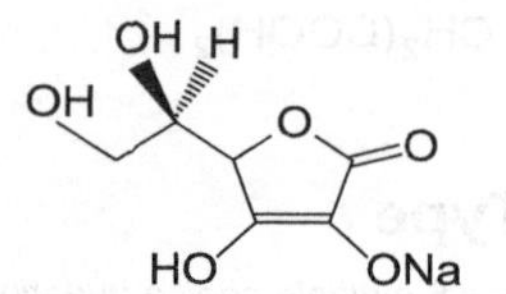

11 Suggest the best combination of reagent for the transformation.

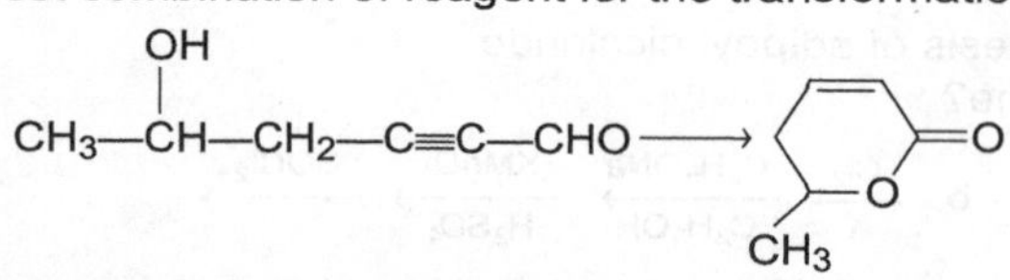

a. $Ag_2O/NH_3(aq)$ then $Pd/BaSO_4/H_2$/ Heat

b. CrO_3-H_2SO_4 then $Pd/BaSO_4/H_2$/Heat

c. $Pd/BaSO_4/H_2$/Heat then CrO_3-H_2SO_4

d. $Ag_2O/NH_3(aq)$ then $Na/NH_3(l)$

12 What is formed as the major organic product in the following reaction?

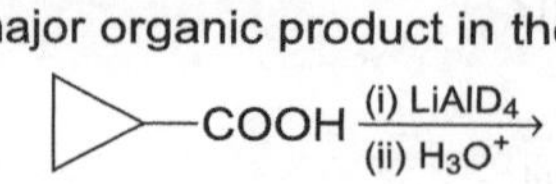

a. 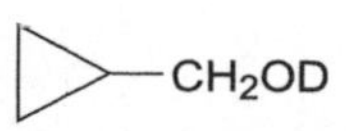

b. 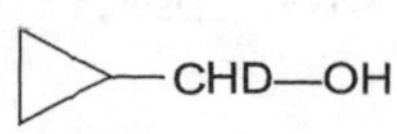

c. 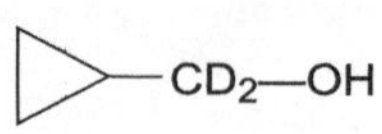

d. 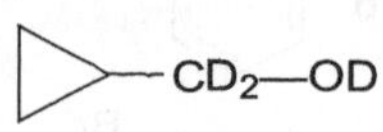

13 What is the major organic product in the following reaction?

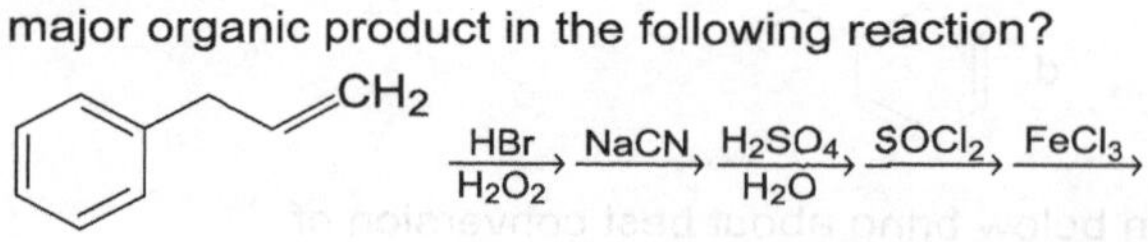

a. 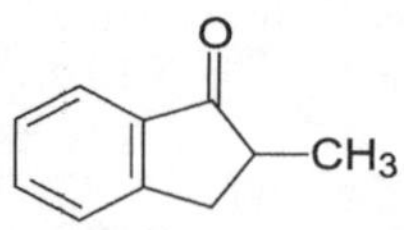

b. 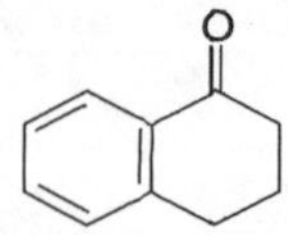

c. 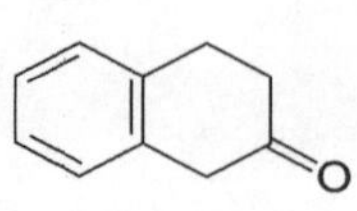

d. 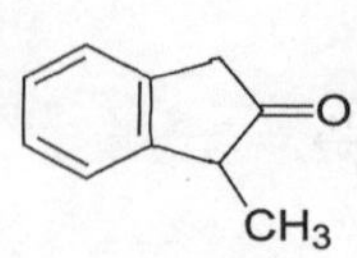

14 What is the major organic product formed in the following reaction?

a. 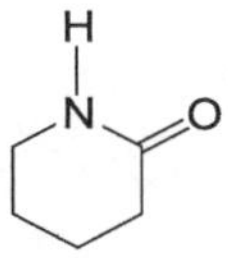

b. 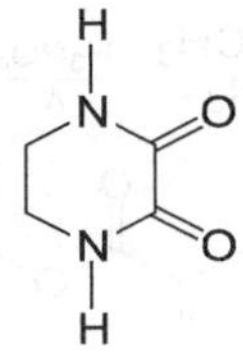

c. 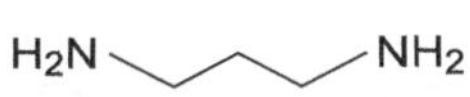

d. 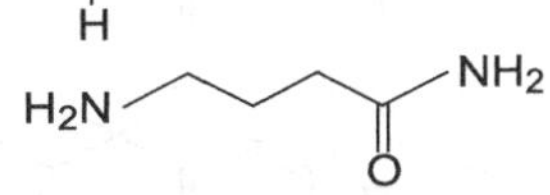

One or More than One Options Correct Type

Direction *(Q. Nos. 15-19) This section contains 5 multiple choice questions. Each question has four choices (a), (b), (c) and (d), out of which ONE or MORE THAN ONE are correct.*

15 Consider the following synthesis of lactone,

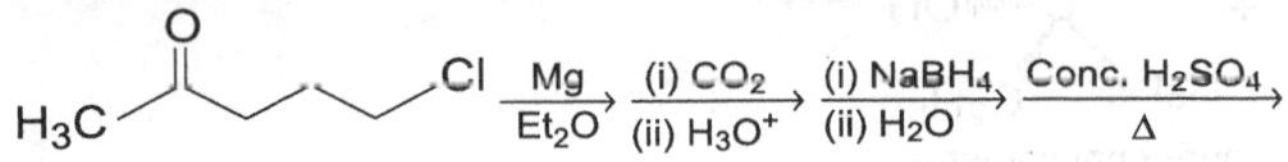

Expected product(s) is/are

a. 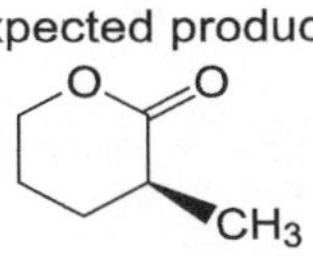

b. 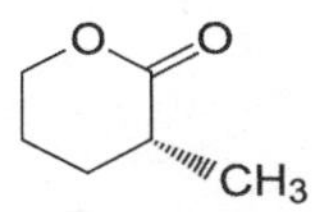

c. 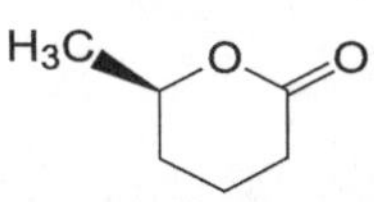

d. 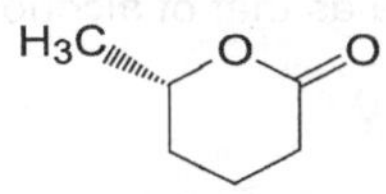

16 Compound *A* given below on treatment with aqueous acid undergoes a rearrangement into indicated product *via* some intermediate product (s).

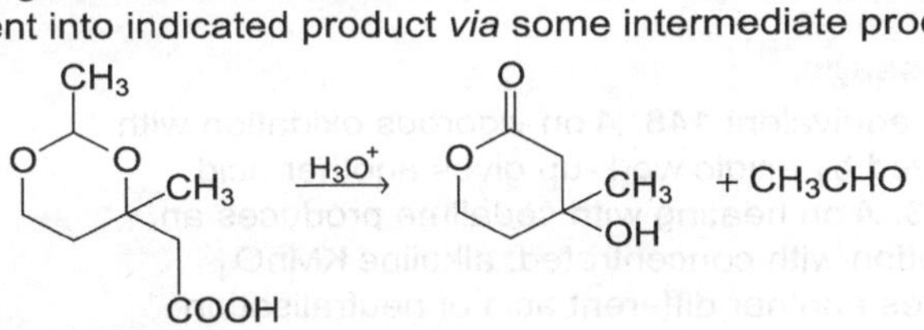

What is/are the intermediate product(s)?

a. CH_3CHO

b. CH_3COOH

c. 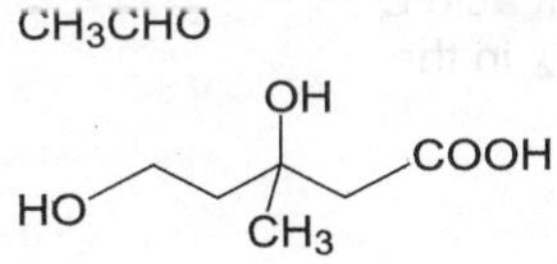

d. 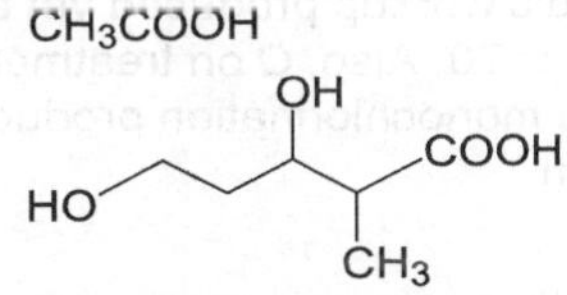

17 Organic product(s) formed in the reaction given below is/are

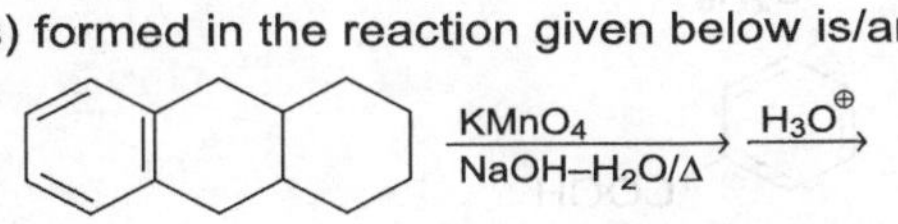

a. 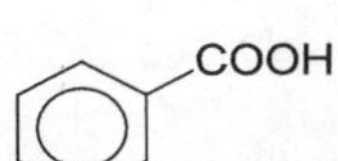

b. 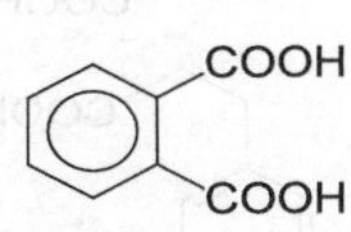

c. 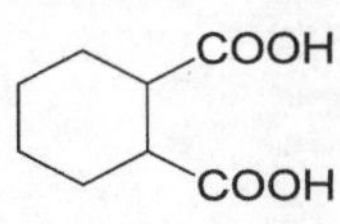

d. 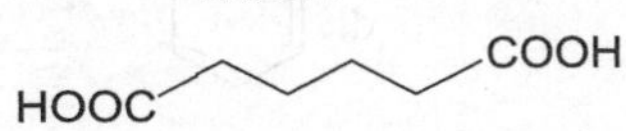

Working Space

Working Space

18 What is/are possible product(s) in the following reaction?

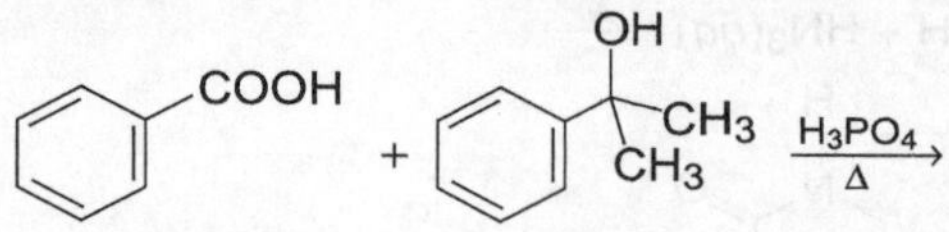

a. 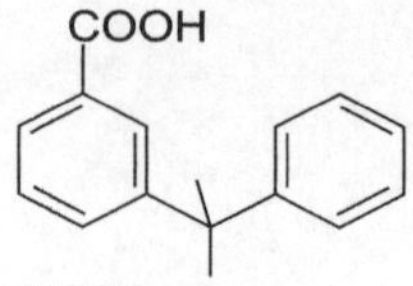

b. 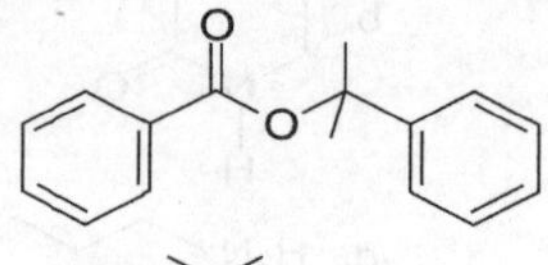

c. 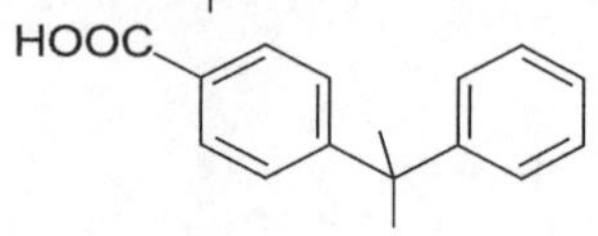

d. 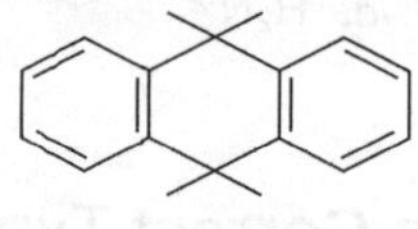

19 Consider the following Fischer esterification reaction.

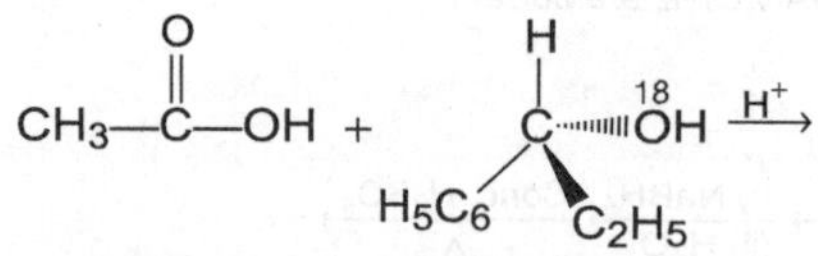

A pure enantiomer

The correct statement is/are

a. O^{18} will be a part of ester

b. alcohol will retain its configuration

c. O^{18} will be sp^2-hybridised in ester

d. sign of specific rotation of ester is same as that of alcohol

Comprehension Type

***Direction** (Q. Nos. 20-22) This section contains a paragraph, describing theory, experiments, data, etc. Three questions related to the paragraph have been given. Each question has only one correct answer among the four given options (a), (b), (c) and (d).*

Passage

An organic acid *A* has neutralisation equivalent 148. *A* on vigorous oxidation with concentrated, alkaline $KMnO_4$ followed by acidic work-up gives another acid whose neutralisation equivalent is 83. *A* on heating with sodalime produces an organic liquid *B*. *B* on vigorous oxidation with concentrated, alkaline $KMnO_4$ solution followed by acid workup gives another different acid of neutralisation equivalent 122. *P*, an isomer of *A*, on vigorous oxidation with concentrated, alkaline $KMnO_4$ followed by acidic workup produced yet another different acid *Q* whose neutralisation equivalent is 70. Also, *Q* on treatment with Cl_2/CCl_4 in the presence of $AlCl_3$ gives a single monochlorination product *via* aromatic electrophilic substitution reaction.

20 The structure of *A* is

a. 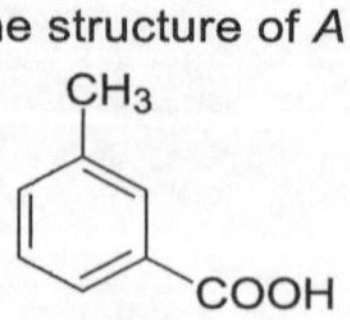

b. 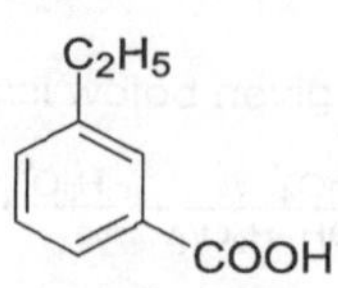

c. 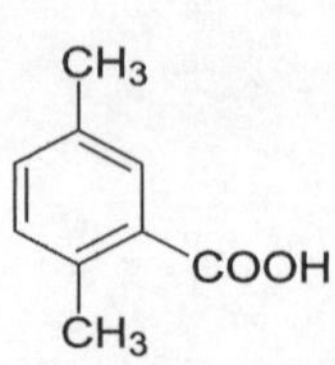

d. 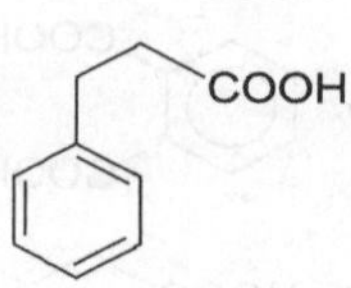

Working Space

21 Which satisfies the criteria of *B*?

a. 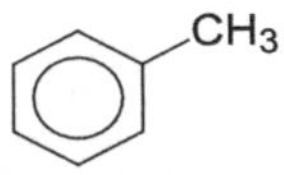

b. 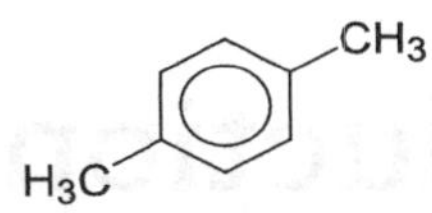

c. 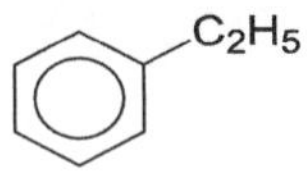

d. 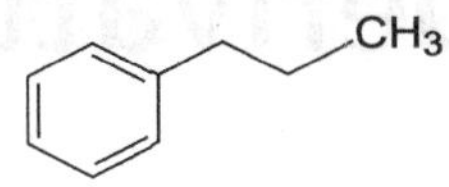

22 Which of the following can be predicted as most suitable structure of *P* ?

a. 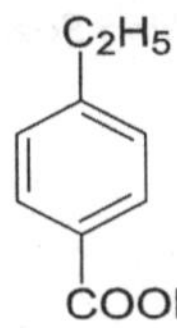

b. 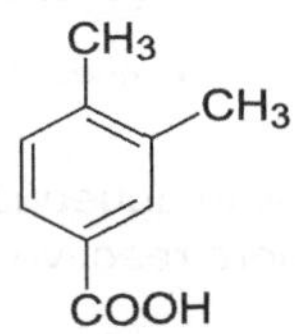

c. 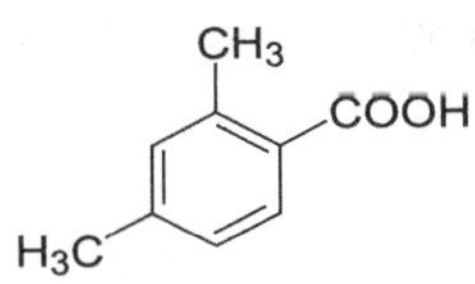

d. 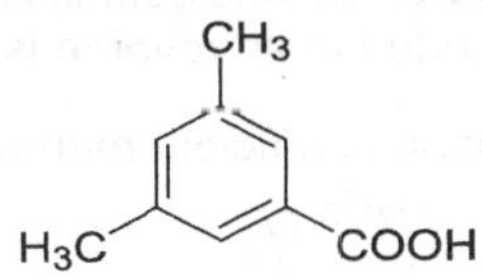

One Integer Value Correct Type

***Direction** (Q. Nos. 23-26) This section contains 4 questions. When worked out will result in an integer from 0 to 9 (both inclusive).*

23 If succinic acid is heated with excess of hydrazoic acid (HN_3), how many molecules of gases are produced per molecule of succinic acid?

24 An aromatic carboxylic acid has some of hydrogen of benzene replaced by —COOH group. Neutralisation equivalent of this acid is exactly 63.5. How many carboxyl groups are present on the benzene ring of original aromatic acid?

25 If a racemic mixture of 2-methyl butanedioic acid is treated with Br_2 in the presence of phosphorus, how many different monobromination products result as an outcome of α-halogenation reaction?

26 Consider the following reaction,

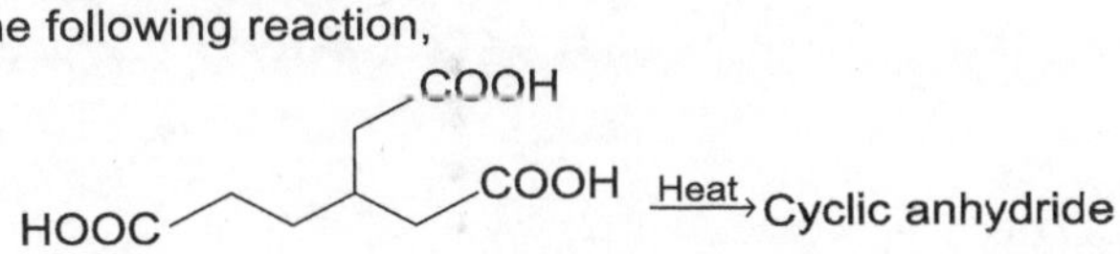

How many carbon atoms are present in the ring of cyclic anhydride?

DPP-3 Introduction to Acid Derivatives

Subjective Questions

Working Space

Direction *(Q. Nos. 1-5) These questions are subjective in nature, need to be solved completely on notebook.*

1 Both CH_3CH_2Cl and CH_3COCl, when treated with aqueous NaOH, undergo nucleophilic substitution reaction. Which is more reactive and why?

2 An ester can spontaneously be prepared by nucleophilic acyl substitution on either an acid chloride or acid anhydride but its reverse is not possible, explain?

3 Bring about the following transformation in efficient manner :

i. 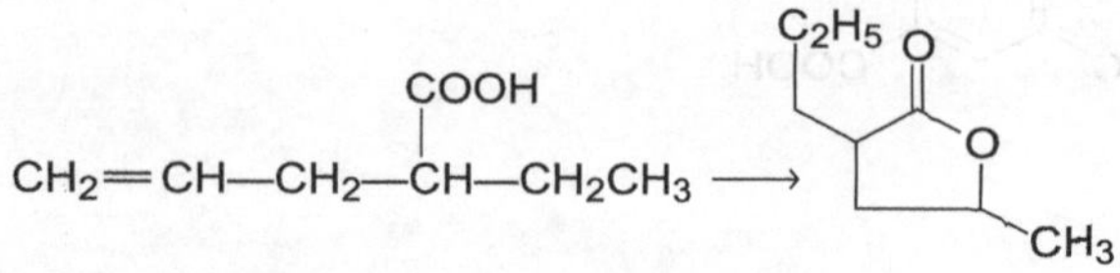

ii. $CH_3—CH{=}CH_2 \longrightarrow CH_3—CN$

iii. $CH_3CH_2OH \longrightarrow CH_3CH_2CONHCH_2CH_3$

4 Convert cyclopentane carboxylic acid into :

i. 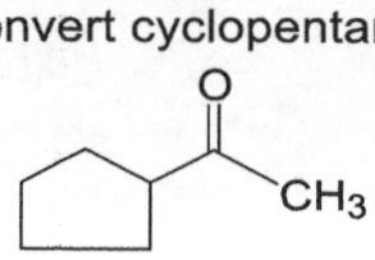

ii. 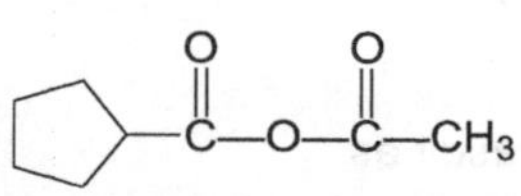

iii. 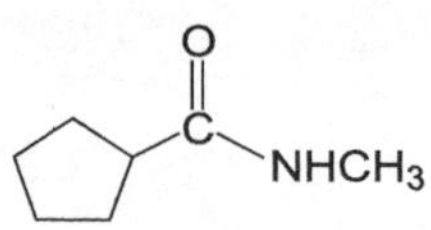

iv. 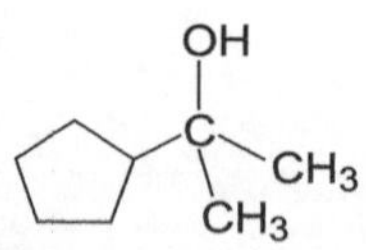

5 Using cyclopentyl methanol as the only organic reagent, synthesise the following compound in good yield.

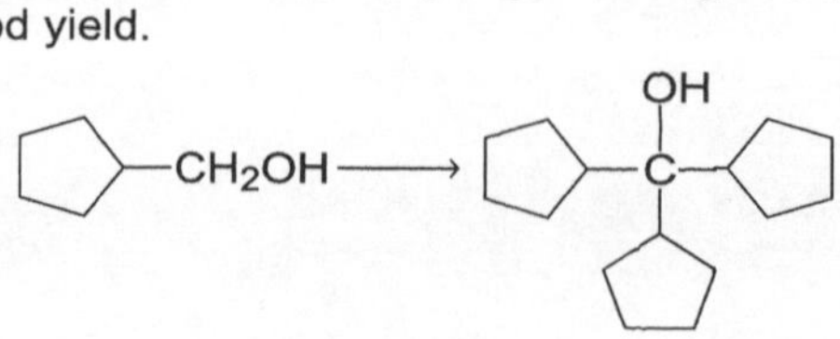

Working Space

Only One Option Correct Type

Direction (Q. Nos. 6-11) *This section contains 6 multiple choice questions. Each question has four choices (a), (b), (c) and (d), out of which ONLY ONE is correct.*

6 Which of the following is most easily decarboxylated on heating?

a. $CH_3CH_2CH_2COOH$

b. $CH_3CH_2—CH(CH_3)—COOH$

c. $CH_3—C(=O)—CH_2COOH$

d. $CH_3—CH_2—C(=O)—COOH$

7 Which of the following reaction fail to occur spontaneously as written?

a. $CH_3COCl + H_2O \longrightarrow CH_3COOH + HCl$

b. $CH_3—C(=O)—O—C(=O)—CH_3 + NH_3 \longrightarrow CH_3CONH_2 + CH_3COOH$

c. 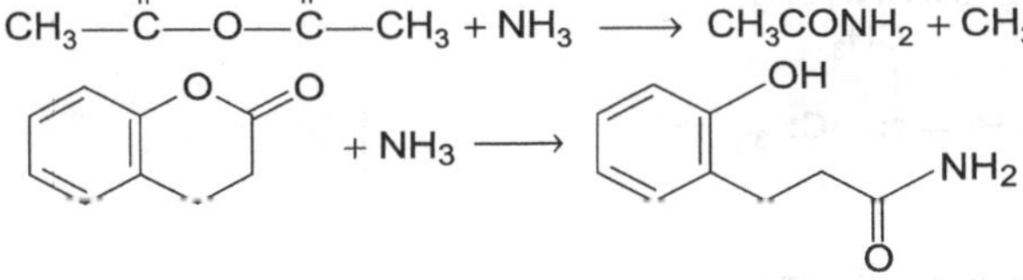

d. 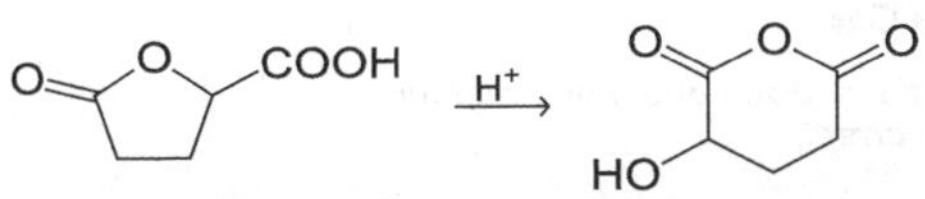

8 Which of the following gives a cyclic anhydride on heating?

a. 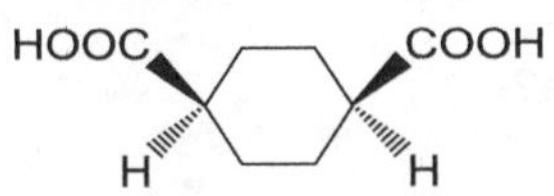

b. 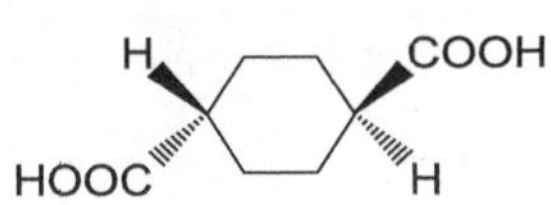

c. $HOOC—CH_2—COOH$

d. 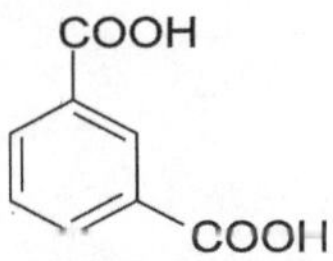

9 What is the increasing order of reactivity of the followings towards base catalysed hydrolysis reaction?

I. CH_3CONH_2

II. CH_3COOCH_3

III. 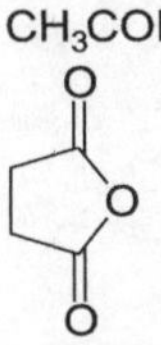

IV. C_6H_5COCl

a. I < II < III < IV
b. IV < III < II < I
c. I < III < II < IV
d. I < IV < II < III

Working Space

10 In the reaction given below, the major organic product formed is

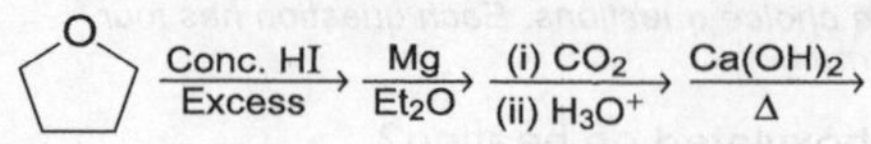

a. 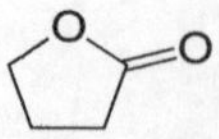

b. 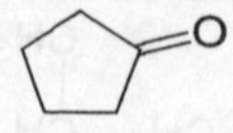

c. 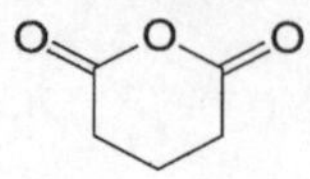

d. 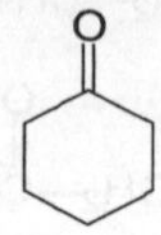

11 In the reaction given below, the major organic product is

$$C_6H_5COOH \xrightarrow{SOCl_2} \xrightarrow{CH_3NH_2} \xrightarrow{LiAlH_4}$$

a. $C_6H_5-CONHCH_3$

b. $C_6H_5CH_2NHCH_3$

c. $C_6H_5NHCH_2CH_3$

d. $C_6H_5-\underset{}{N}(CH_3)-CH_3$

One or More than One Options Correct Type

Direction *(Q. Nos. 12-15) This section contains 4 multiple choice questions. Each question has four choices (a), (b), (c) and (d), out of which ONE or MORE THAN ONE are correct.*

12 Which reagent(s) can convert benzoic acid into corresponding acid chloride?

a. PCl_5 b. $SOCl_2$ c. HCl d. $(COCl)_2$

13 Which reaction condition can convert butanoic acid into methyl butanoate?

a. $\xrightarrow[\Delta]{CH_2N_2}$

b. $\xrightarrow{SOCl_2} \xrightarrow{CH_3ONa}$

c. $\xrightarrow[\Delta]{NH_3} \xrightarrow{CH_3ONa}$

d. $\xrightarrow[\Delta]{P_2O_5} \xrightarrow[H^+/\Delta]{CH_3OH}$

14 What is/are correct regarding the following acyl substitution reaction?

$$C_6H_5COCl + CH_3ONa \xrightarrow{H_2O} C_6H_5COOCH_3 + NaCl$$

a. Order of reaction is two

b. Acylium ion $\left(R-\overset{\oplus}{C}=O\right)$ intermediate is involved

c. Presence of $-NO_2$ at *para* position of phenyl ring increases reactivity

d. Reaction occur at faster rate if carried out in DMSO than in H_2O

15 Which is/are not a suitable nucleophilic substitution reaction?

a. $CH_3COCl + CH_3MgBr/(\text{excess}) \xrightarrow{Et_2O} CH_3-\overset{O}{\overset{\|}{C}}-CH_3$

b. $CH_3-\overset{O}{\overset{\|}{C}}-O-\overset{O}{\overset{\|}{C}}-CH_3 + CH_3NH_2 \longrightarrow CH_3CONHCH_3 + CH_3COOH$

c. $CH_3CONH_2 + CH_3MgBr \xrightarrow{Et_2O} CH_3-\overset{O}{\overset{\|}{C}}-CH_3$

d. 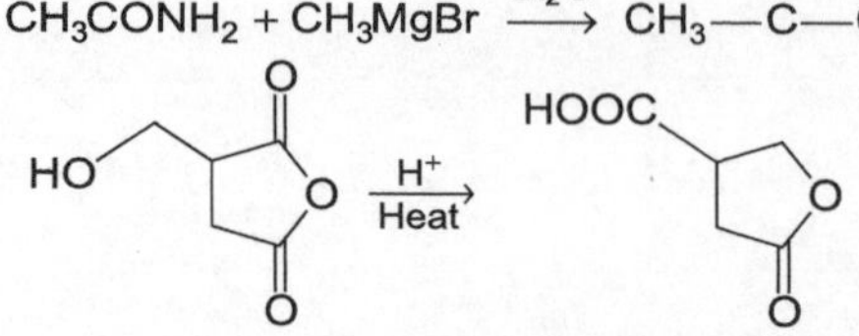

Comprehension Type

Working Space

Direction (Q. Nos. 16-18) This section contains a paragraph, describing theory, experiments, data, etc. Three questions related to the paragraph have been given. Each question has only one correct answer among the four given options (a), (b), (c) and (d).

Passage

In a nucleophilic acyl substitution reaction, a more reactive acid derivative is spontaneously converted into a more stable acid derivative :

$$R-\overset{O}{\overset{\|}{C}}-X + \overset{\ominus}{Nu} \longrightarrow R-\overset{O}{\overset{\|}{C}}-Nu + X^-$$

Here X^- could be Cl^-, Br^-, RO^-, $RCOO^-$ or NH_2^-. The reaction is found to be bimolecular.

16 The leaving ability of X^- is in order of

a. $Cl^- < RCOO^- < RO^- < NH_2^-$
b. $NH_2^- < RO^- < RCOO^- < Cl^-$
c. $NH_2^- < RCOO^- < RO^- < Cl^-$
d. $RCOO^- < RO^- < NH_2^- < Cl^-$

17 If X is Cl, the increasing order of reactivity of the following with CH_3ONa is

I. 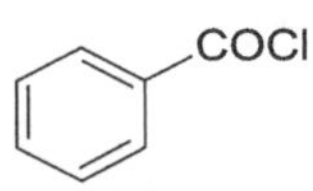

II. 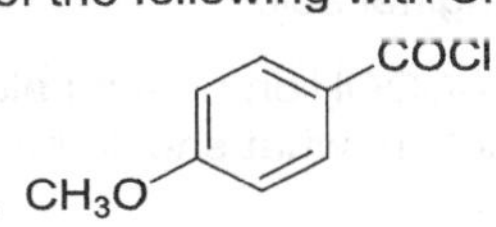

III. 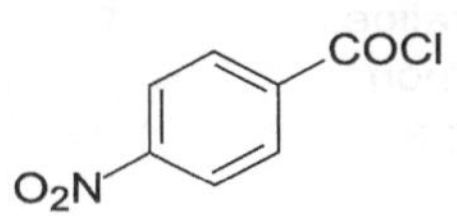

IV. 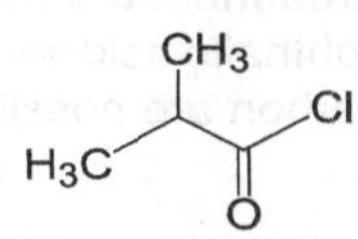

a. I < II < III < IV
b. IV < III < II < I
c. IV < II < I < III
d. II < I < IV < III

18 How benzoic acid X be most efficiently be converted into 2-phenyl propene?

a. $X \xrightarrow[H^{\oplus}/\Delta]{C_2H_5OH} \xrightarrow[Et_2O]{CH_3MgBr} \xrightarrow[\Delta]{H_2SO_4}$
b. $X \xrightarrow{SOCl_2} \xrightarrow[Et_2O]{CH_3MgBr(excess)} \xrightarrow[\Delta]{H_2SO_4}$
c. $X \xrightarrow[\Delta]{NH_3} \xrightarrow[Et_2O]{CH_3MgBr\ (excess)} \xrightarrow[\Delta]{H_2SO_4}$
d. $X \xrightarrow[P_2O_5/\Delta]{CH_3COOH} \xrightarrow[Et_2O]{CH_3MgBr} \xrightarrow[\Delta]{H_2SO_4}$

Statement Type

Direction (Q. Nos. 19-22) This section is based on Statement I and Statement II. Select the correct answer from the codes given below.

a. Both Statement I and Statement II are correct and Statement II is the correct explanation of Statement I
b. Both Statement I and Statement II are correct but Statement II is not the correct explanation of Statement I
c. Statement I is correct but Statement II is incorrect
d. Statement II is correct but Statement I is incorrect

19 **Statement I** Ethanoic anhydride on heating with ammonia gives ethanamide.
Statement II Amide is more stable acid derivative than anhydride.

20 **Statement I** CH_3COOH when treated with $C_2H_5\overset{18}{O}H/H^+$, $CH_3CO\overset{18}{O}C_2H_5$ is formed.
Statement II In Fischer esterification, protonated acid undergo nucleophilic attack by alcohol in the slow, rate determining step.

Working Space

21 **Statement I** 1-butene, when treated with acetic acid in the presence of H_2SO_4, produces racemic mixture of 2-butylpropanoate.

Statement II Acetic acid undergo electrophilic addition at olefinic bond in the presence of an acid catalyst.

22 **Statement I** In the following reaction,

$$CH_3-\overset{\overset{\displaystyle O}{\|}}{C}-CH_2-CH_3 \xrightarrow[\text{(ii) } H_3O^{\oplus}]{\text{(i) } NaBH_4} P \xrightarrow[H_2SO_4/\Delta]{CH_3COOH} \text{Racemic mixture of esters.}$$

Statement II Alcohol (P) undergo nucleophilic addition on protonated acetic acid.

One Integer Value Correct Type

Direction *(Q. Nos. 23-26) This section contains 4 questions. When worked out will result in an integer from 0 to 9 (both inclusive).*

23 A polycarboxylic acid (X) has neutralisation equivalent of 65.5. X when treated with excess of $SOCl_2$ followed by excess of $CH_3ONa(aq)$, gives ester of molar mass 318. What is the basicity of original acid?

24 Dicarboxylic acid $HOOC-(CH_2)_n-COOH$ on heating alone undergo cyclisation giving cyclic anhydride. For this reaction, what should be the minimum values of n?

25 If one of the isomer of a hydrocarbon on treatment with hot, concentrated alkaline, $KMnO_4$ followed by acidic work up gives phthalic acid without loss of any carbon atoms, how many total isomers of hydrocarbon are possible, each containing a phenyl ring?

26 If disodium salt of propanedioic acid is treated with excess of $COCl_2$ a cyclic compound X is formed. How many lone pair of electrons are present in X?

DPP-4 Acid Chlorides and Anhydrides

Subjective Questions

Working Space

Direction (*Q. Nos.* 1-5) *These questions are subjective in nature, need to be solved completely on notebook.*

1 Bring about the following conversion in the efficient manner

i. $CH_3CH_2OH \longrightarrow CH_3CH_2-\underset{14}{\overset{O}{\overset{\|}{C}}}-Cl$

ii. $CH_3CH_2CH_2COOH \longrightarrow CH_3CH_2COCl$

iii. $CH_3COOH \longrightarrow HOC-CH_2-CHO$

iv. $CH_3CH_2COOH \longrightarrow CH_3CH_2-\overset{O}{\overset{\|}{C}}-CH_2CH_2CH_3$

2 Prepare the following anhydride as sole product, starting from any desired reactant.

i. Benzoic anhydride
ii. Phthalic anhydride
iii. Succinic anhydride
iv. Ethanoic propanoic anhydride

3 i. Starting benzene and toluene as the only source of carbon, prepare benzophenone.
ii. Using toluene as only source of carbon, prepare *p*-methyl benzophenone.

4 Complete the following reactions (write only major organic products).

i. 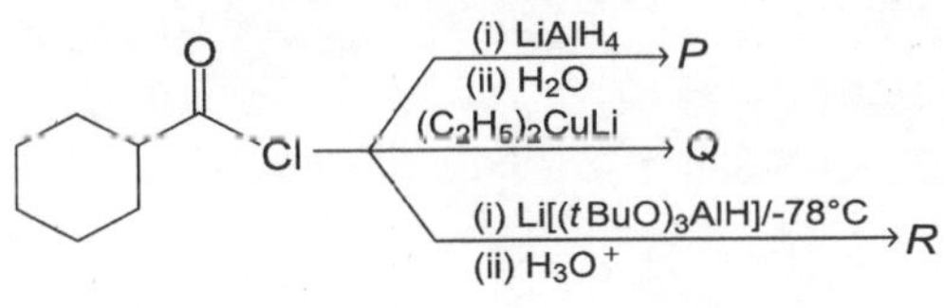

ii. 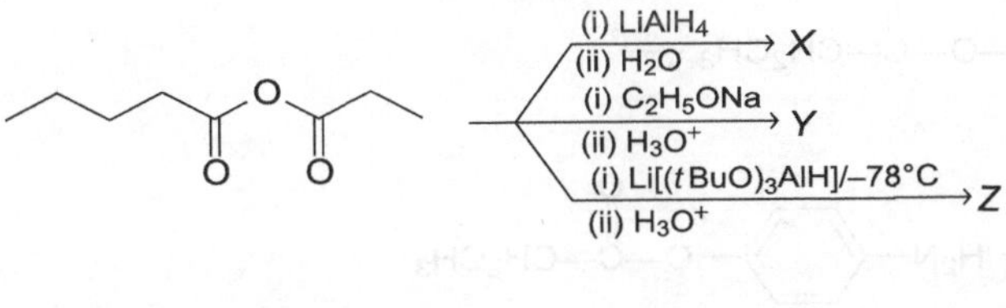

5 Complete the following reactions :

i. 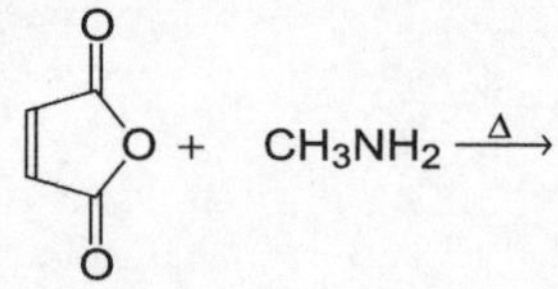

ii. $Cl-\overset{O}{\overset{\|}{C}}-(CH_2)_4-\overset{O}{\overset{\|}{C}}-Cl \xrightarrow{Pd/BaSO_4/H_2} \xrightarrow[\Delta]{NaOH(Dil.)}$

iii. 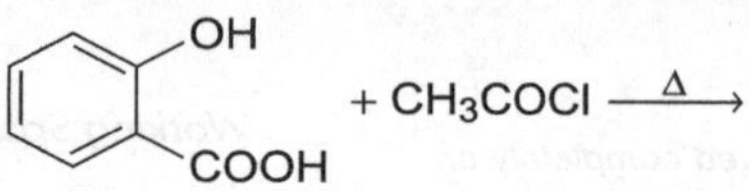

iv. 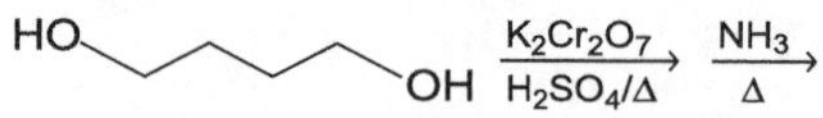

v. $C_6H_5COOH + CH_2{=}C{=}O \xrightarrow[H_2O]{H_2SO_4}$

Working Space

Only One Option Correct Type

Direction (Q. Nos. 6-13) *This section contains 8 multiple choice questions. Each question has four choices (a), (b), (c) and (d), out of which ONLY ONE is correct.*

6 What is the major product formed in the reaction?

$$CH_3-\overset{O}{\overset{\|}{C}}-O-\overset{O}{\overset{\|}{C}}-CF_3 + C_2H_5ONa \xrightarrow{C_2H_5OH}$$

a. $CF_3COOC_2H_5$
b. $CH_3COOC_2H_5$
c. Both (a) and (b) are formed in comparable amount
d. Neither (a) nor (b) is formed as the major product

7 What is the major ester in the following reaction?

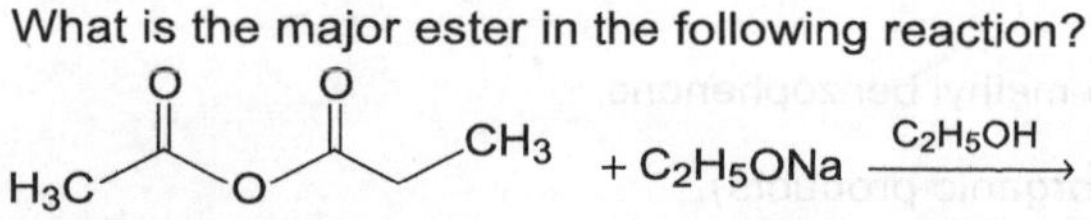

a. $CH_3COOC_2H_5$
b. $CH_3CH_2COOC_2H_5$
c. CH_3COOCH_3
d. $CH_3CH_2COOCH_3$

8 Consider the following reaction,

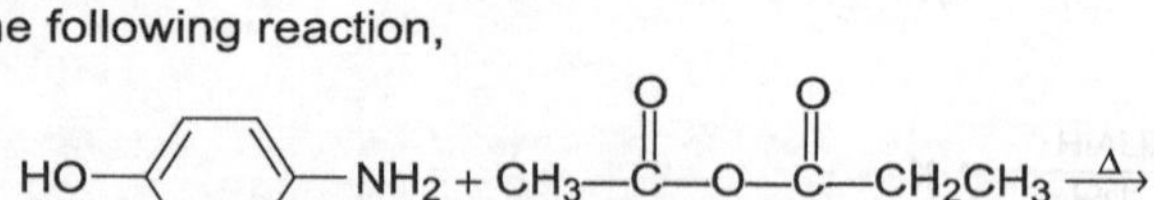

The major product (along with an acid) is

a. 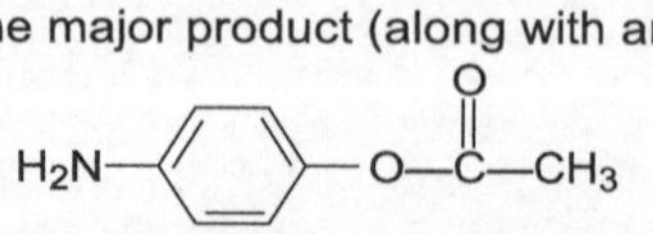
b. 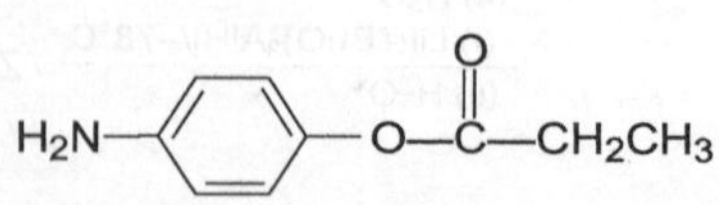

c. 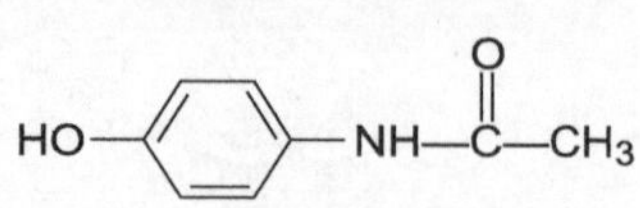
d. 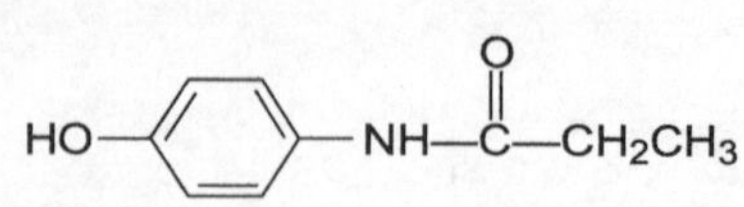

Working Space

9 Which reagent can best bring about the following transformation?

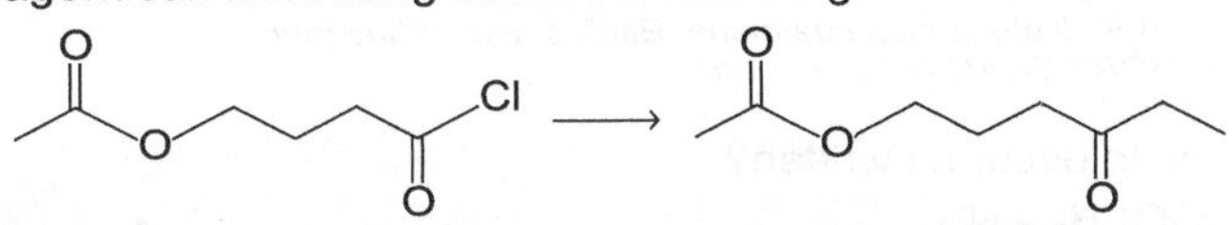

a. C_2H_5MgBr
b. C_2H_5Li
c. $(C_2H_5)_2CuLi$
d. $C_2H_5OH/H^{\oplus}$

10 Which is the best reagent for the following transformation?

$$CH_3-C\equiv C-CH_2-\overset{O}{\overset{\|}{C}}-Cl \longrightarrow CH_3-C\equiv C-CH_2-\overset{O}{\overset{\|}{C}}-H$$

a. $Pd/BaSO_4/H_2$
b. $Li[\{(CH_3)_3CO\}_3AlH]$ / $-78°C$
c. $LiAlH_4$ then CrO_3-H_2SO_4
d. $NaBH_4$ then CrO_3-H_2SO_4

11 Which does not form cyclic anhydride on heating?

a. $HOOC-CH_2-COOH$
b. $HOOC-CH_2CH_2-COOH$
c. *Cis*—$HOOC-CH{=}CH-COOH$
d. 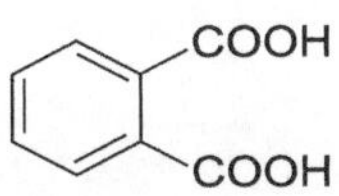

12 Which reaction is the best synthesis of benzoic-ethanoic anhydride?

a. 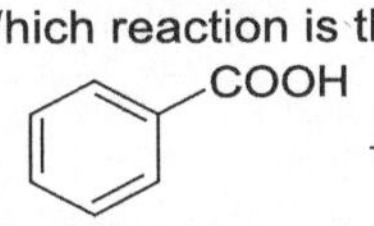 $+ CH_3COOH \xrightarrow[\Delta]{P_2O_5}$

b. 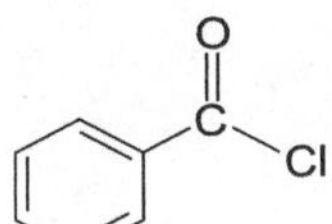 $+ CH_3COOH \xrightarrow{\Delta}$

c. 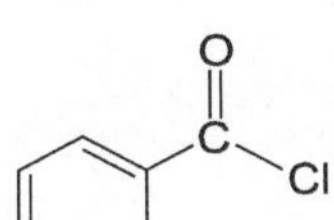 $+ CH_3COONa \longrightarrow$

d. CH_3COCl + 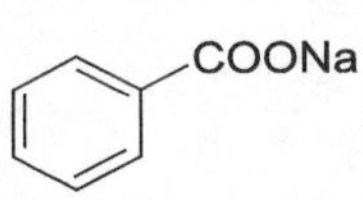 $\longrightarrow$

13 The major organic product formed in the reaction given below is

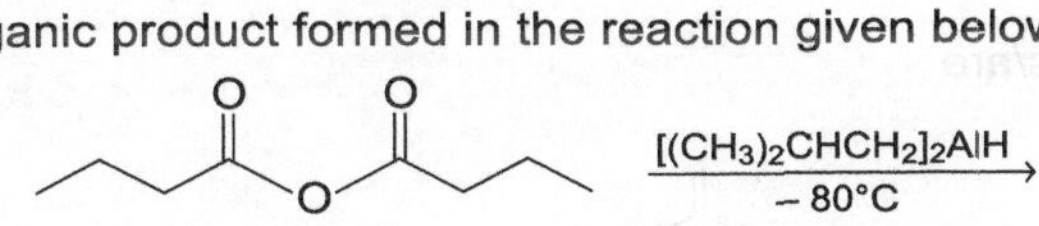

a. 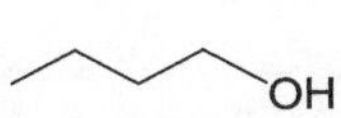
b. 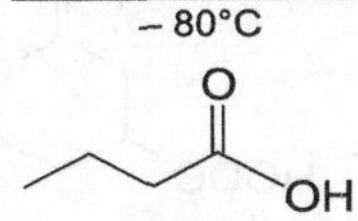
c. 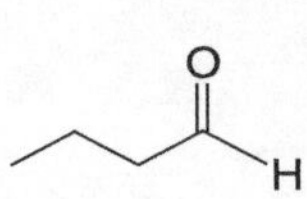
d. 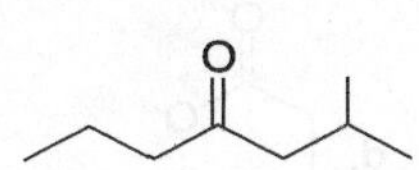

One or More than One Options Correct Type

Working Space

Direction *(Q. Nos. 14-19) This section contains 6 multiple choice questions. Each question has four choices (a), (b), (c) and (d), out of which ONE or MORE THAN ONE are correct.*

14 Which of the following reaction(s) fail to occur as written?

a. $CH_3CONH_2 + CH_3OH \longrightarrow CH_3COOCH_3 + NH_3$

b. $CH_3COOCH_3 + CH_3NH_2 \longrightarrow CH_3CONHCH_3 + CH_3OH$

c. $CH_3—\overset{O}{\overset{\|}{C}}—O—\overset{O}{\overset{\|}{C}}—CH_3 + NH_3 \longrightarrow CH_3CONH_2 + CH_3COOH$

d. $CH_3—\overset{O}{\overset{\|}{C}}—O—\overset{O}{\overset{\|}{C}}—CH_3 + HCl \longrightarrow CH_3COCl + CH_3COOH$

15 If phthalic acid is heated with CH_3COOH in the presence of P_2O_5, possible product(s) formed is/are

a. 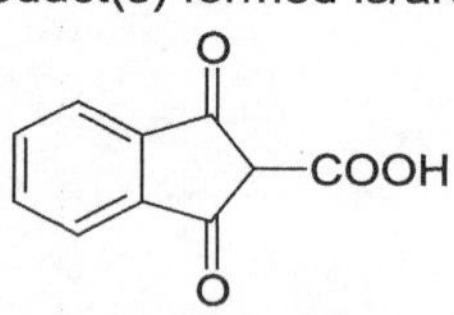

b. 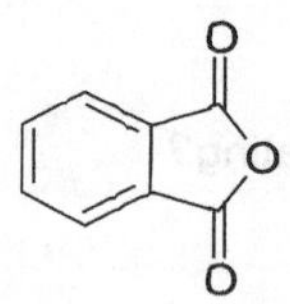

c. 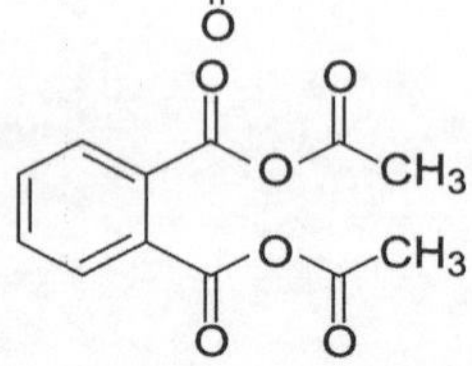

d. $CH_3—\overset{O}{\overset{\|}{C}}—O—\overset{O}{\overset{\|}{C}}—CH_3$

16 Treatment of benzoyl chloride (C_6H_5COCl) with which reagent(s) given below can give 1-phenyl butanone?

a. $\xrightarrow[H_2]{Pd/BaSO_4} \xrightarrow[Et_2O]{CH_3CH_2CH_2MgBr} \xrightarrow{H_3O^+} \xrightarrow[H_2SO_4]{CrO_3}$

b. $(CH_3CH_2CH_2)_2CuLi$

c. $(CH_3CH_2CH_2)MgBr/Et_2O$ then $H_3O^{\oplus}$

d. $\xrightarrow[-80^\circ C]{Li[(tBuO)_3AlH]} \xrightarrow[(ii)\ Et_2O\ (iii)\ H_3O^{\oplus}]{(i)\ CH_3CH_2CH_2Li} \xrightarrow[Pyridine]{CrO_3-HCl}$

17 In the following reaction,

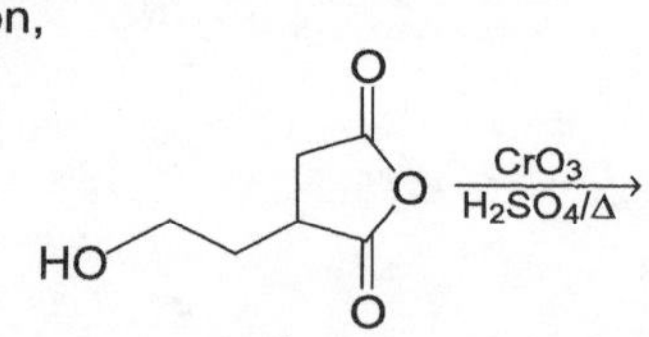

A

Possible product(s) from the following is/are

a. 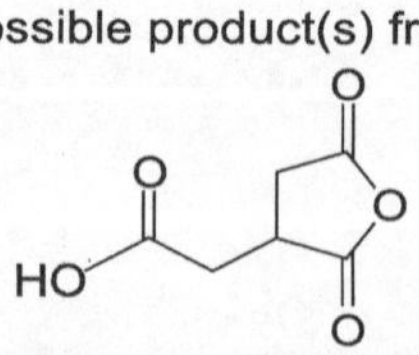

b. 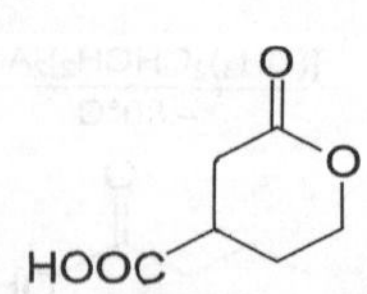

c. 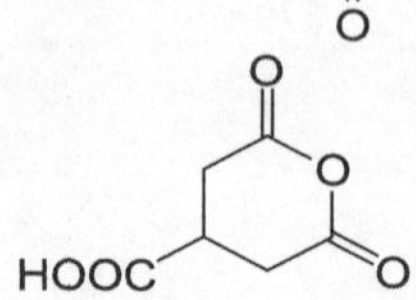

d. 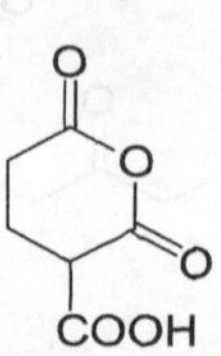

Working Space

18 In the reaction given below, possible product(s) is/are

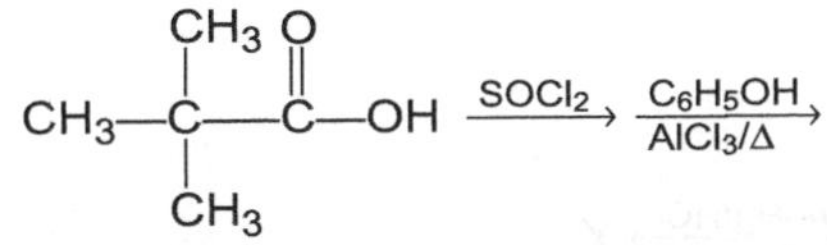

a. 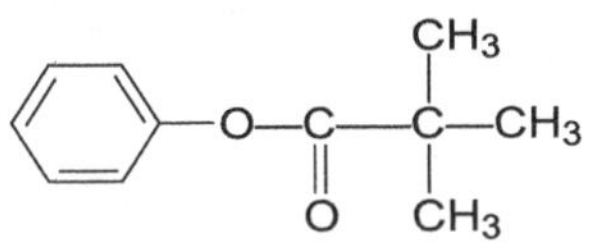

b. 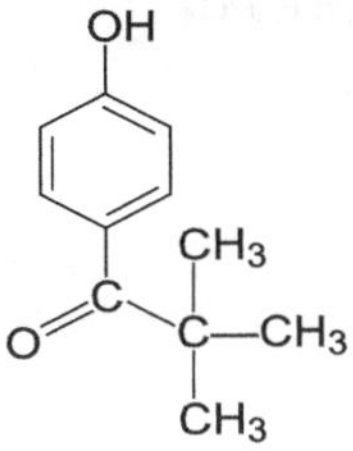

c. 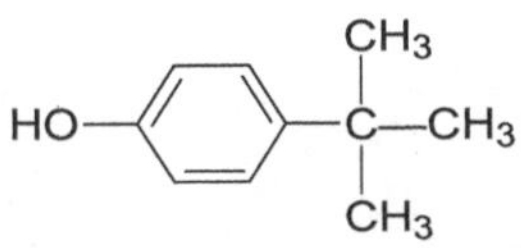

d. 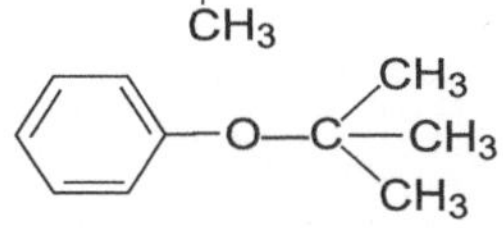

19 Consider the reaction sequence in the following synthesis,

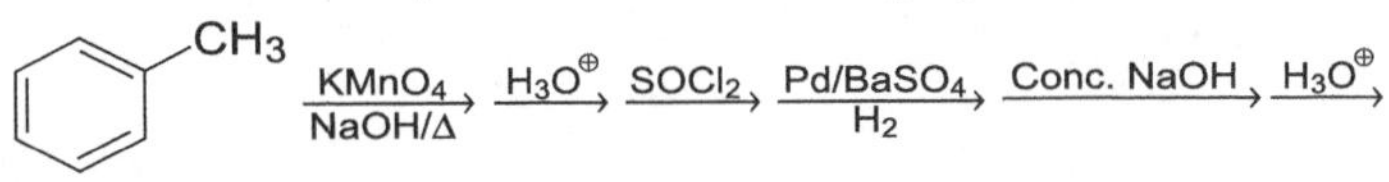

The end product (s) is/are

a. $C_6H_5-\overset{\underset{\|}{O}}{C}-OCH_2C_6H_5$

b. C_6H_5COOH

c. $C_6H_5CH_2OH$

d. 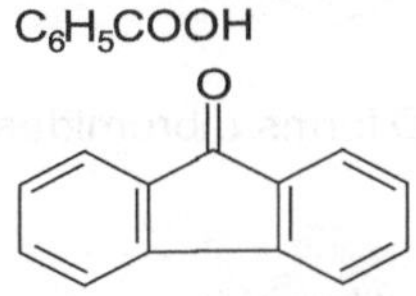

Comprehension Type

Direction *(Q. Nos. 20-22) This section contains a paragraph, describing theory, experiments, data, etc. Three questions related to the paragraph have been given. Each question has only one correct answer among the four given options (a), (b), (c) and (d).*

Passage

P and *Q* are isomers of a compound whose empirical formula is CHO. Both *P* and *Q* decolourises Br_2-H_2O solution. On heating, *P* forms a cyclic anhydride. Compound *P* as well as *Q*, upon treatment with cold, dilute alkaline $KMnO_4$ produces one or more than one of the followings *S*, *T* and *U*.

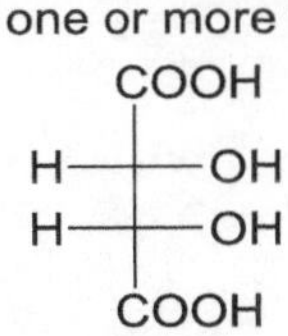

S

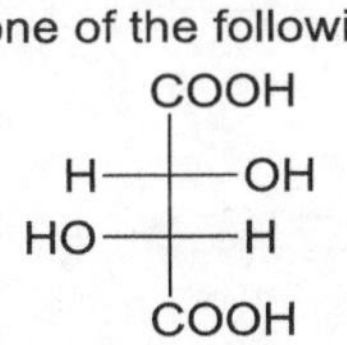

T

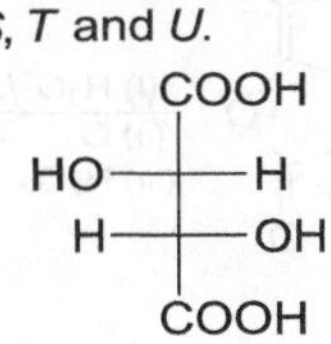

U

20 Compounds formed from *P* and *Q* respectively are

a. optically active *S* and optically active pair (*T*, *U*)
b. optically inactive *S* and optically inactive pair (*T*, *U*)
c. optically active pair (*U*, *T*) and optically active *S*
d. optically inactive pair (*T*, *U*) and optically inactive *S*

Working Space

21 In the reaction given below, *X* and *Y* are

$$Q \xrightarrow[\Delta]{H_2/Ni} X$$

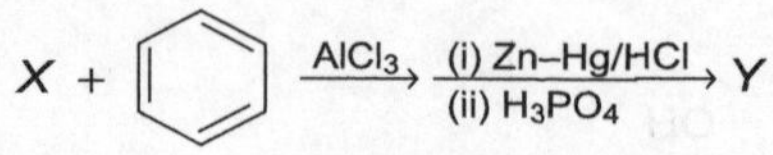

a. 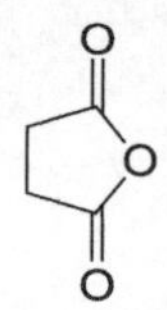 and 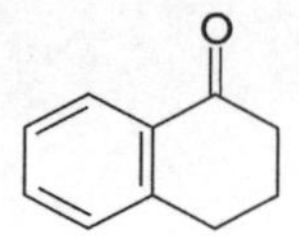

b. 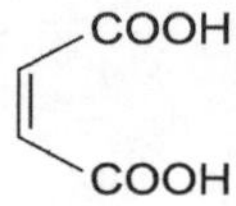 and 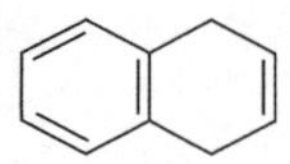

c. 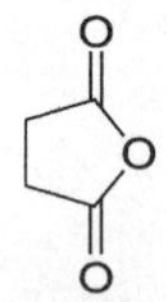 and 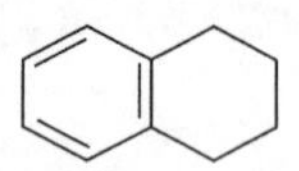

d. 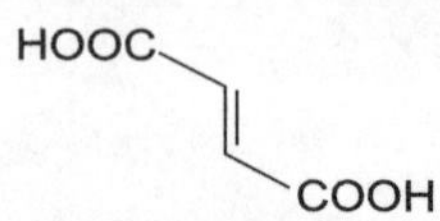 and 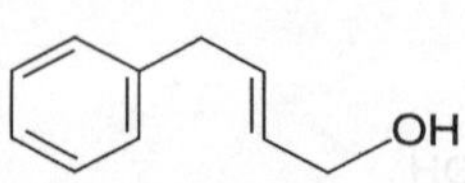

22 Treatment of either *P* or *Q* with Br_2- H_2O forms dibromides. The correct statement concerning product dibromides is

a. Both forms achiral dibromide
b. *P* gives achiral while *Q* forms racemic dibromide
c. *P* forms racemic dibromides while *Q* forms achiral dibromide
d. Both *P* and *Q* form racemic mixture of dibromides

One Integer Value Correct Type

***Direction** (Q. Nos. 23-25) This section contains 3 questions. When worked out will result in an integer from 0 to 9 (both inclusive).*

23 The total number of carboxylic groups in product *P* is/are

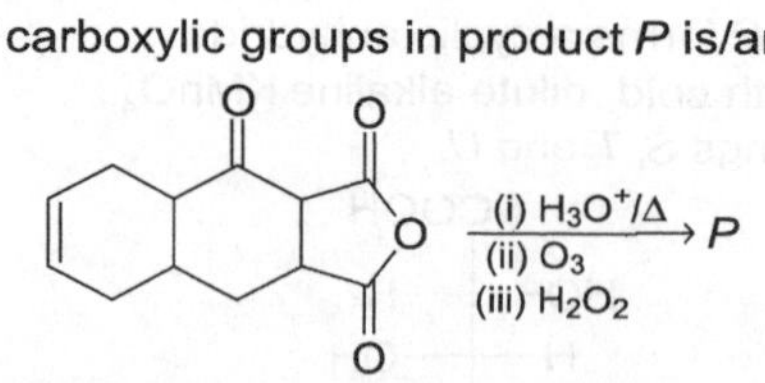

24 Consider the following reaction,

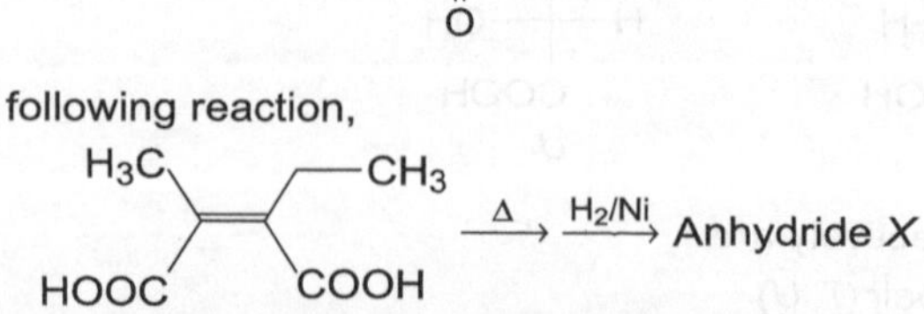

How many different stereoisomers of *X* is/are formed?

25 How many different isomers (including stereoisomers) of cyclohexane dicarboxylic acid on heating with P_2O_5 gives cyclic anhydride?

DPP-5 Esters

Subjective Questions

Working Space

Direction (Q. Nos. 1-5) *These questions are subjective in nature, need to be solved completely on notebook.*

1 Direct ester formation from carboxylic acid and alcohol works well in acidic solution but does not work in alkaline solution. However, its reverse i.e. hydrolysis of ester back to carboxylic acid and alcohol occur in both acidic as well as in alkaline medium. Explain.

2 Explain the formation of products in the following reaction.

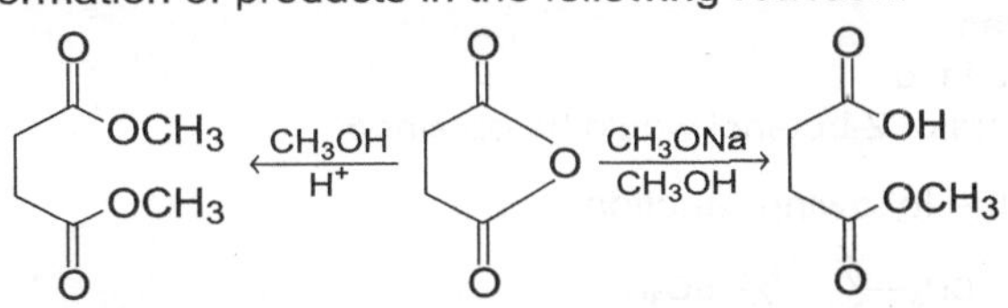

3 A sweet smelling organic compound *A*, on alkaline hydrolysis gave alcohol *B* and cyclohexyl ethanoic acid. *B* on treatment with hot concentrated, alkaline $KMnO_4$, followed by acidic work up gave $C(C_8H_6O_4)$ which did not form anhydride on simple heating. *C* on treatment with $SOCl_2$ followed by the reduction with $Pd/BaSO_4/H_2$ gave $C_8H_6O_2$ which on further reduction with Zn(Hg) in HCl gave $D(C_8H_{10})$. *D* on treatment with 1-chloropropane in the presence of $AlCl_3$ produces *P* and *Q* as principal products with their molecular formula $C_{11}H_{16}$. Deduce the structure of *A*, *B*, *C*, *D*, *P* and *Q*.

4 Organic compounds with composition $C_3H_8O_2$ has five constitutional isomers *A* through *E*. Different isomers give the following reactions,

$$A + (CH_3CO)_2O \xrightarrow{H^+} C_7H_{12}O_4$$

does not give Tollen's or Fehling's tests, does not undergo oxidative cleavage with HIO_4.

$$B + (CH_3CO)_2O \xrightarrow{H^+} C_7H_{12}O_4$$

does not give Tollen's test, Fehling's test but undergoes oxidative cleavage with HIO_4.

$$C + (CH_3CO)_2O \xrightarrow{H^+} C_5H_{10}O_3$$

no Tollen's test, no Fehling test, no oxidative cleavage with HIO_4. After acid hydrolysis gives oxidative cleavage with HIO_4.

$$D + (CH_3CO)_2O \xrightarrow{H^+} C_5H_{10}O_3$$

does not give Tollen's or Fehling's test, does not undergo oxidative cleavage with HIO_4. After acid hydrolysis gives Tollen's test.

$$E + (CH_3CO)_2O \xrightarrow{H^+} \text{No reaction}$$

does not undergo oxidative cleavage with HIO_4. After acid hydrolysis gives Tollen's test.

Identify *A* to *E*.

5 An optically inactive mixture containing all stereoisomers of 2, 5-dimethyl-1, 1-cyclopentane dicarboxylic acids upon fractional crystallisation gave two fractions, *A* and *B*. Fraction *A* on heating produces another compound *C* which is a racemic mixture with molecular formula $C_8H_{14}O_2$. *B* on heating produced isomeric compounds *D* and *E* with their molecular formula $C_8H_{14}O_2$, both of them are optically inactive and can be separated by fractional crystallisation. Identify *A* through *E*.

Working Space

Only One Option Correct Type

Direction *(Q. Nos. 6-13) This section contains 8 multiple choice questions. Each question has four choices (a), (b), (c) and (d), out of which ONLY ONE is correct.*

6 (+)-2-butanol is first treated with tosyl chloride then with C_6H_5ONa. There was obtained secondary butyl benzoate. This ester when subjected to alkaline hydrolysis, alcohol and acid are recovered back. What is true about stereochemistry of 2-butanol obtained after hydrolysis?

a. Pure (−)-2-butanol would be obtained
b. Pure (+)-2- butanol would be obtained
c. Racemic (±)-2-butanol would be obtained
d. Optically active mixture of enantiomers of 2-butanol would be obtained

7 Predict the major addition product of the following reaction

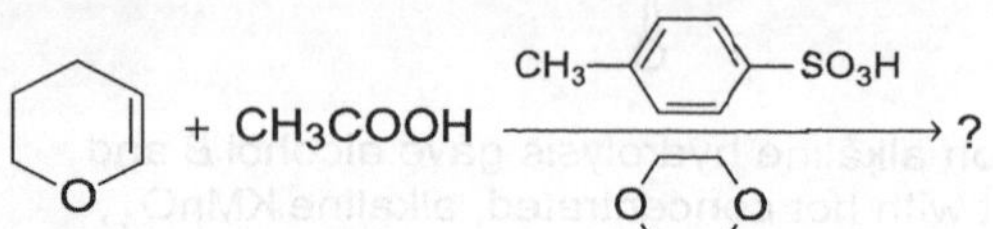

a. 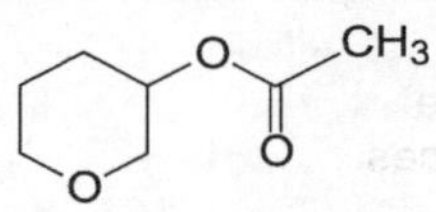

b. 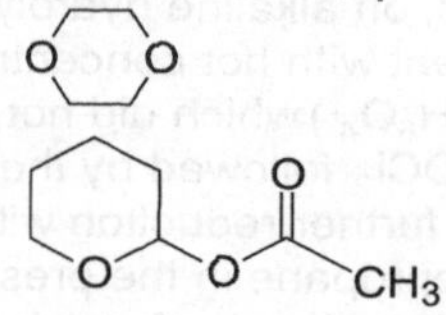

c. 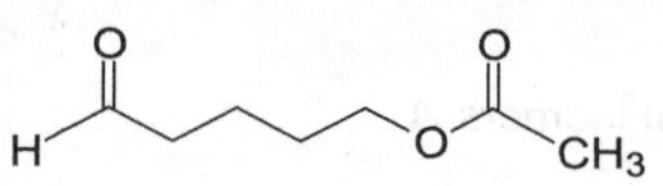

d. 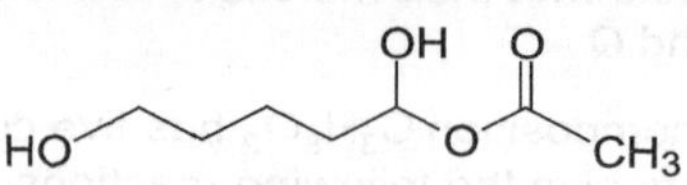

8 What is the major organic product of the following reaction?

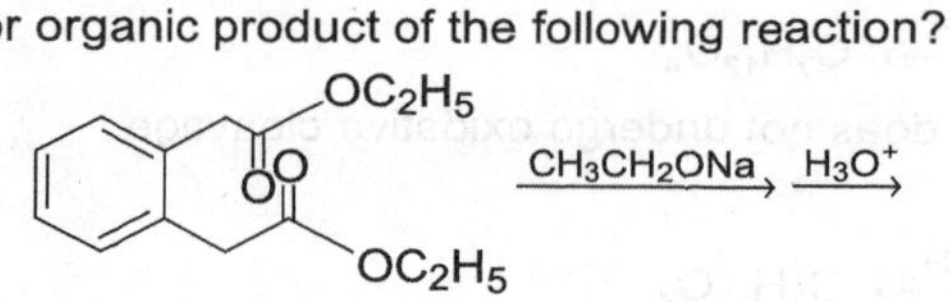

a. 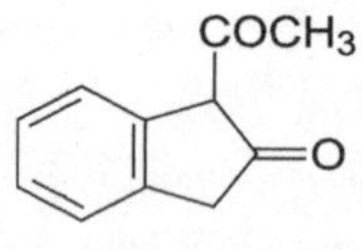

b. 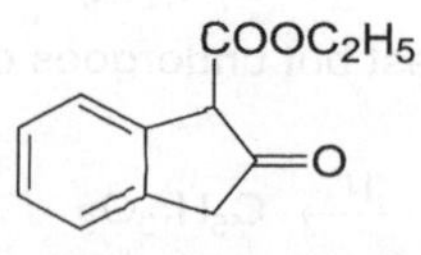

c. 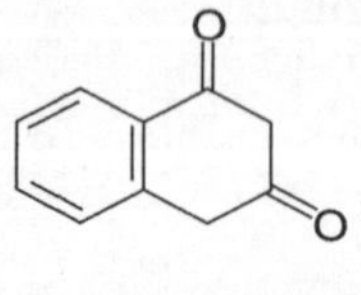

d. 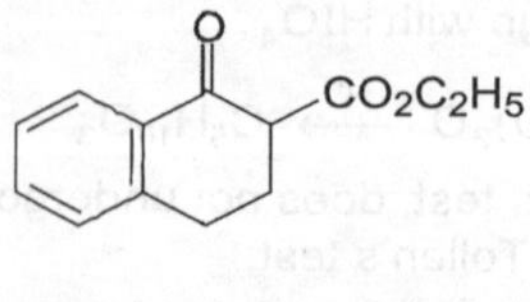

Working Space

9 Choose the best method that could perform the following transformation.

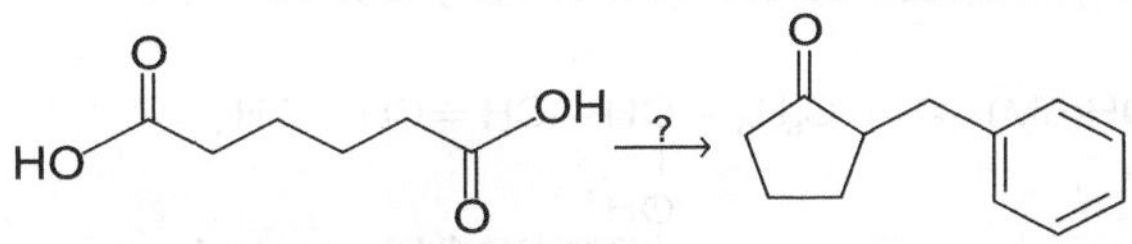

a. $\xrightarrow{HO^-}$ $\xrightarrow{C_6H_5CH_2Br}$ $\xrightarrow[\text{(ii) } H^+ / \text{Heat}]{\text{(i) } HO^-}$

b. $\xrightarrow{HO^-}$ $\xrightarrow[H^+]{EtOH}$ $\xrightarrow{C_6H_5CH_2Br}$

c. $\xrightarrow{LDA}$ $\xrightarrow{C_6H_5CH_2Br}$ $\xrightarrow[\text{(ii) } H^+ / \text{Heat}]{\text{(i) } HO^-}$

d. $\xrightarrow[H^+ / \text{Heat}]{EtOH}$ $\xrightarrow{EtONa}$ $\xrightarrow{C_6H_5CH_2Br}$ $\xrightarrow[\text{(ii) } H^{\oplus} / \text{Heat}]{\text{(i) } HO^-}$

10 Arrange the following esters in the increasing order of reactivity in base catalysed hydrolysis reaction.

I. 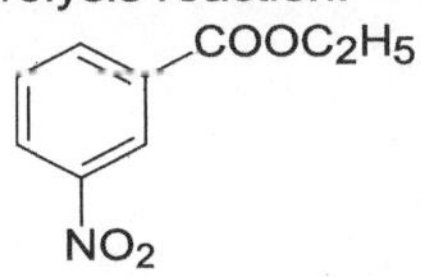

II. 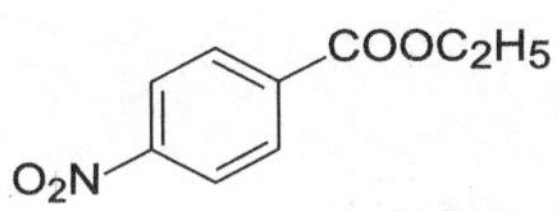

III. 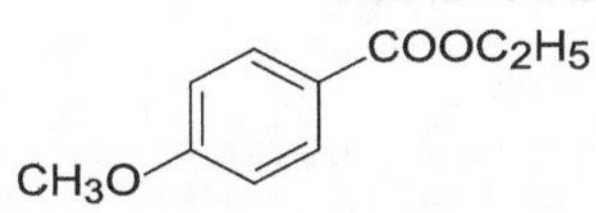

IV. 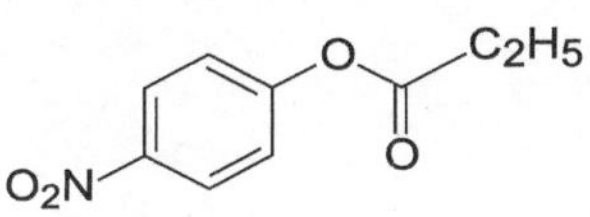

a. I < II < III < IV
b. IV < III < II < I
c. III < I < II < IV
d. IV < II < I < III

11 Which ester given below can most easily be produced by acid catalysed esterification (Fischer's esterification)?

a. $CH_3CH_2COOC_2H_5$
b. $CH_3CH_2COOCH(CH_3)CH_2CH_3$
c. $CH_3CH_2COOC(CH_3)_2CH_2CH_3$
d. $CH_3CH_2COOC_6H_5$

12 Which of the following ester is most likely to undergo unimolecular acid catalysed hydrolysis reaction?

a. $CH_3CH_2COOC_6H_5$
b. $CH_3CH_2COOC_2H_5$

c. 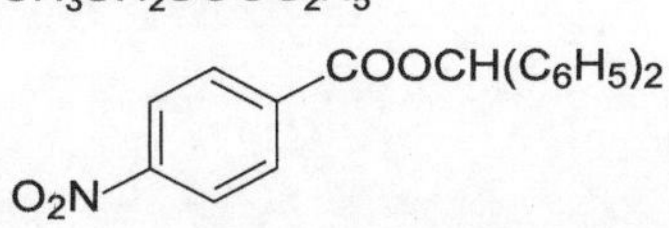

d. 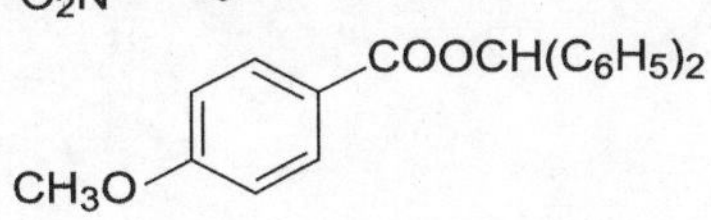

13 Which reaction and indicated product with specified stereochemistry is most accurate?

a. $C_6H_5-\underset{\displaystyle OCOC_2H_5}{\underset{|}{CH}}-CH=CH-CH_3 + NaOH(5M) \longrightarrow C_6H_5-\underset{\displaystyle OH}{\underset{|}{CH}}-CH=CH-CH_3$ (Racemic mixture)

b. $C_6H_5-\underset{\displaystyle OCOC_2H_5}{\underset{|}{CH}}-CH=CH-CH_3 + \text{Dil. NaOH(dil.)} \longrightarrow C_6H_5-\underset{\displaystyle OH}{\underset{|}{CH}}-CH=CH-CH_3$ (Racemic mixture)

c. $C_6H_5-CH=CH-\underset{\displaystyle OCOC_2H_5}{\underset{|}{CH}}-CH_3 + \text{Dil. NaOH} \longrightarrow C_6H_5-CH=CH-\underset{\displaystyle OH}{\underset{|}{CH}}-CH_3$ (Racemic mixture)

d. $C_6H_5-CH=CH-\underset{\displaystyle OCOC_2H_5}{\underset{|}{CH}}-CH_3 + NaOH(5M) \longrightarrow C_6H_5-CH=CH-\underset{\displaystyle OH}{\underset{|}{CH}}-CH_3$ (Retention of configuration)

One or More than One Options Correct Type

Direction (*Q. Nos. 14-19*) *This section contains 6 multiple choice questions. Each question has four choices (a), (b), (c) and (d), out of which ONE or MORE THAN ONE are correct.*

14 What is/are true regarding the following Fischer's ester synthesis?

$$CH_3CH_2COOH + CH_3CH_2OH \xrightarrow{H^+} CH_3CH_2COOC_2H_5$$

a. The reaction involves protonation on sp^2 oxygen of acid in the first step

b. The reaction involves protonation on sp^3 oxygen of acid in the first step

c. The slowest step of reaction is the nucleophilic attack of alcohol on sp^2-carbon of acid

d. Alcohol undergoes protonation in the first step of reaction

15 Consider the following acid catalysed esterification reaction,

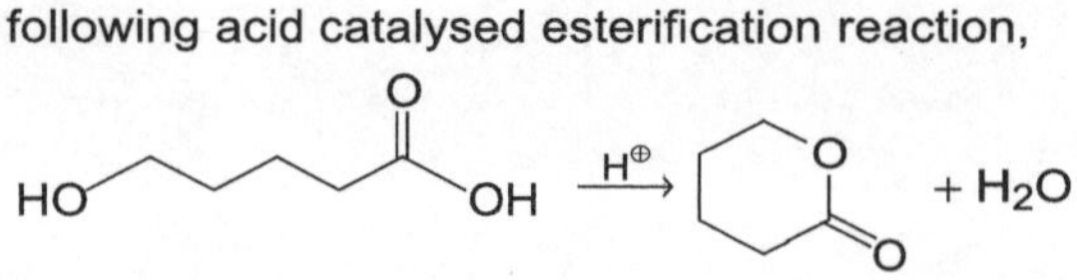

The correct statement concerning the above reaction is/are

a. The above ester is the major product rather than the one formed through intermolecular condensation

b. If the hydroxyl (—OH) group of alcohol contain O^{18}, radioactivity will be retained in the ester

c. If the sp^2 oxygen of acid is O^{18}, it will essentially be absent in the product ester

d. Intramolecular condensation of a hydroxy acid is always favoured over intermolecular condensation

16 In which of the following hydrolysis reaction of esters, reactant and products are correctly matched?

a. $CH_3CH_2COOC_2H_5 \xrightarrow[H_2SO_4]{H_2O^{18}} CH_3CH_2CO\overset{18}{O}H + CH_3CH_2OH$

b. 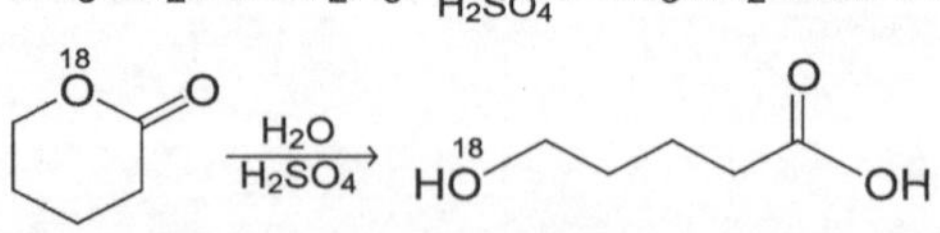

c. $CH_3-\overset{O}{\overset{||}{C}}-O-CH_2-O-\overset{O}{\overset{||}{C}}-CH_3 \xrightarrow[H_2SO_4]{H_2O} CH_3COOH + HCHO$

d. 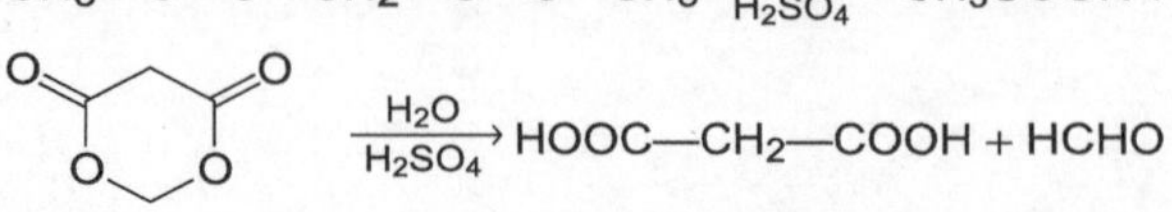

Working Space

17 In which of the following reaction(s), reactant and products are correctly matched?

a. 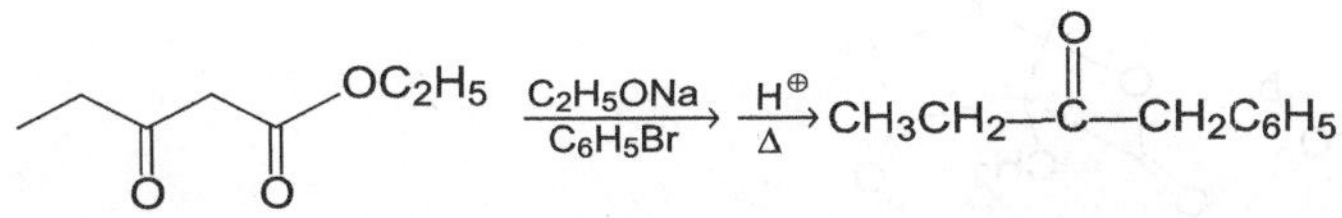

b. 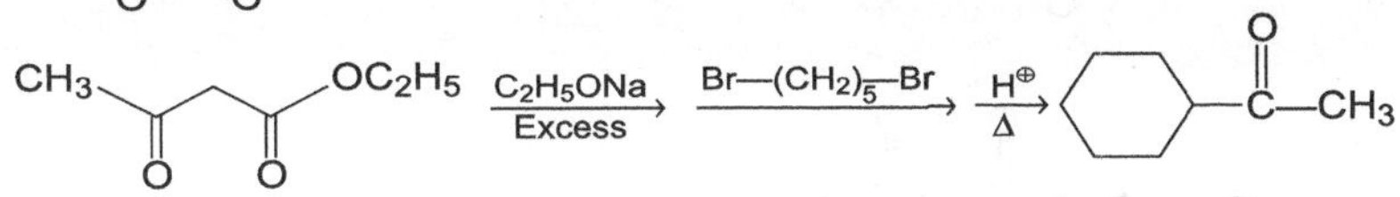

c. 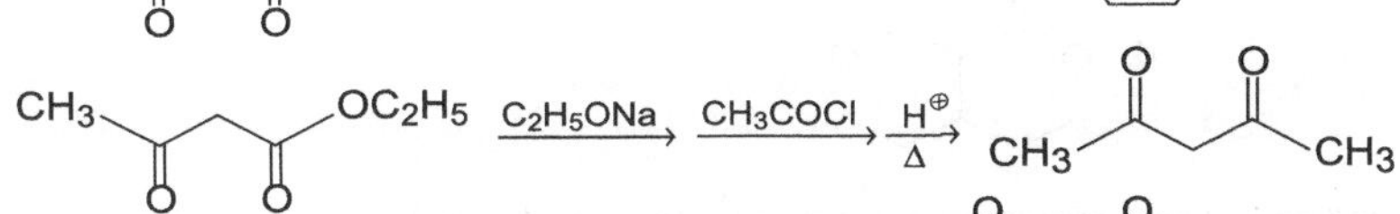

d. 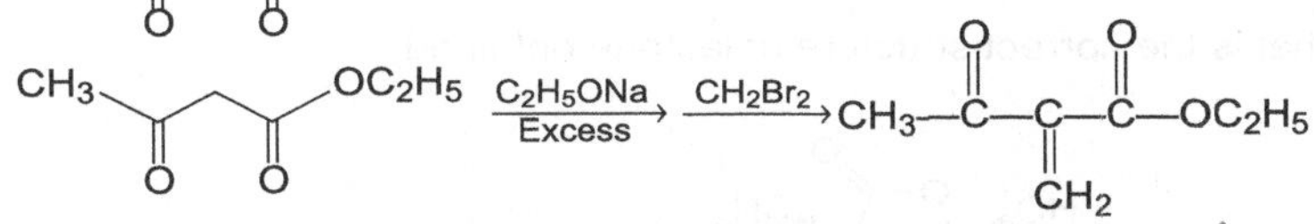

18 If $CH_3-\overset{O^{18}}{\overset{\|}{C}}-OH$ is treated with $CH_3CH_2\overset{18}{O}H$ in the presence of H_2SO_4, possible esters formed is/are

a. $CH_3-\overset{O^{18}}{\overset{\|}{C}}-OC_2H_5$

b. $CH_3-\overset{O}{\overset{\|}{C}}-\overset{18}{O}C_2H_5$

c. $CH_3-\overset{O^{18}}{\overset{\|}{C}}-\overset{18}{O}C_2H_5$

d. $CH_3-\overset{O}{\overset{\|}{C}}-OC_2H_5$

19 The correct statement regarding the following transformation is/are

$$CH_3CH_2-\overset{O}{\overset{\|}{C}}-OCH_2CH_2CH_3 + CH_3CH_2OH \xrightarrow{H^+} CH_3CH_2-\overset{O}{\overset{\|}{C}}-OC_2H_5 + CH_3CH_2CH_2OH$$

a. Presence of excess of ethanol favours the reaction
b. Both acid and base can act as catalyst
c. Smaller alcohols always displace the larger alcohols from ester
d. The reaction is second order in both acid and base catalysed condition

Comprehension Type

Direction *(Q. Nos. 20-22) This section contains a paragraph, describing theory, experiments, data, etc. Three questions related to the paragraph have been given. Each question has only one correct answer among the four given options (a), (b), (c) and (d).*

Passage

Treatment of 2,4-pentanedione with excess of KCN(*aq*) and acetic acid followed by hydrolysis in the presence of H_2SO_4 gives two products *A* and *B*, both having molecular formula $C_7H_{12}O_6$. When heated, *B* first gives a lactonic acid $C_7H_{10}O_5$ and finally a dilactone $P(C_7H_8O_4)$.

20 What is the correct structure of *A*?

a. 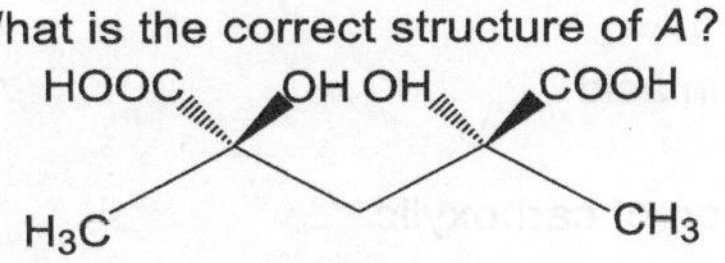

b. 

c. Racemic mixture of (a)

d. Either (a) or (b)

Working Space

21 What is the correct structure of P?

a. 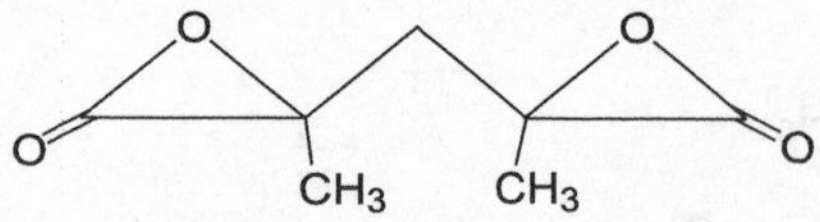 b. 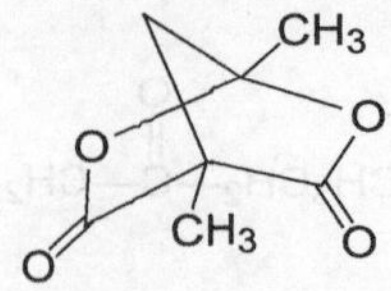

c. 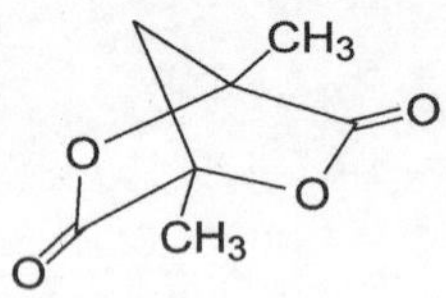 d. 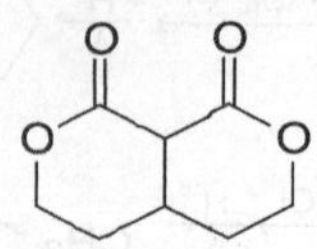

22 A also lactonises on heating. What is the correct structure of lactone obtained from A?

a. 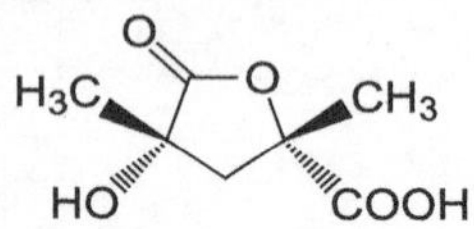 b. 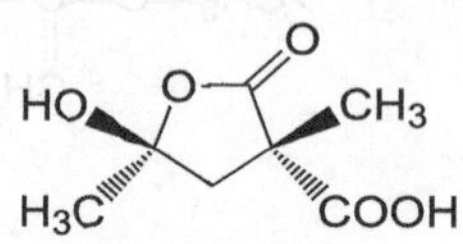

c. 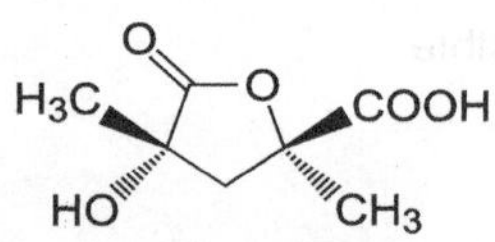 d. 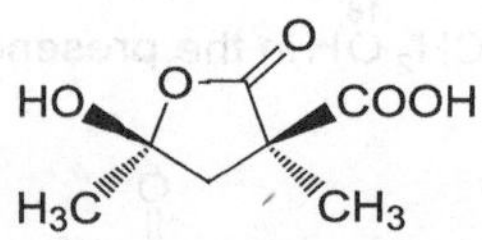

Statement Type

***Direction** (Q. Nos. 23-26) This section is based on Statement I and Statement II. Select the correct answer from the codes given below.*

a. Both Statement I and Statement II are correct and Statement II is the correct explanation of Statement I
b. Both Statement I and Statement II are correct but Statement II is not the correct explanation of Statement I
c. Statement I is correct but Statement II is incorrect
d. Statement II is correct but Statement I is incorrect

23 **Statement I** Ester formation from acid and alcohol occur in acidic medium but not in alkaline medium. However, hydrolysis of esters proceeds in both acidic and alkaline medium.

Statement II In alkaline medium carboxylic acid is neutralised into salt which do not undergo nucleophilic attack by alcohols.

24 **Statement I** When ethyl propanoate is refluxed with butanol in slightly acidic condition at the boiling point of ethanol, butylpropanoate is formed.

Statement II During *trans* esterification, larger alcohols always displaces the smaller one.

25 **Statement I** If $CH_3COOCH(CH_3)C_6H_5$ is heated with dilute NaOH solution, racemic mixture of 1-phenyl ethanol is formed.

Statement II At low hydroxide concentration and alcohol part of ester being capable of forming stable carbocation, hydrolysis proceeds preferably by unimolecular alkyl oxygen cleavage mechanism.

26 **Statement I** p-nitrobenzoic acid is more reactive than benzoic acid in acid catalysed esterification reaction.

Statement II Rate determining step in Fischer's esterification reaction of carboxylic acid in nucleophilic attack alcohols on protonated acid.

Working Space

One Integer Value Correct Type

Direction (Q. Nos. 27-30) *This section contains 4 questions. When worked out will result in an integer from 0 to 9 (both inclusive).*

27 If a mixture containing ethyl acetate and ethyl propanoate is heated in the presence of C_2H_5ONa, how many different condensation products would be formed?

28 In the following lists of reactions, how many of them gives ester as the major organic product?

I. $CH_3COOH \xrightarrow{NaOH} \xrightarrow{CH_3CH_2Br}$

II. $CH_3CH_2COOH + CH_3CH_2Li \longrightarrow$

III. $CH_3CH_2-\overset{O}{\overset{\|}{C}}-C_6H_5 + CF_3CO_3H \longrightarrow$

IV. $CH_3COOH + CH_2N_2 \xrightarrow{\Delta}$

V. $C_6H_5-CH=CH_2 + CH_3COOH \xrightarrow{\Delta}$

VI. 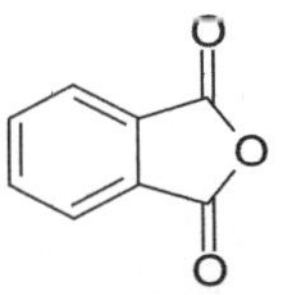 $+ CH_3CH_2OH \xrightarrow[\Delta]{H_2SO_4}$

VII. $CH_3CH_2COOH \xrightarrow[C_2H_5OH]{C_2H_5ONa}$

VIII. $CH_3CH_2CHO + Al(OC_2H_5)_3 \xrightarrow[\Delta]{H_2O}$

IX. $C_6H_5 \cdot COCl + C_2H_5ONa \longrightarrow$

X. $CH_3CH_2CONH_2 + CH_3CH_2OH \xrightarrow[\Delta]{H^+}$

29 How many different isomers exist for $C_3H_6O_2$ which reduces Tollen's reagent as well as forms $C_5H_8O_3$ upon treatment with acetic anhydride?

30 A hydroxy acid has molecular formula $C_5H_{10}O_3$ and it lactonises (forms cyclic esters) on heating. If only five and six membered lactones are considered, how many different isomers of lactones are possible?

DPP-6 Acid Amides

Subjective Questions

***Direction** (Q. Nos. 1-5) These questions are subjective in nature, need to be solved completely on notebook.*

Working Space

1 Provide suitable condition for controlled hydrolysis of nitriles to amide in both acid and base catalysed condition.

2 Arrange the following in the increasing order of reactivity in Hofmann's degradation reaction.

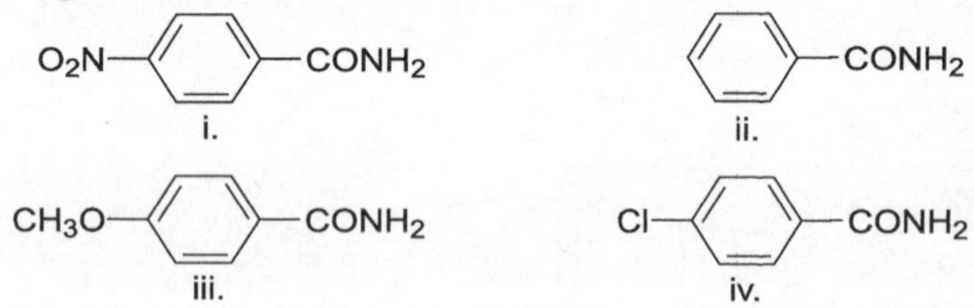

3 Identify the compounds *A*, *B* and *C* in the following reaction.

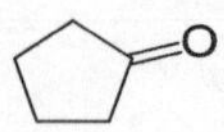
$+ H_2NOH \xrightarrow{\Delta} A \xrightarrow{CH_2SO_4} B \xrightarrow{H_3O^+} C$ (an amino acid)

4 Convert CH_3CONH_2 into

i. CH_3COCl
ii. $(CH_3CO)_2O$
iii. $CH_3COOC_2H_5$

5 An amide is neutral to litmus paper while an amine changes colour of moistened litmus paper from red to blue, explain.

Only One Option Correct Type

***Direction** (Q. Nos. 6-13) This section contains 8 multiple choice questions. Each question has four choices (a), (b), (c) and (d), out of which ONLY ONE is correct.*

6 What is formed as the major organic product in the reaction?

$$CH_3CH_2-\overset{\overset{O}{\parallel}}{C}-NH_2 + P_4O_{10} \xrightarrow{\text{Heat}}$$

a. $CH_3CH_2NH_2$
b. CH_3CH_2CN
c. CH_3CH_2COOH
d. $CH_3CH_2CH_2NH_2$

7 What is formed as the major organic product of the reaction?

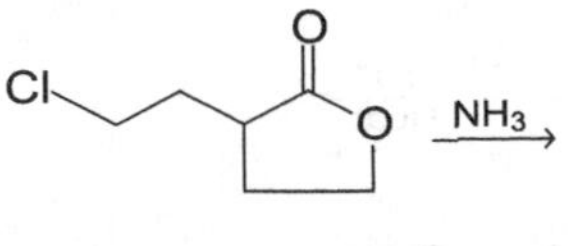

a. 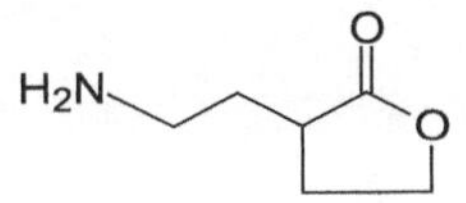

b. 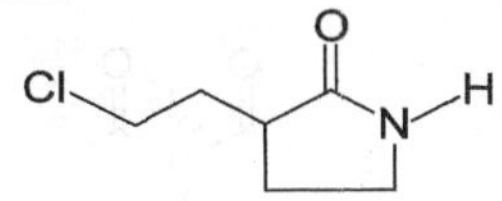

c. 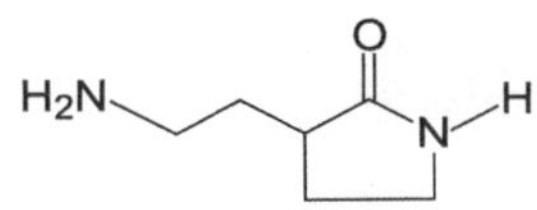

d. 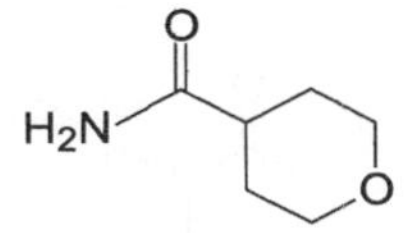

8 Which is most reactive in base catalysed hydrolysis reaction?

a. 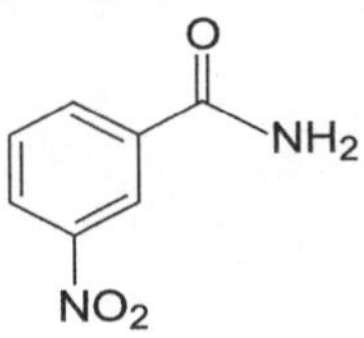

b. 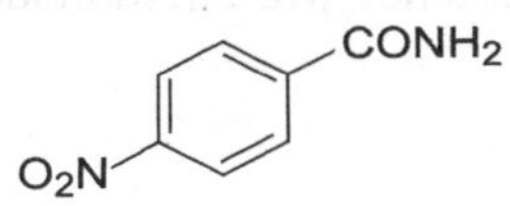

c. 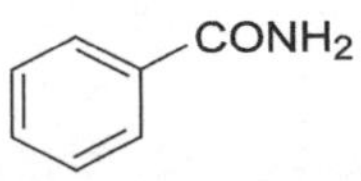

d. 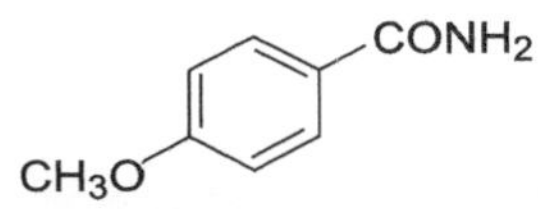

9 Which reaction given below produces a secondary amide?

a. $CH_3-\overset{O}{\overset{\|}{C}}-CH_3 + H_2NOH \xrightarrow{\Delta} \xrightarrow{PCl_5} \xrightarrow[H_2O]{NaOH}$

b. $(CH_3CO)_2O + (CH_3)_2NH \xrightarrow{\Delta}$

c. $(CH_3)_2CHCOCl + CH_3NH_2 \longrightarrow$

d. $(CH_3)_2CHCOOC_2H_5 + CH_3NH_2 \longrightarrow$

10 Which reaction gives a polymeric amide?

a. $CH_3-CH(COOH)_2 + NH_3 \xrightarrow{\Delta}$

b. 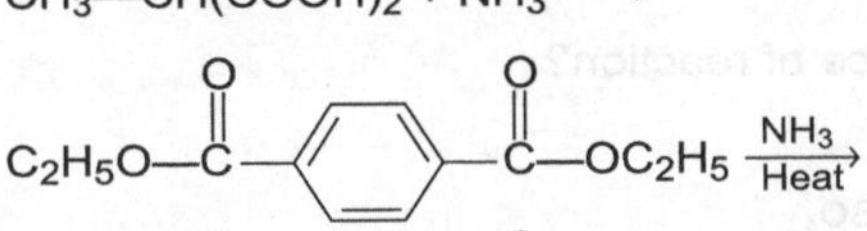

c. 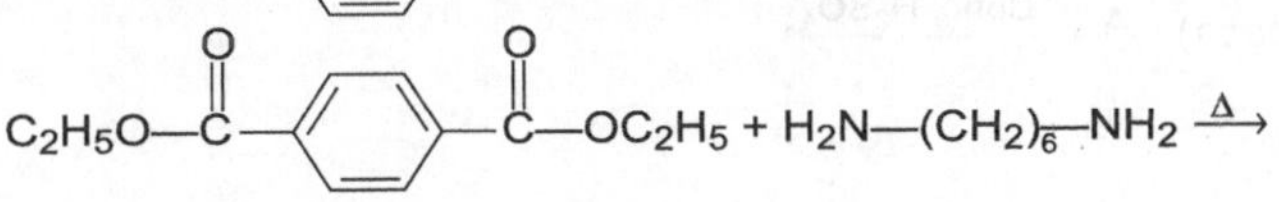

d. $Cl-\overset{O}{\overset{\|}{C}}-(CH_2)_4-\overset{O}{\overset{\|}{C}}-Cl + CH_3NH_2 \longrightarrow$

11 Which reaction given below gives a lactum as the major organic product?

a. 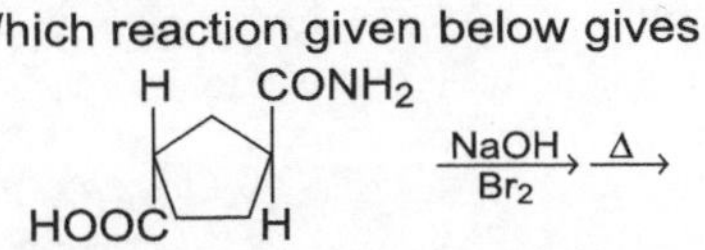

b. 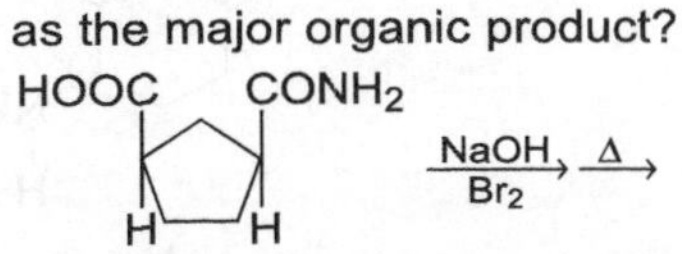

c. 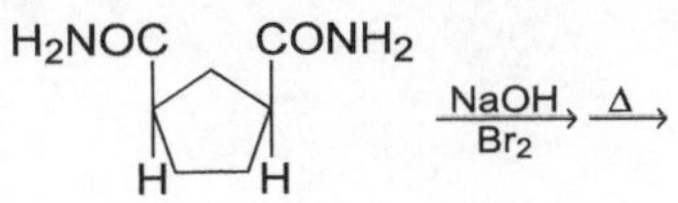

d. 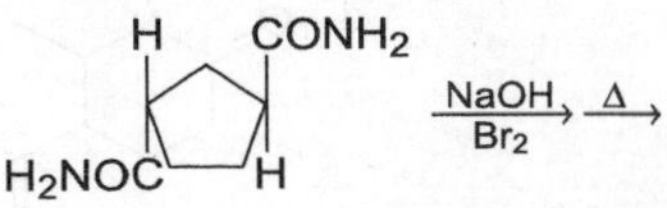

Working Space

12 What is the major product of the reaction?

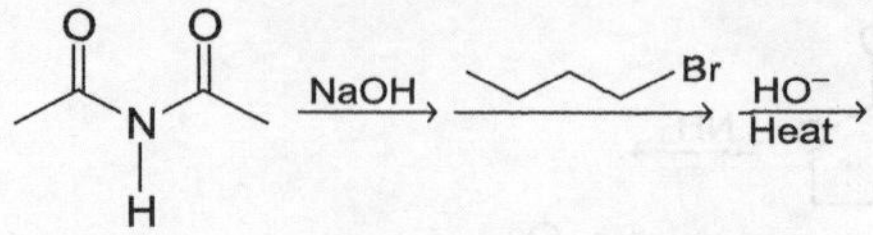

a. 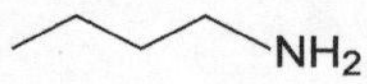 b. 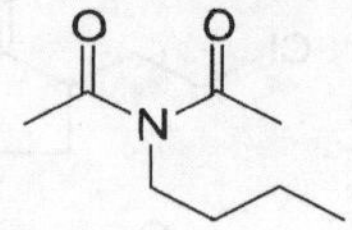

c. 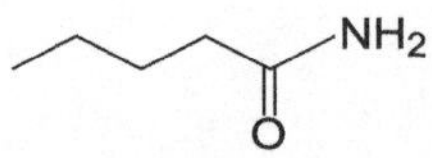 d. 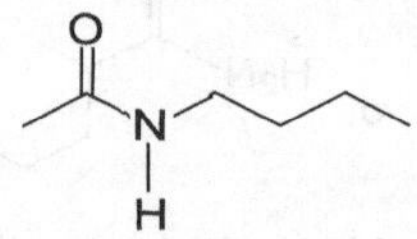

13 Which of the following reactions does not give ethanamide?

a. $(CH_3CO)_2O + NH_3 \xrightarrow{\text{Heat}}$

b. $CH_3CHO + NH_3 \xrightarrow{\text{Heat}}$

c. $CH_3COBr + NH_3 \longrightarrow$

d. $CH_3COOC_2H_5 + NH_3 \xrightarrow{\text{Heat}}$

One or More than One Options Correct Type

***Direction** (Q. Nos. 14-19) This section contains 6 multiple choice questions. Each question has four choices (a), (b), (c) and (d), out of which ONE or MORE THAN ONE are correct.*

14 Which reagent given below can bring about the following transformation?

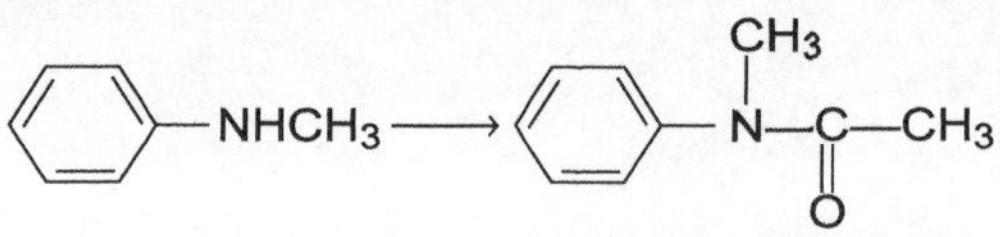

a. CH_3COOH/Δ b. CH_3COCl

c. $(CH_3CO)_2O/\Delta$ d. $CH_3COOC_2H_5/\Delta$

15 Which is/are the expected product in the following sequence of reaction?

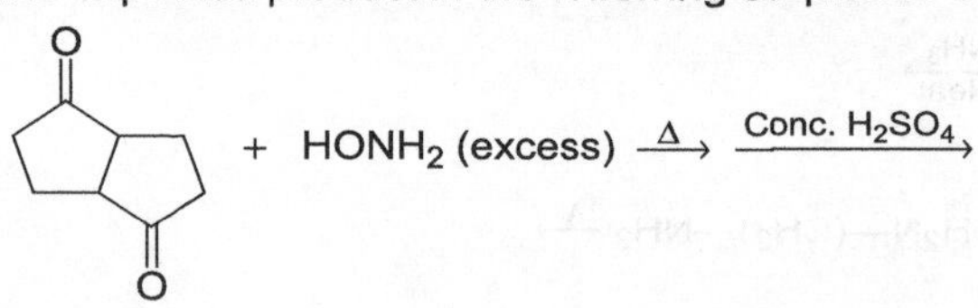

a. 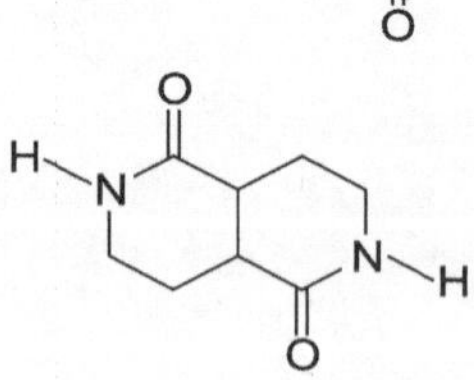 b. 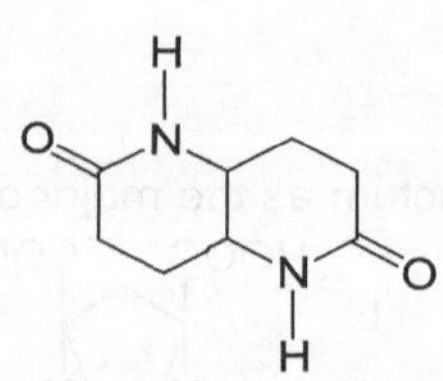

c. 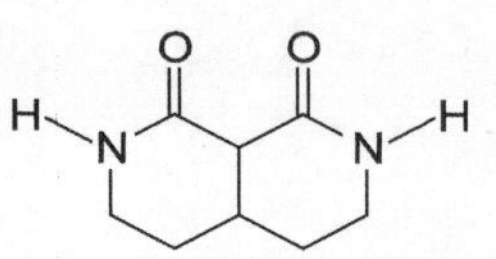 d. 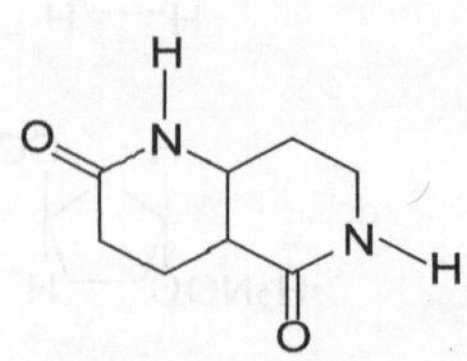

Working Space

16 Consider the following Hofmann's degradation reaction,

$$CH_3-\overset{O}{\overset{\|}{C}}-\overset{15}{N}H_2 + CH_3CH_2-\overset{O}{\overset{\|}{C}}-NH_2 \xrightarrow[Br_2]{NaOH}$$

The expected product(s) is/are

a. CH_3NH_2 b. $CH_3-\overset{15}{N}H_2$

c. $CH_3CH_2-\overset{15}{N}H_2$ d. $CH_3CH_2NH_2$

17 Which reaction(s) below gives primary amine?

a. $CH_3CH_2CH_2CONH_2 + NaOH + Br_2 \longrightarrow$

b. $CH_3CH_2COOH + HN_3 \xrightarrow{\Delta}$

c. $CH_3CH_2CONH_2 + LiAlH_4 \longrightarrow$

d. $CH_3CH_2CONH_2 + NaBH_4 \longrightarrow$

18 In the reaction given below, the intermediates formed is/are

$CH_3CH_2CONH_2 + NaOH + Br_2 \longrightarrow CH_3CH_2NH_2$

a. $CH_3CH_2CONHBr$ b. $CH_3CH_2-\overset{O}{\overset{\|}{C}}-\ddot{N}:$

c. $CH_3CH_2CONHOH$ d. CH_3CH_2NCO

19 Consider the following hydrolysis reaction of a pure enantiomer of amide,

$$CH_3-\overset{O}{\overset{\|}{C}}-NH-\underset{CH_3}{\underset{|}{CH}}-CH_2CH_3 + H_2O \longrightarrow CH_3COOH + CH_3CH_2-\underset{CH_3}{\underset{|}{CH}}-NH_2$$

The correct statement regarding the above hydrolysis reaction is/are

a. Both acid and base can act as catalyst

b. In both acid and base catalysed condition, a pure enantiomer of amine with retention of configuration is obtained

c. Base catalysed hydrolysis is bimolecular while acid catalysed is unimolecular

d. Base catalysed hydrolysis is unimolecular while acid catalysed is bimolecular

Comprehension Type

Direction (Q. Nos. 20-22) This section contains a paragraph, describing theory, experiments, data, etc. Three questions related to the paragraph have been given. Each question has only one correct answer among the four given options (a), (b), (c) and (d).

Passage

Hofmann's bromamide degradation is used to convert a primary amide into primary amine. The proposed mechanism of reaction are :

$$RCONH_2 + Br_2 \xrightarrow{NaOH} RNH_2$$

Step-I $R-\overset{O}{\overset{\|}{C}}-NH_2 + OBr^- \longrightarrow R-\overset{O}{\overset{\|}{C}}-NHBr + HO^-$

Step-II $R-\overset{O}{\overset{\|}{C}}-NHBr + HO^- \longrightarrow R-\overset{O}{\overset{\|}{C}}-\overset{-}{\ddot{N}}-Br + H_2O$

Step-III 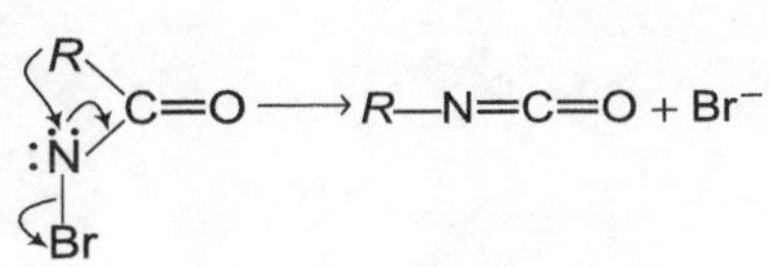

Working Space

Step-IV $R-\overset{O}{\overset{\|}{C}}-\underset{Br}{\underset{|}{\ddot{N}}}: \longrightarrow R-\overset{O}{\overset{\|}{C}}-\ddot{N}: + Br^-$ (Nitrene)

or **Step-V** 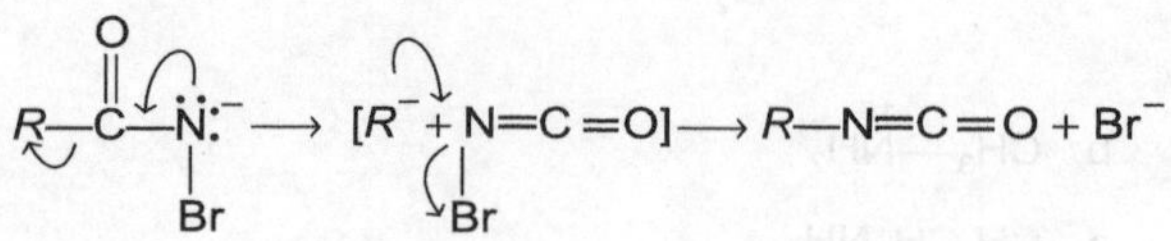

Step-VI $R-N=C=O + H_2O \xrightarrow{HO^-} R-NH_2 + CO_3^{2-}$

20 Which is the most important intermediate formed in the slow, rate determining step?

a. RCONHBr
b. $R-N=C=O$
c. $R-\overset{O}{\overset{\|}{C}}-\ddot{N}:$
d. $Br-N=C=O$

21 In the following reaction,

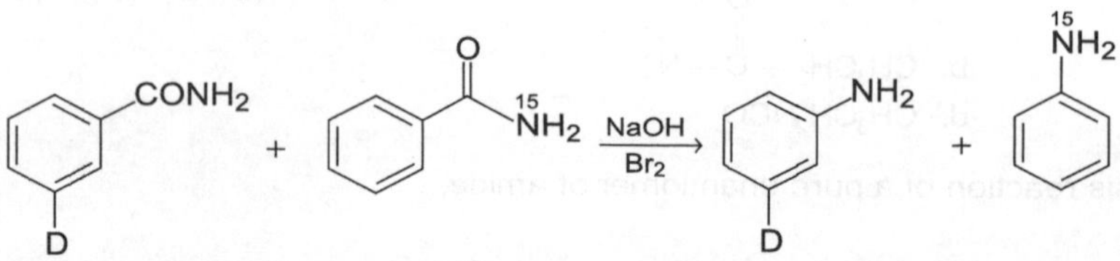

Formation of the above shown products and not any cross-products eliminates completely which step of the reaction mechanism?

a. III b. IV c. V d. VI

22 If a pure enantiomer of amide given below is used

$$C_6H_5-\underset{}{\overset{CH_3}{\overset{|}{C}H}}-CONH_2 + NaOH + Br_2 \longrightarrow$$

What can be predicted regarding product?

a. Product amine will have inverted configuration at α-carbon
b. Racemic mixture of product amine will be formed
c. Product amine has the same sign of specific rotation as of starting amide
d. Product amine has same configuration at α-carbon as of starting amide

One Integer Value Correct Type

Direction *(Q. Nos. 23-26) This section contains 4 questions. When worked out will result in an integer from 0 to 9 (both inclusive).*

23 How many amide isomers exist for C_4H_9ON that do not form amine on treatment with Br_2-NaOH?

24 How many amide isomers exist for $C_5H_{11}ON$ that on treatment with NaOH/ Br_2 give amines?

25 Nylon-6 is formed by the acid catalysed hydrolysis of a cyclic amide. How many carbon atoms are present in this cyclic amide?

26 How many total amide isomers exist for C_3H_7ON?

DPP-7 Preparation of Amines

Subjective Questions

Working Space

Direction (Q. Nos. 1-5) *These questions are subjective in nature, need to be solved completely on notebook.*

1 How butanamine, N-methyl butanamine and N, N-dimethyl butanamine can be prepared from

i. butanoic acid?

ii. pentanoic acid?

2 Starting from phthalic anhydride and 1-bromo butane as only source of carbon, synthesize 1-butanamine as single amine product. In this Gabriel phthalimide synthesis, contrast the difference between alkaline hydrolysis and hydrazine treatment in the final step of synthesis.

3 Complete the following reactions.

i. 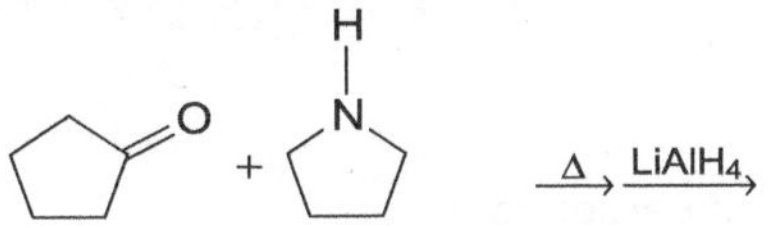

ii. 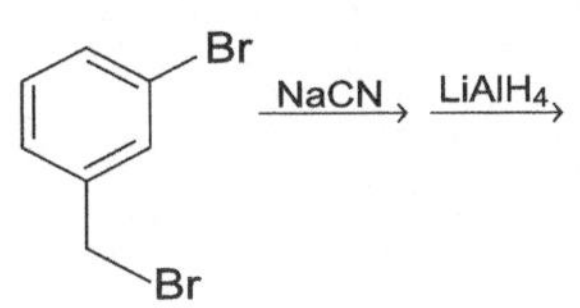

iii. 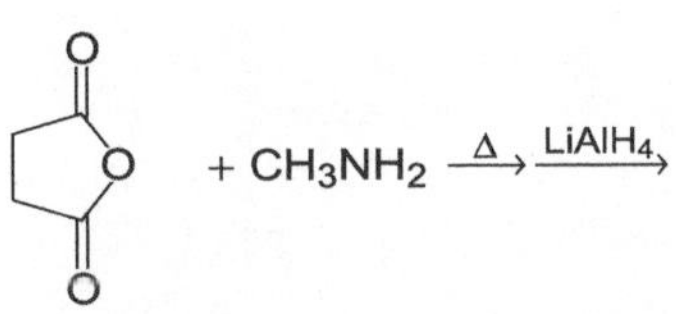

iv. 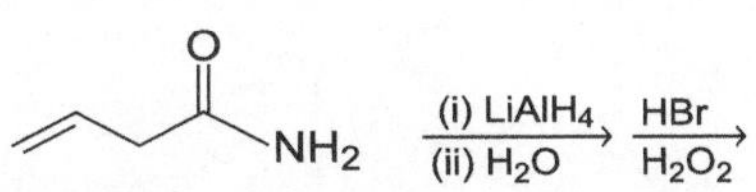

4 An amine *A* ($C_6H_{15}N$) on treatment with CH_3I and then KOH gives *B* ($C_8H_{20}N^+OH^-$). *B* on heating produced *C* (C_4H_8) and *D* ($C_4H_{11}N$). *C* on treatment with aqueous $Hg(CH_3COO)_2$ followed by the reduction with $NaBH_4$ yielded *E* ($C_4H_{10}O$) which does not change the colour of CrO_3-H_2SO_4 solution. Deduce the structures of *A* to *E*.

5 Identify intermediates and final products in the following reaction.

i. $BrCH_2CH_2Br \xrightarrow{NaCN} P \xrightarrow[Na]{C_2H_5OH} Q$

ii. 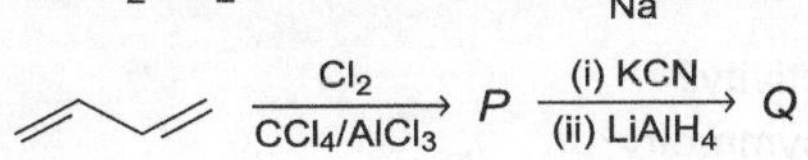

Only One Option Correct Type

Working Space

Direction (*Q. Nos. 6-13*) *This section contains 8 multiple choice questions. Each question has four choices (a), (b), (c) and (d), out of which ONLY ONE is correct.*

6 Which amine given below cannot be obtained by Hofmann's degradation of an amide?

a. $CH_3CH_2NH_2$ b. $C_6H_5NH_2$

c. $C_6H_5-\underset{\underset{NH_2}{|}}{CH}-C_6H_5$ d. 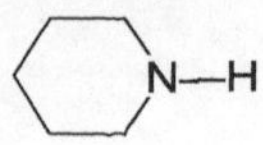

7 A secondary amine cannot be prepared directly by

a. reduction of amide b. reduction of cyanide

c. reduction of imine d. reduction of isocyanide

8 Which is the best preparation of butanamine?

a. 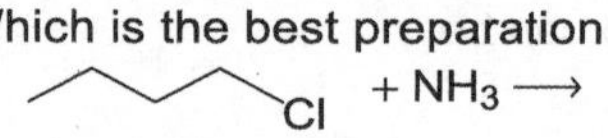

b. 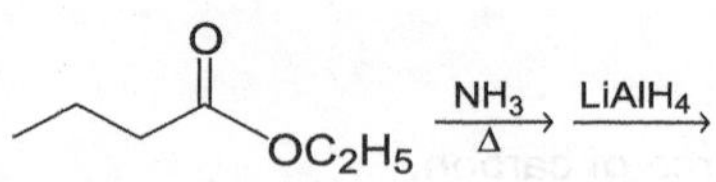

c. 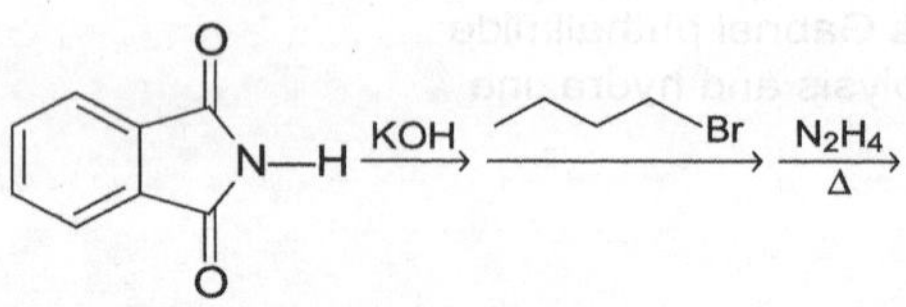

d. 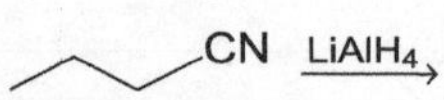

9 In which of the following reaction a primary amine without loss of carbon is produced?

a. $R-CONH_2 + NaOH \xrightarrow{Br_2}$

b. $R-CONH_2 \xrightarrow{LiAlH_4}$

c. $R-CONH_2 \xrightarrow{NaBH_4}$

d. $RCOOH \xrightarrow{HN_3}$

10 Which gives secondary amine as the major product?

a. $CH_3-CH{=}CH_2 \xrightarrow[CH_3CN]{H^+/H_2O} \xrightarrow{LiAlH_4}$

b. $CH_3CN + CH_3MgBr \xrightarrow{Et_2O} \xrightarrow{LiAlH_4}$

c. $CH_3-\overset{\overset{O}{||}}{C}-CH_3 + NH_3 \xrightarrow{\Delta}$

d. $CH_3COOH + (CH_3)_2NH \xrightarrow{\Delta} \xrightarrow{LiAlH_4}$

11 Which reagent can be used for the distinction between a 1° aliphatic amine and $C_6H_5NH_2$?

a. $C_6H_5SO_2Cl$ b. HCl

c. Br_2-H_2O d. $NaNO_2$/HCl

12 Why N-ethyl-N-methyl propanamine does not show optical activity?

a. It has no stereocenter b. It has a center of symmetry

c. It shows fluxional inversion d. It is a *meso* compound

13 The compound which on treatment with nitrous acid at low temperature produces an oily, liquid is

a. CH_3NH_2 b. $C_2H_5NH_2$
c. $(C_2H_5)_2NH$ d. $(CH_3)_3N$

Working Space

One or More than One Options Correct Type

Direction (Q. Nos. 14-17) *This section contains 4 multiple choice questions. Each question has four choices (a), (b), (c) and (d), out of which ONE or MORE THAN ONE are correct.*

14 A primary amine can be prepared by

a. reduction of nitriles b. reduction of $RCONH_2$
c. reduction of nitro alkanes d. reaction of RX with NH_3

15 Which amines given below, when treated with CH_3I forms racemic mixture of salts?

a. $CH_3CH_2NHCH_3$
b. $C_6H_5CH—NHC_2H_5$
c. $C_6H_5CH_2—N(CH_3)—C_6H_5$
d. $C_6H_5—CH_2—NH—C_6H_5$

16 In Gabriel phthalimide synthesis of primary amine, which amine given below is formed with greater difficulty than $(CH_3)_2CHNH_2$?

a. 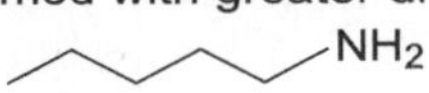
b. 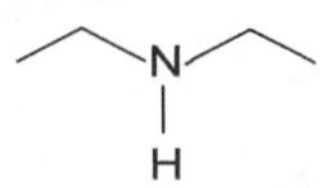
c. $C_6H_5—NH_2$
d. 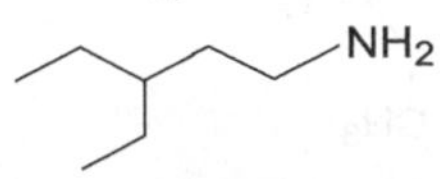

17 Which amine(s) given below forms precipitate with tosyl chloride that is insoluble in KOH solution?

a. $(CH_3)_3N$ b. $CH_3CH_2NHCH_3$
c. $CH_3CH_2NH_2$ d. $C_6H_5NHCH_3$

Comprehension Type

Direction (Q. Nos. 18-20) *This section contains a paragraph, describing theory, experiments, data, etc. Three questions related to the paragraph have been given. Each question has only one correct answer among the four given options (a), (b), (c) and (d).*

Passage

An organic compound *A* has molecular formula $C_9H_{13}NO$ and it can be resolved into enantiomers. *A* does not decolourise Br_2 - H_2O solution. *A* on hydrolysis with dil. H_2SO_4 gives $B(C_9H_{14}O_3)$ which gives effervescence with $NaHCO_3$. *B* on treatment with $NaBH_4$ followed by heating with concentrated H_2SO_4 yielded a sweet smelling liquid $C(C_9H_{14}O_3)$. Also, *A* on reduction with $LiAlH_4$ yields $C_9H_{19}ON$ which on further heating with concentrated H_2SO_4 produces the following compound.

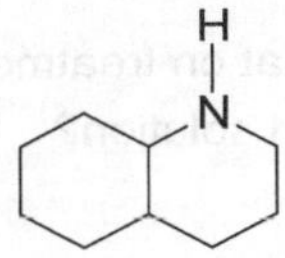

Working Space

18 What is the structure of *A*?

a. 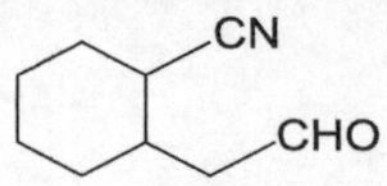

b. 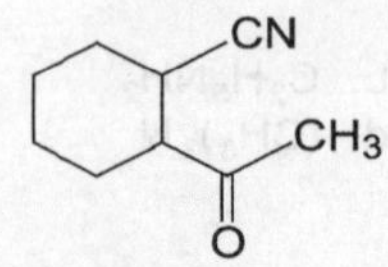

c. 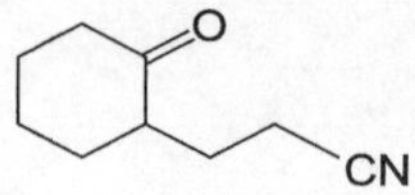

d. 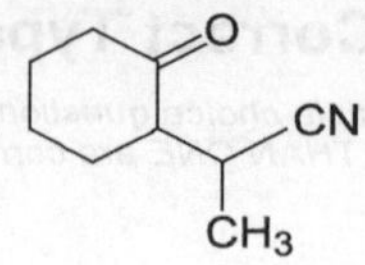

19 *B* and *C* respectively are

a. 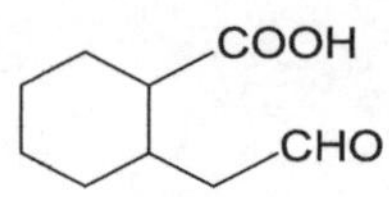 and 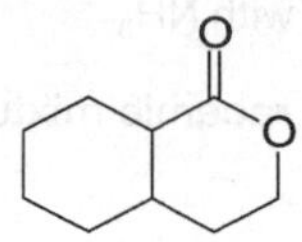

b. 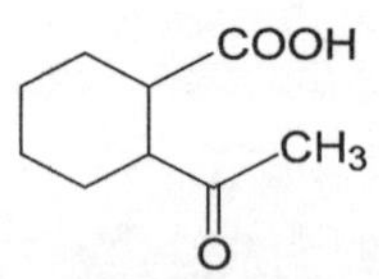 and 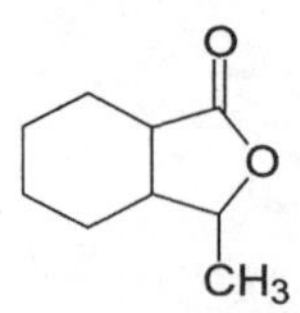

c. 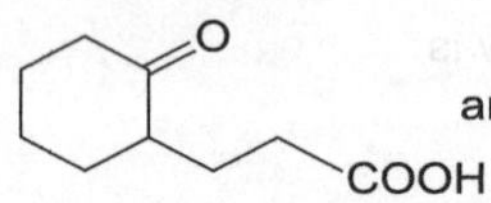 and 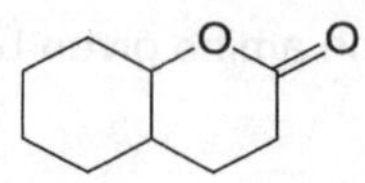

d. 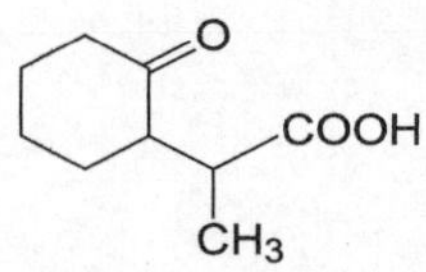 and 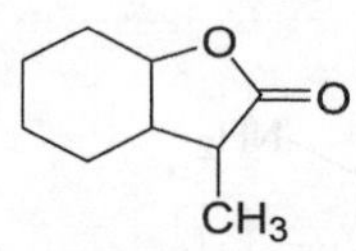

20 What is formed at the end of following reaction $C \xrightarrow[\Delta]{NH_3} \xrightarrow[Br_2]{NaOH} ?$

a. 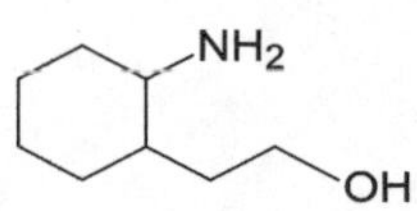

b. 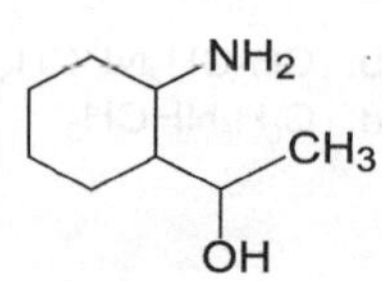

c. 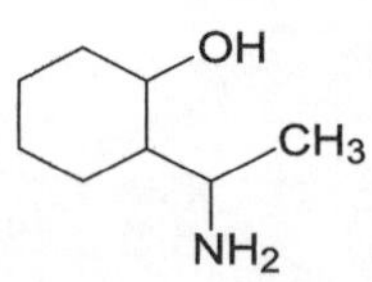

d. 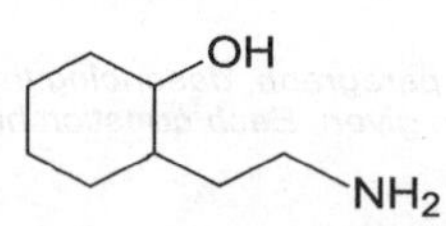

One Integer Value Correct Type

***Direction** (Q. Nos. 21 and 22) This section contains 2 questions. When worked out will result in an integer from 0 to 9 (both inclusive).*

21 How many tertiary amines are possible for $C_6H_{15}N$?

22 How many different isomers of amine exist for $C_5H_{13}N$ that on treatment with tosyl chloride form precipitate which does not dissolve in KOH solution?

DPP-8 Reactions of Amines

Subjective Questions

Working Space

Direction *(Q. Nos. 1-5) These questions are subjective in nature, need to be solved completely on notebook.*

1 How primary, secondary and tertiary amines are differentiated by Hinsberg's test?

2 Bring about the following transformation.

i. $CH_3CH_2NH_2 \longrightarrow CH_3CH_2NHCH_3$

ii. $CH_3CH_2CH_2NH_2 \longrightarrow CH_3CH_2COOH$

3 How a mixture of $CH_3CH_2NH_2$ and CH_3CONH_2 be separated using HCl solution?

4 In aqueous solution, $(CH_3)_3N$ is a weaker base than $(CH_3)_2NH$ but its reverse is true in gas phase, explain.

5 Write the reactions involved in distinction between $CH_3CH_2NH_2$ and $(CH_3CH_2)_2NH$ using CS_2 and $HgCl_2$.

Only One Option Correct Type

Direction *(Q. Nos. 6-13) This section contains 8 multiple choice questions. Each question has four choices (a), (b), (c) and (d), out of which ONLY ONE is correct.*

6 Carbyl amine test is performed in alcoholic KOH by heating a mixture of

a. $CHCl_3$ and Ag powder
b. $CHCl_3$ and R—NH_2
c. R—Cl and R—NH_2
d. R—CN and R—NH_2

7 Tetraethyl ammonium bromide is produced by the reaction of

a. triethyl amine and $(CH_3)_3CBr(l)$
b. triethyl amine and $Br_2(l)$
c. diethyl amine and ethyl bromide
d. triethyl amine and HBr

8 A primary aliphatic amine is converted to alkyl chloride by treatment with

a. HCl
b. PCl_5
c. CH_3COCl
d. NOCl

9 Which of the following reagent is useful for separating aniline from mixture of aniline and nitrobenzene?

a. NaOH(*aq*)
b. H_2O
c. $NaHCO_3(aq)$
d. HCl(*aq*)

10 Which of the following reagent gives a white precipitate with an aqueous solution of ethyl amine?

a. $FeSO_4(aq)$
b. $CuSO_4(aq)$
c. $ZnSO_4(aq)$
d. $[Ag(NH_3)_2]^+(aq)$

11 When a solution of aliphatic amine is treated with $NaNO_2$ in HCl, effervescence is observed due to

a. CO_2
b. NO_2
c. N_2
d. H_2

Working Space

12 Which of the following is not true regarding a primary amine?

a. It forms liquid oxamic ester
b. It forms N-alkyl benzene sulphonamide
c. It forms mustard oil
d. It is prepared by Gabriel phthalimide synthesis

13 The compound $C_5H_{13}N$ is optically active and reacts with HNO_2 to give an alcohol $C_5H_{11}OH$. Hence, the compound is

a. 2-aminopentane
b. N-methyl butanamine
c. N, N-dimethyl propanamine
d. 1-amino pentane

One or More than One Options Correct Type

Direction *(Q. Nos. 14-19) This section contains 6 multiple choice questions. Each question has four choices (a), (b), (c) and (d), out of which ONE or MORE THAN ONE are correct.*

14 Amine(s) that form precipitate with $p\text{-}CH_3—C_6H_4—SO_2Cl$ is/are

a. $C_6H_5NH_2$ b. $C_6H_5CH_2NH_2$
c. $C_6H_5NHCH_3$ d. $C_6H_5N(CH_3)_2$

15 Amine (s) that gives visible change with $NaNO_2/HCl$ is/are

a. 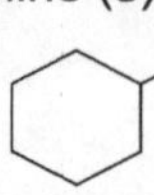
b. 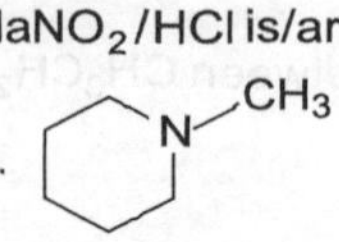
c. 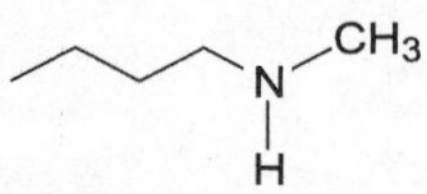
d. 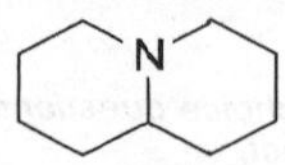

16 In which reaction(s) reactants and products are correctly matched?

a. $C_6H_5NH_2 + CHCl_3 \xrightarrow{NaOH} C_6H_5NC + NaCl$

b. $CH_3CH_2NH_2 + CHCl_3 \xrightarrow{NaOH} CH_3CH_2NC + NaCl$

c. $(CH_3)_2NH + CHCl_3 \xrightarrow{NaOH} H—C(=O)—N(CH_3)_2 + NaCl$

d. $(CH_3)_3N + CHCl_3 \xrightarrow{NaOH} (CH_3)_3C—NH_2 + NaCl$

17 In which of the following reaction(s), reactants and products are correctly matched?

a. $CH_3CH_2NH_2 \xrightarrow[\Delta]{CS_2} \xrightarrow{HgCl_2} CH_3CH_2NCS + HgS + 2HCl$

b. $(CH_3)_2NH \xrightarrow[\Delta]{CS_2} (H_3C)_2N—C(=S)—SH$

c. $C_6H_5NH_2 + CS_2 \xrightarrow{\Delta}$ No reaction

d. $(CH_3)_2N + CS_2 \xrightarrow{\Delta} CH_3—N(CH_3)—C(=S)—SCH_3$

Working Space

18 In which of the following pair(s), the first member is more basic than second one in aqueous solution?

a. $CH_3CH_2NH_2$; $(CH_3CH_2)_2NH$
b. $(CH_3CH_2)_2NH$; $(CH_3CH_2)_3N$
c. $CH_2(NH_2)_2$; $CH_3CH_2NH_2$
d. $C_6H_5NH_2$; $CH_3CH_2NH_2$

19 In the following reaction,

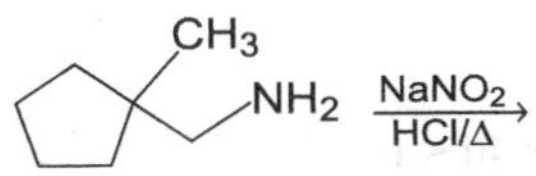

The expected product(s) is/are

a. 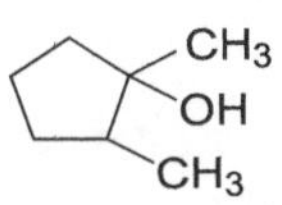

b. 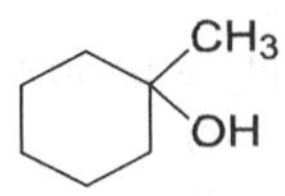

c. 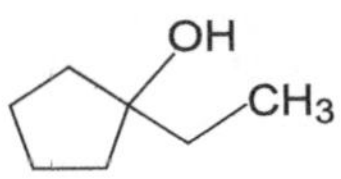

d. 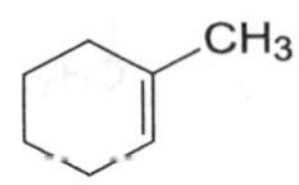

Comprehension Type

Direction *(Q. Nos. 20-22) This section contains a paragraph, describing theory, experiments, data, etc. Three questions related to the paragraph have been given. Each question has only one correct answer among the four given options (a), (b), (c) and (d).*

Passage

Carbyl amine reaction is used to differentiate a primary amine from secondary amines. Primary amine form a foul smelling isocyanide (*R*NC) which serves for its detection. Both aliphatic and aromatic primary amines give this test. A widely accepted mechanism of this reaction is

Step-I $CHCl_3 + OH^- \longrightarrow :CCl_2 + H_2O + Cl^-$
Dichlorocarbene

Step-II 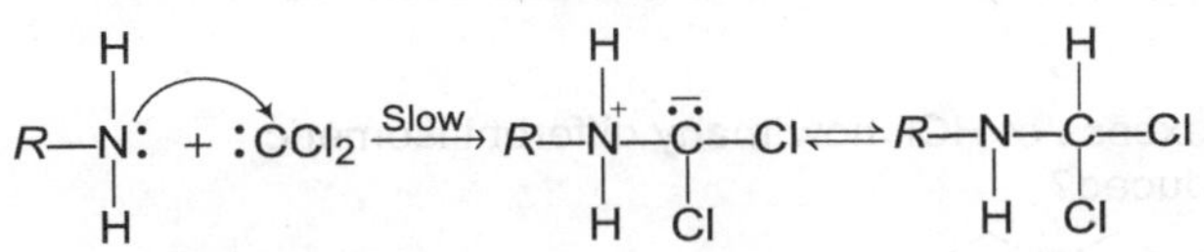

Step-III 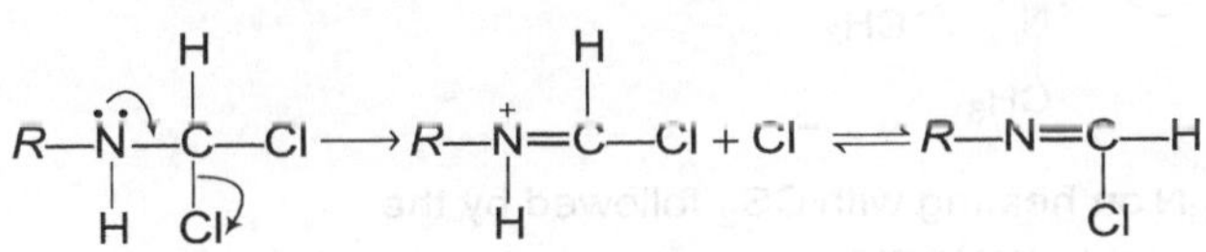

Step-IV 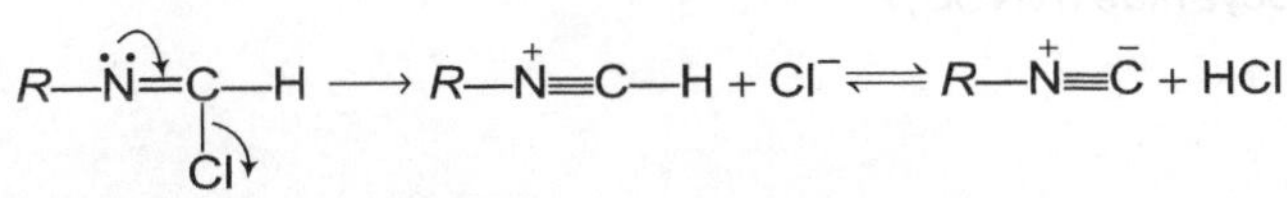

20 What is true regarding the first step of mechanism?

a. Nucleophile $\ddot{C}Cl_2$ is formed by α-elimination

b. Nucleophile $\ddot{C}Cl_2$ is formed by β-elimination

c. Electrophile $\ddot{C}Cl_2$ is formed by α-elimination

d. Electrophile $\ddot{C}Cl_2$ is formed by β-elimination

Working Space

21 What is the increasing order of reactivity of the following amines in the above carbyl amine reaction?

a. I < II < III < IV
b. II < IV < III < I
c. II < IV < I < III
d. III < I < IV < II

22 What would be formed if amine $CH_3CH_2NHCH_3$ is used?

a. It does not react with CCl_2

b. $H-\overset{\overset{\displaystyle O}{\|}}{C}-\underset{\underset{\displaystyle CH_3}{|}}{N}-CH_2CH_3$

c. $CH_3CH_2-\overset{\overset{\displaystyle CH_3}{|}}{N}-CHCl_2$

d. $CH_3CH_2-\underset{\underset{\displaystyle CH_3}{|}}{N}-COOH$

One Integer Value Correct Type

Direction *(Q. Nos. 23-26) This section contains 4 questions. When worked out will result in an integer from 0 to 9 (both inclusive).*

23 If adipic acid (hexane dioic acid) is heated with excess of hydrazoic acid (HN_3), how many gaseous molecules are formed per molecule of diacid?

24 If chloroform is treated with KOH followed by the addition of 1, 6-diamino hexane, hexane-1,6-di isocyanide is formed. How many formula units of KCl are formed per molecule of diamine?

25 If the following amine is treated with excess of HCl, how many different isomeric quarternary ammonium salts are produced?

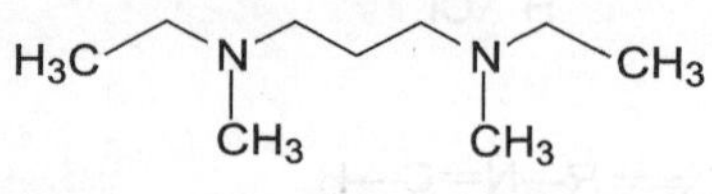

26 How many different isomers of $C_4H_{11}N$ on heating with CS_2 followed by the addition of $HgCl_2$ gives alkyl isothiocyanide ($RNCS$)?

SET - 1

Revisal Problems for JEE Main

Only One Option Correct Type

Working Space

***Direction** (Q. Nos. 1-26) This section contains 26 multiple choice questions. Each question has four choices (a), (b), (c) and (d), out of which ONLY ONE option is correct.*

1 Which of the following statements is not generally true?

a. The boiling point of a carboxylic acid is higher than that of its methyl ester
b. Methyl esters are more reactive acylating agents than their amide counterparts
c. Amide hydrolysis may be carried out with either strong acid or base catalyst
d. Fischer esterification of acids with alcohols requires a strong base catalyst

2 An equimolar mixture of benzoic acid and benzyl alcohol is dissolved in equal volumes of ether and 5% aqueous NaOH. The resulting mixture separates into two immiscible liquid layers. Which of the following is approximately correct?

a. Both organic solutes are largely in the ether layer
b. The benzyl alcohol is in the ether layer and the benzoic acid is in the water layer
c. Both organic solutes are largely in the water layer
d. The benzyl alcohol is in the water layer and benzoic acid is in the ether layer

3 Fischer esterification of phenyl acetic acid with 1-propanol gave a mixture of 93% of the ester, propyl phenylacetate, contaminated with 7% unreacted acid. Which of the following treatments would be best used to purify the ester?

a. Reduce the unwanted acid with $LiAlH_4$ in ether
b. Wash an ether solution of the crude product with concentrated brine (*aq*. NaCl)
c. Wash an ether solution of the crude product with 5% aqueous sulphuric acid
d. Wash an ether solution of the crude product with 5% aqueous sodium carbonate

4 Which of the following would not be a useful reaction for preparing isobutyric acid, $(CH_3)_2CHCO_2H$?

a. 2-methyl-1-propanol + Jone's reagent
b. 2-bromopropane + CO_2; followed by hydrolysis.
c. *cis*-2, 5 dimethyl-2-hexene + O_3; followed by H_2O_2
d. 2-bromopropane + NaCN; followed by acid-catalysed hydrolysis

5 Two C_8H_9Br isomers form Grignard reagents which on carboxylation give isomeric carboxylic acids. Oxidation of each isomeric acid with hot $KMnO_4$ produces the same $C_9H_6O_6$ tricarboxylic acid. Which of the following compounds are the original isomeric bromides?

I. 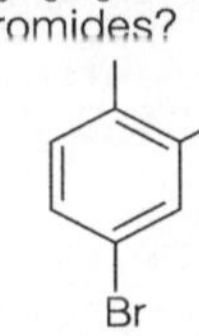

II. 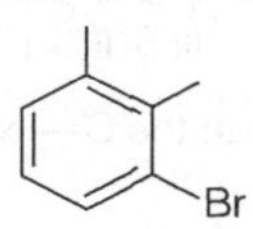

III. 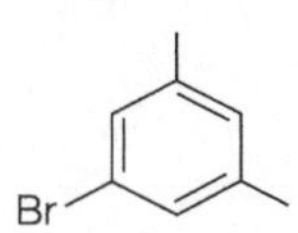

IV. 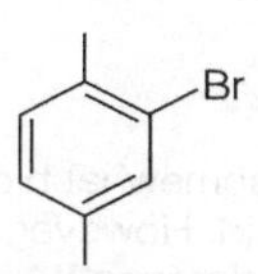

a. Both I and II
b. Both III and IV
c. Both I and IV
d. Both II and III

Working Space

6 Which sequence of reactions would be best for the conversion of toluene into *para*-bromophenylacetic acid? (NBS = N-bromosuccinimide)

a. (i) Br_2 and $FeBr_3$; (ii) $KMnO_4$ and heat; (iii) NaCN in ethanol; (iv) H_3O^+ and heat
b. (i) Br_2 and $FeBr_3$; (ii) NBS in CCl_4; (iii) Mg in ether followed by CO_2; (iv) H_3O^+
c. (i) Br_2 and $FeBr_3$; (ii) NBS in CCl_4; (iii) NaCN in ethanol; (iv) H_3O^+ and heat
d. (i) Br_2 and $FeBr_3$; (ii) $KMnO_4$ and heat; (iii) Mg in ether followed by CO_2; (iv) H_3O^+

7 Which of the following reagents are suitable for reducing an acyl chloride to an aldehyde?

I. $LiAlH_4$ in ether
II. $NaBH_4$ in *tert*-butanol
III. $LiAlH(t-C_4H_9O)_3$ in ether at – 78° C
IV. H_2 and Pd / $BaSO_4$ catalyst

a. Both III and IV
b. Both I and IV
c. Only I
d. All are suitable

8 What is the expected product from the double reaction drawn below?

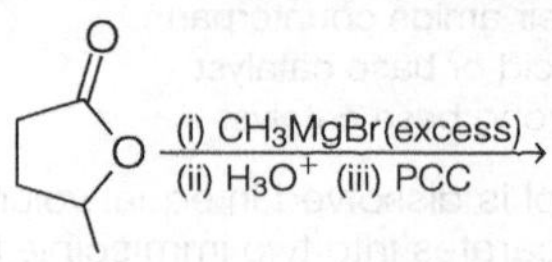

a. 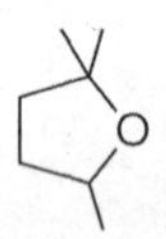

b. 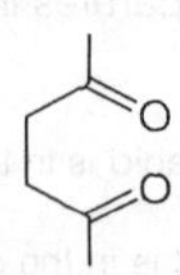

c. 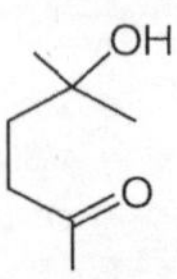

d. 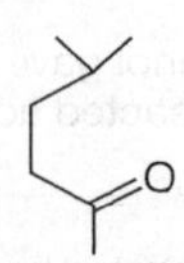

9 Which of the following reactions is most likely to produce ethyl propanoate?

a. Sodium ethoxide + propanoic acid
b. Propanol + acetyl chloride
c. Sodium propanoate + acetic anhydride
d. Potassium propanoate + ethyl iodide

10 If diethyl amine is treated separately with the following derivatives of isobutyric acid, what order of reactivity is expected?

I. Isobutyronitrile (2-methylpropanenitrile)
II. Isobutyryl chloride (2-methylpropanoyl chloride)
III. Ethyl isobutyrate
IV. Isobutyric anhydride (2-methylpropanoic anhydride)

a. I > II > III > IV
b. II > IV > III > I
c. IV > II > III > I
d. IV > III > II > I

11 The C—NH_2 bond in acetamide is 0.1 Å shorter than the C—NH_2 bond in ethylamine. Why?

a. N : N repulsion in ethylamine
b. Hyperconjugation in ethylamine
c. Dipole interactions in acetamide
d. $p-\pi$ resonance in acetamide

12 Methyl esters of carboxylic acids, RCO_2CH_3, have somewhat higher molecular masses than 1°-amide, $RCONH_2$, derivatives of the same acid. However, the amides have much higher boiling points. What is responsible for this difference?

a. Hydrogen bonding
b. Resonance conjugation of N with C$=$O
c. The lower electronegativity of N *versus* O
d. Rapid pyramidal inversion of the nitrogen

Working Space

13 Ethyl acetate undergoes the following sequence of reactions,

I. treatment with excess phenylmagnesium bromide in ether
II. heating with Conc. H_3PO_4
III. treatment B_2H_6 in ether, followed by alkaline H_2O_2
IV. treatment with Jones' reagent (CrO_3 in aqueous acid + acetone)

Which of the following is the expected product? Note that C_6H_5 = phenyl

a. 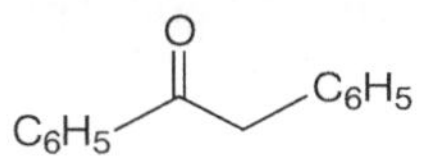

b. 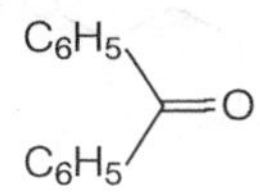

c. 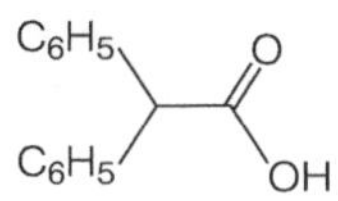

d. 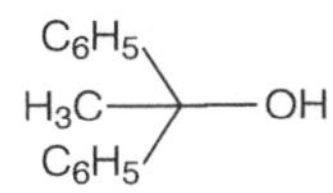

14 Which reactions would best accomplish in the following transformation?

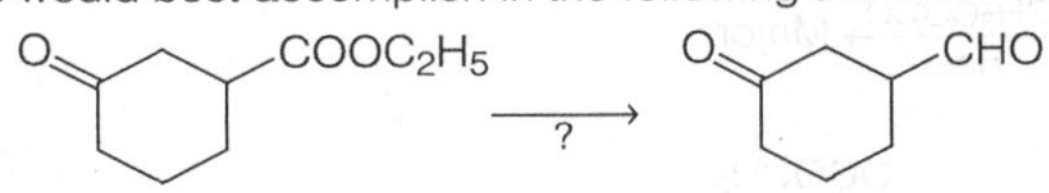

a. (i) $LiAlH_4$, (ii) Jones' reagent
b. (i) $LiAlH_4$, (ii) PCC in CH_2Cl_2
c. (i) $NaBH_4$; (ii) Jones' reagent
d. (i) $NaBH_4$; (ii) PCC in CH_2Cl_2

15 Adipic acid (hexanedioic acid) is converted to its diacid chloride by reaction with $SOCl_2$, and this then reacts with 2 equivalents of sodium azide in ether solution. Addition of conc. HCl, followed by heating, results in considerable gas evolution and the formation of a crystalline water soluble solid. What is this product?
(addition of *aq*. NaOH to the solid produces a foul smelling liquid.)

a. 1, 4-diaminobutane, dihydrochloride salt
b. 1, 5-diaminopentane, dihydrochloride salt
c. 1, 6-diaminohexane, dihydrochloride salt
d. diammonium adipate

16 2, 2-dimethyl-1, 3-cyclohexanedione is refluxed several hours in an aqueous dioxane solution of sodium hydroxide (10%). The resulting solution is adjusted to pH = 5 by the addition of dilute HCl and then extracted with ether. What compound has been prepared by this procedure?

a. 2-cyclohexenone
b. 5-oxohexanoic acid
c. 6-methylheptanoic acid
d. 5-oxo-6-methylheptanoic acid

17 Which set of reaction conditions is best suited for the preparation of 5-oxo-hexanoic acid from 5-bromo-2-pentanone?

a. (i) NaCN in ethanol; (ii) H_3O^+ and heat
b. (i) NaC≡CH in ether; (ii) aqueous $KMnO_4$ and heat
c. (i) Mg in ether; (ii) CO_2 then H_3O^+
d. (i) NaOH in ethanol; (ii) Jones' reagent

18 A C_7H_9N base reacts with sodium nitrite and hydrochloric acid at 0° C, giving a clear solution. On heating with KCN and $Cu_2(CN)_2$ a gas evolves, and continued heating with conc. HCl yields a $C_8H_8O_4$ crystalline acid. Heating this acid with aqueous $KMnO_4$ produces a $C_8H_6O_4$ product, which dehydrates on strong heating to give a crystalline $C_8H_4O_3$ compound. What is the C_7H_9N base?

a. Benzylamine
b. N-methylaniline
c. *para*-toluidine (the toluidines are aminotoluenes)
d. *ortho*-toluidine

19 Which of the following acids does not decarboxylate on heating?

a. 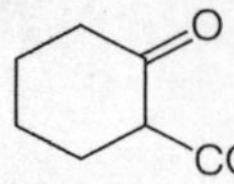

b. 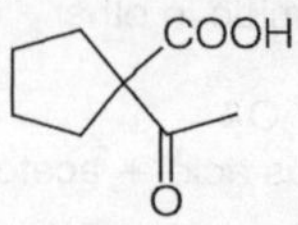

c. 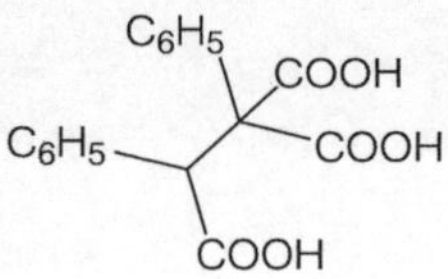

d. 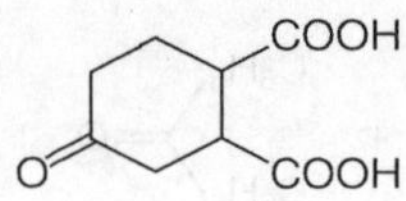

20 What is the major product of the following reaction?

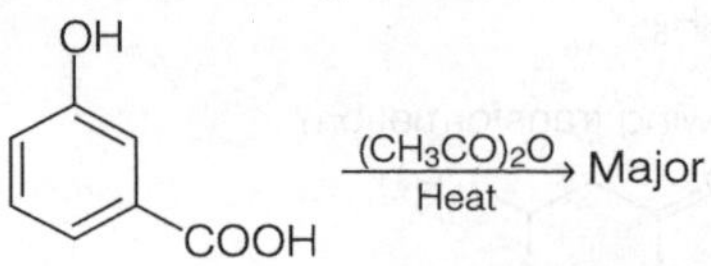

a. 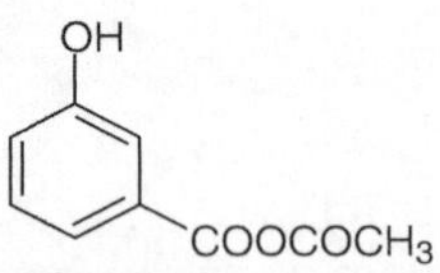

b. 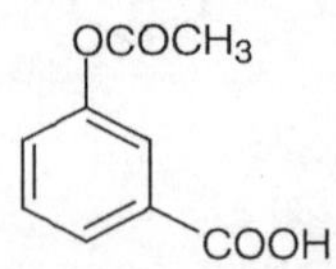

c. 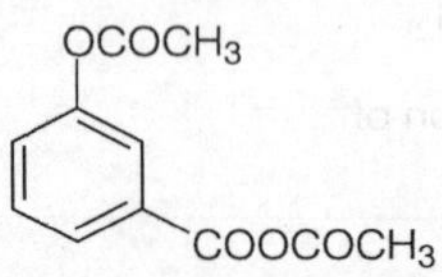

d. 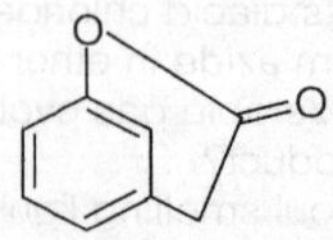

21 Treatment of chiral ester I with one equivalent of LDA followed by allyl bromide II affords the alkylation product III.

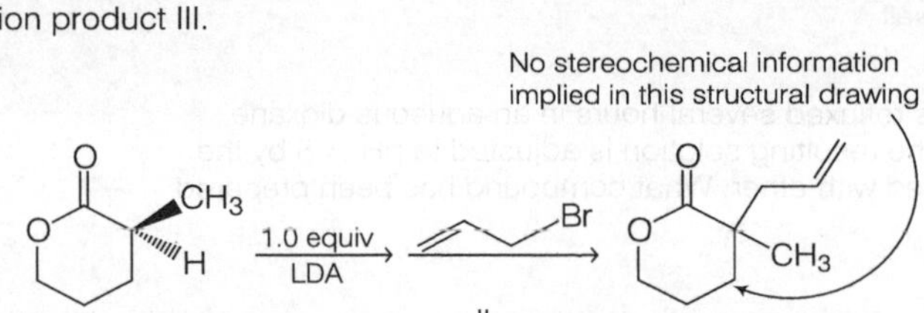

Which of the following statement is true?

a. III is chiral
b. III is achiral
c. III is a *meso* compound
d. III is racemic

22 What is the most likely product in the following reaction?

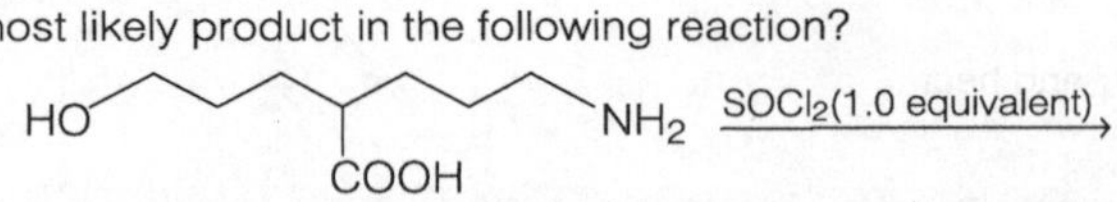

a. 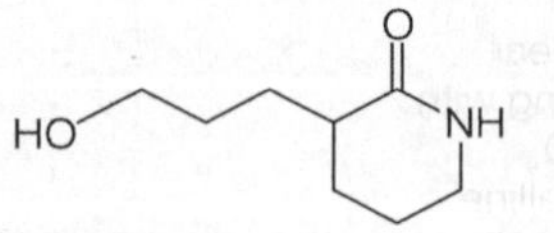

b. 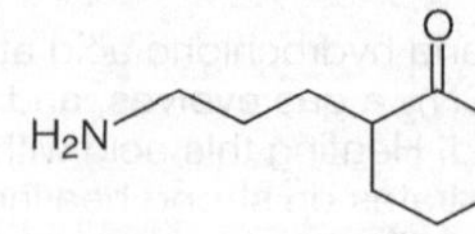

c. 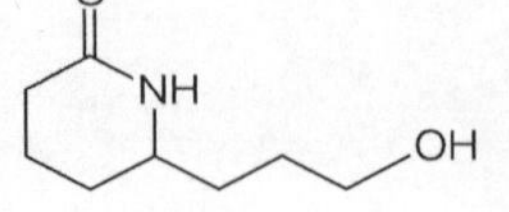

d. 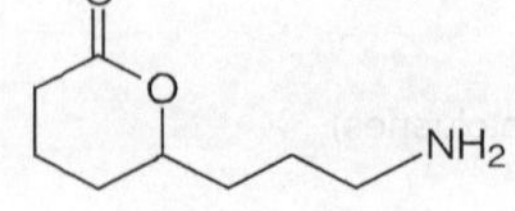

Working Space

Working Space

23 Arrange the following esters in the increasing order of reactivity towards base catalysed hydrolysis reaction.

I. 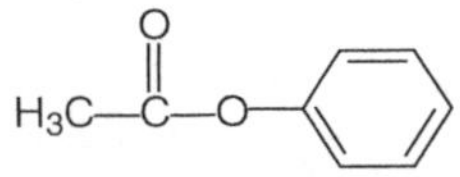

II. 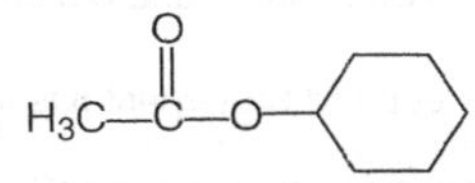

III. 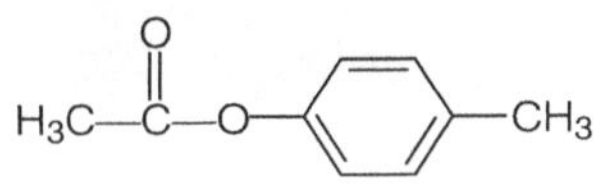

IV. 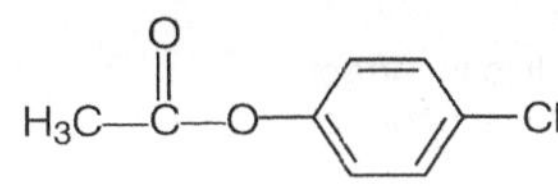

a. II < I < III < IV
b. IV < I < III < II
c. II < III < I < IV
d. IV < II < I < III

24 The compound given below can be acetylated in a similar manner to acetylation of hydroxy benzoic acid to give salicylate ester derivatives. Which of the following product would result upon acetylation of this compound with acetic anhydride?

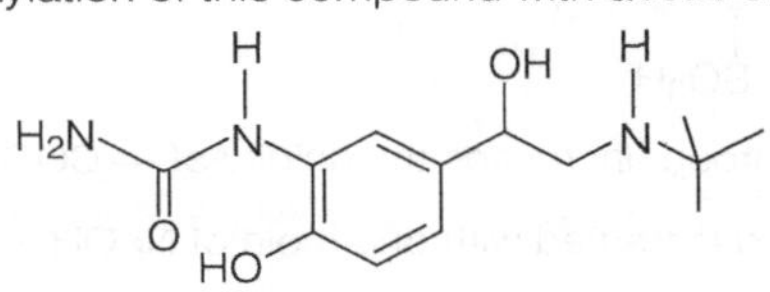

a. 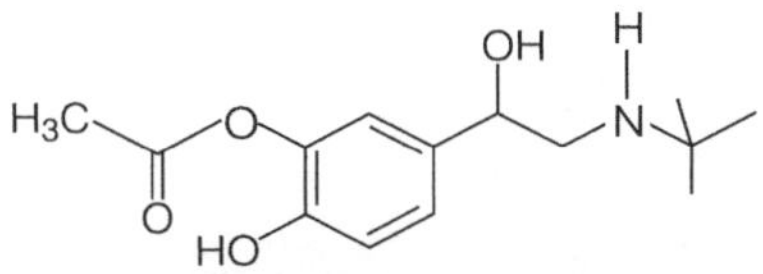

b. 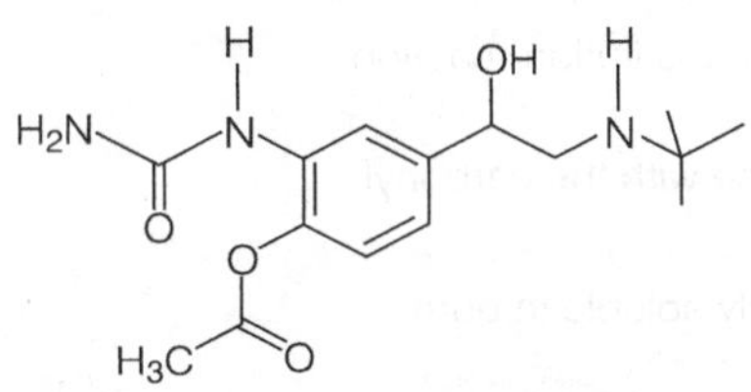

c. 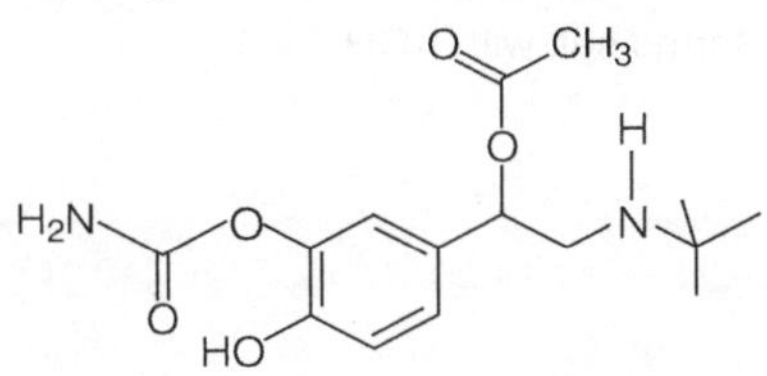

d. 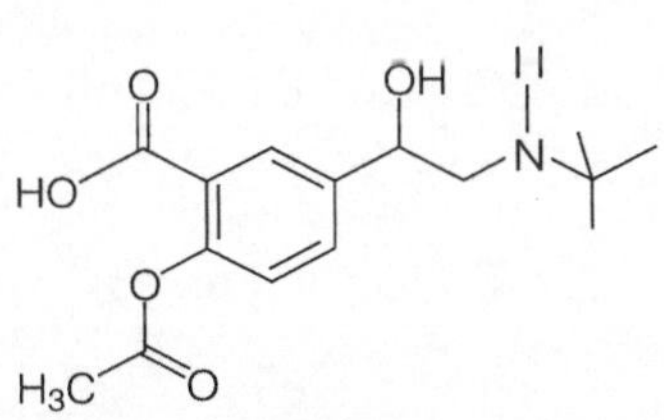

25 A $C_5H_{13}N$ compound gives a base soluble derivative in the Hinsberg's test. Which of the following best satisfy this condition?

a. N, N-diethyl propyl amine
b. Isopropyl dimethyl amine
c. 2, 2-dimethyl propyl amine
d. N-methyl-2-methyl propyl amine

26 Which of the following reagent cannot be used for differentiating propanamine from N-methyl butanamine?

a. $CHCl_3/KOH$
b. *p*-$CH_3—C_6H_4—SO_2Cl$
c. $CS_2/HgCl_2$
d. CH_3COOH

Statement Type

Working Space

Direction *(Q. Nos. 27-30) This section is based on Statement I and Statement II. Select the correct answer from the codes given below.*

a. Both Statement I and Statement II are correct and Statement II is the correct explanation of Statement I
b. Both Statement I and Statement II are correct but Statement II is not the correct explanation of Statement I
c. Statement I is correct but Statement II is incorrect
d. Statement II is correct but Statement I is incorrect

27 **Statement I** Consider the following neutralisation reaction,

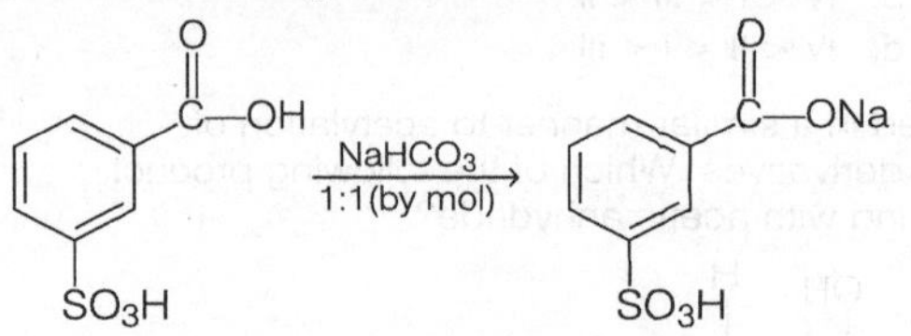

Statement II —SO_3H is an electron withdrawing group, increases the acidity of —COOH.

28 **Statement I** If one mole of the following compound is treated with one mole of NaOH, one mole of phenyl propanoate is formed.

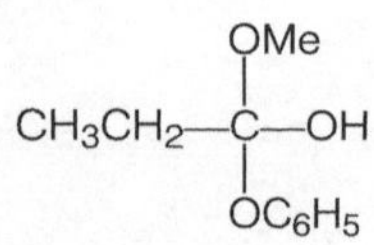

Statement II Phenoxide ion ($C_6H_5O^-$) has resonance stability.

29 **Statement I** Acetyl chloride (CH_3COCl) undergoes faster nucleophilic substitution reaction (S_N2) than chloroethane.
Statement II In acetyl chloride, the leaving group (—Cl) is in resonance with the carbonyl ($>C=O$) group.

30 **Statement I** Ethyl acetoacetate ($CH_3COCH_2COOC_2H_5$) is very sparingly soluble in pure water and highly soluble in aqueous KOH solution.
Statement II It's methylenic (—CH_2—) hydrogen is sufficiently acidic, forms salt with KOH quantitatively.

SET - 2

Revisal Problems
for JEE Advanced

Only One Option Correct Type

Working Space

Direction *(Q. Nos. 1-10) This section contains 10 multiple choice questions. Each question has four choices (a), (b), (c) and (d), out of which ONLY ONE is correct.*

1 Reduction of 4-oxobutyric acid, also called succinic semialdehyde, by sodium borohydride followed by aqueous acid gives a $C_4H_6O_2$ product. This product is which of the following compounds?

a. 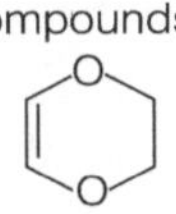

b. 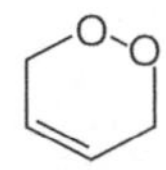

c. 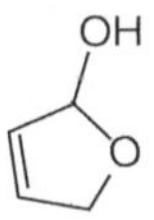

d. 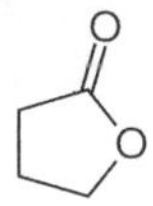

2 Which of the following would not be a useful method for converting a carboxylic acid into an ester derivative?

a. $RCO_2H + CH_2N_2$ in ether
b. $RCO_2H + (CH_3)_2C{=}CH_2$ and acid catalyst
c. $RCO_2H + C_2H_5OH$ and acid catalyst + Heat ($—H_2O$)
d. $RCO_2^-Na^+ + (CH_3)_3CBr$

3 On heating in the presence of bromine, 2, 2-dimethyl-3-oxobutyric acid produces 3-bromo-3-methyl-2-butanone. What unstable intermediate is involved in this reaction?

a. 1, 2, 2-trimethylcyclopropanol
b. 2, 2-dimethyl β-butyrolactone
c. 2-hydroxy-3-methylbutene
d. dimethylketene

4 Fischer esterification of mesitoic acid (2, 4, 6-trimethylbenzoic acid) is extremely slow compared with benzoic acid itself. What is the major factor accounting for this difference in reactivity?

a. Steric hindrance by the *ortho* methyl groups
b. The electron donating inductive effect of methyl substituents
c. Acid-catalysed rearrangement of the methyl substituents
d. Rapid acid-catalysed decarboxylation of mesitoic acid

5 The major product in the following reaction is :

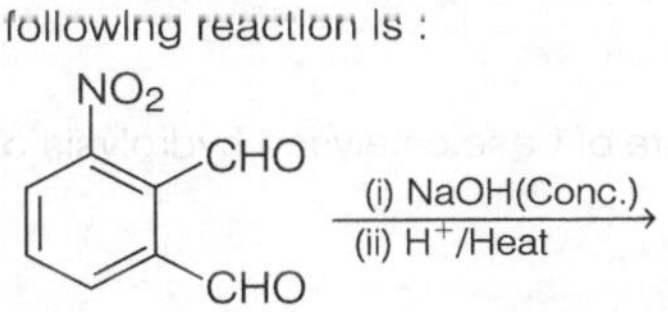

a. 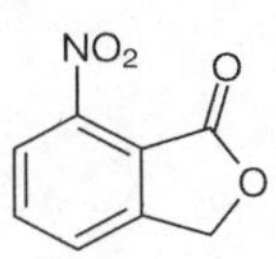

b. 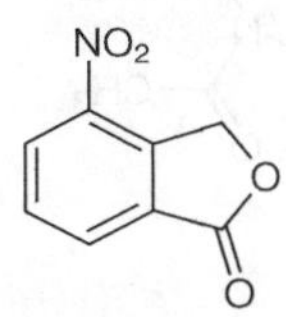

c. 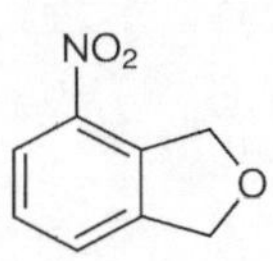

d. 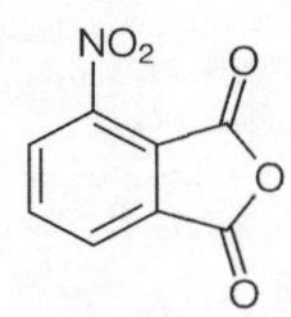

Working Space

6 What donor and acceptor reactants should be used to prepare the following compound by Claisen condensation?

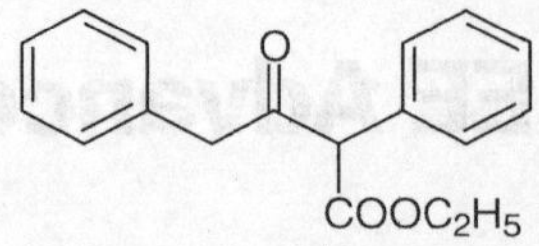

a. Acceptor = ethyl benzoate; donor = ethyl phenylacetate
b. Acceptor = ethyl phenylacetate; donor = ethyl benzoate
c. Acceptor = ethyl benzoate; donor = ethyl benzoate
d. Acceptor = ethyl phenylacetate; donor = ethyl phenylacetate

7 Device a series of reactions to convert ethyl 3-oxobutyrate to ethyl 4-oxopentanoate. Select the reagents and conditions from the following table, listing them in the order of use.

I. sodium ethoxide in ethanol
II. ethanol + acid catalyst and heat
III. H_3O^+ and heat
IV. CO_2 then H_3O^+
V. Mg in ether
VI. PBr_3
VII. $NaBH_4$ in alcohol
VIII. CH_2I_2 in ether Zn(Cu)
IX. $BrCH_2CO_2C_2H_5$
X. $(CH_3CO)_2O$ + pyridine

a. I then IX then III then II
b. VII then VI then V then X then II
c. III then VII then VI then V then X then II
d. VIII then III then II

8 Dibutyl amine $(C_4H_9)_2NH$ and anisole $(C_6H_5OCH_3)$, have similar boiling points, and are relatively insoluble in water. How might a mixture of these compounds be separated into pure components?

a. (i) dissolve mixture in ether; (ii) extract the anisole into 10% NaOH solution
b. (i) dissolve mixture in ether; (ii) extract the amine into 10% aqueous HCl solution
c. (i) dissolve mixture in ether; (ii) extract the amine into 10% NaOH solution
d. (i) dissolve mixture in ether; (ii) extract the anisole into 10% HCl solution

9 Amines are well known to be stronger bases and nucleophiles than alkenes. Why do enamines, such as 1-dimethylamino cyclopentene, preferentially reacts with electrophiles at a double bond carbon rather than at nitrogen?

a. The nitrogen is sterically hindered by alkyl groups
b. Nitrogen is more electronegative than carbon
c. The carbocation formed by electrophilic attack at C-2 is stabilised by pi-bonding with the lone pair at nitrogen
d. Ammonium cations are less stable than carbocation

10 Which of the following substitution would increase rate of base catalysed hydrolysis of the compound shown below?

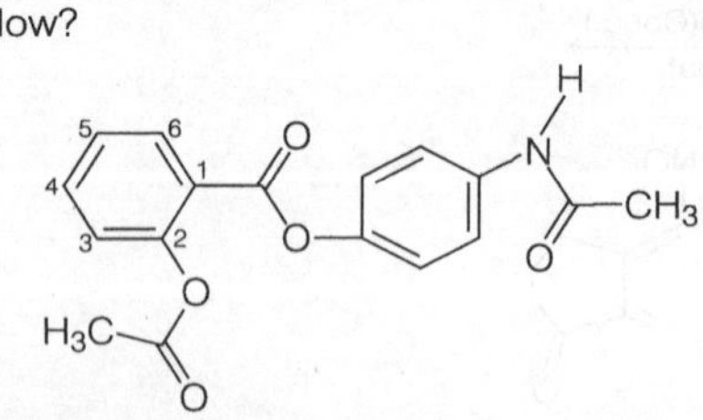

a. A nitro group at C_4 position
b. An amino group at C_6 position
c. A methoxy group at C_3 position
d. A methoxy group at C_2 position

One or More than One Options Correct Type

Direction *(Q. Nos. 11-16) This section contains 6 multiple choice questions. Each question has four choices (a), (b), (c) and (d), out of which ONE or MORE THAN ONE are correct.*

11 A carboxylic acid *X* is treated with bromine (1 mole) in the presence of a little phosphorus. The isolated product is heated with a strong base and then oxidised with O_3 / Zn - H_2O. One of the two oxidation products is a neutral compound which reacts with hydroxylamine to give an oxime; *X* can be

a. $CH_3CH_2CH(CH_3)CH_2CO_2H$
b. $CH_3CH_2CH(CH_3)CO_2H$
c. $CH_3CH_2CO_2H$
d. $(CH_3)_2CHCH_2CO_2H$

12 Which one of the following statements is/are true?

a. Acetamide is a liquid with an ammoniacal smell, which reacts with mineral acids to form stable salts
b. Acetamide is neutral to litmus, reacts with sodium hydroxide to liberate ammonia
c. Acetamide reacts with benzene sulphonyl chloride in alkaline solution to give a crystalline sulphonamide
d. Acetamide reacts on warming with phosphorus pentaoxide to give acetonitrile

13 Which of the following reaction will produce isopropyl acetate in the best yield?

a. Acetyl chloride + 2-propanol $\longrightarrow$
b. Silver acetate + 2-chloropropane $\longrightarrow$
c. Acetic acid + 2-propanol/H_2SO_4 $\longrightarrow$
d. 3-methylbutanone + $C_6H_5CO_3H$ $\longrightarrow$

14 When acetic acid is dissolved in H_2O^{18}, which of the following species is(are) expected to exist in solution?

a. 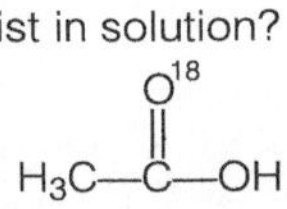
b. 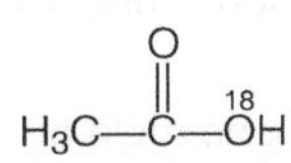
c. 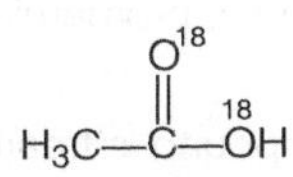
d. 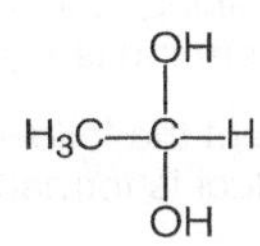

15 The expected product(s) in the following reaction is(are)

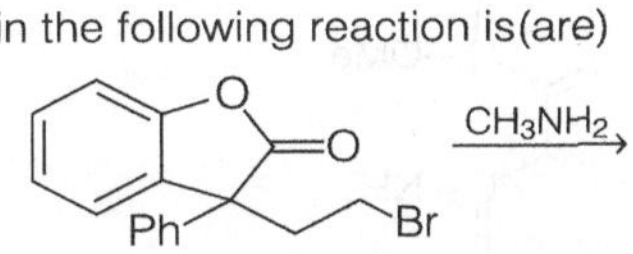

a. 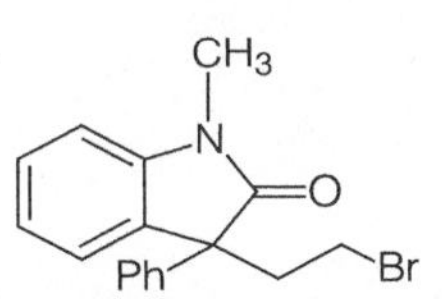
b. 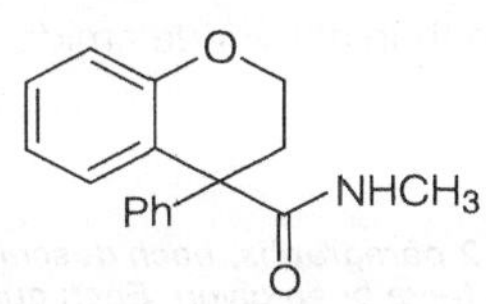
c. 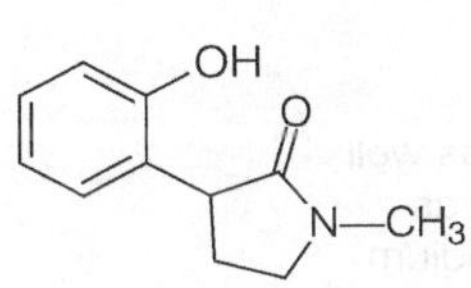
d. 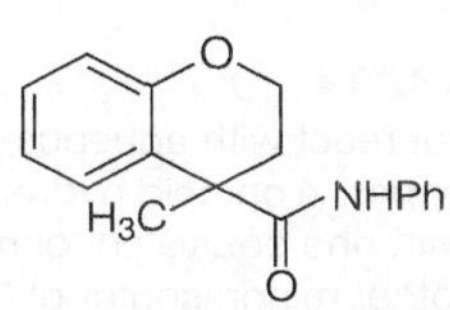

16 The expected product(s) in the following Hofmann's bromamide reaction is/are

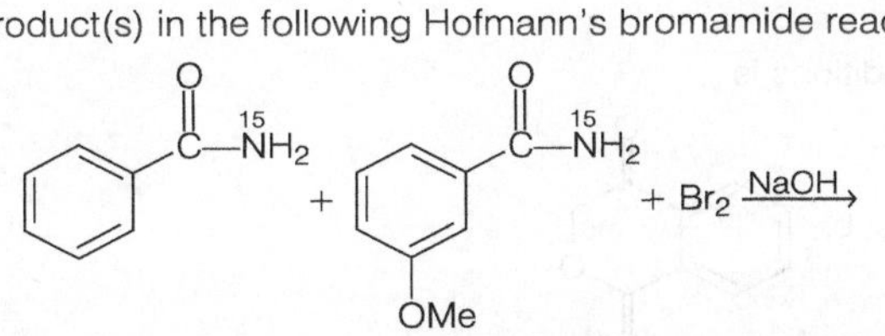

a. 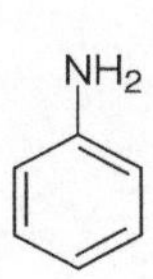
b. 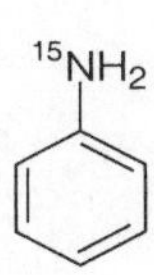
c. 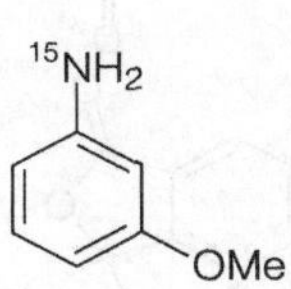
d. 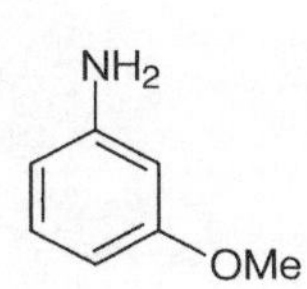

Working Space

Statement Type

Working Space

Direction (*Q. Nos. 17 and 18*) ***This section is based on Statement I and Statement II. Select the correct answer from the codes given below.***

a. Both Statement I and Statement II are correct and Statement II is the correct explanation of Statement I
b. Both Statement I and Statement II are correct but Statement II is not the correct explanation of Statement I
c. Statement I is correct but Statement II is incorrect
d. Statement II is correct but Statement I is incorrect

17 **Statement I** Consider the following two amides undergoing Hofmann's bromamide reaction,

I. 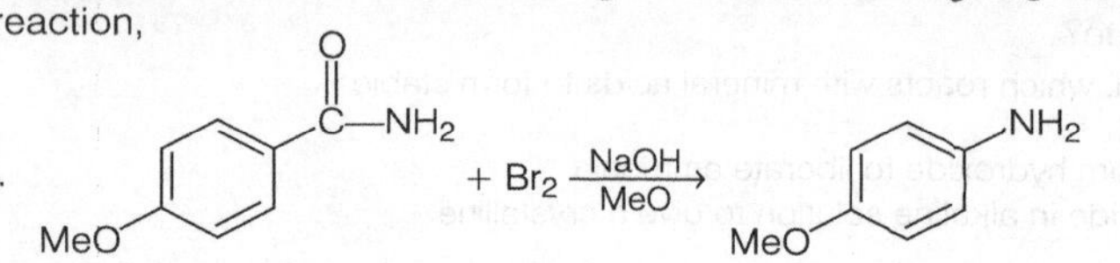

and

II. 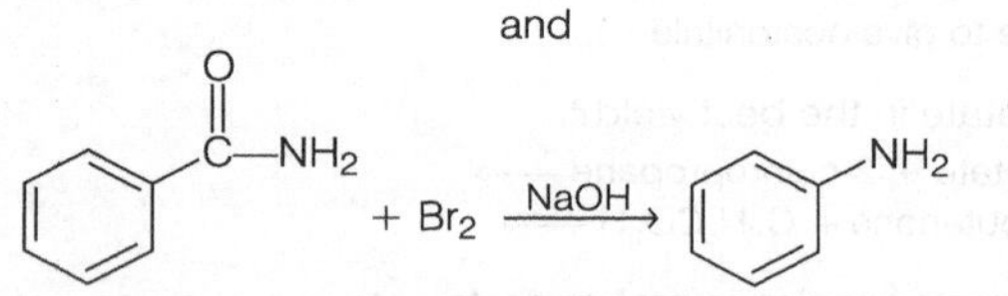

Reaction (I) occur more easily than (II).

Statement II The rate determining step in Hofmann's bromamide reaction is unimolecular elimination of bromide (Br^-) forming isocyanate.

18 **Statement I** When one mole of the following compound is treated with one mole of ammonia, the indicated product is formed as the major product.

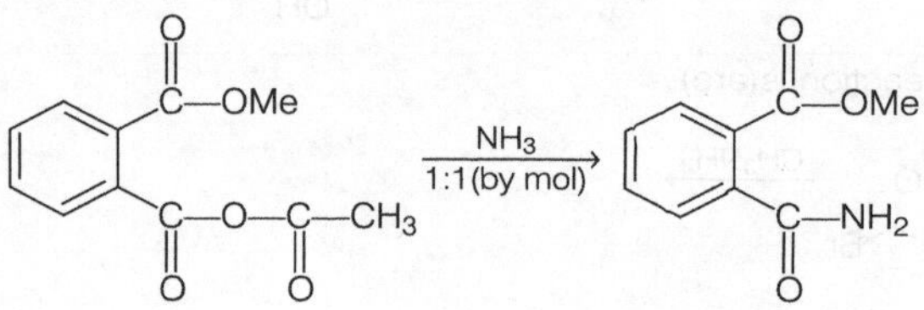

Statement II Ester amide is more stable than anhydride amide.

Comprehension Type

Direction (*Q. Nos. 19-24*) ***This section contains 2 paragraphs, each describing theory, experiments, data, etc. Six questions related to the paragraphs have been given. Each question has only one correct answer among the four given options (a), (b), (c) and (d).***

Passage I

An organic compound $A(C_9H_6O_3)$ does not react with aqueous solution of $NaHCO_3$ as well as does not change the colour of litmus paper. *A* on acid hydrolysis gives $B(C_9H_8O_4)$ as major product whereas, *A* on treatment with one equivalent of methanol in acidic medium gives $C(C_{10}H_{10}O_4)$ as the major product (other minor isomer of *C* is also formed). *B* on heating with soda lime gives toluene.

19 The structure of *A* satisfying the above conditions is

a. 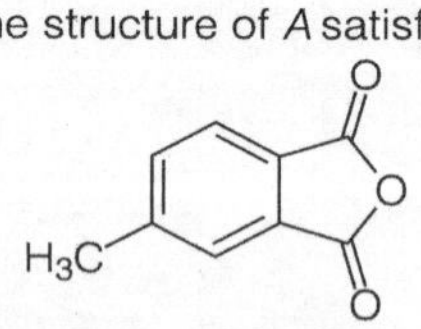

b. 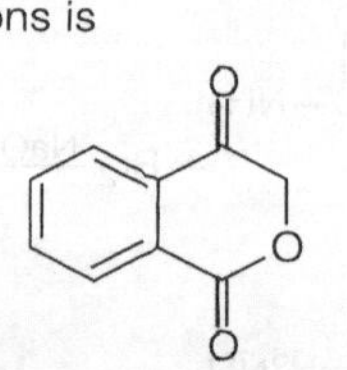

c. 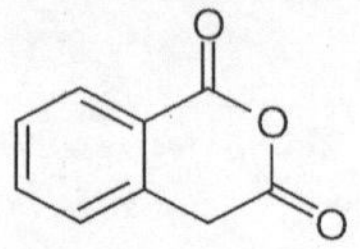

d. 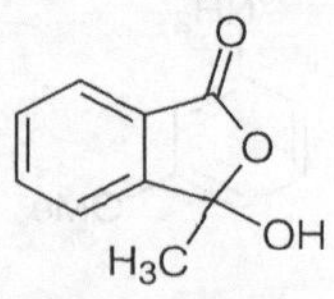

Working Space

20 The most likely structure of *B* is

a. 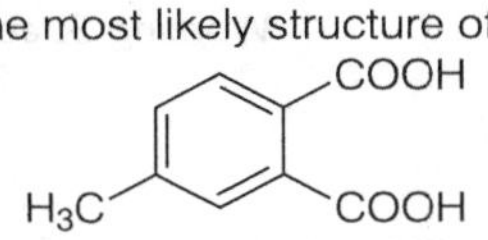

b. 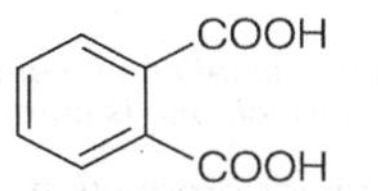

c. 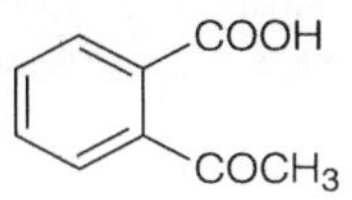

d. 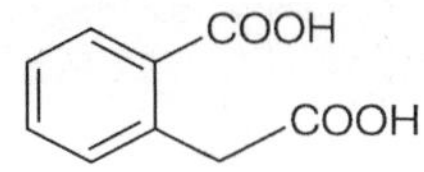

21 The structure of product *C* is

a. 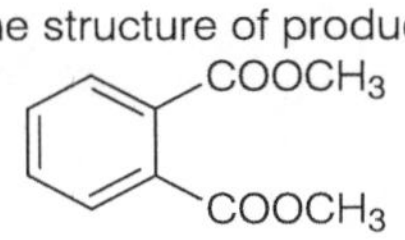

b. 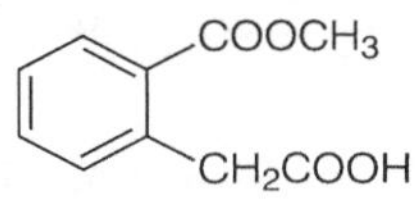

c. 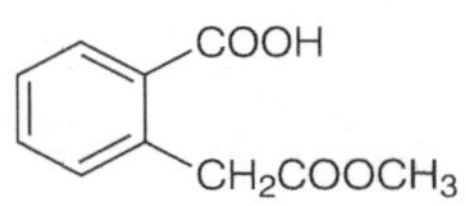

d. 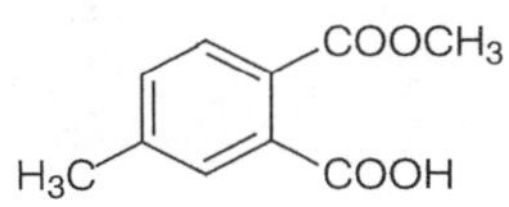

Passage II

Consider the following reaction scheme.

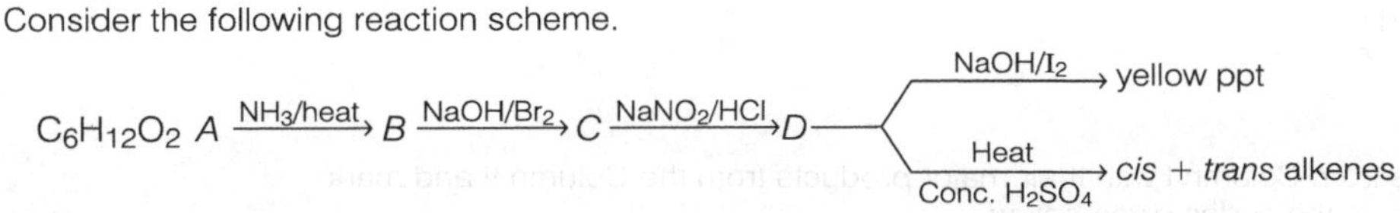

Chiral gives of CO_2 with $NaHCO_3$

22 The most likely structure of *A* is

a. 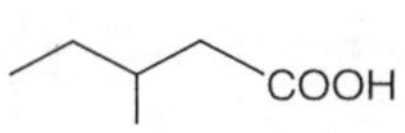

b. 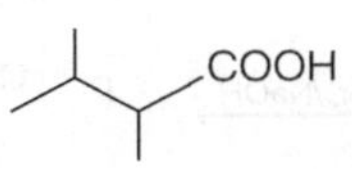

c. 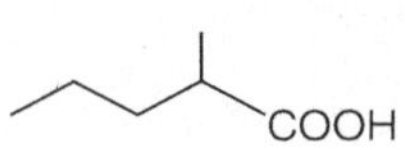

d. 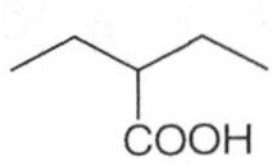

23 Which of the following regarding *B* is not true?

a. It is an amide
b. It has the same configuration as that of *A*
c. It is an achiral compound
d. It on heating with P_2O_5 gives a chiral alkyl nitrile

24 The structure of *C* and *D* respectively are

a. 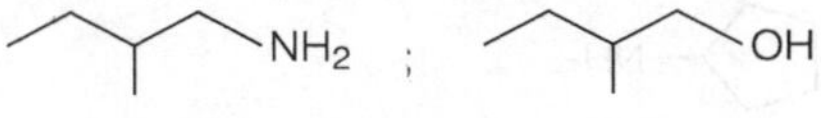

b. 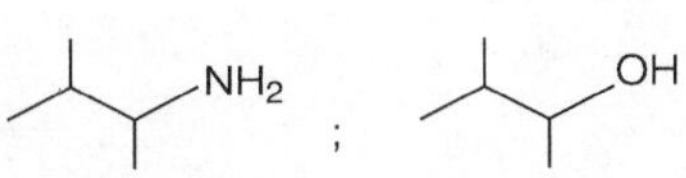

c. 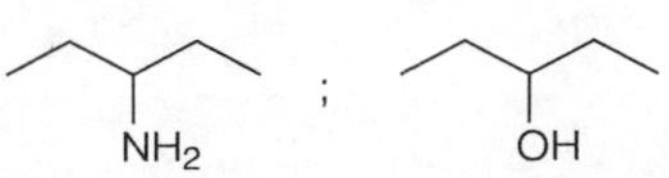

d. 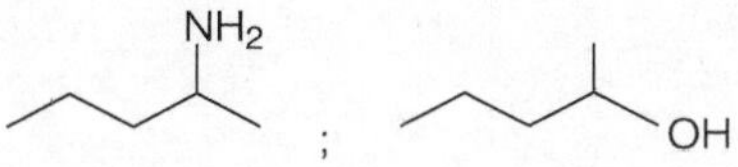

Matching List Type

Working Space

***Direction** (Q. Nos. 25 and 26) Choices for the correct combination of elements from Column I and Column II are given as options (a), (b), (c) and (d), out of which one is correct.*

25 Match the reactions from Column I with their major products from Column II and mark the correct option from the codes given below.

	Column I		Column II
i.	$CH_3CH_2-\overset{O}{\overset{\|\|}{C}}-\overset{18}{O}-\overset{CH_3}{\overset{\|}{CH}}-C_6H_5$ + dil. H_2SO_4 $\xrightarrow{\text{Heat}}$	p.	$^{13}CO_2$
ii.	$CH_3CH_2Br \xrightarrow[\text{(ii) dil. } H_2SO_4]{\text{(i) Na}-^{13}CN}$	q.	CH_3CH_2COOH
iii.	$CH_3CH_2MgBr \xrightarrow[\text{(ii) dil. } H_2SO_4]{\text{(i) } ^{13}CO_2}$	r.	$CH_3CH_2-\overset{\overset{18}{O}}{\overset{\|\|}{C}}-OH$
iv.	$HOOC-CH_2-\overset{13}{C}OOH \xrightarrow{\text{Heat}}$	s.	$CH_3CH_2\overset{13}{C}OOH$

Codes

	i	ii	iii	iv
a.	p,q	r,s	s	q,r
b.	r,s	p	q	r,s
c.	r	s	s	q,r,s
d.	s	p	q,r	r

26 Match the reactions from Column I with their major products from the Column II and mark the correct option from the codes given below.

	Column I		Column II
i.	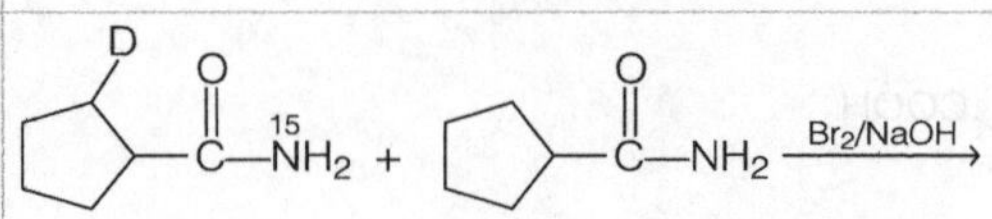	p.	$CH_3COCH_2COOC_2H_5$
ii.	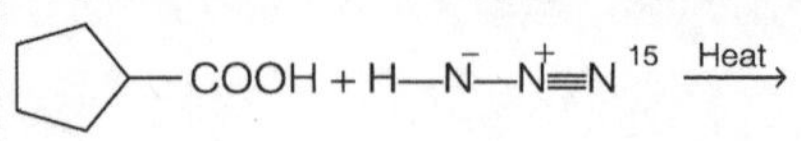	q.	$CH_3CH_2COCH_2COOC_2H_5$
iii.	$CH_3CH_2COOC_2H_5 + CH_3COOC_2H_5 \xrightarrow[C_2H_5OH]{C_2H_5ONa}$	r.	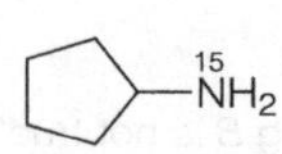
iv.	$H_3C-\overset{O}{\overset{\|\|}{C}}-CH_3 + C_2H_5O-\overset{O}{\overset{\|\|}{C}}-OC_2H_5 \xrightarrow[C_2H_5OH]{C_2H_5ONa}$	s.	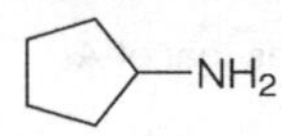
		t.	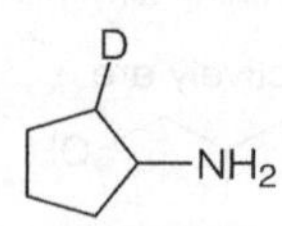

Codes

	i	ii	iii	iv
a.	p,q	r	s,t	q
b.	s,t	r,s	p,q	p
c.	p	r	s	t
d.	r,t	q	p	s

One Integer Value Correct Type

Working Space

Direction *(Q. Nos. 27-30) This section contains 4 questions. When worked out will result in an integer from 0 to 9 (both inclusive).*

27 An organic acid *X* has molecular formula $C_8H_{16}O_2$. *X* on heating with NH_3 forms *Y*. *Y* on treatment with alkaline Br_2 solution forms *Z*. *Z* on treatment with nitrous acid followed by heating with concentrated H_2SO_4 solution gives 2, 4-dimethyl-2-pentene. How many different acids *X* can satisfy the above mentioned condition?

28 How many different isomers exist for $C_4H_{11}N$?

29 If a mixture containing ethyl acetate and ethyl propanoate is refluxed with C_2H_5ONa / C_2H_5OH, ester condensation takes place. How many different condensation would be formed?

30 How many of the following gives visible change when treated with nitrous acid ($NaNO_2$ + HCl)?

i. $CH_3CH_2CH_2NH_2$

ii. $CH_2CH_2NHCH_3$

iii. 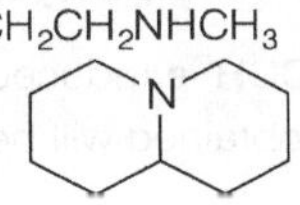

iv. 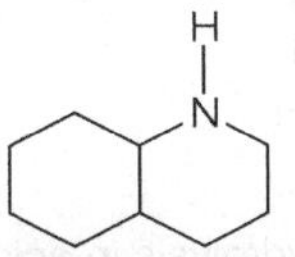

v. $CH_3CH_2CONH_2$

vi. $CH_3CH_2CH(CH_3)NO_2$

vii. $CH_3CH_2N(CH_3)_2$

viii. 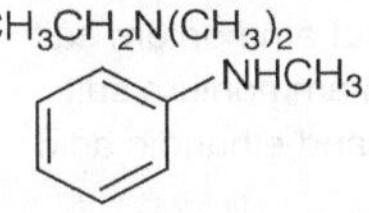

ix. $CH_3COOC_2H_5$

JEE Main & AIEEE Archive

(Compilation of Last 13 Years' Questions)

1 In the reaction,

$$CH_3COOH \xrightarrow{LiAlH_4} A \xrightarrow{PCl_5} B \xrightarrow{alc.\ KOH} C$$

the product C is **(JEE Main 2014)**

a. acetaldehyde b. acetylene
c. ethylene d. acetylchloride

2 A compound with molecular mass 180 is acylated with CH_3COCl to get a compound with molecular mass 390. The number of amino groups present per molecule of the former compound is **(JEE Main 2013)**

a. 2 b. 5
c. 4 d. 6

3 An organic compound A upon reacting with NH_3 gives B. On heating, B gives C. C in the presence of KOH reacts with Br_2 to give $CH_3CH_2NH_2$. A is **((JEE Main 2013)**

a. CH_3COOH
b. $CH_3CH_2CH_2COOH$
c. $CH_3-CH(CH_3)-COOH$
d. CH_3CH_2COOH

4 The strongest acid amongst the following compounds is **(AIEEE 2011)**

a. CH_3COOH
b. $HCOOH$
c. $CH_3CH_2CH(Cl)CO_2H$
d. $ClCH_2CH_2CH_2COOH$

5 Sodium ethoxide has reacted with ethanoyl chloride. The compound that is produced in the above reaction is **(AIEEE 2011)**

a. diethyl ether
b. 2-butanone
c. ethyl chloride
d. ethyl ethanoate

6 A liquid was mixed with ethanol and a drop of concentrated H_2SO_4 was added. A compound with a fruity smell was formed. The liquid was **(AIEEE 2009)**

a. CH_3OH b. $HCHO$
c. CH_3COCH_3 d. CH_3COOH

7 The general formula $C_nH_{2n}O_2$ could be for open chain **(AIEEE 2003)**

a. diketones b. carboxylic acids
c. diols d. dialdehydes

8 When $CH_2=CH-COOH$ is reduced with $LiAlH_4$, the compound obtained will be **(AIEEE 2003)**

a. CH_3-CH_2-COOH
b. $CH_2=CH-CH_2OH$
c. $CH_3-CH_2-CH_2OH$
d. CH_3-CH_2-CHO

9 Ethyl isocyanide on hydrolysis in acidic medium generates **(AIEEE 2003)**

a. ethylamine salt and methanoic acid
b. propanoic acid and ammonium salt
c. ethanoic acid and ammonium salt
d. methylamine salt and ethanoic acid

10 End product of the following reaction is

$$CH_3CH_2COOH \xrightarrow[\text{Red P}]{Cl_2} \xrightarrow{\text{Alcoholic KOH}}$$ **(AIEEE 2002)**

a. $CH_3CH(OH)COOH$
b. $CH_2(OH)CH_2COOH$
c. $CH_2=CHCOOH$
d. $CH_2(Cl)CH(OH)COOH$

JEE Advanced & IIT JEE Archive

(Compilation of Last 10 Years' Questions)

1 Different possible thermal decomposition pathways for peroxyesters are shown below. Match each pathway from Column I with an appropriate structure from Column II and select the correct answer using the code given below the lists.

(2014 Adv., Matching Type)

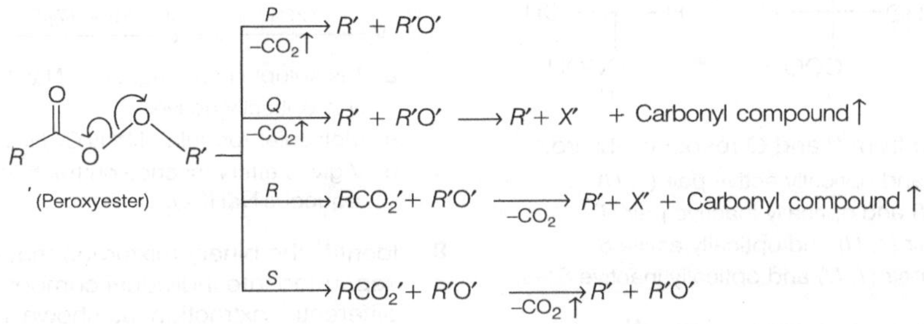

	Column I		Column II
i.	Pathway *P*	p.	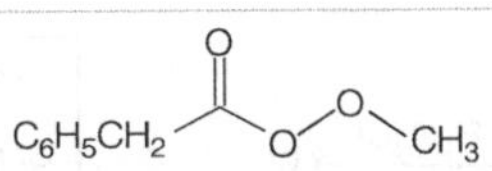
ii.	Pathway *Q*	q.	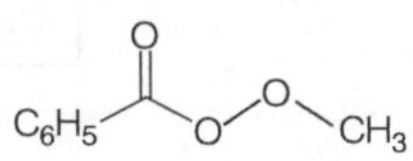
iii.	Pathway *R*	r.	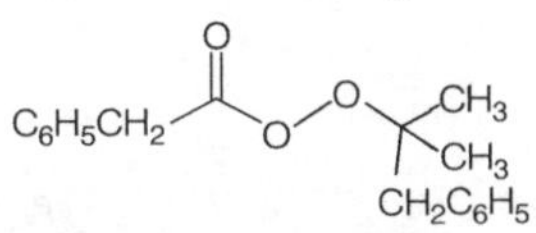
iv.	Pathway *S*	s.	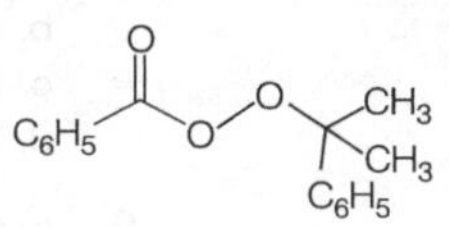

Codes

	i	ii	iii	iv
a.	p	r	s	q
b.	q	s	r	p
c.	s	p	q	r
d.	r	q	p	s

2 The total number of carboxylic acid groups in the product *P* is **(2013 Adv., Integer Type)**

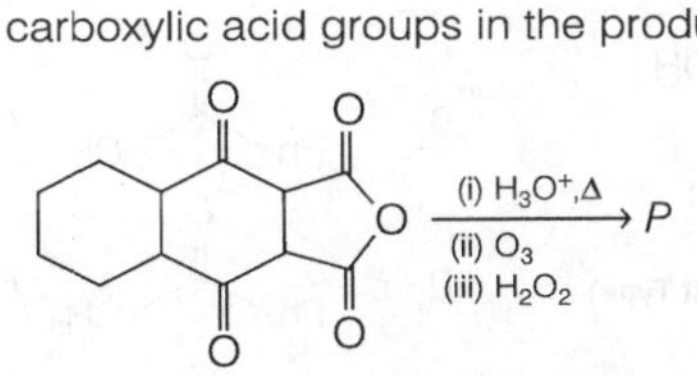

3 The compound that does not liberate CO_2, on treatment with aqueous sodium bicarbonate solution, is **(2013 Adv., Only One Option Correct Type)**

a. benzoic acid

b. benzenesulphonic acid

c. salicylic acid

d. carbolic acid (Phenol)

JEE Archive

Passage for Q. Nos. (4-5)

P and *Q* are isomers of dicarboxylic acid $C_4H_4O_4$. Both decolourise Br_2 / H_2O. On heating, *P* forms the cyclic anhydride.

Upon treatment with dilute alkaline $KMnO_4$. *P* as well as *Q* could produce one or more than one form *S*, *T* and *U*. **(2013 Adv., Comprehension Type)**

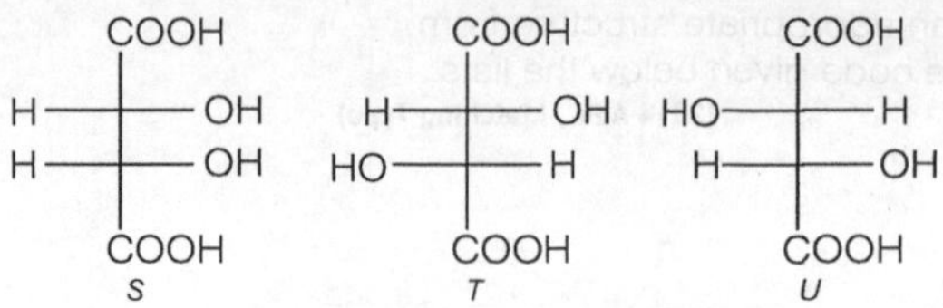

4 Compounds formed from *P* and *Q* respectively are,

a. Optically active *S* and optically active pair (*T*, *U*)
b. Optically inactive *S* and optically inactive pair (*T*, *U*)
c. Optically active pair (*T*, *U*) and optically active *S*
d. Optically inactive pair (*T*, *U*) and optically inactive *S*

5 In the following reaction sequences *V* and *W* respectively are

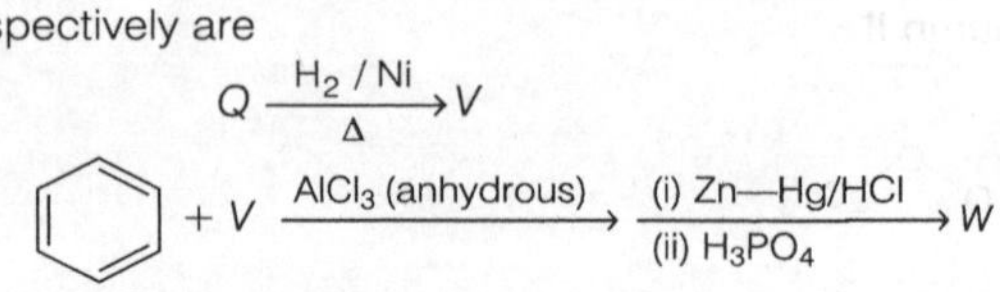

a. 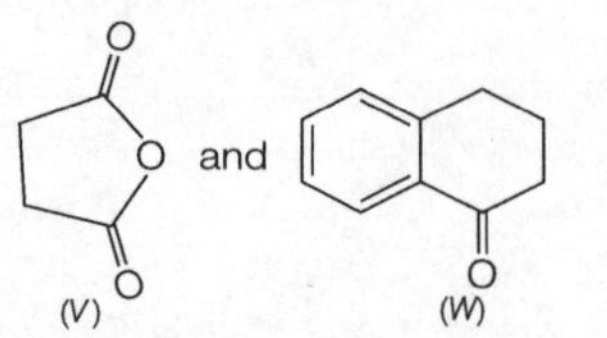

b. 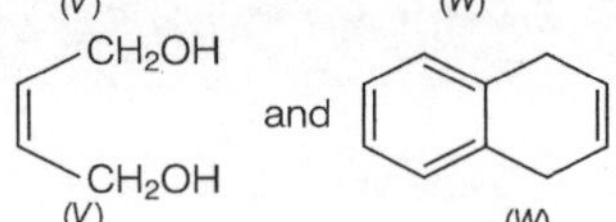

c. 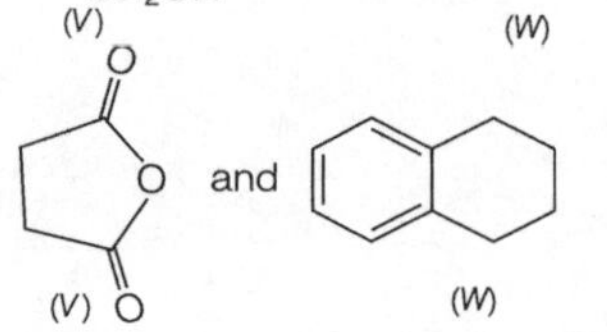

d. 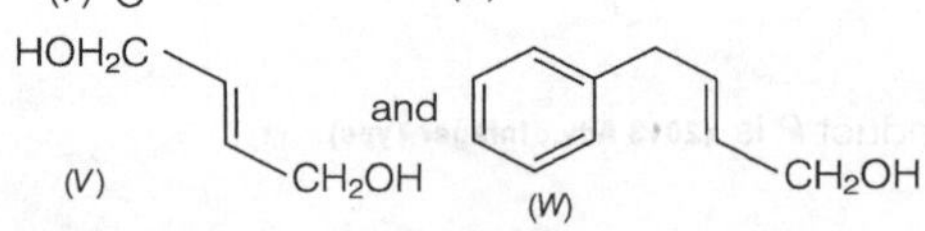

6 The compound that undergoes decarboxylation most readily under mild condition is **(2012, Only One Option Correct Type)**

a. 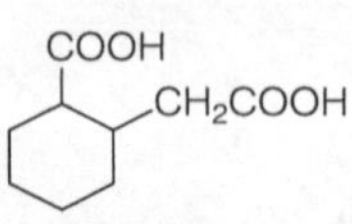
b. 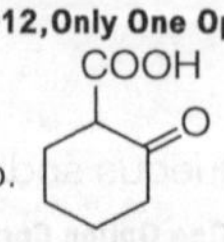
c. 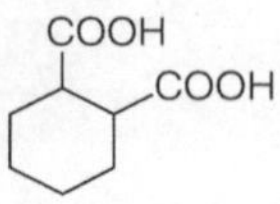
d. 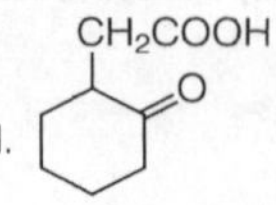

7 With reference to the scheme given, which of the given statement (s) about *T*, *U*, *V* and *W* is (are) correct? **(2012, One or More than One Options Correct Type)**

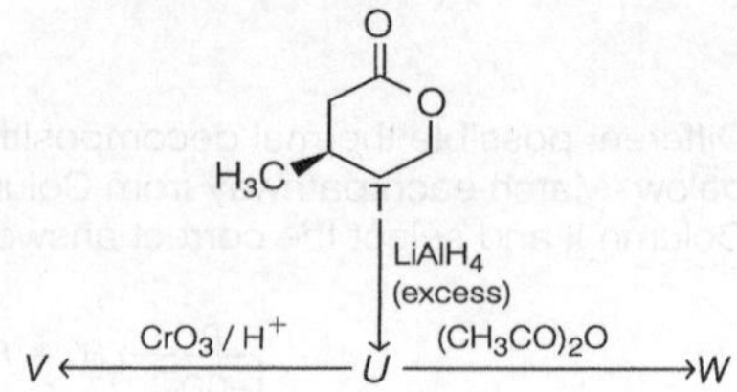

a. *T* is soluble in hot aqueous NaOH
b. *U* is optically active
c. Molecular formula of *W* is $C_{10}H_{18}O_4$
d. *V* gives effervescence on treatment with aqueous $NaHCO_3$

8 Identify the binary mixture(s) that can be separated into individual compounds, by differential extraction, as shown in the given scheme. **(2012, One or More than One Options Correct Type)**

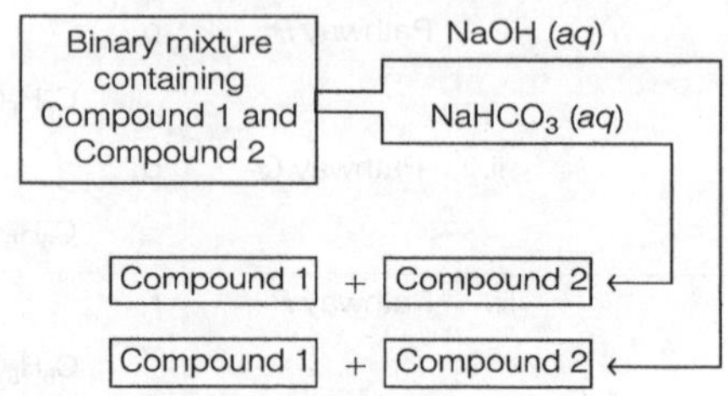

a. C_6H_5OH and C_6H_5COOH
b. C_6H_5COOH and $C_6H_5CH_2OH$
c. $C_6H_5CH_2OH$ and C_6H_5OH
d. $C_6H_5CH_2OH$ and $C_6H_5CH_2COOH$

9 In the following reaction sequence, the correct structures of *E*, *F* and *G* are

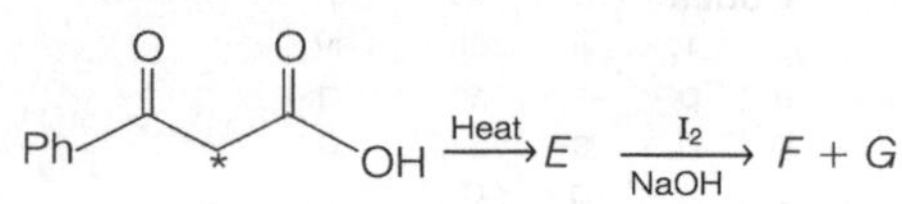

(* implies ^{13}C labelled carbon) **(2008, Only One Option Correct Type)**

a. 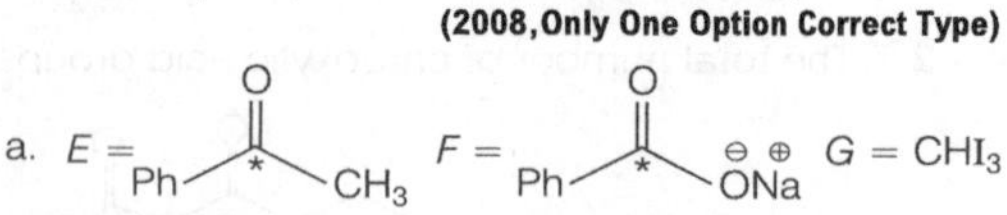
b. 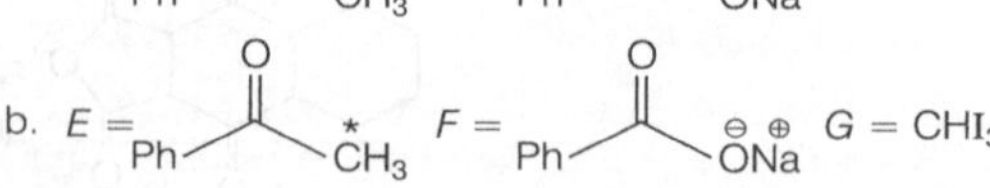
c. 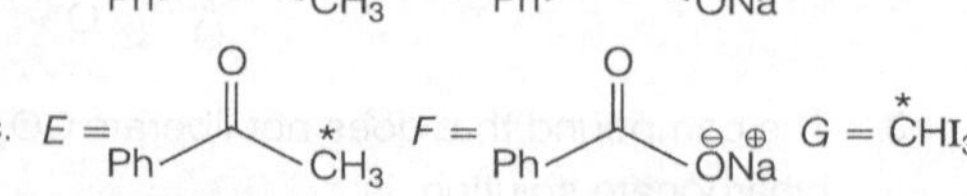
d. 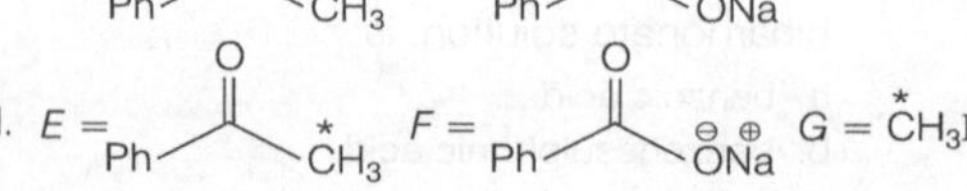

JEE Archive

Answers with Explanations

Answers

1. Alkyl Halides

DPP-1 Preparation of Alkyl Halides

6. *b* **7.** *c* **8.** *a* **9.** *d* **10.** *d* **11.** *b* **12.** *b* **13.** *b,c* **14.** *b,c,d*
15. *a,b,c,d* **16.** *a,c* **17.** *b,c* **18.** *c* **19.** *b* **20.** *b* **21.** 4 **22.** 6 **23.** 3 **24.** 4

DPP-2 Physical and General Properties of Alkyl Halides

6. *b* **7.** *d* **8.** *b* **9.** *a* **10.** *a* **11.** *b* **12.** *d* **13.** *c* **14.** *d*
15. *b,c,d* **16.** *a,c,d* **17.** *a,b* **18.** *c,d* **19.** *b,c* **20.** *a* **21.** *d* **22.** *b* **23.** 3 **24.** 7
25. 6 **26.** *b* **27.** *a*

DPP-3 S_N2 Reaction Basics

1. (i) $CH_3ONa + CH_3I$, (ii) $NaSH + CH_3I$, (iii) $PH_3 + CH_3CH_2CH_2I$, (iv) $NaCN + CH_3CH_2I$, (v) $(CH_3)_3CCH_2—SNa + CH_3I$, (vi) CH_3I (excess) $+ NH_3$, (vii) $C_6H_5COONa + C_6H_5CH_2Cl$, (viii) $(CH_3)_2CHONa + CH_2=CH_2—I$, (ix) $CH_2=CH—CH_2—ONa + CH_2=CH—CH_2—I$

3. (i) CH_3CH_2Br, (ii) CH_3CH_2I, (iii) 1 - bromopropane, (iv) neopentyl iodide, (v) CH_3S^-, (vi) CH_3S^-, (vii) $CH_3CH_2O^-$, (viii) CH_3NH_2, (ix) CH_3NH_2, (x) $CH_3CH_2S^-$, (xi) $CH_3CH_2S^-$, (xii) $HOCH_2CH_2CH_2CH_2Br$

6. *a* **7.** *d* **8.** *d* **9.** *c* **10.** *d* **11.** *d* **12.** *b* **13.** *b* **14.** *a,b,c*
15. *a,b,d* **16.** *a,c* **17.** *a,c,d* **18.** *a,c,d* **19.** *b,c,d* **20.** *b* **21.** *c* **22.** *c* **23.** 9 **24.** 6
25. 3

DPP-4 S_N2 Reaction Reactivity

1. (i) $HO^- < H_2O < H_3O^+$, (iii) $NH_2^- < NO_2^- < NO_3^-$, (iii) $F^- < Cl^- < Br^- < I^-$, (iv) $F_3CCH_2CH_2COO^- < F_3CCH_2COO^- < F_3CCOO^-$, (v) $CH_3CH_2^- < CH_3CH_2O^- < CH_3COO^-$

6. (i) HO^-, (ii) HO^-, (iii) CH_3S^-, (iv) Cl^-, (v) I^- **10.** *d* **11.** *d* **12.** *a* **13.** *d* **14.** *b*
15. *c* **16.** *c* **17.** *c* **18.** *b* **19.** *b* **20.** *c* **21.** *b* **22.** *a,d* **23.** *c,d* **24.** *a,d*
25. *a,b,c,d* **26.** *a,c,d* **27.** *b* **28.** *a* **29.** *c* **30.** *d* **31.** *c* **32.** *d* **33.** 5
34. 4

DPP-5 Ambident Nucleophiles and NGP in S_N2 Reaction

1. (i) CH_3S^-, (ii) $(CH_3)_3N$, (iii) CH_3Cl, (iv) $(CH_3)_3P$, (v) HS^-, (vi) CH_3SCH_3 **7.** *c* **8.** *a* **9.** *d*
10. *c* **11.** *c* **12.** *c* **13.** *c* **14.** *c* **15.** *a,b,c* **16.** *a,b,c,d* **17.** *b,d* **18.** *a,c*
19. *a,c* **20.** *b* **21.** *b* **22.** *c* **23.** 4 **24.** 6 **25.** *a*

DPP-6 S_N1 Reaction

1. (i) S_N2, (ii) S_N2, (iii) S_N2, (iv) S_N2, (v) S_N2, (vi) S_N1, (vii) S_N1, (viii) S_N1, (ix) S_N2, (x) S_N2

6. *b* **7.** *d* **8.** *a* **9.** *a* **10.** *c* **11.** *a* **12.** *c* **13.** *d* **14.** *d* **15.** *d*
16. *c* **17.** *b* **18.** *b* **19.** *c* **20.** *d* **21.** *a* **22.** *b* **23.** *a,b,c* **24.** *a,b,c,d*
25. *b,d* **26.** *a,b,d* **27.** *a,d* **28.** *c* **29.** *a* **30.** *c* **31.** *b* **32.** *b* **33.** 6 **34.** 5
35. 6

DPP-7 E2 Reaction Basics

2. (i) Both S_N1 and E1, (ii) S_N2 reaction, (iii) Both S_N2 and S_N1, (iv) Only S_N1, (v) S_N1 reaction, (vi) E1 reaction, (vii) E2 reaction, (viii) S_N1 and E1, (ix) S_N1 and E1, (x) S_N1, E1 and E2 reactions (xi) S_N2 reactions.

6. *b*	7. *a*	8. *c*	9. *c*	10. *d*	11. *a*	12. *b*	13. *c*	14. *b*	15. *d*
16. *b,c*	17. *b,c*	18. *b,c*	19. *a,b,d*	20. *b,c*	21. *c*	22. *b*	23. *b*	24. *c*	25. *d*
26. *d*	27. 5	28. 5	29. 6	30. 7					

DPP-8 E2 Reaction Advanced, E1 cb and Hofmann's Elimination

4. *c*	5. *b*	6. *d*	7. *c*	8. *d*	9. *a,c,d*	10. *a,c*	11. *a,b,c*	12. *a*	13. *c*
14. *d*	15. *b*	16. *a*	17. *c*	18. *c*	19. *d*	20. *d*	21. *b*	22. *c*	23. 4
24. 3									

DPP-9 E1 Reaction and Related Reactions

6. *d*	7. *a*	8. *c*	9. *b*	10. *b*	11. *c*	12. *b*	13. *c*	14. *a,b,d*
15. *a,b,d*	16. *a,c*	17. *d*	18. *b*	19. *b*	20. 8	21. 7		

DPP-10 Dihalides and Polyhalides

5. *a*	6. *c*	7. *b*	8. *c*	9. *a*	10. *b*	11. *a*	12. *d*	13. *a*	14. *b*
15. *c*	16. *b*	17. *b*	18. *b,c,d*	19. *a,b,d*	20. *a,b,d*				

Revisal Problems (JEE Main)

1. *d*	2. *d*	3. *d*	4. *c*	5. *a*	6. *d*	7. *c*	8. *a*	9. *a*	10. *b*
11. *c*	12. *c*	13. *c*	14. *c*	15. *d*	16. *d*	17. *d*	18. *a*	19. *c*	20. *c*
21. *c*	22. *d*	23. *c*	24. *c*	25. *b*	26. *b*	27. *b*	28. *b*	29. *c*	30. *a*

Revisal Problems (JEE Advanced)

1. *b*	2. *c*	3. *a*	4. *b*	5. *d*	6. *d*	7. *a*	8. *b*	9. *b*	10. *c*
11. *b*	12. *b*	13. *a,b,d*	14. *a,b,c,d*	15. *a,b,d*	16. *a,b,c,d*	17. *c,d*	18. *c,d*	19. *a,c*	20. *a,c*
21. *a,d*	22. *a,b*	23. *a*	24. *d*	25. *b*	26. 3	27. 5	28. 6	29. 6	30. *a*
31. *a*									

JEE Main & AIEEE Archive

1. *c,d*	2. *c*	3. *d*	4. *c*	5. *b*	6. *d*	7. *a*	8. *d*	9. *a*	10. *a*
11. *b*	12. *c*	13. *c*	14. *c*						

JEE Advanced & IIT JEE Archive

1. *a*	2. *b*	3. 5	4. *a*	5. *a*	6. *a*

2. Aliphatic Alcohols, Ethers and Epoxides

DPP-1 Preparation of Alcohols

6. *c*	7. *d*	8. *a*	9. *d*	10. *b*	11. *b*	12. *b*	13. *b*	14. *b,c*	15. *b,c*
16. *a,c,d*	17. *a,c,d*	18. *a,b*	19. *b,d*	20. *a,b,c,d*		21. *d*	22. *b*	23. *d*	24. 3
25. 5	26. 6	27. 3							

DPP-2 Reactions of Alcohols-I

2. (i) H_2 CO_3 (ii) CH_3 COOH (iii) CH_3 CH_2OH (iv) CH_3 CH_2 SH
3. (i) Equilibrium lies towards left because CH_3CH_2OH is stronger acid.
   (ii) Equilibrium lies towards right because HCl is stronger acid.
   (iii) Equilibrium lies towards right because phenol is stronger acid.
   (iv) Equilibrium lies towards right because 1-butanol is stronger acid.

6. *b* 7. *d* 8. *d* 9. *c* 10. *b* 11. *d* 12. *d* 13. *c* 14. *a,b* 15. *a,d*
16. *a,b,c,d* 17. *a,c,d* 18. *d* 19. *c* 20. *a* 21. 5 22. 5 23. 3 24. 5

DPP-3 Reactions of Alcohols-II

6. *d* 7. *d* 8. *a* 9. *d* 10. *c* 11. *d* 12. *d* 13. *b* 14. *a,b,d* 15. *a,b*
16. *b,c,d* 17. *a,b,c* 18. *a,b,c* 19. *c,d* 20. *d* 21. *c* 22. *c* 23. *d* 24. *a* 25. 9
26. 7 27. 3 28. 4

DPP-4 Tests and Distinctions of Alchohols

5. (i) Heating with sodium metal : 1-butanol gives effervescence but 1-octene does not.
   (ii) Treatment with $AgNO_3$ solution : 1-bromo pentane gives brown precipitate of AgBr, 1-propanol does not.
   (iii) Treatment with Br_2 - H_2O : 3-butene-1-ol decolourises brown colour of bromine water but 1-butanol does not.
   (iv) Treatment with conc. $HCl/ZnCl_2$: 3-pentanol gives white turbidity in cold, 1-pentanol does not.
   (v) Treatment with I_2 + NaOH : α-phenyl ethanol gives yellow precipitate of CHI_3, β-phenyl ethanol does not.

6. *d* 7. *c* 8. *d* 9. *c* 10. *c* 11. *b* 12. *d* 13. *c* 14. *a,b* 15. *a,c*
16. *a,b,c* 17. *a,d* 18. *b* 19. *c* 20. *b* 21. 8 22. 4 23. 4 24. 5

DPP-5 Ethers

7. *c* 8. *b* 9. *d* 10. *b* 11. *c* 12. *d* 13. *b* 14. *b* 15. *b,c*
16. *b,c,d* 17. *c,d* 18. *a,b,c* 19. *a,b,c* 20. *a,c* 21. *c* 22. *a* 23. *d* 24. 6 25. 8
26. 4 27. 8 28. 6

DPP-6 Epoxides

6. *d* 7. *c* 8. *b* 9. *a* 10. *b* 11. *c* 12. *c* 13. *b* 14. *c,d* 15. *c,d*
16. *a,c,d* 17. *b,c* 18. *b,c,d* 19. *b,d* 20. *b* 21. *c* 22. *a* 23. 4 24. 3 25. 2
26. 4 27. 3

DPP-7 Diols and Polyols

1. *b* 2. *c* 3. *b* 4. *c* 5. *d* 6. *b* 7. *c* 8. *c* 9. *a,c,d*
10. *a,b,c* 11. *a,d* 12. *a,b* 13. *a,b,d* 14. *a,c* 15. *a* 16. *a* 17. *c* 18. *a* 19. *a*

Revisal Problems (JEE Main)

1. *b* 2. *d* 3. *a* 4. *d* 5. *d* 6. *d* 7. *a* 8. *d* 9. *d* 10. *d*
11. *d* 12. *a* 13. *b* 14. *a* 15. *a* 16. *a* 17. *b* 18. *b* 19. *a* 20. *c*
21. *a* 22. *c* 23. *a* 24. *d* 25. *c* 26. *d* 27. *c* 28. *a* 29. *a* 30. *d*

Revisal Problems (JEE Advanced)

1. c 2. a 3. c 4. b 5. b 6. a 7. c 8. b 9. c 10. c
11. b 12. b 13. a,b,c 14. a,c,d 15. a,b,d 16. a,b,c 17. b,c,d 18. a,b,c 19. b 20. d
21. b 22. b 23. b 24. a 25. a 26. a 27. a 28. 6 29. 6 30. 8

JEE Main & AIEEE Archive

1. d 2. b 3. c 4. b 5. a 6. c 7. d 8. c 9. c 10. d
11. a 12. b

JEE Advanced & IIT JEE Archive

1. c 2. d 3. c 4. b 5. b 6. d 7. c 8. b 9. b

3. Aliphatic Aldehydes and Ketones

DPP-1 Preparation of Aldehydes and Ketones

6. c 7. c 8. c 9. d 10. b 11. b 12. d 13. b 14. b 15. c
16. c 17. c 18. a,c,d 19. a,c 20. a,b,c 21. c,d 22. a,c 23. b 24. c 25. d
26. 5 27. 7 28. 5 29. 5 30. 4

DPP-2 Physical Properties and General Reactions

7. b 8. c 9. c 10. a 11. b 12. d 13. d 14. c 15. d 16. b,c
17. a,b,c,d 18. a,b 19. a,c,d 20. b 21. c 22. c 23. 3 24. 3 25. 8 26. 3
27. 8 28. 5 29. a 30. a

DPP-3 Reactions with Ammonia, Amines, Hydrazine and Its Derivatives

6. b 7. c 8. d 9. d 10. c 11. a 12. d 13. c 14. d 15. b
16. a,b,c 17. b,c 18. a,b,c,d 19. a,c 20. b 21. a 22. c 23. 6 24. 3
25. 4 26. 6

DPP-4 Keto-Enol Tautomerism and Related Reactions

7. d 8. d 9. d 10. b 11. c 12. c 13. b 14. c 15. d 16. b
17. d 18. d 19. b 20. a 21. b,c 22. a,b,c,d 23. a,b,c 24. a,b 25. a,c
26. c 27. b 28. d 29. 6 30. 8 31. 7 32. 6

DPP-5 Aldol Condensation

6. b 7. a 8. c 9. c 10. c 11. c 12. a 13. b 14. a 15. b
16. c,d 17. a,b,c 18. b,d 19. a,c,d 20. a,c,d 21. a 22. c 23. a 24. d 25. c
26. d 27. d 28. 6 29. 4 30. 3 31. 4 32. 5

DPP-6 Cannizaro Reaction

6. c 7. a 8. b 9. d 10. d 11. d 12. a 13. d 14. b 15. b
16. a,c,d 17. a,b,c 18. b,d 19. a,b,c,d 20. a,c 21. c 22. d 23. b 24. a
25. b 26. 6 27. 5 28. 4

DPP-7 Beckmann Rearrangement

4. d 5. a 6. b 7. a 8. c 9. c 10. b 11. a,b,c 12. a,c,d 13. a,c
14. b,c,d 15. c 16. c 17. d 18. 4 19. 4 20. 6

DPP-8 Unsaturated Aldehydes and Ketones

6. *c* 7. *d* 8. *a* 9. *b* 10. *c* 11. *a,b* 12. *c,d* 13. *a,b,d* 14. *a,b* 15. *b*
16. *b* 17. *c* 18. 4 19. 8 20. 8

DPP-9 Reformatsky, Witting Benzil-Benzilic Acid Rearrangements, etc.

1. *d* 2. *b* 3. *d* 4. *d* 5. *c* 6. *c* 7. *c* 8. *b* 9. *a,b,c*
10. *a,c,d* 11. *a,b,c* 12. *a,b,d* 13. *a,c* 14. *b,c* 15. *a* 16. *c* 17. *a* 18. 4 19. 4
20. 3

DPP-10 Tests and Distinctions

1. *c* 2. *c* 3. *b* 4. *d* 5. *a* 6. *c* 7. *a* 8. *b* 9. *a,b,c,d*
10. *a,b,d* 11. *b,d* 12. *a,d* 13. *a,d* 14. *c,d* 15. *b* 16. *c* 17. *b* 18. 4 19. 9
20. 4 21. *b*

Revisal Problems (JEE Main)

1. *b* 2. *b* 3. *a* 4. *c* 5. *a* 6. *c* 7. *a* 8. *a* 9. *c* 10. *c*
11. *b* 12. *b* 13. *b* 14. *d* 15. *c* 16. *d* 17. *d* 18. *c* 19. *d* 20. *a*
21. *d* 22. *c* 23. *b* 24. *c* 25. *c* 26. *c* 27. *d* 28. *b* 29. *c* 30. *c*

Revisal Problems (JEE Advanced)

1. *b* 2. *a* 3. *d* 4. *c* 5. *b* 6. *d* 7. *d* 8. *b* 9. *a,d* 10. *a,c*
11. *b,d* 12. *a,b,c* 13. *b,c* 14. *a,b,d* 15. *a* 16. *d* 17. *d* 18. *b* 19. *c* 20. *c*
21. *b* 22. *c* 23. *c* 24. *b* 25. *d* 26. 4 27. 6 28. 4 29. 4 30. 4

JEE Main & AIEEE Archive

1. *a* 2. *a* 3. *a* or *c* 4. *d* 5. *c* 6. *b* 7. *d*

JEE Advanced & IIT JEE Archive

1. *a* 2. *d* 3. 5 4. *a* 5. *c* 6. 1 7. *b* 8. *a* 9. *d* 10. *b*
11. *b* 12. *a* 13. *d* 14. *a* 15. *c* 16. *b* 17. *b* 18. *c*

4. Carboxylic Acids and Acid Derivatives

DPP-1 Preparation and General Properties

6. *c* 7. *b* 8. *d* 9. *c* 10. *b* 11. *a* 12. *c* 13. *c* 14. *a,c,d* 15. *c,d*
16. *a,b,c* 17. *a,b,c* 18. *b,c* 19. *b,c,d* 20. *c* 21. *b* 22. *b* 23. 2 24. 4 25. 3
26. 3

DPP-2 Reactions of Carboxylic Acids

7. *b* 8. *b* 9. *b* 10. *c* 11. *a* 12. *c* 13. *b* 14. *c* 15. *c,d* 16. *c*
17. *b,d* 18. *a,b,d* 19. *a,b* 20. *b* 21. *c* 22. *d* 23. 4 24. 4 25. 6 26. 5

DPP-3 Introduction to Acid Derivatives

6. *c* 7. *d* 8. *a* 9. *a* 10. *b* 11. *b* 12. *a,b,c* 13. *a,b,d* 14. *a,c,d* 15. *a,c*
16. *b* 17. *d* 18. *b* 19. *a* 20. *a* 21. *b* 22. *b* 23. 4 24. 2 25. 4
26. 10

DPP-4 Acid Chlorides and Anhydrides

6. *b* 7. *a* 8. *c* 9. *c* 10. *b* 11. *a* 12. *d* 13. *c* 14. *a,d*
15. *b,c,d* 16. *a,b,d* 17. *a,b,c* 18. *a,b,c,d* 19. *b,c* 20. *b* 21. *a* 22. *c* 23. 3
24. 2 25. 3

DPP-5 Esters

6. *a* 7. *b* 8. *b* 9. *d* 10. *c* 11. *a* 12. *d* 13. *d* 14. *a,c* 15. *a,b*
16. *a,b,c,d* 17. *b,c,d* 18. *b,c* 19. *a,b,d* 20. *b* 21. *c* 22. *c* 23. *a* 24. *c*
25. *a* 26. *a* 27. 6 28. 7 29. 4 30. 7

DPP-6 Acid Amides

6. *b* 7. *d* 8. *b* 9. *b* 10. *c* 11. *b* 12. *a* 13. *b* 14. *a,b,c,d*
15. *a,b,d* 16. *b,d* 17. *a,b,c* 18. *a,d* 19. *a,b* 20. *b* 21. *c* 22. *d* 23. 6 24. 5
25. 6 26. 4

DPP-7 Preparation of Amines

6. *d* 7. *b* 8. *c* 9. *b* 10. *a* 11. *c* 12. *c* 13. *c* 14. *a,b,c* 15. *b,d*
16. *b,c* 17. *b,d* 18. *c* 19. *c* 20. *d* 21. 8 22. 7

DPP-8 Reactions of Amines

6. *b* 7. *c* 8. *d* 9. *d* 10. *c* 11. *c* 12. *a* 13. *a* 14. *a,b,c* 15. *a,c*
16. *a,b,c* 17. *a,b* 18. *b* 19. *b,c,d* 20. *c* 21. *c* 22. *b* 23. 4 24. 6 25. 3
26. 5

Revisal Problems (JEE Main)

1. *d* 2. *b* 3. *d* 4. *c* 5. *c* 6. *c* 7. *a* 8. *c* 9. *d* 10. *b*
11. *d* 12. *a* 13. *c* 14. *b* 15. *a* 16. *d* 17. *a* 18. *d* 19. *d* 20. *b*
21. *d* 22. *a* 23. *c* 24. *b* 25. *c* 26. *d* 27. *d* 28. *d* 29. *b* 30. *a*

Revisal Problems (JEE Advanced)

1. *d* 2. *d* 3. *c* 4. *a* 5. *b* 6. *d* 7. *a* 8. *b* 9. *c* 10. *a*
11. *c,d* 12. *b,d* 13. *a,b,c,d* 14. *a,b,c* 15. *a,b,c* 16. *b,d* 17. *b* 18. *a* 19. *c*
20. *d* 21. *c* 22. *c* 23. *c* 24. *d* 25. *c* 26. *b* 27. 4 28. 9 29. 6
30. 4

JEE Main & AIEEE Archive

1. *c* 2. *b* 3. *d* 4. *c* 5. *d* 6. *d* 7. *b* 8. *b* 9. *a* 10. *c*

JEE Advanced & IIT JEE Archive

1. *a* 2. 2 3. *d* 4. *a* 5. *a* 6. *b* 7. *b* 8. *b,d* 9. *c*

Answers

1. Alkyl Halides

DPP-1 Preparation of Alkyl Halides

1. Alkyl fluoride is best synthesised by heating an alkyl chloride or bromide in the presence of a metallic fluoride such as AgF, Hg_2F_2, CoF_2 or SbF_3. The reaction is known as Swarts reaction.

$$CH_3-Br + AgF \longrightarrow CH_3-F + AgBr$$

2. A → (II)

B → (III)

C → (I)

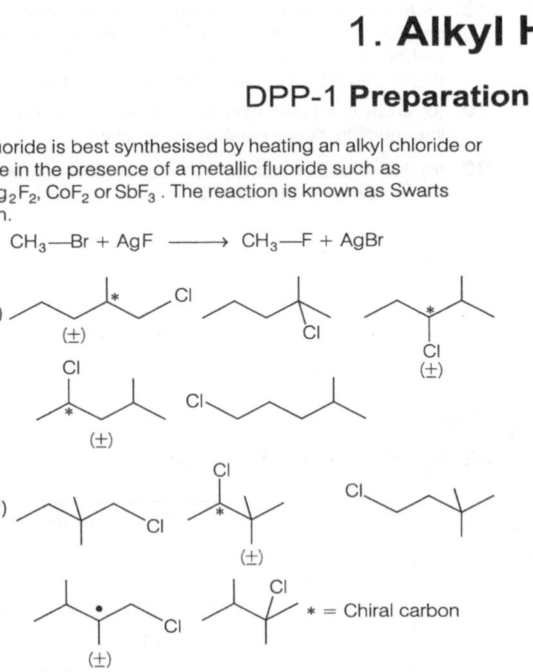

* = Chiral carbon

3. $$CH_3-CH_2-\ddot{O}-H + H^+ \longrightarrow CH_3-CH_2-\overset{+}{O}H_2 \xrightarrow[S_N2]{Br^-} CH_3-CH_2-Br + H_2O$$

4. $R-H + I_2 \xrightarrow{\Delta} R-I + HI$

The above reaction fail to occur because the byproduct HI is a strong reducing agent, reduces iodoalkane back to alkane. Using some oxidising agent makes the reaction possible as it reacts with HI.

5. $R-Cl + NaI \xrightarrow{\text{Acetone}} R-I + NaCl$

In the above reaction, NaCl due to its greater ionic character than NaI, is less soluble in acetone, phased out. Phasing out of NaCl is the driving force of this reaction.

However the above reaction proceeds by S_N2 mechanism, alkyl chloride should be preferably less substituted at α-carbon i.e. it should be primary or secondary but not tertiary.

6. (b)

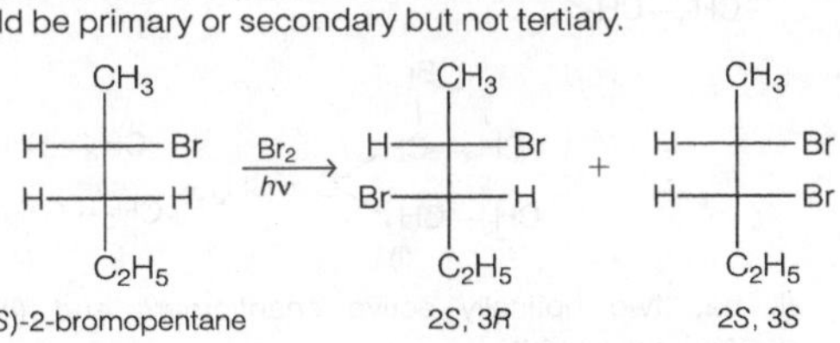

(S)-2-bromopentane → 2S, 3R + 2S, 3S

Compound with *R*-configuration at C-2 would not be formed.

7. (c) The reaction follows free radical mechanism and alkyl free radical is formed in the propagation step as

$$R-\overset{O}{\overset{\|}{C}}-O-Ag + Br_2 \longrightarrow R-\overset{O}{\overset{\|}{C}}-O-Br + AgBr$$

$$R-\overset{O}{\overset{\|}{C}}-O-Br \longrightarrow R-\overset{O}{\overset{\|}{C}}-\dot{O} + \dot{Br}$$

$$R-\overset{O}{\overset{\|}{C}}-\dot{O} \longrightarrow \dot{R} + CO_2$$

Hence, stability of alkyl free radical (3° > 2° > 1°) determine the reactivity.

8. (a) Except F_2, almost all halogens react with *R*COOAg giving alkyl halide *via* Hunsdiecker reaction. With I_2 if *R*COOAg is in excess, *R*—I formed in first step reacts further with unreacted salt to give ester as

$$R-COOAg + R-I \longrightarrow R-COOR + AgI$$

9. (d) Free radical bromination occur. Preferably at highest degree carbon where most stable free radical is formed.

10. (d) If free radical halogenation generate a chiral carbon, racemic mixture of halides are always formed due to equal probability of halogenation from both sides of planar free radical.

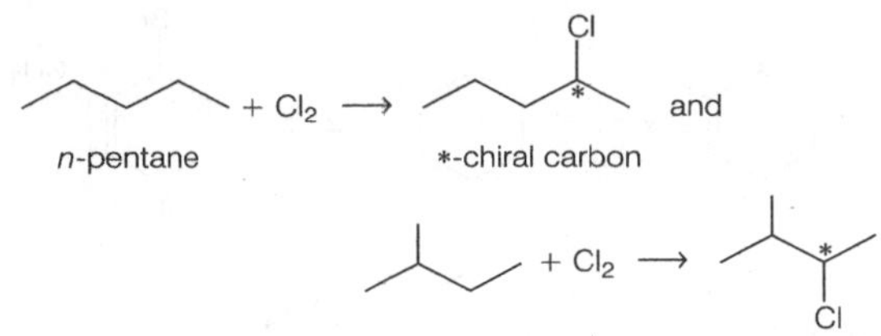

11. (b) $R-OH + SOCl_2 \longrightarrow R-Cl\,(l) + HCl\,(g) + SO_2(g)$

The gaseous byproducts escape out on its own continuously driving the reaction in forward direction.

12. (b)

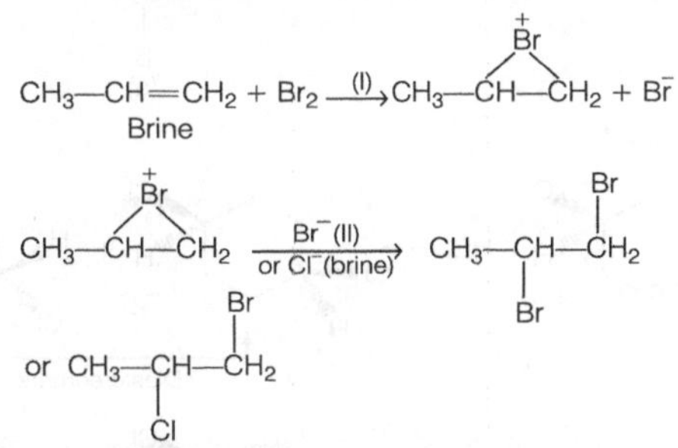

Nucleophilic attack in step-II occur at the carbon atom which can better accommodate the positive charge. Hence, attack of Br^- or Cl^- in second step occur at 2° carbon rather that at 1° carbon.

13. (b,c) Both SO_2Cl_2 and Cl_2 undergo homolytic bond fission when heated or irradiated with light.

14. (b,c,d)

(A)

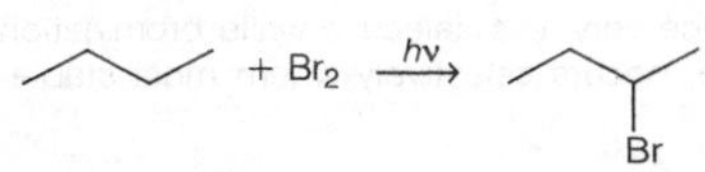

Secondary (major)

(B)

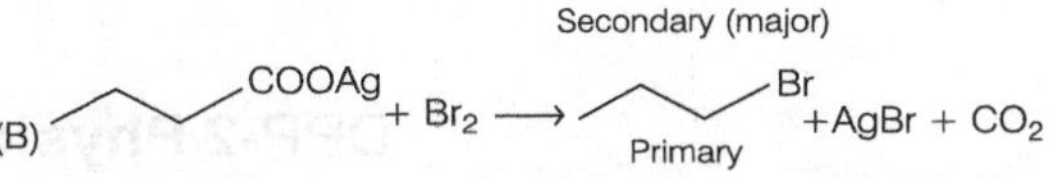

(C)

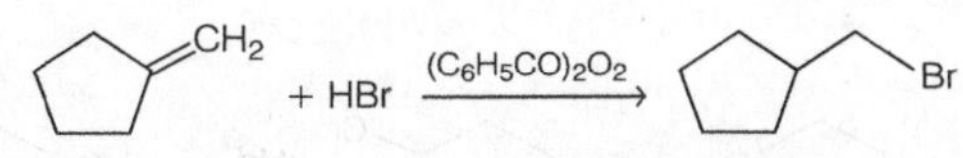

Anti-Markownikoff's addition (Primary)

(D)

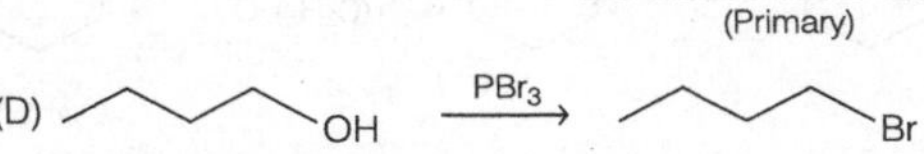

15. (a,b,c,d) NBS in CCl_4 brings about allylic bromination by free radical mechanism:

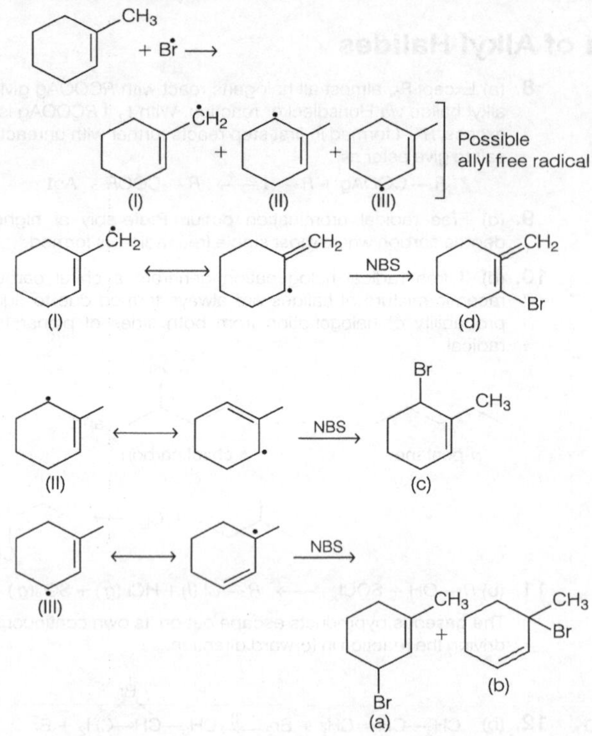

16. (a,c)

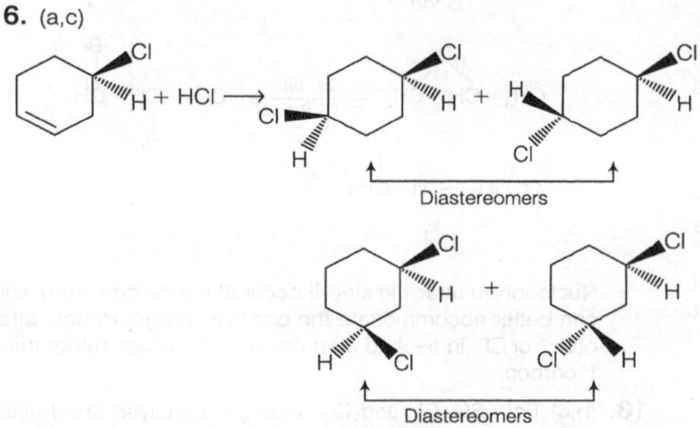

17. (b,c) In free radical halogenation of alkane, the first step of propagation is exothermic when Cl_2 is used while it is endothermic when Br_2 is used. Also chlorination occur at very fast rate, hence very less selective while bromination occur at very slow rate, occurs selectively where most stable free radical is formed.

18. (c) As mentioned in mechanism, reaction proceed *via* carbocation intermediate. Hence, alcohol forming most stable carbocation reacts most easily. Alcohol (c) forms aromatic (highly stable) carbocation, hence most reactive.

19. (b) $ZnCl_2$ is a Lewis acid, helps in the formation of carbocation intermediate, hence catalyse the reaction.

20. (b) The order of stability of carbocation intermediates formed follows the order of reactivity.

21. (4)

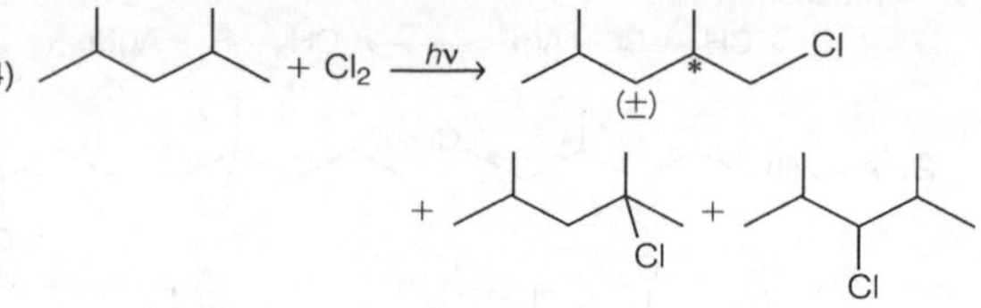

22. (6)

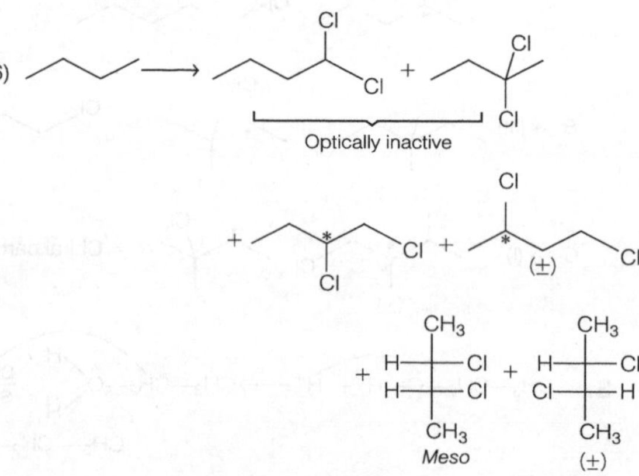

Only three pairs of enantiomers are formed.

23. (3)

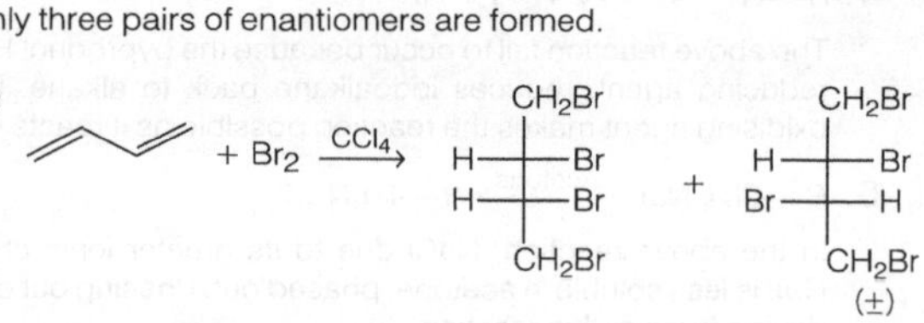

24. (4)

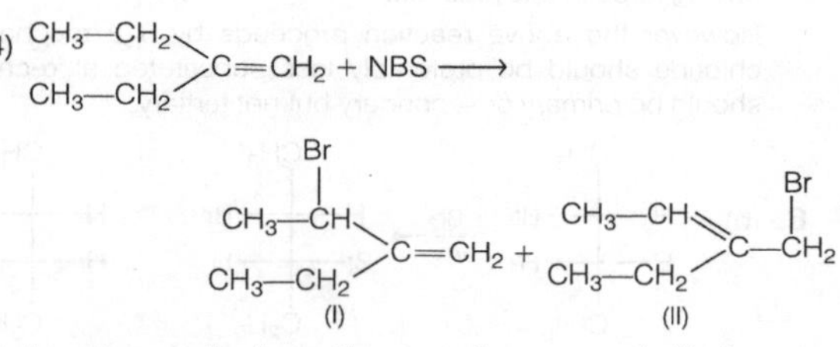

(I) has two optically active enantiomers and (II) has two geometrical isomers.

DPP-2 **Physical and General Properties of Alkyl Halides**

1.

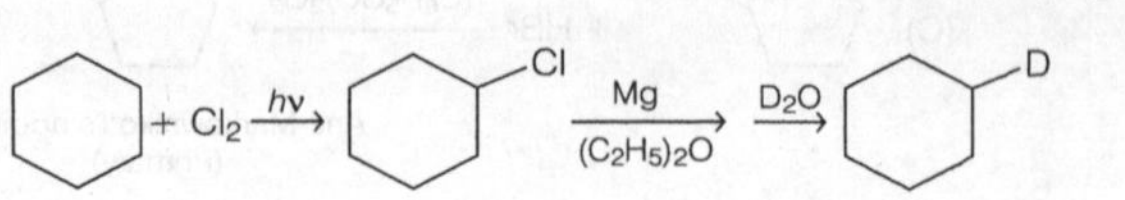

2. CH_3Cl, although polar but gas while CCl_4 is liquid at room temperature although it is non-polar. It is due to very large van der Waals' force in CCl_4 than in CH_3Cl.

3. 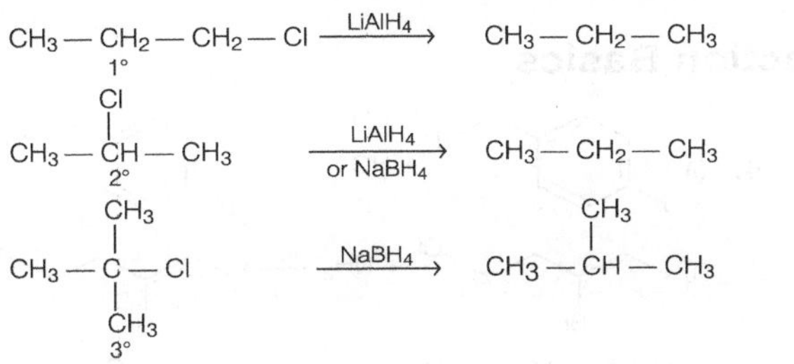

With $LiAlH_4$, 3° alkyl halide gives alkene. Triphenyl Tin hydride $(C_6H_5)_3SnH$ reduces all three types of halides into alkane.

4. 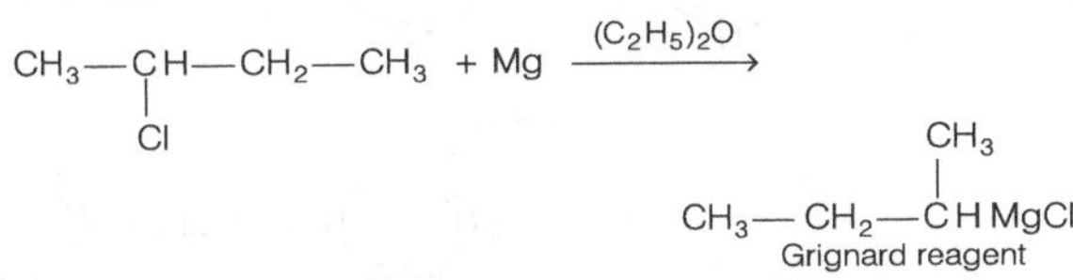

Grignard reagent consist of very strong base (R^-) which brings about some elimination reaction giving alkene as

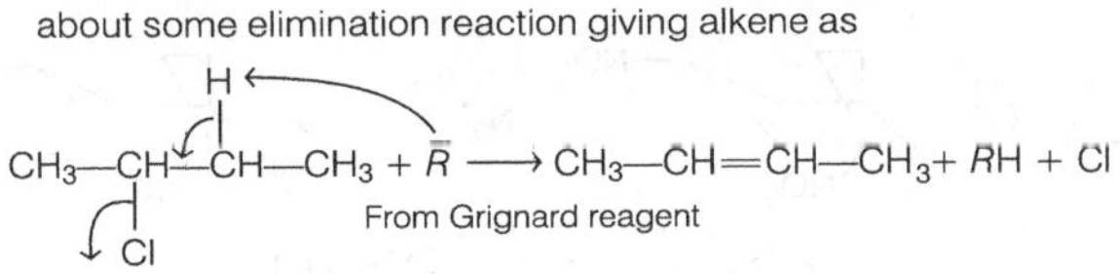

5. RLi is more nucleophilic than RMgX. This is because Mg^{2+} has greater polarising power than Li^+, R—Mg bond is more covalent, a poorer source of R^- nucleophilic than RLi.

6. (b)

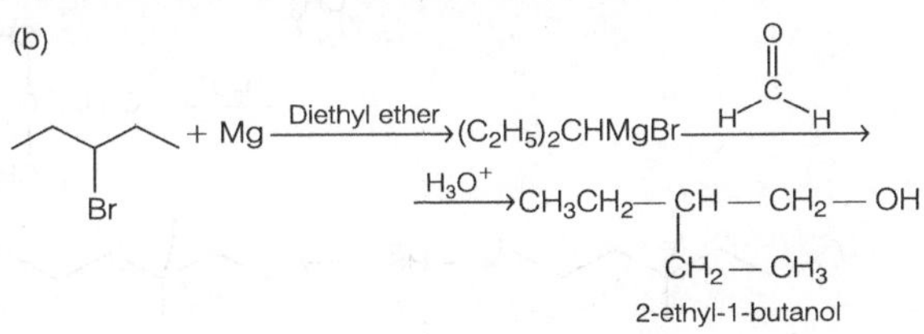

7. (d)

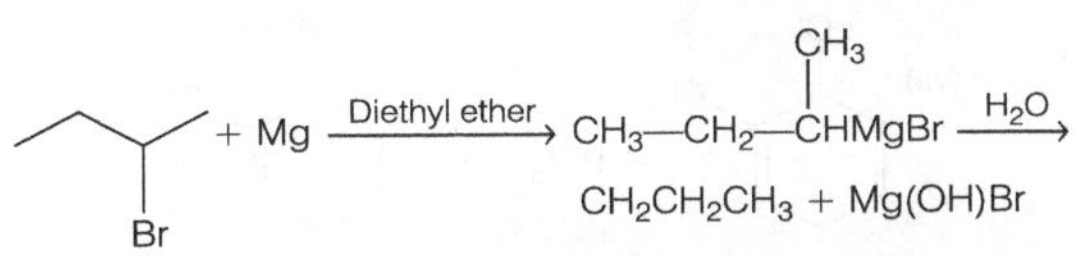

8. (b) 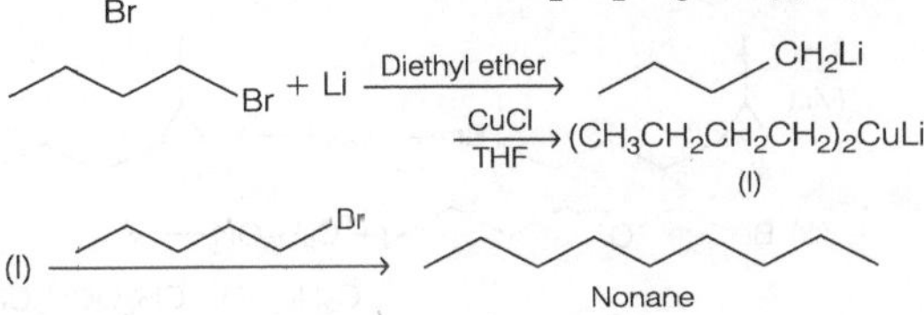

9. (a) 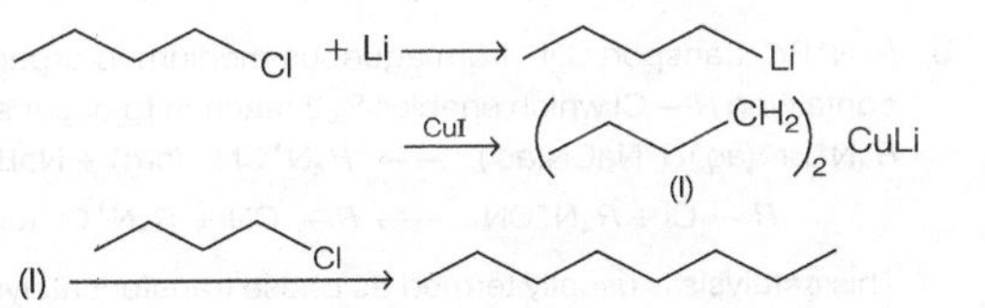

10. (a) 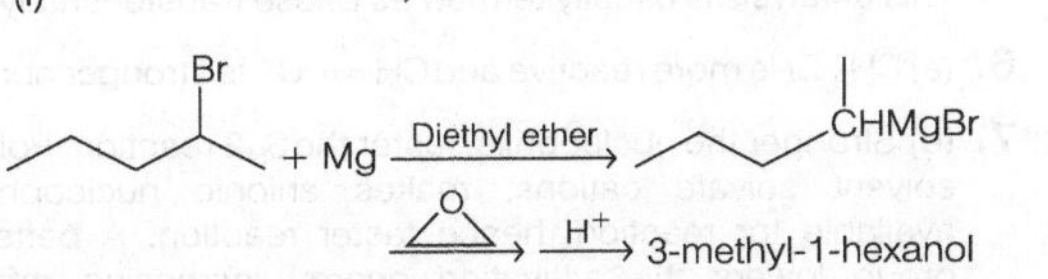

11. (b) 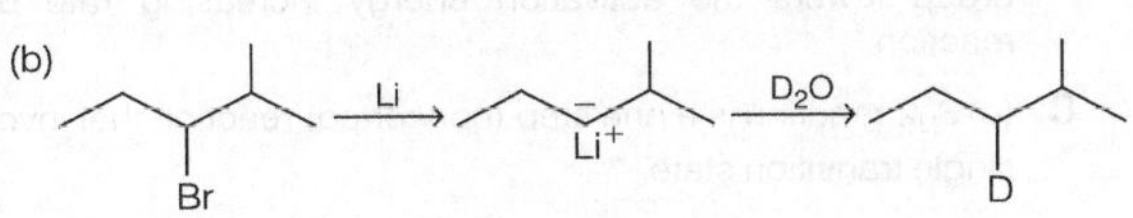

12. (d) All of these are true regarding chlorinated hydrocarbons.

13. (c) Halides containing acidic hydrogen (here —OH) cannot be used in Grignard's synthesis as Grignard reagent consist of very basic carbanion (R^-).

14. (d) All others are gas at room temperature.

15. (b,c,d) 1-chlorobutane is a primary halide, $LiAlH_4$, Zn, $(C_6H_5)_3SnH$ all reduces it to butane but $NaBH_4$ does not reduce primary halide.

16. (a,c,d) $LiAlH_4$ fails to reduce tertiary halide, rather elimination reaction takes place giving alkene.

17. (a,b) Chloro and fluoro alkanes are less dense than water while bromo and iodoalkane are more denser than water.

18. (c,d) Both contain acidic hydrogen, not suitable for preparation of Grignard reagent.

19. (b,c) Cyclic bromonium ion formed in the first step undergo nucleophilic attack by both Br^- and Cl^- ions, giving both option (b) and option (c).

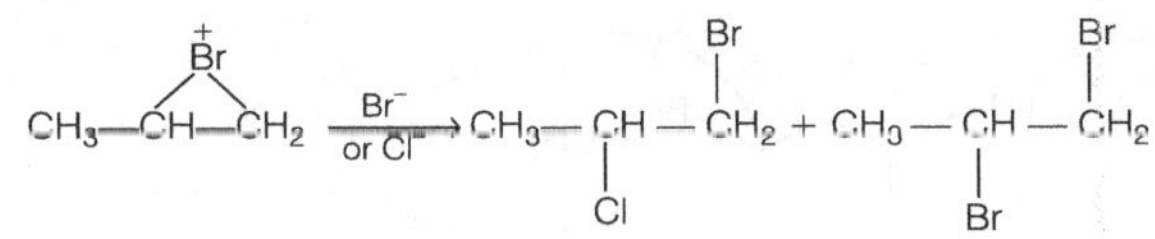

20. (a) As shown in mechanism, reduction occur by hydride transfer (S_N2) mechanism. Hence, primary halide would be most reactive here.

21. (d) Grignard's reagent contain alkyl carbanion (R^-). Hence, Grignard's reagent with 3° carbanions are very less likely to be formed.

22. (b) If two similar alkyl halides join by Wurtz's reaction alkane containing even number of carbons are obtained. If mixed alkyl halides with one containing even and other containing odd number of carbons are used, alkane with odd number of carbons are obtained.

23. (3) 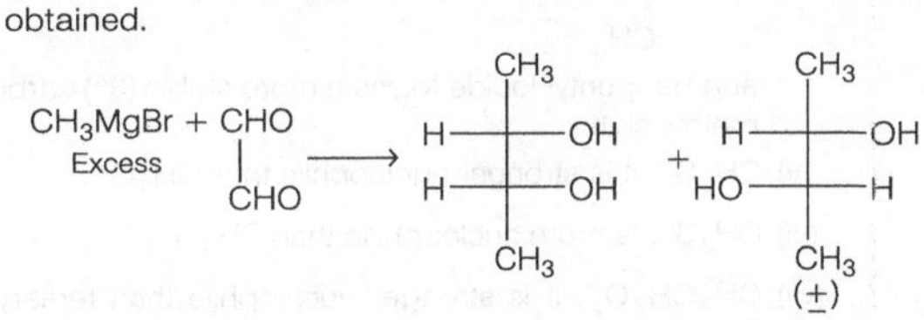

24. (7) 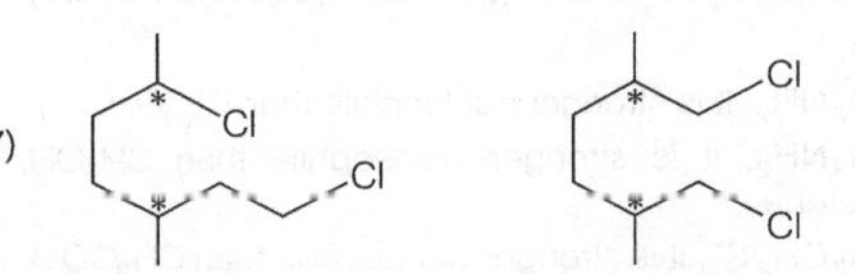

Two chiral carbons, unsymmetrical, 4 stereoisomers

Two chiral carbons, symmetrical, one *meso* and two optically active isomers

Both of the above dichloride on heating with alkali metal gives the same 1,4-dimethyl cyclohexane.

25. (6) Only (ii), (iv), (vii) and (ix) are less dense than water. Alkyl bromides, alkyl iodides and all alkyl halides containing two or more halogen atoms.

26. (b) Although F is more electronegative than Cl, CH_3F, is less polar.

Dipole moment (μ) is defined as

$$\mu = q.d$$

Due to smaller radius of F than Cl, $q \cdot d$ in CH_3Cl is greater than in CH_3F although q is greater in CH_3F.

27. (a) Tertiary carbanion is very unstable, difficult to form.

DPP-3 S_N2 Reaction Basics

1. (i) $CH_3ONa + CH_3I$

(ii) $NaSH + CH_3I$

(iii) $PH_3 + CH_3CH_2CH_2I$

(iv) $NaCN + CH_3CH_2I$

(v) $(CH_3)_3CCH_2{-}SNa + CH_3I$

(vi) CH_3I (excess) + NH_3

(vii) $C_6H_5COONa + C_6H_5CH_2Cl$

(viii) $(CH_3)_2CHONa + CH_2 = CH_2{-}I$

(ix) $CH_2{=}CH{-}CH_2{-}ONa + CH_2 = CH{-}CH_2{-}I$

2. (i)

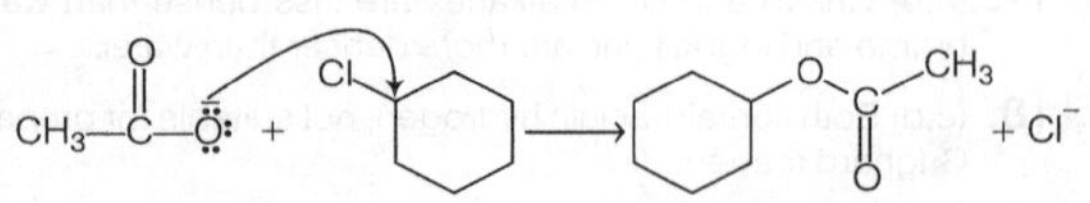

(ii)

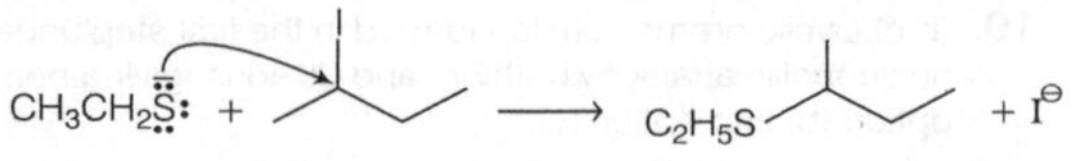

(iii)

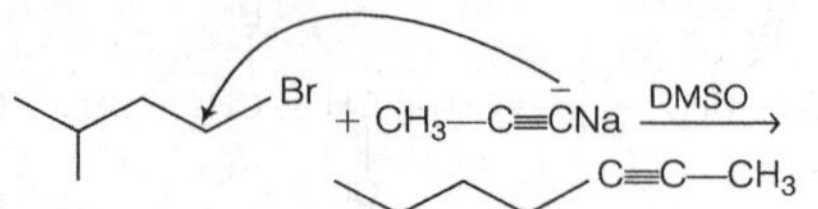

3. (i) CH_3CH_2Br, NaBr is less soluble in acetone, phased out driving S_N2 reaction in forward direction.

(ii) CH_3CH_2I, in hydroxylic solvents, leaving group ability determine the reactivity. I^- is better leaving group than Br^-.

(iii) 1-bromopropane, S_N2 reaction occur more rapidly at less substituted α-carbon.

(iv) $H_3C{-}C(CH_3)_2{-}CH_2{-}I$, reaction proceeds by S_N1 mechanism and neopentyl iodide forms a more stable (3°) carbocation by methyl shift.

(v) CH_3S^-, it is stronger nucleophile than CH_3O^-.

(vi) CH_3S^-, is more nucleophilic than CH_3O^-.

(vii) $CH_3CH_2O^-$, it is stronger nucleophile than tertiary butoxide ion.

(viii) CH_3NH_2, it is stronger nucleophile than CH_3OH.

(ix) CH_3NH_2, it is stronger nucleophile than CH_3OH, even in acetone.

(x) $CH_3CH_2S^-$, it is stronger nucleophile than CH_3COO^-.

(xi) $CH_3CH_2S^-$, it is stronger nucleophile than CH_3COO^-, irrespective of solvents.

(xii) $HOCH_2CH_2CH_2CH_2Br$ as it forms a more stable, five membered and cyclic ether.

(xiii) Anti side of the leaving group has less bulky groups, less steric hindrance to the attack of nucleophile is observed, hence more reactive.

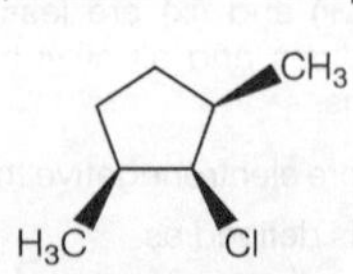

(xiv) It is more nucleophilic than phenoxide ion.

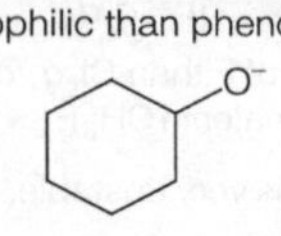

4. (i)

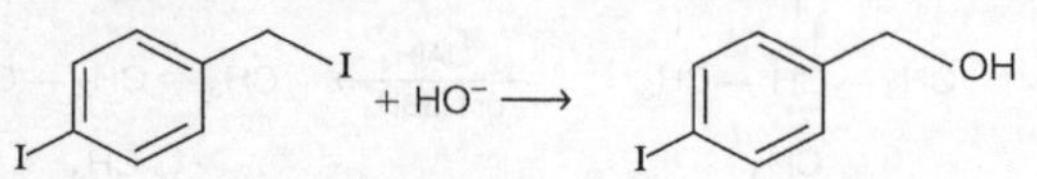

(ii)

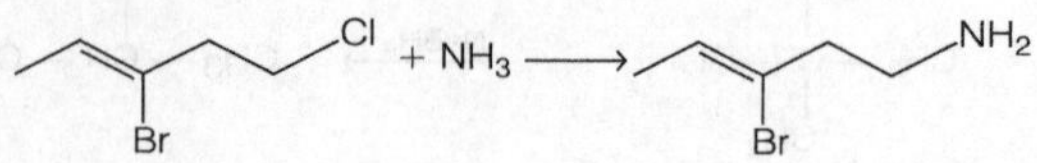

(iii)

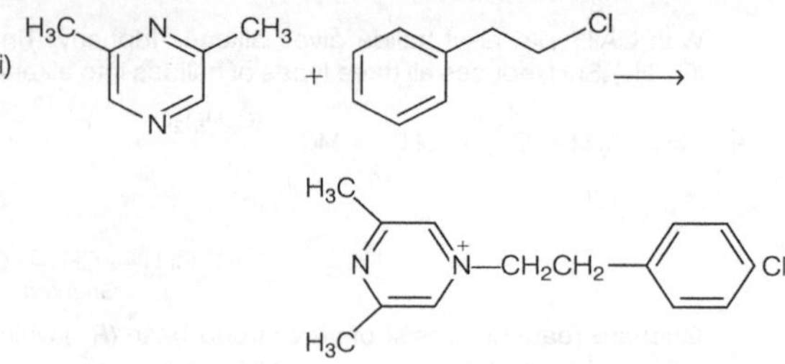

(iv)

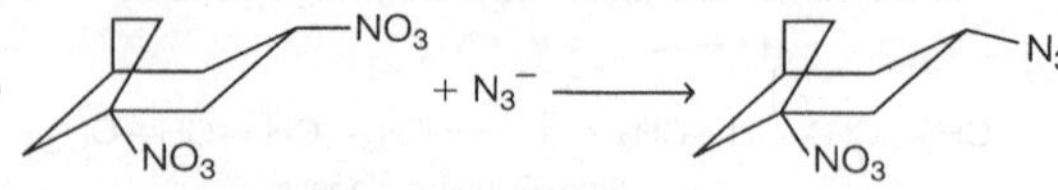

(v)

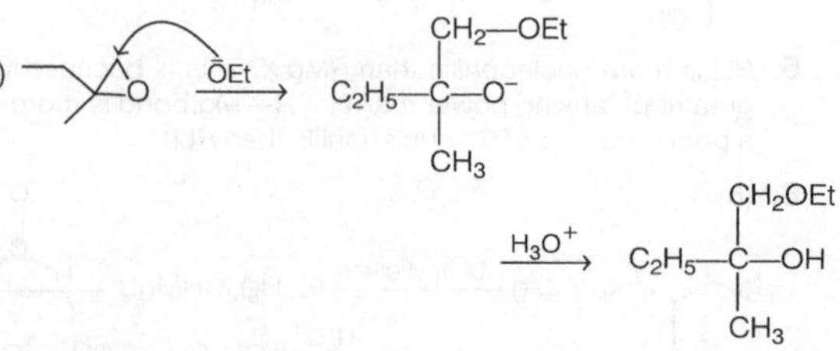

(vi)

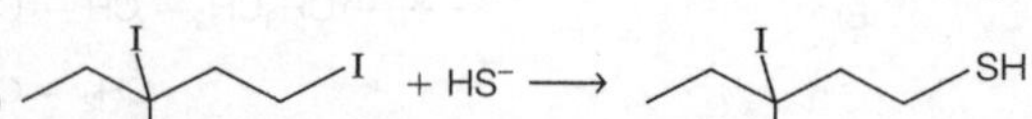

(vii)

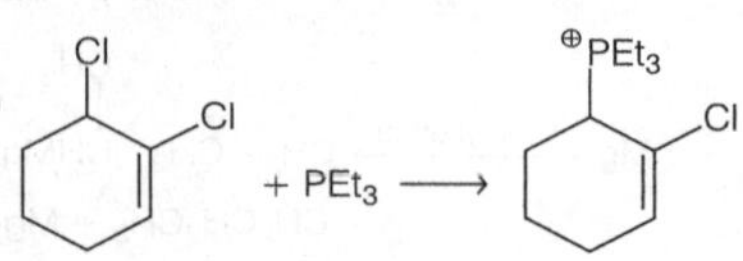

(viii)

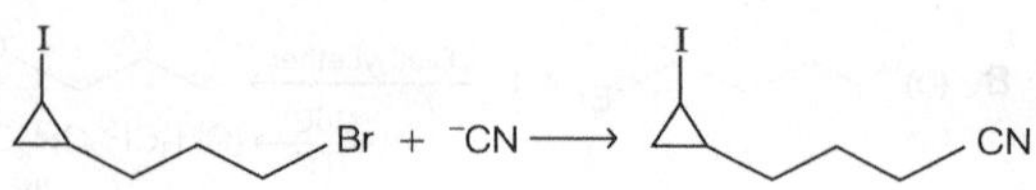

(ix) 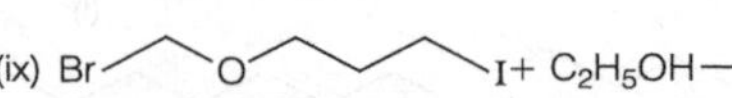 + $C_2H_5OH \longrightarrow$

$C_2H_5{-}O{-}CH_2OCH_2CH_2CH_2{-}I$

5. $R_4N^+Br^-$ transport CN^- from aqueous medium to organic phase containing $R{-}Cl$ which enables S_N2 reaction to occur as

$$R_4N^+Br^-(aq) + NaCN(aq) \longrightarrow R_4N^+CN^- \text{ (org)} + NaBr\ (aq)$$

$$R{-}Cl + R_4N^+CN^- \longrightarrow R{-}CN + R_4N^+Cl^-(aq)$$

This catalysis is usually termed as phase transfer catalysis.

6. (a) CH_3Br is more reactive and $CH \equiv C^-$ is stronger nucleophile.

7. (d) Stronger the nucleophile, faster the S_N2 reaction. Polar aprotic solvent solvate cations, makes anionic nucleophile more available for reaction, hence faster reaction. A better leaving group lowers the activation energy increasing rate of S_N2 reaction.

8. (d) S_N2 reaction is a one step (concerted) reaction that involves a single transition state.

9. (c) Allyl bromide (I) is most reactive among the given halides as pi bonds from allylic position stabilises the transition state. Vinyl iodide (II) is least reactive due to partial double bond character. Electron withdrawing inductive effect of CH_3O^- increases reactivity of (IV) over (III).

10. (d) (IV) is most reactive as it is benzylic as well as electron withdrawing effect of $-NO_2$ further increases the reactivity. (III) is least reactive due to resonance effect resulting in partial double bond character between carbon and chlorine. (I) is less reactive in S_N2 reaction than II, due to greater steric hindrance in (I).

11. (d) Substrates are same as they are at same potential energies. Nucleophiles are different as magnifested by different activation energies and different products with different potential energies.

12. (b) HO^- is stronger nucleophile than CH_3COO^-. Hence, lower activation energy of (I) than (II).

13. (b) The transition state in S_N2 reaction is pentavalent as indicated here.

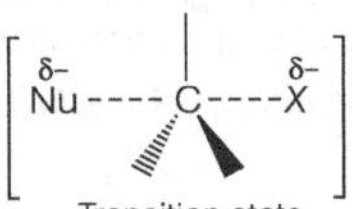

Transition state

14. (a,b,c) In a S_N2 reaction, Rate = k[substrate] [nucleophile]

Hence, rate depends both on concentration of substrate as well as nucleophile. Rate of reaction also depends on nature of leaving group. A better leaving group gives lower activation energy hence, greater rate of reaction. Number of lone pair has no role to play in nucleophilicity hence reactivity.

15. (a,b,d) Increasing concentration of nucleophile (NaOH) increases the reactivity. Adding some NaI also catalyse reaction because of following equilibrium.

$$CH_3-Br + I^- \rightleftharpoons CH_3-I + Br^-$$

Now reaction occur with CH_3I which has better leaving group. Heavier isotope forms stronger covalent bond. Hence, replacing CH_3Br by $^{14}CH_3Br$ stabilises substrate, decreases rate whereas replacing NaOH by $NaO^{18}H$ stabilises transition state increases rate of reaction.

16. (a,c) Both nucleophile and leaving group are electron rich species, nucleophile attacks from backside (most remote position) to have minimum repulsion with leaving group. Also, due to the presence of leaving group on front side there is less physical space for attack of nucleophile.

17. (a,c,d) $CH_3-CH_2-CH_2-Cl$ 1-chloro propane (1°) $\quad CH_3-C(CH_3)_2-Cl$ 2-chloro-2-methylpropane (2°)

As indicated in the given reaction, α-carbon is less hindered in 1-chloropropane. Also $+I$ effect of three methyl groups in 2-chloro-2-methyl propane decreases electropositive character of α-carbon, decreases reactivity further in S_N2 reaction.

18. (a,c,d) Steric hindrance plays the most important role in S_N2 reaction. Hence, although Cl is poorer leaving group than I, Cl is substituted predominantly in the above reaction due to less steric hindrance at α-carbon. Addition of NaI replace Cl by I and substitution becomes easier.

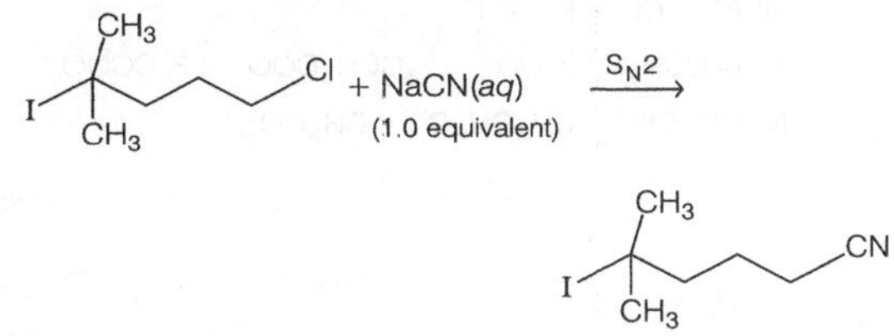

19. (b,c,d) Stronger bond formation at α-carbon affects the stability of product (thermodynamics) not the rate (kinetics) of reaction. Change of isotope at α-position or in halogens affect the rate, hence show kinetic isotopic effect. Presence of electron withdrawing group in substrate increases electrophilicity of α-carbon hence, increases its reactivity towards nucleophiles.

20. (b) Products are same in both reactions, hence same potential energies of products are shown. Also I^- is better leaving group, has lower activation energy in S_N2 reaction as indicated by curve-I in diagram.

21. (c) Since I^- is better leaving group, $k_I > k_{II}$

22. (c) Following neutralisation reaction is preferred over S_N2 reaction :

$$CH_3-CH_2-OH + NaNH_2 \longrightarrow CH_3-CH_2-O^-Na^+ + NH_3$$

23. (9) Rate $= k\,[C_2H_5I]\,[OH^-]$

Rate is linear function of both alkyl halide and nucleophile concentration both will decrease by same factor. Hence, if after 1.0 h concentration of iodoethane decreases to $\left(\frac{1}{3}\right)^{rd}$ concentration of hydroxide will also decrease by the same factor and rate by $\left(\frac{1}{3}\right)^2$, i.e. by 9 times.

24. (6) Compounds (II), (III), (IV), (V), (VIII) and (X) react at faster rate than 2-bromobutane in S_N2 reaction.

25. (3) Only the circled groups are substituted in S_N2.

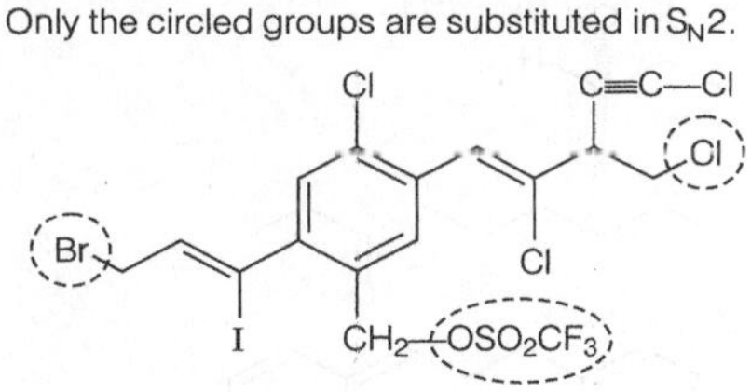

DPP-4 S_N2 Reaction Reactivity

1. (i) $HO^- < H_2O < H_3O^+$

(iii) $NH_2^- < NO_2^- < NO_3^-$

(iii) $F^- < Cl^- < Br^- < I^-$

(iv) $F_3CCH_2CH_2COO^- < F_3CCH_2COO^- < F_3CCOO^-$

(v) $CH_3CH_2^- < CH_3CH_2O^- < CH_3COO^-$

(vi) 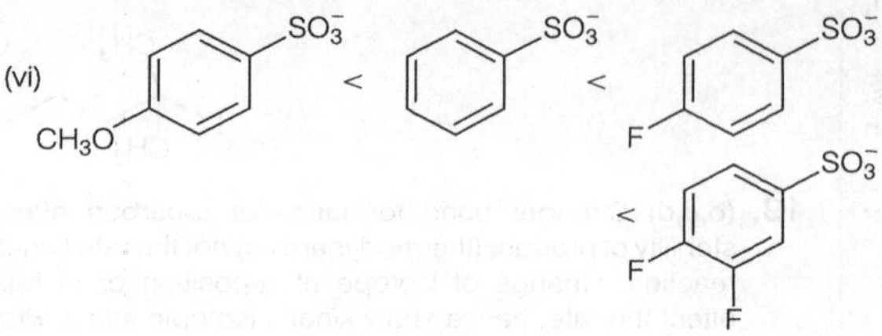

2. (i) 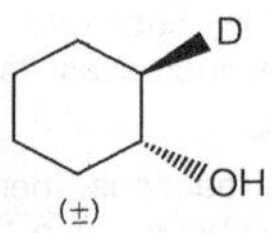

(ii) 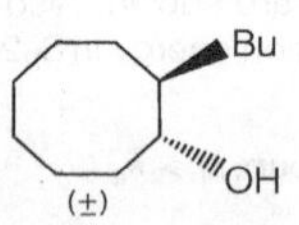

(iii) 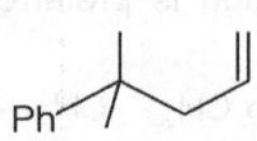

(iv) 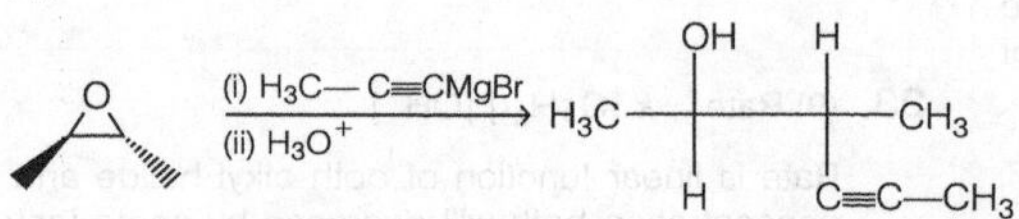

(v) 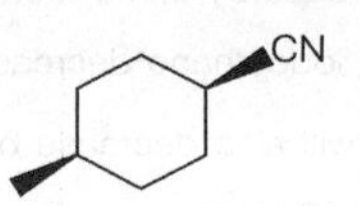

(vi) 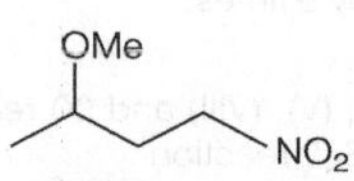

(vii) 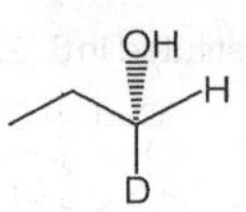

(viii) 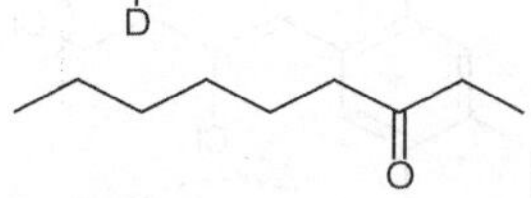

(ix) 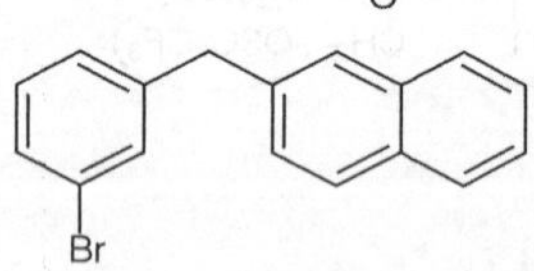

(x) 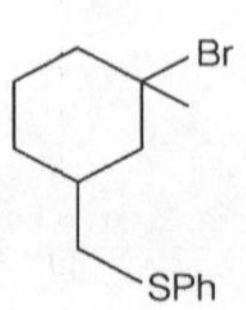

(xi) 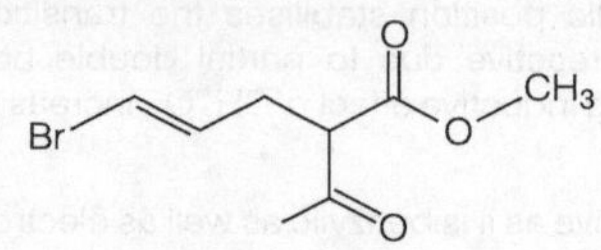

(xii) 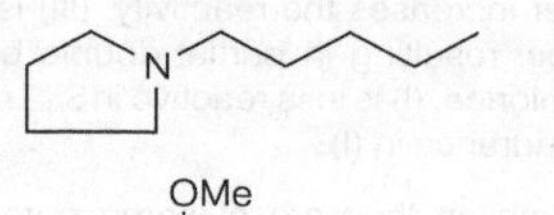

3. 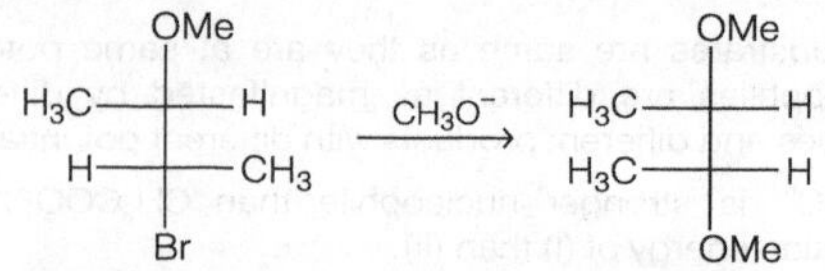

Inversion of configuration at only α-C takes place.

4. Chloroacetic acid is stronger acid than acetic acid hence, acetate ion (CH_3COO^-) is stronger base as well as stronger nucleophile than chloroacetate ion ($Cl—CH_2COO^-$).

5. 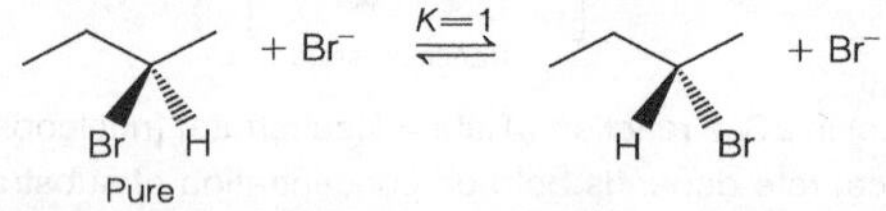

Establishment of above equilibrium is responsible for racemisation. No such substitution equilibrium is possible with alcohol because — OH is a very poor leaving group.

6. (i) HO^- (ii) HO^- (iii) CH_3S^- (iv) Cl^- (v) I^-

7. Polar aprotic solvents like DMSO, THF, DMF etc., increases rate of S_N2 reaction by solvating cationic fragments of nucleophilic reagent.

8. It is due to establishment of the following equilibrium.

$$R—Br + I^- \rightleftharpoons R—I + Br^-$$

9. In the presence of 0.5 mole of Na_2CO_3 and 4-hydroxyphenol, the nucleophile is 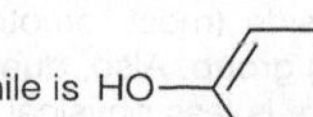 (*MW* = 109) Now let molar mass of leaving group be *X*. Then

$$171 - X + 109 = 200$$

$$\Rightarrow \quad X = 80 \text{ (Br)}$$

Therefore, in substrate, molar mass of remaining group is

$$171 - 80 = 91$$

As given in question, *A* has a phenyl ring ($C_6H_5 = 77$), remaining part has molecular weight = 14(CH_2)

Hence, $C_6H_5CH_2Br$ (*A*)

and 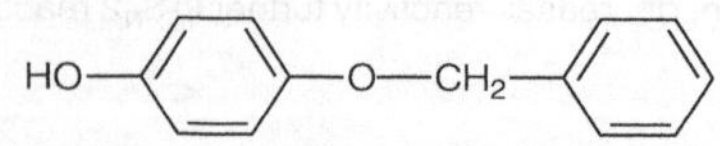

B

10. (d) Nucleophilicity increases down the group in hydroxylic solvents.

11. (d) Inversion of configuration takes place at chiral carbon.

12. (a) $(CH_3CH_2\bar{C}HCH_3)\overset{+}{Mg}Br$ attack at less substituted carbon of epoxide ring from the side opposite to — CH_3 group.

13. (d)

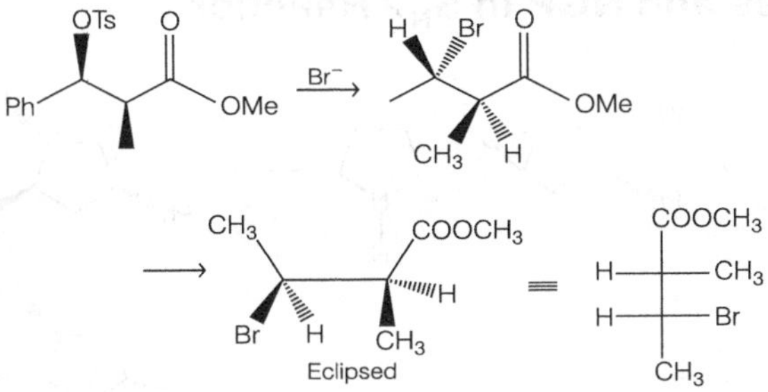

14. (b) α-carbon is achiral, hence retention of configuration at β-carbon.

15. (c) Both CH_3S^- and I^- are weak bases but strong nucleophiles.

16. (c) (S)-2-butanol is formed with specific rotation + 13.52°.

17. (c) Although CH_3O^- is weaker base than $(CH_3)_3CO^-$, former is stronger nucleophile due to smaller size.

18. (b) In halogen-exchange (Finkelstein reaction), reactivity is determined by solubility of Na *X* formed in acetone. Lower the solubility of Na *X* in acetone, easier the reaction. NaCl is less soluble than NaBr in acetone.

19. (b) CH_3S^- (I) is strongest nucleophile and $C_6H_5O^-$ (IV) is weakest due to resonance stabilisation in it.

20. (c) $CH_3ONa + CH_3COOH \longrightarrow CH_3OH + \underset{\text{(Nucleophile)}}{CH_3COO^-Na^+}$

21. (b)

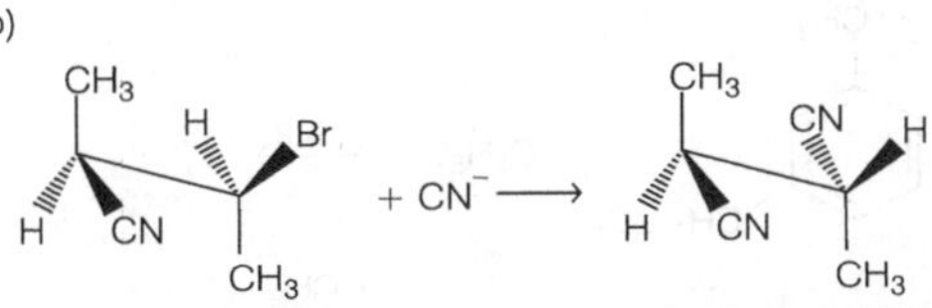

meso achiral

22. (a,d) CH_3O^- is stronger nucleophile (smaller size) than $(CH_3)_3CO^-$. HS^- is stronger nucleophile than HO^- as S is more polarisable than O.

23. (c,d) Both $(CF_3SO_3^-)$ and $p\text{-}O_2N—C_6H_4—SO_3^-$ are weaker base than its other counterpart.

24. (a,d) In both option (a) and option (d) α-carbon is chiral, hence S_N2 reaction will lead to inversion of configuration. In option (b), α-carbon is achiral, configuration of chiral β-carbon will not be affected. In option (c) S_N2 reaction occur twice at the same α-carbon hence, double inversion will give net retention of configuration.

25. (a,b,c,d) All the mentioned solvents are polar and aprotic, hence increases rate of S_N2 reaction.

26. (a,c,d) In both option (a) and option (c) products have inverted configuration but at α-carbon only. In option (b) product has inverted configuration at both α and β-carbons, hence not true in S_N2 reaction. In case of option (d), an equilibrium will be established with $K = 1$. :

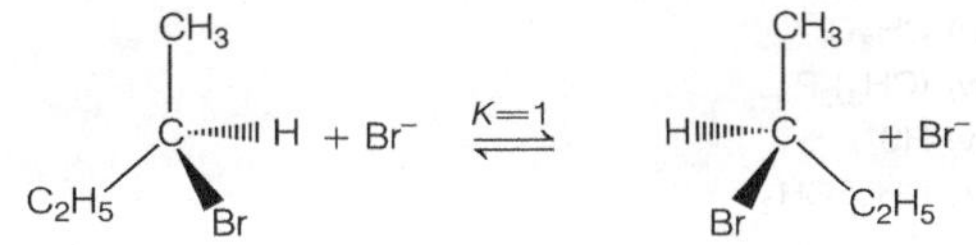

At equilibrium both enantiomers exist in equal amounts giving racemic mixture.

27. (b) F^- is most hydrated, least nucleophilic and I^- is most polarisable, strongest nucleophile among halides in H_2O.

28. (a) In polar aprotic solvent, basicity parallels nucleophilicity.

29. (c) Hydroxylic solvents like H_2O, stabilises both nucleophile and transition state, but to a greater extent to nucleophile. As a result, activation energy increases and rate of reaction decreases.

30. (d) The product would have opposite configuration at α-carbon, but reactants and products are not isomers, they are different compounds, have their characteristic stereochemical properties not inter-related.

31. (c) Both Cl can act as leaving group and there will be equal probability of the two chlorides to leave, giving racemic mixture of products.

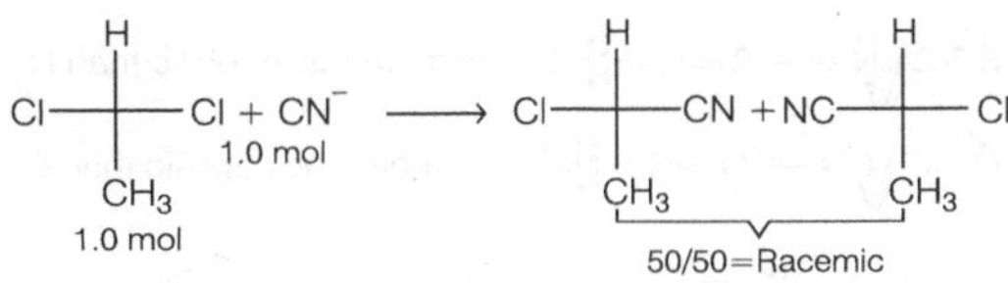

32. (d)

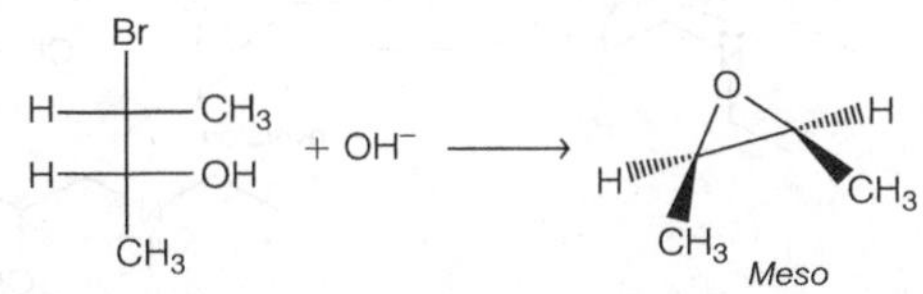

33. (5) In (iii), (iv), (v) and (ix); Neutralisation will be preferred over S_N2 reaction. In (vi), due to tertiary α-carbon, S_N2 is very difficult, however E2 elimination reaction will be preferred in this case.

34. (4) Any of the two Cl can be substituted by CN^- as:

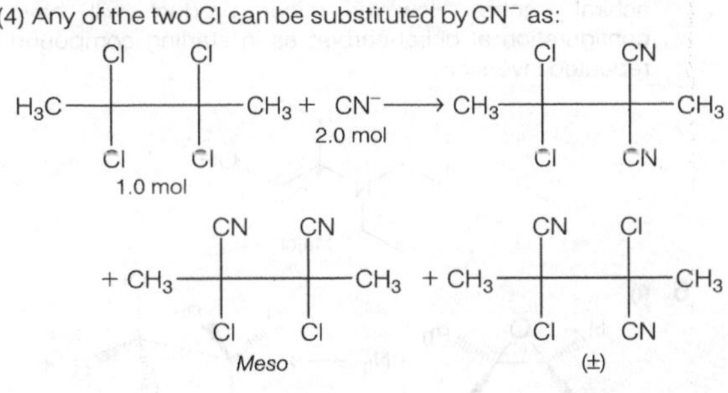

DPP-5 Ambident Nucleophiles and NGP in S_N2 Reaction

1. (i) CH_3S^-
(ii) $(CH_3)_3N$
(iii) CH_3Cl
(iv) $(CH_3)_3P$
(v) HS^-
(vi) CH_3SCH_3

2. Due to neighbouring group participation.

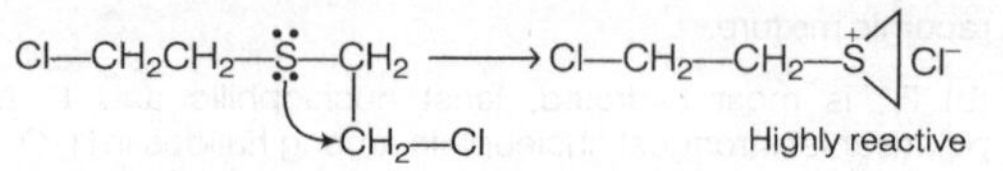

3.

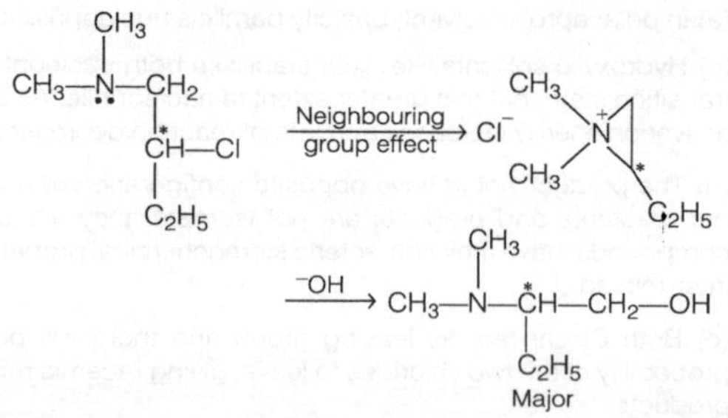

4.

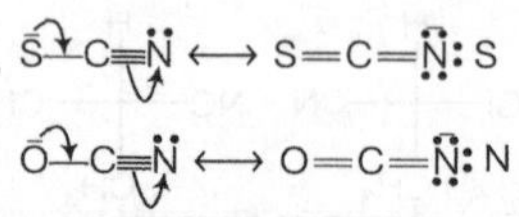

S is more nucleophilic than N

N is more nucleophilic than O.

5.

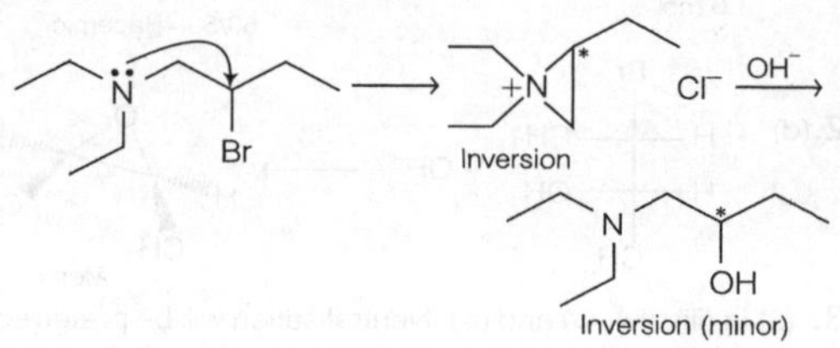

Major product will have inverted configuration at chiral carbon because in the second final step nucleophilic attack occurs at achiral centre. However minor product will have same configuration at chiral carbon as in starting compound due to repeated inversion.

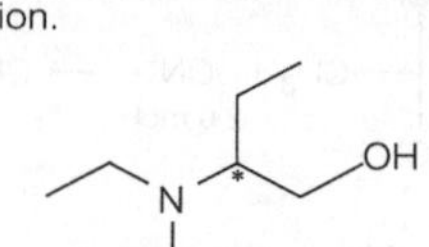

Major

6. (i)

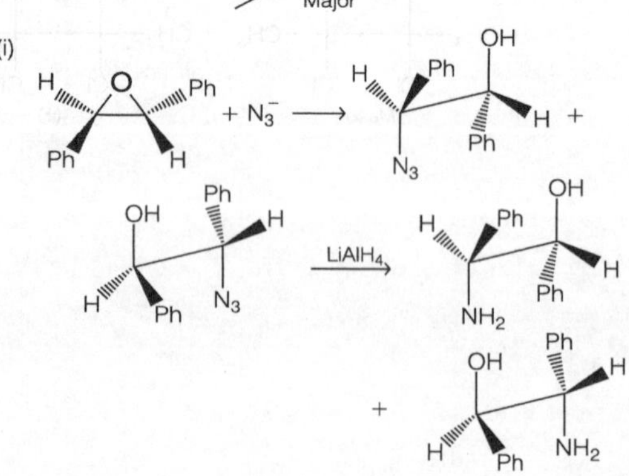

(ii)

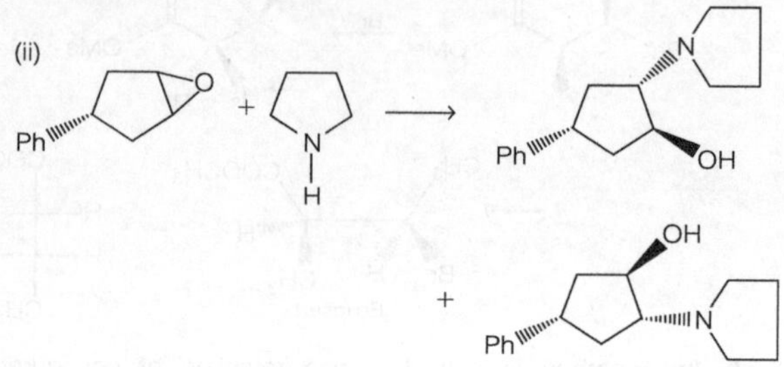

(iii)

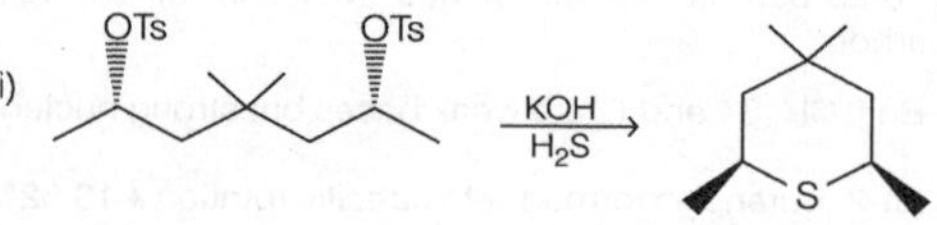

7. (c) $:\!C\equiv N\!:$ Both carbon and nitrogen are capable of donating electron pair but carbon is better electron donor than nitrogen, carbon donates electron pair in most of the cases.

8. (a) By the following equilibrium

$$CN^-\,(aq) + C_6H_5CH_2\overset{+}{N}(CH_3)_3Cl^-\,(aq) \rightleftharpoons Cl^-(aq) + \underset{(org)}{C_6H_5CH_2\overset{+}{N}(CH_3)_3CN^-}$$

Cyanide ion is transferred to organic phase otherwise KCN is insoluble in organic solvents.

9. (d)

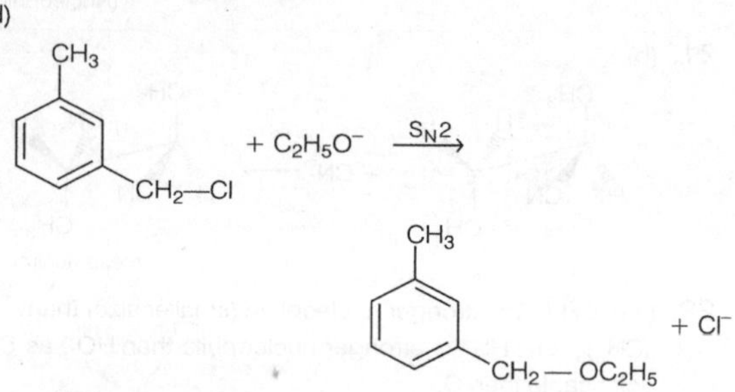

In all other cases leaving group is on phenyl ring, not substituted. usually by nucleophiles under ordinary condition.

10. (c)

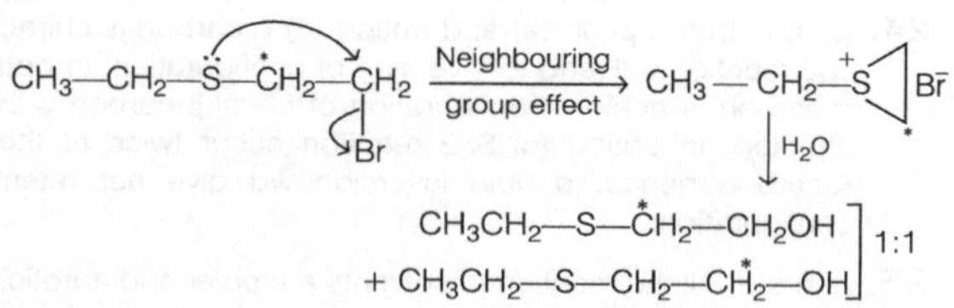

11. (c)

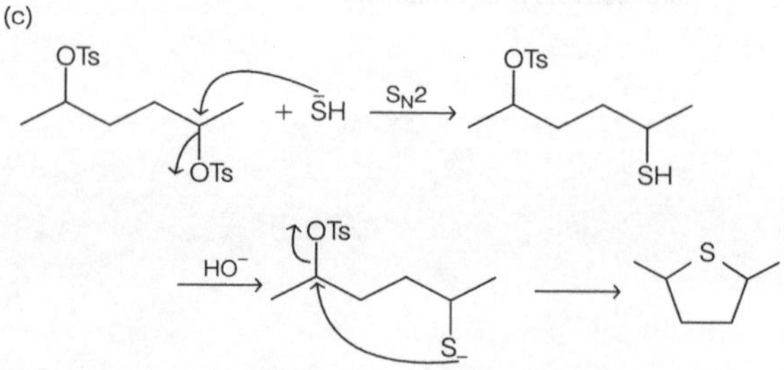

12. (c) 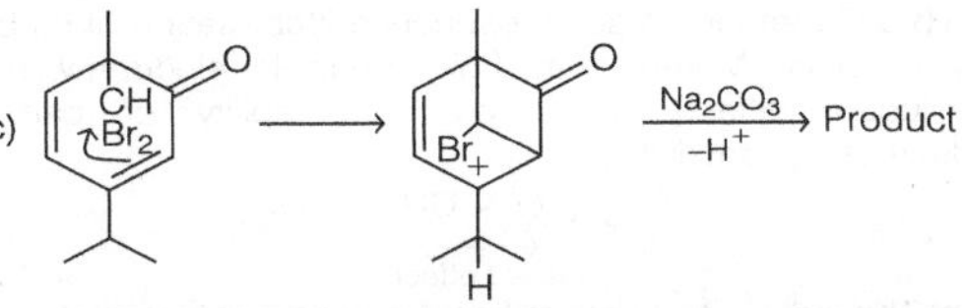

13. (c) 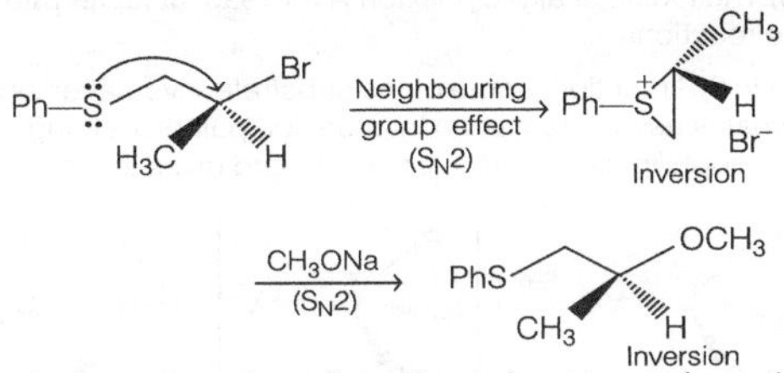

Due to inversion of configuration twice at α-carbon, there will be overall retention of configuration.

14. (c) 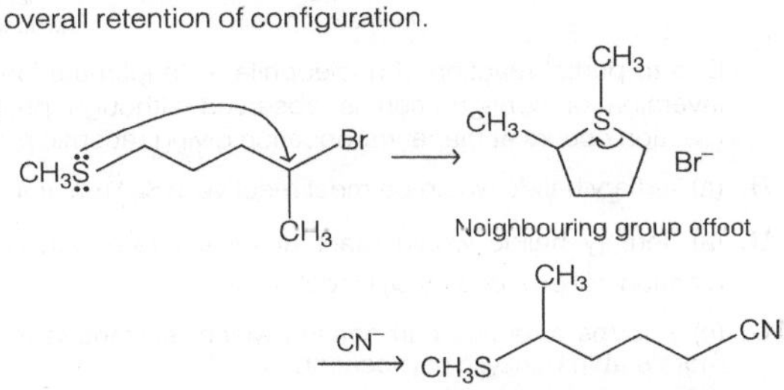

15. (a,b,c) All will react *via* cyclic sulphonium ion intermediate involving S_N2 reaction twice giving over all retention of configuration.

16. (a,b,c,d) All have more than one donor atoms within the same donor group, hence ambident nucleophiles.

17. (b,d) Both donate lone pair from better donor nitrogen atom.

18. (a,c) $CH_3CH_2—Br + NaCN \longrightarrow CH_3CH_2CN + NaBr$

$CH_3CH_2—Br + AgCN \longrightarrow CH_3CH_2CN + AgBr$

Similarly,

$CH_3CH_2—Br + NaSCN \longrightarrow CH_3CH_2SCN + NaBr$

$CH_3CH_2—Br + AgSCN \longrightarrow CH_3CH_2NCS + AgBr$

Silver salts have covalent character, hence lone pair of free atom i.e. nitrogen in both cases donate lone pair. With sodium salts, nucleophiles are free ions in solution, better donor donate lone pair.

19. (a,c) Due to neighbouring group participation, reaction occurs at much faster rate than 2-bromo-3, 3-dimethylpentane. Also it involves S_N2 reaction twice, overall retention of configuration is observed.

20. (b) With NaCN free cyanide (CN^-) is obtained in solution, hence better donor atom carbon donate lone pair. In AgCN, due to covalent character of Ag — C bond CN^- is not free in solution hence, lone pair of 'N' is donated.

21. (b) In both salts, lone pair on nitrogen (better donor atom than oxygen) is free for donation to alkyl halide.

22. (c) With free NCO^-, better donor atom (nitrogen) donate its lone pair. If AgNCO is used, due to covalent character of Ag—N bond lone pair of oxygen is donated giving $CH_3CH_2—OCN$ as major product.

23. (4)

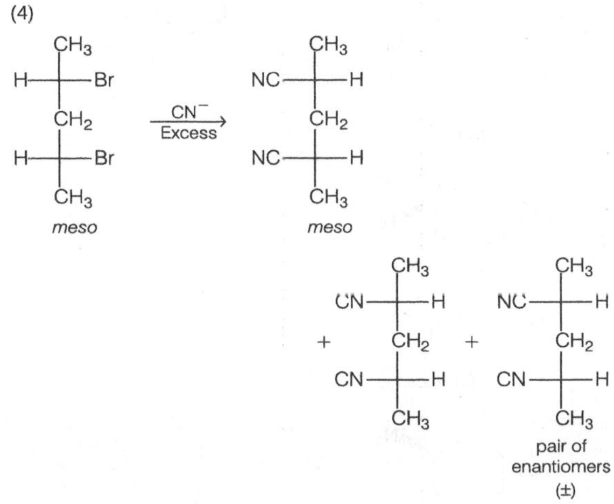

24. (6)

$$k_{II}/k_I = 5 \times 10^4$$

$$R_I = k_I\,(0.05)\,(0.01) \quad \ldots(i)$$

$$R_{II} = k_{II}\,(0.1)\,(0.1) \quad \ldots(ii)$$

Dividing the above two rates gives

$$\frac{R_{II}}{R_I} = \frac{k_{II}}{k_I}\left(\frac{0.01}{0.05 \times 0.01}\right)$$

$$= 5 \times 10^4 \times \frac{100}{5} = 10^6$$

25. (a)

(i) Due to 'S' at β-position, neighbouring group participation occur giving net retention (twice inversion).

(ii) S_N2 at α-carbon gives inversion and product is *meso* diol.

(iii) Only S_N2 hence inversion

(iv) Show neighbouring group effect

Hence, (i) → (q,s), (ii) → (p,r), (iiii) → (p), (iv) → (r,s)

DPP-6 S_N1 Reaction

1. (i) S_N2, (ii) S_N2, (iii) S_N2, (iv) S_N2, (v) S_N2 (vi) S_N1, (vii) S_N1, (viii) S_N1, (ix) S_N2, (x) S_N2

2. (i) $CH_3CH_2CH_2—OCH_3$ S_N2

(ii) 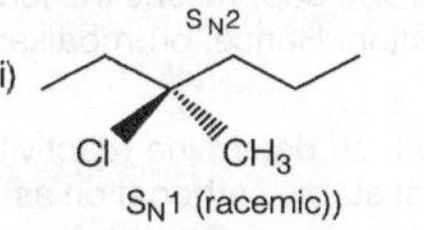

S_N1 (racemic))

(iii) $(CH_3CH_2CH_2)_4\overset{+}{N}\overset{-}{Br}$ S_N2

(iv) 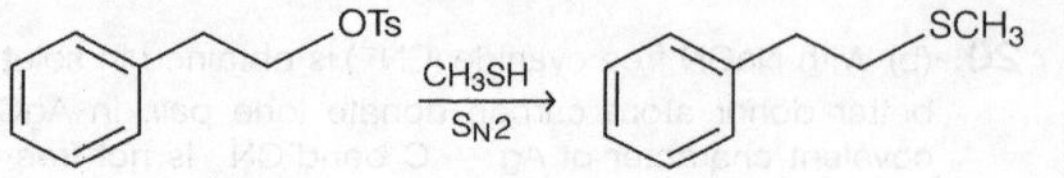

(v) 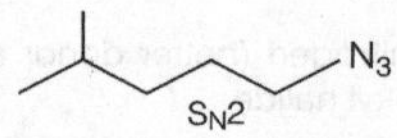

S_N2

(vi) 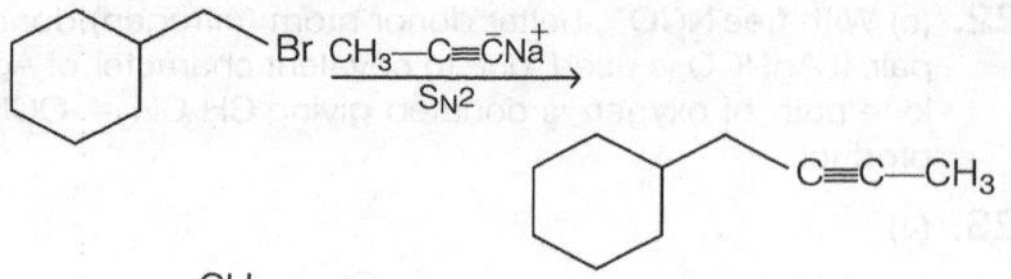

(vii) 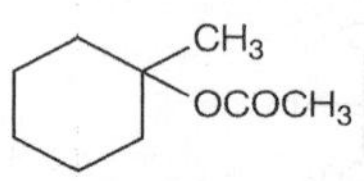

S_N1

(viii) 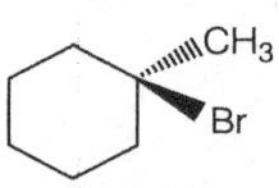

S_N1

(ix) 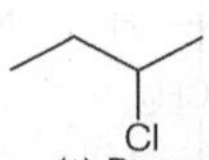

(±)-Racemic

S_N1

(x) 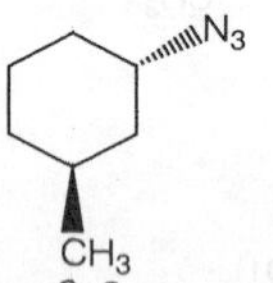

S_N2

(xi) 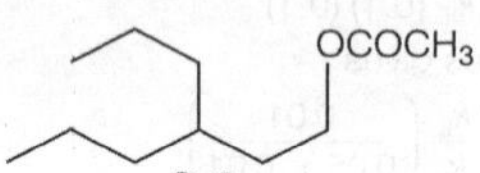

S_N2

(xii) 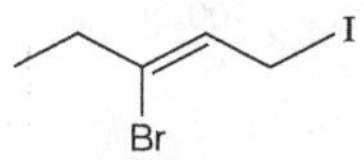

S_N2

3. In the presence of moisture NaI will dissolve in water, will not be available in acetone for halogen exchange. Also, the tertiary halide will start reacting with water *via* S_N1 reaction.

4. Tertiary butyl alcohol reacts with HCl or HBr by S_N1 reaction in which nucleophile is not involved in the slow, rate determining step. Therefore, it does not matter whether nucleophile is Cl^- or Br^-, reaction proceeds at same rate with both nucleophiles.

When equimolar mixture of HCl and HBr is used.

$$HCl \longrightarrow H^+ + Cl^-$$

$$HBr \longrightarrow H^+ + Br^-$$

HBr being a stronger acid than HCl, it suppresses the ionisation of HCl, decreasing Cl^- concentration. Hence, bromoalkane is formed predominantly.

5. It is the stability of carbocation which determine reactivity of S_N1 reaction. α-chloro ether form most stable carbocation as

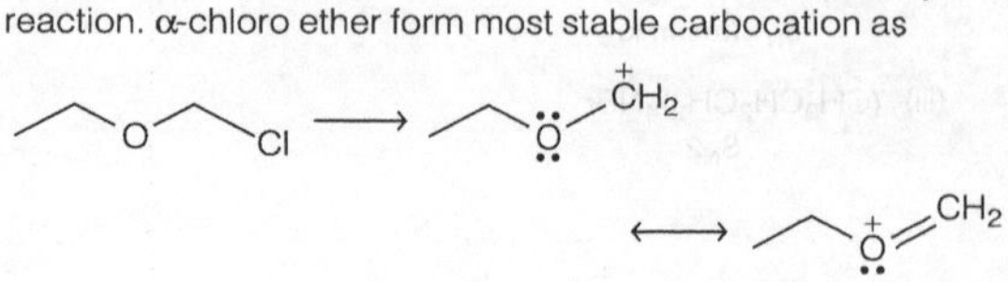

In β-chloroether, no such resonance stabilisation of carbocation takes place. Moreover, methoxy group, by electron withdrawing inductive effect, decreases the stability of carbocation, decreasing reactivity

$$CH_3-O-CH_2-\overset{+}{C}H_2$$

$\longleftarrow$ − I effect

6. (b) The iodine at allylic position *A* will react at faster rate in both of the reactions.

7. (d) In S_N1 reaction, some of the substrate, even after ionisation, remains associated as an intimate ion pair preventing nucleophilic attacks from front side and gives inverted product.

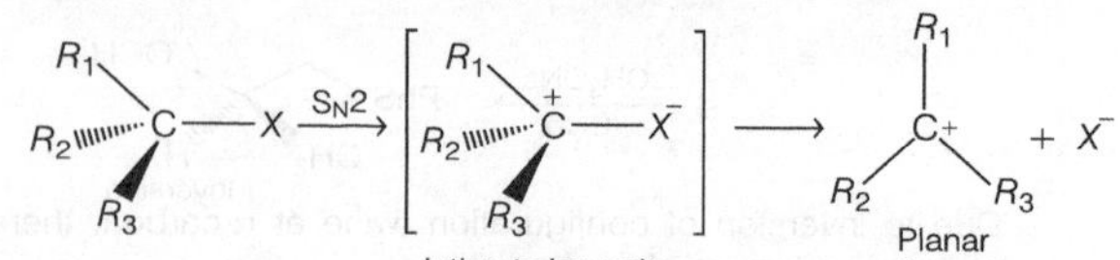

Intimate ion pair

Planar carbocation

Due to partial reaction of nucleophile with intimate ion pair, net inversion of configuration is observed although predominant reaction occurs at planar carbocation giving racemic products.

8. (a) Tertiary halide would be most reactive in S_N1 reaction.

9. (a) Tertiary halide would react at fastest rate with $AgNO_3$ as reaction will proceed by S_N1 mechanism.

10. (c) *Y* forms a tertiary carbocation which rearranges to a further more stable benzylic carbocation.

11. (a) Rate of S_N1 reaction is directly proportional to concentration of alkyl halide but independent of concentration of nucleophile. Non-polar solvents play no role in S_N1 reaction.

12. (c) S_N1 reaction proceeds *via* carbocation intermediate which is being indicated here by two transition states and a small trough between reactant and product. Also the second transition state must be below first transition state in S_N1 reaction.

13. (d) Electron donating —CH_3 group stabilises benzylic carbocation, hence increases reactivity of corresponding substrate. Electron withdrawing —NO_2 decreases stability of carbocation, decreases reactivity of corresponding substrate in S_N1 reaction.

14. (d) It forms a secondary carbocation which, by hydride shift, rearranges to a tertiary carbocation.

15. (d) Increasing temperature increases rate of any reaction. Also, rate of S_N1 reaction is directly proportional to concentration of alkyl halide but independent of nucleophile and aprotic solvent.

16. (c) (III) forms tertiary carbocation, hence most reactive. (I) is least reactive as highly unstable carbocation is formed at bridge head carbon of bicyclic compound.

17. (b) Protic solvents solvate carbocation, promotes S_N1 reaction.

18. (b) (V) forms most stable, tertiary allylic, carbocation. (II) is vinylic halide, least reactive.

19. (c)

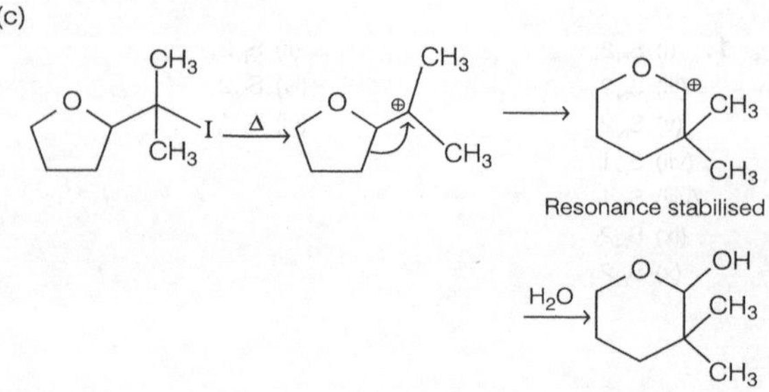

20. (d) (II) is most reactive as it forms aromatic carbocation (III) is next most reactive as it forms allylic carbocation. (I) is least reactive as it forms least stable cyclopropyl carbocation.

21. (a) (II) forms more stable carbocation than (I).

22. (b) Both statements are true but the two different products are due to resonance in carbocation intermediate as

$$^{13}CH_2=CH-CH_2-Cl \longrightarrow {}^{13}CH_2=CH-\overset{+}{C}H_2 \longleftrightarrow {}^{13+}CH_2-CH=CH_2$$

23. (a,b,c) H_2O has higher solvating power than CH_3CH_2OH,hence faster S_N1 reaction occur in H_2O. Reaction proceeds *via* carbocation intermediate, it passes through more than one transition states. Due to the presence of some intimate ion pair, S_N1 reaction occur resulting in partial racemisation and net inversion of configuration.

Since, nucleophile is not involved in rate determining step, reaction occur at same rate with both $CH_3-{}^{18}OH$ and CH_3OH.

24. (a,b,c,d)

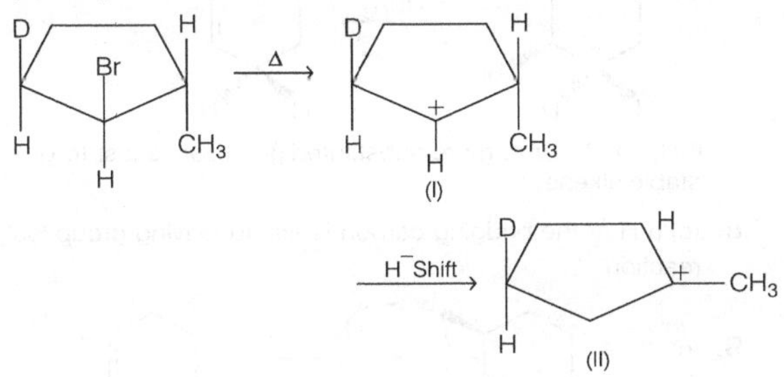

(I) gives (a) and (b) while (II) gives (c) and (d).

25. (b,d) S_N1 reaction takes place giving partial racemisation but net inversion of configuration at chiral α-carbon.

26. (a,b,d)

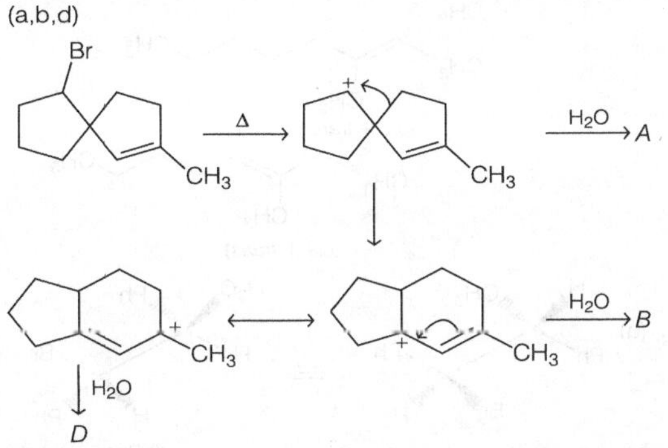

27. (a,d) $ZnCl_2$ is a Lewis acid, helps in formation of carbocation. $AgNO_3$ forms AgCl precipitate, derive heterolysis reaction below in forward direction :

$$(CH_3)_3C-Br + Ag^+ \longrightarrow AgBr\downarrow + (CH_3)_3C^+$$

From the given information structure of *A* can be derived as:

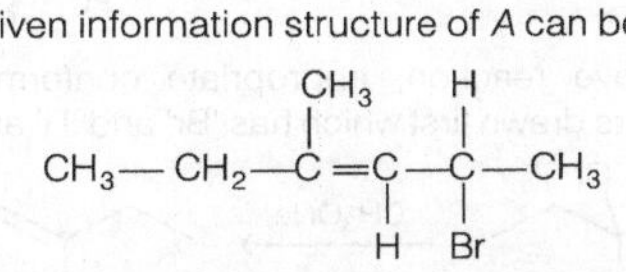

28. (c) It shows geometrical isomerism as well as it has a chiral carbon. Hence, *A* will have four stereoisomers.

29. (a) Alcohols *B* and *C* are:

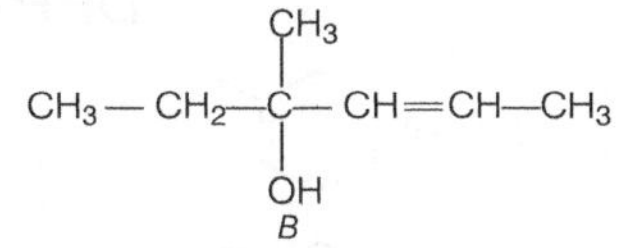

and

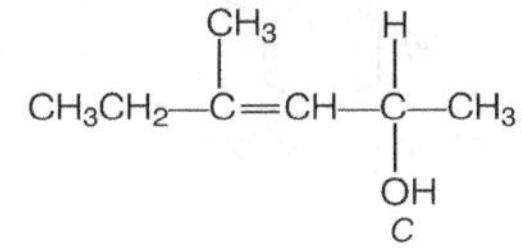

Both *B* and *C* have one chiral carbon each, both show optical isomerism.

30. (c) It has three chiral carbon, hence eight stereoisomers.

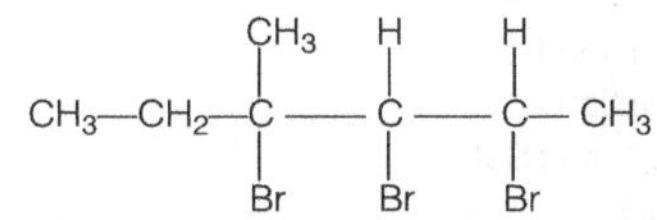

31. (b)

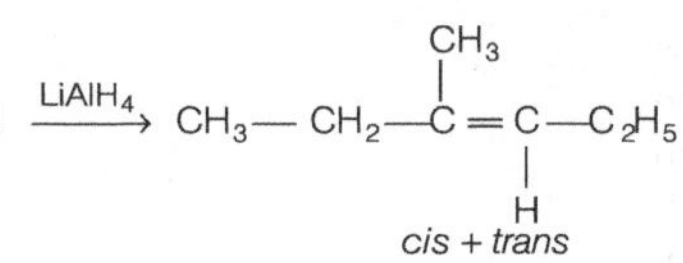

32. (b)

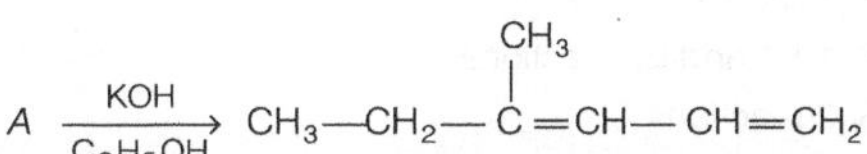

Only two geometrical isomers

33. (6)

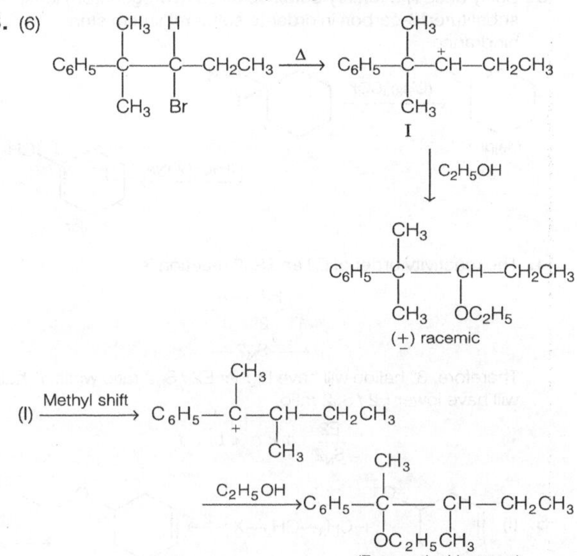

(Four optical isomers)

34. (5) Only (iii), (iv), (v), (vi) and (viii) form carbocations more stable than benzyl carbocation.

35. (6)

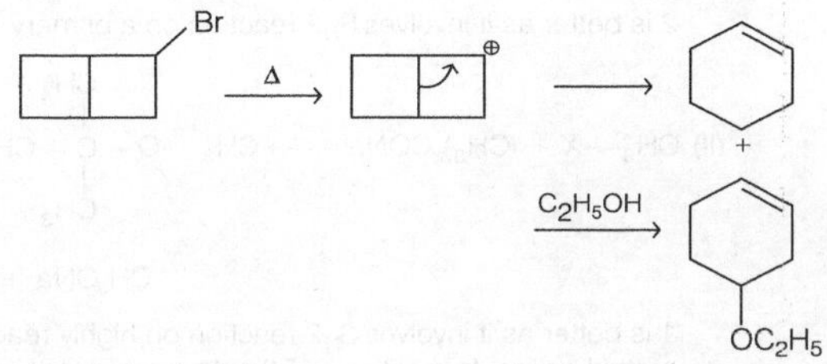

DPP-7 E2 Reaction Basics

1. (i) 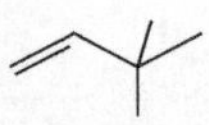 (ii) 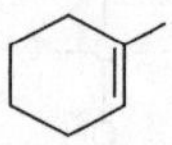

(iii) 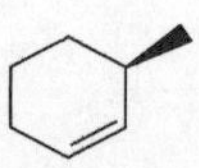 (iv) 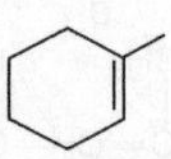

No *anti* H on 3° β-carbon is avaiable.

(v) 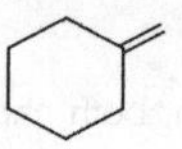

(vi) 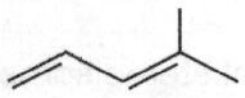

2. (i) Both S_N1 and E1
(ii) S_N2 reaction
(iii) Both S_N2 and S_N1
(iv) Only S_N1
(v) S_N1 reaction
(vi) E1 reaction
(vii) E2 reaction
(viii) S_N1 and E1
(ix) S_N1 and E1
(x) S_N1, E1 and E2 reactions
(xi) S_N2 reactions

3. Bulky base like tertiary butoxide takes hydrogen from least substituted β-carbon in order to suffer minimum steric hindrance.

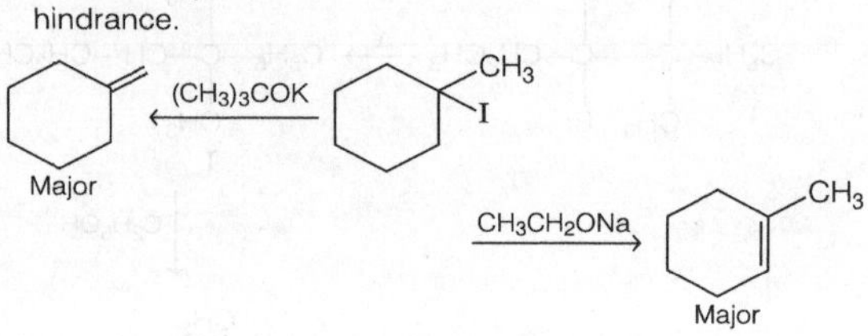

4. The reactivity order of E2 and S_N2 reaction is

$$\xrightarrow{\text{E2}}$$
$$1° \quad 2° \quad 3°$$
$$\xleftarrow{S_N2}$$

Therefore, 3° halide will have higher E2 / S_N2 ratio while 1° halide will have lower E2 / S_N2 ratio.

$$\Rightarrow \quad \frac{E2}{S_N2} : a < c < b < d$$

5. (I) 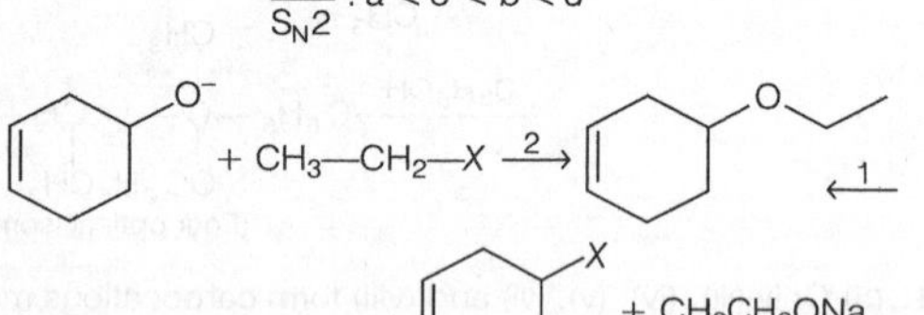

2 is better as it involves S_N2 reaction on a primary halide.

(II) $CH_3—X + (CH_3)_3CONa \xrightarrow{1} CH_3—O—C(CH_3)_3 \xleftarrow{2} CH_3ONa + (CH_3)_3CX$

1 is better as it involves S_N2 reaction on highly reactive methyl halide. In reaction 2, E2 will be preferred.

(III) $CH_2{=}CH—CH_2—X + (CH_3)_3C—CH_2ONa \xrightarrow{1} CH_2{=}CH—CH_2—O—CH_2—C(CH_3)_3$

This is preferred over reaction 2, as it involves reaction at more reactive allyl halide.

6. (b) 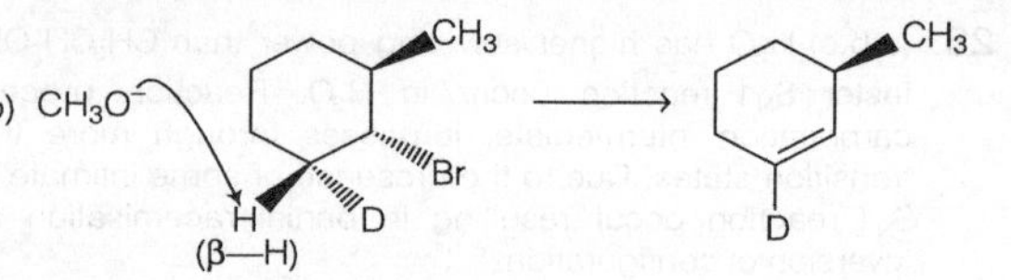

β – H *anti* to leaving halide is taken by base.

7. (a) 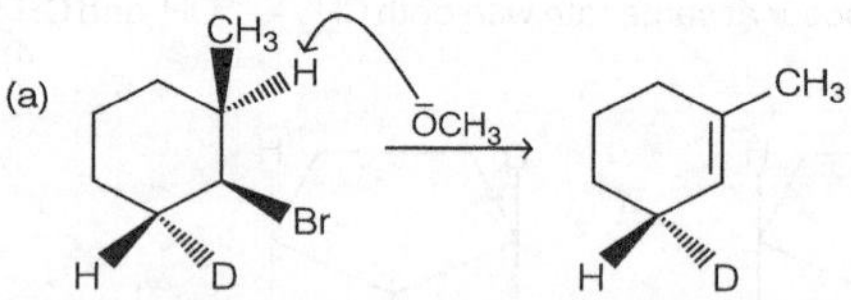

β-H, *anti* to Br at more substituted β-carbon is lost to give more stable alkene.

8. (c) β-H at the bridging carbon is *anti* to leaving group,lost in E2 reaction.

9. (c) 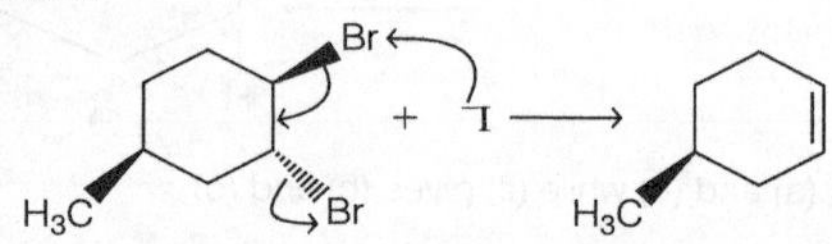

10. (d) 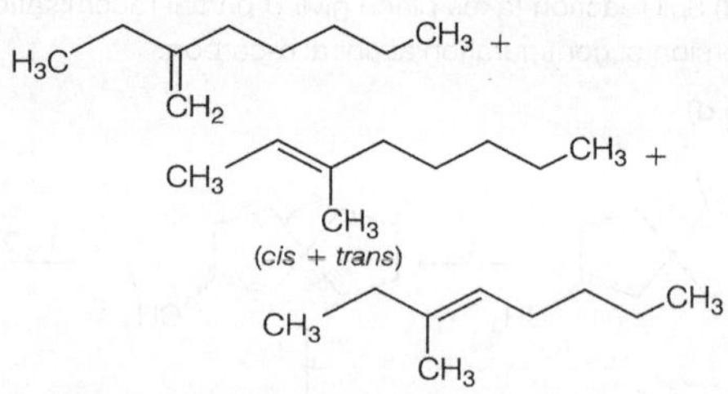

11. (a) 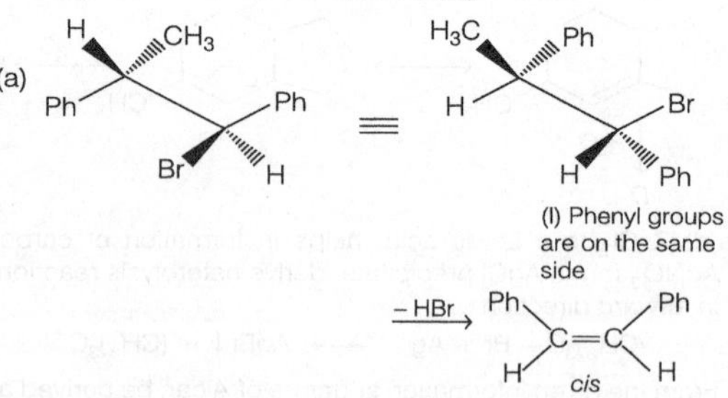

In the above reaction, appropriate conformer of the given compound is drawn first which has 'Br' and 'H' an *anti* position (I).

12. (b) 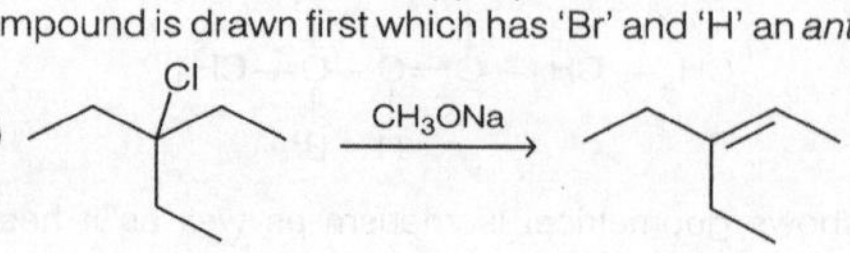

In the above reaction, all β-H are equivalent. Hence, a single alkene is formed in E2 reaction.

13. (c)

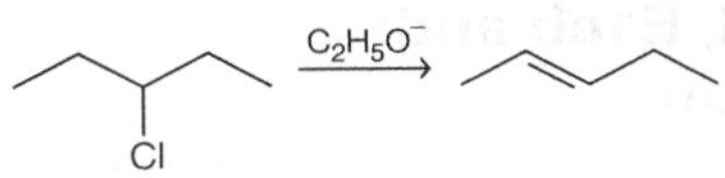

14. (b) Fluorine is most electronegative, increases acidity of β-H to the most.

15. (d) (III) is most reactive as it has most acidic β-H as well as it gives resonance stabilised alkene. Next to (III) is (IV) as it is tertiary. (I) is less reactive than (II) because (I) gives less substituted alkene.

16. (b,c) Reaction proceed by E2 mechanism, hence there would be only one transition state in the activation energy diagram. Moreover, since very bulky base (tertiary butoxide) is being used, least substituted alkene, 1-butene in the present case, would be formed as the major product.

17. (b,c)

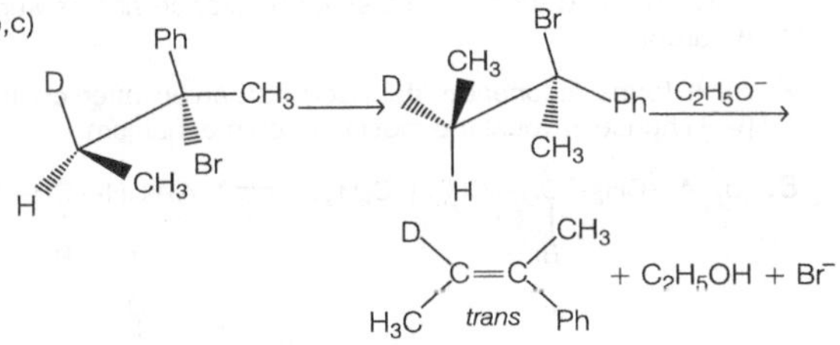

C—H bond is weaker than C—D bond hence, hydrogen is taken preferably by base. If diastereomers of the given substrate is taken, *cis*-alkene would be formed.

18. (b,c)

19. (a,b,d) With ethoxide base, most substituted alkene (I) is formed as the major product. In the formation of (II), $C_2H_5O^-$ takes proton from less hindered β-carbon, hence less activation energy and greater rate of reaction although stability of product determines it content at equilibrium. Also, since E2 reaction is an elementary reaction in which halogen leaves in the rate determining step, iodide leaves most easily and fluoride with maximum difficulty.

20. (b,c)

(I) Iodide ion attacks as nucleophile at α-carbon, brings about S_N2 reaction.

(II)

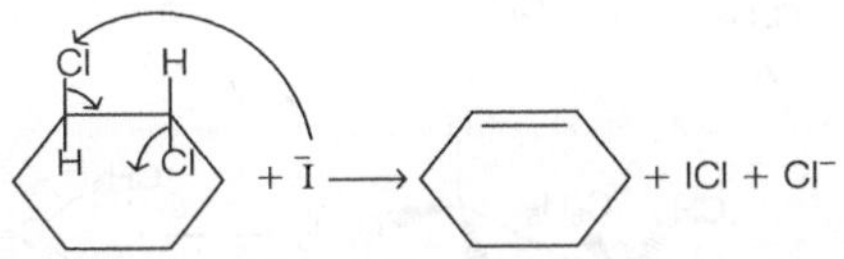

21. (c) Both option (a) and option (b) can be resolved into enantiomers. Option (c) and option (d) are non-enantiomeric, but show geometrical isomerism. However, option (d) is a 3° halide, it is almost inert in S_N2 reaction but highly reactive in E2 reaction.

22. (b)

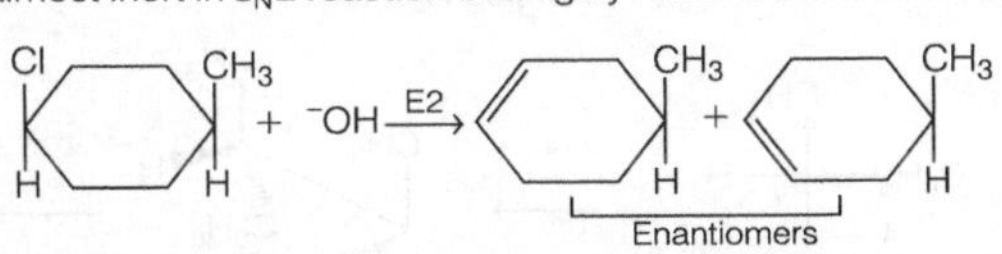

23. (b)

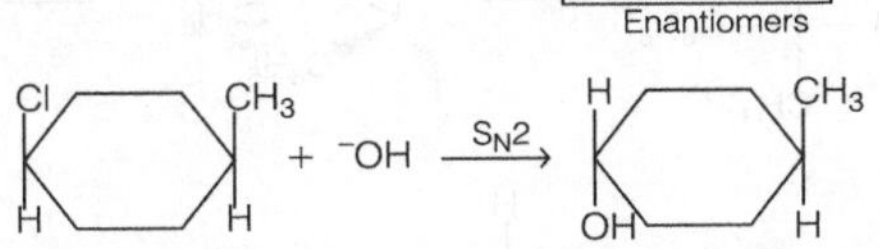

Has plane of symmetry

24. (c) Only deuterium is present at *anti* position of a β-carbon, it is abstracted by base, although giving less substituted alkene as the major product. On tertiary β-carbon, hydrogen is not available in *anti* position to leaving group.

25. (d)

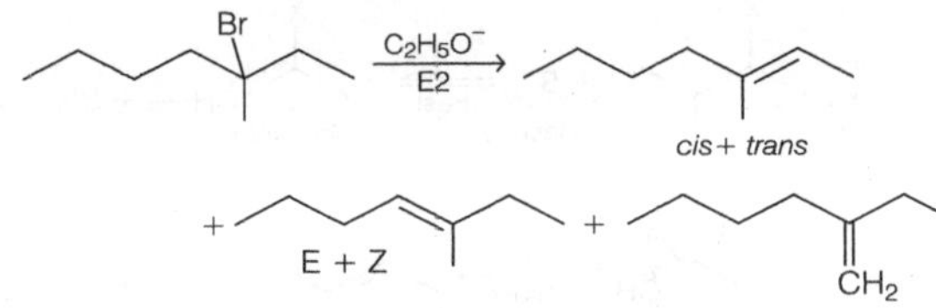

26. (d) Bulky base takes β-H from least hindered β-carbon, giving least substituted alkene as the major product.

27. (5)

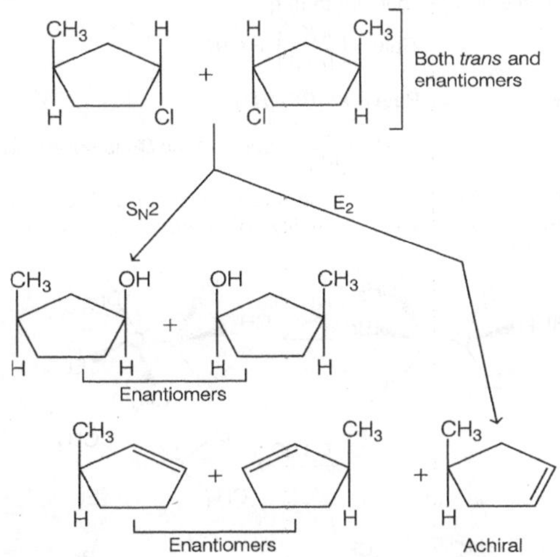

28. (5)

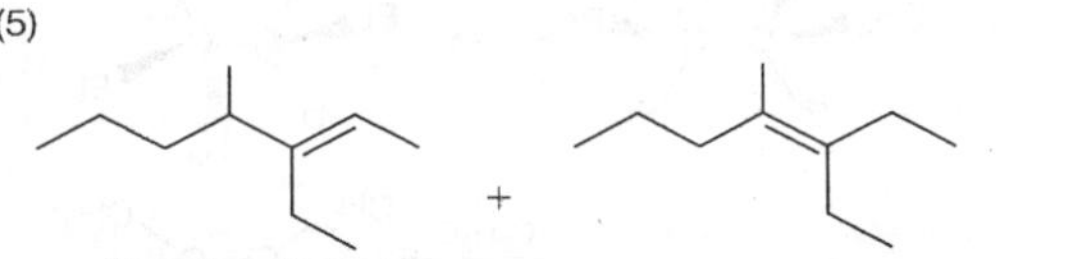

(Four stereoisomers)

29. (6)

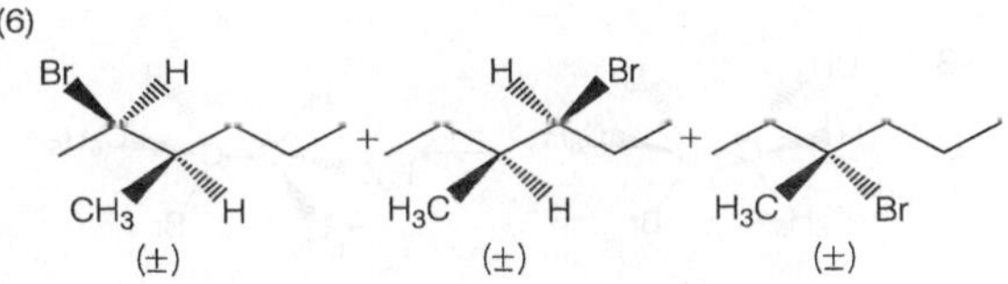

30. (7) *X* is:

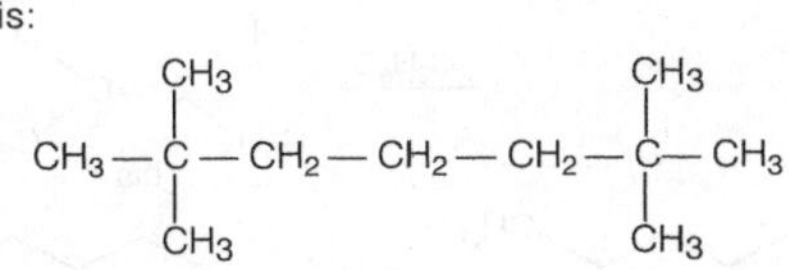

DPP-8 E2 Reaction Advanced, E1cb and Hofmann's Elimination

1. E1 cb mechanism is a two step process which proceeds *via* a carbanion mechanism.

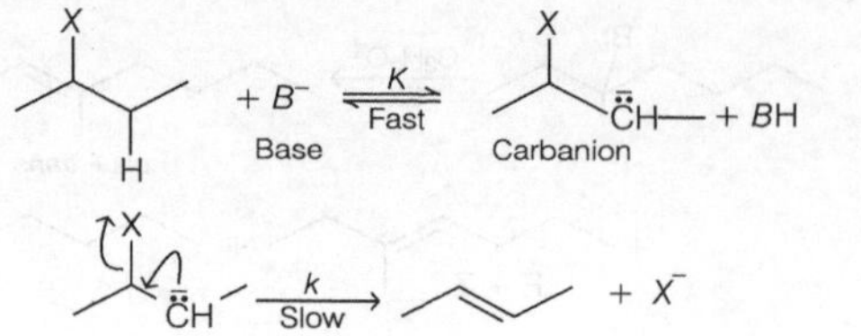

$$\text{Rate} = k\,[\text{carbanion}] \qquad \ldots\text{(i)}$$

Also $$k = \frac{[\text{carbanion}][BH]}{[RX][B^-]}$$

Substituting [carbanion] in (i) gives

$$\text{Rate} = \left(\frac{kK}{[BH]}\right)[RX][B^-]$$

$$\Rightarrow \qquad \text{Rate} = k'\,[RX]\,[B^-];$$

$$k' = \frac{Kk}{[BH]} = \text{constant as } BH \text{ is solvent in excess}$$

Hence, rate expression is same as that of normal E 2 reaction. Any factor that favours stability of carbanion, favours E1 cb mechanism.

2. (*A*)

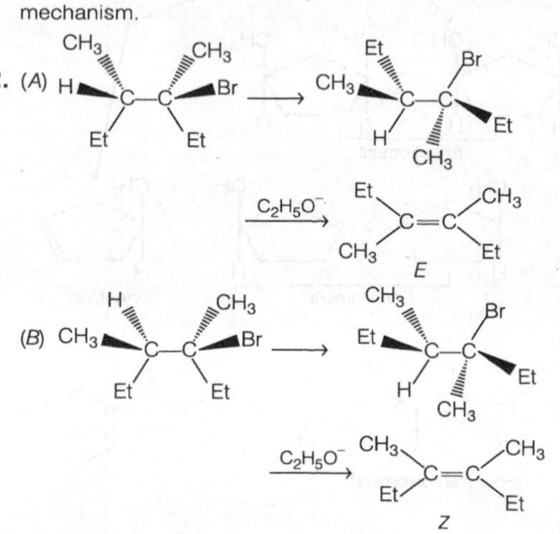

3.

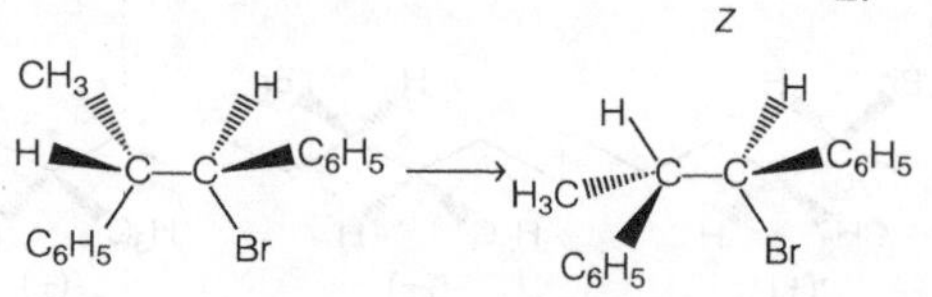

H and Br are in *anti* position

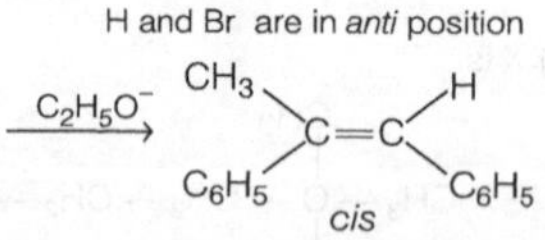

4. (c)

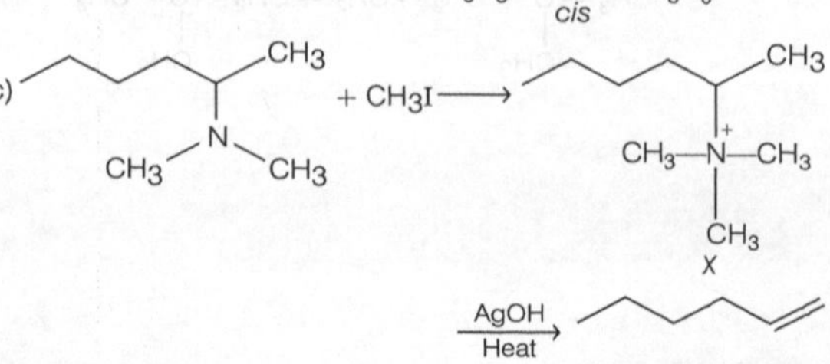

Hofmann's elimination

5. (b)

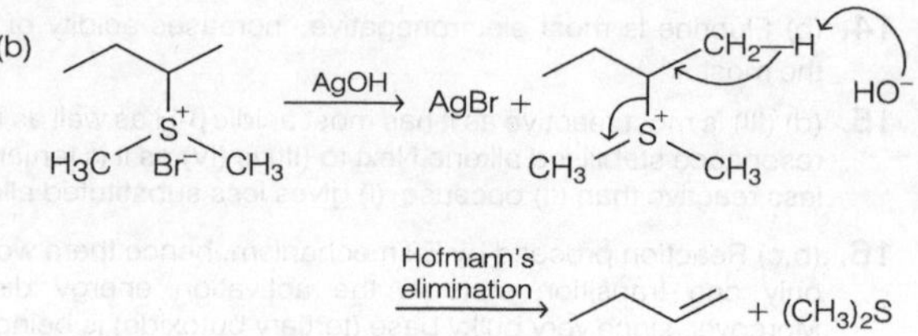

6. (d) Rate of reaction in E2 reaction depends on acidity of β—H as well as steric hindrance at β-carbon. When *X* is F, acidity is maximum and there is less steric hindrance at less substituted β-carbon.

7. (c) It forms resonance stabilised carbanion after abstraction of β—H hence, favours the most to E1cb mechanism.

8. (d)

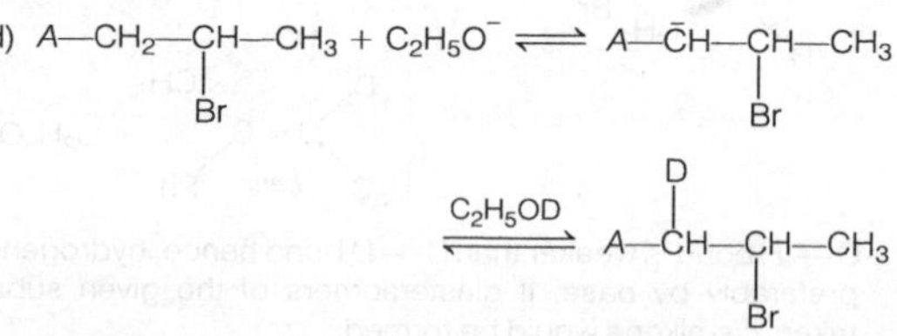

Above equilibrium can explain the formation of given product, hence reaction must proceed by E1cb mechanism.

9. (a,c,d) All these lacks β—H undergo S_N2 reaction giving methanol. b has β—H undergoes Hofmann's elimination giving ethene.

10. (a,c)

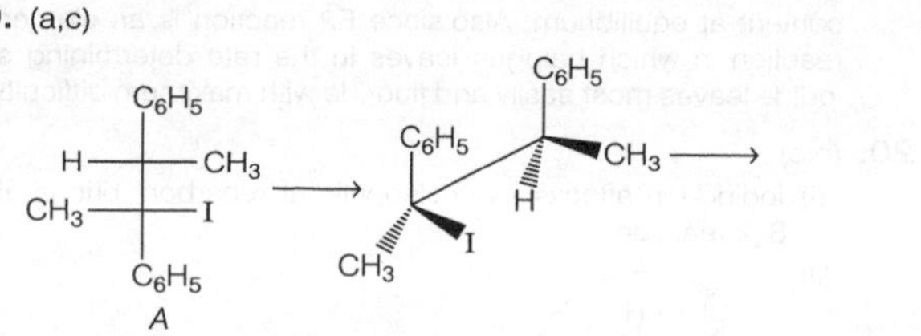

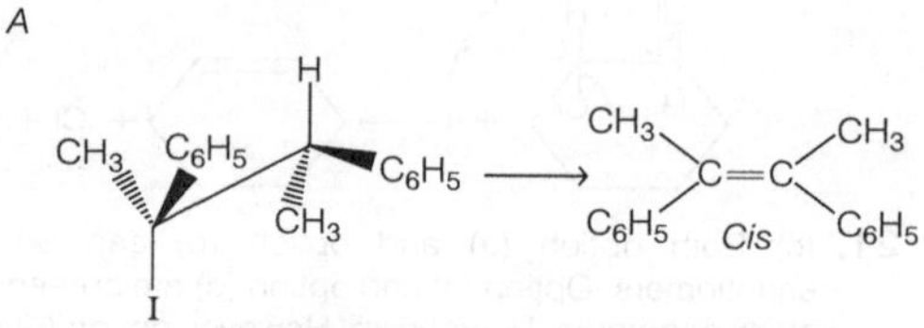

H and I are in *anti*-position

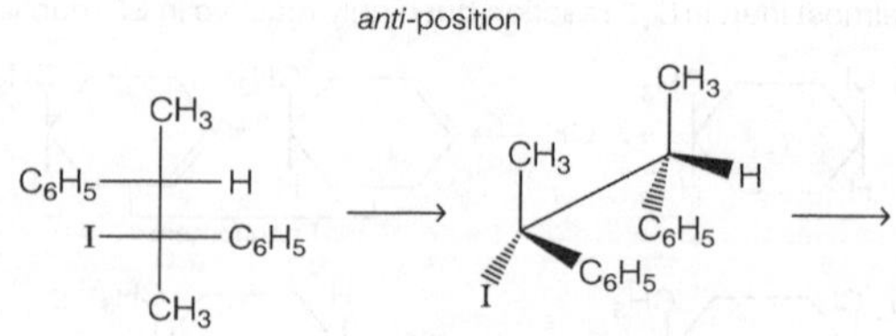

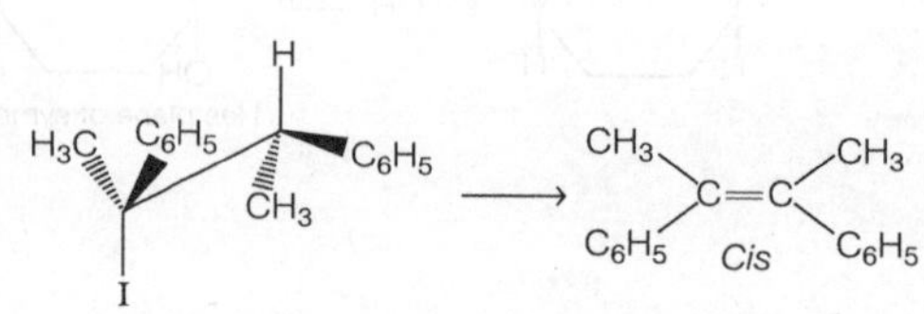

11. (a,b,c)

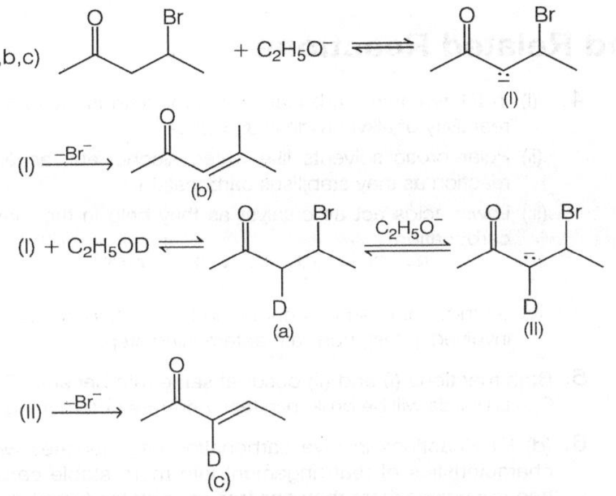

12. (a) Maximum elimination products would be formed in flask 1 and minimum elimination products would be formed in flask 4. Hence, even the 19% of total products in flask 1 would be much greater than 70% of total products in flask 4.

13. (c) Fluorine being the poorest leaving group, major product is one that is formed at fastest rate.

14. (d)Fluorine is poorest leaving group and formation of 2-hexene involve greater steric hindrance hence, highest energy transition state.

15. (b)

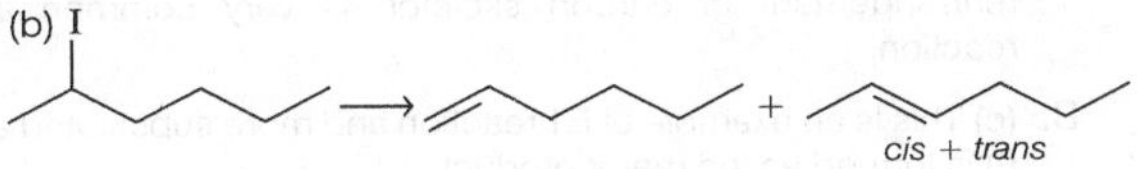

16. (a) Iodine is best leaving group and 2-hexene is more substituted product, hence major product.

17. (c) It forms the highest energy transition state.

18. (c)

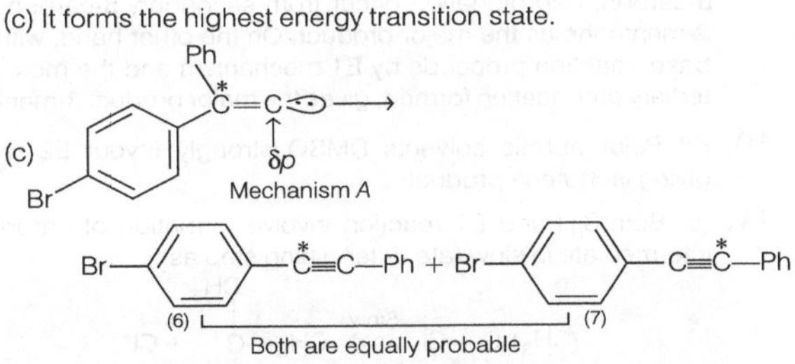

19. (d) 2 and 6 are different compounds, can be separated by gas chromatography. They have different mass, can be distinguished by mass-spectrometry. 2 and 6 have different polarity, can be distinguished by dipole moment measurement. However both are achiral, cannot be distinguished by polarimetry.

20. (d)

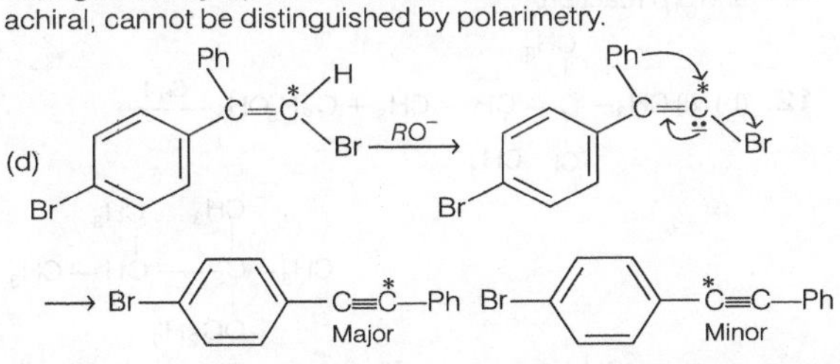

21. (b)

(i) Given halide is a primary, predominantly undergo S_N2 reaction.

(i) → (q)

(ii) Given halide is secondary, can undergo S_N2 reaction. Also, E2 reaction leads to a conjugated system. Also it may react by E1cb mechanism because it forms resonance stabilised carbanion.

(ii) → (q,r,s)

(iii) It is a 3° halide and in the presence of weak base H_2O, weak nucleophile H_2O, it may undergo unimolecular substitution (S_N1) and elimination (E1) reaction.

(iii) → (p)

(iv) It may form a stable benzylic carbocation after hydride shift, hence may react by S_N1 mechanism. Also it is a secondary halide, may undergo S_N2 reaction. It may also react by E2 reaction at it gives conjugated system. Carbanion. formed at β-C will be stabilised by resonance from ring, hence may undergo E1cb mechanism.

(iv) → (p,q,r,s)

22. (c)

(i)

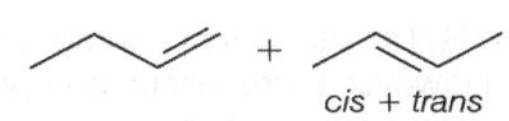

(i) → (s)

(ii)

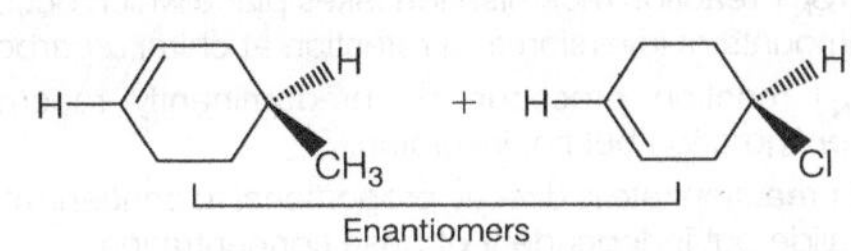

(ii) → (r)

(iii) $CH_2=C(CH_3)-CH_2CH_3$

and $CH_3-C(CH_3)=CH-CH_3$

(iii) → (q)

(iv) Substrate has no β-H.

(iv) → (p)

23. (4)

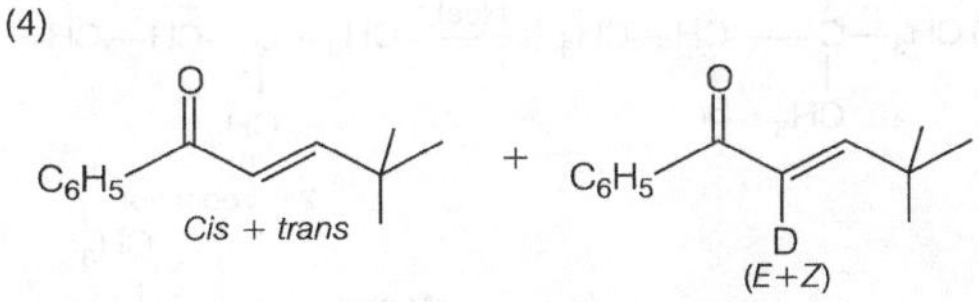

24. (3) The major elimination product is

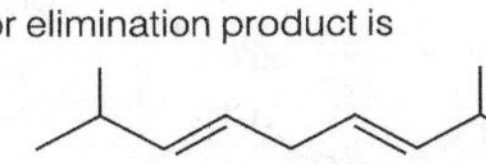

Since, the diene above is symmetrical, only three geometrical isomers exist.

DPP-9 **E1 Reaction and Related Reactions**

1. 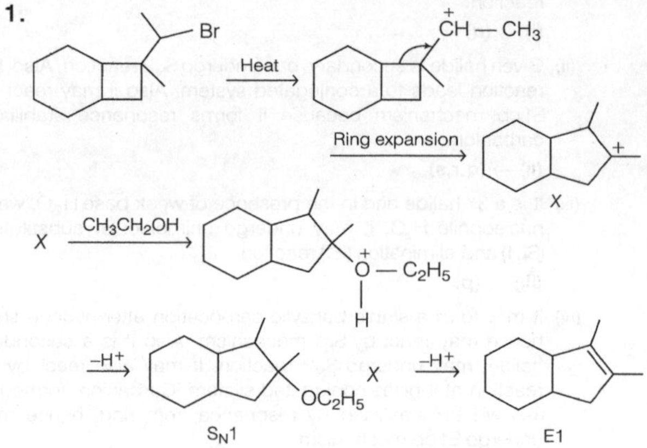

2. (i) Both S_N1 and E1 reaction mechanism involve carbocation reactive intermediates.

(ii) S_N2 reaction is first order with respect to each alkyl halide and nucleophile.

(iii) Both S_N2 and S_N1 reactions involve inversion of configuration at α-carbons. However there occur complete inversion in S_N2 (Walden's inversion) whereas partial racemisation and net inversion in S_N1 reaction.

(iv) In S_N1 reaction racemisation takes place which occur by equal amounts of inversion and retention at chiral α-carbon.

(v) S_N1 reaction proceeds by predominantly racemisation of configuration but net inversion.

(vi) E1 reaction rate is directly proportional to concentration of alkyl halide but independent of base concentration.

$$R_{E_1} = k[R - X]$$

(vii) E2 reaction is second order overall, first-order in alkyl halide and first order in base concentration.

(viii) Both S_N1 and E1 reactions are accelerated by protic solvents like water, alcohol, etc.

(ix) Rearrangements of carbocation intermediate is very common in both S_N1 and E1 reaction.

(x) This is the order of reactivity in E2, E1 and S_N1 reactions.

(xi) This is the order of reactivity in S_N2 reaction.

3. 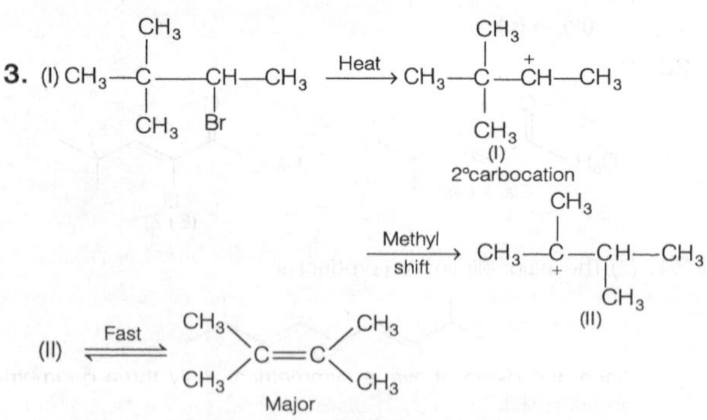

Activation energy diagram of the above reaction is

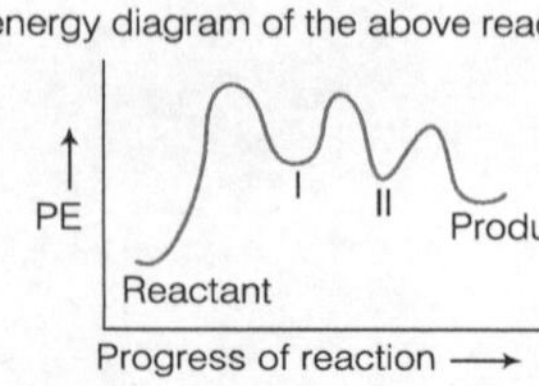

4. (i) In E1 reaction, carbocation intermediate is involved. Hence, reactivity of alkyl halide is 3° > 2° > 1°.

(ii) Polar protic solvents like water, alcohol etc. accelerate E1 reaction as they stabilises carbocation.

(iii) Lewis acids act as catalyst as they help in the formation of carbocation.

$$R{-}Cl + \underset{\text{L.A.}}{AlCl_3} \longrightarrow R^+ + AlCl_4^{\ominus}$$

(iv) Strength of base has no role in E1 reaction because base is involved in fast, non-rate determining step.

5. Both reactions (i) and (ii) occur at same rate because C—H and C—D bonds will be broken in fast, non-rate determining step.

6. (d) E1 Reactions involve carbocation intermediates which has characteristics of rearrangement into more stable carbocation, hence involves more than one transition states (usually).

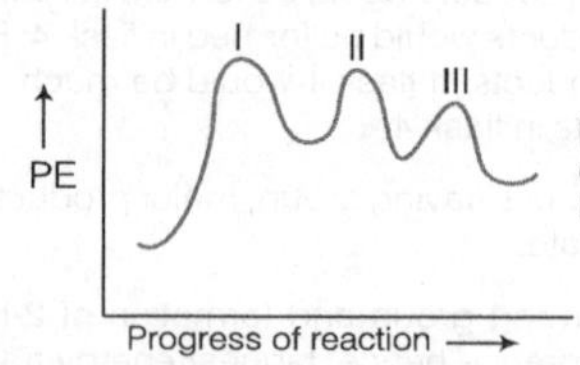

Potential energy diagram of a typical E1 reaction. I, II, III are transition states.

7. (a) E1 reaction proceeds *via* carbocation intermediate, rearrangement of carbon skeleton is very common in this reaction.

8. (c) This is an example of E1 reaction and more substituted alkene (I) is formed as the major product.

9. (b) With strong base C_2H_5ONa, reaction proceeds by E2 mechanism. Since, there is no H in the *anti* position at tertiary β-carbon, deprotonation occur from secondary β-carbon giving 2-menthene as the major product. On the other hand, with weak base, reaction proceeds by E1 mechanism and the most stable tertiary carbocation formed, gives the major product 3-menthene.

10. (b) Polar aprotic solvents DMSO strongly favour E2 reaction giving isobutene product.

11. (c) Both S_N1 and E1 reaction involve formation of carbocation intermediate in slow rate determining step as :

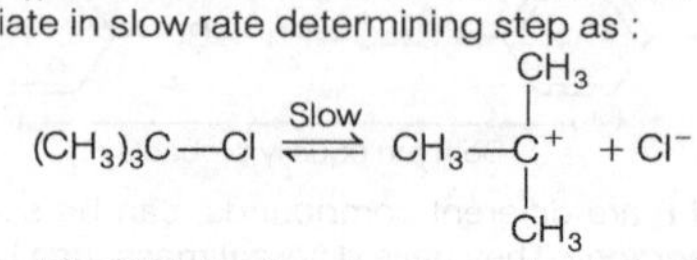

Presence of NaCl derive the above reaction in backward direction by Le-Chatelier's principle, hence has adverse effect on both E1 and S_N1 reaction.

12. (b) 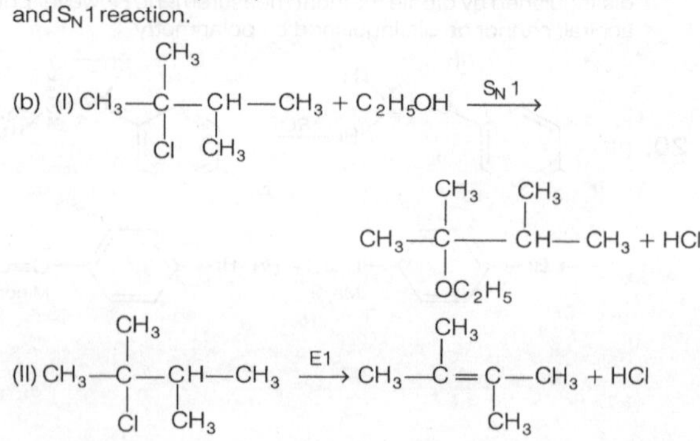

Reaction (II) leads to greater increase in entropy ($\Delta n = 1$). Hence, favoured at higher temperature due to greater decrease in ΔG than for reaction (I).

13. (c) Carbocation at bridge head carbon of bicyclic compound is highly unstable due to lack of planarity.

14. (a,b,d) Reaction proceeds by E1 mechanism:

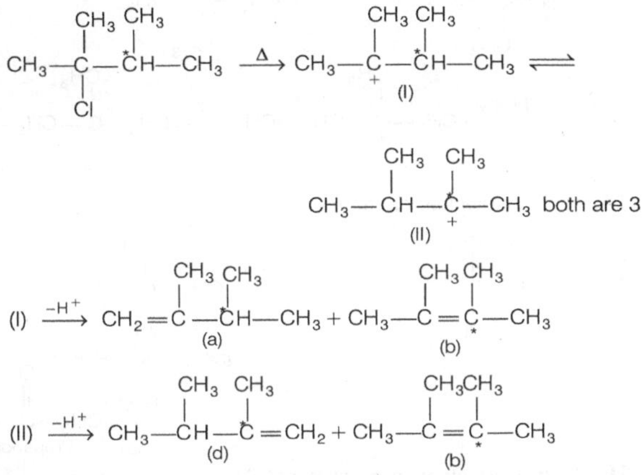

15. (a,b,d) Reaction proceeds by E1 mechanism.

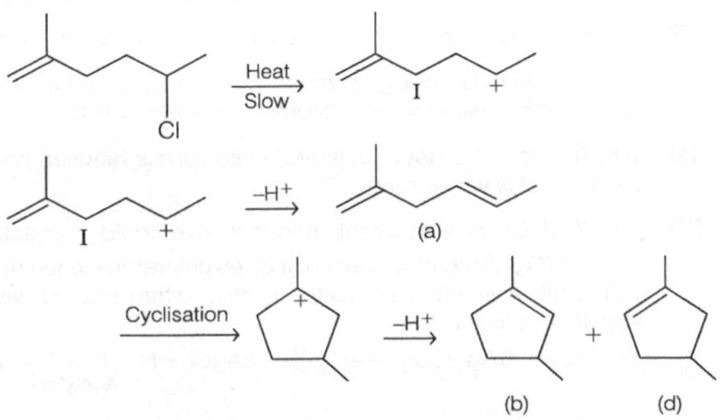

16. (a,c) In E1 reaction, carbocation intermediate undergo deprotonation in the fast, non-rate determining step, hence C—H and C—D bonds react at same rate. However, C—*X* bond is broken in first, slow, non-rate determining step, hence $R—I^{127}$ reacts at faster rate than $R—I^{131}$.

17. (d) The reaction involves E2 elimination mechanism in which base approaches to β-H from side *anti* to leaving halide in the rate determining step. Hence, if tertiary butyl group is *cis* to tosylate, it would not give steric hindrance to the base approaching from opposite side to β-H.

18. (b) When tertiary butyl group is *trans* to tosylate, E2 reaction will be difficult and E1 reaction would be preferred. Since, E1 reaction occur slowly and hydrogen is lost in fast step, both H and D will be lost with equal ease.

19. (b) When tertiary butyl group is *trans* to leaving halide, E2 reaction will be highly sterically hindered and E1 reaction would be preferred.

20. (8) Total 8 alkenes

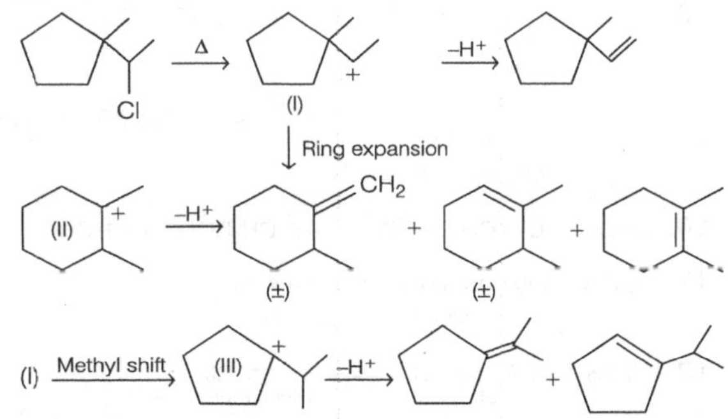

21. (7) Total seven isomers.

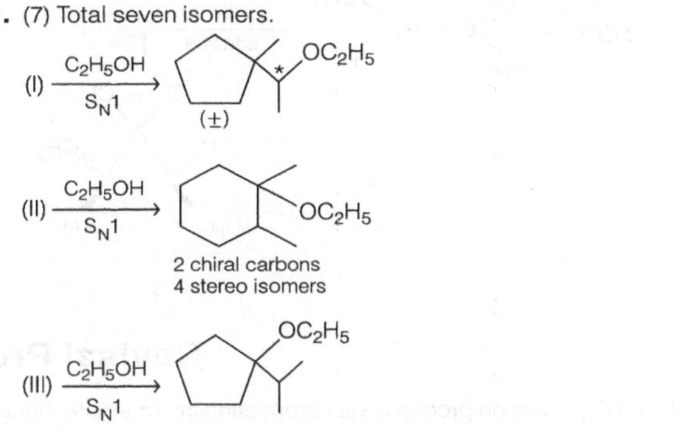

DPP-10 **Dihalides and Polyhalides**

1. $CH_4 + 4Cl_2 \xrightarrow{hv} CCl_4 + 4HCl$ Use of large excess of chlorine gas will favour formation of CCl_4, otherwise less halogenated product would be formed.

2. $CH_3-C\equiv C-CH_3 \xrightarrow{Br_2}$

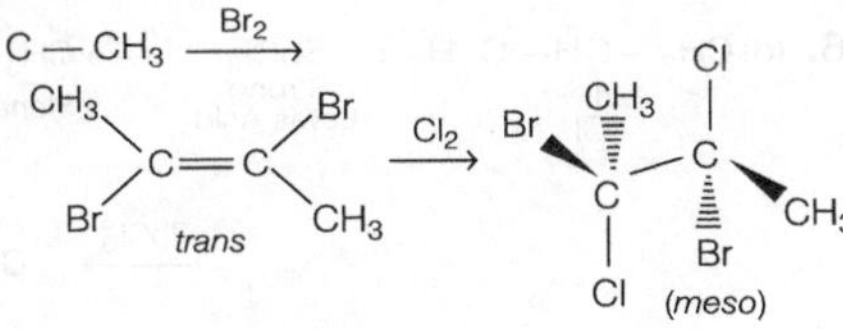

3.

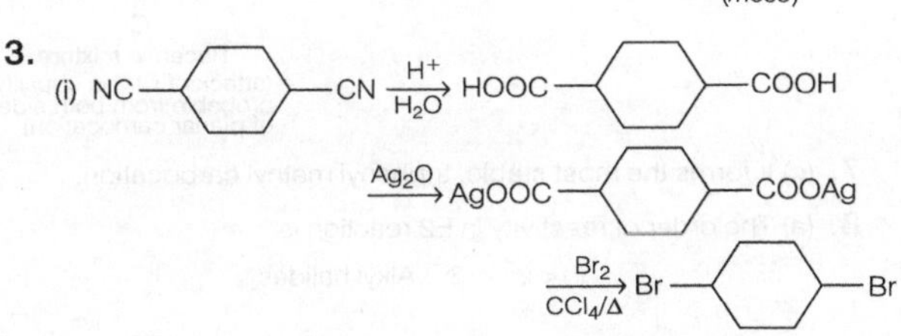

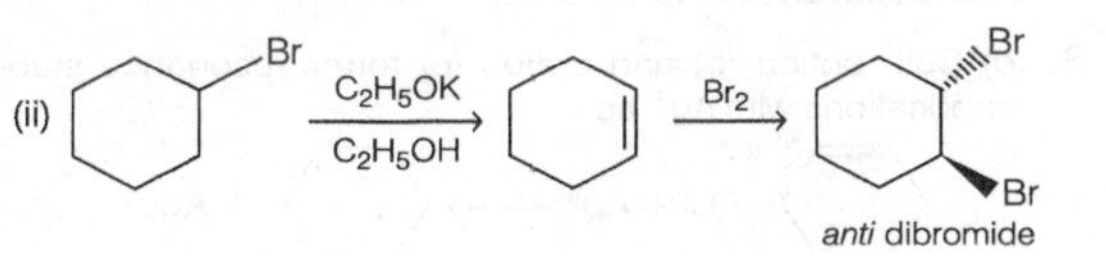

(iii) $CH_3-\overset{O}{\overset{\|}{C}}-CH_3 + Cl_2 \xrightarrow{NaOH} CH_3COONa + CHCl_3$

4. $C_2H_6 + 6F_2(g) \longrightarrow C_2F_6 + 6HF$

Upon direct fluorination of alkane, all hydrogens are simultaneously substituted by fluorine.

For monofluorination purpose, alkane is first mono chlorinated or mono brominated conveniently. These monohalogen intermediate is then heated with fluoride salts like AgF, HgF_2, CoF_3 etc.

(Swart's reaction)

$$R—Cl + AgF \xrightarrow{\Delta} R—F + AgCl$$

5. (a) $CCl_4 + KOH \longrightarrow [H_2CO_3] \longrightarrow K_2CO_3$

6. (c) $CHCl_3 + \text{Conc. } HNO_3 \xrightarrow{\Delta} \underset{\text{Nitro chloropicrin}}{O_2N—CCl_3}$

7. (b) $\begin{matrix} CH_2COH \\ | \\ CH—OH \\ | \\ CH_2OH \end{matrix} \xrightarrow{HI} CH_3—\underset{I}{\underset{|}{CH}}—CH_3$

8. (c) $CH_2{=}CH_2 + Cl_2 \longrightarrow \underset{vicinal}{Cl—CH_2—CH_2—Cl}$

9. (a) $CH_3—\underset{OH}{\underset{|}{CH}}—CH_3 + OBr^- \xrightarrow[\text{O.A.}]{} CH_3—\overset{O}{\overset{||}{C}}—CH_3$

$\xrightarrow{OBr^-} CH_3COO^- + \underset{\text{Bromoform}}{CHBr_3}$

10. (b) $CH_3—\overset{O}{\overset{||}{C}}—CH_3 + OCl^- \longrightarrow CH_3COO^- + CHCl_3$

11. (a) Reducing power of Zn increases as :

alcohol $< H_2O < H_2O/H^+$

12. (d) $CHCl_3 + \underset{\text{St. base}}{(CH_3)_3CO^-} \longrightarrow \underset{\text{Dichlorocarbene}}{:CCl_2}$

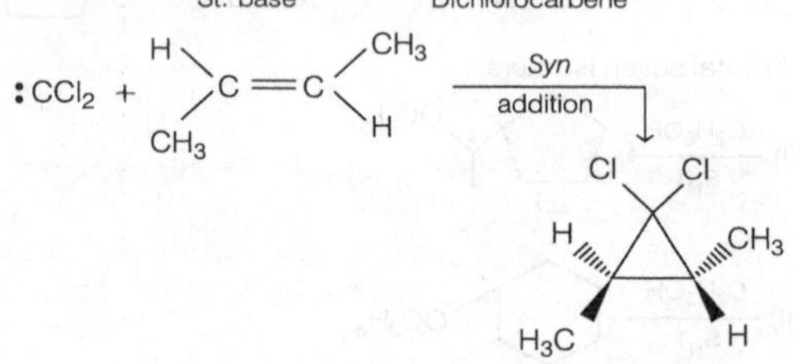

13. (a) Alkyl halides on heating undergo rearrangement into higher degree alkyl halide as :

$CH_3—\underset{CH_3}{\overset{CH_3}{C}}—CH_2—Cl \xrightarrow{300°C} CH_3—\underset{CH_3}{\overset{CH_3}{C}}—\overset{+}{C}H_2$

$\xrightarrow[\text{shift}]{\text{Methyl}} CH_3—\underset{+}{\overset{CH_3}{C}}—CH_2—CH_3 \xrightarrow{Cl^-} CH_3—\underset{Cl}{\overset{CH_3}{C}}—CH_2—CH_3$

14. (b) Hunsdiecker reaction.

15. (c) $CH_3—\underset{Cl}{\overset{Cl}{C}}—CH_3 + 2OH^- \xrightarrow{-2Cl^-} \underset{\textit{gem}\text{ diol (unstable)}}{CH_3—\underset{OH}{\overset{OH}{C}}—CH_3}$

$\xrightarrow{-H_2O} \underset{\text{Propanone}}{CH_3—\overset{O}{\overset{||}{C}}—CH_3}$

16. (b) It is industrial preparation of CCl_4.

$$CS_2 + 3Cl_2 \longrightarrow CCl_4 + S_2Cl_2$$

17. (b) Swarts reaction-both chlorides are replaced here by fluoride.

18. (b,c,d) option (a) will give *meso* dibromide, all other will give racemic dibromide by *anti* dibromination of double bond.

19. (a,b,d) Fluorine is not substituted in solvolysis reaction because C—F bond is very strong.

20. (a,b,d) $CHCl_3$ is a covalent, non-ionisable liquid. It reacts with conc. HNO_3 giving chloropicrin, it gives chloretone when refluxed with alkaline solution of acetone. Also, when heated with Ag, acetylene is formed.

$$H—CCl_3 + 6Ag + Cl_3C—H \xrightarrow{\Delta} 6AgCl + \underset{\text{Acetylene}}{H—C{\equiv}C—H}$$

Revisal Problems (JEE Main)

1. (d) S_N1 reaction proceeds *via* carbocation intermediate, hence 3° halide (forms most stable tertiary carbocation) is most reactive. Also, polar protic solvent stabilises carbocation, increases reactivity of S_N1 reaction.

2. (d) Reactivity follow the order of stability of carbocation. Stability of carbocation follows the order 1° < 2° < 3°.

3. (d) Both option (b) and option (c) forms resonance stabilised carbocations with Ag^+ as

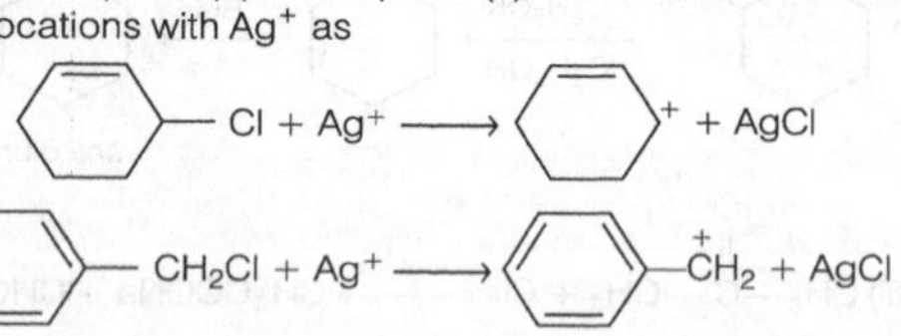

4. (c) It forms the most stable carbocation.

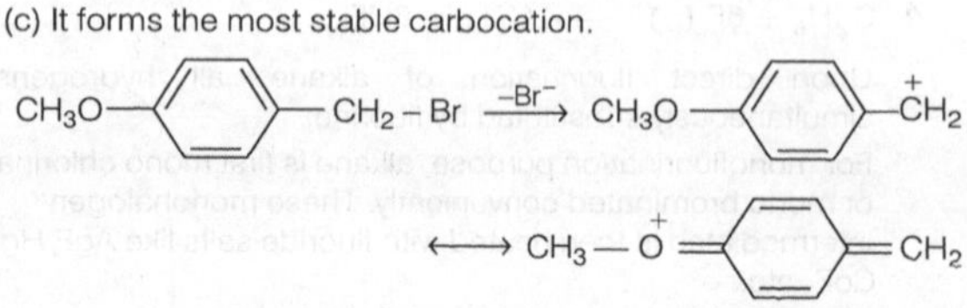

Highly stable

5. (a) Iodide is best leaving group among halides and it forms the most stable carbocation.

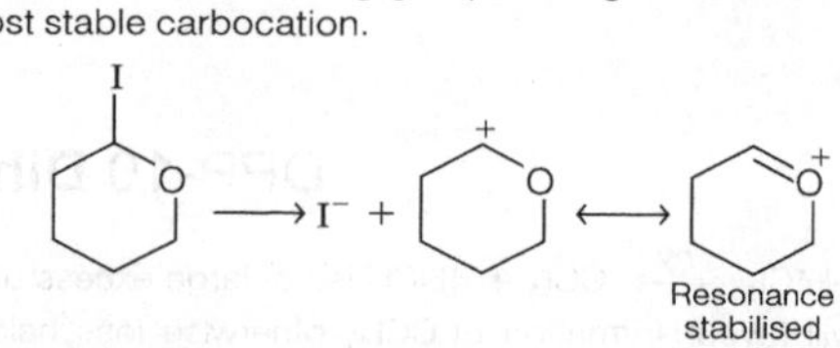

Resonance stabilised

6. (d) $CH_3—\underset{Cl}{\underset{|}{CH}}—C_6H_5 + \underset{\text{Lewis Acid}}{\underset{\text{Strong}}{SbCl_5}} \longrightarrow \underset{\text{Planar carbocation}}{CH_3—\overset{H}{\underset{+}{C}}—C_6H_5}$

$\xrightarrow{SbCl_6^-} CH_3—\underset{Cl}{\overset{H}{C}}—C_6H_5$

Racemic mixture (attack of Cl^- is equally probable from both sides of planar carbocation)

7. (c) It forms the most stable, triphenyl methyl carbocation.

8. (a) The order of reactivity in E2 reaction is :

1° < 2° < 3° : Alkyl halides

9. (a) It is resonance stabilised, most stable diene.

10. (b) Bromide is a better leaving group and in E2 elimination reaction, more substituted, more stable product is formed as major product.

11. (c) E1 reaction proceeds *via* carbocation intermediate and reactivity follow the stability of carbocation.

12. (c) In E2 reaction, strong base takes β-H from side *anti* to the leaving group. In the present case, on tertiary β-carbon, hydrogen atom is not available in *anti* position to leaving group, hence elimination occur from other β-carbon although it produces less substituted, less stable alkene.

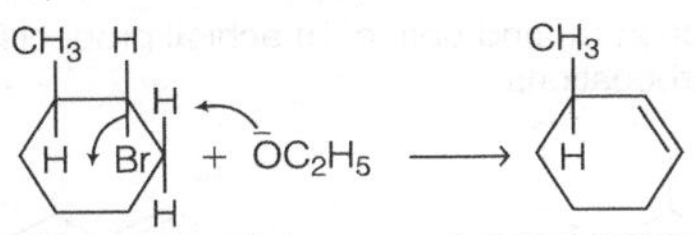

trans-1-bromo-2-methyl cyclohexane → 3-methyl cyclohexane

13. (c) Carbon, being less electronegative than nitrogen, a better electron donor and stronger nucleophile.

14. (c) In aprotic solvents, nucleophiles are not solvated. Hence, basicity parallels nucleophilicity in case of halides.

15. (d) Rate of S_N2 reaction is directly proportional to concentration of both substrate and nucleophile. Also, polar aprotic solvents increases reactivity as they solvates cationic part of nucleophile. A better leaving group increases reactivity of S_N2 reaction because leaving group departs in the rate determining step.

16. (d) S_N2 reaction leads to inversion of configuration at the chiral α-carbon.

17. (d) Electron withdrawing nitro group increases electrophilicity at the α-carbon.

18. (a) HS^- is the strongest amongst the given nucleophiles.

19. (c) In S_N2 reaction, the order of reactivity is

$$1° > 2° > 3° \quad : \text{Alkyl halides}$$

20. (c) Steric hindrance in substrate to the attack of nucleophile decreases reactivity in S_N2 reaction.

21. (c) Both SCN^- and OCN^- are ambident nucleophiles. In SCN^-, more nucleophilic 'S' acts as donor while in OCN^-, more nucleophilic nitrogen acts as donor.

22. (d) A more polar solvents have greater effect on stability of nucleophiles (due to fully developed negative charge on it) than on transition state (has partially developed negative charge).

23. (c) Only the primary aliphatic bromide is substituted here in S_N2 reaction.

24. (c) Nitro group exert strong electron withdrawing resonance effect from *para* position, hence destabilises the stability of free radical formed at *para* methyl group. Therefore, bromination occur preferably at *meta* methyl group.

25. (b) Greater the stability of leaving group as base, better the leaving group.

26. (b) Grignard's reagent consists of very basic carbanion, reacts with water at room temperature as :

$$RMgX + H_2O \longrightarrow RH + Mg(OH)X$$

27. (b) Fluorine is highly reactive, it is difficult to control the reaction at monofluorination level.

28. (b) Ag—C bond is covalent in AgCN, hence nucleophilic attacks occur from nitrogen atom.

29. (c) It is the lower solubility of NaCl in acetone that enables substitution to take place successfully.

30. (a) In E2 reaction, either enantiomer.

Revisal Problems (JEE Advanced)

1. (b) H and Br from *anti* position eliminated.

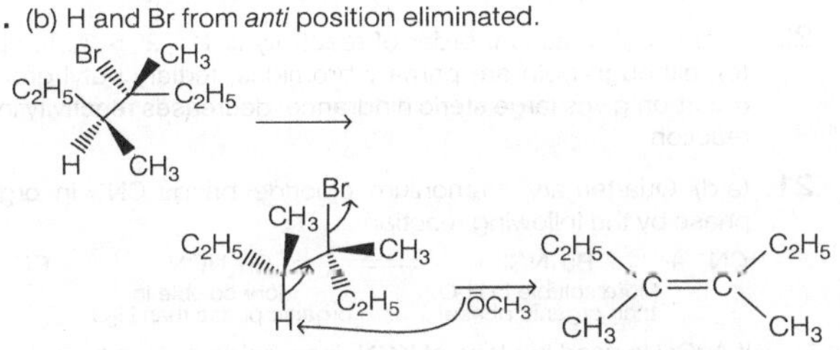

2. (c) Drawing Sawhorse projection of the given compound

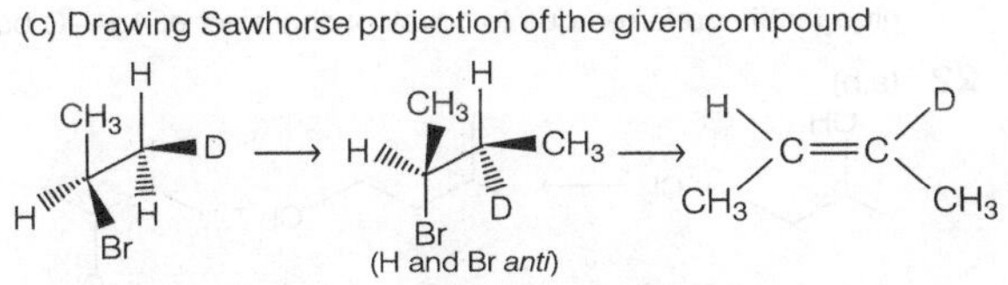

3. (a) E2 elimination first gives cyclopentene. In the next step, addition of bromine gives *anti* vicinal dibromide.

4. (b) S_N2 reaction gives inversion of configuration at α-carbon.

5. (d) S_N2 reaction gives inversion of configuration at α-carbon. If the compound also has chiral carbon other than α-carbon, configuration at those chiral carbons remains unchanged after the reaction.

6. (d) S_N1 reaction involves planar carbocation, giving racemic products predominantly. However same carbocation is not completely detached from leaving group, exists as intimate ion pair. Such ion pairs prevent attacks from front side favouring backside attacks.

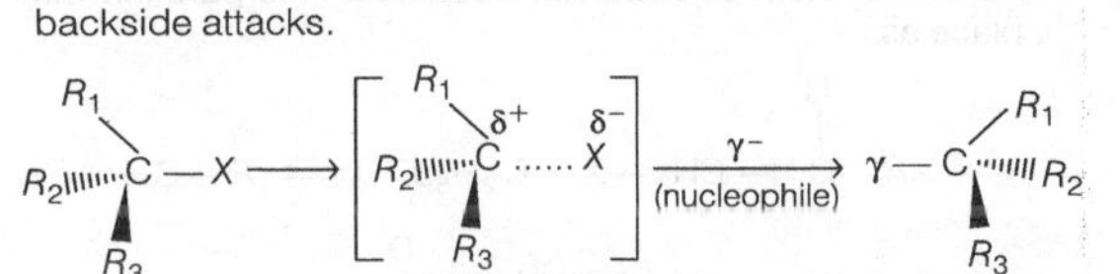

Due to above reason, in S_N1 reaction, partial racemisation and net inversion is usually observed.

7. (a) S_N1 reaction proceeds *via* carbocation intermediate, hence reactivity follow the order of stability of carbocation. In the present case, besides resonance stability of benzylic carbocations, hyperconjugation by alkyl group at *para* position is also affecting stability. Methyl group from *para* position will stabilise the most by hyperconjugation effect due to three α-H followed by ethyl group with two α-H and least by isopropyl group with only one α-H.

8. (b) Hydrolysis will proceed by S_N1 mechanism. III will be least reactive as it will involves the least stable, *anti* aromatic carbocation. IV will be most reactive as its carbocation will involve ring expansion of highly strained cyclopropyl ring.

9. (b) (III) is most reactive as it forms most stable, aromatic carbocation:

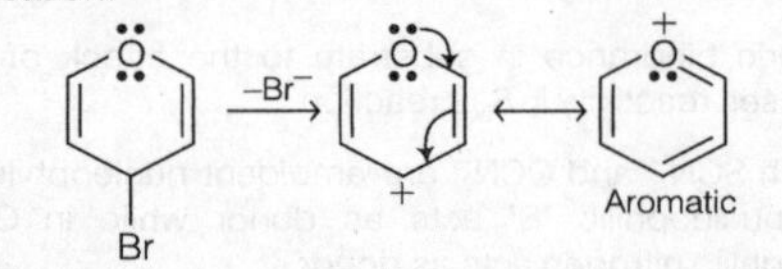

(II) is least reactive as it involves least stable, *anti* aromatic carbocation.

10. (c) It involves least stable, *anti* aromatic carbocation.

11. (b) (III) is most reactive as it proceeds *via* most stable aromatic carbocation (as shown in Q. 9). (II) is least reactive as it proceeds *via* least stable *anti* aromatic carbocation.

12. (b) (IV) is most reactive as it involves the most stable carbocation.

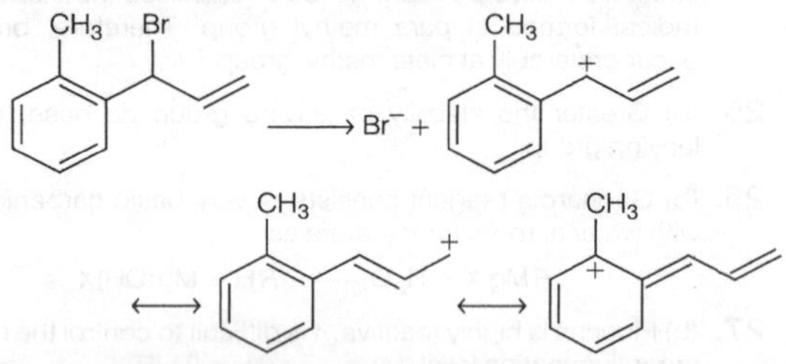

(III) is least reactive as it is vinylic halide. (II) is more reactive than (I) because (II) proceeds *via* benzylic carbocation.

13. (a,b,d) The reaction may proceed either by E2 or S_N2 mechanism. The only elimination product possible here is 1-pentene. 2-pentene is not possible because its formation will involve rearrangement which is not possible in E2 reaction.

Also due to following equilibrium :

$$(CH_3)_3COK + C_2H_5OH \rightleftharpoons (CH_3)_3COH + C_2H_5OK$$

Both tertiary butoxide and an ethoxide undergo S_N2 reaction forming (b) and (d).

14. (a,b,c,d) E1cb is a modification of E2 mechanism that involve formation of a carbanion intermediate in the first, fast step followed by unimolecular elimination of halide ion in the second slow rate determining step. Overall order is two. Electron withdrawing group stabilises carbanion, favours the mechanism. Since first step is reversible, deuterium incorporation may take place as :

$$R-\overset{H}{\overset{|}{C}}H-CH_2-X + C_2H_5O^- \rightleftharpoons R-\overset{-}{C}H-CH_2-X \xrightleftharpoons{C_2H_5OD} R-\overset{D}{\overset{|}{C}}H-CH_2-X$$

15. (a,b,d)

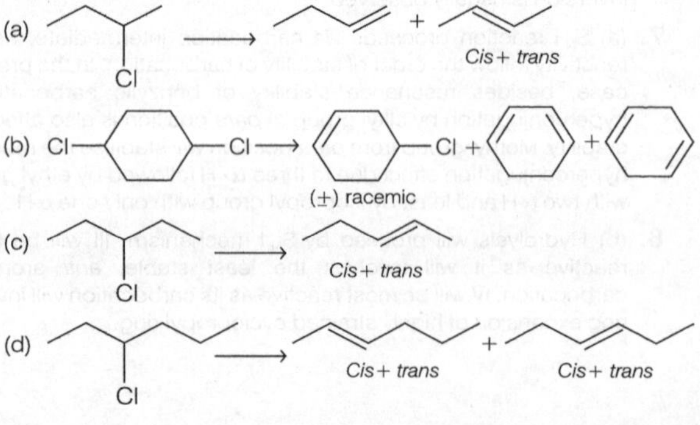

16. (a,b,c,d) *Cis* and *trans* 2-butenes are pair of diastereomers. C—H bond is broken in the rate determining slow step, stronger C—D bond will slow down the reaction. C—X bond is broken in slow, ate determining step, C—Cl bond will break slowly than C—Br bond.

Bulky base like tertiary butoxide results in less substituted Hofmann's elimination product.

17. (c,d) Reaction-I has involved rearrangement, must have proceeded by S_N1 mechanism. Reaction-II involves strong nucleophile and devoids of rearrangement, proceeded by S_N2 mechanism.

18. (c,d) In option (a) and option (b) achiral products are formed *via* tertiary carbocations.

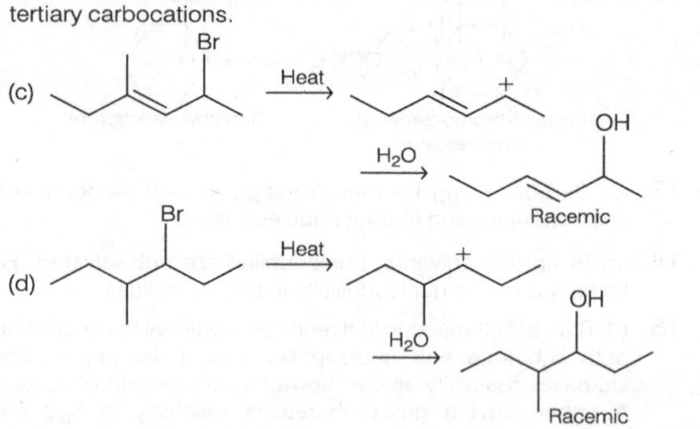

19. (a,c)

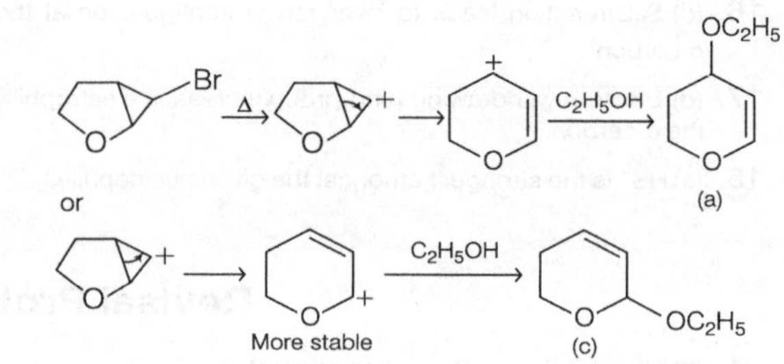

20. (a,c) In S_N2 reaction, order of reactivity is 1° > 2° > 3°. In option (c), although both are primary bromides, tertiary butyl group at α-carbon gives large steric hindrance, decreases reactivity in S_N2 reaction.

21. (a,d) Quarternary ammonium chloride brings CN^- in organic phase by the following reaction:

$$CN^- + \underset{\text{More soluble in } H_2O \text{ than organic phase}}{Bu_4NCl} \rightleftharpoons \underset{\text{More soluble in organic phase than } H_2O}{Bu_4NCN} + Cl^-$$

If AgCN is used in place of KCN, isocyanide (linking from nitrogen) is preferred due to covalent character of Ag—C bond.

22. (a,b)

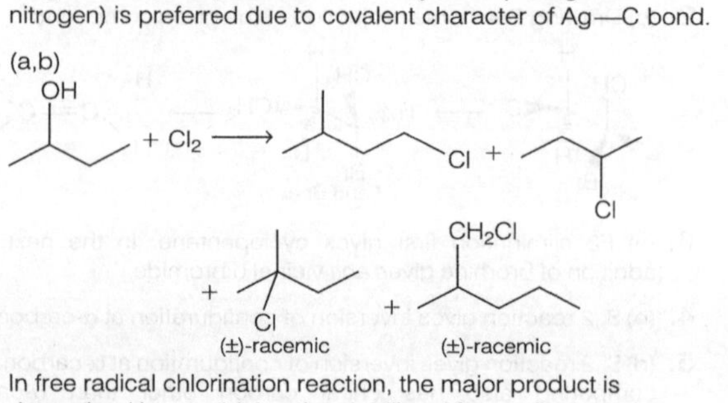

In free radical chlorination reaction, the major product is determined by combining the stability of free radicals and probability of reaction at a particular carbon both. Hence, only the basis of stability of tertiary free radical alone, tertiary chloride can't be predicted as major product.

Explanations for Q. Nos. (23-25)

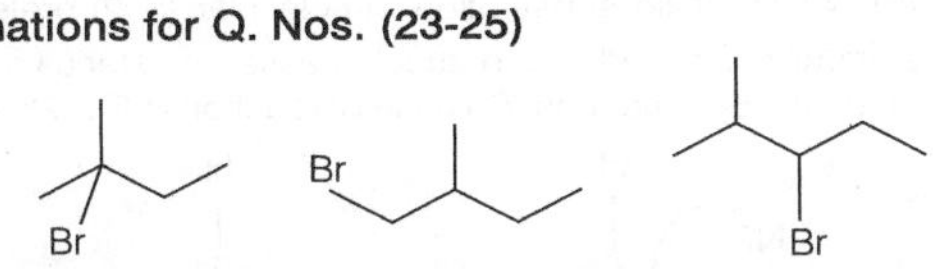

A *B* *C*

B being 1° halide, undergoes faster S_N2 reaction than 2° halide *C*.
Both *B* and *C* are chiral while *A* is achiral.
Grignard's reagent ($R^- Mg X^+$) with most stable carbanion (1°) is formed with *B*.

23. (a) **24.** (d) **25.** (b)

26. (3) Only the following chlorination products, upon E2 reaction can give 1-methyl cyclopentene as one of the product.

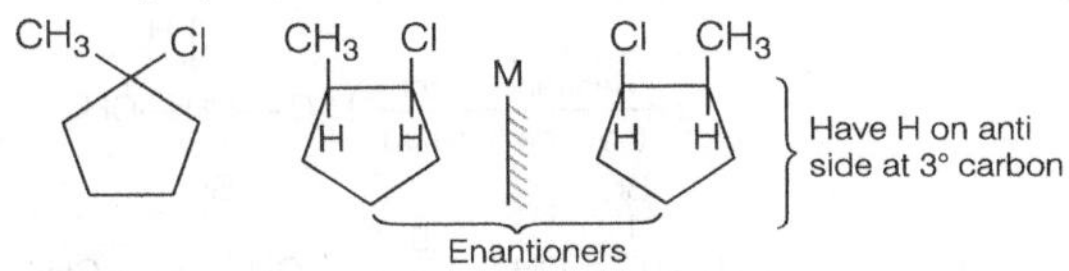

27. (5)

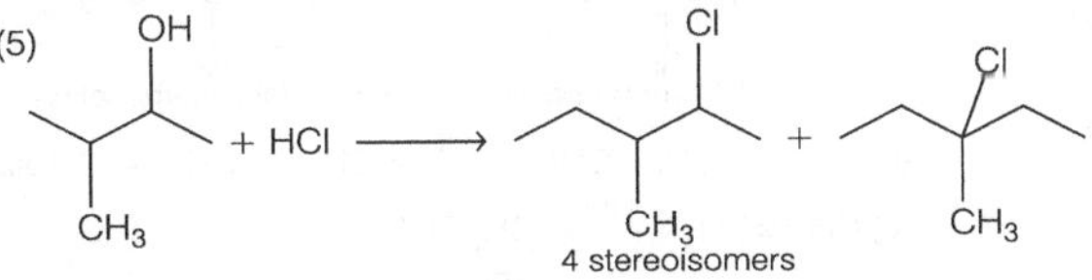

28. (6)

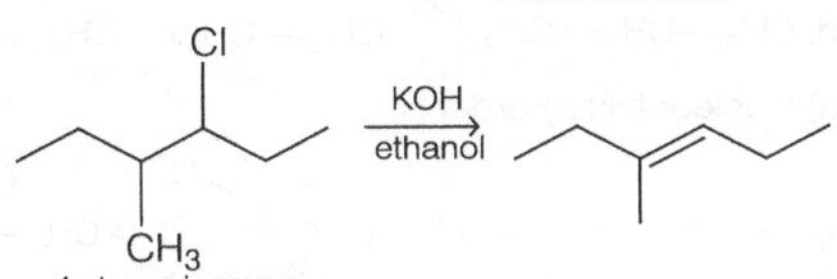

4 stereo isomers

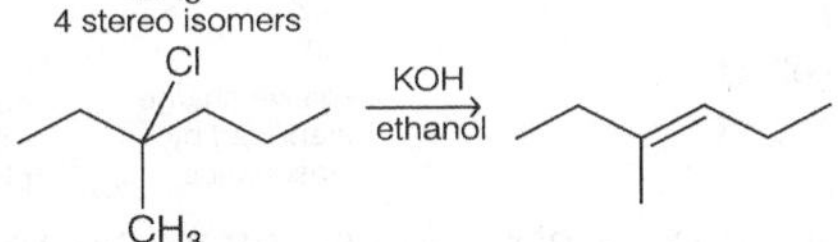

Pair of enantiomers

29. (6)

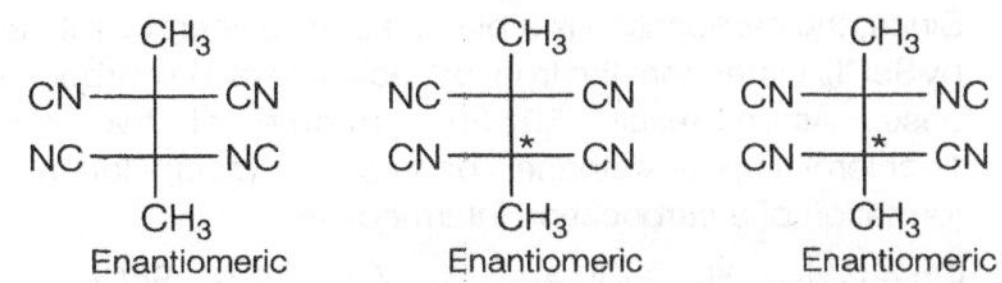

30. (a)

(i) With $SOCl_2$, an intramolecular nucleophilic substitution (S_N1) reaction takes place. In absence of base like pyridine, complete retention of configuration takes place.

(i) → (p,s)

(ii) With $SOCl_2$, in the presence of pyridine, S_N2 reaction takes place with inversion of configuration at chiral α-carbon.

(ii) → (r)

(iii) S_N1 reaction takes place giving racemic mixture of product.

(iii) → (q

(iv)

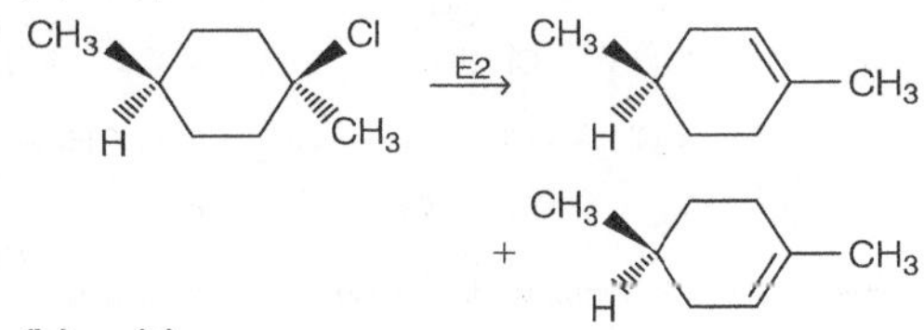

(iv) → (q)

31. (a)

(i) Deuterium incorporation indicates E1cb mechanism which follow 2nd order reaction rate law.

(i) → (q,s)

(ii) E2 reaction giving major product, which follow 2nd order rate law.

(ii) → (q,r)

(iii) Loss of fluoride indicates that carbanion intermediate is involved.

(iii) → (s)

(iv) Rearrangement of carbocation intermediate gives this product, indicating E1 mechanism.

(iv) → (p)

JEE Main & AIEEE Archive

1. (c,d) Alkyl fluorides can be prepared by action of mercurous fluoride or antimony trifluorides (inorganic fluorides) on corresponding alkyl halide.

But, when action of NaI/acetone takes place on alkyl chloride bromide, alkyl iodide forms. This reaction is called 'Finkelstein reaction'

$$C_2H_5Cl \xrightarrow[\text{Acetone}]{\text{NaI}} CHI + NaCl$$

Free radical fluorination is highly explosive reaction. So not preferred for the preparation of fluoride.

2. (c) $2Cl-\overset{Cl}{\underset{Cl}{C}}-CH_3 \xrightarrow{Ag} CH_3-C\equiv C-CH_3 + 6AgCl$
(2-butyne)
(1, 1, 1–trichloroethane)

3. (d) Compound *A* gives a precipitate with alcoholic $AgNO_3$ (here white is misprinting because the colour of ppt is light yellow), so it must contain Br in side chain.

On oxidation, it gives $C_8H_6O_4$, which shows the presence of two alkyl chains attached directly with the benzene nucleus.

Since compound *B* gives anhydride on heating, the two alkyl substituent must occupy adjacent (1, 2) position.

Thus, *A* must be

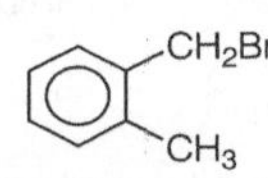

and the reactions are as follows

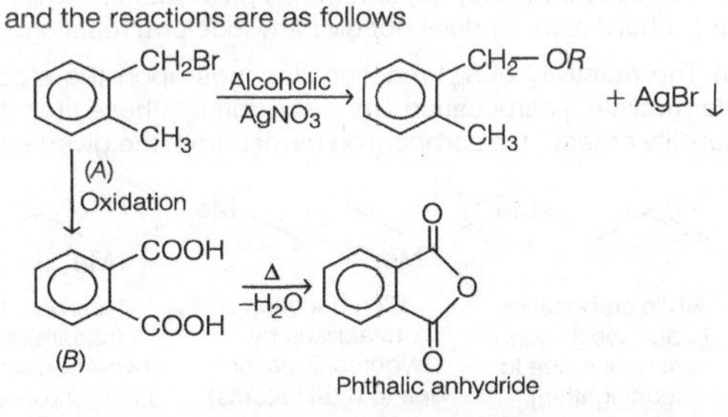

4. (c) The given compound, (–I)– chloro-1-phenylethane in the presence of $SbCl_5$ forms a carbocation.

$$Cl-\underset{Ph}{CH}-CH_3 \xrightarrow[\text{Toluene}]{SbCl_5} [Ph-\overset{+}{C}H-CH_3]\,SbCl_6^-$$

Since, the carbocation is a planar species, therefore it can be attacked by $SbCl_6^-$ either from the front or back side of the carbocation with equal ease. As a result, 50 : 50 mixture of two enantiomers of 1-chloro-1-phenylethane undergoes racemisation due to the formation of a carbocation intermediate.

$$[Ph-\overset{+}{C}H-CH_3]\,SbCl_6^- \longrightarrow Ph-\underset{Cl}{\underset{|}{CH}}-CH_3 + SbCl_5$$

[*d* and *l* forms]

5. (b) 2-methylbutane on monochlorination gives 4 isomers, among which I and III are chiral in nature.

$$CH_3-CH_2-\overset{*}{\underset{CH_3}{\underset{|}{C}H}}-CH_3 \xrightarrow{Cl_2} \underset{(I)}{CH_3CH_2-\overset{*}{\underset{CH_3}{\underset{|}{C}H}}-CH_2Cl} + \underset{(II)}{CH_3-CH_2-\overset{Cl}{\overset{|}{\underset{CH_3}{\underset{|}{C}}}}-CH_3}$$

$$+\underset{(III)}{CH_3-\overset{Cl}{\overset{|}{\overset{*}{C}H}}-\underset{CH_3}{\underset{|}{CH}}-CH_3} + \underset{(IV)}{\underset{Cl}{\underset{|}{CH_2}}\,CH_2-\underset{CH_3}{\underset{|}{CH}}-CH_3}$$

Hence, 2 chiral compounds are formed in the above reaction.

6. (d) Iodoform reaction is given by alcohols and ketones containing $CH_3-\underset{OH}{\underset{|}{CH}}-$ and $CH_3-\overset{O}{\overset{||}{C}}-$ group, respectively. Thus, among the given compounds, isobutyl alcohol does not contain $CH_3-\underset{OH}{\underset{|}{CH}}-$ group.

Hence, it does not give iodoform reaction on treatment with I_2 / NaOH.

$CH_3-\underset{OH}{\underset{|}{CH}}-CH_3$; Isopropyl alcohol (b)

$CH_3-\overset{O}{\overset{||}{C}}-CH_2CH_3$ Ethyl methyl ketone (a)

$CH_3-\overset{O}{\overset{||}{C}}-\underset{CH_3}{\underset{|}{CH}}-CH_3$ 3-methyl-2-butanone (c)

$CH_3-\overset{CH_3}{\overset{|}{CH}}-CH_2OH$ Isobutyl alcohol (d)

Hence, compounds (a), (b) and (c) will give iodoform while compound (d) (isobutyl alcohol) does not give any iodoform reaction.

7. (a) The reactivity of S_N1 reaction depends upon the stability of the intermediate, carbocation formed during these reactions. The stabililty order of the carbocation formed from the given species is

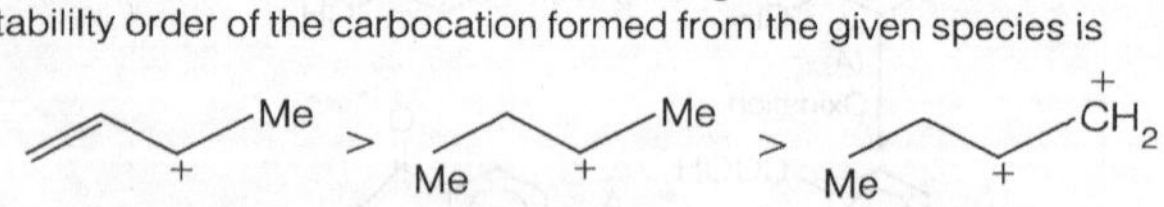

Allylic carbocation (stabilises through resonance due to conjugation) > 2°carbocation (stabilises by hyperconjugation due to 6 α-H atoms) > 1°carbocation (stabilises by hyperconjugation due to 2 α-H-atoms)

Hence, the reactivity order of the given bromide towards S_N1 reaction is

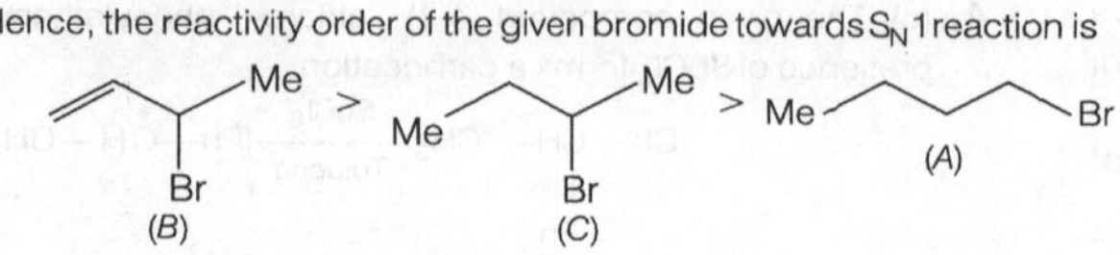

(B) > (C) > (A)

8. (d) Nucleophilic substitution bimolecular (S_N2) prefers less sterically hindered site to attack. Lesser the steric hindrance faster the S_N2 reaction. So ease of reaction is 1° > 2° > 3°.

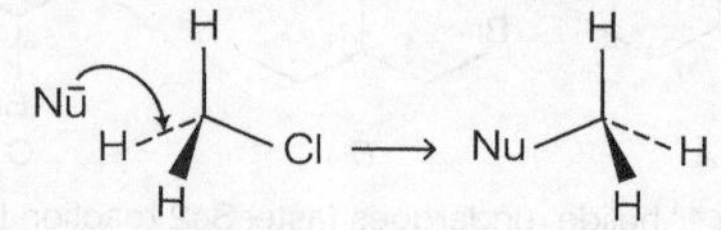

S_N2 involves inversion of configuration stereo, chemically. Since, 1° alkyl halides are prone to S_N2 reactions, therefore CH_3Cl undergoes complete strerochemical inversion.

9. (a)

(a) $CH_3-C\equiv CH + 2HBr \longrightarrow CH_3-\underset{Br}{\underset{|}{C}}=CH_2 \xrightarrow{H^+} H_3C-\underset{Br}{\underset{|}{CH}}-\overset{+}{C}H_2$

$\xleftarrow[\text{(Hydride transfer)}]{\text{Rearrangement}}$ $CH_3-\overset{+}{\underset{Br}{\underset{|}{C}}}-CH_3$ (More stable carbocation) $\xrightarrow{Br^-}$ $CH_3-\overset{Br}{\overset{|}{\underset{Br}{\underset{|}{C}}}}-CH_3$ (2, 2-dibromopropane)

(b) $CH_3-CH{=}CHBr \xrightarrow{HBr} CH_3-CH_2-CHBr_2$

(c) $CH\equiv CH \xrightarrow{2\,HBr} CH_3CHBr_2$

(d) $CH_3-CH=CH_2 \xrightarrow{HBr} CH_3-CHBr-CH_3$

10. (a) Nucleophilicity order is

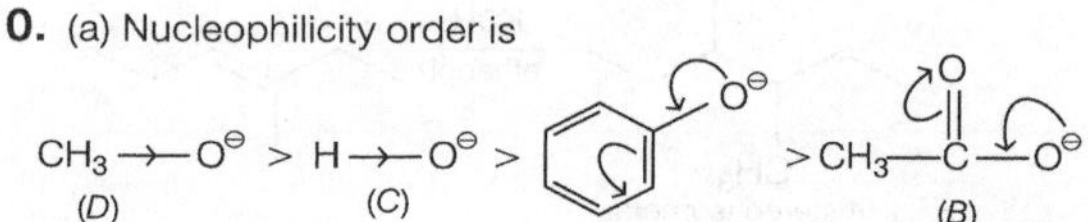

$CH_3\rightarrow O^\ominus$ (D) + *I* effect > $H\rightarrow O^\ominus$ (C) > phenoxide (A) Negative charge stabilised by resonance > $CH_3-\overset{O}{\overset{||}{C}}-O^\ominus$ (B) Negative charge comparatively stable than phenoxide ion

11. (b) $R_2CuLi + R'X \longrightarrow R-R' + R-Cu + LiX$

(Dialkyl copper lithium)

12. (c) $H_3C-\overset{H}{\overset{|}{\underset{H}{\underset{|}{C}}}}-\overset{H}{\overset{|}{\underset{CH_3}{\underset{|}{C}}}}-CH_3 \xrightarrow{Br_2} CH_3-CH_2-\overset{Br}{\overset{|}{\underset{CH_3}{\underset{|}{C}}}}-CH_3$

2-bromo-2-methyl butane

13. (c) $CH_3CH_2\underset{Br}{\underset{|}{C}H}CH_3 \rightarrow CH_3CH_2\overset{\oplus}{C}HCH_3$ (2°carbocation) $\xrightarrow{-H^+}$

$\underset{\text{(II)}}{\underset{\text{Less substituted}}{CH_3CH_2CH=CH_2}} + \underset{\text{(I)}}{\underset{\text{More substituted}}{CH_3CH=CHCH_3}}$

Stability of (I) > (II), hence (I) is predominant.

14. (c) $CH_2=CHCH=CH_2 + HBr \longrightarrow CH_3\underset{Br}{\underset{|}{C}H}CH=CH_2$ (1, 2-addition product)

$+ CH_3CH=CHCH_2Br$ (1, 4-addition product)

Addition is through the formation of allylic carbocation.

$\underset{\text{(more stable)}}{\underset{\text{(2° allylic)}}{CH_2=CH\overset{\oplus}{C}HCH_3}} \longleftrightarrow \underset{\text{(less stable)}}{\underset{\text{(1° allylic)}}{CH_3CH=CH\overset{\oplus}{C}H_2}}$

Under mild conditions (temperature ≈ − 80°C) kinetic product is the 1, 2-addition product and under vigorous conditions, (temp. ≈ 40°C) thermodynamic product is the 1, 4-addition product.

Thus, 1-bromo-2-butene is the major product under given condition.

JEE Advanced & IIT JEE Archive

1. (a)

	Column I	Column II	Explanation
i.	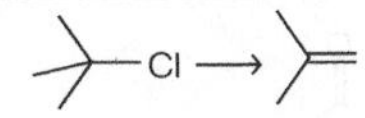	NaOEt(q)	$\bar{O}$ Et (strong nucleophile) causes dehydrohalogenation of 3° alkyl halide
ii.	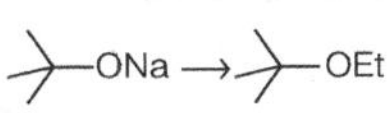	EtBr (r)	3° butoxide undergoes S_N reaction with 1° alkyl halide
iii.	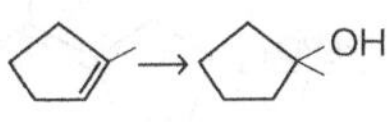	(i) $Hg(OAC)_2$ (ii) $NaBH_4$ (p)	Mercuration-demercuration adds H_2O by Markownikoff's rule without rearrangement
iv.	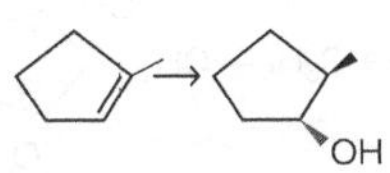	(i) BH_3 (ii) H_2O_2 / OH^- (s)	Hydroboro-oxidation adds H_2O by anti-Markownikoff's rule

2. (b) Acetone is an aprotic solvent and can dissolve both the nucleophile and the substrate and thus S_N2 reaction is favoured. Also

$$S_N2 \xleftarrow{\quad 1^\circ \quad 2^\circ \quad 3^\circ \quad} \text{Alkyl halides} \xrightarrow{\qquad} S_N1$$

S.	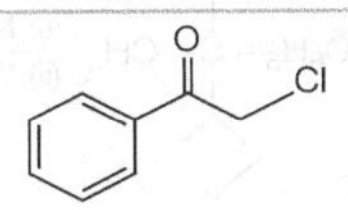	1° alkyl halide but (C—Cl). BE is decreased by electron withdrawing [C_6H_5CO] group, (a case of I-effect). Thus, maximum rate in S_N2 reaction
Q.	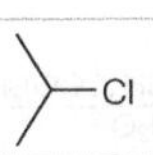	2° alkyl halide, rate is minimum
P.	CH_3 — Cl	1° alkyl halide
R.	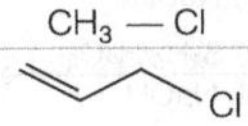	1° allylic halide but allylic 1° carbocation is resonance stabilised in S_N1 reaction

Thus, reactivity order is $S > P > R > Q$

3. (5) The substrate has three different types of B—H, therefore, first, three structural isomers of alkenes are expected as

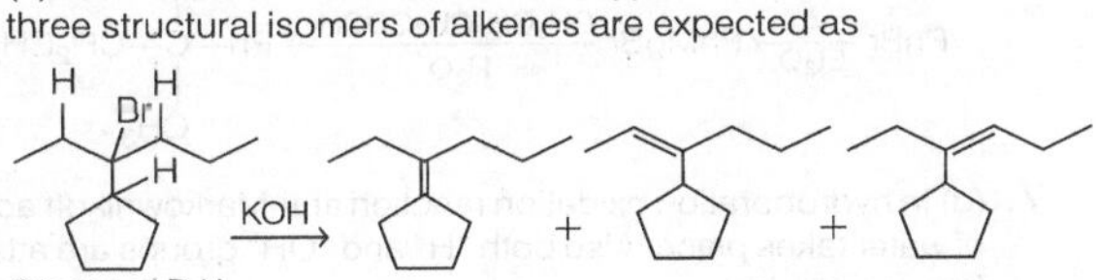

3 types of B-H (I) (II) (III)

The last two alkenes (II) and (III) are also capable of showing geometrical isomerism hence, two geometrical isomers for each of them will be counted giving a total of five isomers.

4. (a) Nucleophile PhS^- substitute the Br^- through S_N2 mechanism with inversion of conjiguration at α-C.

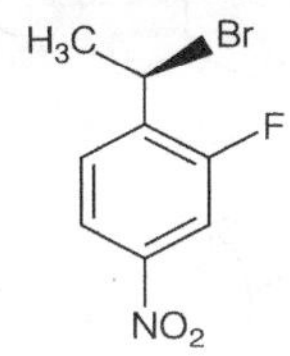

5. (a)

(i) $CH_3—CHBr—CD_3 \xrightarrow[E2]{\text{Alc. KOH}} CH_2 = CH—CD_3$

E2 reaction is a single-step reaction in which both deprotonation from β-C and loss of leaving group from α-C occur simultaneously in the rate-determining step. C—D bond is stronger than C—H bond, C—H is preferably broken in elimination.

(ii) Ph—CHBr—CH_3 reacts faster than Ph—CHBr—CD_3 in E2 reaction because in later case, stronger C—D bond is to be broken in the rate determining step.

(iii) $Ph—CH_2—CH_2Br \xrightarrow[C_2H_5O^-]{C_2H_5OD} Ph—CD = CH_2$

(iv) Both $PhCH_2CH_2Br$ and $PhCD_2CH_2Br$ will react at same rate in E1 reaction because C—H bond is broken in fast non-determining.

Thus, (i) → (q), (ii) → (q), (iii) → (r), (iv) → (p,s)

6. (a) Reaction proceed through carbocation intermediate

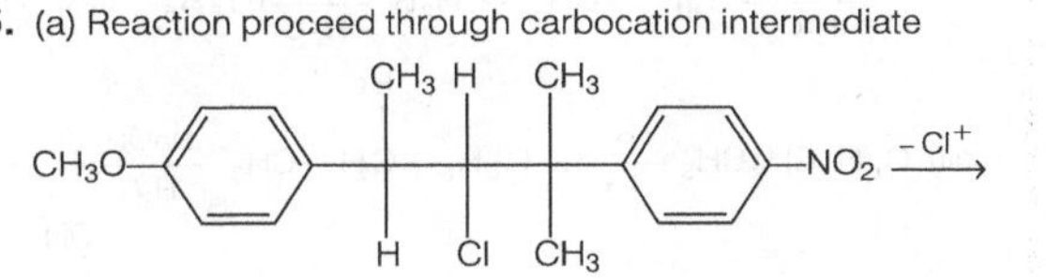

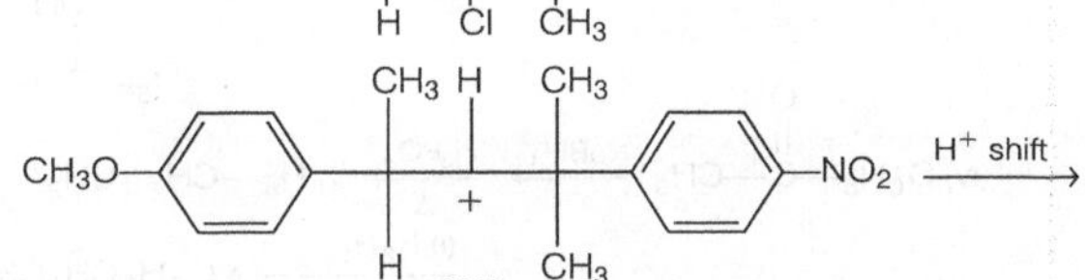

2°(I)

(I) $\xrightarrow{H_2O} K$

(II) $\xrightarrow{H_2O} L$

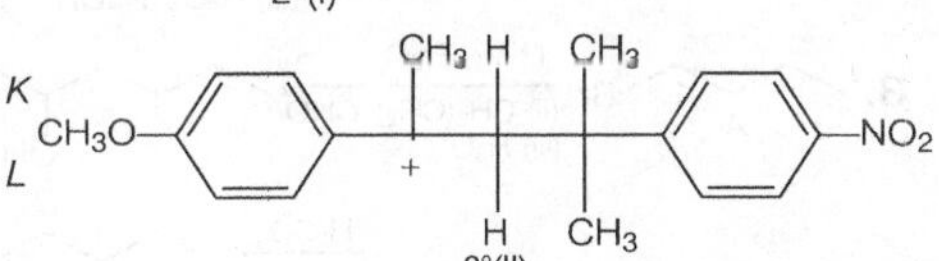

3°(II)

stabilised by + R effect of – OCH_3

2. Aliphatic Alcohols, Ethers and Epoxides

DPP-1 Preparation of Alcohols

1. Various alcohol isomers of $C_5H_{12}O$ are as follows:

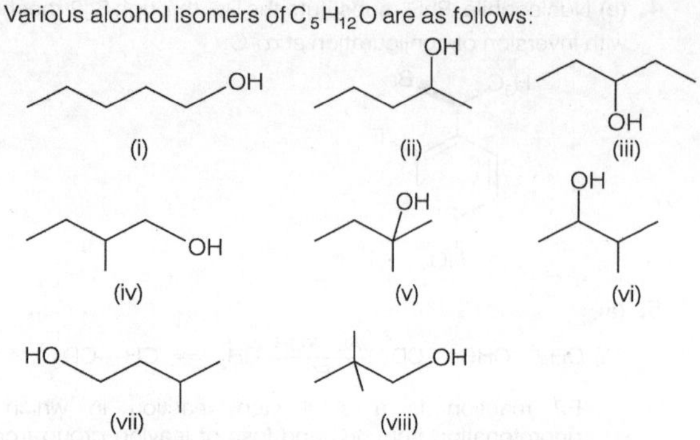

Aldehydes, carboxylic acids and esters on reduction give primary alcohols while ketone on reduction gives secondary alcohols. Hence, alcohols (i), (iv), (vii) and (viii) can be prepared by reduction of either aldehydes or acid of esters while alcohols (ii), (iii) and (vi) can be prepared by reduction of ketones only. Alcohol (v) is tertiary, cannot be prepared by reduction of any of these three types of compound.

2. (i) $C_6H_6 \xrightarrow[Fe]{Br_2} C_6H_5Br \xrightarrow[Et_2O]{Mg} \xrightarrow[(ii)\ H_3O^+]{(i)\ CH_3CHO} C_6H_5CH(OH)CH_3$

(ii) $C_6H_5CH_2CH_3 \xrightarrow[\Delta]{Br_2} C_6H_5-\underset{Br}{CH}-CH_3 \xrightarrow[C_2H_5OH]{KOH} C_6H_5-CH=CH_2$

$\xrightarrow[H_2O_2]{HBr} C_6H_5-CH_2-CH_2Br \xrightarrow[H_2O]{NaOH} C_6H_5-CH_2CH_2OH$

(iii) $C_6H_5CH_2CH_3 \xrightarrow[\Delta]{Br_2} C_6H_5-\overset{Br}{CH}-CH_3 \xrightarrow[H_2O]{NaOH} C_6H_5-\overset{OH}{CH}-CH_3$

(iv) $C_6H_5-\overset{O}{\overset{\|}{C}}-CH_3 \xrightarrow{NaBH_4} \xrightarrow[\Delta]{H_3PO_4} C_6H_5-CH=CH_2 \xrightarrow[(ii)\ H_2O_2/NaOH]{(i)\ B_2H_6} C_6H_5CH_2CH_2OH$

3.

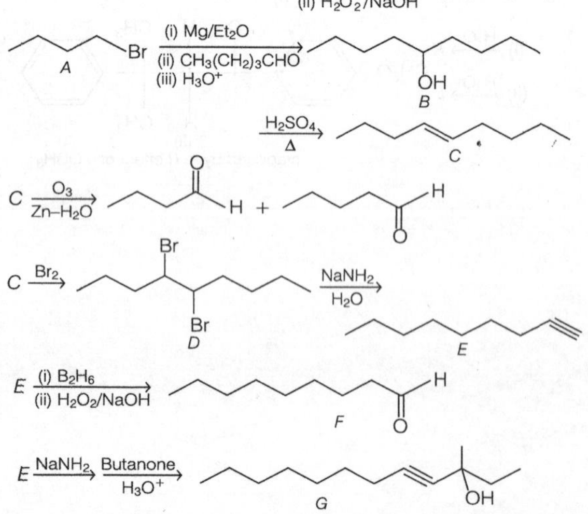

4. (i)

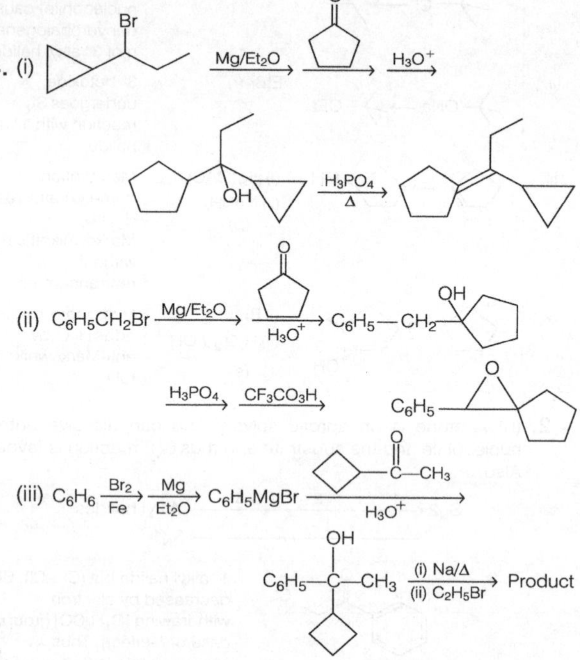

5.

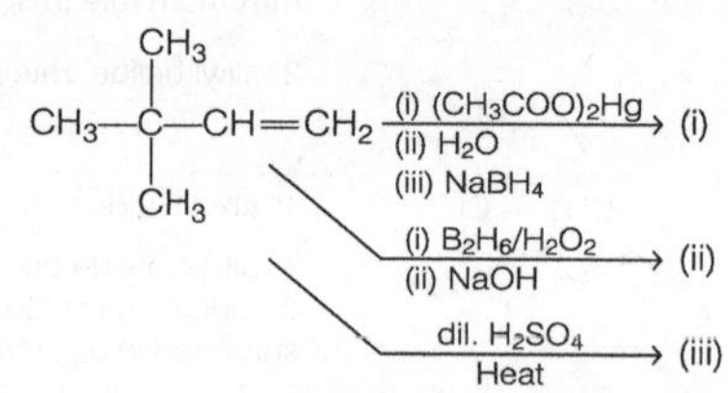

6. (c)

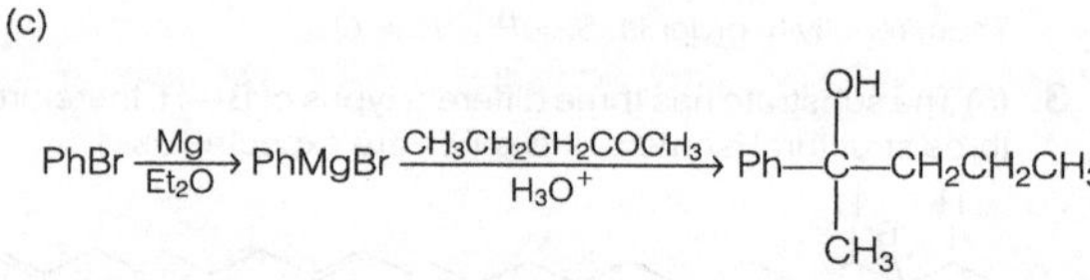

7. (d) In hydroboration oxidation reaction anti-Markownikoff addition of water takes place. Also both "H" and "OH" groups are attached from same side.

8. (a) (I)

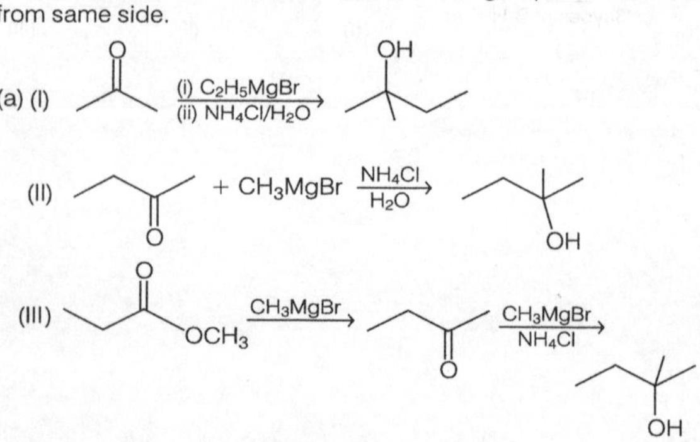

All of the above reactions give 2-methyl-2-butanol.

9. (d) It gives Markownikoff's addition of water without rearrangement.

10. (b) Reaction involves rearrangement of carbocation.

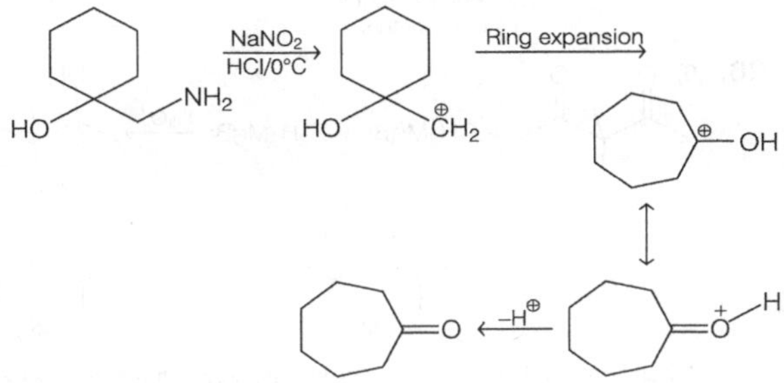

11. (b) It is an oxo process of synthesis of alcohols.

$$CH_3—CH=CH—CH_3 + CO + H_2 \xrightarrow{[CoH(CO)_4]} CH_3—CH_2—\underset{\underset{CHO}{|}}{CH}—CH_3 \xrightarrow{Zn\text{-}Cu/H_2} CH_3CH_2—\underset{\underset{CH_3}{|}}{CH}—CH_2OH$$

12. (b)

$$(CH_3)_3CBr + CH_3COOAg(aq) \longrightarrow \underset{\text{Ester}}{CH_3—\overset{\overset{O}{||}}{C}—O—C(CH_3)_3} \xrightarrow{H_3O^{\oplus}} CH_3COOH + CH_3—\underset{\underset{CH_3}{|}}{\overset{\overset{CH_3}{|}}{C}}—OH$$

13. (b) In oxymercuration-demercuration reaction, Markownikoff's addition of H_2O takes place with HO^- added to carbon where more stable carbocation could be formed.

14. (b,c) It is a S_N1 reaction.

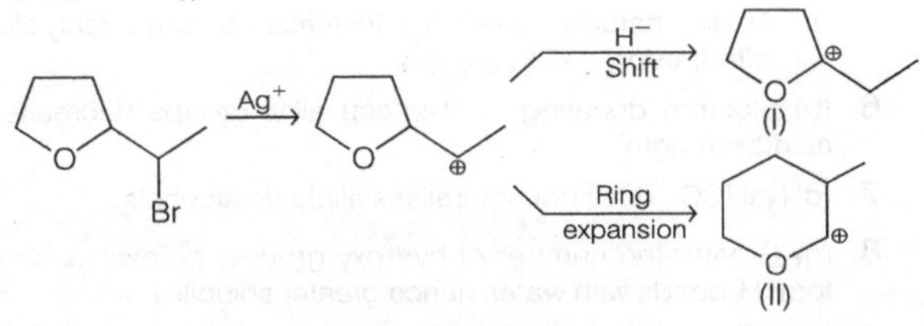

(I) and (II) undergo nucleophilic attack by H_2O giving the desired products.

15. (b,c) Oxo process

$$CH_3CH_2—CH=CH_2 + CO + H_2 \xrightarrow{[CoH(CO)_4]} \xrightarrow[H_2/\Delta]{Cu\text{-}Zn}$$

$$CH_3CH_2—\underset{\underset{CHO}{|}}{CH}—CH_3 + CH_3CH_2CH_2CH_2CHO$$

The aldehydes formed above undergo reduction to give the desired products.

16. (a,c,d) Option (b) does not give 2-cyclopentyl-2-butanol as

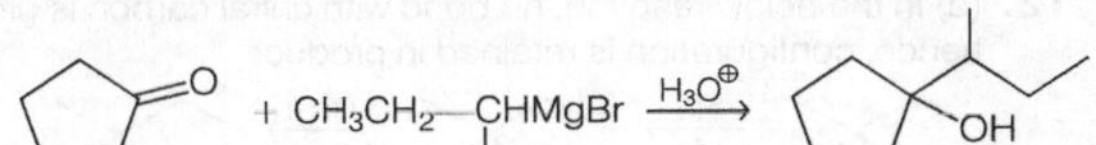

1-(1-methyl propyl) cyclopentanol

17. (a,c,d) Isomerisation occur *via* carbocation intermediates.

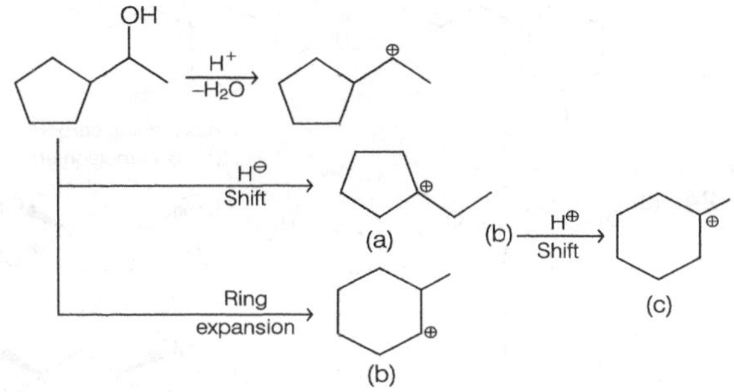

The above carbocations upon nucleophilic attack by water gives the desired product.

18. (a,b)

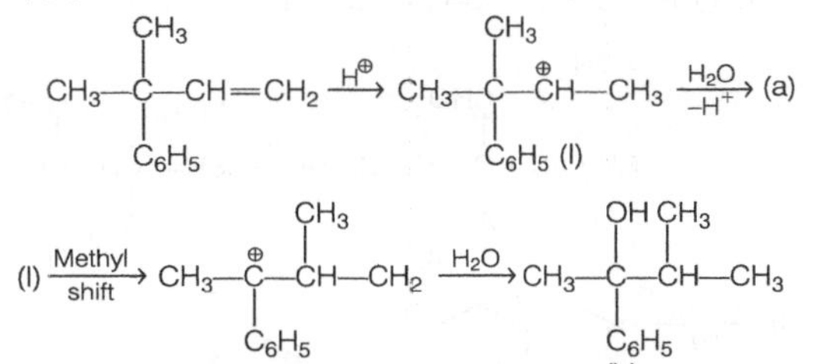

More stable benzylic carbocation is formed

19. (b,d) Deutride (D^-) addition at planar carbonyl carbon occur from both side of plane, with equal probability giving racemic mixture of alcohols. Also deuterium is attached to carbonyl carbon atom only.

20. (a,b,c,d) Reaction proceed *via* carbocation intermediates.

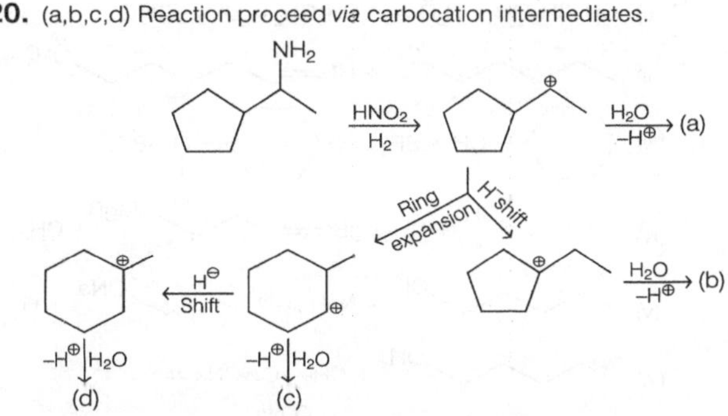

21. (d) *B* is an alcohol formed by the attack of Grignard's reagent on acetone. Hence, alcohol must have the skeleton

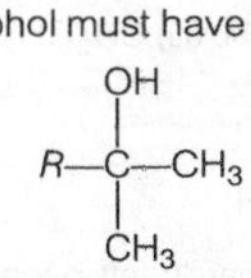

Also, *R* is C_7H_{11} with two degree of unsaturation it must be cyclohexenyl methyl not cyclohexyl methyl.

22. (b) Hydroboration oxidation at double bond gives anti-Markownikoff's addition of H_2O in *syn* orientation.

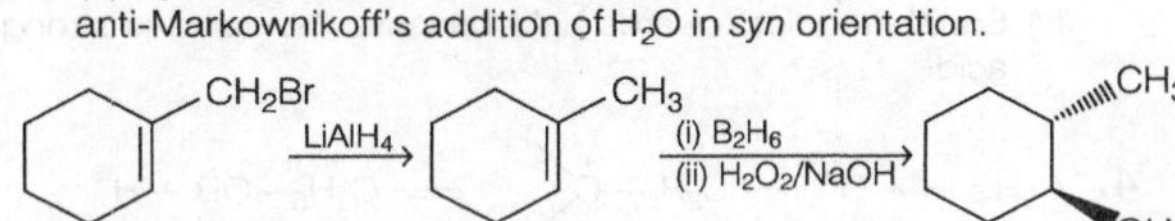

A

23. (d)

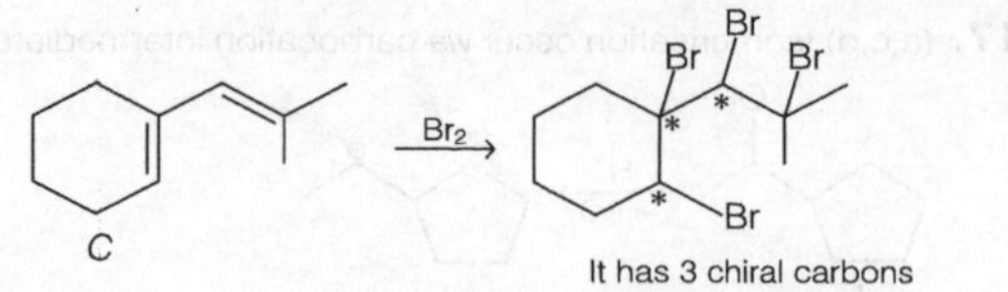

It has 3 chiral carbons
$2^3 = 8$ stereoisomers

24. (3)

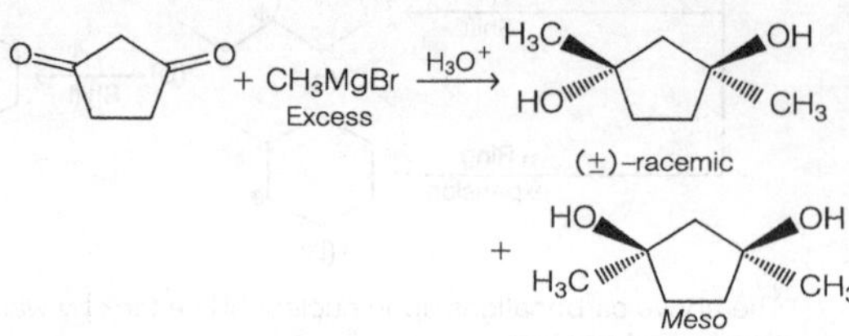

25. (5)

$$CH_3CH_2-\underset{\substack{|\\CH_3}}{C}=CH-CH_3 + CO + H_2 \xrightarrow{Co_2(CO)_8}$$

$$CH_3CH_2-\underset{\substack{|\\CH_3}}{\overset{\substack{CHO\\|}}{C}}-CH_2CH_3 \; (I) + CH_3CH_2-\underset{\substack{|\\CH_3}}{CH}-\overset{\substack{CHO\\|}}{CH}-CH_3 \; (II)$$

(I) $\xrightarrow[H_2]{Cu\text{-}Zn}$

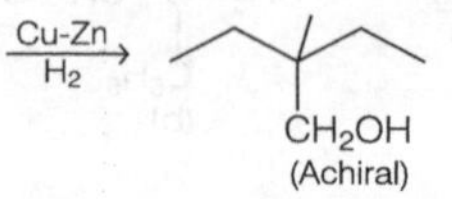

(II) $\xrightarrow[H_2]{Cu\text{-}Zn}$

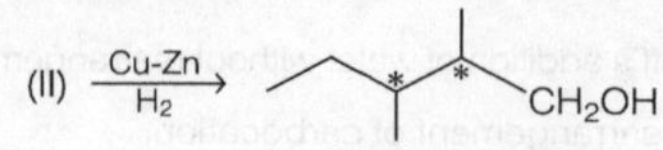

Has 2 chiral carbons
4 stereoisomers

26. (6)

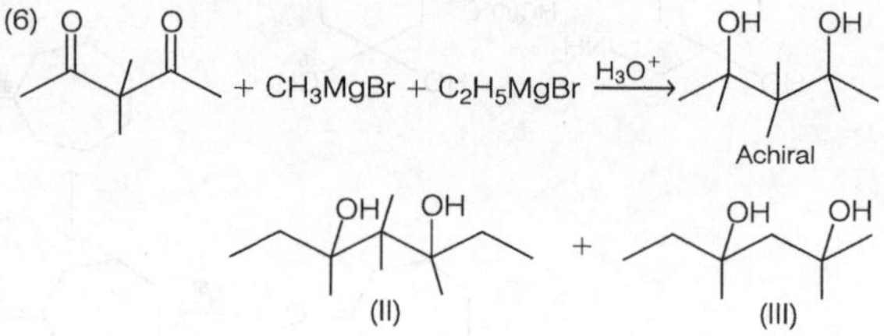

(II) has two chiral carbon but symmetrical, hence, three stereoisomers. (III) has only one chiral carbon hence, two stereoisomers (pair of enantiomers).

27. (3)

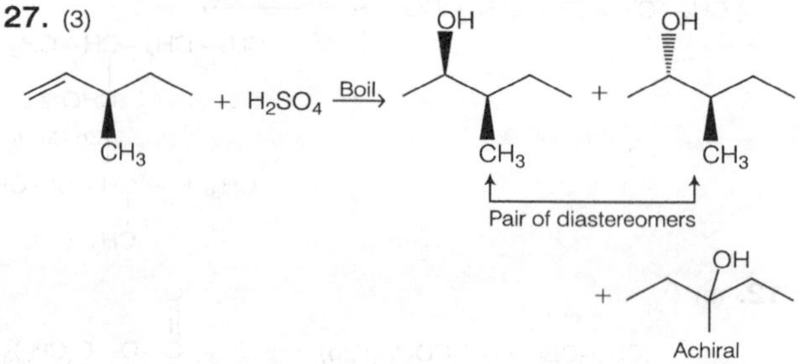

DPP-2 Reactions of Alcohols-I

1. (i) $CH_3CH_2OH + H_3\overset{\oplus}{O} \rightleftharpoons CH_3CH_2-\overset{\oplus}{O}H_2 + H_2O$

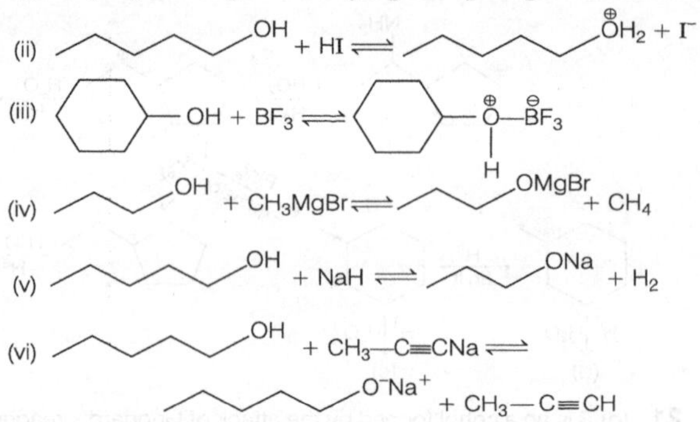

2. (i) H_2CO_3, (ii) CH_3COOH, (iii) CH_3CH_2OH, (iv) CH_3CH_2SH

(v)

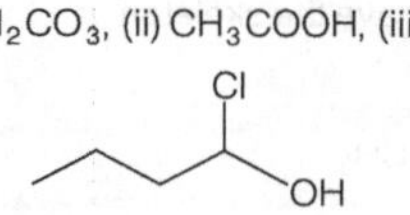

3. (i) Equilibrium lies towards left because CH_3CH_2OH is stronger acid.

(ii) Equilibrium lies towards right because HCl is stronger acid.

(iii) Equilibrium lies towards right because phenol is stronger acid.

(iv) Equilibrium lies towards right because 1-butanol is stronger acid.

4. $C_2H_5OH + D^+ \longrightarrow C_2H_5-\overset{+}{O}\begin{smallmatrix}H\\D\end{smallmatrix} \rightleftharpoons C_2H_5-OD + H^{\oplus}$

O—D bond is stronger than O—H bond, hence equilibrium lies predominantly towards right.

5. (i) The two strong electron withdrawing (—CF_3) groups in (II) operating in opposite direction decreases the polarity of C—O bond hence, decreases its H-bonding ability. Hence, it has less boiling point than (I).

(ii) The dipole character of C—O bond is cancelled by the resultant of two —CF_3 groups, hence lower dipole moment than (I).

(iii) Acidic character of —OH is increased by two strong electron withdrawing —CF_3 groups.

6. (b) Electron donating (+ I-effect) alkyl groups decreases the acidic strength.

7. (d) $NaHCO_3$ does not neutralises aliphatic alcohols.

8. (d) Greater the number of hydroxy groups, greater its ability to form H-bonds with water, hence greater solubility.

9. (c) Both (I) and (II) are 2° alcohols, less acidic than (I) and (IV) where there is electron withdrawing chloride group is present. Between (I) and (II), (I) is less acidic as ring is considered here equivalent to two ethyl groups attached to α-carbon. (III) is more acidic than (IV) because in (III) Cl is at α-carbon while it is at β-carbon in (IV).

10. (b) Dihydric alcohols are always more soluble in water than monohydric alcohol. Between (I) and (II), (I) is more soluble as it forms intermolecular H-bonds with water while (II) forms intramolecular H-bonds which decreases its ability to form intermolecular H-bonds with water.

11. (d) Phenol forms salt with NaOH while aliphatic alcohols do not.

12. (d) In the below reaction, no bond with chiral carbon is disturbed, hence, configuration is retained in product.

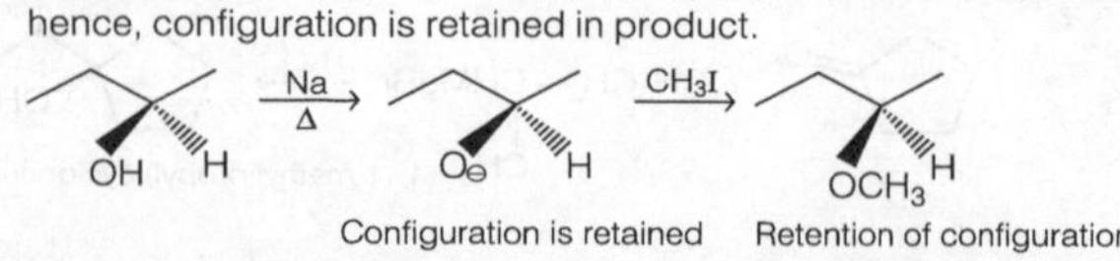

Configuration is retained Retention of configuration

13. (c) Phenolic — OH is neutralised here

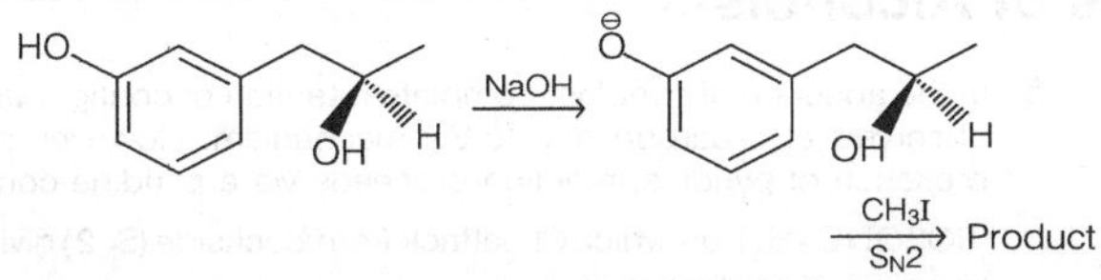

14. (a,b) The conjugate base of given acid is highly stabilised by resonance.

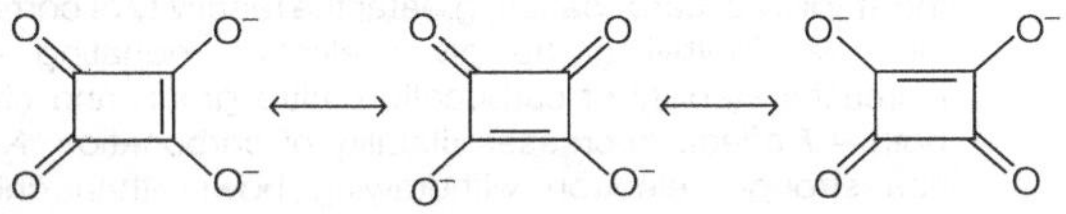

15. (a,d) $CH_3CH_2OH + NaOH \xrightarrow{I_2} \underset{\text{Yellow}}{CHI_3\downarrow} + HCOONa$

No such reaction occur with CH_3OH.

Acidic $KMnO_4$ forms CO_2 with methanol but CH_3COOH with ethanol.

16. (a,b,c,d) All these groups increases the stability of *gauche* conformer due to stable intramolecular H-bonding.

17. (a,c,d)

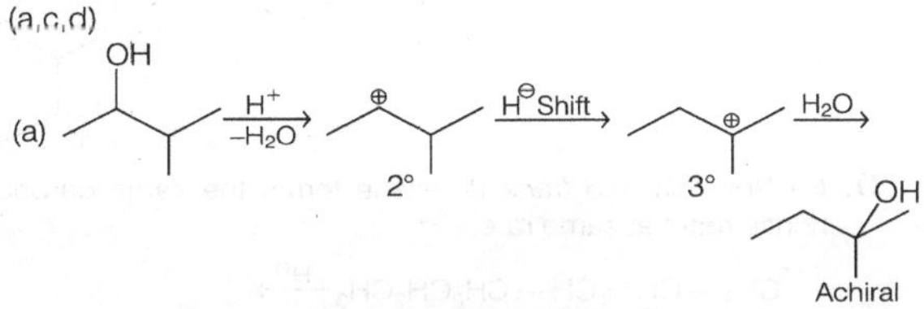

(c) PCl_5 brings about S_N2 reaction, chloro alkane with inverted configuration would be produced.

(d) 3-methyl-2-butanol is a secondary alcohol, will be oxidised to ketone with acidic dichromate solution.

Explanations for Q. Nos. (18-20)

The key point is formation of *Y* from *ortho* methoxy phenol by Reimer-Tiemann reaction.

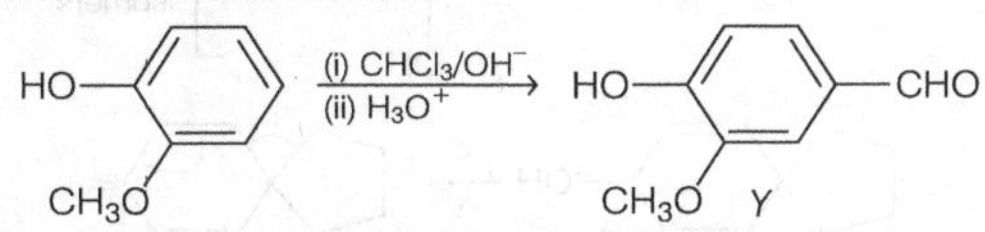

Also, *Y* is an ozonolysis product of *X*.

18. (d)

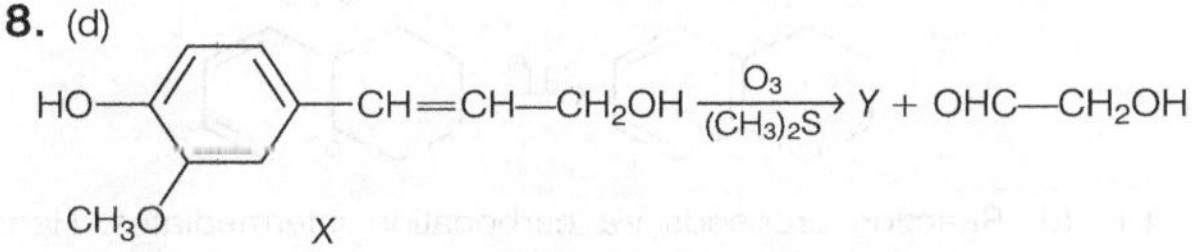

19. (c)

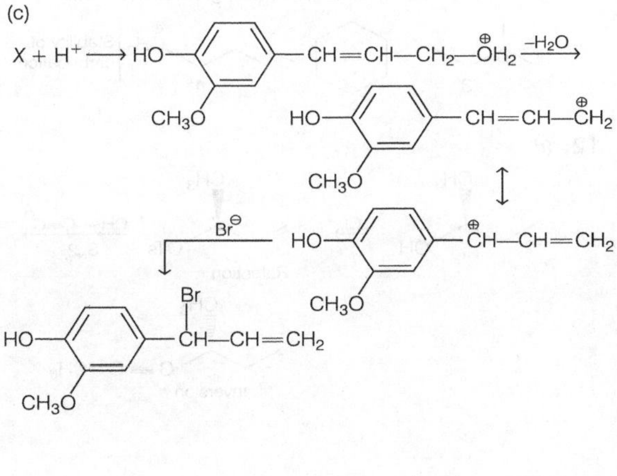

20. (a)

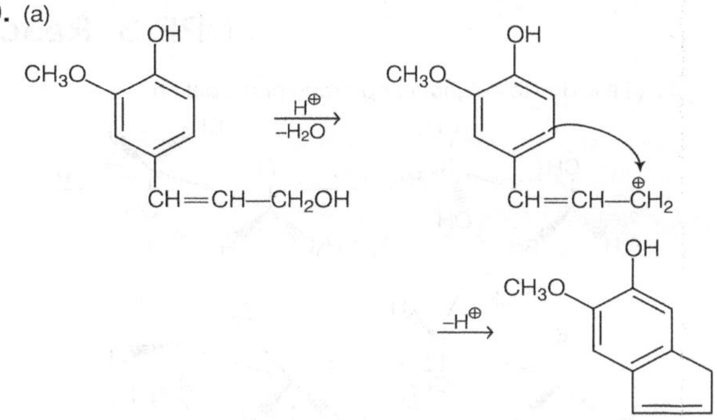

21. (5)

(i) $HCl/ZnCl_2$ (Lucas reagent) gives white turbidity.

(ii) NaOH does not give any visible change.

(iii) $\underset{\text{Orange solution}}{K_2Cr_2O_7} + H_2SO_4 + CH_3—CH(OH)—CH_3 \longrightarrow CH_3—\overset{O}{\overset{\|}{C}}—CH_3 + \underset{\text{Green}}{Cr^{3+}}$

(iv) MnO_2 does not oxidises 2-propanol.

(v) $NaNH_2$ gives effervescence of NH_3.

(vi) C_2H_5MgBr gives $C_2H_6(g)$

(vii) NaH gives $H_2(g)$

(viii), (ix) and (x) all react with 2-propanol but no visible change would be observed.

Hence, only (i), (iii), (v), (vi) and (vii) gives visible change.

22. (5)

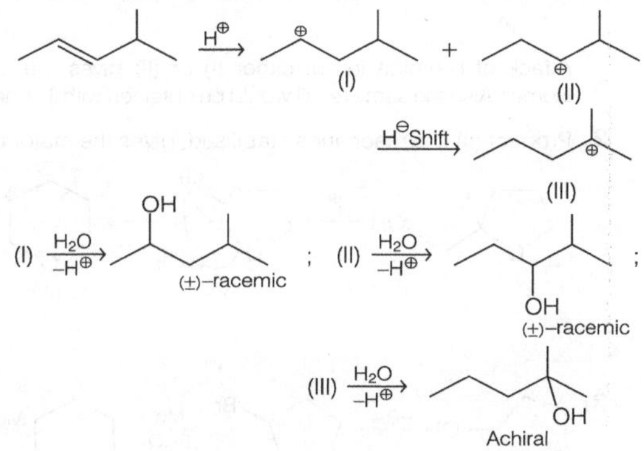

23. (3)

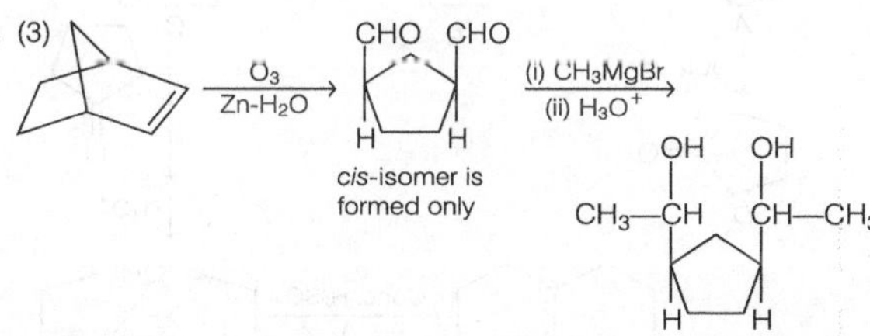

Three stereomers of final product are formed, a *meso* and a pair of enantiomers.

24. (5)

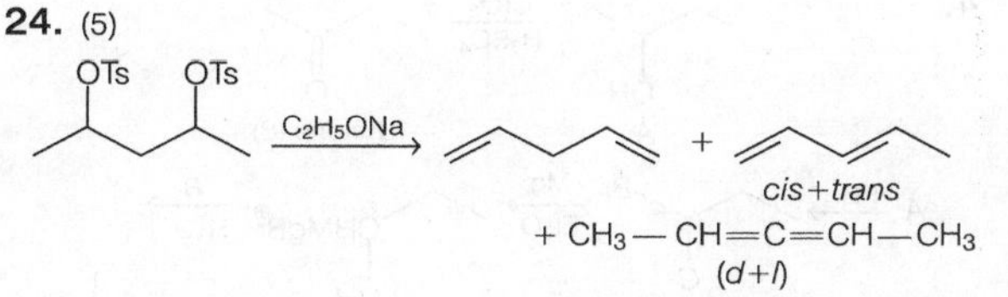

DPP-3 Reactions of Alcohols-II

1. (i) It is due to neighbouring group participation.

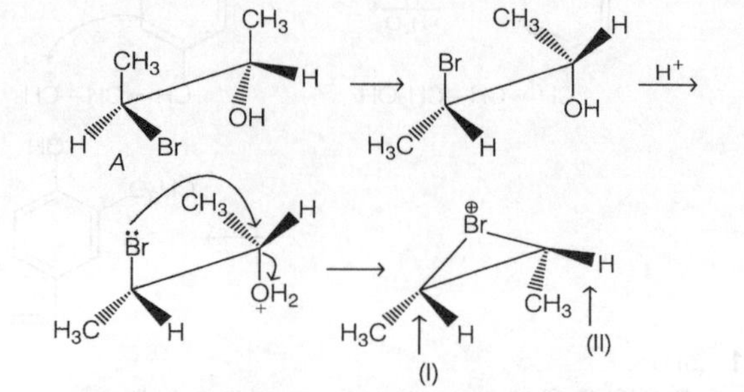

Now attack of bromide ion at (I) and (II) above gives pair of enantiomer in equal amount, hence racemic mixture. Same will be true for its enantiomer *B*.

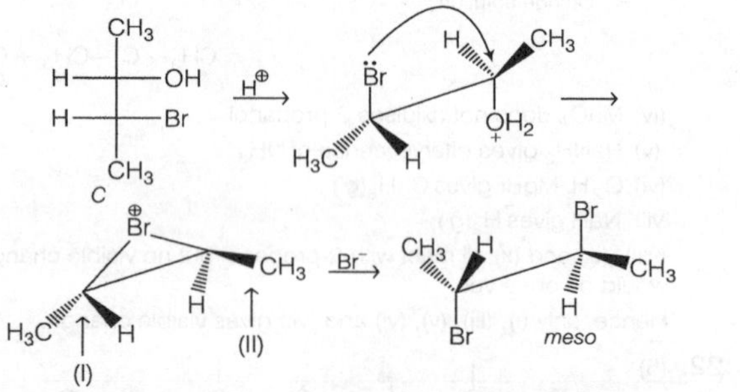

Attack of bromide ion at either (I) or (II) gives the same *meso* isomer. Also the same result would be obtained with its enantiomer *D*.

2. Product (II) It is resonance stabilised, gives the major product.

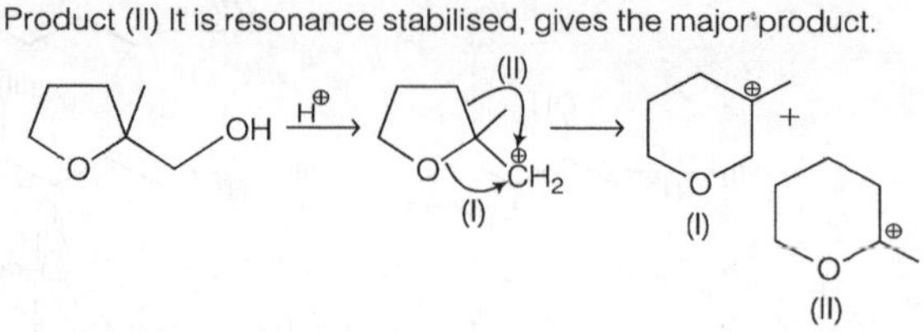

3.

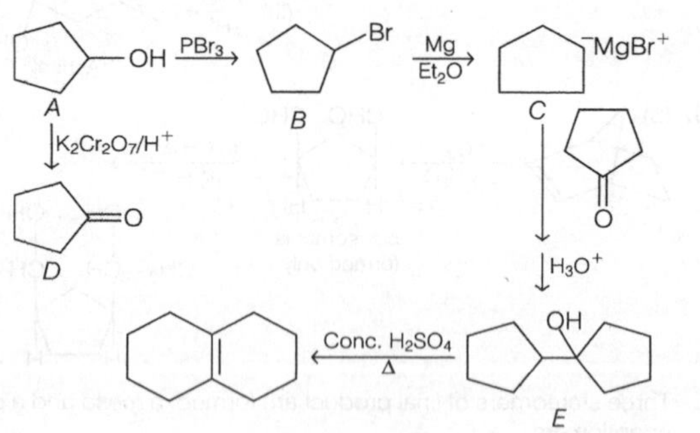

4.

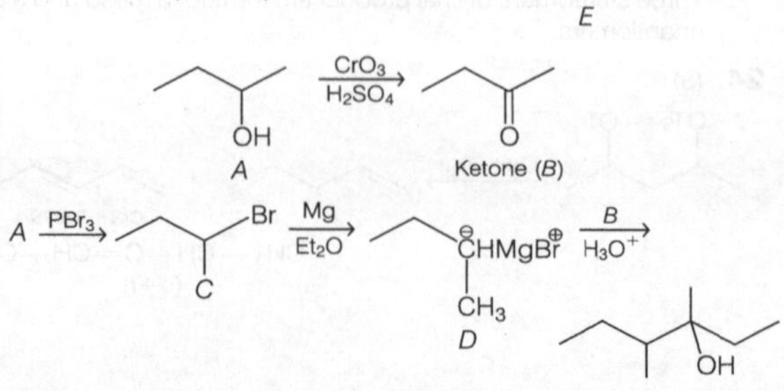

5. In the absence of pyridine, complete retention of configuration is observed at α-carbon due to S_Ni mechanism. However, in the presence of pyridine, reaction proceeds *via* a pyridine complex ($R-OSO\overset{+}{N}C_5H_5$) on which Cl^- attack from backside (S_N2) giving inversion of configuration.

6. (d) Reactive intermediate carbocation is involved, hence greater the stability of carbocation, greater the reactivity of corresponding alcohols. Methyl group gives electron donating $+I$-effect, increases stability of carbocation, nitro group and chloro have both $-I$-effect, decreases stability of carbocation. Nitro group has stronger electron withdrawing power than chloro, nitro derivative is further less reactive than chloro derivative.

7. (d) It is Jone's reagent which does not affect the olefinic bonds.

8. (a)

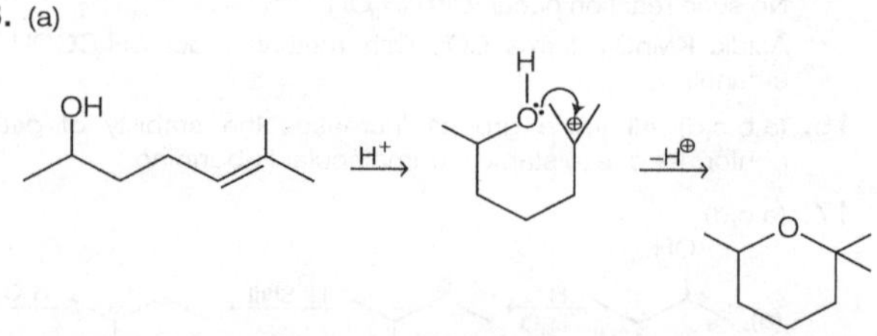

9. (d) Both *cis* and *trans* 2-hexene forms the same carbocation, hence react at same rate.

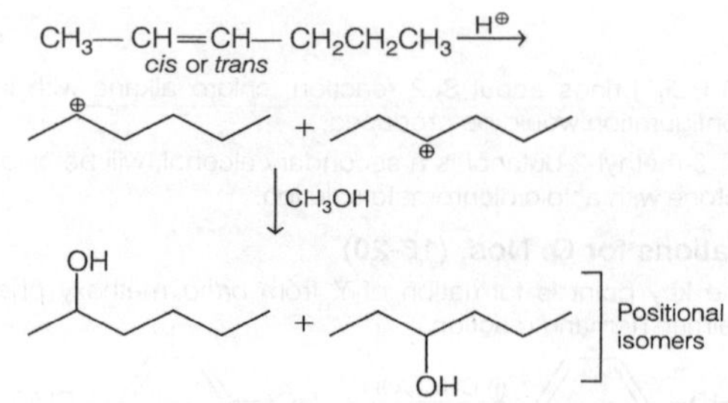

10. (c)

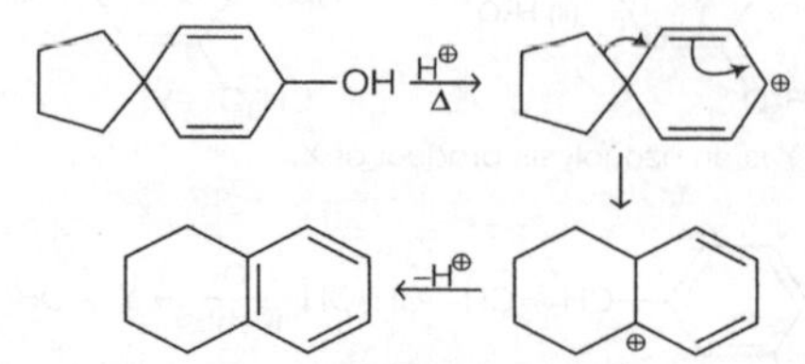

11. (d) Reaction proceeds *via* carbocation intermediates. Hence, greater the stability of carbocation, greater the reactivity.

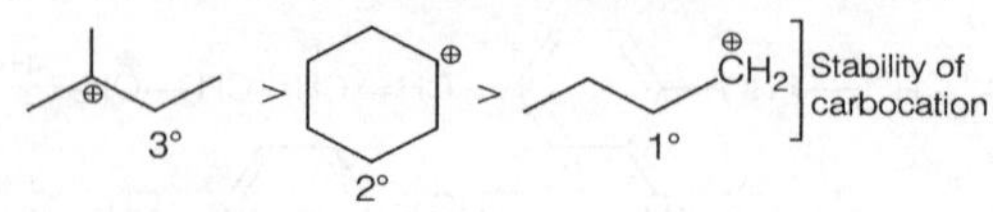

12. (d)

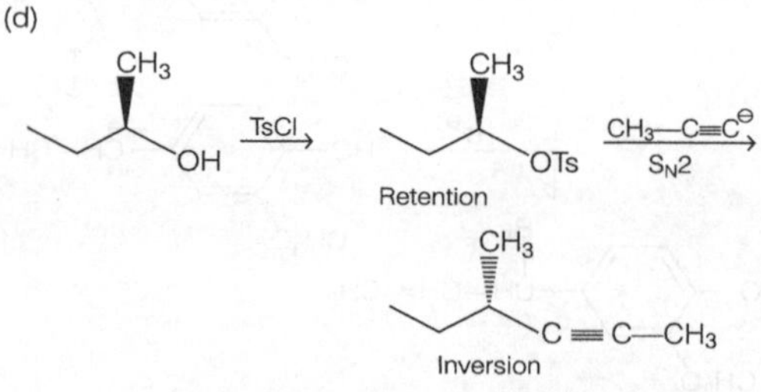

13. (b)

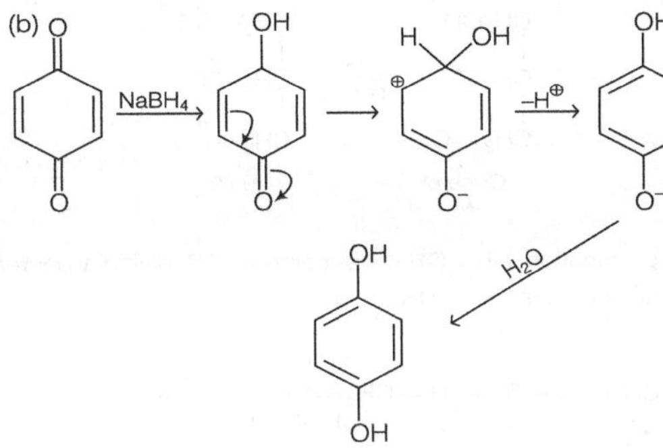

14. (a,b,d) Reaction proceeds *via* carbocation intermediates.

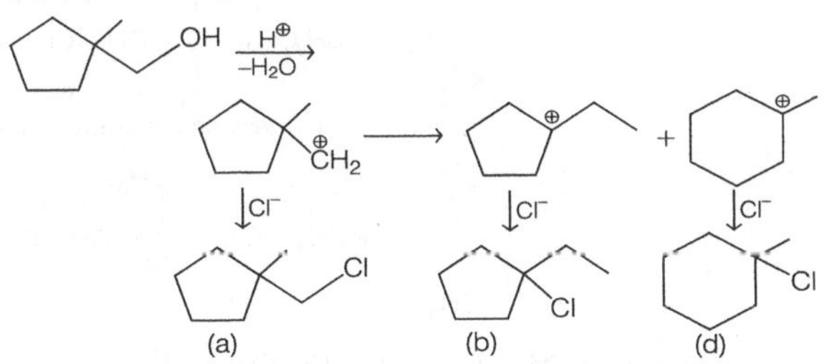

15. (a,b)

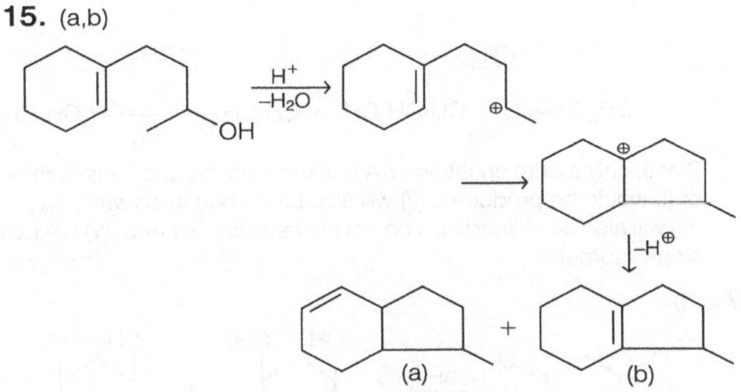

16. (b,c,d)

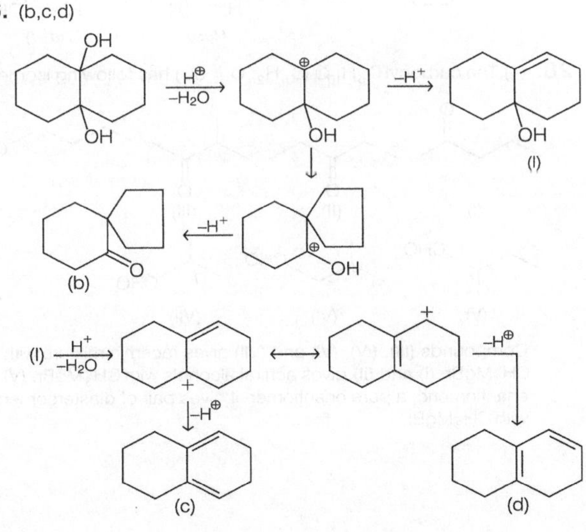

17. (a,b,c)

(a)

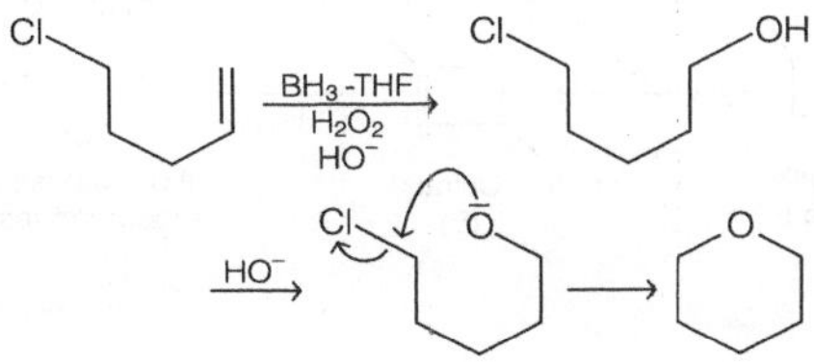

(b)

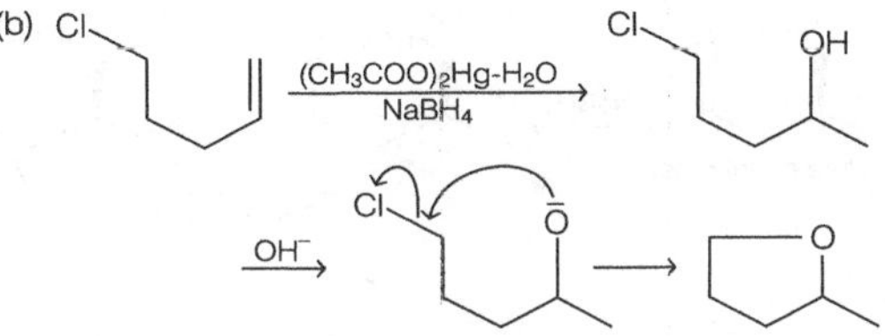

(c)

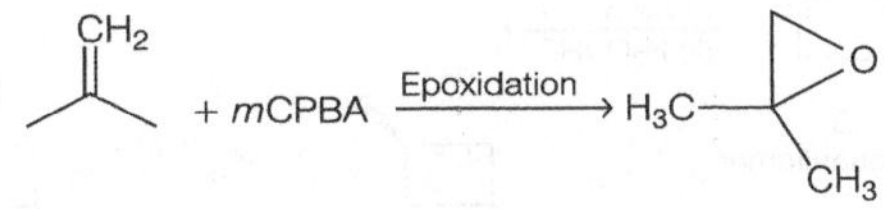

(d) No S_N2 reaction on C_6H_5Br.

18. (a,b,c)

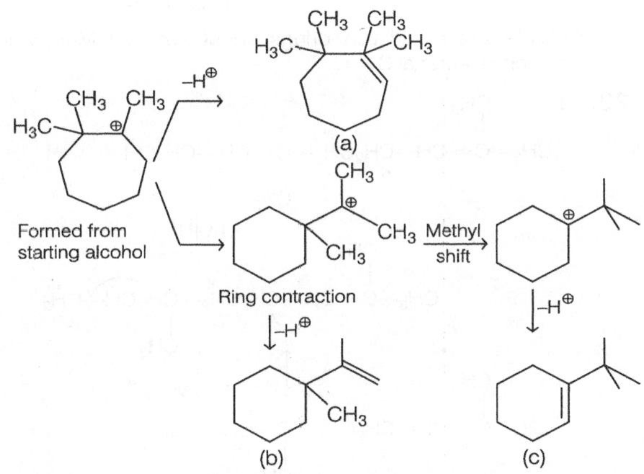

19. (c,d)

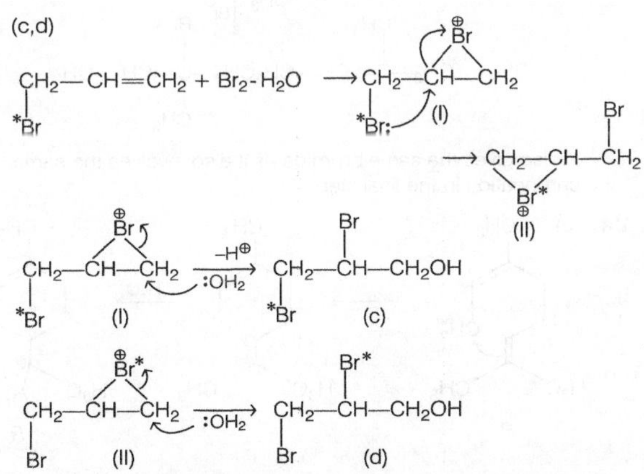

Nucleophilic attack of H_2O occur at α-C of cyclic bromonium ion where positive charge is more easily accommodated.

Explanations for Q. Nos. (20-22)

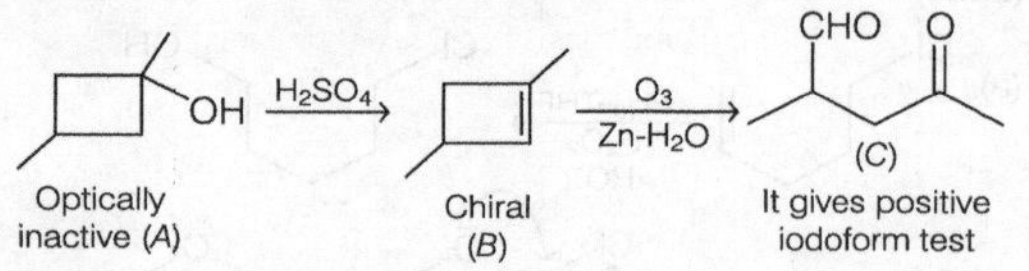

20. (d)

21. (c)

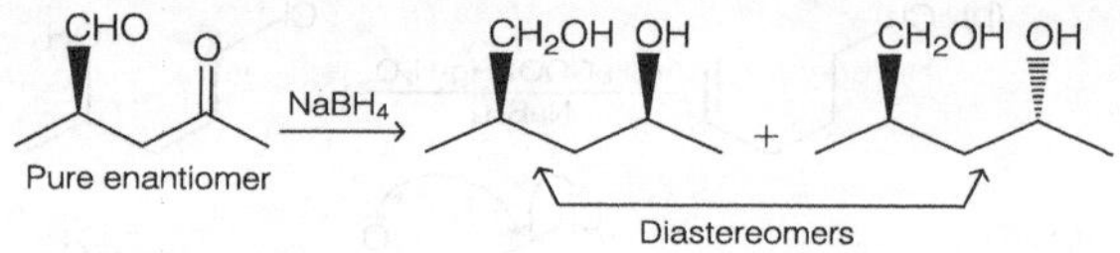

22. (c)

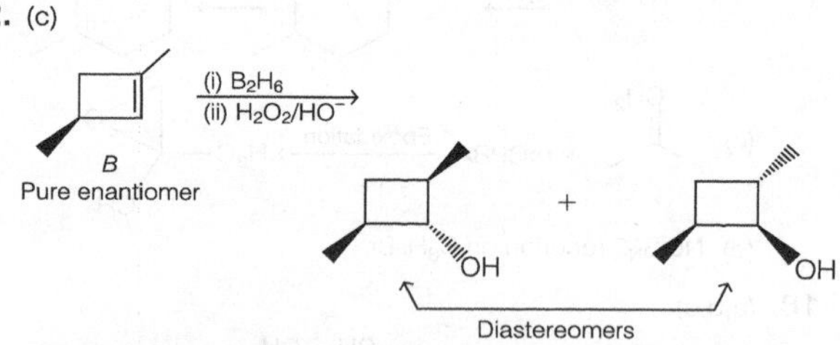

Hydroboration oxidation brings about *syn*, anti-Markownikoff's addition of water at C=C.

23. (d)

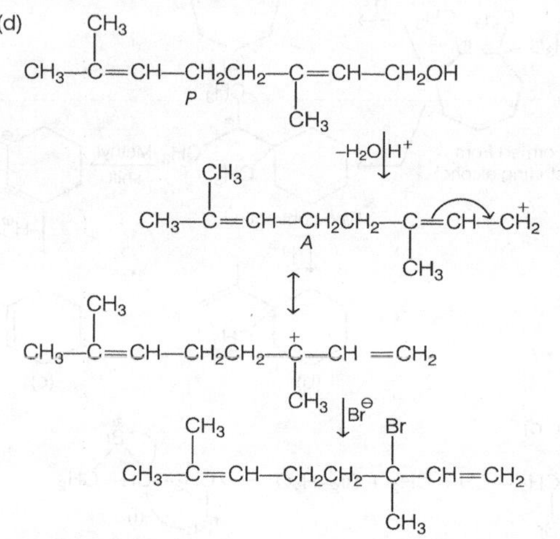

X also gives the same bromide as it also involves the same carbocation in the final step.

24. (a)

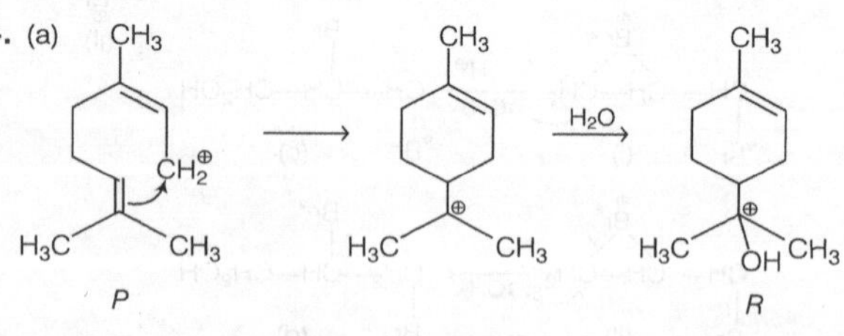

25. (9)

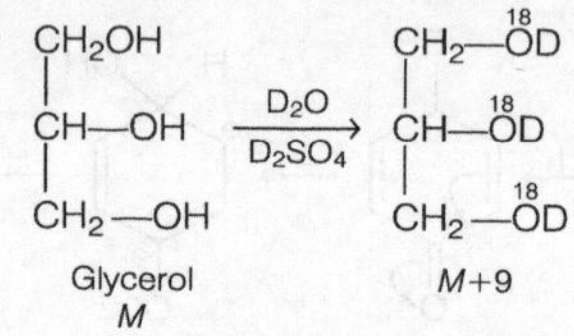

via protonation, all —OH will become $-O^{18}D$, giving an increase in 3 units of mass per —OH group.

26. (7)

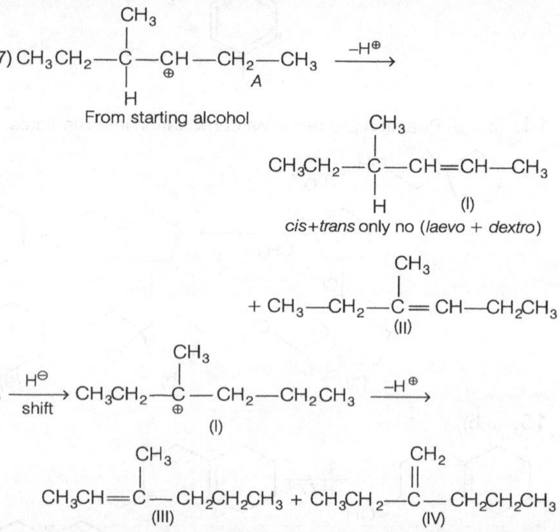

Since, only a pure enantiomer *A* is taken, only *cis* and *trans*-isomers of (I) would be produced. (II) will also be formed in *cis-trans* pair. (III) will also be formed as a pair of *cis-trans* isomer and (IV) will be a single isomer.

27. (3)

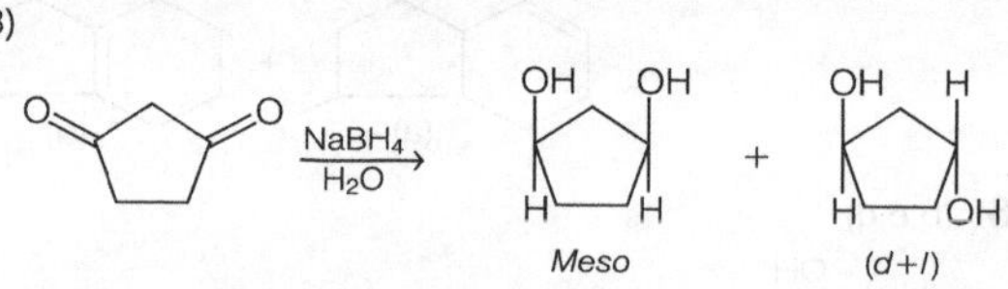

28. (4) The carbonyl $C_5H_{10}O$ ($C_nH_{2n}O$ = 86) has following isomers.

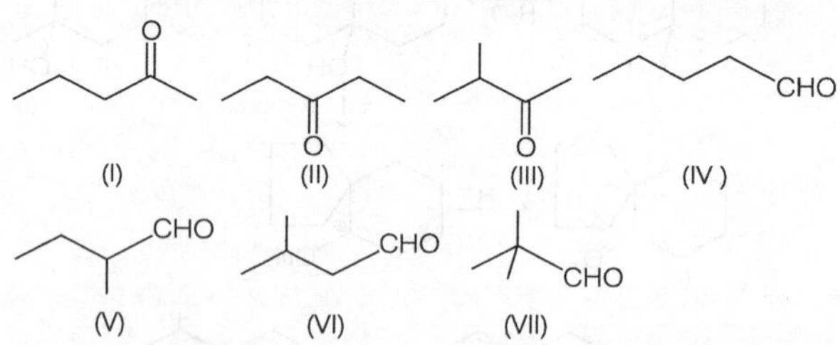

Compounds (III), (IV), (VI) and (VII) gives racemic mixture with CH_3MgBr. (I) and (II) gives achiral alcohols with CH_3MgBr. (V) is enantiomeric, a pure enantiomer it gives pair of diastereomers with CH_3MgBr.

DPP-4 Tests and Distinctions of Alcohols

1. $\underset{\text{Primary}}{R\text{—}CH_2OH} + \underset{\text{Orange}}{CrO_3\text{-}H_2SO_4} \xrightarrow{\text{Cold}} R\text{—}COOH + \underset{\text{Blue-green}}{Cr^{3+}}$

$\underset{\text{Secondary}}{R\text{—}CH(R)\text{—}OH} + \underset{\text{Orange}}{CrO_3\text{-}H_2SO_4} \xrightarrow{\text{Cold}} R\text{—}\overset{O}{\overset{\|}{C}}\text{—}R + \underset{\text{Blue-green}}{Cr^{3+}}$

$\underset{\text{Tertiary}}{R_3C\text{—}OH} + \underset{\text{Orange}}{CrO_3\text{-}H_2SO_4} \xrightarrow{\text{Cold}} \text{No reaction}$

Turning of orange CrO_3 - H_2SO_4 solution into blue-green solution indicates 1° or 2° alcohols otherwise 3° alcohol. After adding chromic acid solution and removing unreacted chromic acid, if effervescence of CO_2 is observed on adding $NaHCO_3$, $RCOOH$, hence RCH_2OH is confirmed. If no effervescence is observed in the above step, R_2CHOH is confirmed.

2. (i) 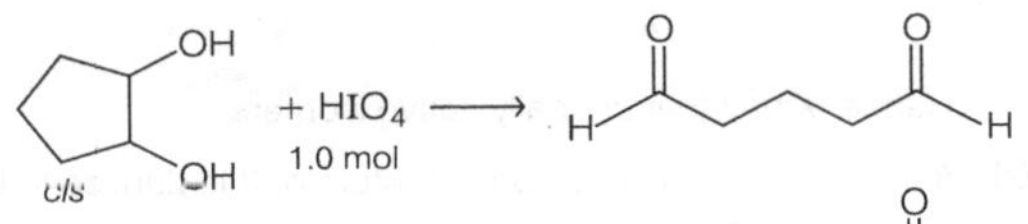

(ii) $CH_3\text{—}CH(OH)\text{—}CHO + \underset{1.0\text{ mol}}{HIO_4} \longrightarrow CH_3\text{—}CHO + H\text{—}\overset{O}{\overset{\|}{C}}\text{—}OH$

(iii) $CH_2OH\text{—}CHOH\text{—}CHO + \underset{2.0\text{ mol}}{HIO_4} \longrightarrow CH_2O + CO_2 + H\text{—}\overset{O}{\overset{\|}{C}}\text{—}OH$

(iv) $CH_2(OH)\text{—}CH(OCH_3)\text{—}CH_2(OH) + HIO_4 \longrightarrow \text{No reaction}$

3. $CH_3\text{—}C(CH_3)_2\text{—}OH + HCl \xrightarrow{ZnCl_2} CH_3\text{—}C(CH_3)_2\text{—}Cl$

Immediate white turbidity

$CH_3\text{—}CH(OH)\text{—}CH_2\text{—}CH_3 + HCl \xrightarrow{ZnCl_2} CH_3\text{—}CH(Cl)\text{—}CH_2CH_3$

Delayed white turbidity

$CH_3\text{—}CH_2\text{—}CH_2\text{—}CH_2OH + HCl \xrightarrow{ZnCl_2}$ No turbidity in cold, turbidity on heating

4.

$R\text{—}CH_2OH \xrightarrow{HI} R\text{—}CH_2I \xrightarrow{AgNO_2} R\text{—}CH_2\text{—}NO_2 \xrightarrow{HNO_2} R\text{—}\underset{\underset{N\text{—}OH}{\|}}{C}\text{—}NO_2 \xrightarrow{OH^-} \text{Red}$

Nitrolic acid (blue)

$R_2CHOH \xrightarrow{HI} R_2CHI \xrightarrow{AgNO_2} R_2CH\text{—}NO_2 \xrightarrow{HNO_2} R\text{—}\underset{\underset{NO}{|}}{\overset{\overset{R}{|}}{C}}\text{—}NO_2 \xrightarrow{OH^-}$ No red Colouration

Pseudonitrol (blue)

3° (tertiary) alcohols do not give positive Victor Meyer test.

5. (i) Heating with sodium metal : 1-butanol gives effervescence but 1-octene does not.

(ii) Treatment with $AgNO_3$ solution : 1-bromo pentane gives brown precipitate of AgBr, 1-propanol does not.

(iii) Treatment with Br_2 -H_2O : 3-butene-1-ol decolourises brown colour of bromine water but 1-butanol does not.

(iv) Treatment with conc. $HCl/ZnCl_2$: 3-pentanol gives white turbidity in cold, 1-pentanol does not.

(v) Treatment with I_2 + NaOH : α-phenyl ethanol gives yellow precipitate of CHI_3, β-phenyl ethanol does not.

6. (d) Chromic acid oxidises 3-pentanol but does not oxidise 3-methyl-3-pentanol (3°). Hence, 3-pentanol would change colour of chromic acid from orange to blue green but 3-methyl-3-pentanol does not.

7. (c) X is a primary, chiral alcohol.

$\underset{\text{Chiral}}{CH_3\text{—}CH_2\text{—}C(CH_3)_2\text{—}CH_2OH} \xrightarrow{HBr} \underset{\text{Achiral}}{CH_3\text{—}CH_2\text{—}C(CH_3)(Br)\text{—}CH_3}$

8. (d) X must be a primary alcohol as indicated by Victor Meyer's test. Both (a) and (b) are primary satisfy the condition.

9. (c) 

Chiral, 3° alcohol — Turbidity

10. (c) H_2CO forms cyclic acetal with ethylene glycol but the change cannot be observed visually. In all other cases, change can be observed visually-effervescence with $NaNH_2$ and NaH while colour change with CrO_3 -H_2SO_4.

11. (b) Although it is secondary alcohol, it forms a very unstable carbocation (Cl is electron withdrawing group), hence it does not give turbidity with Lucas reagent in cold condition.

12. (d) All alcohols react (1°, 2° and 3°) with cerric nitrate and a colour change from yellow to red is observed.

13. (c) 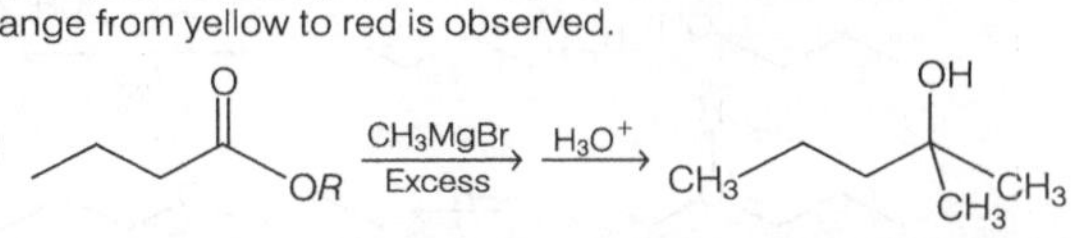

14. (a,b) 2-pentanol forms yellow precipitate of CHI_3 with I_2/NaOH but 1-pentanol does not. 2-pentanol give immediate turbidity with Lucas reagent but 1-pentanol does not gives turbidity with Lucas reagent at room temperature. Both alcohols (1° and 2°) change colour of chromic acid solution from orange to blue green and both give effervescence of $NH_3(g)$ with $NaNH_2$, hence these reagents cannot be used for distinction between 1° and 2° alcohols.

15. (a,c) Both have electron withdrawing groups, destabilises carbocation, do not form turbidity with Lucas reagent at room temperature like primary alcohols. Option (b) and option (d) have electron donating groups, stabilises benzylic carbocation, forms immediate turbidity with Lucas reagent like 2° and 3° alcohols.

16. (a,b,c) Both primary and secondary alcohols form blue solution in Victor Meyer test due to the formation of respectively nitrolic acid and pseudonitrol. Tertiary alcohols form neither, hence no blue colouration is observed.

17. (a,d) $\underset{X\text{(Chiral)}}{CH_2{=}CH\text{—}C(CH_3)(OH)\text{—}CH_2\text{—}CH_3} \xrightarrow{H_2/Ni} \underset{Y\text{(Achiral)}}{CH_3\text{—}CH_2\text{—}C(CH_3)(OH)\text{—}CH_2\text{—}CH_3}$

X neither oxidised by chromic acid nor gives iodoform.

Explanation for Q. Nos. (18–20)

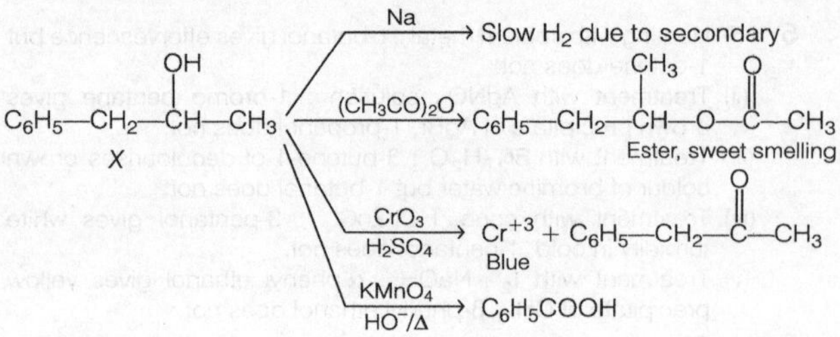

Also X, has CH_3—CH(OH)—, gives iodoform test and it is chiral.

18. (b)

19. (c)

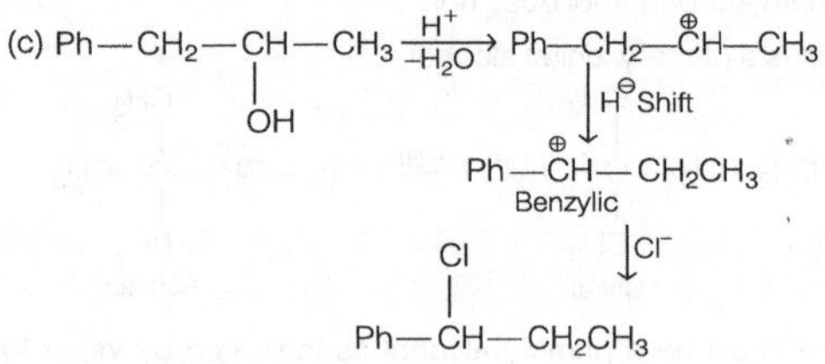

20. (b)

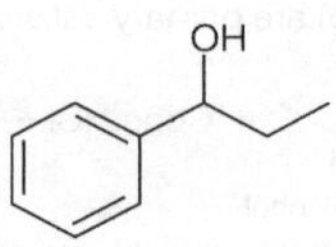

does not give iodoform test but gives all other reaction similar to X.

21. (8) X is a primary alcohol as indicated by Victor Meyer test as well as it has a chiral carbon, hence

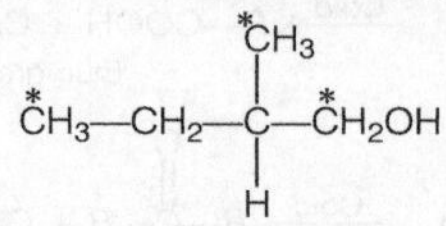

22. (4) Compound X satisfying the given criteria is

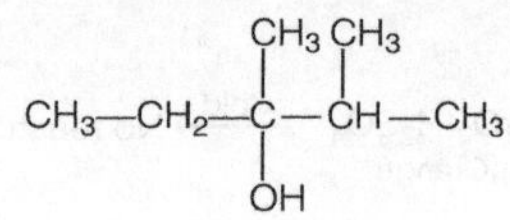

X : 3°, chiral, has a 3° H.

23. (4) X satisfying the given criteria is

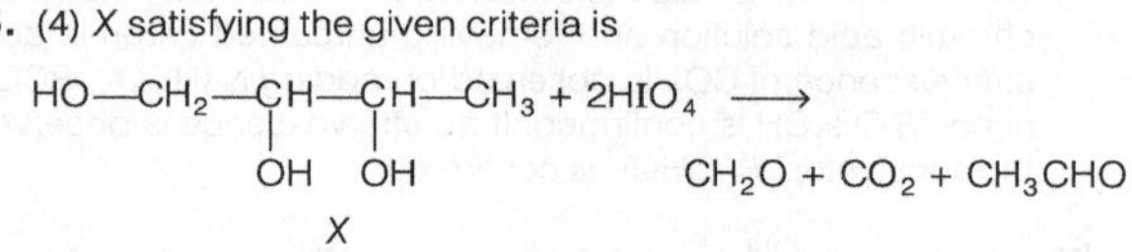

Hence, X has four optically active isomers.

24. (5) From the give condition, structure of X is derived to be

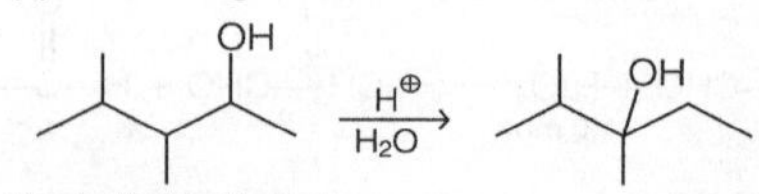

Has both enantiomers and diastereomers

Gives neither iodoform test nor oxidised by chromic acid

DPP-5 Ethers

1. Ether formation proceeds by S_N2 mechanism as :

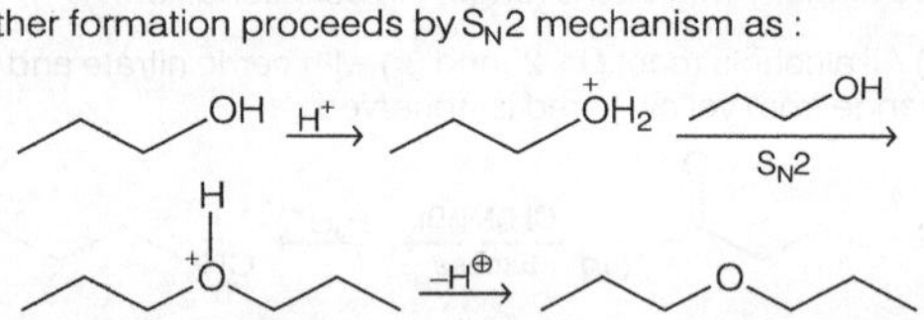

If 2-propanol (2° alcohol) is used, it is less reactive than 1°alcohol in S_N2 reaction. Moreover, 2-propanol is more likely to undergo E1 reaction as it forms 2° carbocation more easily than 1-propanol which forms less stable 1°carbocation.

$$CH_3-\underset{OH}{CH}-CH_3 \xrightarrow[-H_2O]{H^\oplus} CH_3-\overset{+}{C}H-CH_3 \;(2°) \xrightarrow{-H^\oplus} CH_3-CH{=}CH_2$$

2. (i)

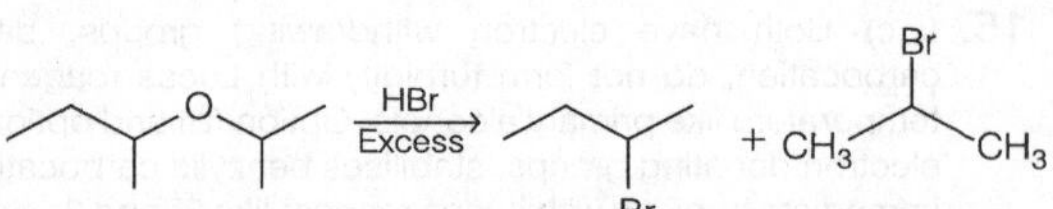

(ii)

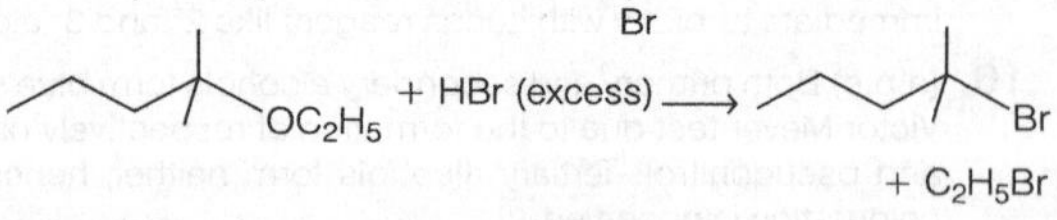

(iii)

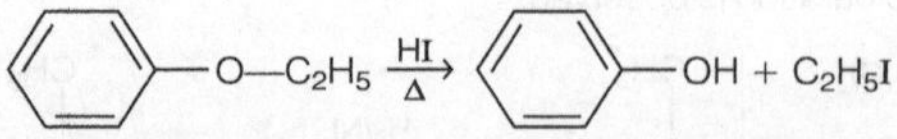

Phenol does not react further with HI to give C_2H_5I.

(iv)

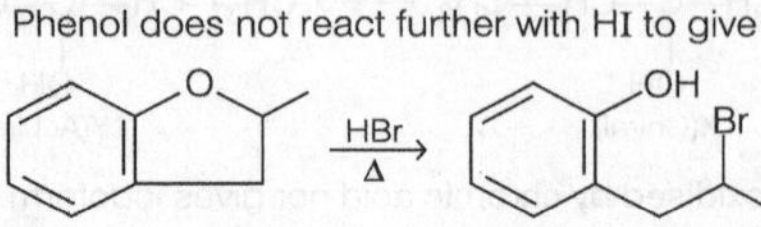

3. It is due to oxidation of ether partially into peroxide as

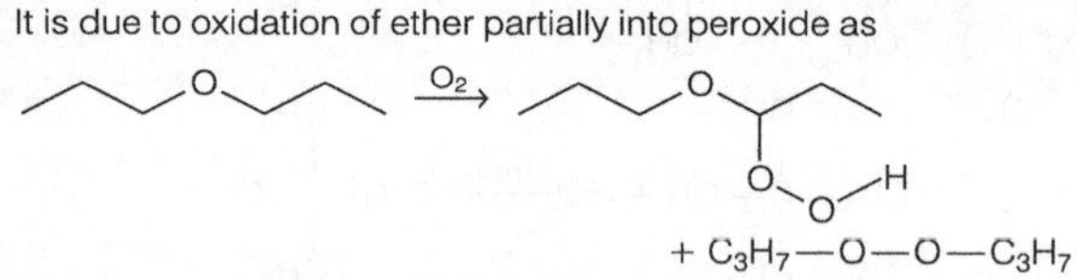

4.

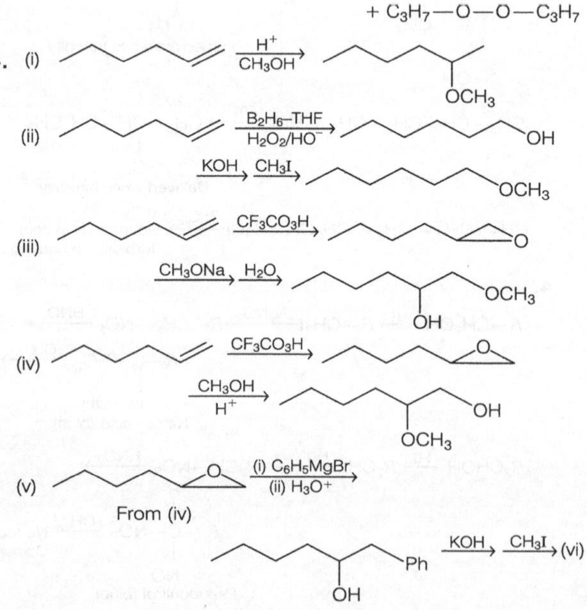

5. (i)

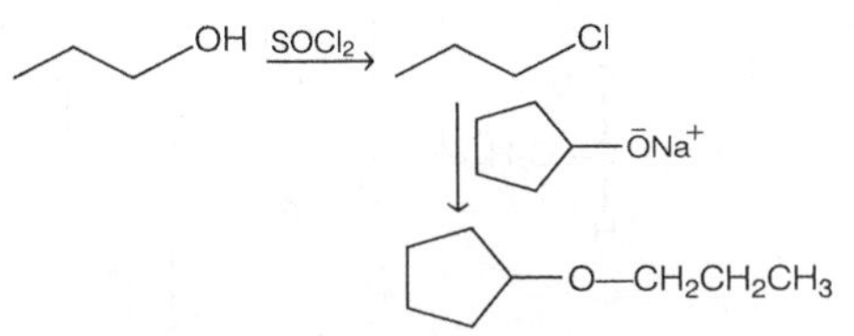

(ii)

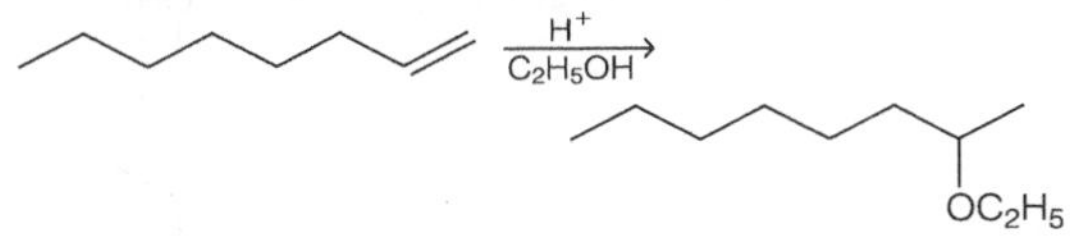

(iii)

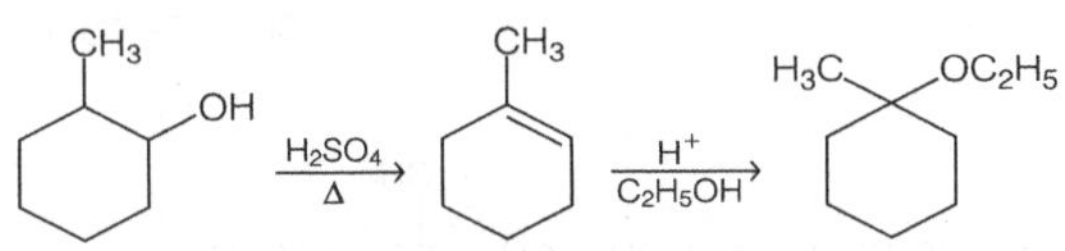

(iv)

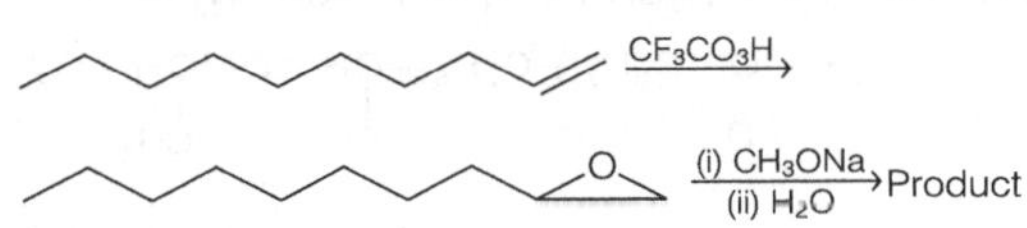

6.

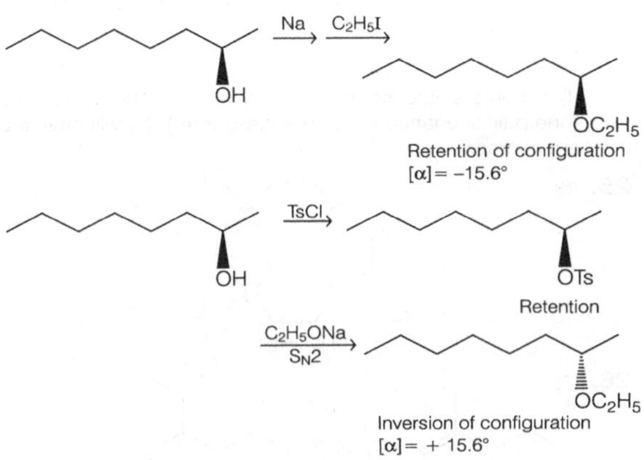

7. (c)

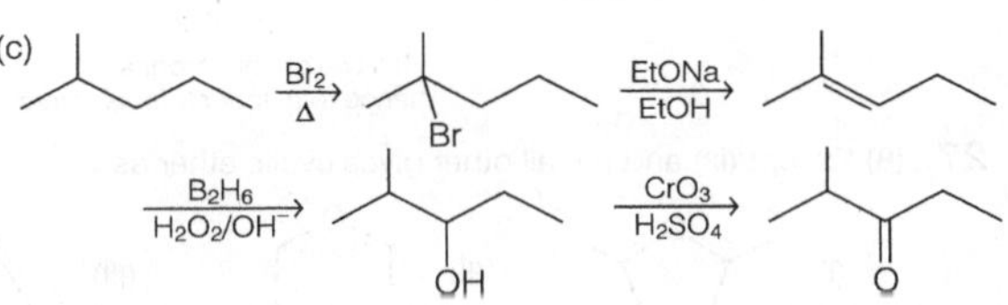

8. (b)

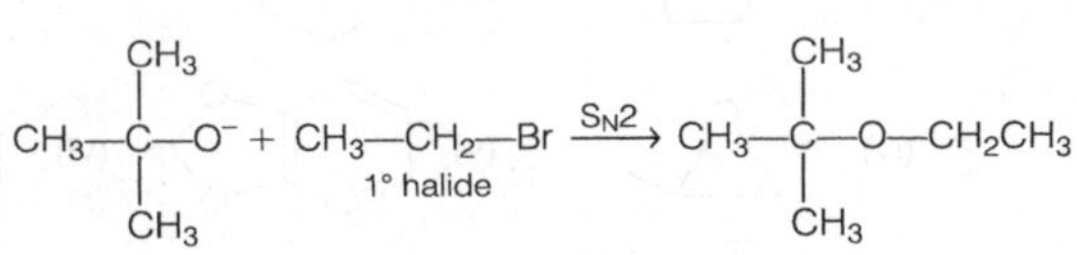

9. (d)

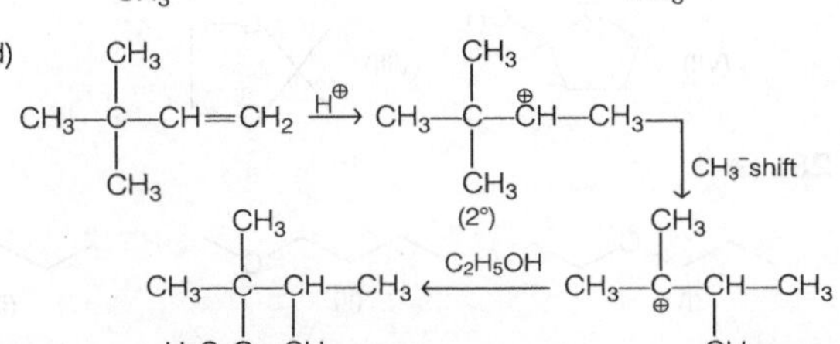

10. (b)

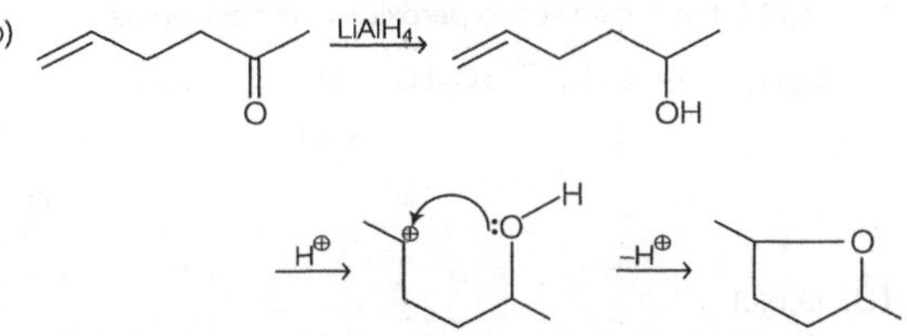

11. (c) Alkoxy mercuration-demercuration brings about Markownikoff's addition of alcohols at C = C without any rearrangement.

12. (d)

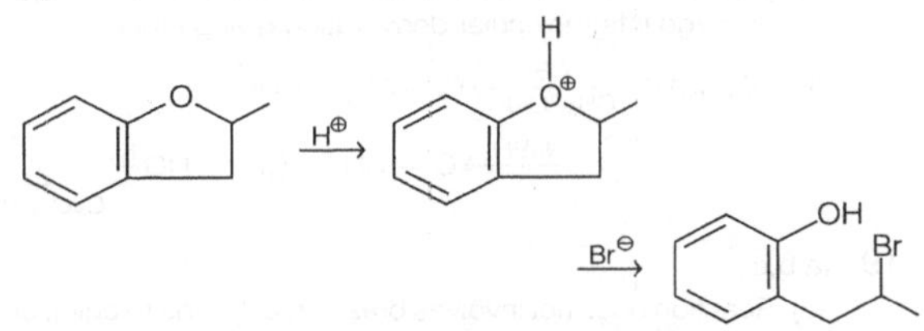

Phenolic —OH does not undergo further substitution.

13. (b) It is a crown ether which traps K^+ ion and makes $KMnO_4$ soluble.

14. (b)

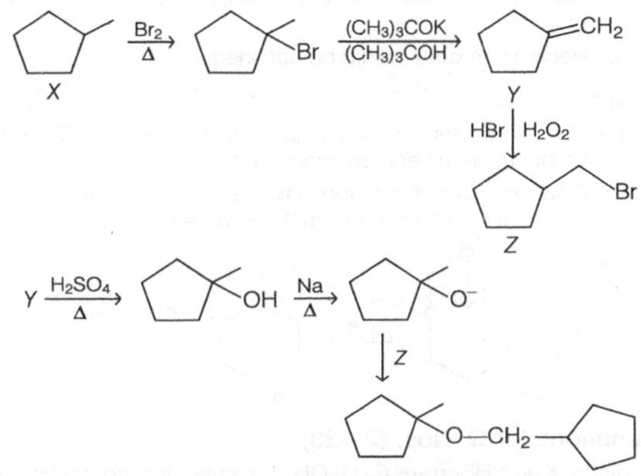

15. (b,c) Reaction proceeds by S_N1 mechanism.

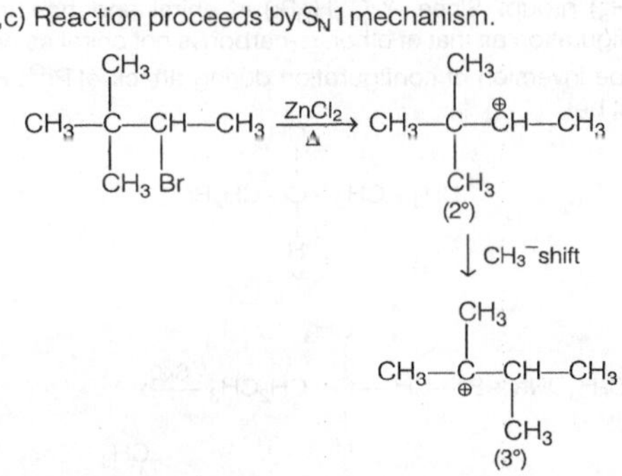

Above carbocations give the desired products.

16. (b,c,d) For preparation of ether by intermolecular dehydration of alcohol, the alcohols must by primary aliphatic because S_N2 reaction has to take place.

17. (c,d) Ether is oxidised to peroxide at α-position as

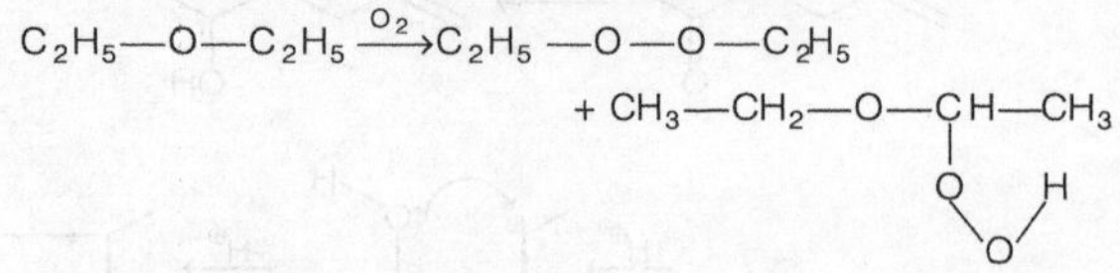

18. (a,b,c)

(a) $(CH_3)_2SO_4$ brings about methylation of alcohol giving ether.

(b) Diazomethane reacts with alcohol to form methyl ether.

$$R-OH + CH_2N_2 \longrightarrow R-O-CH_3 + N_2\uparrow$$

(c) Primary alcohols on heating with conc. H_2SO_4 at 140°C undergo intermolecular dehydration giving ether.

(d)

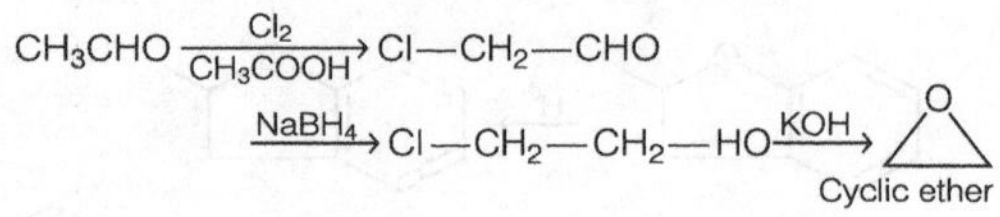

19. (a,b,c)

(a) Reaction does not involves breaking of bonds to chiral carbon, hence retention of configuration.

(b) With TsCl, —OTs is formed with retention of configuration. Subsequent reaction with CH_3ONa involves S_N2 reaction, hence inversion of configuration takes place.

(c) With conc. H_2SO_4, alkene is formed. Alkene in the next step reacts *via* carbocation intermediate, hence racemic product is obtained.

(d) Racemic mixture would be obtained.

20. (a,c)

(a) The first step involves protonation at C=C forming carbocation rather than oxonium ion.

(c) Final product formation involves nucleophilic attack at carbocation, racemic product would be formed.

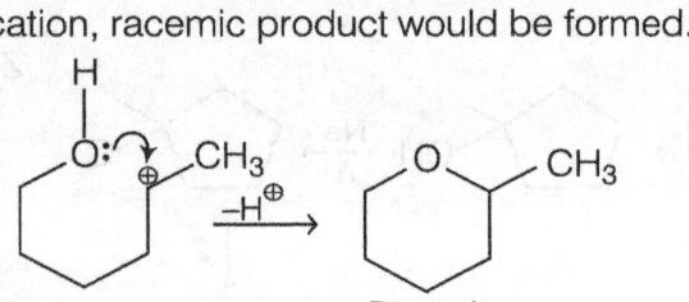

Racemic

Explanations for Q. Nos. (21-23)

Since, *X* with HBr gives C_6H_5OH, the ether must contain a phenyl (C_6H_5) group. Since, $Y\,(C_5H_{11}Br)$ is chiral and has the same configuration as that of ether, α-carbon is not chiral as here there will be inversion of configuration during attack of $Br^{\ominus}$. Hence, *Y* must be

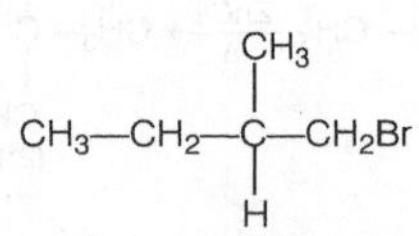

Y

21. (c)

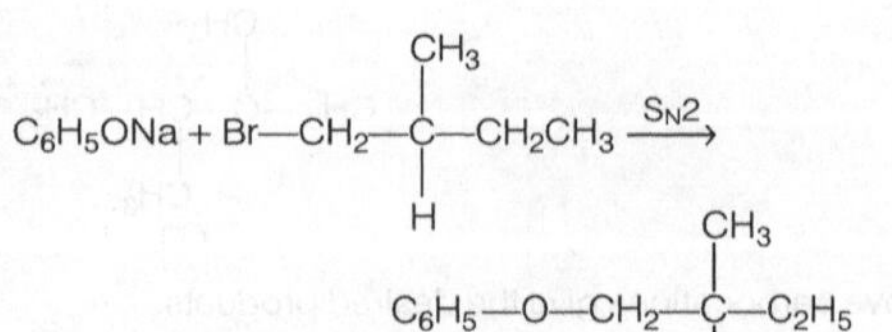

22. (a)

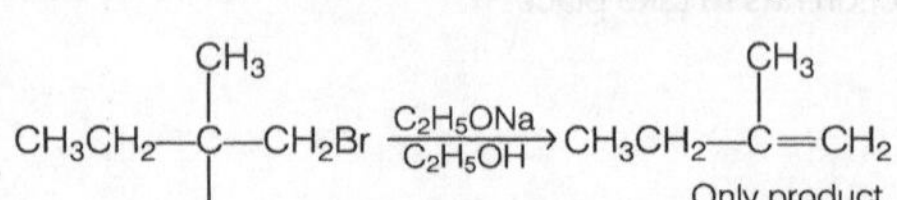

23. (d)

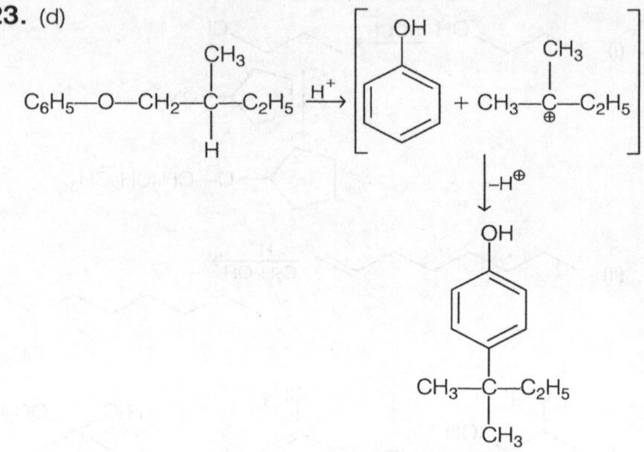

24. (6)

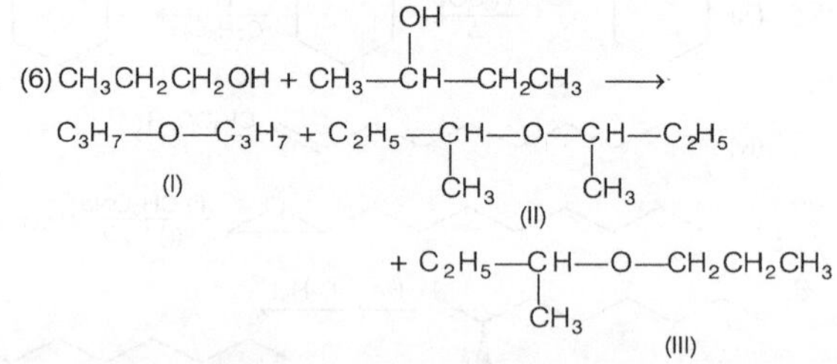

(I) will be a single isomer while (II) will have three stereoisomers (one pair of enantiomer and a *meso* form). (II) will have a pair of enantiomer.

25. (8)

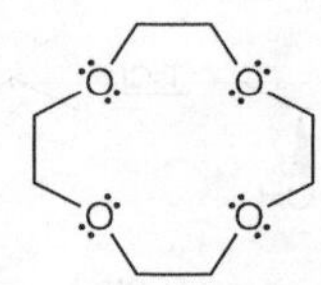

12-crown-4

26. (4)

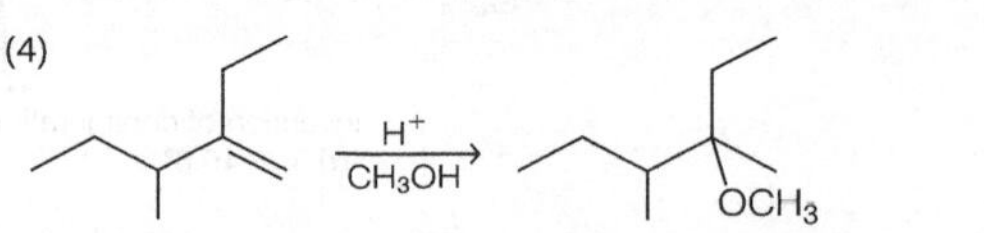

Has two chiral carbons, hence total four stereoisomers

27. (8) Except (ix) and (x), all other gives cyclic ether as

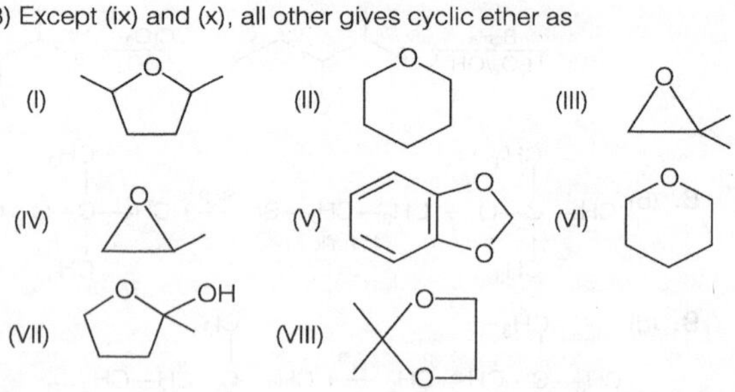

28. (6)

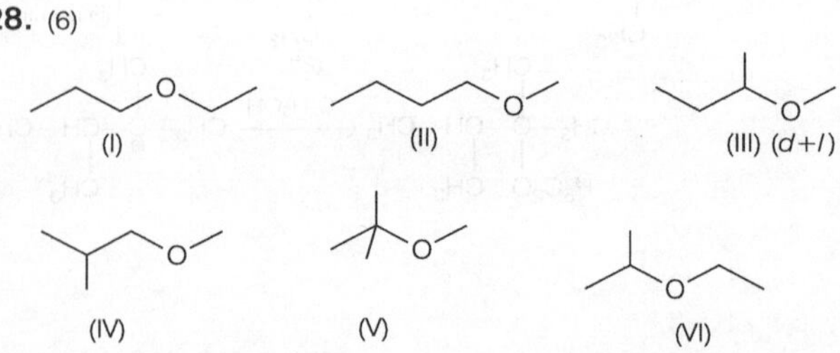

DPP-6 Epoxides

1.

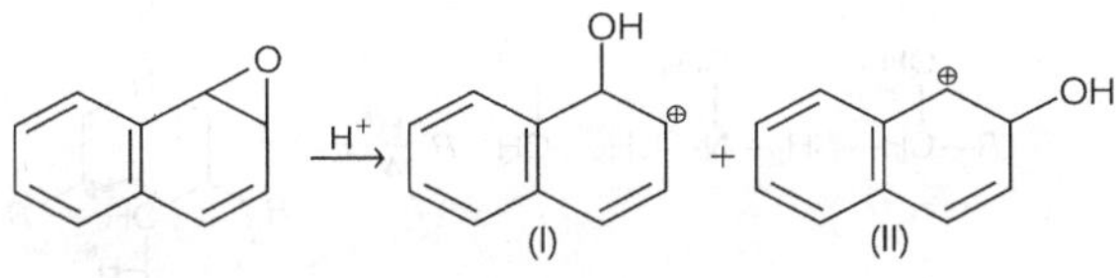

(I) has greater resonance stability than (II).

2. (i)

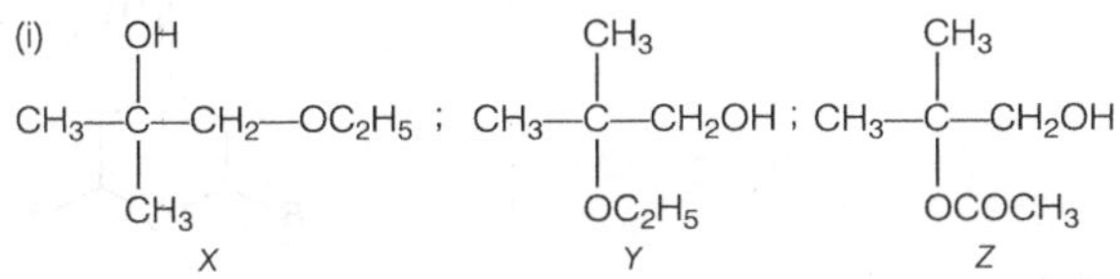

(ii)

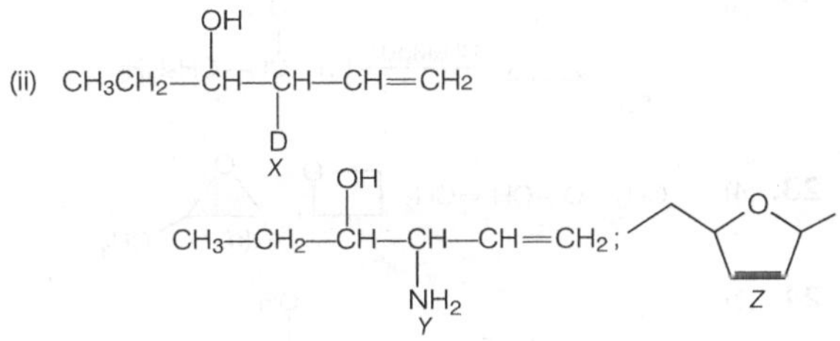

3.

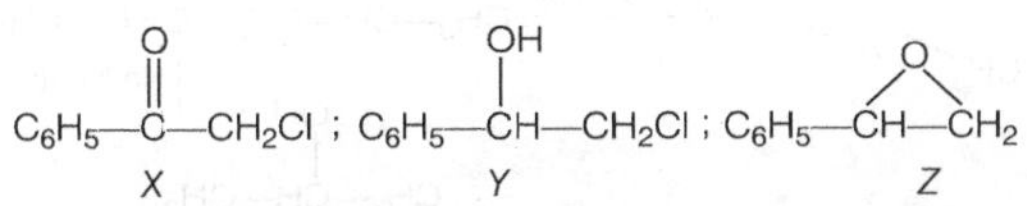

4.

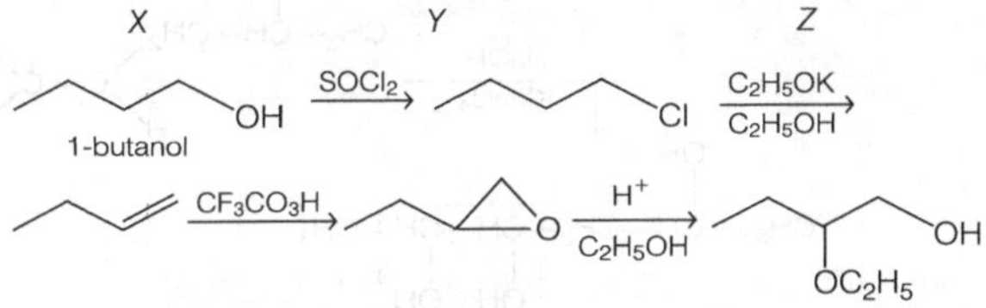

5.

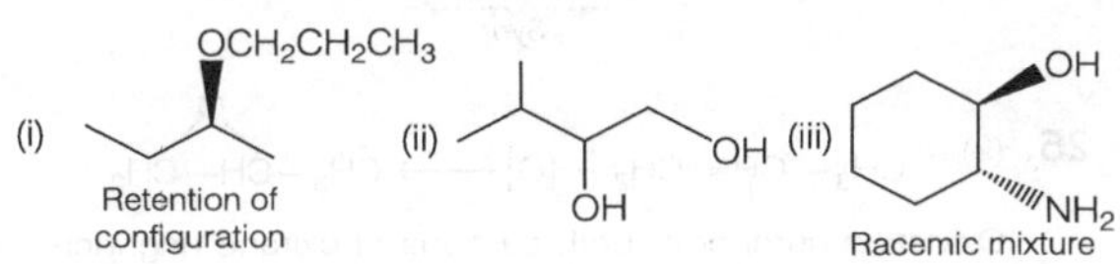

6. (d) Epoxide upon hydrolysis gives *anti*, *vicinal* diols.

7. (c) Epoxide undergo nucleophilic attack of NH_3 at less substituted α-carbon giving *anti* hydroxy amine.

8. (b)

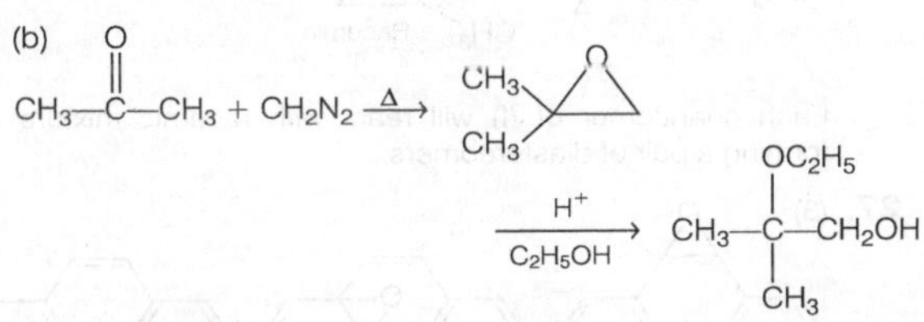

9. (a)

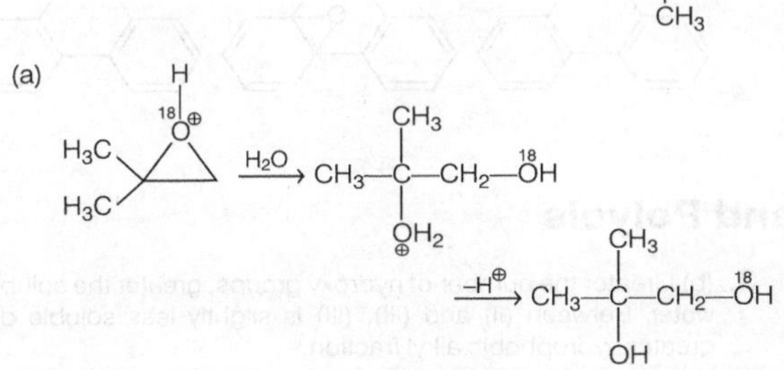

On protonated epoxide, nucleophilic attack occur at α-carbon Where more stable carbocation could be formed.

10. (b)

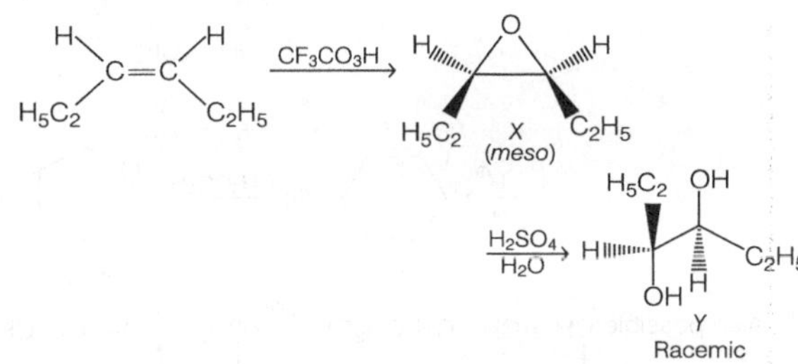

11. (c) Epoxidation is an example of electrophilic addition at olefinic bonds. Therefore, greater the electron density at olefinic bond, greater its reactivity. Electron donating groups increases reactivity while electron withdrawing groups like —CF_3 decreases reactivity.

12. (c) It gives racemic mixture of oxiranes.

13. (b)

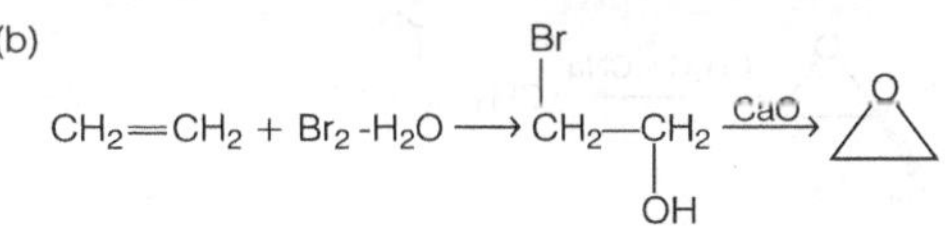

14. (c,d)

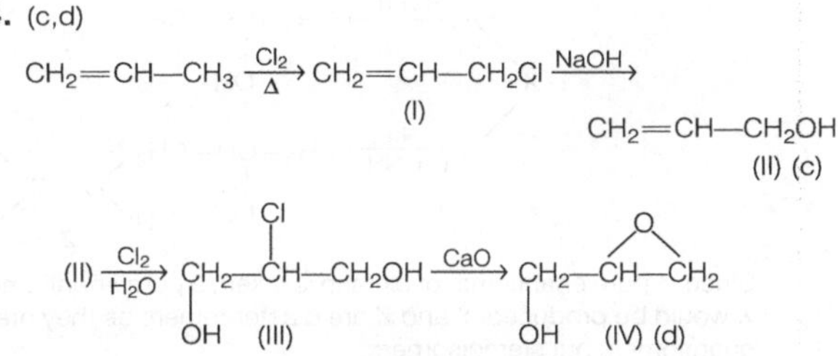

15. (c,d) (c) Epoxide on hydrolysis gives *anti* diol.

(d) Halohydrin is formed in first step which gives epoxide with CaO. Epoxide is then hydrolysed in H_2SO_4 giving *anti* diol.

16. (a,c,d)

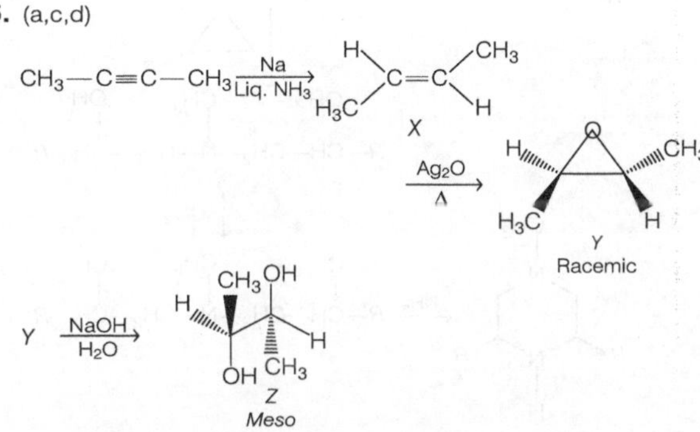

If *trans* X is taken, Y will be *meso* by Z will be racemic.

17. (b,c) The major product would be the only enantiomer formed from a given enantiomer of starting compound.

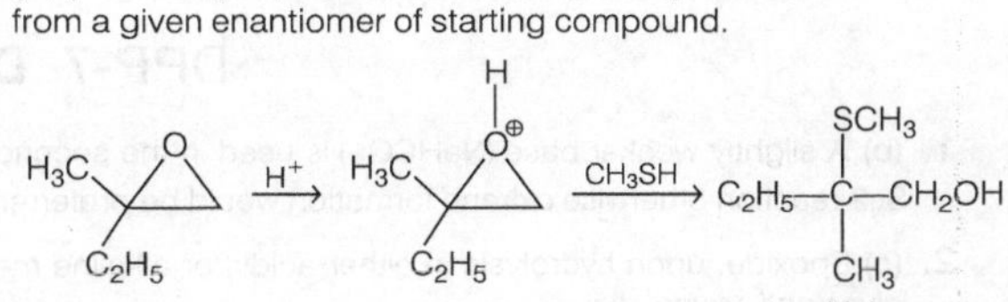

18. (b,c,d)

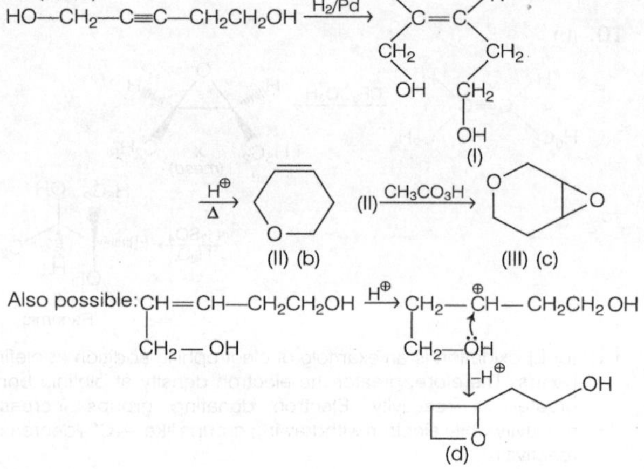

19. (b,d)

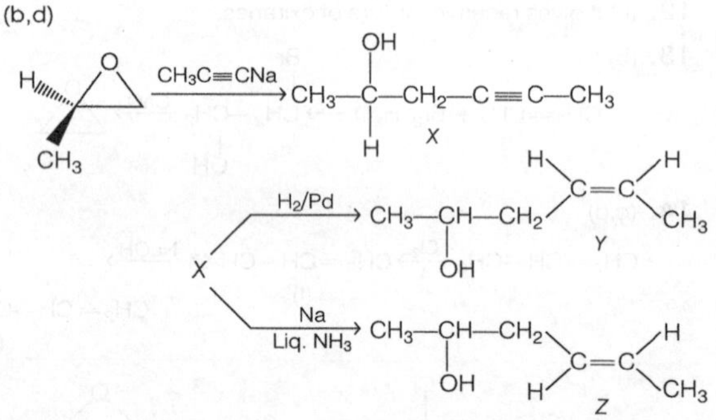

Since, a pure enantiomer of oxirane is taken, a pure enantiomer of *X* would be produced. *Y* and *Z* are diastereomers as they are not enantiomers but stereoisomers.

20. (b)

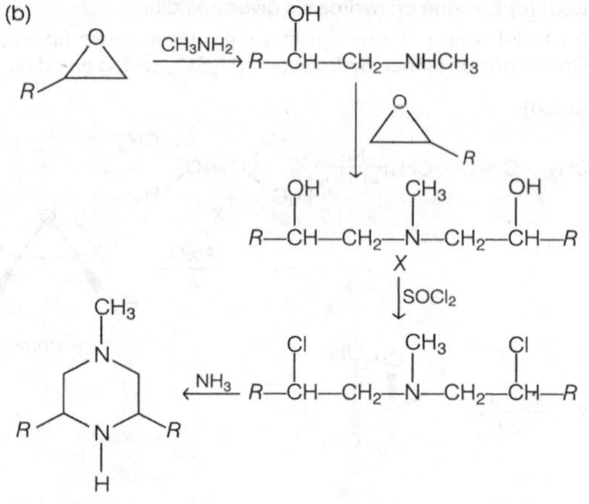

21. (c)

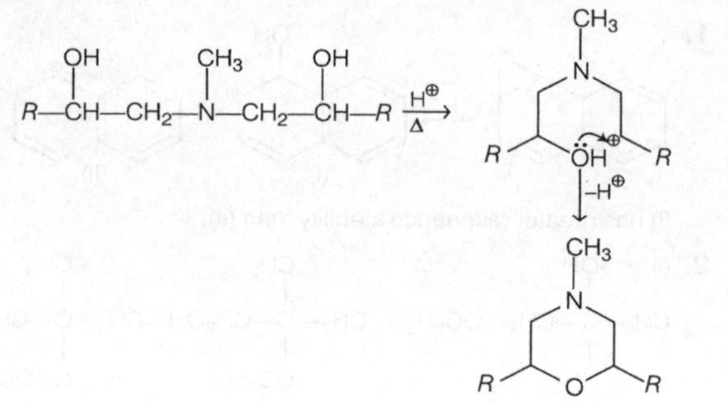

22. (a)

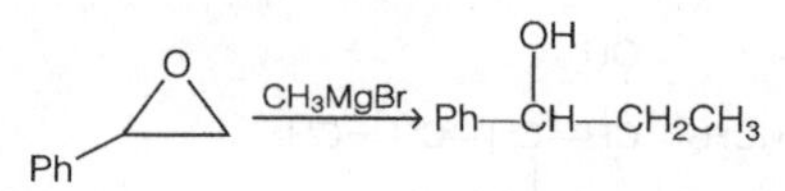

23. (4)

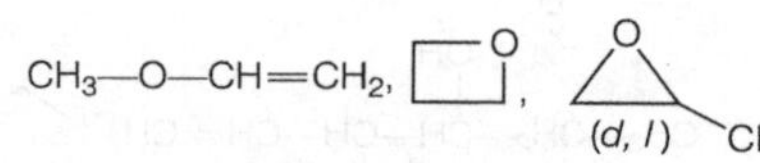

24. (3)

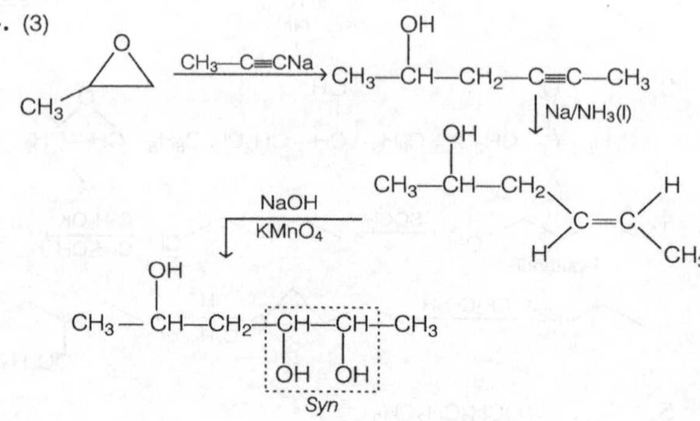

25. (2)

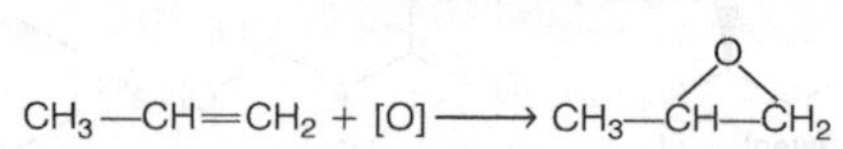

Oxidation number of both carbons of oxirane ring increases by one unit each.

26. (4)

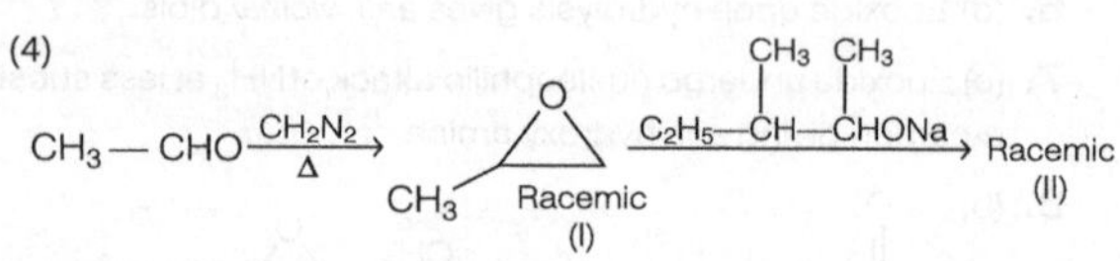

Each enantiomer of (I) will react with racemic mixture of (II) forming a pair of diastereomers.

27. (3)

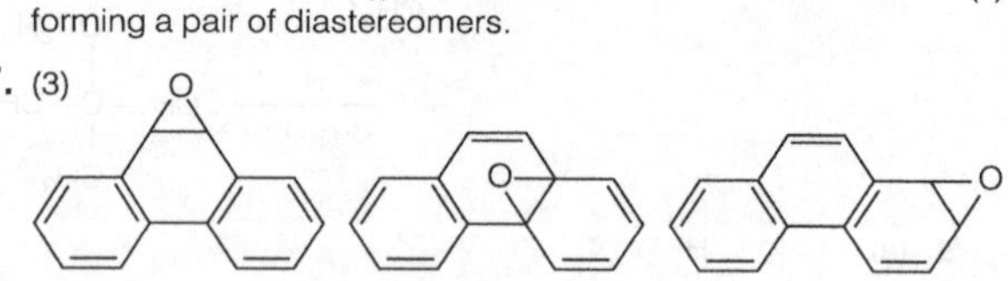

DPP-7 **Diols and Polyols**

1. (b) A slightly weaker base ($NaHCO_3$) is used in the second step S_N2 reaction otherwise oxirane formation would be preferred.

2. (c) Epoxide, upon hydrolysis in either acidic or alkaline medium gives *anti*, *vicinal* diol.

3. (b) Greater the number of hydroxy groups, greater the solubility in water. Between (II) and (III), (III) is slightly less soluble due to greater hydrophobic alkyl fraction.

4. (c) At planar carbonyl carbons, hydride ion attacks from both sides giving both stereomers. Hence, the product diols would be all stereomers of 2, 3-butanediol.

5. (d) PI_3 does not form di iodide with *vicinal* diols.

6. (b)

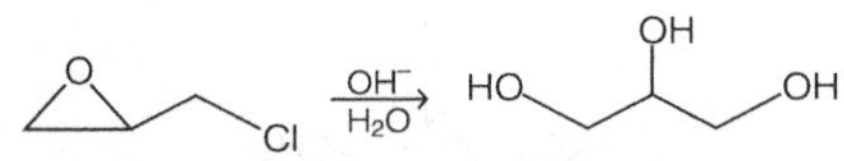

Both oxirane hydrolysis and halide substitution (S_N2) take place.

7. (c)

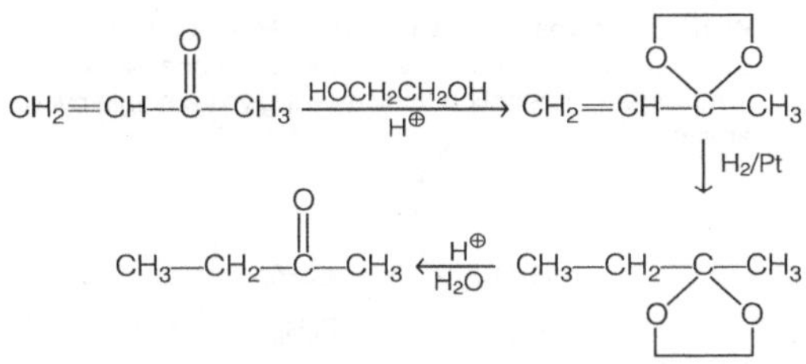

8. (c) Ethylene glycol on heating with concentrated H_2SO_4 gives 1, 4-dioxane.

9. (a,c,d)

(a) Alkaline $KMnO_4$ gives *syn*, *vicinal* diol with alkene.

(b) Here, oxirane is formed as the major product.

(c) Epoxidation followed by hydrolysis gives diol.

(d) Dihalide formed in the first step undergoes S_N2 reaction with $NaHCO_3$ giving diol.

10. (a,b,c)

(a) *Vicinal* diols, on heating with concentrated H_2SO_4 gives dioxane as the major product.

(b)

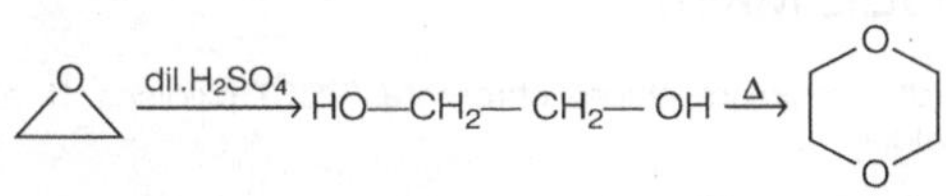

(c)

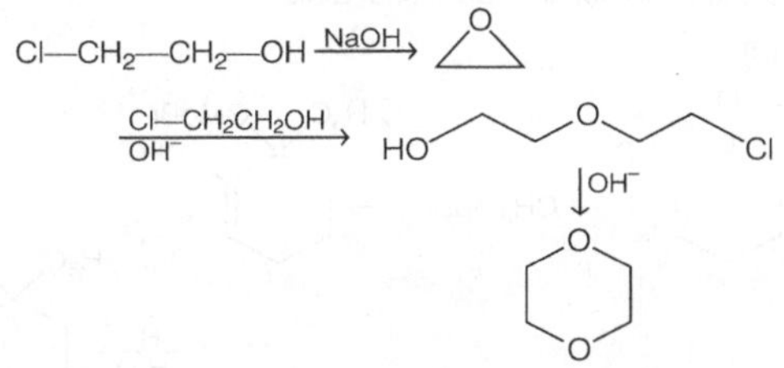

(d) Diols do not give dioxane in alkaline medium.

11. (a,d) *Syn vicinal* diols undergo oxidative cleavage with HIO_4 giving diols. Option (b) and option (c) are *anti* diol and have restricted rotation.

12. (a,b) It is a Pinacol-pinacolone rearrangement.

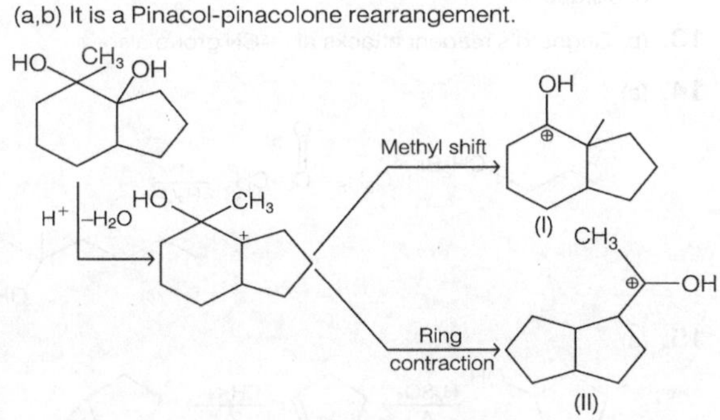

(I) on deprotonation gives (a) while (II) on deprotonation gives (b).

13. (a,b,d)

(a)

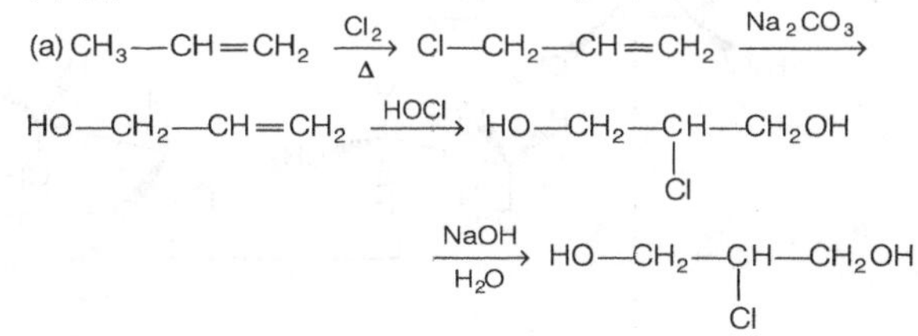

(b)

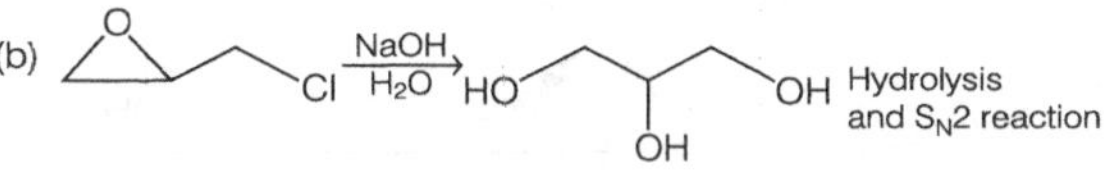

(d)

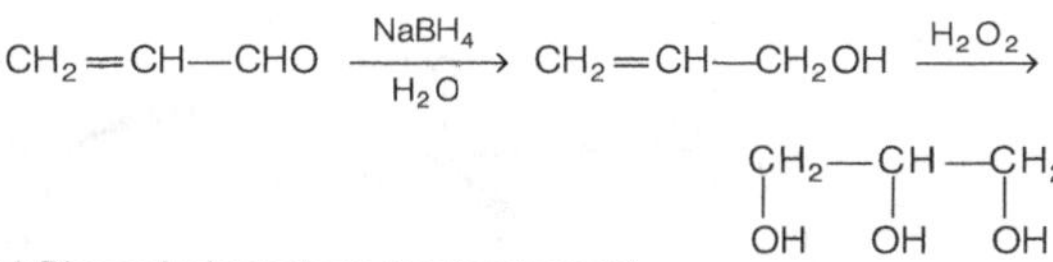

14. (a,c) Pinacol-pinacolone rearrangement.

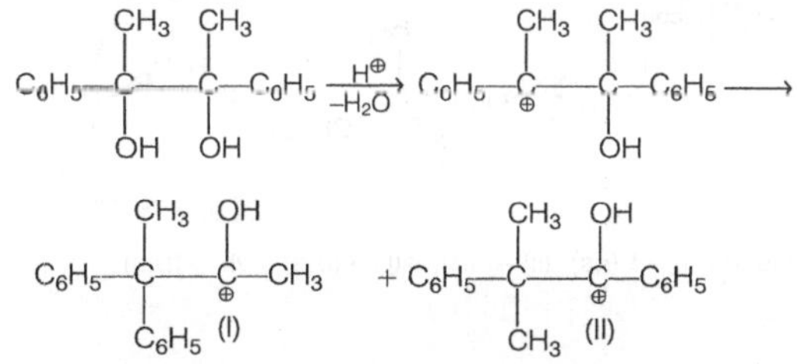

Deprotonation of (I) gives (a) while deprotonation of (II) gives (c).

Explanations for Q. Nos. (15-17)

Vicinal diols which is not tertiary, prefer to form dioxane while tertiary diols prefer to undergo Pinacol-pinacolone rearrangement.

15. (a) The diol is not tertiary, hence more likely to form dioxane.

16. (a) It is tertiary diol, undergo Pinacol-pinacolone rearrangement.

17. (c)

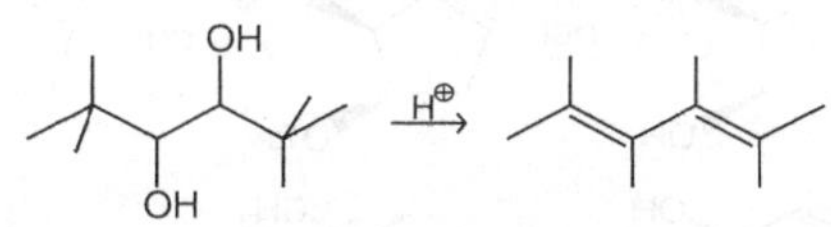

18. (a) (i)

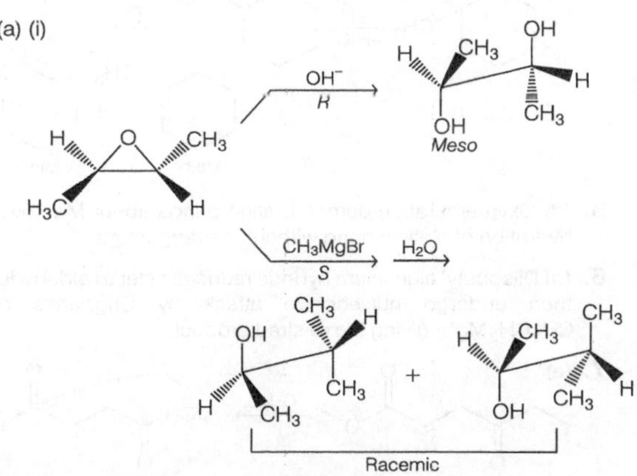

(ii)

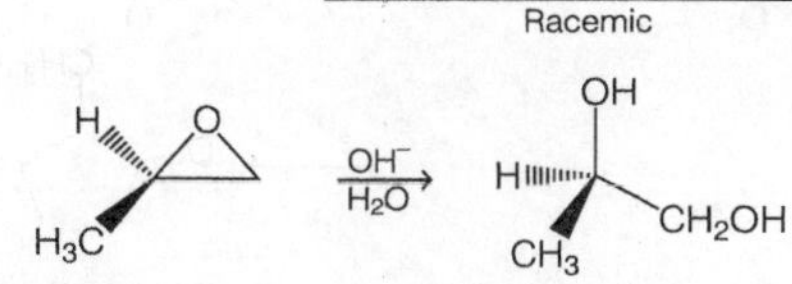

Pure enantiomer Pure enantiomer

(iii)

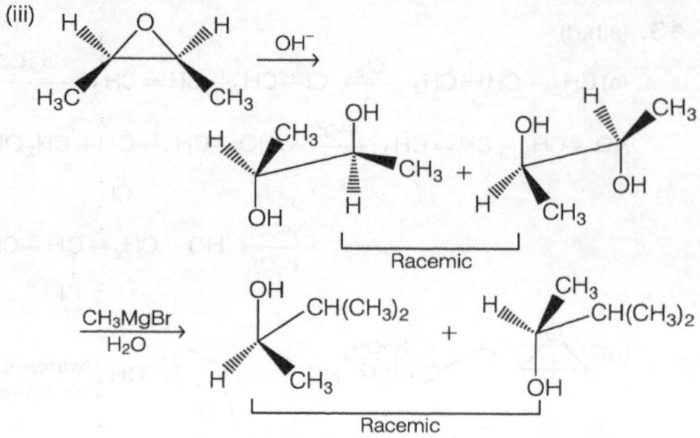

(iv)

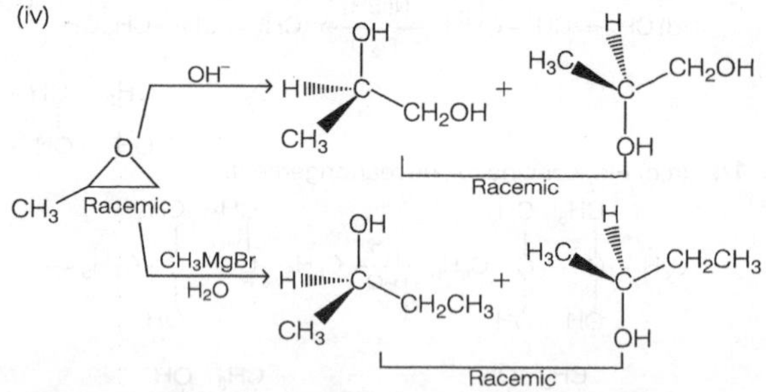

Hence, (i) → (r,s), (ii) → (q), (iii) → (p,s), (iv) → (p,s)

19. (a)

(i) It is a symmetrical, 3° diol, gives $(CH_3)_2CO$ with HIO_4, a single pinacolone $(CH_3)_3C—COCH_3$ and a single cyclic amine :

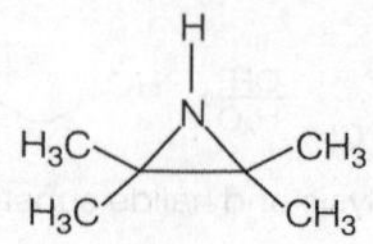

(i) → (p,r,t)

(ii) With HIO_4 gives $(CH_3)_2CO$ and $(C_6H_5)_2CO$. (ii) is unsymmetrical, 3° diol, gives more than one pinacolone in rearrangement reaction. However it gives a single cyclic amine.

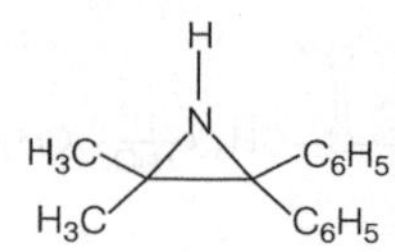

(ii) → (q,s,t)

(iii) (iii) with HIO_4 gives a single ketone-acetophenone. It gives two different pinacolone in rearrangement reaction.

(iii) → (p,s)

(iv) It is unsymmetrical, 3° diol, gives two different ketones with HIO_4 and more than one pinacolone in rearrangement reaction.

(iv) → (q,s)

Revisal Problems (JEE Main)

1. (b) Nucleophilic attack of I^- occur at $—CH_2$ of protonated ether giving phenol and benzyl iodide.

2. (d) Nucleophilic attack of CH_3OH on protonated oxirane occur at more substituted α-carbon.

3. (a)

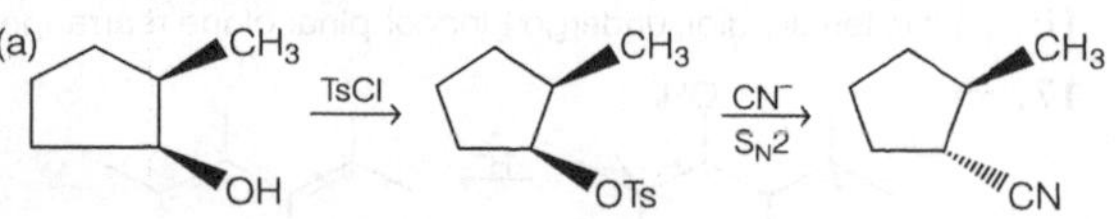

4. (d)

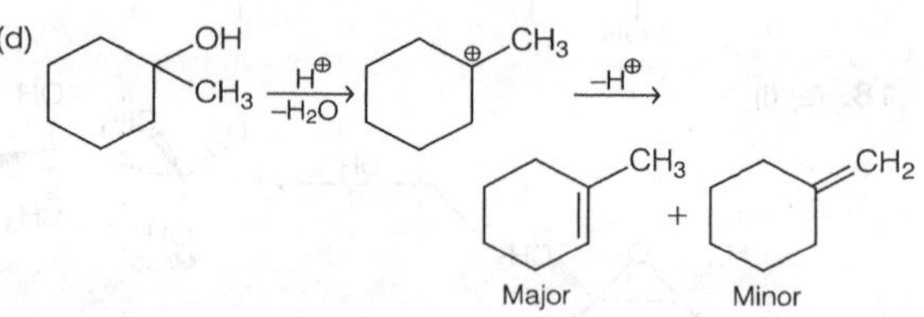

5. (d) Oxymercuration-demercuration brings about Markownikoff's hydration of olefinic bond without rearrangement.

6. (d) Diisobutyl aluminium hydride reduces ester to aldehyde which then undergo nucleophilic attack by Grignard's reagent CH_3CH_2MgBr giving the desired product.

7. (a)

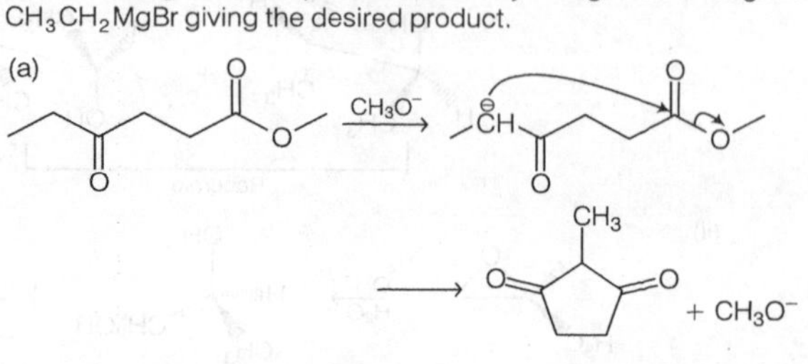

8. (d) Pyridinium chloro chromate (PCC) oxidises 1° alcohols to aldehyde.

9. (d) (II) and (III) are non-*vicinal* diols.

10. (d)

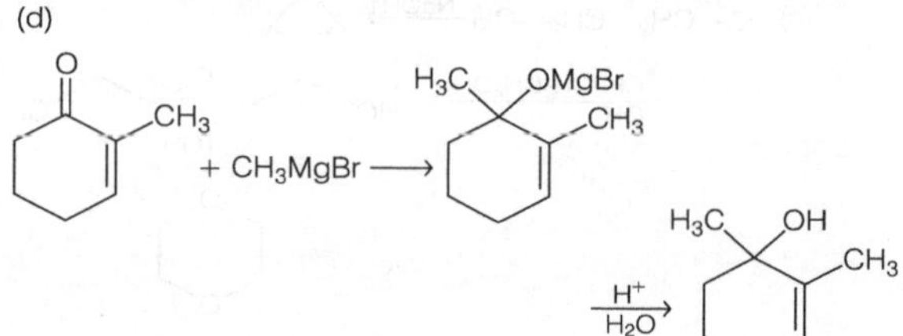

11. (d) Oxo anion is not formed in acidic medium.

12. (a) Nucleophilic attack of $C_6H_5^-$ occur at less hindered α-carbon of oxirane.

13. (b) Grignard's reagent attacks at —CN group also.

14. (a)

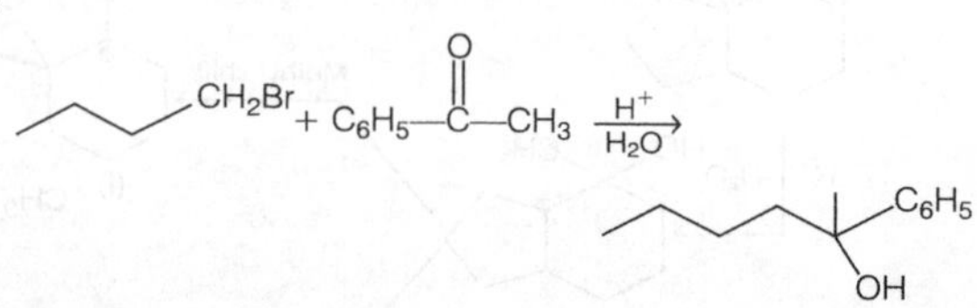

15. (a)

16. (a)

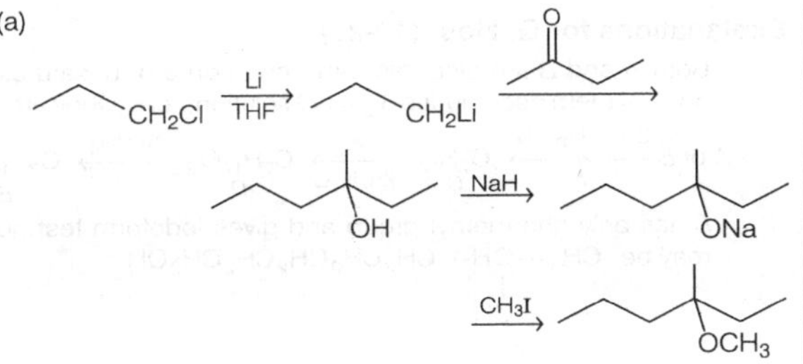

17. (b)

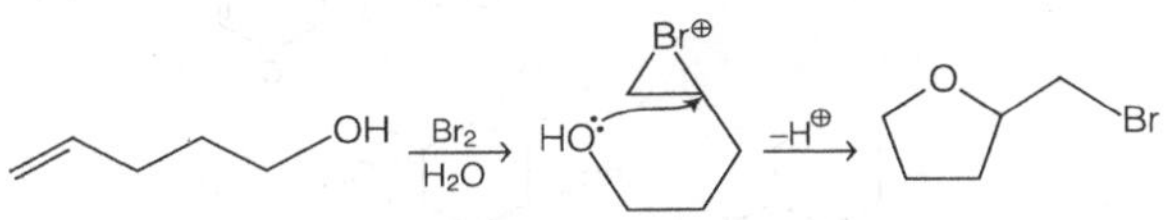

18. (b) *Vicinal* diols undergo oxidative cleavage with HIO_4.

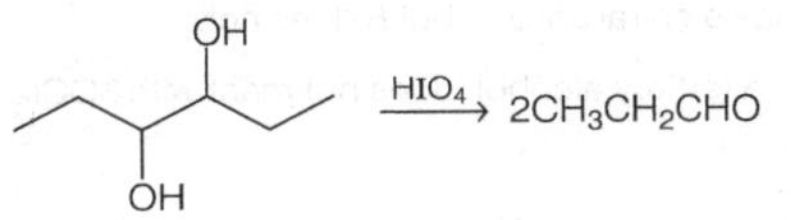

19. (a)

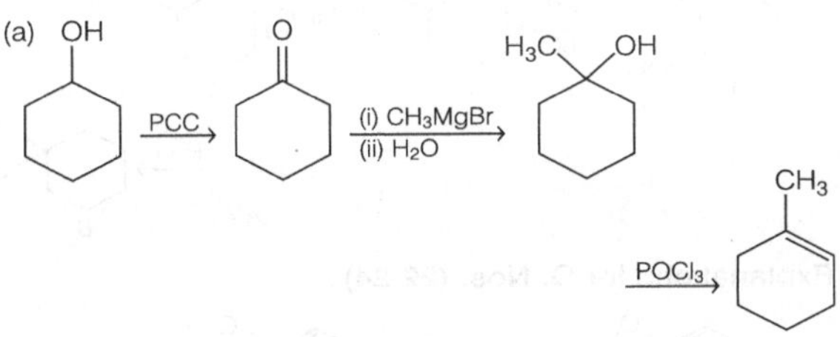

20. (c) E2 elimination occur from *anti*-position. Also the better leaving group (tosylate) is eliminated.

21. (a)

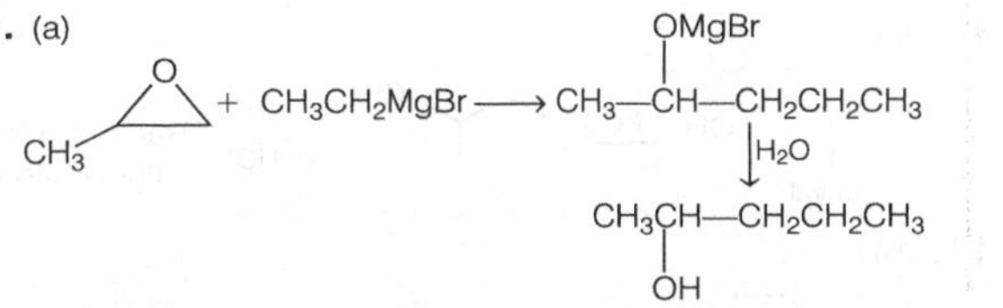

22. (c) Alkene does not react with alkali.

23. (a)

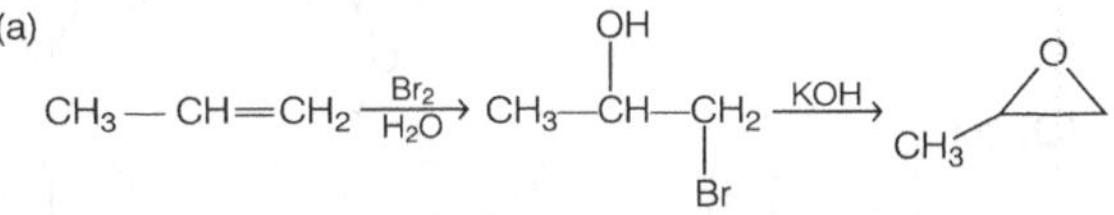

24. (d) $CH_3CH_2Cl \xrightarrow{AgNO_3} AgCl\downarrow + CH_3\overset{\oplus}{C}H_2 \xrightarrow{H_2O} CH_3CH_2OH.$

25. (c) S_N2 reaction is preferred at 1° alkyl halide.

26. (d)

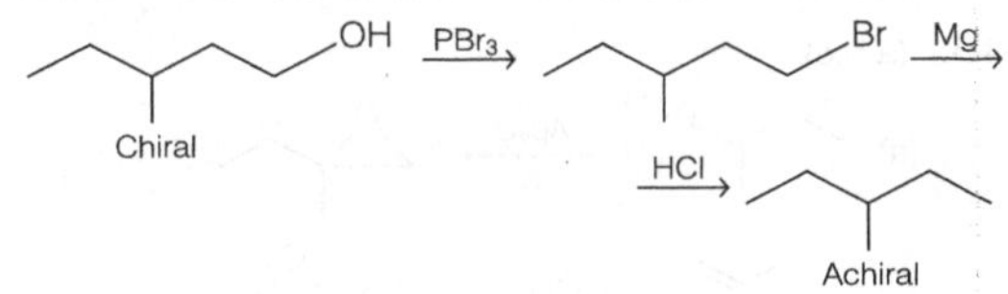

27. (c) Isomerisation occur *via* carbocation intermediate product stability is not the driving force here.

28. (a) Resonance of oxygen with phenyl rings on both side increases bond order and hence, bond energy of O—C bonds, not hydrolysed.

29. (a) Acid catalysed dehydration of alcohols proceed *via* carbocation intermediates. Hence, greater the stability of carbocation, greater the reactivity of corresponding alcohols.

30. (d) S_N2 reaction would occur more easily with primary alkyl halide chloro ethane rather than with secondary 2-chloropropane.

Revisal Problems (JEE Advanced)

1. (c) The major product is a tertiary alcohol, gives immediate turbidity with Lucas reagent.

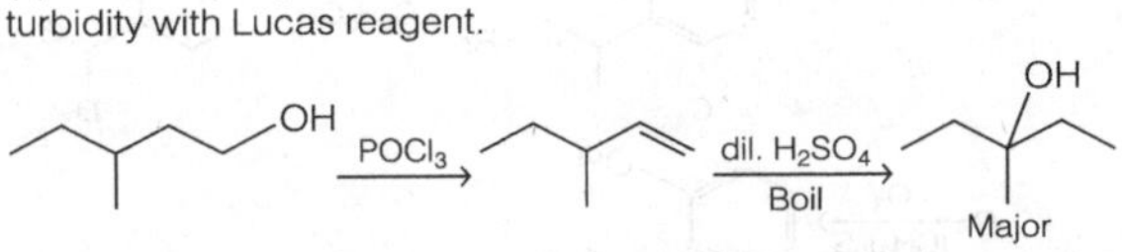

2. (a)

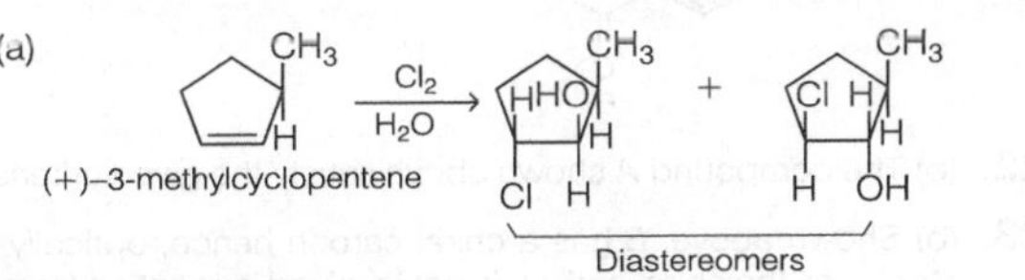

The two anti halohydrins above are diastereomers.

3. (c) A primary alkyl halide and alkoxide would give the indicated product by S_N2 reaction.

4. (b)

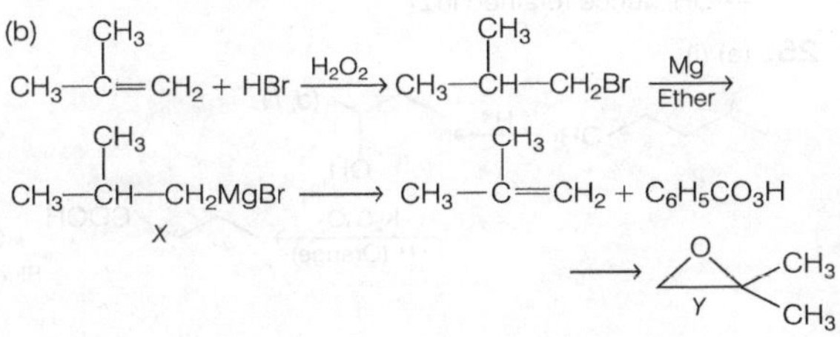

Reaction of *X* and *Y* gives the desired product.

5. (b) It is a Pinacol-pinacolone rearrangement.

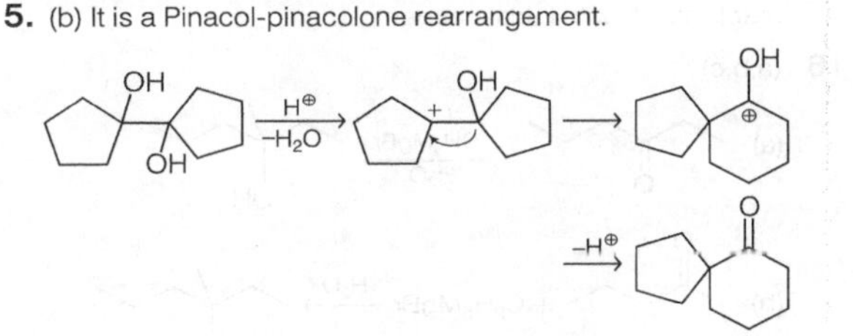

6. (a)

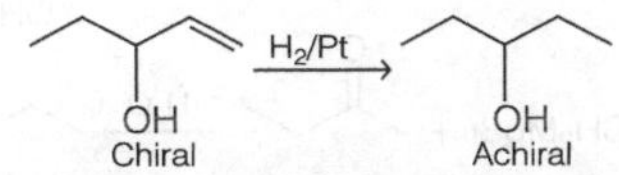

Also, allyl alcohol is oxidised by MnO_2 to corresponding carbonyls.

7. (c) White turbidity of alkyl halide is formed rather than gaseous alkene is formed.

8. (b)

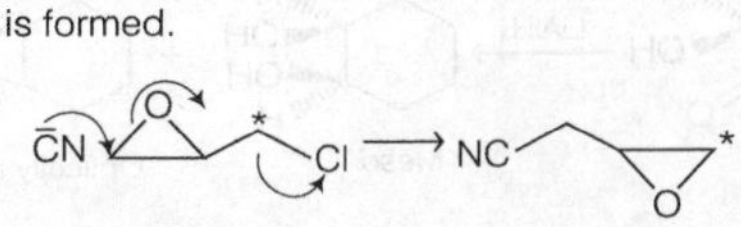

9. (c)

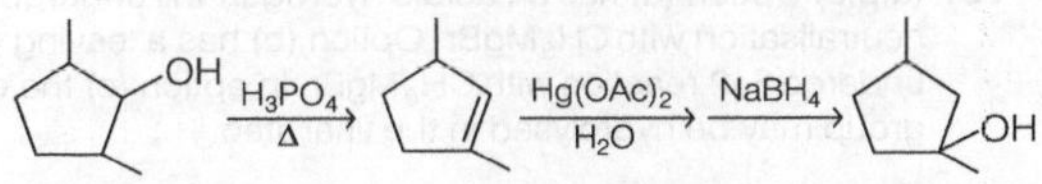

10. (c)

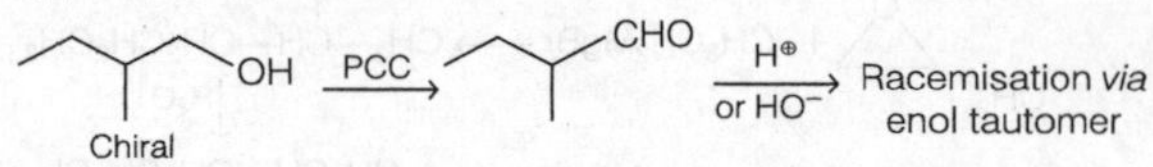

11. (b)

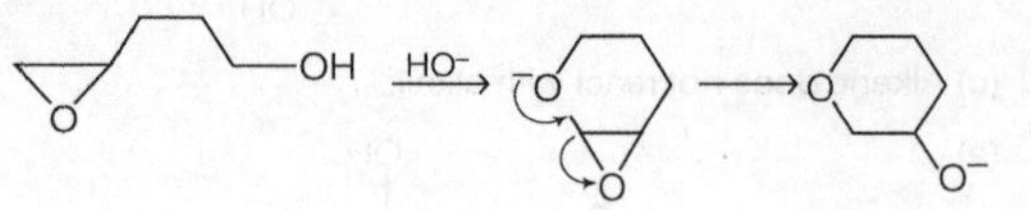

12. (b)

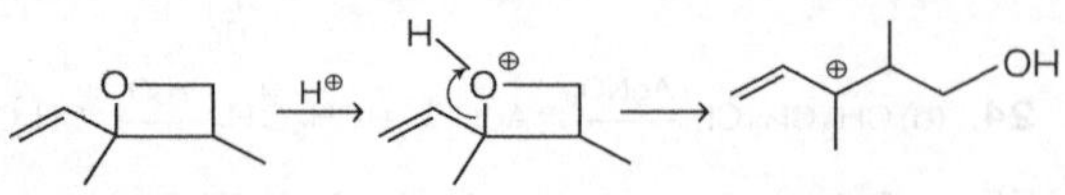

The above carbocation can also be formed by protonation of "b".

13. (a,b,c)

(a) 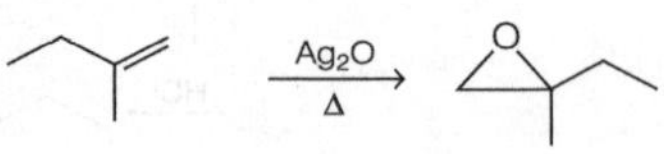

(b) 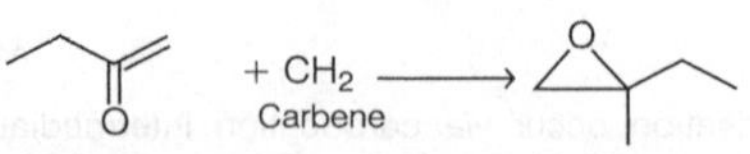

(c) 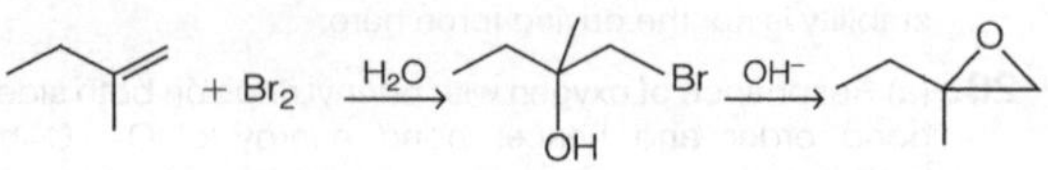

(d) 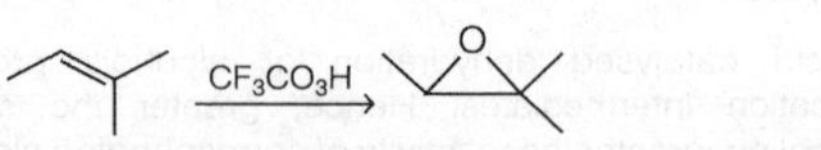

14. (a,c,d) All of these on reduction produces alcohols while amide gives amines on reduction.

15. (a,b,d) The first step is protonation of alcohol base. Weaker acid (greater pK_a) is more substituted (higher degree) alcohols, hence reacts faster in Lucas test. Tertiary alcohols are not oxidised by dichromate but reacts very rapidly with Lucas reagent. Alcohols reacting faster with Na are stronger acid hence, less substituted, reacts slowly with Lucas reagent.

16. (a,b,c)

(a) 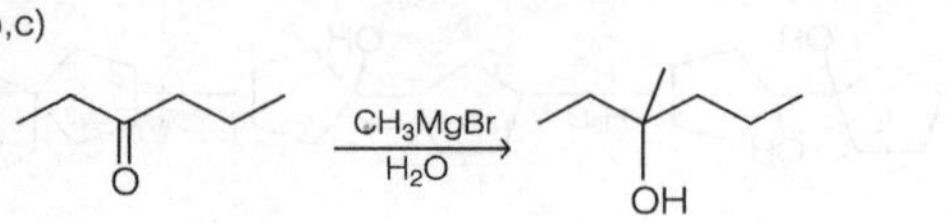

(b) 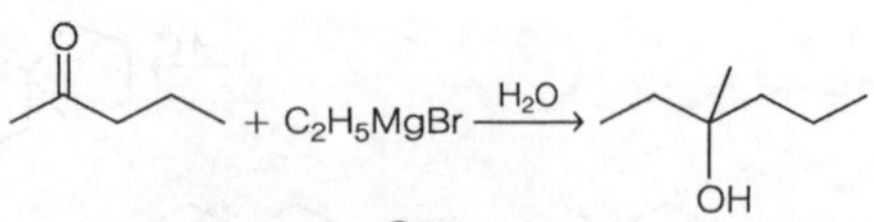

(c) 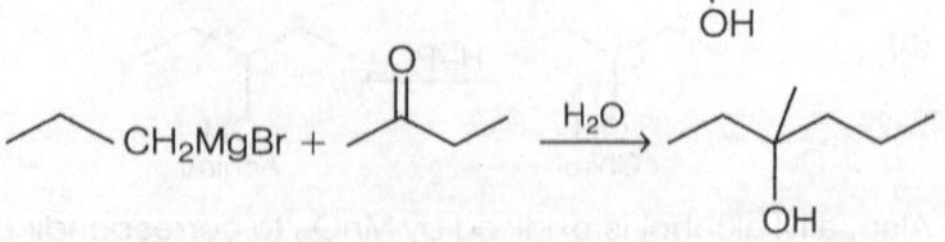

17. (b,c,d) Reduction with $LiAlH_4$ gives a pair of diastereomers.

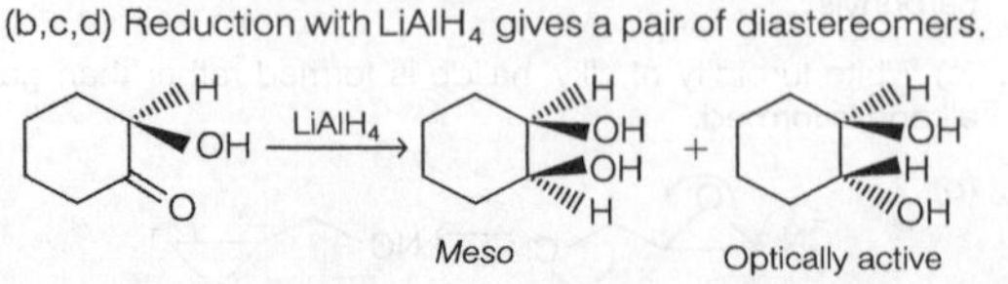

18. (a,b,c) Option (a) has an acidic hydrogen will undergo neutralisation with CH_3MgBr. Option (b) has a leaving group will undergo S_N2 reaction with CH_3MgBr. In option (c) the ether group may be hydrolysed in the final step.

Explanations for Q. Nos. (19-21)

Both *A* and *B* are alcohols with one degree of unsaturation but none of them decolourise Br_2 - H_2O solution, hence they are cyclic.

$$A \text{ or } B \xrightarrow[\Delta]{\text{Conc·}H_2SO_4} \underset{C}{C_7H_{12}} \xrightarrow[Zn/H_2O]{O_3} \underset{D}{C_7H_{12}O_2} \xrightarrow{NaBH_4} \underset{E}{C_7H_{16}O_2}$$

E has only one methyl group and gives iodoform test hence, *E* may be : $CH_3—\underset{\displaystyle OH}{\underset{|}{C}H}—CH_2CH_2CH_2CH_2CH_2OH$

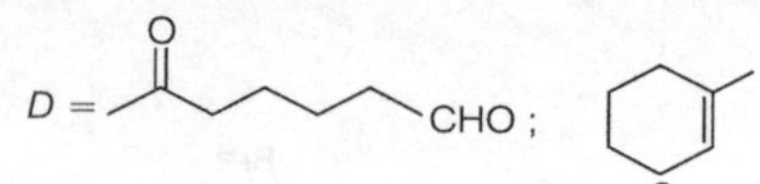

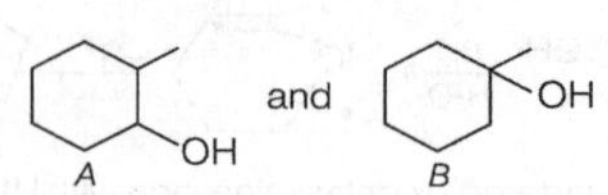

19. (b) *A* has a chiral carbon but *B* does not.

20. (d) *B* is a tertiary alcohol, does not react with $SOCl_2$.

21. (b)

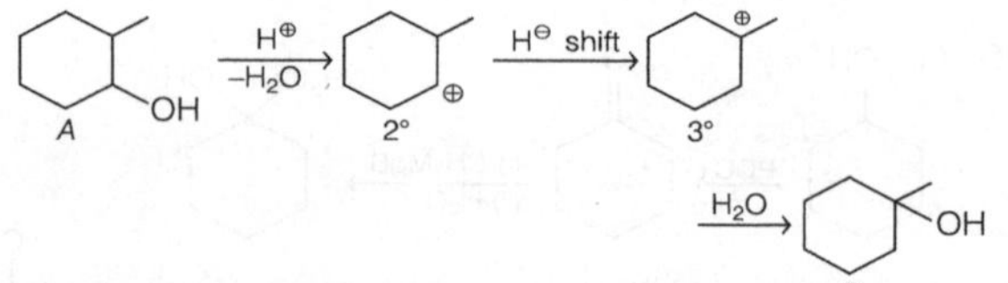

Explanations for Q. Nos. (22-24)

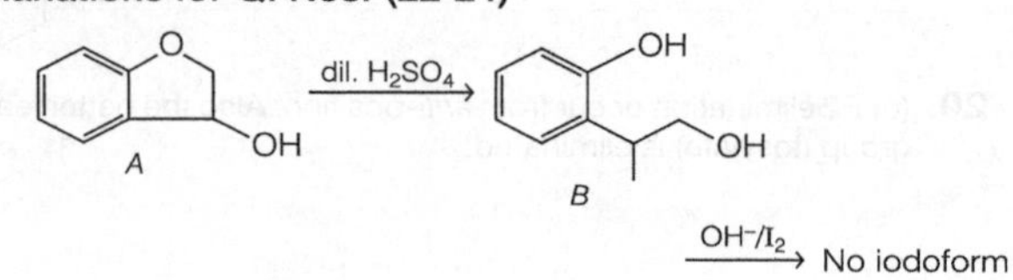

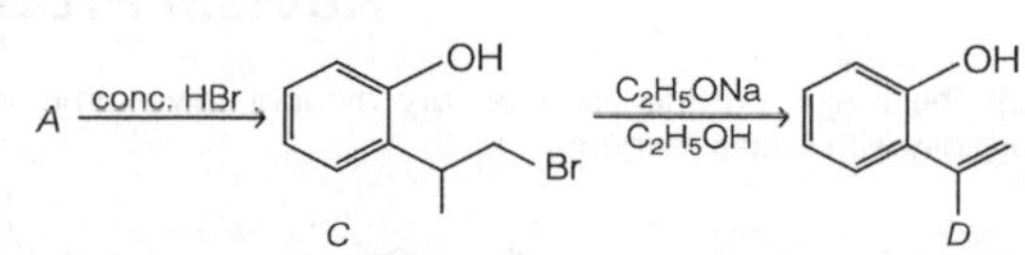

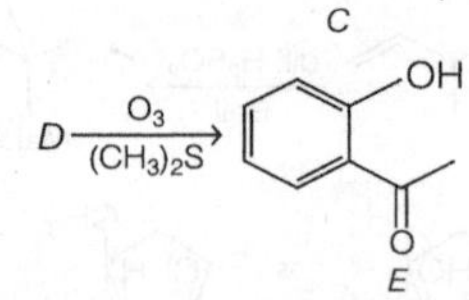

22. (b) The compound *A* shown above satisfy the given criteria.

23. (b) Shown above, *B* has a chiral carbon hence, optically active. Moreover, the chiral carbon is not involved in reaction from *A* to *B*, configuration would be retained in *B*.

24. (a) If oxygen of *A* is labelled as O^{18}, it will be part of phenolic —OH, hence retained in *D*.

25. (a) (i)

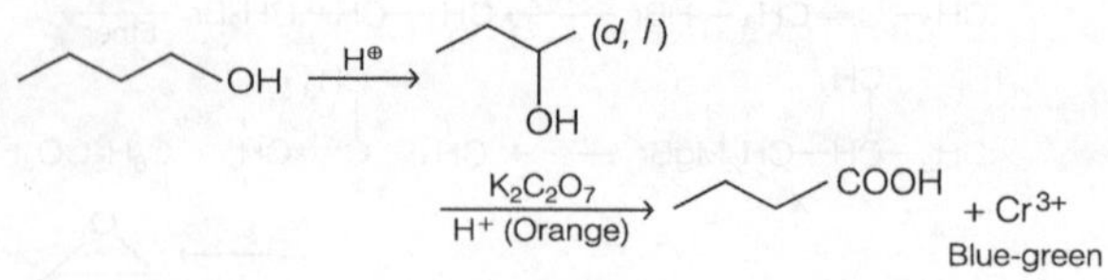

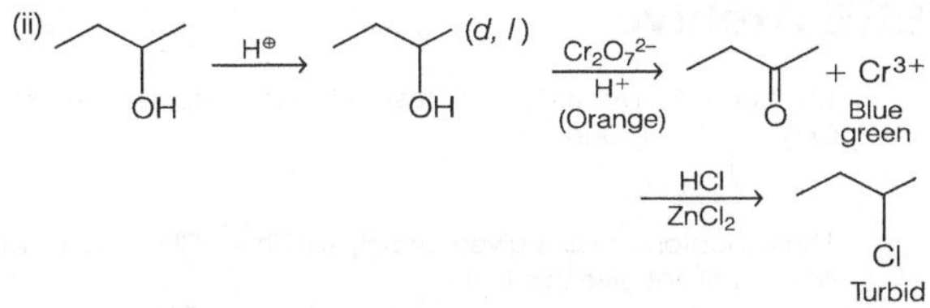

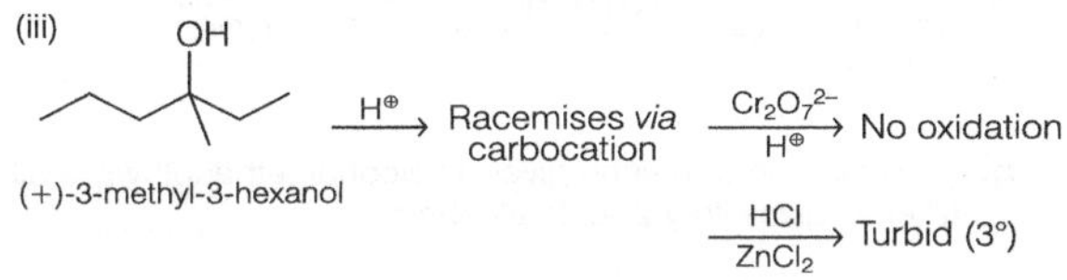

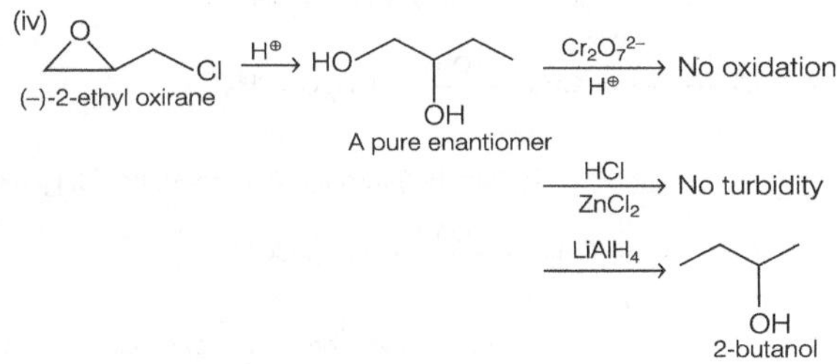

Hence, (i) → (p,q), (ii) → (p,q,r), (iii) → (p,r), (iv) → (s)

26. (a)

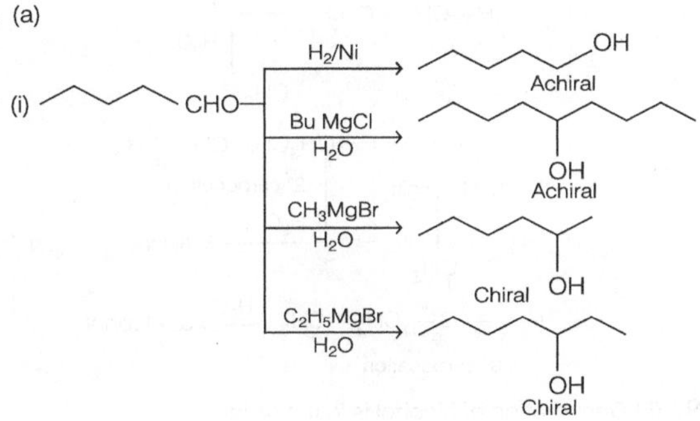

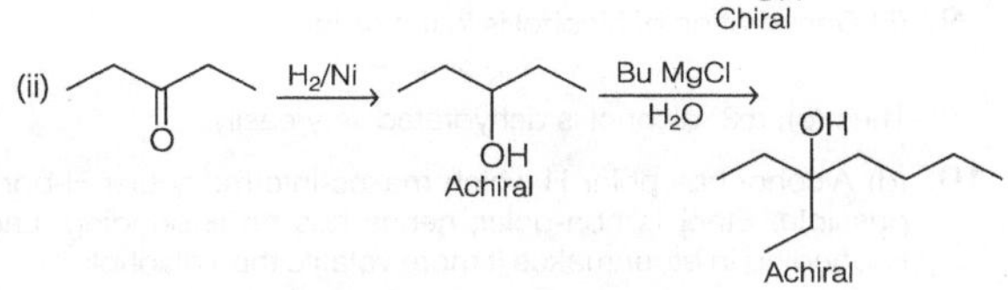

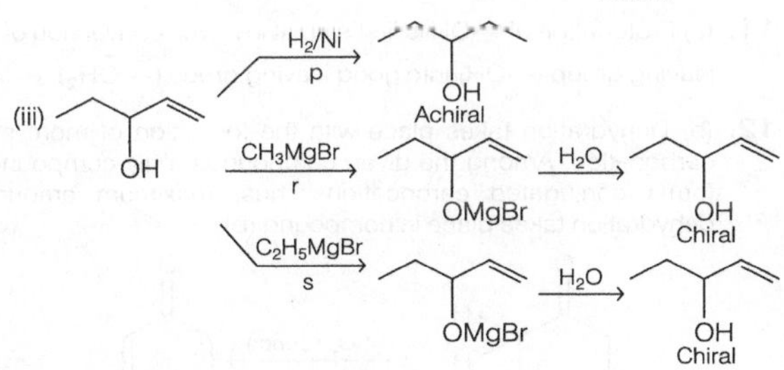

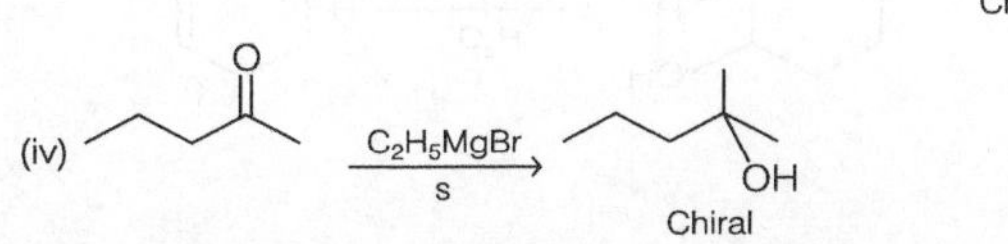

Hence, (i) → (p,q,r,s), (ii) → (p,q), (iii) → (p,r,s), (iv) → (s)

27. (a) (i)

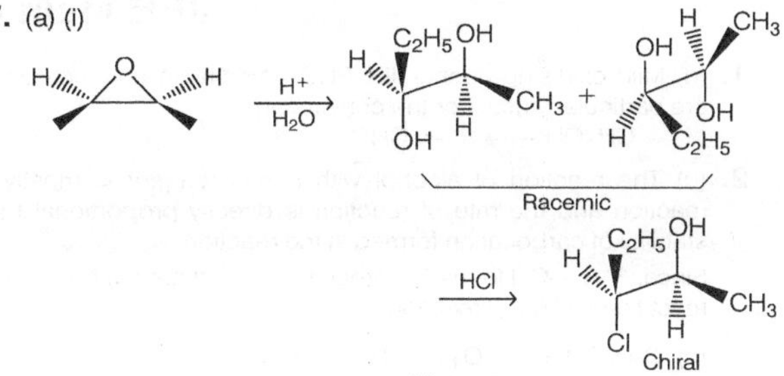

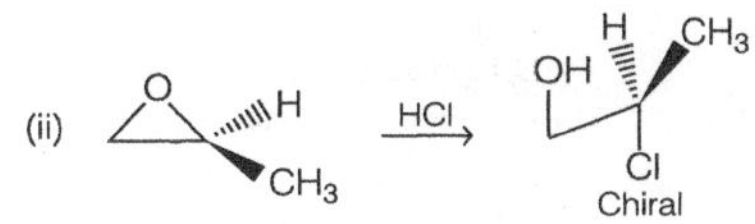

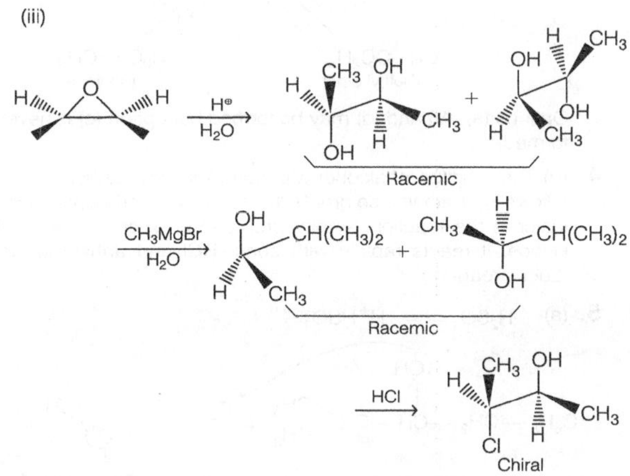

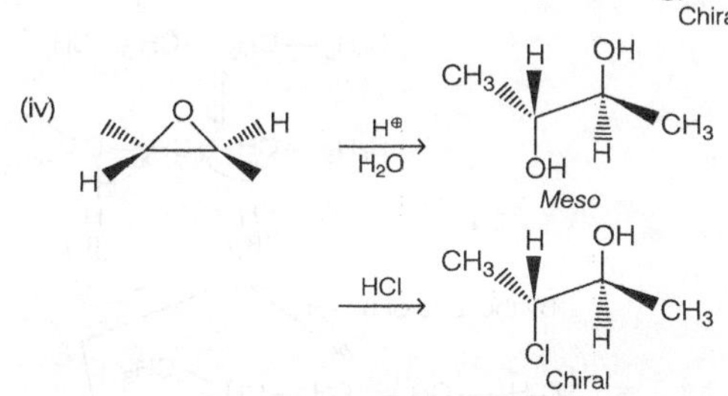

Hence, (i) → (p,s), (ii) → (s), (iii) → (p,r,s), (iv) → (q,s)

28. (6) From % of iodine, formula of iodoalkane is derived as C_3H_7I.

Hence, ether is $C_3H_7OC_3H_7$.

29. (6) $\underset{17}{(-OH)} + CH_3COCl \longrightarrow \underset{59}{\left(-O-\overset{O}{\overset{\|}{C}}-CH_3\right)} + HCl$

Mass gain due to incorporation of one acetyl group $= 59 - 17 = 42$

Net mass gain due to acetylation $= 518 - 266 = 252$

Hence, six hydroxy groups $(6 \times 42 = 252)$ were present.

30. (8)

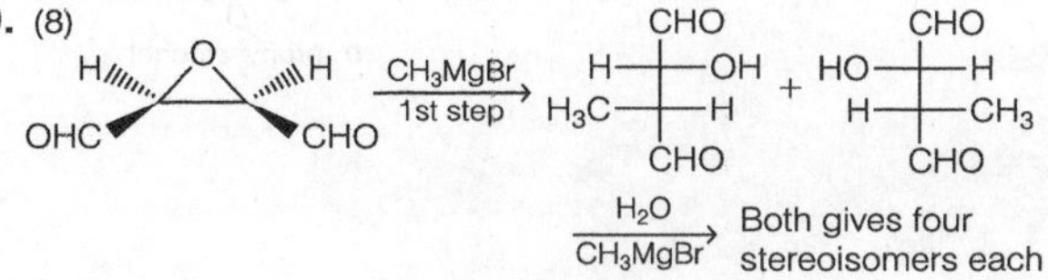

JEE Main & AIEEE Archive

1. (d) Mild oxidising agents like PCC (Pyridinium chlorochromate) are particularly used for the conversion of
$R - CH_2OH \longrightarrow R - CHO$.

2. (b) The reaction of alcohol with Lucas reagent is mostly S_N1 reaction and the rate of reaction is directly proportional to the stability of carbocation formed in the reaction.

Since, 3° R—OH forms 3° carbocation (most stable) hence, it will react fastest by S_N1 reaction.

3. (c) $C_2H_5OH + H_2SO_4$ Room temp.

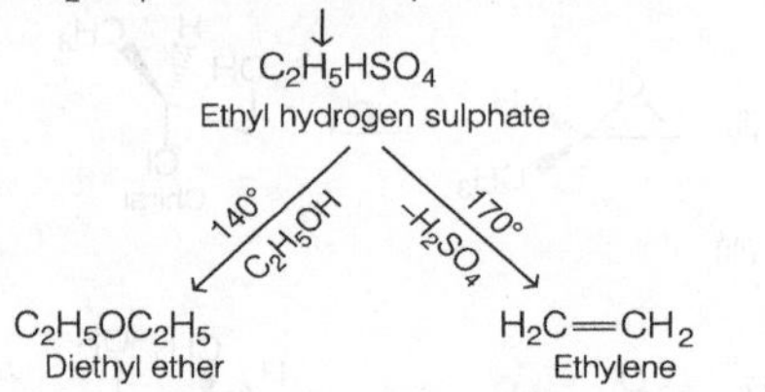

Options (a), (b) and (d) may be formed but option (c) is never formed.

4. (b) The reaction of alcohol with conc. HCl and anhydrous $ZnCl_2$ follows S_N1 pathway, so greater the stability of carbocation formed, faster is the reaction.2-methylpropan-2-ol gives 3° carbocation. Hence, it reacts rapidly with conc. HCl and anhydrous $ZnCl_2$ (Lucas reagent).

5. (a) $H_2SO_4 \rightleftharpoons H^+HSO_4^-$

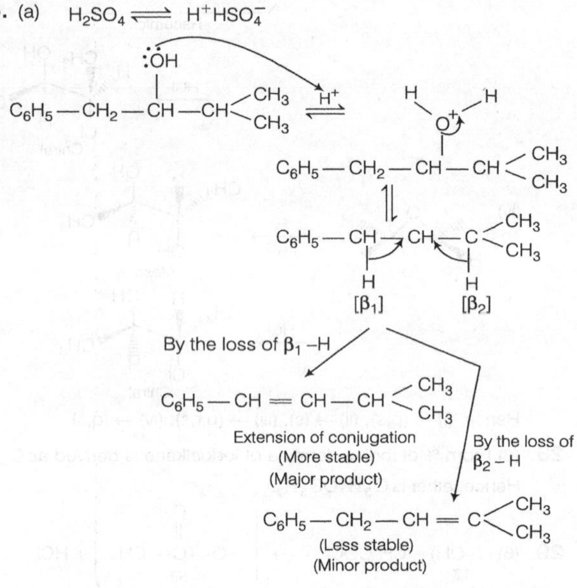

6. (c) $CH_3CH_2OH \xrightarrow[-H_3PO_3]{P + I_2} \underset{A}{CH_3CH_2—I} \xrightarrow{Mg,\ ether} \underset{B}{CH_3CH_2MgI}$

$H—\overset{O}{\overset{\|}{C}}—H \longrightarrow \underset{C}{CH_3CH_2—\overset{OMgI}{\overset{|}{CH_2}}} \xrightarrow{H_2O} \underset{D}{CH_3CH_2\overset{OH}{\overset{|}{CH_2}}} + Mg(OH)I$

n-propyl alcohol

7. (d) For positive iodoform test, alcohol molecule must have $CH_3—\underset{OH}{\underset{|}{CH}}—$ group.

Thus, iodoform test is given by only (d) $Ph—\underset{OH}{\underset{|}{CH}}—CH_3$ while others will not give this test.

$Ph—\underset{OH}{\underset{|}{CH}}—CH_3 \xrightarrow{I_2 + NaOH} CHI_3 + Ph—COO^-$

8. (c) Hydration of ethene gives 1° alcohol (ethanol) while all other alkenes give either 2° or 3° alcohols.

$\underset{\text{Ethene}}{CH_2{=}CH_2} \xrightarrow{H_2O/H^+} CH_3CH_2OH$

$CH_3—CH{=}CH_2 \xrightarrow{H_2O/H^+} CH_3\underset{OH}{\underset{|}{CH}}CH_3$

[(2°alcohol) through 2° carbocation $CH_3\overset{\oplus}{C}HCH_3$]

$CH_3—\underset{CH_3}{\underset{|}{C}}{=}CH_2 \xrightarrow{H_2O/H^+} (CH_3)_3COH$

[(3° alcohol) through 3° carbocation $(CH_3)_3\overset{\oplus}{C}$]

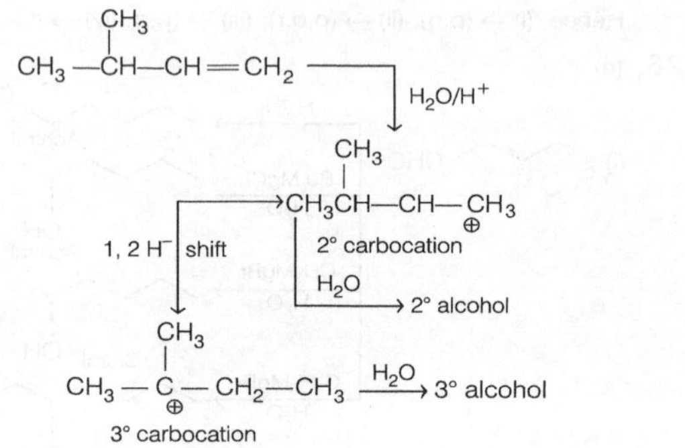

9. (c) Dehydration of alcohol is in the order

$1° < 2° < 3°$

Thus (c), a 3° alcohol is dehydrated very easily.

10. (d) Alcohol has polar H which makes intermolecular H-bonding possible. Ether is non-polar, hence has no H-bonding. Lack of H-bonding in ether makes it more volatile than alcohol.

11. (a) Protonation of —OH is first step. It involves conversion of poor leaving group (—OH) into good leaving group ($—\overset{+}{O}H_2$).

12. (b) Dehydration takes place with the formation of more stable carbocation. Among the given compounds, only compound (b) forms conjugated carbocation. Thus, maximum amount of dehydration takes place in compound (b).

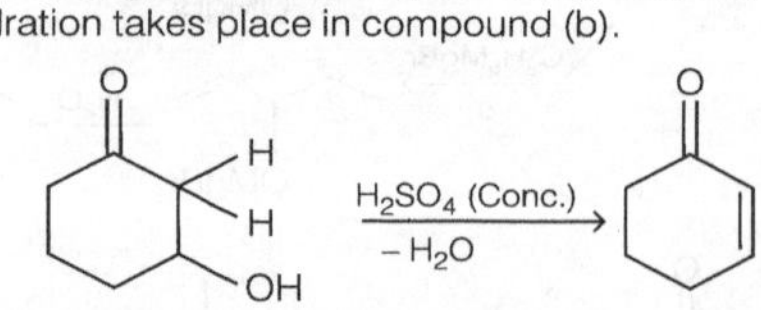

JEE Advanced & IIT JEE Archive

1. (c) The reaction condition indicates that starting compound is phenyl acetylene.

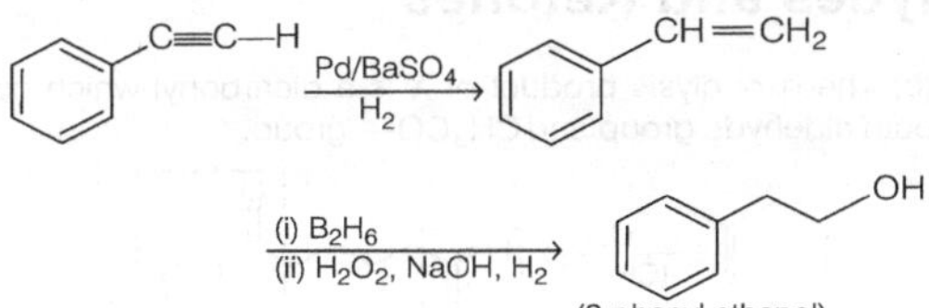

(2-phenyl ethanol)
(X)

Hydroboration oxidation brings about anti-Markonikoff's hydration of alkene.

2. (d)

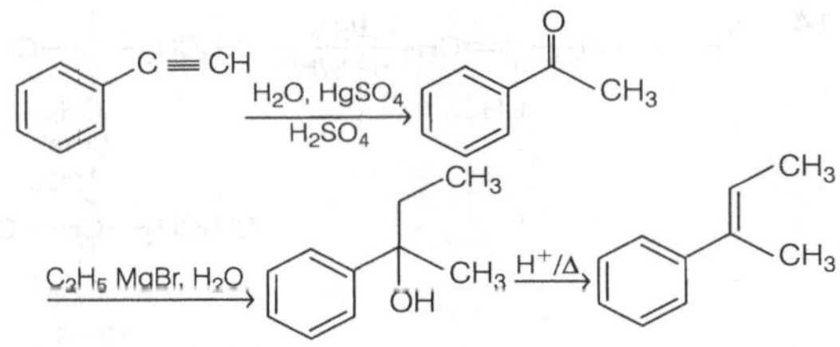

3. (c) This problem can be solved by using the concept of stability of carbocation and S_N1 reaction.

When two phenyl groups are replaced by two *para* methoxy group, carbocation formed will be more stable.

As the stability of carbocation formed increases, rate of acidic hydrolysis increases.

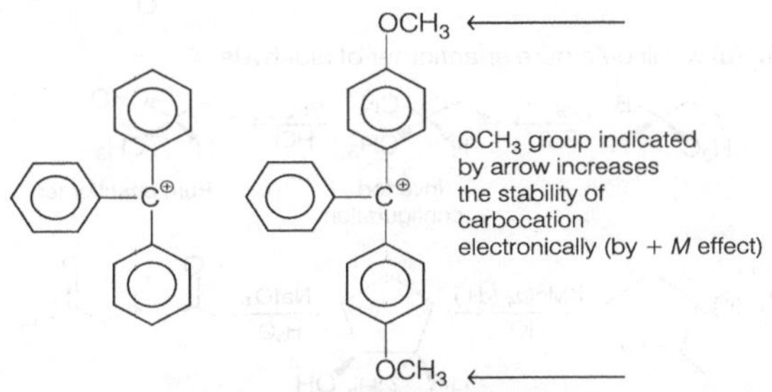

More stable carbocation

4. (b)

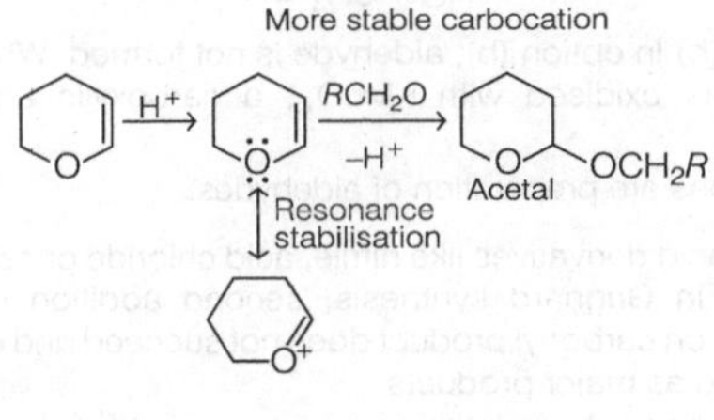

5. (b) $Ph-\overset{O}{\overset{\|}{C}}-CH_3 + Ph-CH_2MgBr \xrightarrow{H_2O} Ph-\overset{OH}{\overset{|}{\underset{CH_3}{\underset{|}{C}}}}-CH_2-Ph$

H

6. (d) I = $\overset{Ph}{H_3C}\rangle C=CH-Ph$

7. (c) All dihydroxy benzene will have higher boiling points than monohydroxy benzene. Also, among dihydroxy benzenes, 1, 2,-di-hydroxy benzene has lowest boiling point due to intramolecular H-bonding.

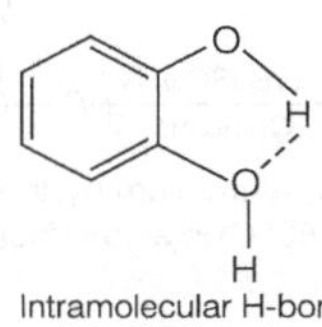

Intramolecular H-bonding in 1,2-dihydroxy benzene

8. (b)

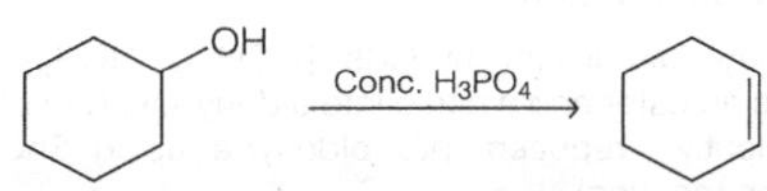

Concentrated H_3PO_4 solution does not involve any substitution product while with others, substitution product are also formed.

9. (b) $C_6H_5MgBr + (CH_3)_3COH \longrightarrow C_6H_6 + Mg[(CH_3)_3CO]Br$

10.

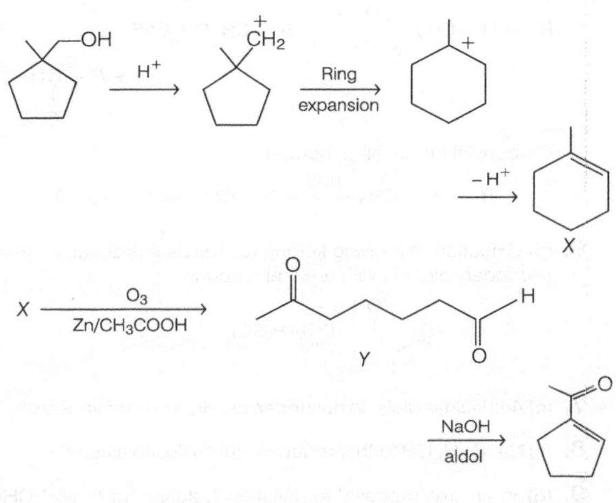

3. Aliphatic Aldehydes and Ketones

DPP-1 Preparation of Aldehydes and Ketones

1. Primary alcohols ($R—CH_2OH$) can conveniently be oxidised into aldehydes ($RCHO$) using pyridinium chlorochromate ($CrO_3/HCl/Pyridine = PCC$).

$$R—CH_2OH \xrightarrow[\text{Pyridine}]{CrO_3/HCl} R—CHO$$

Conventional oxidising agents like CrO_3/H_2SO_4 or $KMnO_4$ further oxidises aldehydes into carboxylic acids.

2. Rosenmund's reduction is a convenient method of reducing an acid chloride into aldehyde.

$$R—\overset{O}{\overset{\|}{C}}—Cl \xrightarrow[\text{Quinoline}]{Pd/BaSO_4/H_2} R—\overset{O}{\overset{\|}{C}}—H$$

Lithium tri(tertiary butoxy) aluminium hydride $Li[\{(CH_3)_3CO\}_3AlH]$ at low temperature ($\sim -80°C$) is also effective in reducing acid chloride into aldehyde.

3. Both $Li[\{(CH_3)_3CO\}_3AlH]$ and $[\{(CH_3)_2CH \cdot CH_2\}_2AlH]$ at low temperature ($\sim -80°C$) are effective in reducing acid chloride and acid anhydride into aldehyde.

4. Di-isobutyl aluminium hydride $[\{(CH_3)_2CHCH_2\}_2AlH]/-80°C$ reduces an ester and a nitrile into aldehydes. Nitriles can also be conveniently reduced into aldehyde using $SnCl_2$/conc. HCl (Stephen reaction).

$$R—CN \xrightarrow[\text{conc. HCl}]{SnCl_2} R—CHO + NH_4Cl$$

5. Oxo process:

$$R—CH{=}CH_2 \xrightarrow[\text{[Co(CO)}_4\text{H]}]{CO + H_2} R—CH_2CH_2CHO + R—\underset{CHO}{\underset{|}{CH}}—CH_3$$

Carbonylation of alkyl borane :

$$R—CH{=}CH_2 \xrightarrow[CO]{R_2BH} R—CH_2—CH_2—CHO$$

6. (c) Oxidation of *X* giving ketone as well as *X* is chiral, it must be a secondary alcohol with α-chiral carbon.

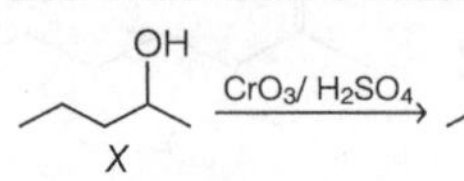

7. (c) *Anti vicinal*-diols do not undergo oxidative cleavage with HIO_4.

8. (c) Na/C_2H_5OH further reduces aldehydes to alcohols.

9. (d) In an oxo process, formylation (addition of H and CHO) of double bond takes place, hence ketones cannot be synthesised in direct oxo process.

10. (b) It gives aldehydes as major product.

11. (b) It gives a ketone.

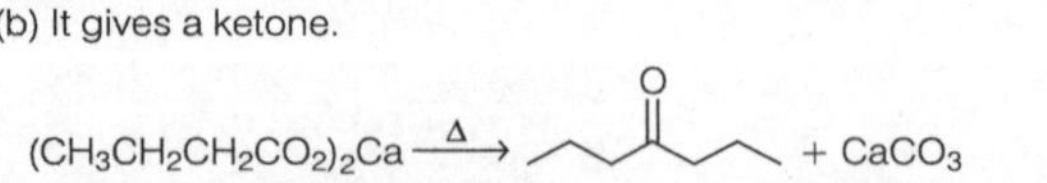

12. (d) Alkaline permanganate (Baeyer's reagent) also oxidises olefinic bonds to *syn vicinal*-diols.

13. (b) The ozonolysis product of *X* is a dicarbonyl which contains both aldehyde group and $CH_3CO—$ group.

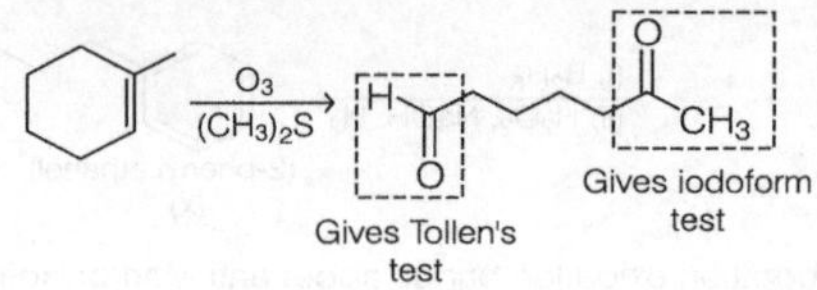

14. (b)

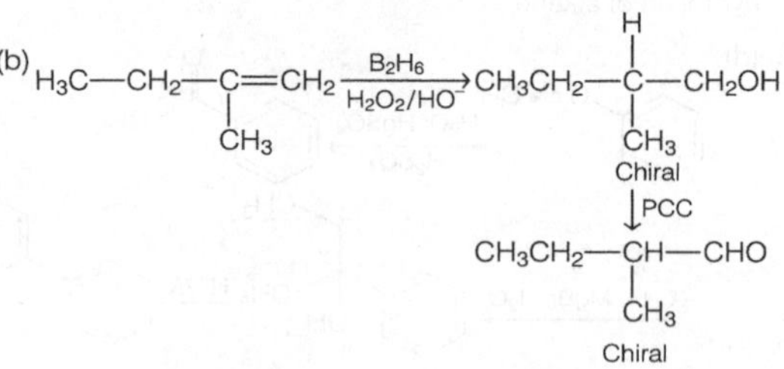

15. (c)

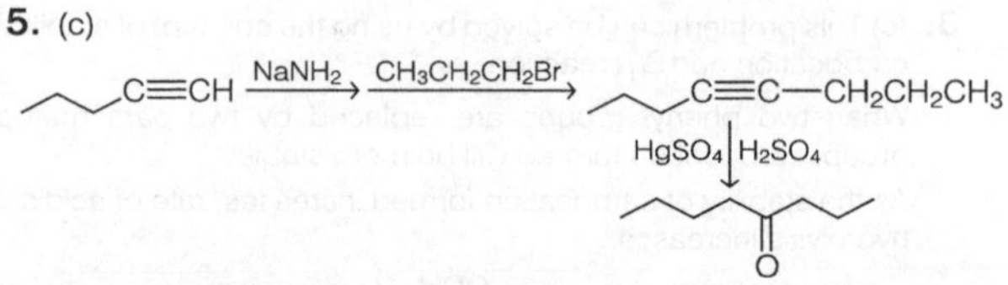

16. (c) *X* will be a pure enantiomer of aldehyde.

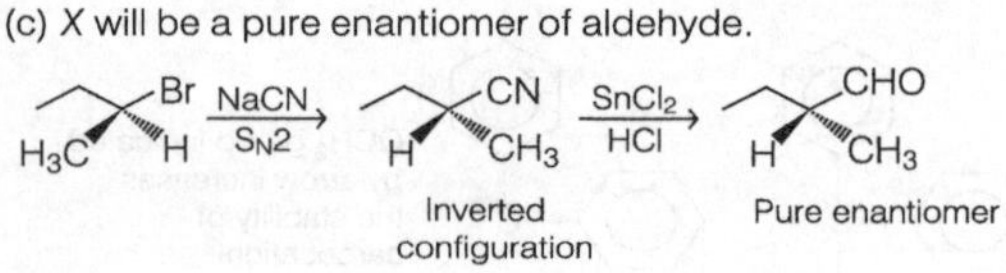

17. (c)

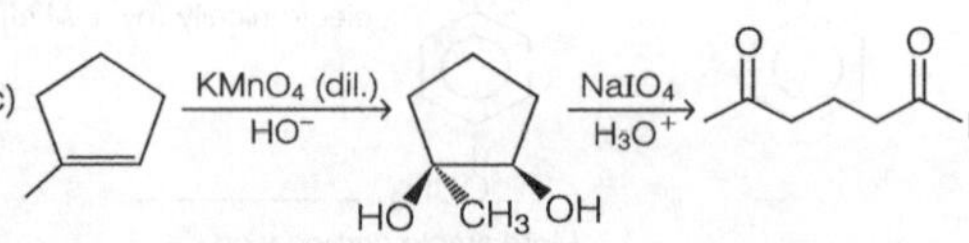

18. (a,c,d) (b) In option (b), aldehyde is not formed. When a terminal alkyne is oxidised with $KMnO_4$, a carboxylic acid is always formed.

All options are preparation of aldehydes.

19. (a,c) If acid derivatives like nitrile, acid chloride or ester is taken in excess in Grignard synthesis, second addition of Grignard's reagent on carbonyl product does not succeed and carbonyls are obtained as major products.

In option (b), carboxylic acids and in option (d), an aldehyde is formed.

20. (a,b,c)

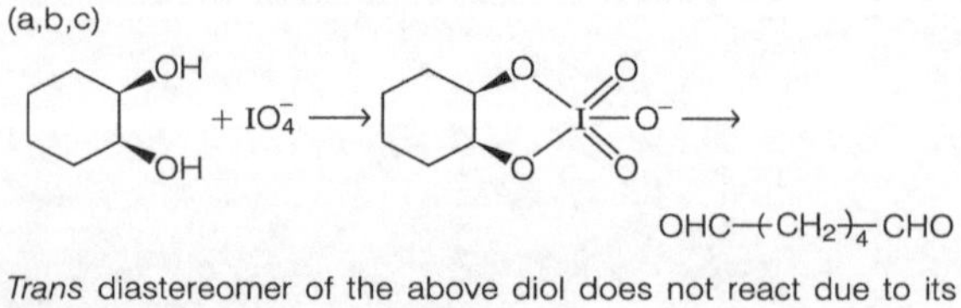

Trans diastereomer of the above diol does not react due to its inability to form cyclic intermediate.

21. (c,d) Formylation of double bond (oxo process) occurs. However, the reaction is not regioselective, hence both isomers of aldehyde are formed.

22. (a,c) Both Rosenmund reduction (Pd/$BaSO_4$) and reduction with Li[{$(CH_3)_3CO$}$_3$AlH] at −80°C are effective in reducing acid chlorides into aldehydes.

Explanations for Q. Nos. (23-25)

Compound *A* has a triple bond and it is symmetrical because its partially reduced product *B* gives single ozonolysis product *C*. Also, *A* shows both enantiomerism and diastereomerism, it must be

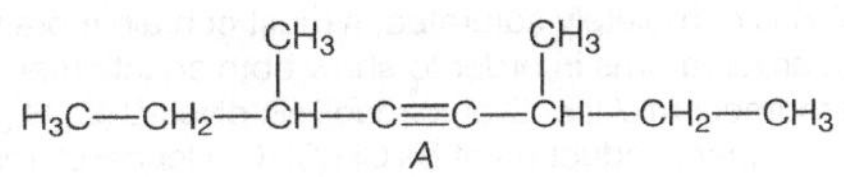

As shown above, *A* has two chiral carbons but simultaneously, it is symmetrical. Hence, it has both *meso* and a pair of enantiomers as stereoisomers.

23. (b)

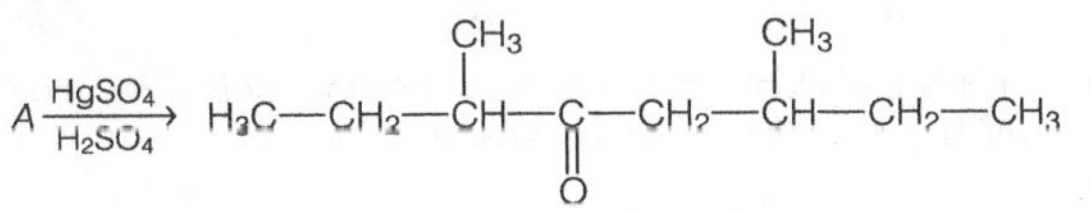

24. (c)

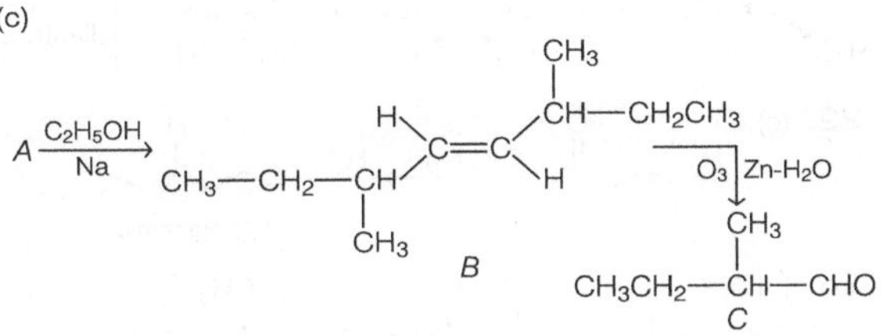

25. (d)

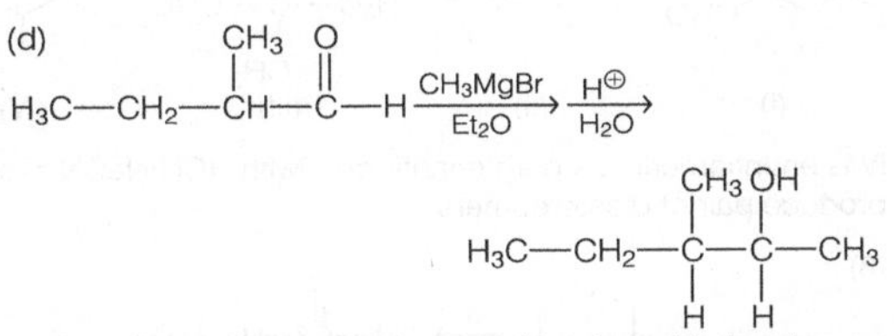

Has two chiral carbons, i.e. 4 stereoisomers

26. (5)

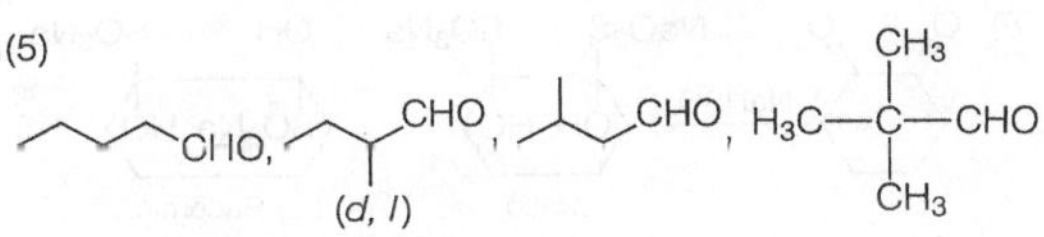

27. (7)

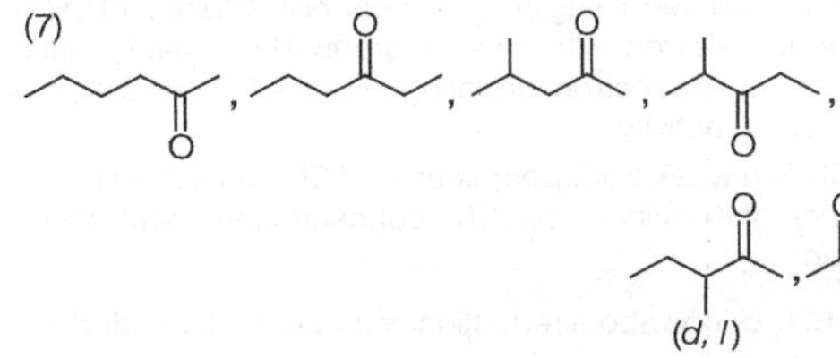

28. (5) All secondary alcohol isomers can be oxidised to ketones.

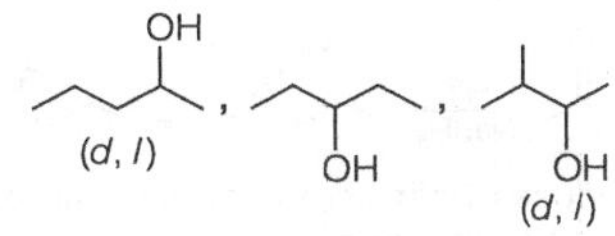

29. (5)

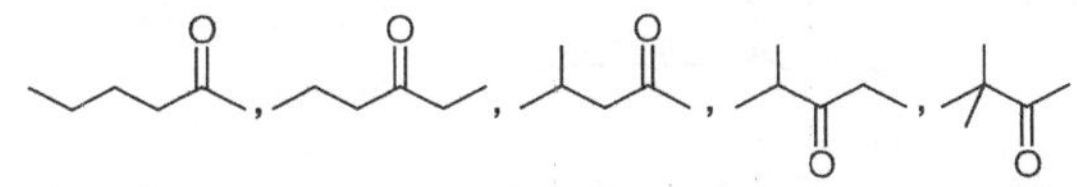

(I) (II) (III) (IV) (V)

All the above shown ketones on reduction with $NaBH_4$ will produce racemic mixtures of alcohols. The isomer below already has a chiral carbon, show enantiomerism. A pure enantiomer of this ketone, on reduction with $NaBH_4$, will produce pair of diastereomers.

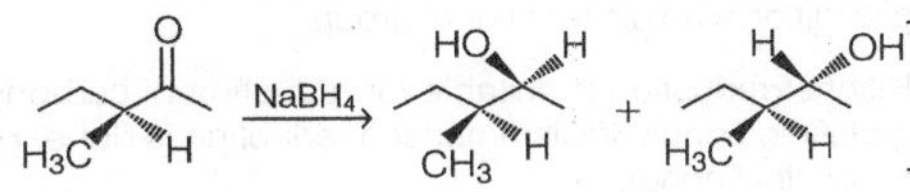

Pair of diastereomers

30. (4)

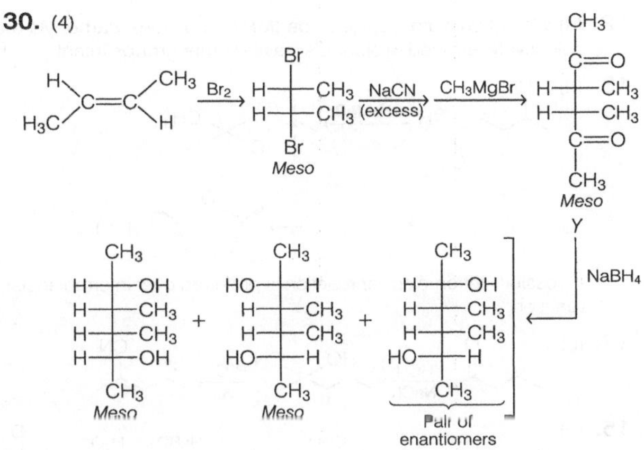

DPP-2 Physical Properties and General Reactions

1. The boiling points is in order of (I) < (II) < (III) < (IV). Alcohols are capable of forming intermolecular hydrogen bond. Alkane has the weakest intermolecular forces of attraction, hence lowest boiling point. Greater charge separation (due to +I effect of two methyl groups) in acetone makes it more polar than propanal, hence higher boiling point.

2. (i) H_2/Pt-Very strong reducing agent, reduces all reducible groups.

(ii) $LiAlH_4$–Strong reducing agent, reduces carbonyls and carboxylic groups but fail to reduce olefinic bond.

(iii) $NaBH_4$–Reduces carbonyls but fail to reduce olefinic bond and carboxylic group.

3. Propanal, an aldehyde, forms precipitate of bisulphite salt with $NaHSO_3$ but ethanol does not.

4. Trioxane is a cyclic trimer while paraformaldehyde is a polymer of formaldehyde. Both are used to store formaldehyde which is otherwise unstable.

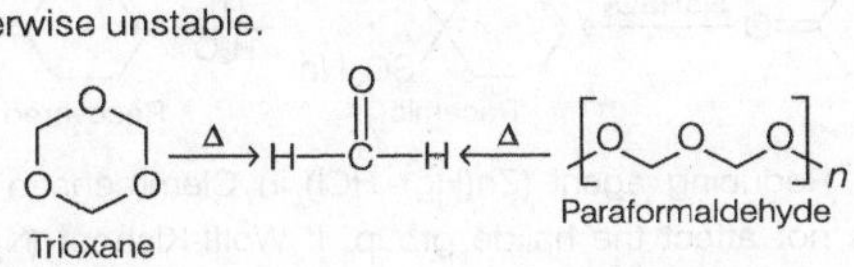

5. Formaldehyde is stored as 40% aqueous solution (formalin) because pure formaldehyde is gas at room temperature.

6. (i) In the rate determining step, nucleophilic addition of CN^- occur. Therefore, electron withdrawing group like $—NO_2$ increases reactivity while electron donating groups like $—CH_3$, $—OCH_3$ decreases reactivity.

(ii) HCN is a weak acid, poor source of CN^- nucleophile. Presence of KCN increases CN^- concentration, catalyses reaction.

7. (b) $NaBH_4$ brings about reduction of carbonyls by hydride transfer mechanism at planar sp^2 carbon. Hence, if a chiral carbon is generated, racemic mixture is always produced.

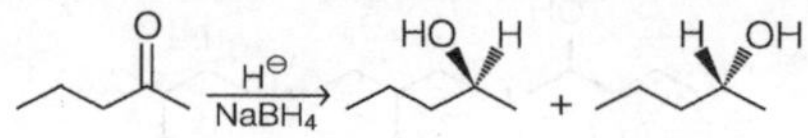

8. (c) $LiAlH_4$ reduces both carbonyls and carboxylic groups but does not reduce olefinic bond.

9. (c)

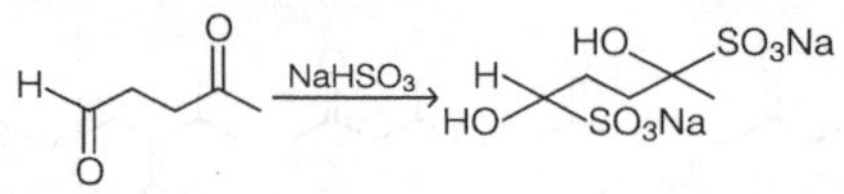

In the above reaction, four stereoisomers, two pairs of enantiomers are formed. Each member of a pair of enantiomer is diastereomer of each member of other pair of enantiomer. Hence, fractional crystallisation would give two fractions, each containing racemic mixture.

10. (a) Clemmensen reduction is suitable for reduction of carbonyls containing additional acidic functional group.

11. (b) Wolff-Kishner reduction is suitable for reduction of carbonyls containing olefinic bonds. If Clemmensen reduction is done, HCl also attacks olefinic bonds.

12. (d) With aluminium isopropoxide (MPV reduction), carbonyls are selectively reduced to alcohols leaving other groups intact.

13. (d)

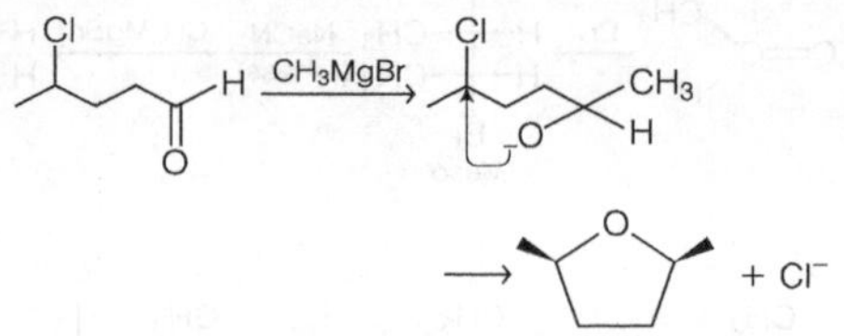

If possible, intramolecular reaction is preferred over intermolecular reaction.

14. (c)

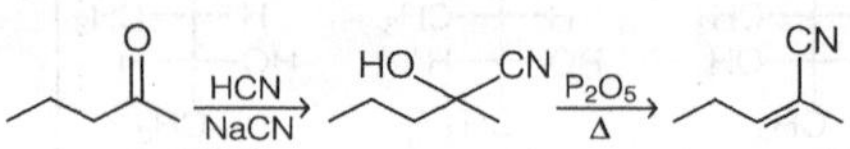

15. (d)

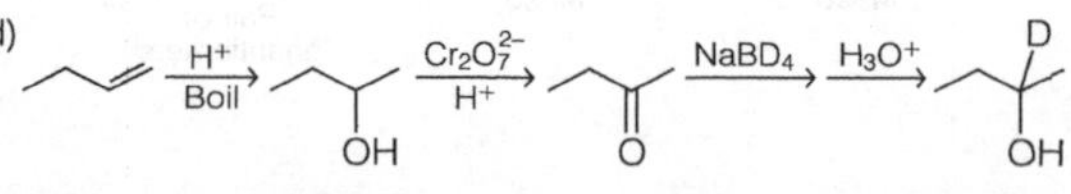

16. (b,c) In MPV reduction, carbonyl is reduced to alcohol while isopropoxide fraction of reducing agent is oxidised to acetone.

17. (a,b,c,d) Aldehydes and ketones form bisulphite salt with $NaHSO_3$ which is insoluble in concentrated $NaHSO_3$ solution due to common ion effect. Hence, if $NaHSO_3$ is in limited quantity, precipitation may not take place.

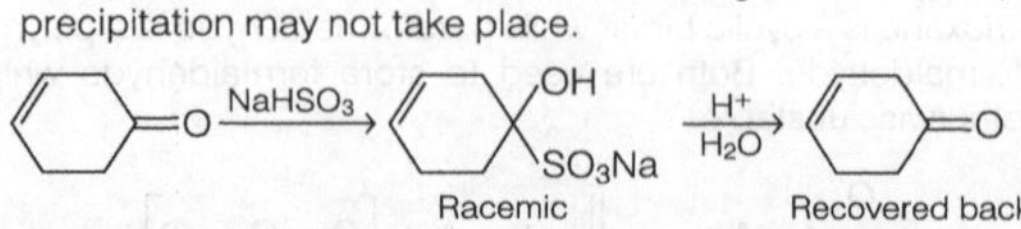

18. (a,b) Reducing agent (Zn(Hg)-HCl) in Clemmensen reduction does not affect the halide group. If Wolff-Kishner (N_2H_4-OH^-) reduction is done, OH^- may bring about S_N2 and E2 reaction with halide group.

19. (a,c,d) H^-(hydride) ion undergoes nucleophilic addition at sp^2 carbon of carbonyls. Aldehydes are more reactive than ketones due to less steric hindrance in former case. Since, nucleophilic attack of hydride ion occur at planar carbonyl carbon, there is equal probability of attack from both side of plane giving racemic mixture if chiral carbon is generated.

Also,

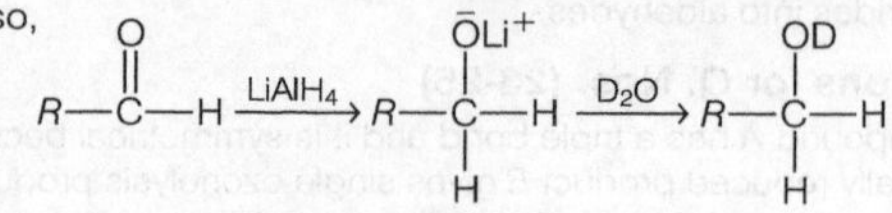

Explanations for Q. Nos. (20-22)

Since, *A* is completely saturated, it must contain more than one chiral carbon atoms in order to show both enantiomerism and diastereomerism. Also, *C* on ozonolysis gives *D* ($C_5H_{10}O$) as one product, other product must be CH_3CHO. Hence, *C* must be $CH_3CH = C_5H_{10}$.

20. (b)

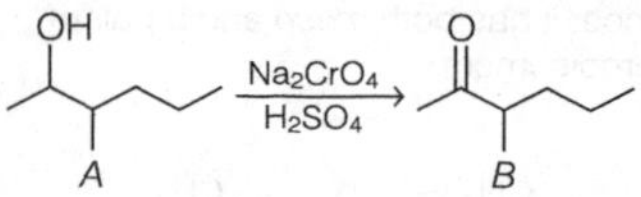

21. (c) *B* is enantiomeric. If a pure enantiomer of *B* is reduced with $LiAlH_4$, pair of diastereomers would be formed.

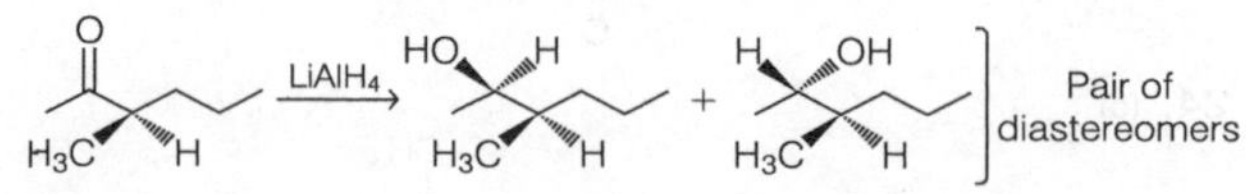

22. (c)

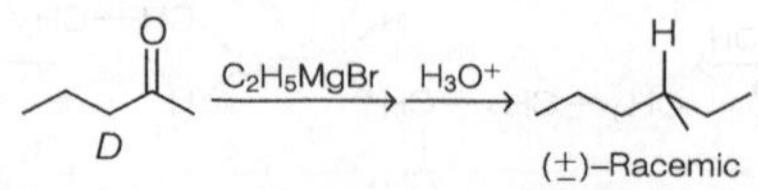

23. (3)

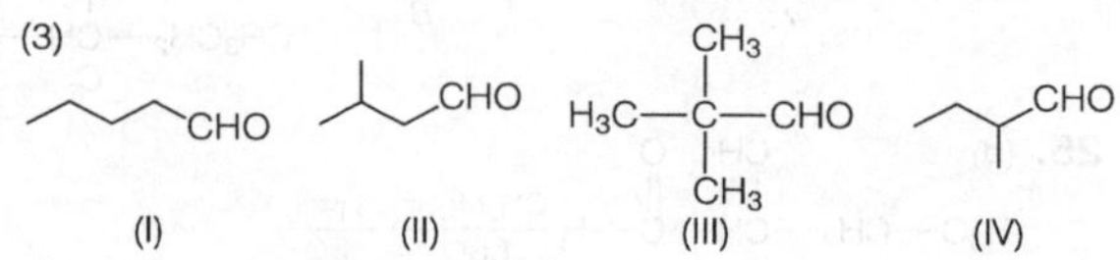

IV is enantiomeric. Its pure enantiomer, with HCN/NaCN, would produce pair of diastereomers.

24. (3)

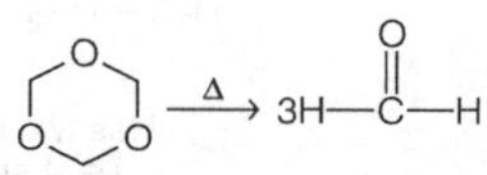

25. (8)

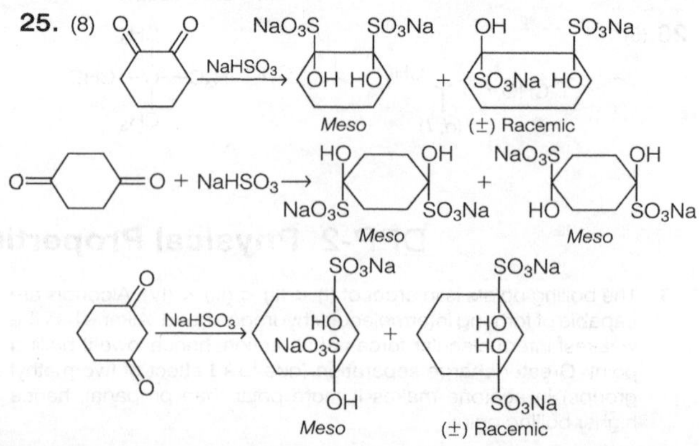

26. (3)

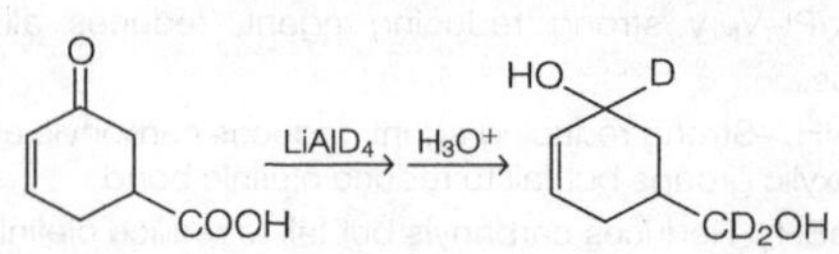

27. (8)

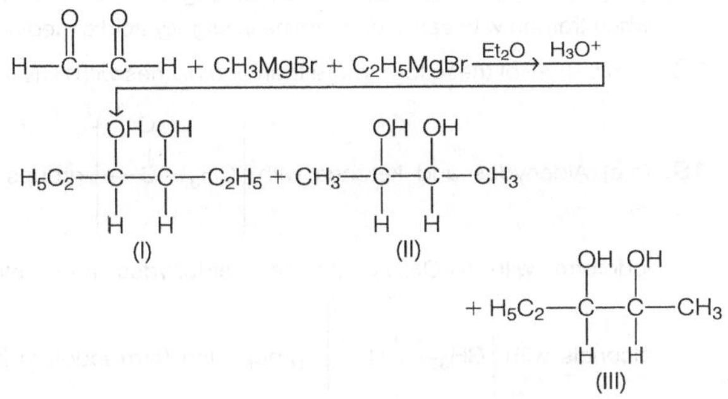

One pair of enantiomers for each (I) and (II) while two pairs of enantiomers for (III).

28. (5)

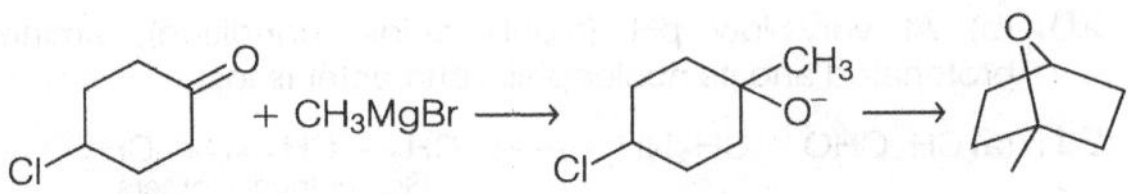

29. (a)

(i) It has a carbonyl carbon that turns chiral on reduction with hydrides. Hence, with $LiAlH_4$ or aluminium isopropoxide, gives racemic mixture. Also, it has no other functional groups, either Clemmensen reduction or Wolff-Kishner reduction can be used.

(ii) It has an olefinic double bond, Clemmensen reduction would not be suitable for selective reduction of carbonyl group.

(iii) $LiAlH_4$ also reduces primary halide but aluminium isopropoxide does not. Wolff-Kishner reduction would not be suitable because HO^- reacts with halide group (S_N2 or E2).

(iv) Same reasons as with (i).

Hence, (i) → (p,q,r,s), (ii) → (p,q,s), (iii) → (q,r) (iv) → (p,q,r,s).

30. (a)

(i)

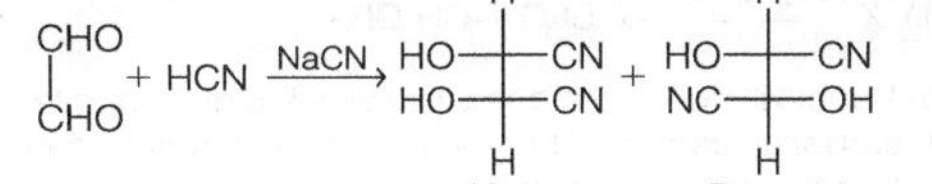

Meso Racemic
Diastereomers

(ii)

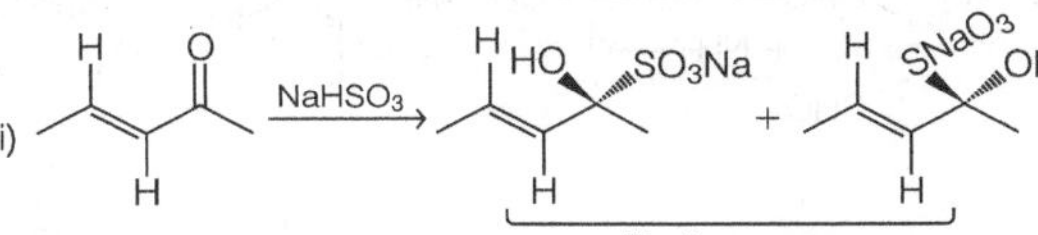

Diastereomers

(iii)

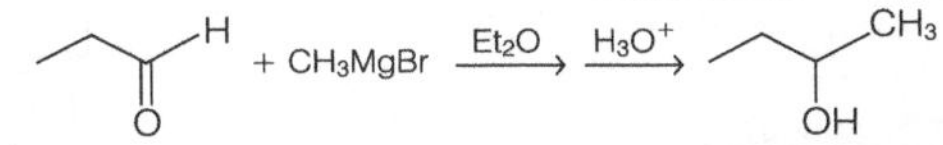

(+) Racemic

(iv)

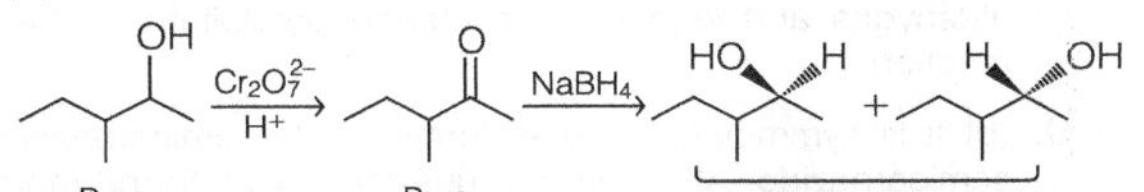

Pure enantiomer Pure enantiomer Diastereomers

Hence, (i) → (p,q,r,s), (ii) → (q,s), (iii) → (p), (iv) → (q,s).

DPP-3 Reactions with Ammonia, Amines, Hydrazine and Its Derivatives

1. (i) $6CH_2O + 4NH_3 \longrightarrow \underset{\text{Tetramine}}{(CH_2)_6N_4} + 6H_2O$

(ii) $3CH_3CHO + NH_3 \longrightarrow$

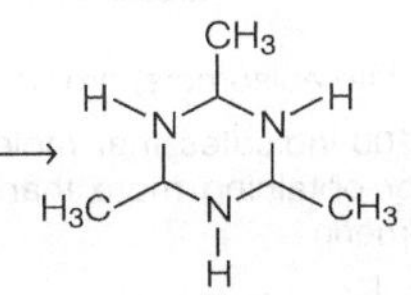

(iii) $CH_3CH_2\overset{O}{\overset{\|}{C}}CH_3 + NH_3 \longrightarrow CH_3—CH_2—\overset{NH}{\overset{\|}{C}}—CH_3$

(iv) $CH_3CH_2CHO + CH_3NH_2 \longrightarrow CH_3CH_2—CH{=}N—CH_3$

2. In slightly acidic medium, nucleophilic addition of amine occur very fast at protonated acetone.

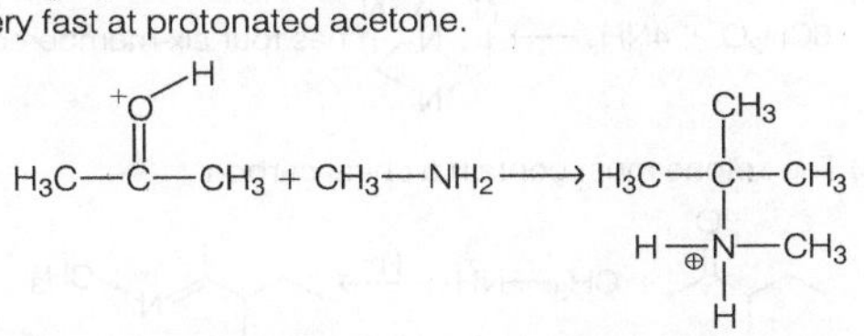

However, if solution is highly acidic, amine gets protonated and its nucleophilicity is lost.

3. (i) $CH_3CH_2CHO + NH_2OH \longrightarrow CH_3CH_2—CH{=}N—OH$

(ii) $CH_3CH_2CHO + N_2H_4 \longrightarrow CH_3CH_2—CH{=}N—NH_2$

(iii) $CH_3CH_2CHO + \underset{\text{Semicarbazide}}{H_2N—\overset{O}{\overset{\|}{C}}—NH—NH_2} \longrightarrow \underset{\text{Semicarbazone}}{CH_3H_2C—CH{=}N—NH—\overset{O}{\overset{\|}{C}}—NH_2}$

(iv)

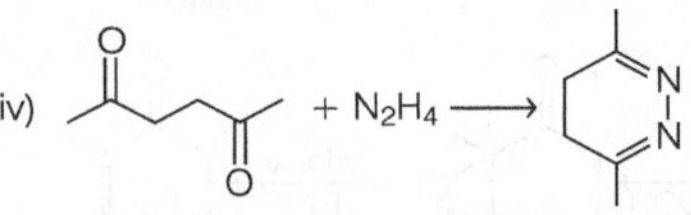

(v)

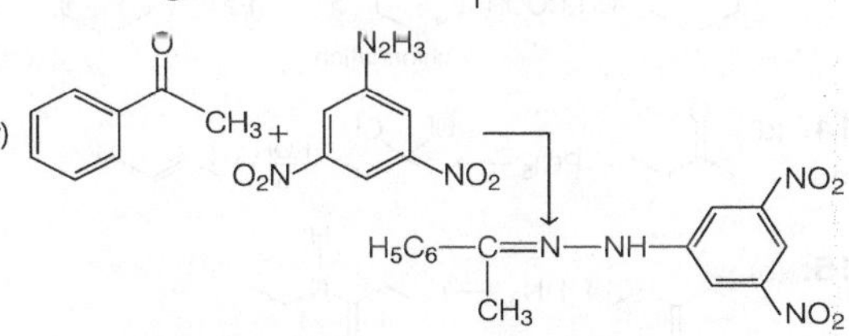

4. (i)

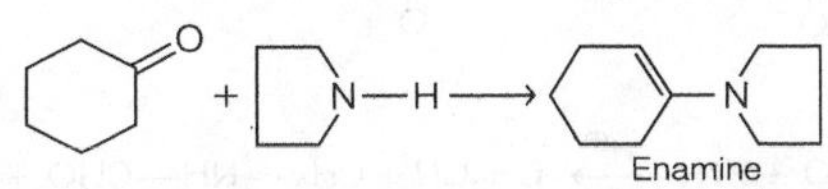

Enamine

(ii) $CH_3—CH_2—CHO + SeO_2 \longrightarrow CH_3—\overset{O}{\overset{\|}{C}}—CHO$

(iii) $CH_3CH_2CHO + Br_2 \longrightarrow CH_3CHBr—CHO$

(iv) $CH_3CH_2COCH_3 + Cl_2 \xrightarrow{NaOH} CH_3CH_2—COONa + CHCl_3$

5. $CH_3CH_2OH \xrightarrow{Cl_2} \underset{X}{Cl_3C—CHO}$

(i) $X \xrightarrow{KOH} CHCl_3 + HCOOK$

(ii) $X \xrightarrow{HNO_3} Cl_3C—COOH$

(iii) $X \xrightarrow{(C_2H_5O)_3Al} Cl_3C—CH_2OH$

6. (b) Reactivity follow the ease with which a nucleophile can attack at carbonyl carbon. Greater the steric hindrance at carbonyl carbon, smaller the reactivity.

7. (c)

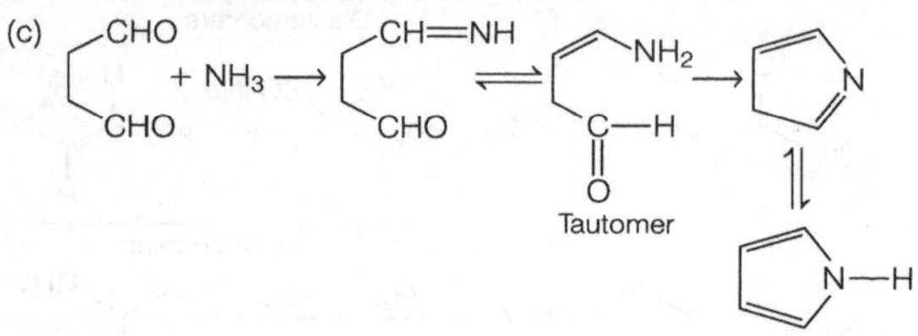

8. (d) Conjugated aldehydes and ketones give reddish orange precipitate with 2,4-dinitrophenyl hydrazine. Non-conjugated aldehydes and ketones give orange precipitate in the same reaction.

9. (d) It is symmetrical ketone, forms single semicarbazone with semicarbazide. All others are unsymmetrical, forms more than one stereoisomers of semicarbazones in the same reaction.

10. (c) Ozonolysis of X gives $C_6H_{12}O$ which has one carbon less than X. Hence, X must contain double bond at terminal carbon.

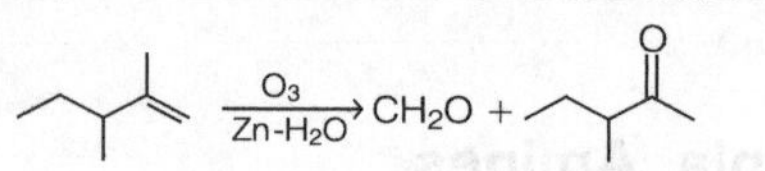

Gives iodoform test

11. (a) It is unsymmetrical aldehyde, forms *syn* and *anti* oximes with hydroxylamine. All others are symmetrical, forms single oxime.

12. (d) Secondary amine forms enamine with aldehydes or ketones.

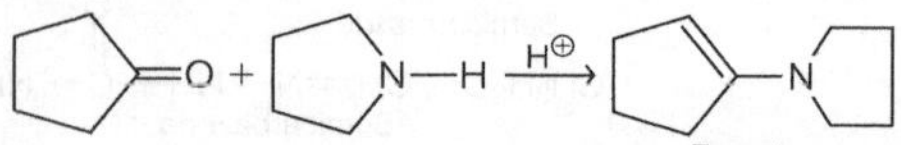

Enamine

13. (c)

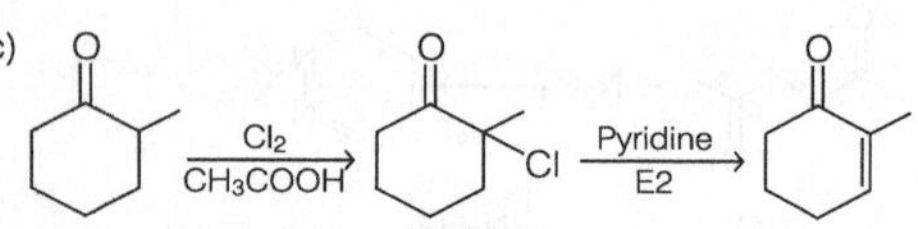

α-chlorination

14. (d)

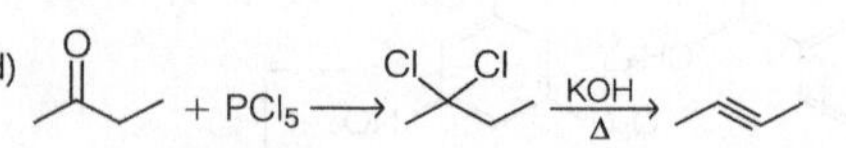

15. (b)

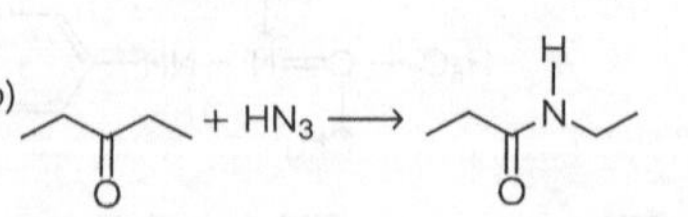

16. (a,b,c)

$CH_3CHO + HN_3 \xrightarrow[\Delta]{H^{\oplus}} CH_3CN + CH_3—NH—CHO + N_2$

17. (b,c) Aldehydes and ketones containing α-H form enamines when treated with secondary amine in slightly acidic medium.

18. (a,b,c,d) All of these form more than one oximes with $H_2N—OH$.

19. (a,c) Aldehydes and ketones with $\left(CH_3—\overset{O}{\overset{\|}{C}}—\right)$ groups form iodoform with $NaOH/I_2$. Besides, aldehydes and ketones, alcohols with $\left(CH_3—\overset{OH}{\overset{|}{C}}H—\right)$ group also form iodofom in the same reaction.

Explanations for Q. Nos. (20-22)

20. (b) At very low pH (highly acidic condition), amines get protonated and its nucleophilic character is lost.

21. (a) $CH_3CHO + CH_3NH_2 \longrightarrow \underset{\text{Geometrical isomers}}{CH_3—CH{=}N—CH_3}$

22. (c)

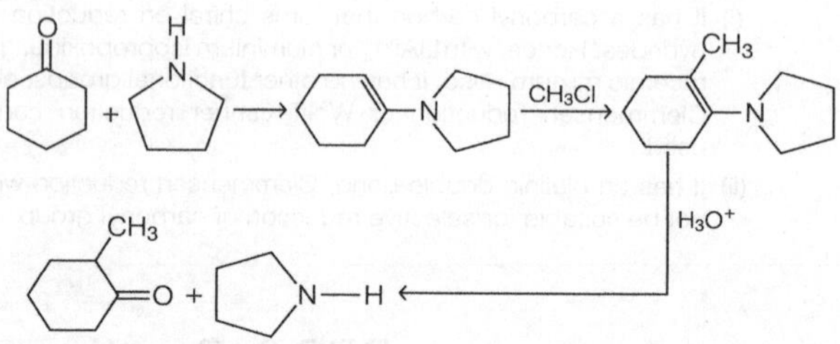

23. (6)

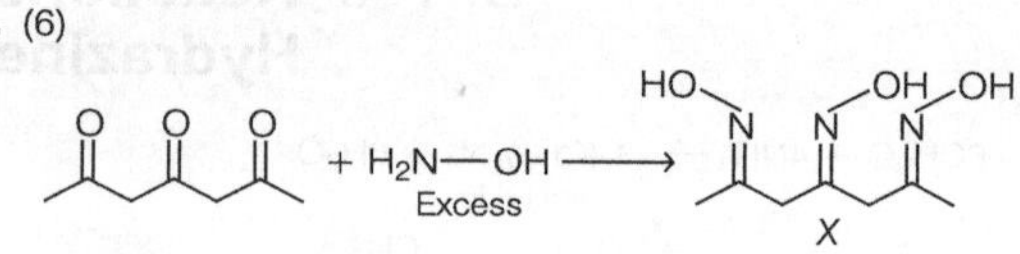

Six isomers (stereoisomers) are possible for X.

24. (3) MW = 100 indicates that molecular formula of aldehyde is $C_6H_{12}O$. For obtaining more than two oximes, aldehyde must exist stereomeric.

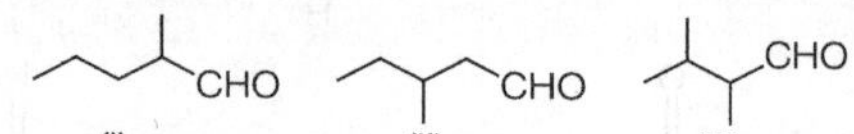

(I) (II) (III)

All the above has one chiral carbon each. Hence, when reacted with H_2NOH, forms more than two oximes.

25. (4)

$6CH_2O + 4NH_3 \longrightarrow$ 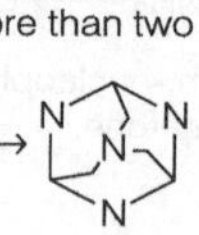 has four six-membered rings.

26. (6) The ketone must contain a chiral carbon.

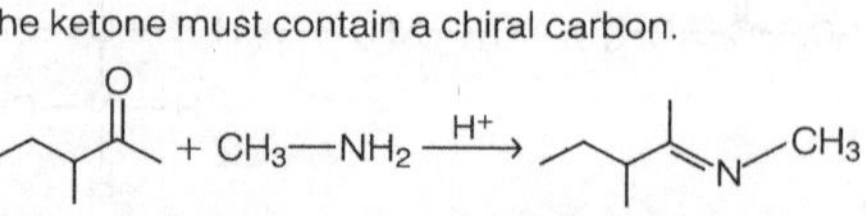

Four stereoisomers

DPP-4 **Keto-Enol Tautomerism and Related Reactions**

1. Ethyl acetoacetate forms a stable enol as:

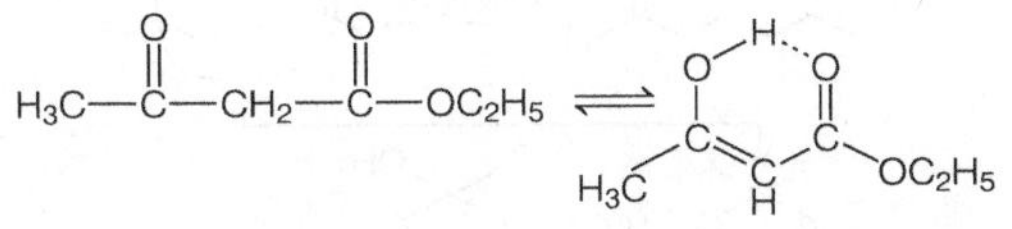

Intramolecular H-bonding

The stability of above enol is due to intramolecular H-bonding. Further, hydroxylic solvents like H_2O, CH_3OH increases the intermolecular H-bonding which decreases the stability of enol form.

2.

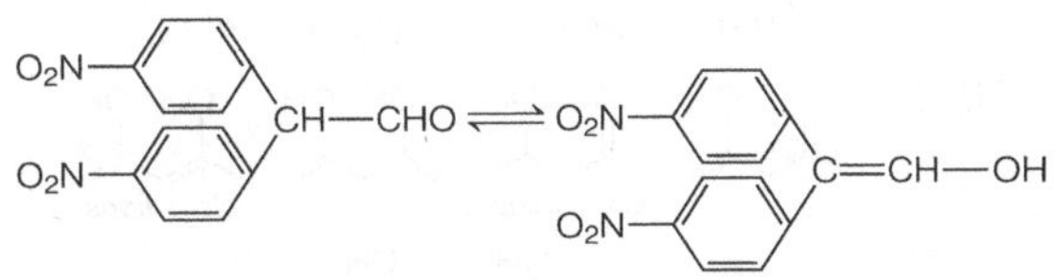

Highly stable enol

3.

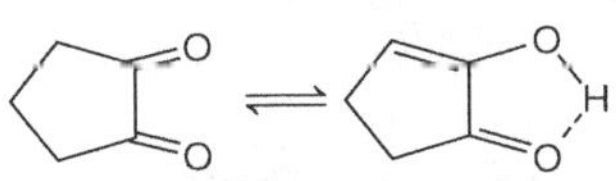

Stable enol

4.

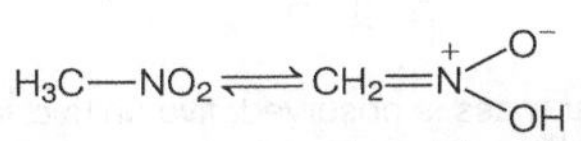

Forms intermolecular hydrogen bond, high boiling point

CH_3ONO does not form enol tautomer, low boiling point.

5. Exchange of O^{16} by O^{18} occur *via* hydrate equilibrium

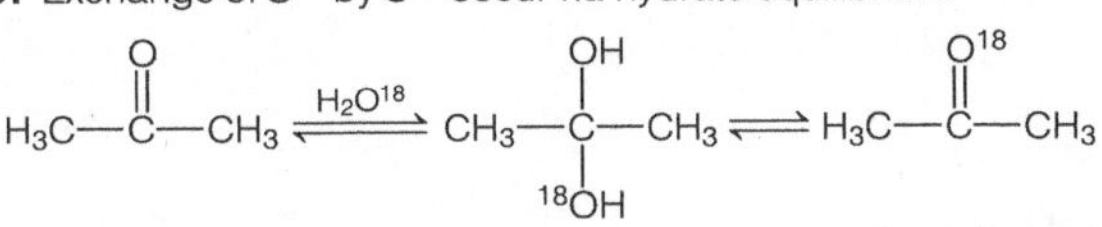

6. (i)

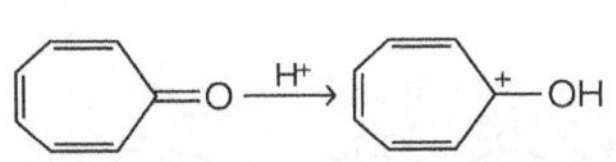

Aromatic

(ii)

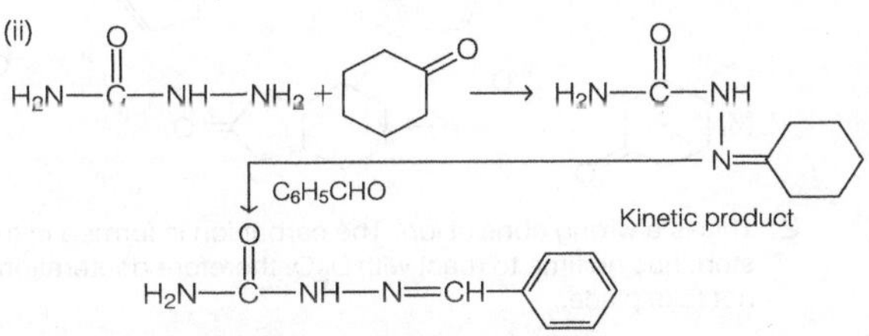

Kinetic product

More stable thermodynamic product

7. (d)

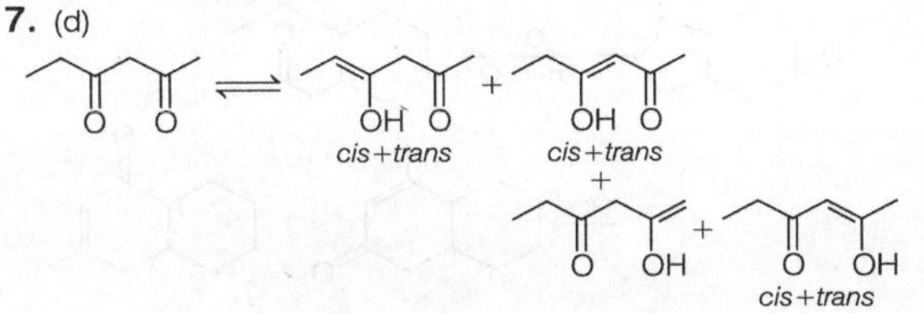

8. (d) A 1,3-diketo compound forms more stable enol than a monocarbonyls. Also ester group forms less stable enol than carbonyls. Hence, III, a 1, 3-diketone forms highest enol content while I (monocarbonyl) forms least enol content at equilibrium.

9. (d) Both option (a) and option (c) forms enol but option (b) does not form enol.

10. (b) sp^2 hybridisation is very less stable at bridgehead carbon of a bicyclic compound.

11. (c) Due to negligible steric hindrance, formaldehyde forms large hydrate content at equilibrium. Also electron withdrawing group in (IV) increases hydrate content compared to III.

12. (c) In the absence of any special stability factor, a hydrate of ketone is highly unstable.

13. (b) Diketo (I) forms highest enol content due to stabilisation of enol by intermolecular H-bonding. Electron donating resonance effect by ester group slightly decreases enol content.

14. (c)

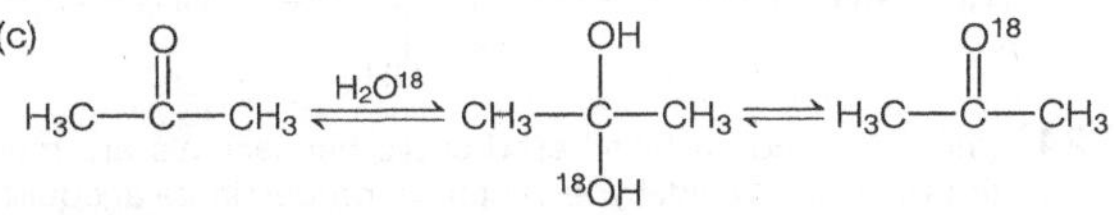

The above exchange occur due to existence of hydrate equilibrium.

15. (d)

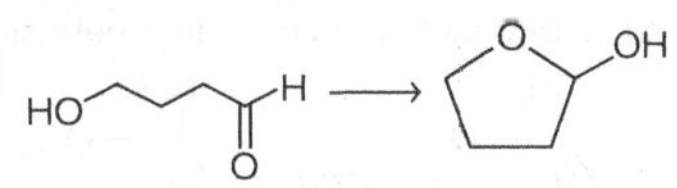

If a hemiacetal is five or six membered cyclic, they are stable.

16. (b)

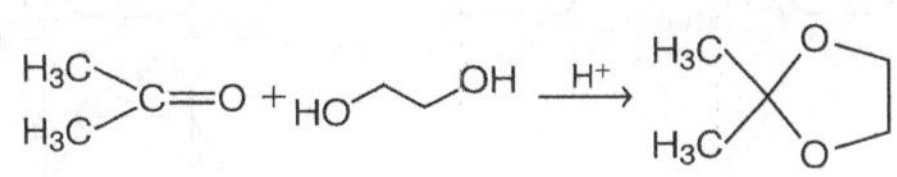

Highly stable

17. (d)

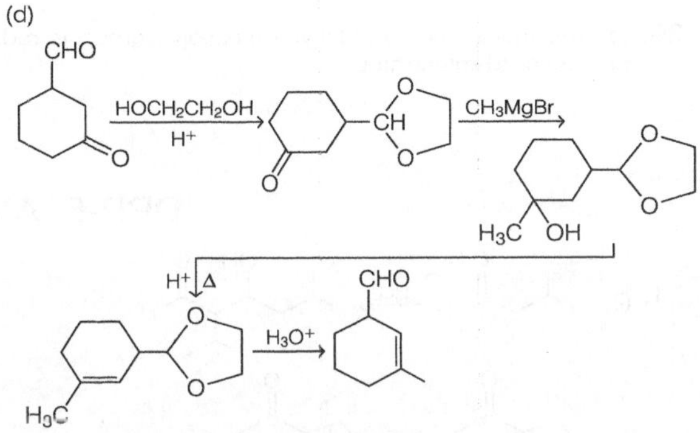

18. (d) Acetal is formed by cyclisation.

19. (b) It is 1, 3-diketo compound, forms stable enol.

20. (a) IV is most acidic as its conjugate base is resonance stabilised by two carbonyl groups, followed by III whose conjugate base is resonance stabilised by a carbonyl group and phenyl ring.

21. (b,c)

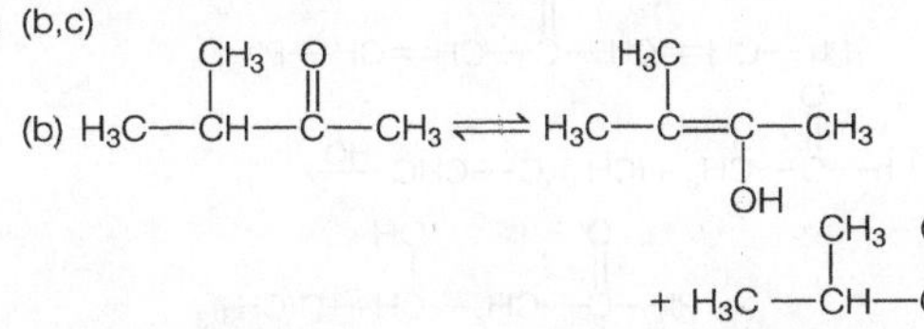

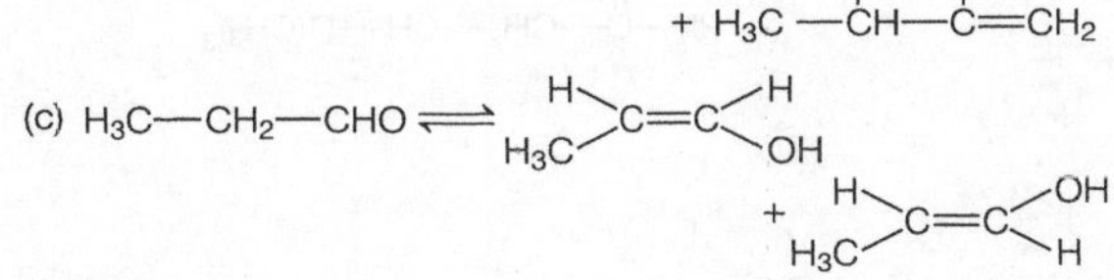

22. (a,b,c,d) Increasing temperature increases equilibrium content of less stable enol tautomers.

Enolisation decreases stability, hence introducing two or more enol groups are further more difficult.

Enols of ketones are more substituted at double bond, hence more stable.

Both acid and base catalyses keto-enol tautomerism.

23. (a,b,c) Hydrates of aldehydes and ketones are less stable than anhydrous form (*gem* diols are unstable).

Hydrates form H-bonds with water, hence hydrate content is more in water than in hexane.

This exchange occur *via* hydrate.

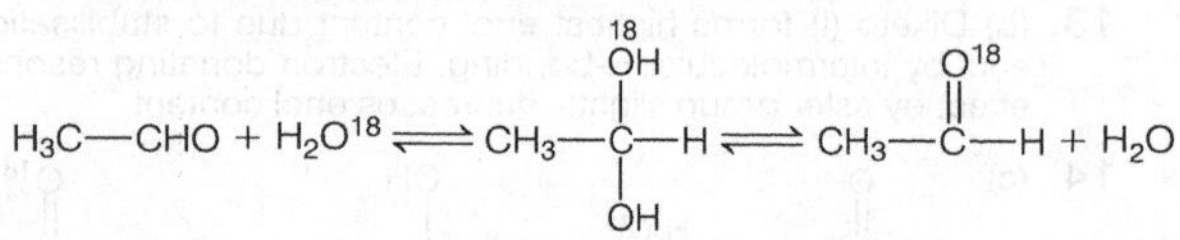

24. (a,b) Five and six-membered cyclic hemiacetals are more stable than its hydroxy aldehydes/ketones, predominate at equilibrium.

Cyclic acetals are highly stable, predominate at equilibrium.

25. (a,c) Product is an acetal, more stable than hemiacetal reactant. Acetal is formed *via* carbocation, both diastereomers are formed.

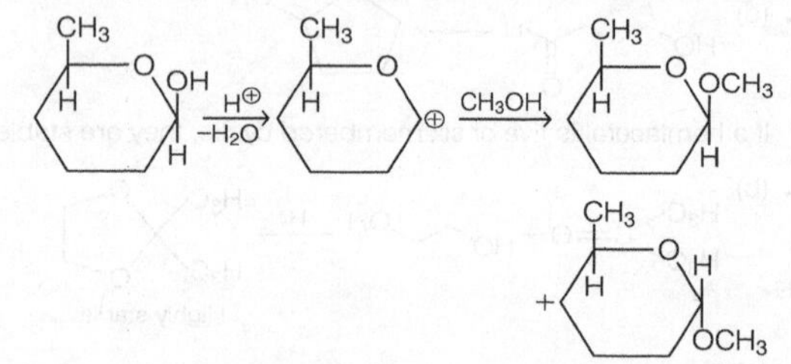

26. (c) The more reactive aldehyde functional group is reduced in preference to ketone group.

27. (b) Aldehyde being more reactive, protected first by acetal formation followed by reduction of ketone group.

28. (d)

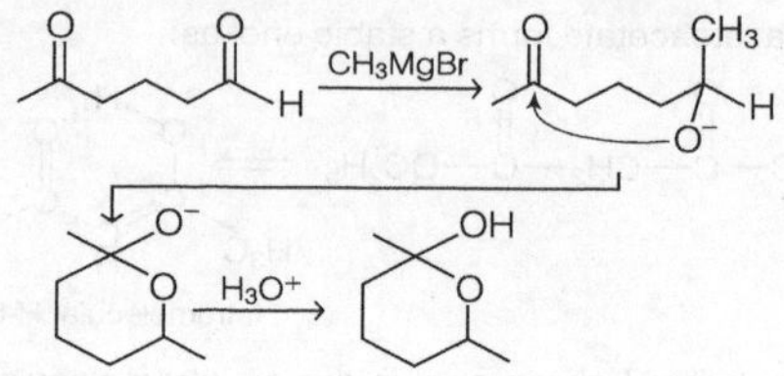

29. (6)

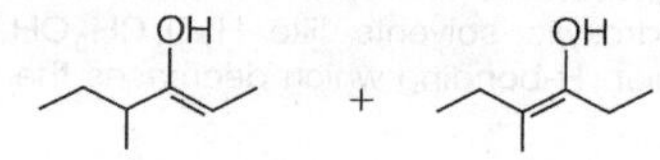

(4 stereoisomers) (2 stereoisomers)

30. (8)

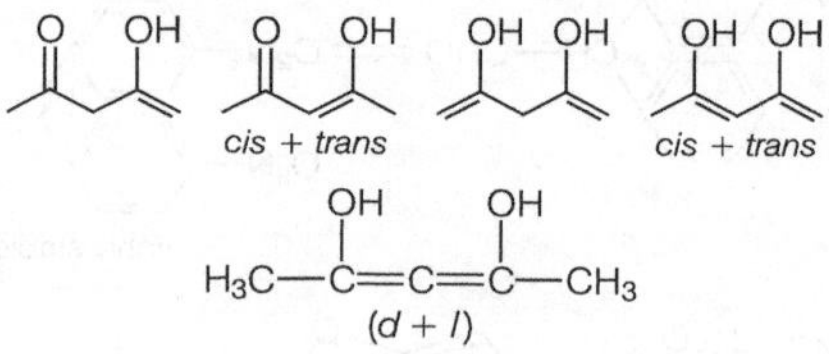

31. (7)

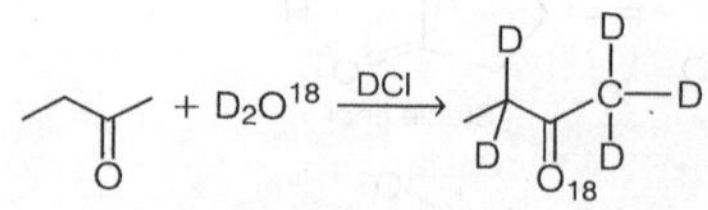

Gain of 7 units in molar mass is observed, five units due to 'D' and two units due to 'O^{18}'.

32. (6)

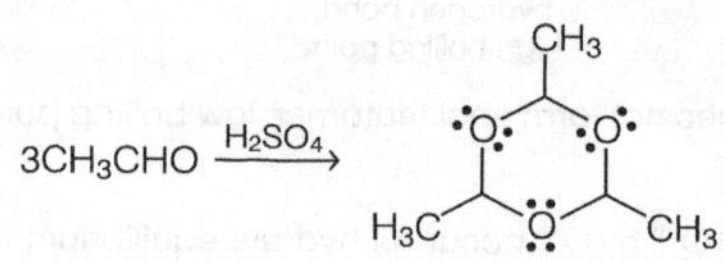

DPP-5 Aldol Condensation

1. (i)

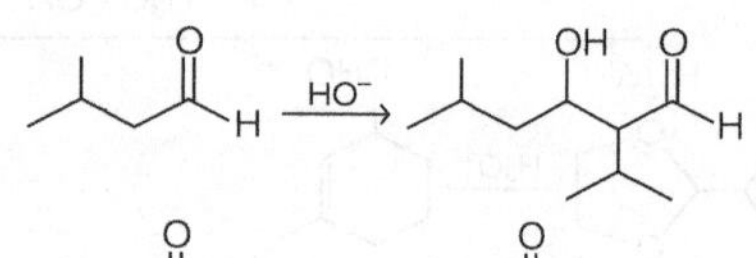

(ii)

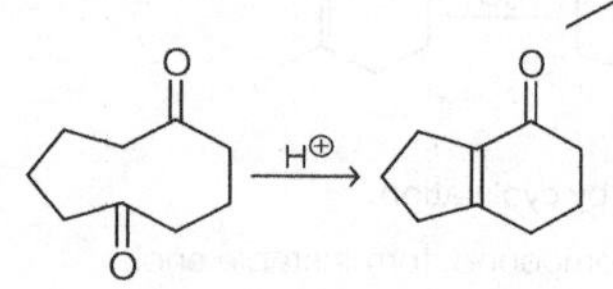

(iii) $2PhCHO + CH_3-\overset{O}{\overset{\|}{C}}-CH_3 \xrightarrow{HO^-}$

$Ph-CH=CH-\overset{O}{\overset{\|}{C}}-CH=CH-Ph$

(iv) $Ph-\overset{O}{\overset{\|}{C}}-CH_3 + (CH_3)_3C-CHO \xrightarrow{HO^-}$

$Ph-\overset{O}{\overset{\|}{C}}-CH_2-\overset{OH}{\overset{|}{C}H}-C(CH_3)_3$

(v)

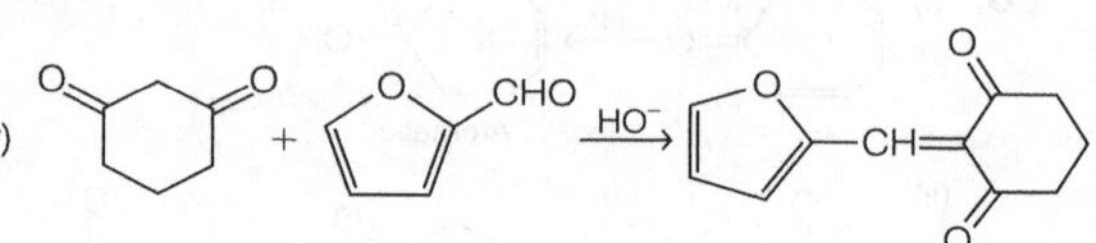

(vi)

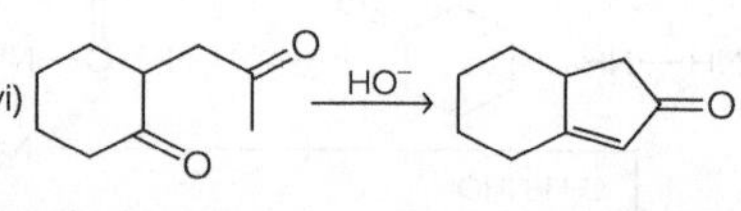

2. This is a wrong conclusion. The carbanion is formed in the fast step, has no time to react with D_2O, therefore deuteration does not take place.

3.

(i)

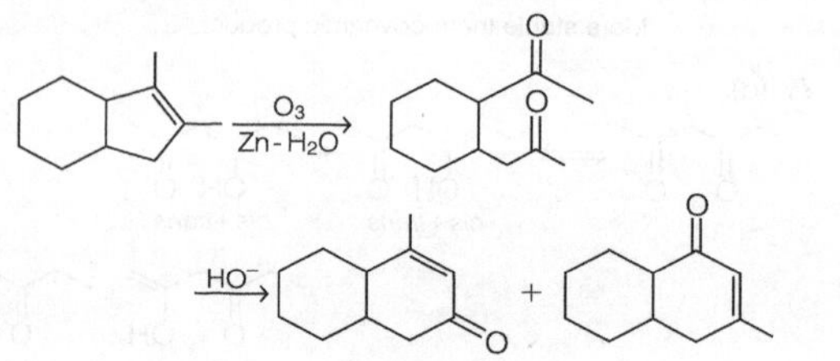

(ii) 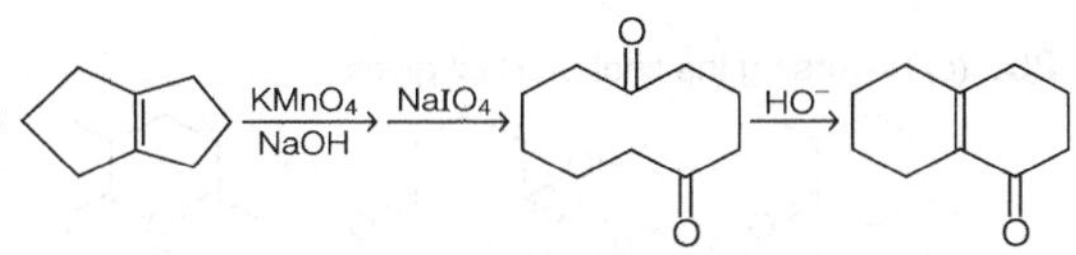

(iii) 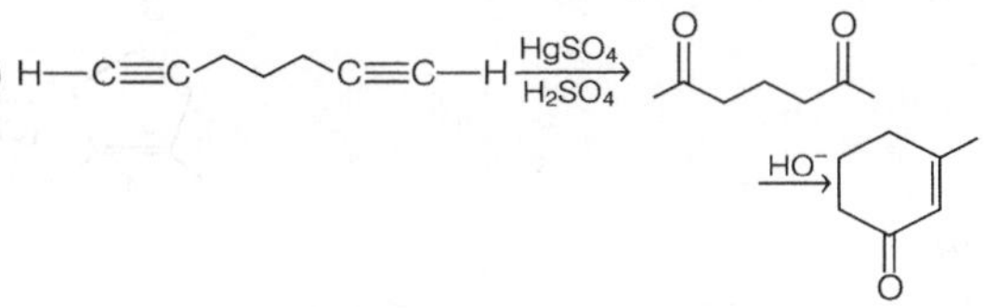

(iv) 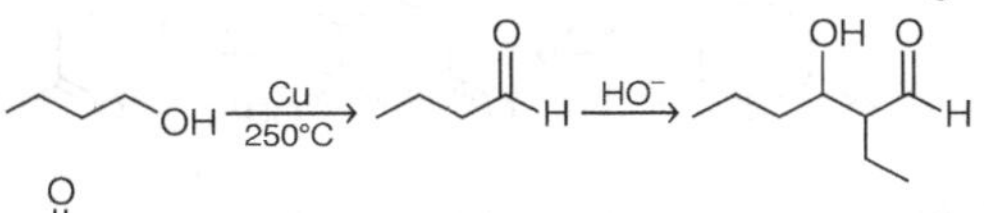

(v) 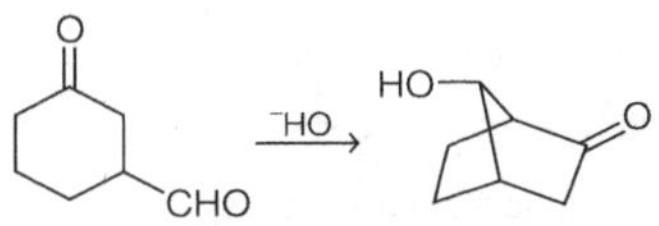

4. (i) 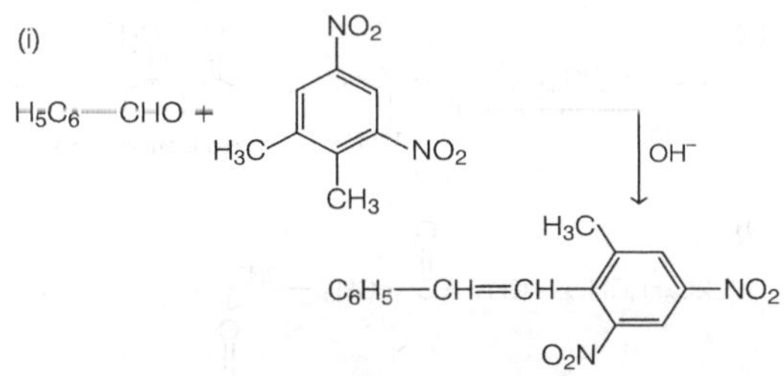

Para methyl group has more acidic hydrogen.

(ii) 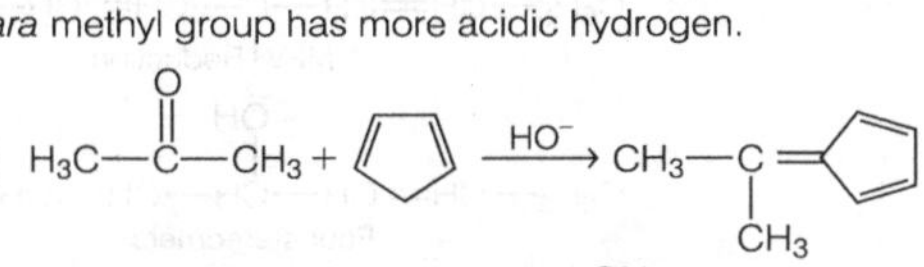

(iii) $CH_3CHO + CHCl_3 \xrightarrow{HO^-} CH_3—CH(OH)—CCl_3$

(iv) $CH_3—CHO + CH_3—C\equiv CH \xrightarrow{NaNH_2} CH_3—CH(OH)—C\equiv C—CH_3$

(v) $CH_3CHO + CH_3NO_2 \xrightarrow[\Delta]{NaOH} CH_3—CH=CH—NO_2$

5. (i) $CH_3CHO + CH_2O \xrightarrow{OH^-} HO—CH_2CH_2CHO \rightarrow HOH_2C—C(CH_2OH)_2—CHO$

(if excess of CH_2O)

(ii) $CH_3—CO—CH_2—CH_3 \xrightarrow{OH^-} CH_3—CH_2—C(OH)(CH_3)—CH_2—CO—CH_2CH_3$

(iii) $CH_3—CH(CH_3)—CHO + CH_3CHO \xrightarrow{HO^-} CH_3—CH(CH_3)—CH(OH)—CH_2—CHO$

(iv) 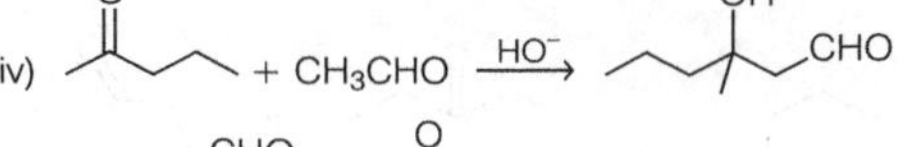

(v) 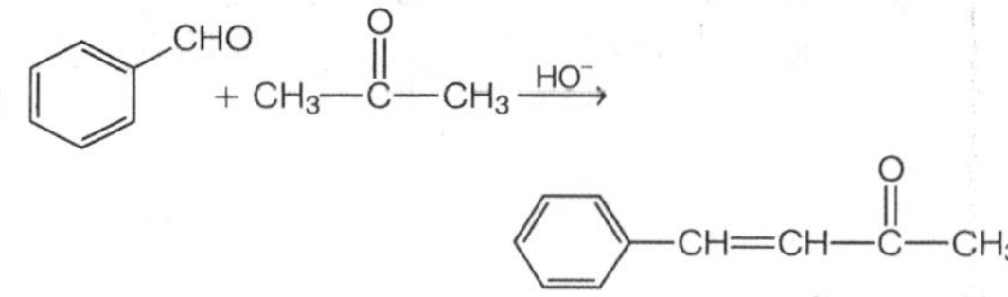

6. (b)

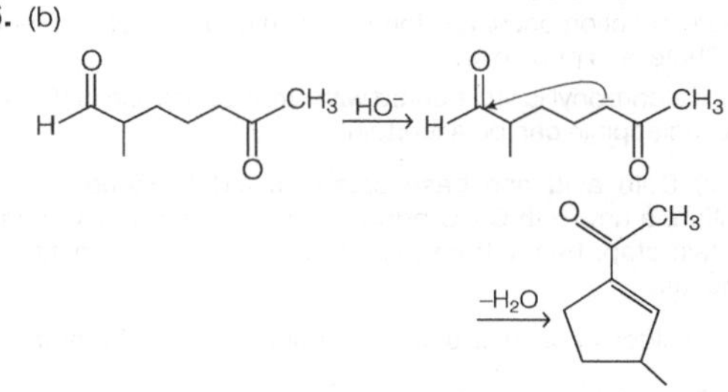

7. (a) An α-β C—C bond is always formed in aldol reaction.

8. (c) Nucleophilic addition at carbonyl carbon of cyclopropane releases angle strain.

9. (c)

$$CH_3NO_2 \xrightarrow{HO^-} \bar{C}H_2—NO_2 \xrightarrow{CH_2O} CH_2(OH)—CH_2—NO_2 \xrightarrow{\Delta} \text{Product}$$

10. (c) In acid catalysed aldol condensation, enol acts as a nucleophile.

11. (c) $CH_2O + CH_3CHO \xrightarrow{HO^-} HOCH_2—CH_2—CHO \xrightarrow{CH_2O \,|\, HO^-} HOCH_2—CH(CH_2OH)—CHO \xrightarrow[CH_2O]{HO^-} HOCH_2—C(CH_2OH)_2—CHO$

Repeated aldol condensation occur due to very high reactivity of formaldehyde towards nucleophile.

12. (a) 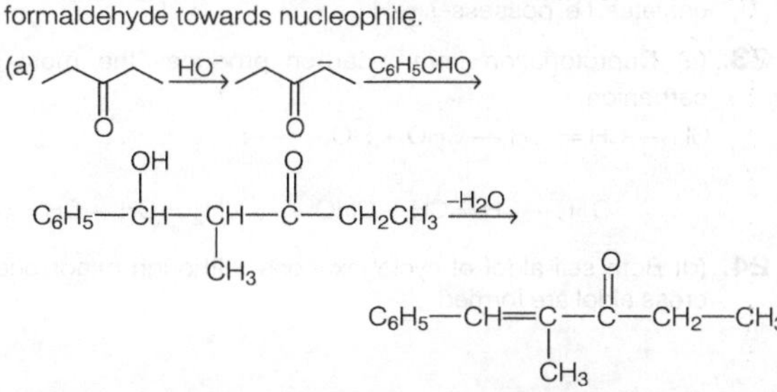

13. (b) Enolate is formed in first fast step which undergo nucleophilic addition in second slow step.

14. (a)

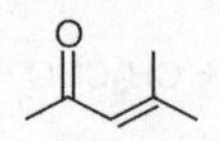

It is an α, β-unsaturated ketone which can be formed in an aldol condensation followed by dehydration.

15. (b)

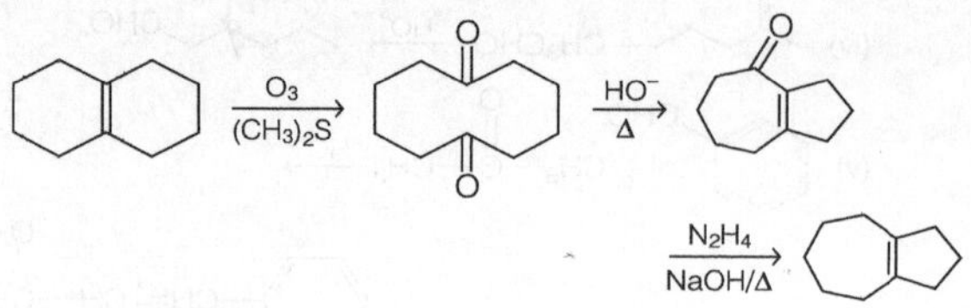

16. (c,d) If an aldol product is continuously removed from the reaction mixture, reaction occur predominantly in direction of that product (Le-Chatelier's principle).

Also, if a carbonyl is first treated with strong base like LDA, its role as a nucleophile can be ascertained.

17. (a,b,c) Both acid and base catalyses aldol reaction. In aldol reaction, a new α-β C—C bond is always formed. α-H is lost in first fast step, hence both CH_3CHO and CD_3CHO react at the same rate.

18. (b,d) In aldol reaction as α-β C—C bond is always formed.

19. (a,c,d) All of these react in aldol reaction giving six-membered ring *via* intramolecular reaction.

Option (b) gives five and seven membered ring intramolecular aldol condensation.

20. (a,c,d)

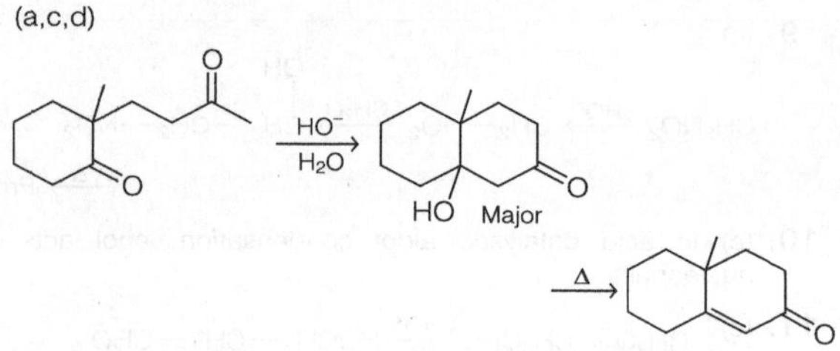

Also, the starting compound has three different types of α-H, three different aldols can be formed in principle. If a pure enantiomer of starting compound is taken, a pair of diastereomers of aldol would be formed.

21. (a) Reversibility of aldol into carbonyls establishes that carbanion nucleophile is formed in first fast reversible step.

22. (c) It would be true only if both carbonyls are capable of forming enolates, i.e. possesses α-H.

23. (a) Deprotonation from γ-carbon produces the most stable carbanion.

$$CH_3—CH{=}CH—CHO + HO^- \longrightarrow$$

$$\bar{C}H_2—CH{=}CH—CHO \longleftrightarrow CH_2{=}CH—CH{=}\overset{O^-}{\overset{|}{C}}—H$$

24. (d) Both self-aldol of cyclohexanone (although minor one) and cross aldol are formed.

25. (c) Reversing the final product gives

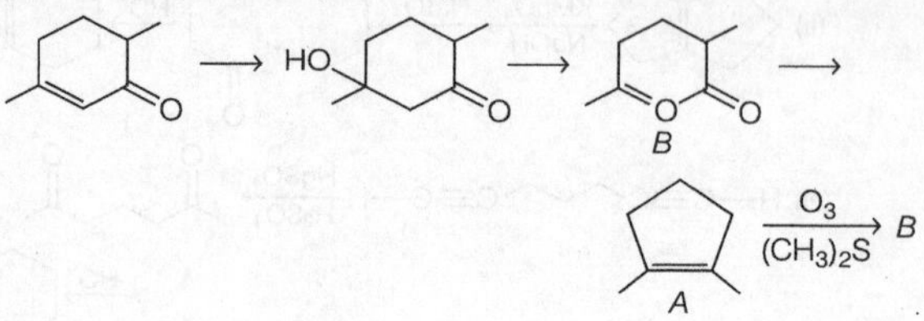

26. (d)

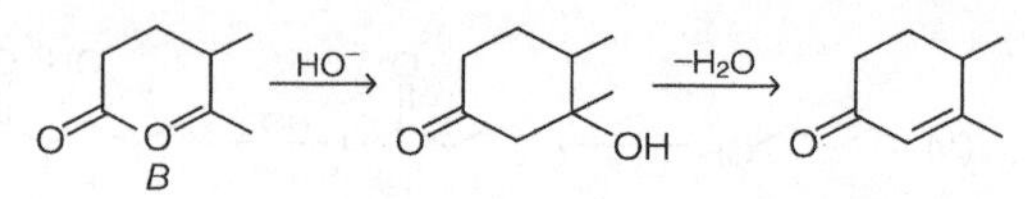

27. (d)

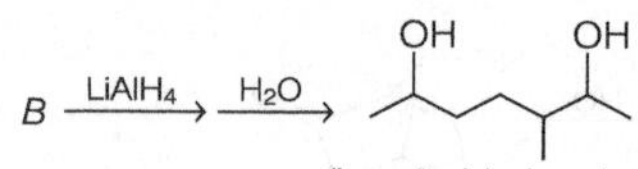

(has 3 chiral carbons)

28. (6)

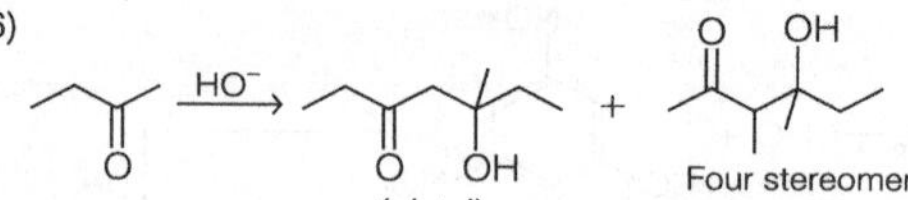

(*d* + *l*) Four stereomers

29. (4)

$$2C_6H_5CHO + CH_3—\overset{O}{\overset{\|}{C}}—CH_3 \xrightarrow{OH^-}$$

$$C_6H_5—CH{=}CH—\overset{O}{\overset{\|}{C}}—CH{=}CH—C_6H_5$$

MPV Reduction

$$C_6H_5—CH{=}CH—\overset{OH}{\overset{|}{C}H}—CH{=}CH—C_6H_5$$

Four stereomers

30. (3) All three hydrogen can be deprotonated from nitromethane giving

$$3CH_3CHO + CH_3NO_2 \xrightarrow{OH^-} CH_3—\overset{OH}{\overset{|}{C}H}—\underset{CH(OH)CH_3}{\underset{|}{\overset{CH(OH)CH_3}{\overset{|}{C}}}}—NO_2$$

31. (4)

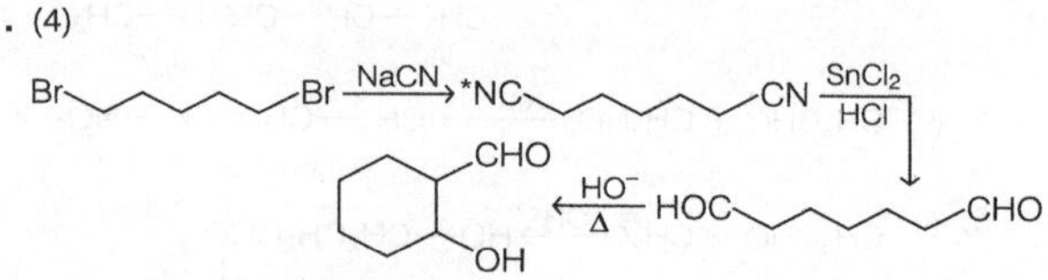

Four stereomers

32. (5)

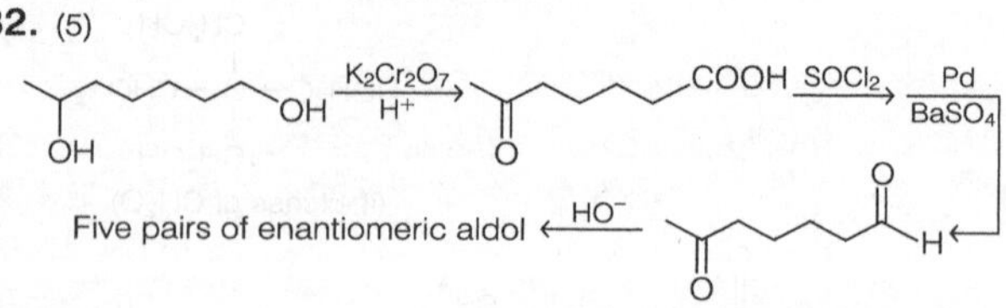

DPP-6 Cannizaro Reaction

1.

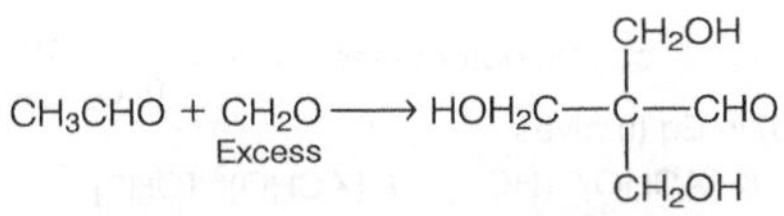

Repeated aldol

The above aldol product undergoes Cannizaro reaction with NaOH but the same does not happen when weak base K_2CO_3 is used.

2. (iii) < (ii) < (i) < (iv). Electron withdrawing group increases reactivity of benzaldehyde in Cannizaro reaction while electron donating groups like CH_3O—, CH_3—, etc., decreases reactivity.

3. In Cannizaro reaction, hydride transfer occur as,

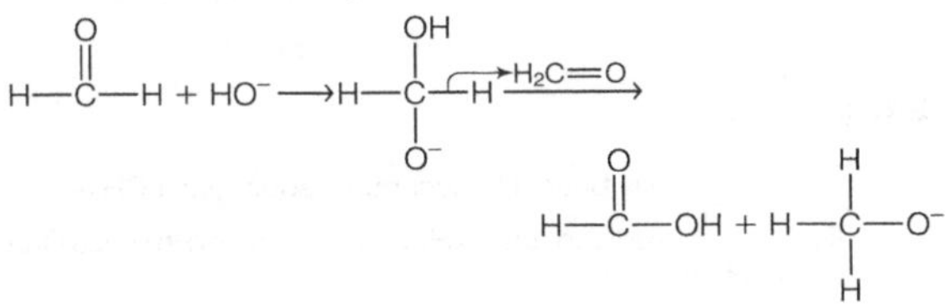

Therefore, there is no scope of D in corporation from solvent as C—D bond.

4. In moderately conc. NaOH solution,

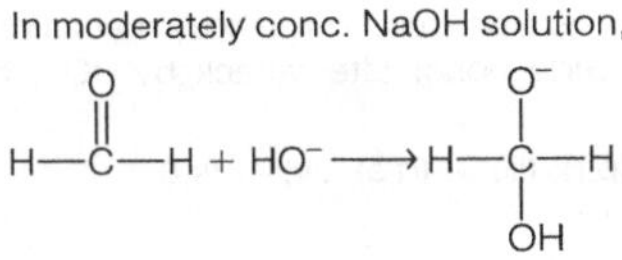

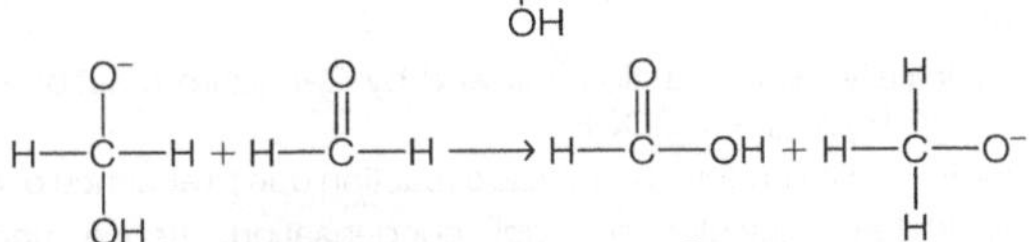

Overall : $2H_2CO + OH^- \longrightarrow H\text{—}\overset{O}{\overset{\|}{C}}\text{—}O^- + CH_3OH$

$R = k\,[H_2CO]^2\,[OH^-]$: 3rd order

In strongly basic medium,

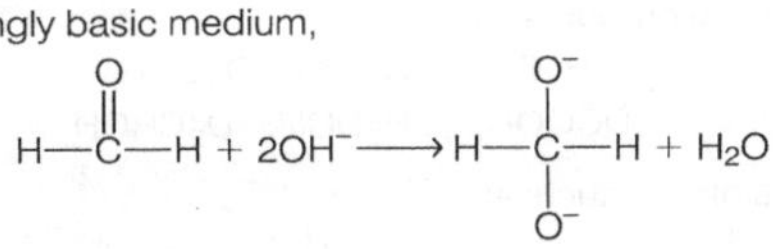

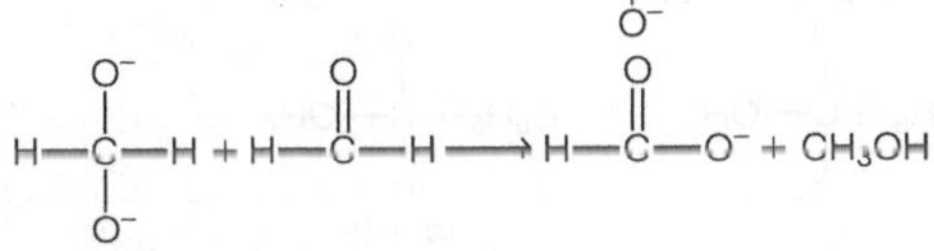

Overall : $2CH_2O + 2OH^- \longrightarrow CH_3OH + HCOO^-$

Rate $= k\,[CH_2O]^2\,[OH^-]^2$: 4th order

5. $CH_3CHO + LDA \longrightarrow \overset{+}{Li}\,\overset{-}{C}H_2CHO$

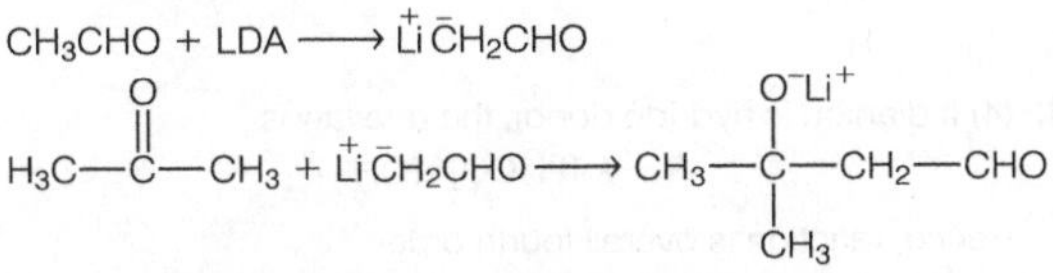

Here, LDA = Lithium diisopropyl amide $\{(CH_3)_2CH\}_2NLi$. It is a very strong base, forms salt with aldehyde and ketone in quantitative amount.

6. (c)

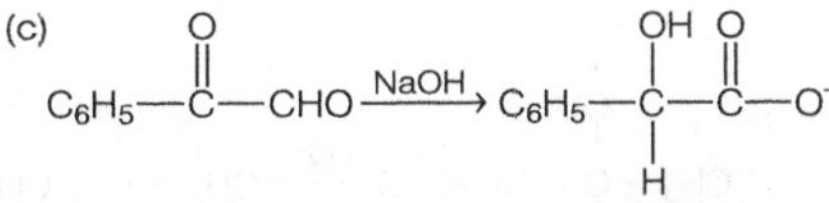

Intramolecular Cannizaro

7. (a)

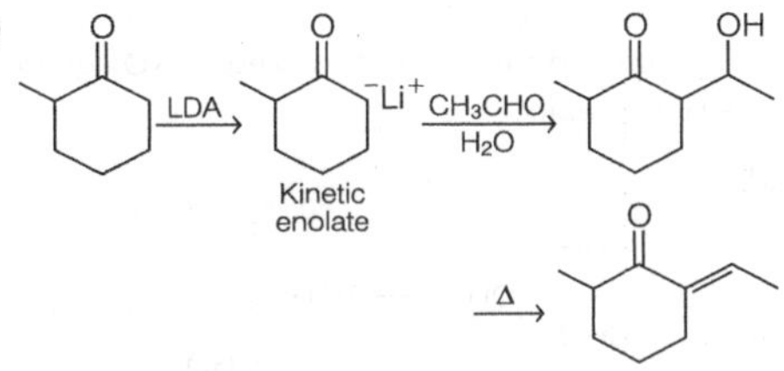

8. (b) Dianion is better hydride donor. Also the electron donating group (CH_3O) increases hydride (H^-) donating ability .

9. (d) It is a simple Tischenko reaction.

10. (d)

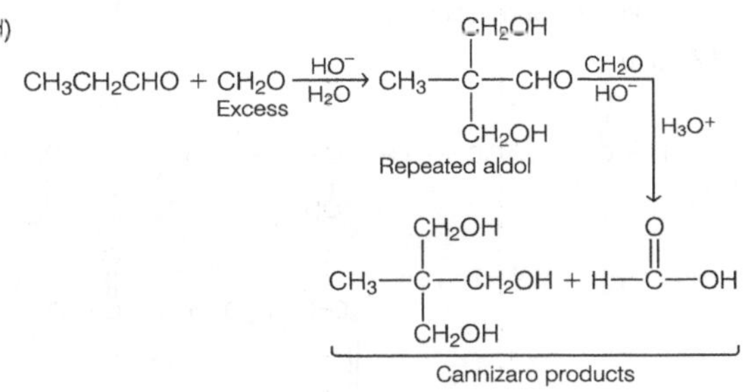

11. (d) Intramolecular Cannizaro reaction occur.

12. (a)

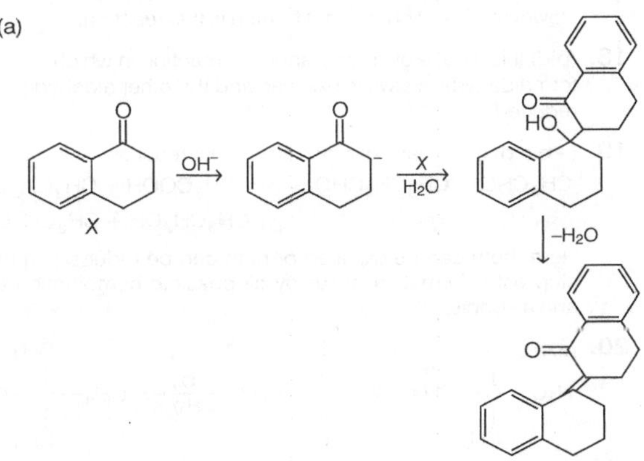

13. (d) For formation of dianion hydride donor, an additional mole of NaOH is consumed.

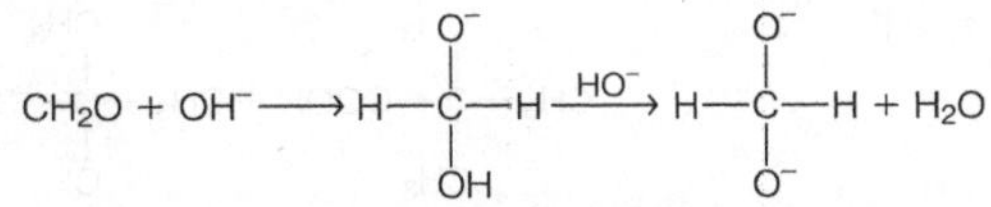

Hence, rate becomes 2nd order with respect to HO^- and fourth order overall.

14. (b)

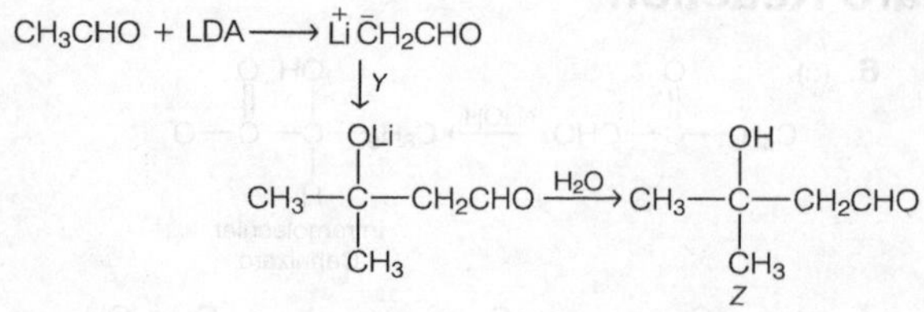

15. (b) —CHO *para* to electron withdrawing $—NO_2$ group is better hydride (H^-) acceptor.

16. (a,c,d)

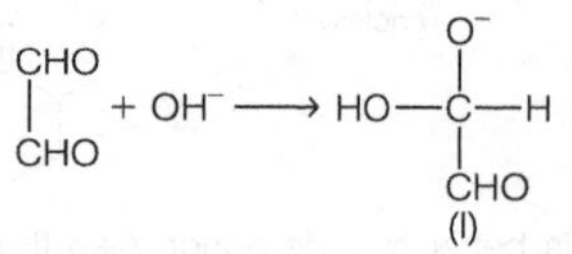

(I) is formed by nucleophilic attack in the first step. (I) may undergo intramolecular hydride transfer giving option (a) or may undergo intermolecular hydride transfer to C_6H_5CHO giving option (c) and option (d).

17. (a,b,c)

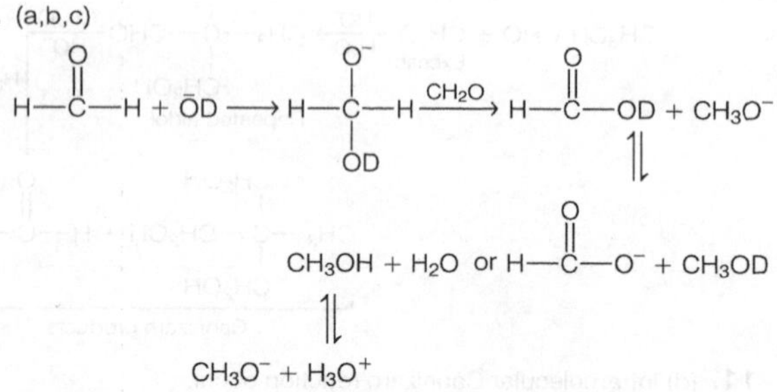

However, C—D bond is not formed in this reaction.

18. (b,d) It is a case of cross Cannizaro reaction in which formaldehyde is always oxidised and the other aldehyde is reduced.

19. (a,b,c,d) It is a case of cross Tischenko reaction.

$$CH_3CHO + CH_3CH_2CHO \longrightarrow CH_3COOH + CH_3CH_2COOH + CH_3CH_2OH + CH_3CH_2CH_2OH$$

Here, both can be oxidised or both can be reduced. In the next step ester formation occur by all possible combination of acid and alcohols.

20. (a,c)

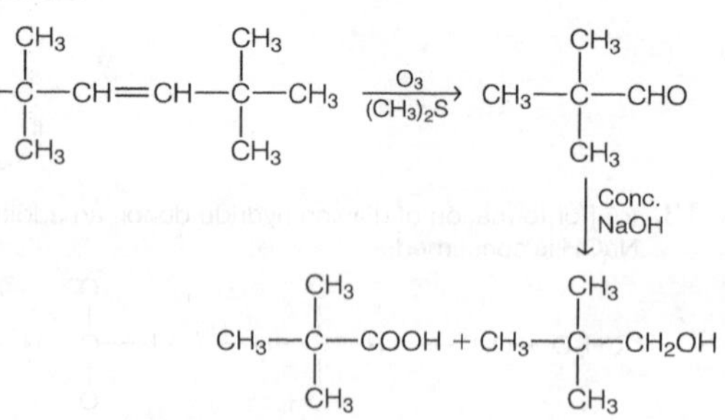

21. (c) $R = k_1 [I][A—\overset{O}{\overset{\|}{C}}—H]$ ...(i)

The intermediate "I" can be determined as : $K = \dfrac{[I]}{[HO^-][ACHO]}$

Substituting [I] in Eq (i) gives

$$R = k_1 K [ACHO]^2 [HO^-] = k [ACHO]^2 [OH^-]$$

22. (d) When reaction is carried out in D_2O, no C—D bond formation occur. It establishes the correctness of above mechanism that reaction proceed by H^- (hydride) transfer in the slow rate determining step.

23. (b) Formation of dianion involves two moles of HO^- per mole of aldehyde

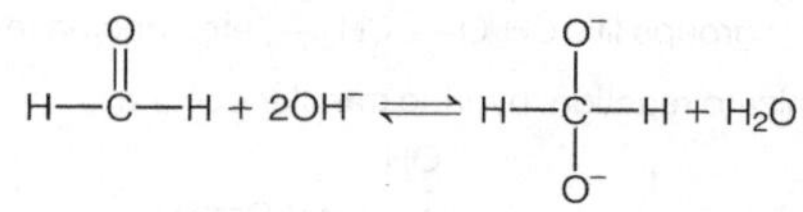

24. (a)

(i) $F—\overset{O}{\overset{\|}{C}}—H$ is better H^- acceptor, reduced to FCH_2OH.

(ii) CH_2O is oxidised and C_6H_5CHO is reduced in cross Cannizaro reaction.

(iii) Electron withdrawing $—NO_2$ makes —CHO a better hydride acceptor, hence *p*-nitrobenzaldehyde is reduced and benzaldehyde is oxidised.

(iv) *p*-nitrobenzaldehyde is reduced. *p*-methoxy benzaldehyde becomes better hydride donor after attack by HO^-, hence oxidised.

Hence, (i) → (p), (ii) → (p,r), (iii) → (q,s), (iv) → (s,t)

25. (b)

(i) Initially, aldol reaction followed by Cannizaro reaction giving $C(CH_2OH)_4$ + HCOOH.

(ii) F—CHO undergo Cannizaro reaction due to absence of α -H.

(iii) It has difficulty in aldol condensation, hence undergo Cannizaro reaction predominantly.

(iv) All aldol, Cannizaro and Claisen reaction occur.

Hence, (i) → (p,q), (ii) → (q), (iii) → (p,q), (iv) → (p,q,r)

26. (6) Possible products are

CH_3OH, HCOOH, D_3COH

DCOOH, CH_2DOH, D_2CHOH

27. (5) Possible products are

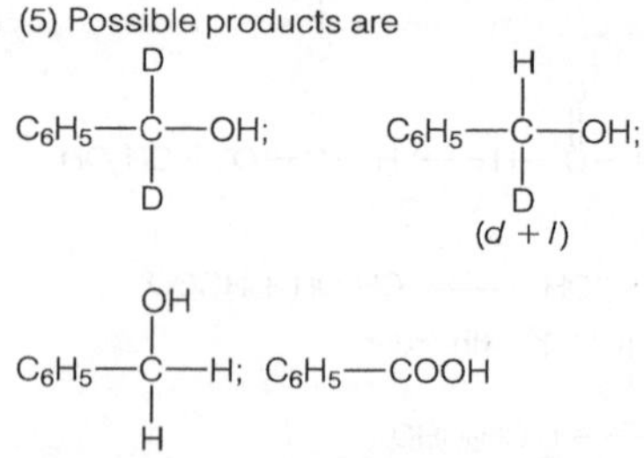

28. (4) If dianion is hydride donor, the rate law is

$$R = k [H_2CO]^2 [OH^-]^2$$

Hence, reaction is overall fourth order.

DPP-7 Beckmann Rearrangement

1. 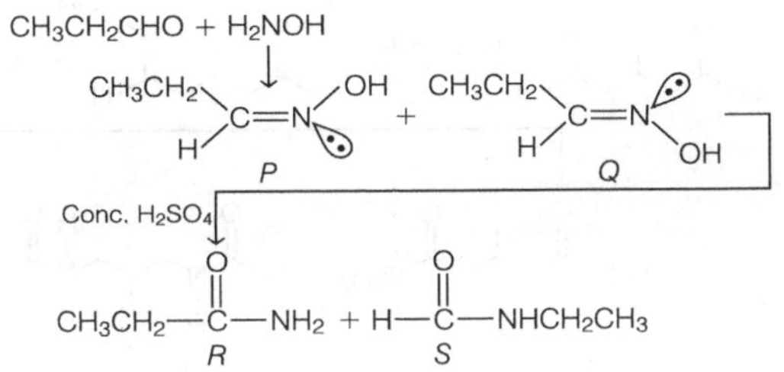

2. (i) 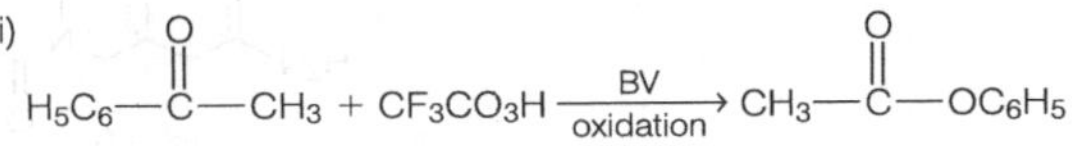

Phenyl group has greater migrating aptitude than methyl group.

(ii) 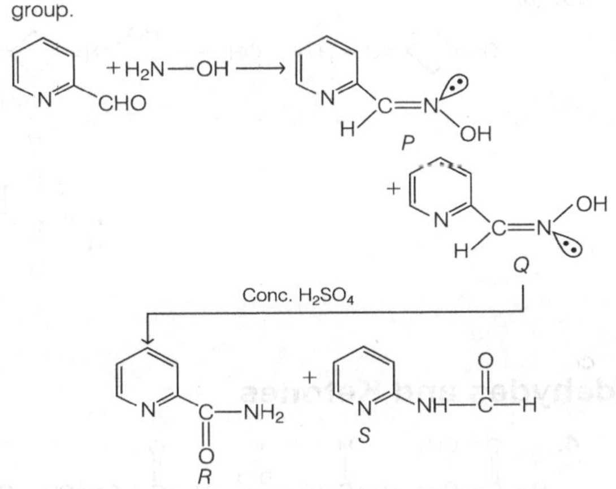

(iii) 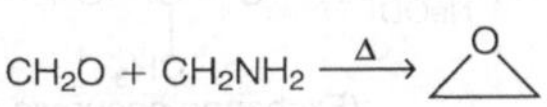

3. (i) 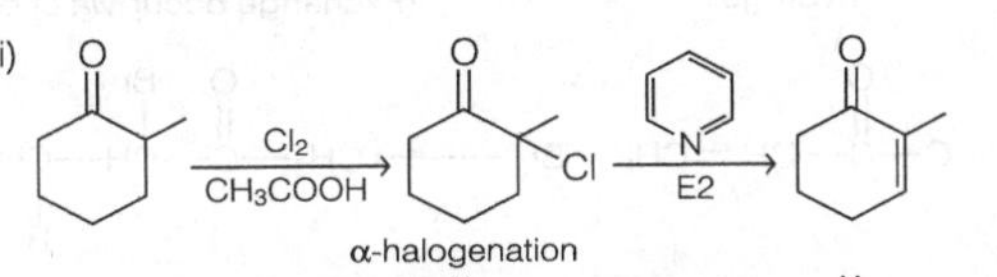

(ii) 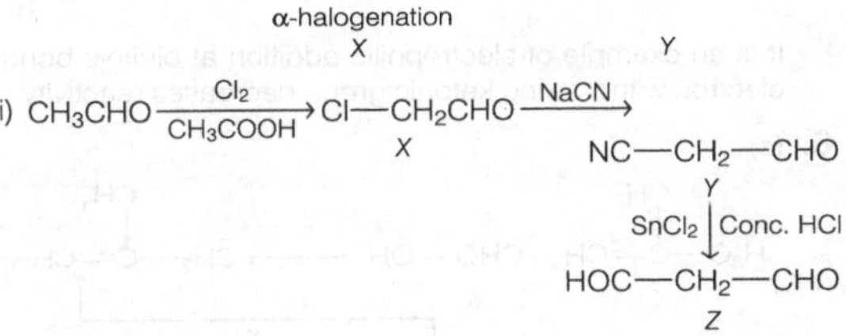

Stephen reaction

(iii) 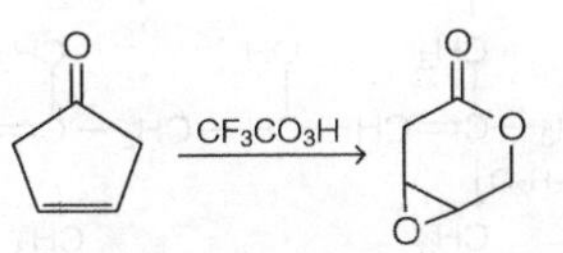

4. (d) Unsymmetrical carbonyls give more than one oximes with hydroxylamine, hence more than one amides in the subsequent Beckmann rearrangement.

5. (a) Symmetrical oximes give single amide in Beckmann rearrangement.

6. (b) In BV oxidation, tertiary alkyl group has greater migrating aptitude than primary and secondary alkyl group.

7. (a) 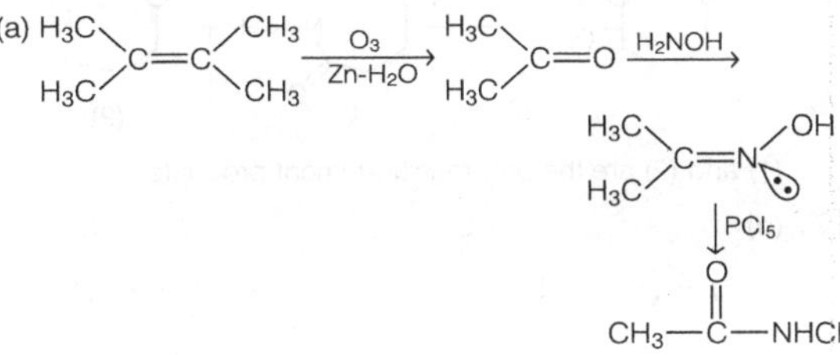

8. (c) 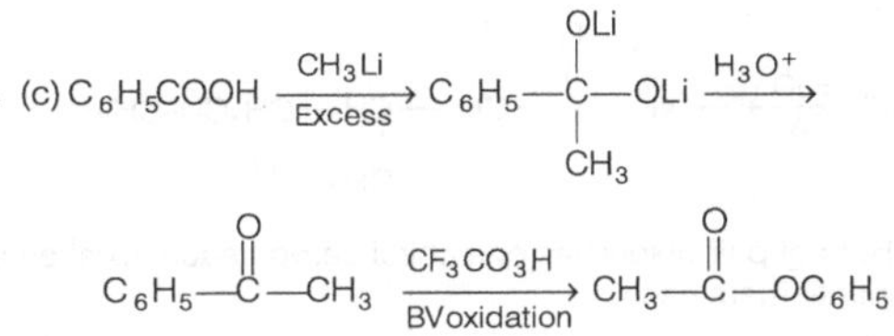

9. (c)

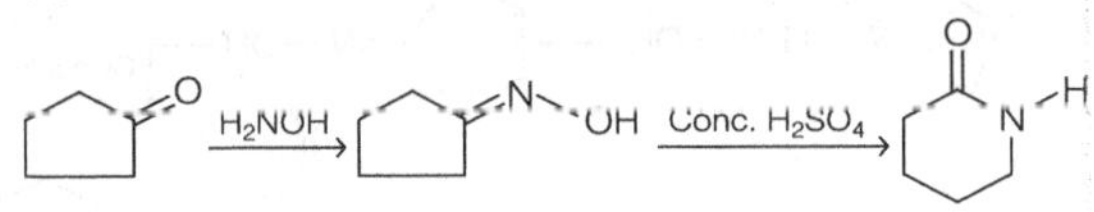

10. (b)

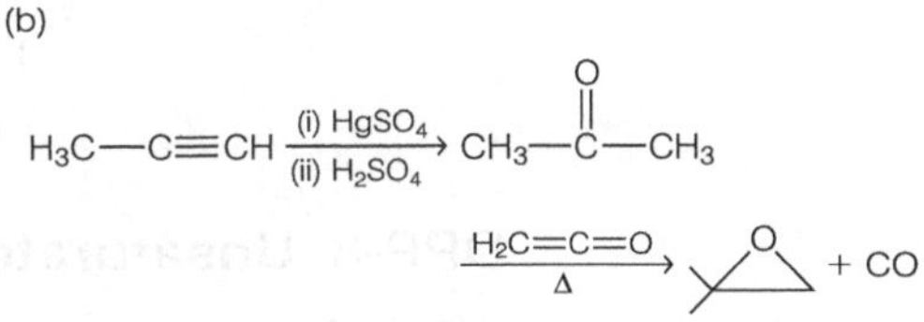

11. (a,b,c) Beckmann rearrangement of oximes can be carried out with either PCl_5 or Conc. H_2SO_4 or $SOCl_2$.

12. (a,c,d)

(a) Addition of carbene at C=O gives oxirane.

(c) Epoxidation of alkene gives oxirane.

(d) 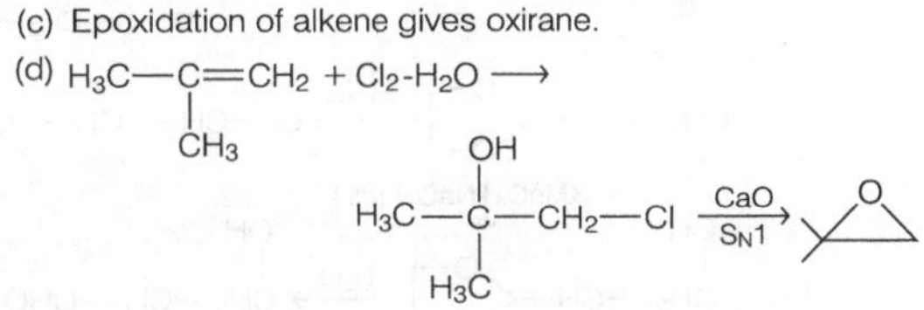

13. (a,c)

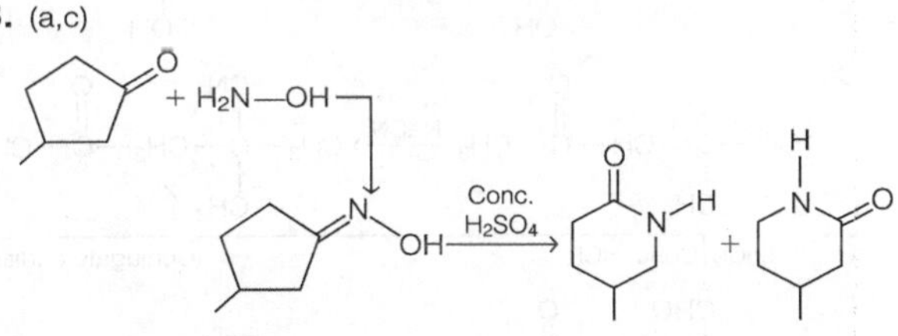

14. (b,c,d) $CH_3-\overset{O}{\overset{\|}{C}}-C_6H_5 \xrightarrow[\text{BV oxidation}]{CF_3CO_3H} CH_3-\overset{O}{\overset{\|}{C}}-OC_6H_5$ Phenyl acetate

15. (c)

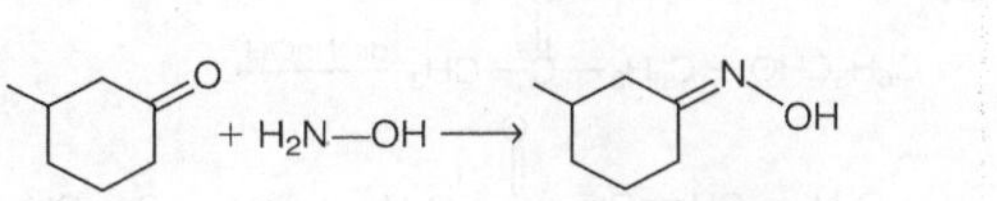

has a chiral carbon (4 Stereoisomers)

16. (c)

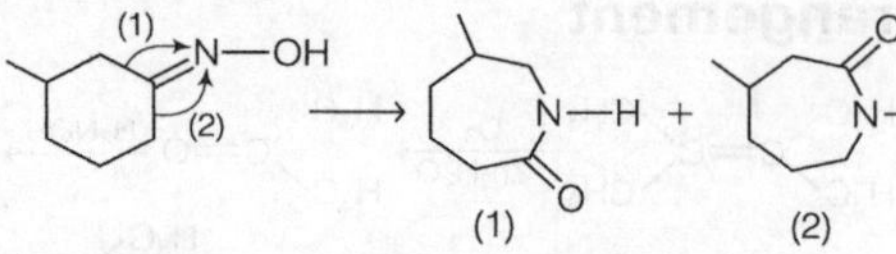

(1) and (2) are the only rearrangement products.

17. (d)

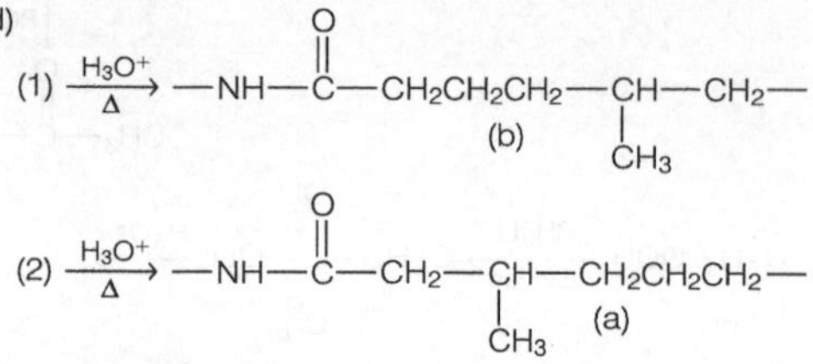

18. (4) Both of pair below has one chiral carbon each, exist as a pair of enantiomers each.

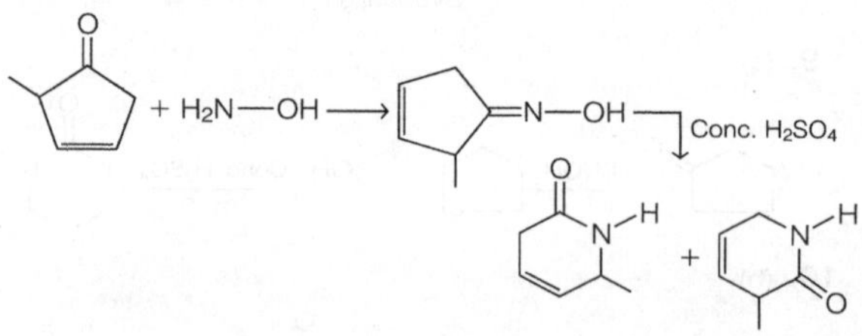

19. (4)

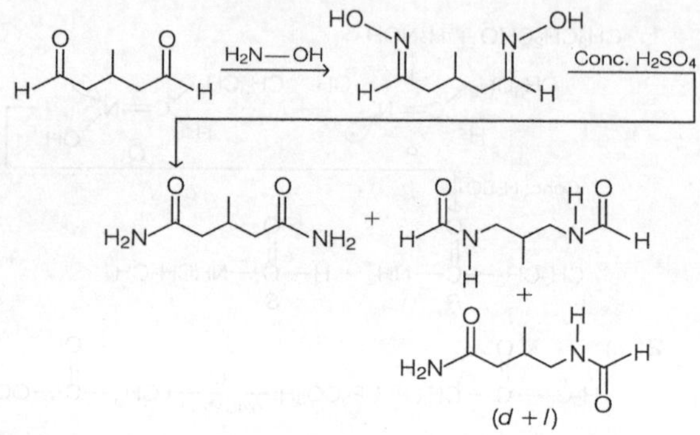

20. (6)

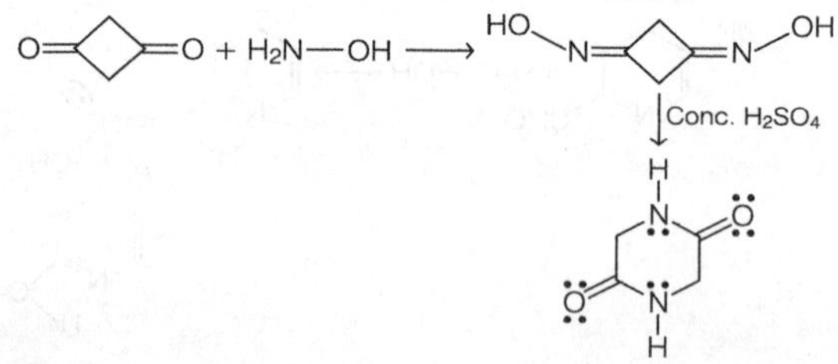

DPP-8 Unsaturated Aldehydes and Ketones

1. Direct hydroxylation with $KMnO_4/H_2O$ will result in oxidation of —CHO, hence it needs to be protected first by acetal formation. However, before acetal formation, double bond must be masked otherwise in acidic medium, alcohol will attack olefinic bond first.

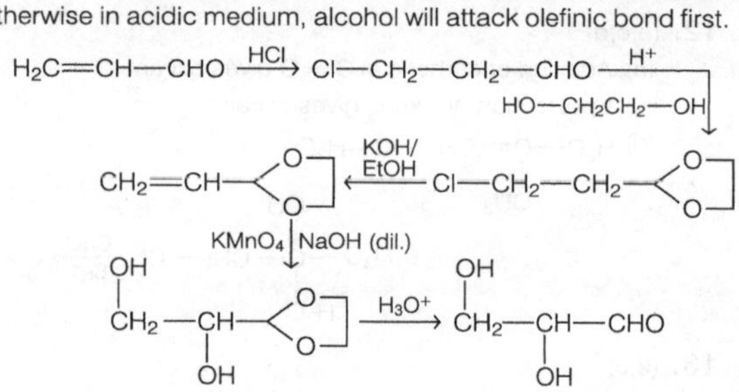

2.

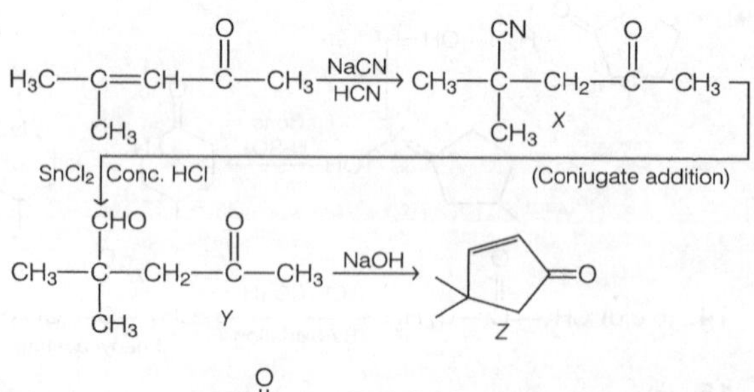

3.

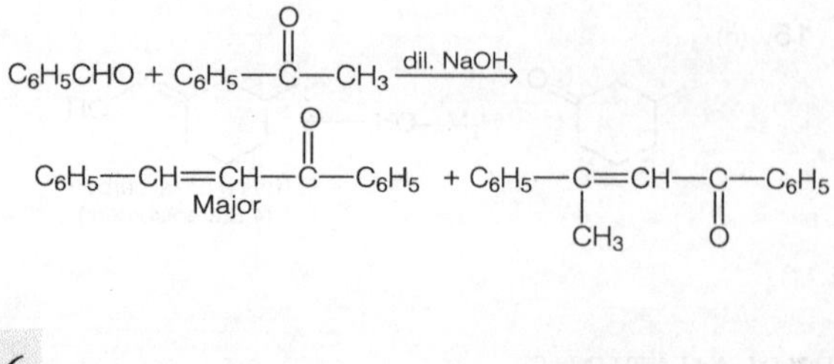

4.

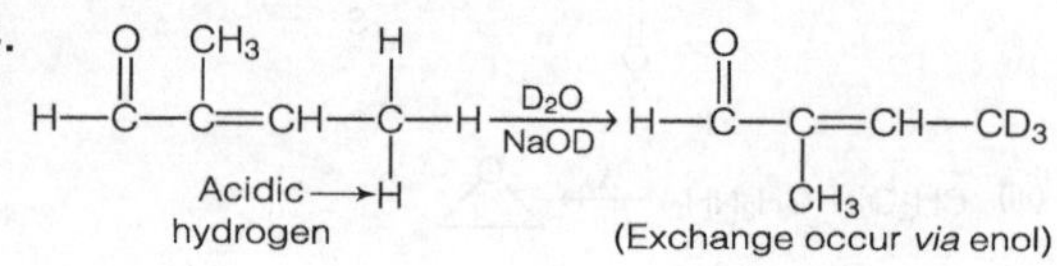

5.

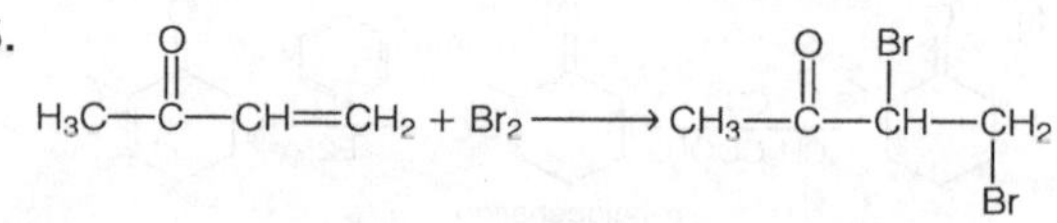

It is an example of electrophilic addition at olefinic bond, hence electron withdrawing ketonic group decreases reactivity.

6. (c)

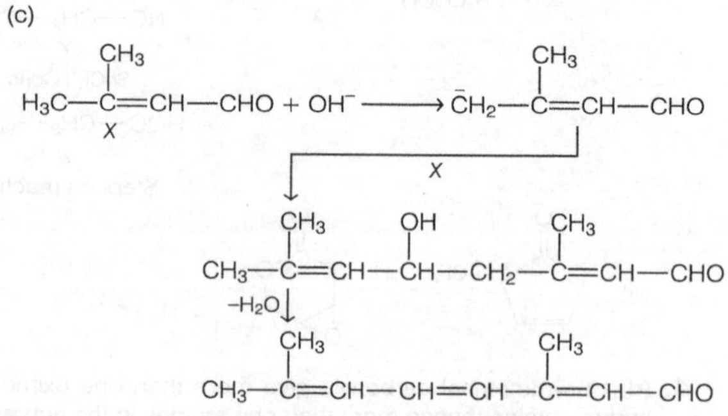

7. (d)

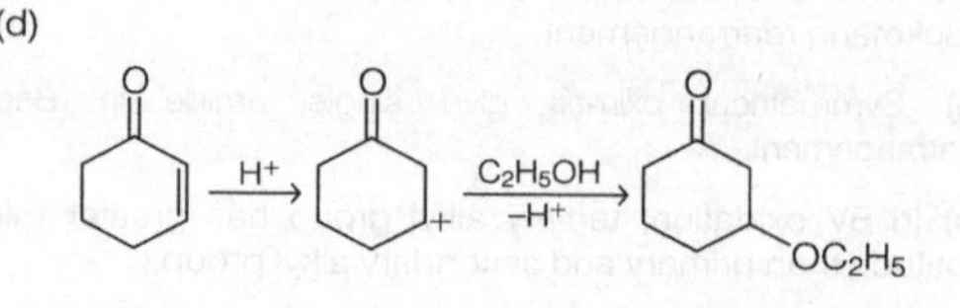

8. (a)

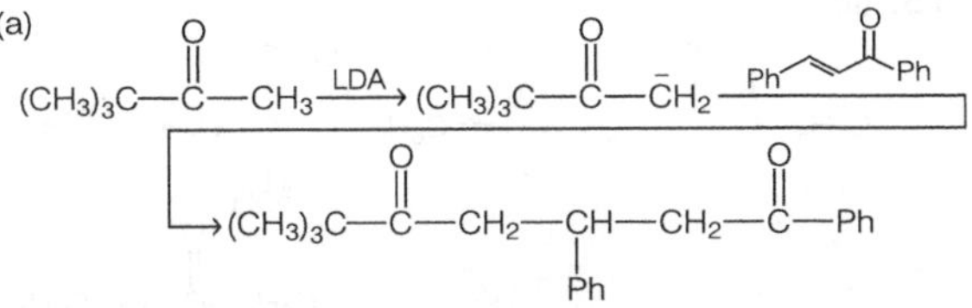

(Conjugate addition)

9. (b)

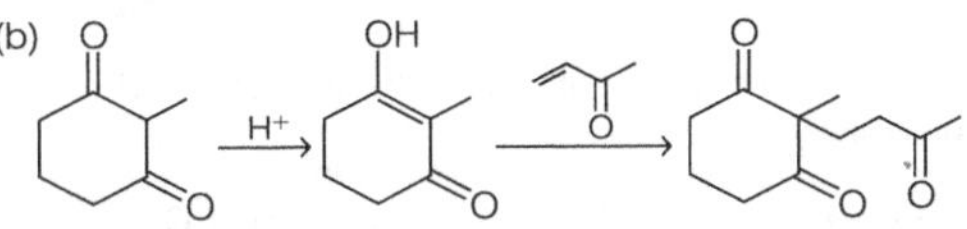

10. (c)

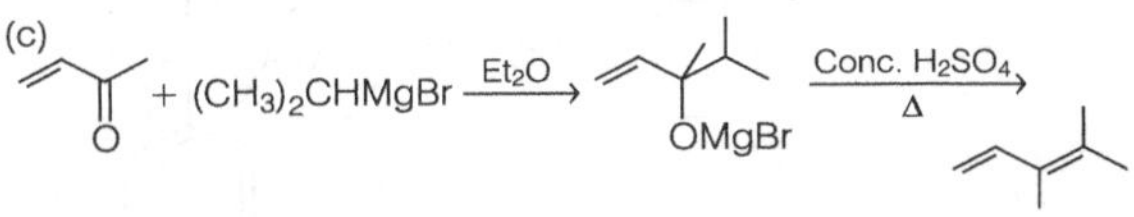

11. (a,b) Weak nucleophiles (Cl^-, NH_3) undergo conjugate addition. Strong nucleophiles (R^-, H^-) undergo direct addition.

12. (c,d) Strong nucleophiles (CH_3MgBr, CH_3Li) undergo direct addition.

13. (a,b,d) Strong nucleophiles (like H^- from $LiAlH_4$) undergo direct addition while weak nucleophiles (like R_2CuLi, RNH_2etc.) undergo conjugate addition at olefinic bond.

14. (a,b) CH_3COOH undergoes nucleophilic addition at conjugate position giving indicated product as major one.

15. (b) *Y* has 2 carbon extra than unsaturated ketone, hence starting compound must be acetylene and nucleophile undergoing conjugate addition is $H—C\equiv C^-Na^+$.

16. (b)

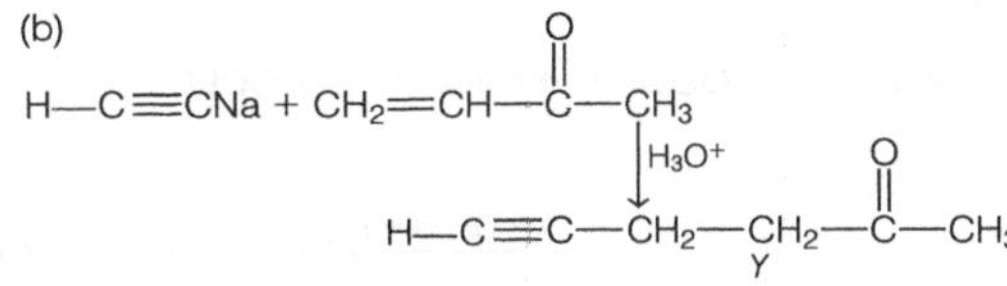

17. (c)

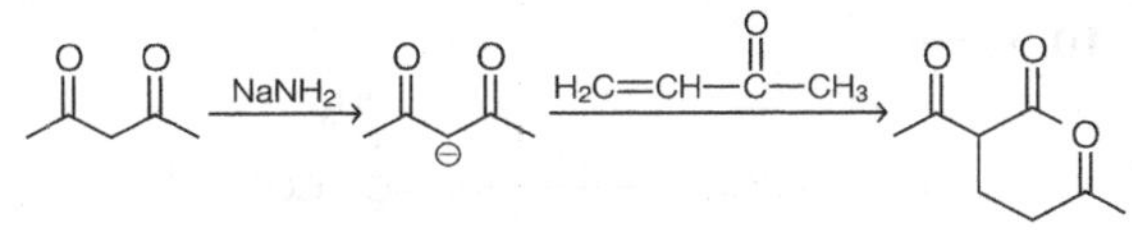

18. (4)

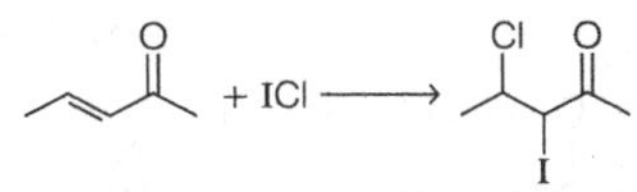

(Two chiral carbons)

19. (8)

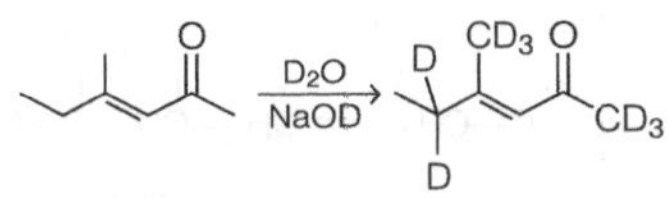

20. (8)

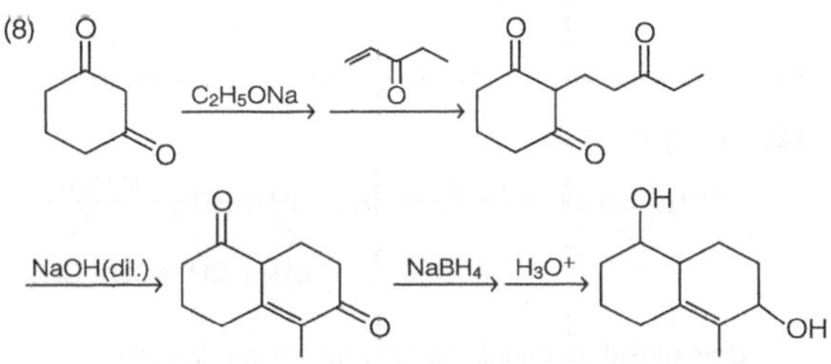

(3 chiral carbons)

DPP-9 **Reformatsky, Witting Benzil-Benzilic Acid Rearrangements, etc.**

1. (d) Free radical coupling reaction of aldehydes and ketones.

2. (b) α-coupling of aldehydes and ketones occur with nitrous acid.

3. (d)

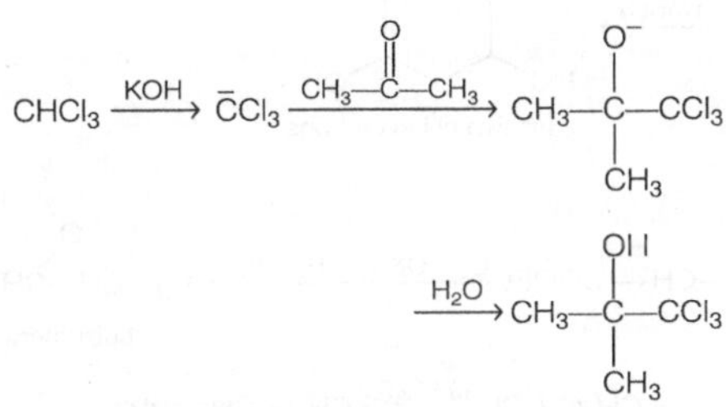

4. (d)

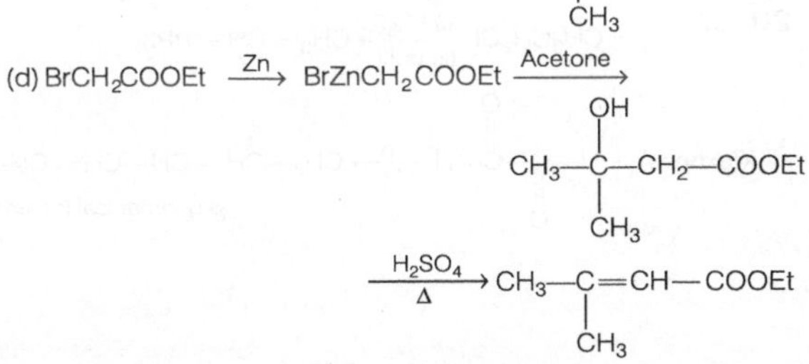

5. (c)

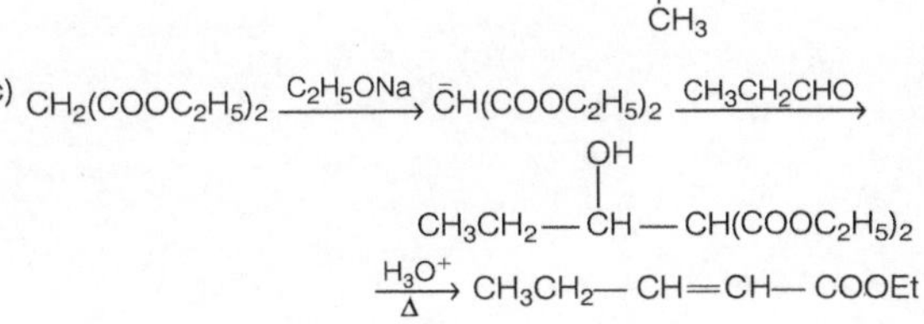

6. (c)

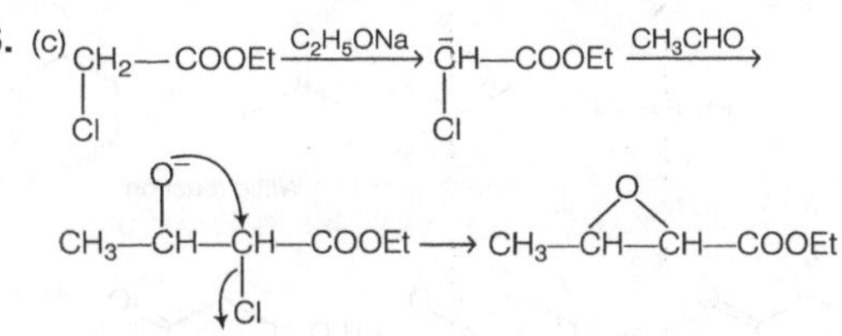

7. (c)

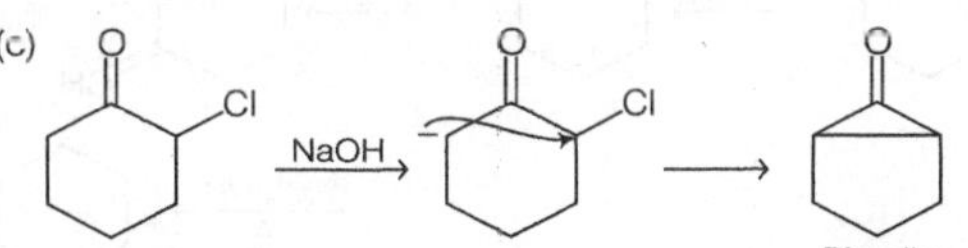

Bicyclic

8. (b)

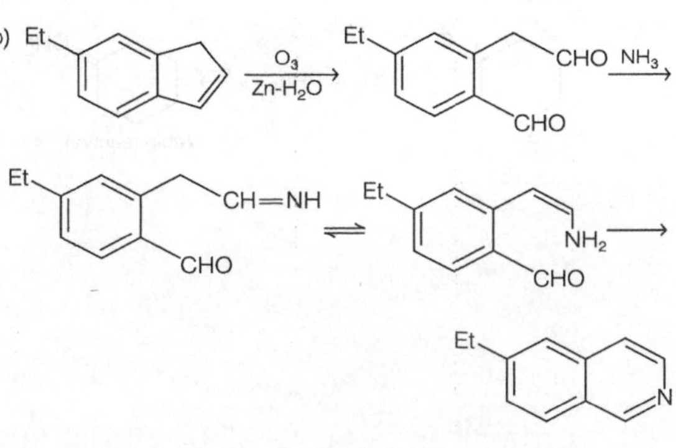

9. (a,b,c)

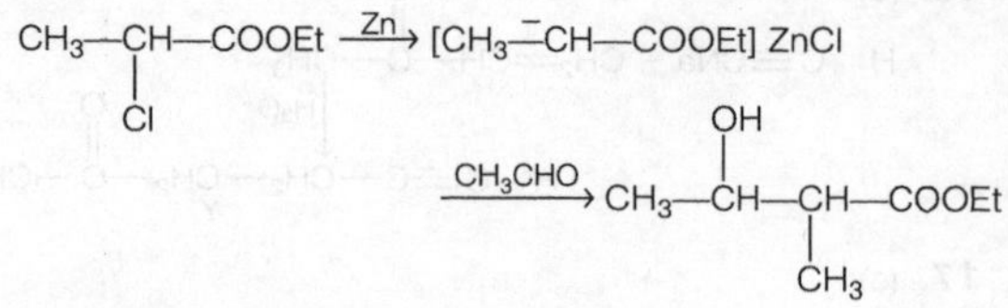

10. (a,c,d)

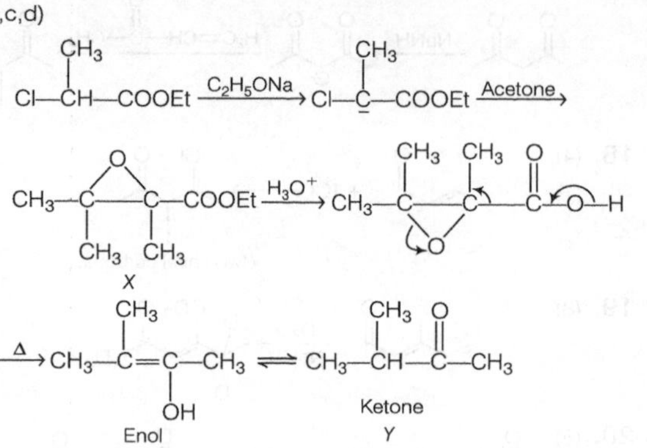

11. (a,b,c) It is an example of Benzil-benzilic acid rearrangement.

12. (a,b,d) Wittig reaction

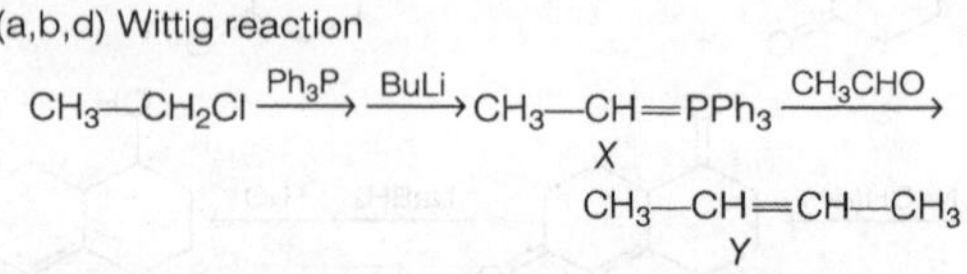

Both diastereomers (*cis* and *trans*) of *Y* are formed.

13. (a,c)

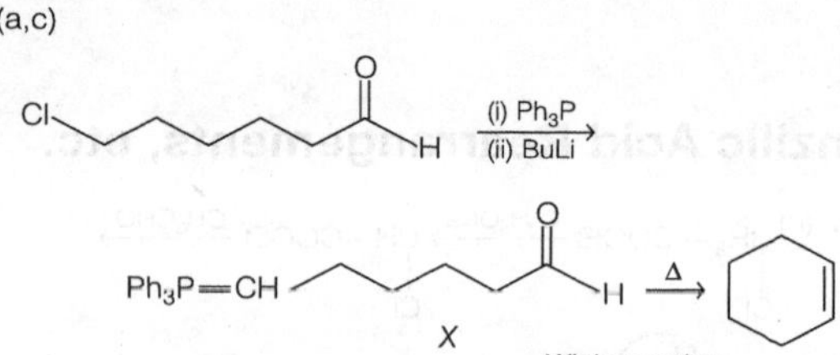

Wittig reaction

14. (b,c)

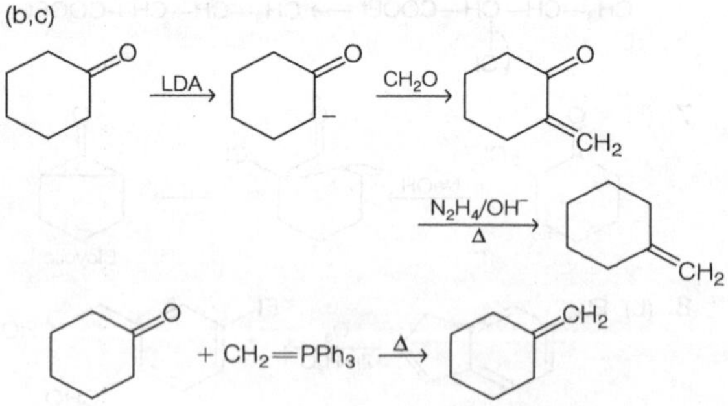

Wittig reaction

15. (a)

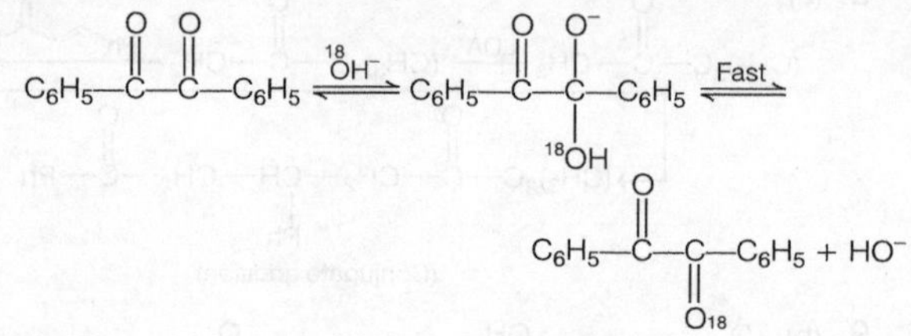

16. (c) Nucleophilic attack at any of the two carbonyl carbons are equally likely.

17. (a)

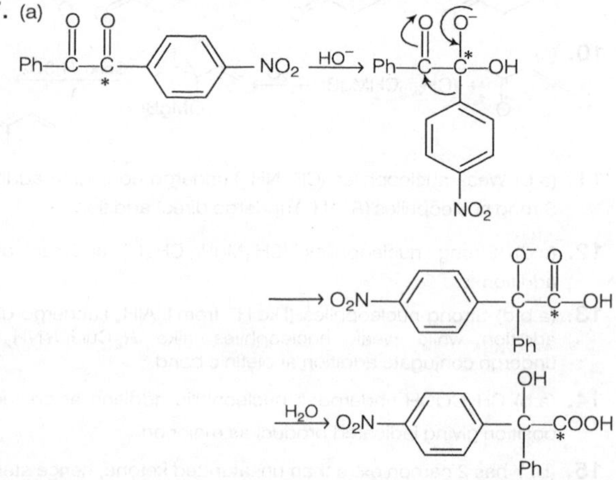

18. (4)

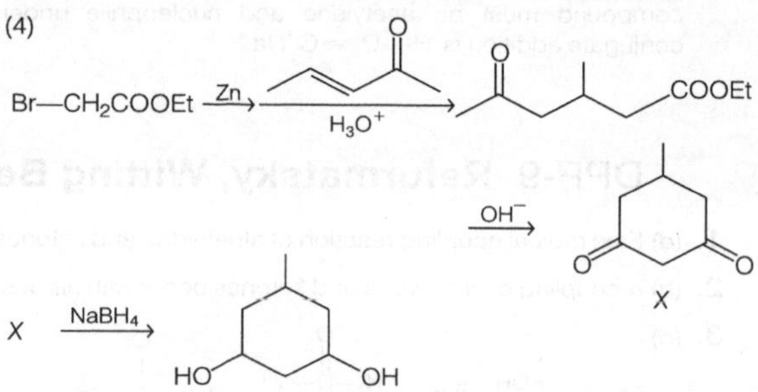

Has two chiral carbons

19. (4)

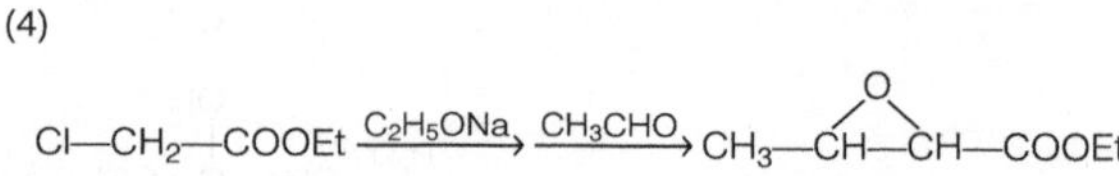

4 Stereoisomers

20. (3)

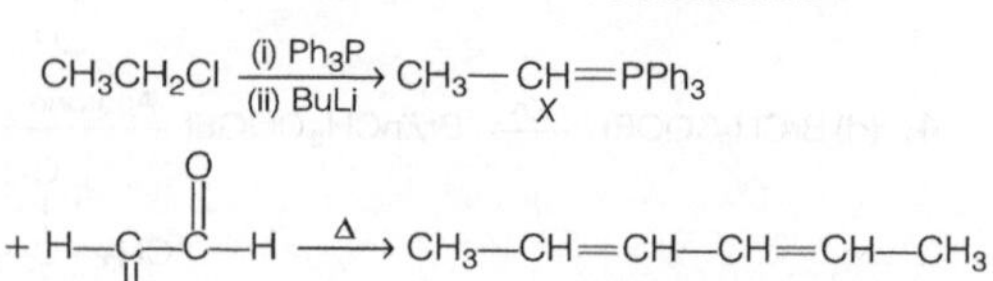

(3 geometrical isomers)

DPP-10 Tests and Distinctions

1. (c) Tollen's reagent is ammoniacal solution of silver nitrate in which silver remains as $[Ag(NH_3)_2]^+$.

2. (c)

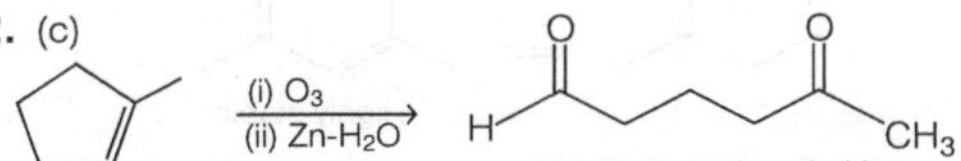

Aldehyde and methyl ketone
Y

The above product has both —CHO and —$COCH_3$ groups required for Tollen's test and iodoform test respectively.

3. (b) Fehling's test is not given by benzaldehyde (aromatic aldehyde).

4. (d) Benzaldehyde gives Tollen's test but acetophenone does not. On the other hand, acetophenone (Ph—$COCH_3$) gives iodoform test but benzaldehyde does not.

5. (a) Butanone forms precipitate with 2, 4-dinitrophenyl hydrazine, which after separation, can be hydrolysed to obtain back the ketone.

6. (c) Cerric nitrate forms coloured solution with alcohols but does not react with ketone.

7. (a) Schiff's test is given by aldehydes but not by ketones.

8. (b) Conjugated aldehydes and ketones form orange coloured precipitate with 2, 4-dinitrophenyl hydrazine. Non-conjugated carbonyls form yellow precipitate in the same reaction.

9. (a,b,c,d) Carbonyls with $\left(CH_3-\overset{\overset{O}{||}}{C}-\right)$ group or an alcohol with a methyl group and a "H" at α-carbon gives iodoform test.

10. (a,b,d) Both aliphatic aldehydes and aromatic aldehydes give Tollen's test. Besides, α-hydroxy ketones are also oxidised by Tollen's reagent.

11. (b,d) Aromatic aldehydes are not oxidised by Fehling's solution. Acid derivatives like ester, anhydride and amide are not oxidised by this reagent. However, α-hydroxy ketones are oxidised by both Fehling's solution and Tollen's reagent.

12. (a,d) Both $NaHSO_3$ and 2,4-dinitrophenyl hydrazine form precipitate with aldehydes and ketones, from which carbonyls can be recovered back easily.

13. (a,d)

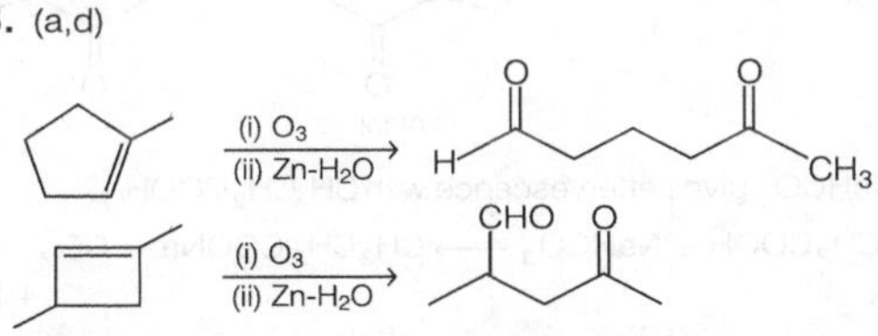

Products in the above reactions have both —CHO and —$COCH_3$ groups required for Tollen's test and iodoform test respectively.

14. (c,d) In acid hydrolysis will give a compound which forms a yellow precipitate with KOH/I_2 solution.

Explanations for Q. Nos. (15-17)

From the given information, it appears that *A* is an ester. Since, *D* on Tischenko reaction gives back *A* indicates that in ester, both acid and alcohol fragments (*B* and *C* respectively) has five carbon atoms each.

Also, *C* is an alcohol which has a chiral carbon and no $CH_3CH(OH)$— grouping (not giving iodoform). Controlled oxidation of *C* gives *D*, a carbonyl which is still chiral. Hence, *D* must be an aldehyde with a chiral carbon.

15. (b)

$$CH_3-CH_2-\overset{*}{\underset{}{C}}H(CH_3)-\overset{O}{\overset{||}{C}}-O-CH_2-\overset{*}{C}H(CH_3)-CH_2-CH_3$$

A

The above ester has two chiral carbons and no symmetry hence, four stereoisomers.

16. (c)

$$\underset{D}{CH_3CH_2-\overset{CH_3}{\overset{|}{C}}H-CHO} \xrightarrow[H_2SO_4]{K_2Cr_2O_7} \underset{B}{CH_3CH_2-\overset{CH_3}{\overset{|}{C}}H-COOH}$$

17. (b)

$$D + Cl_2 \xrightarrow{H^+} CH_3CH_2-\underset{Cl}{\underset{|}{\overset{CH_3}{\overset{|}{C}}}}-CHO \text{ (Racemic)}$$

18. (4) $Ph-\overset{O}{\overset{||}{C}}-CH_3 + 4KOH + 3I_2 \longrightarrow Ph-COOK + CHI_3 + 3KI + 3H_2O$

19. (9) Conjugated aldehydes and ketones give orange precipitate with 2, 4-dinitrophenyl hydrazine.

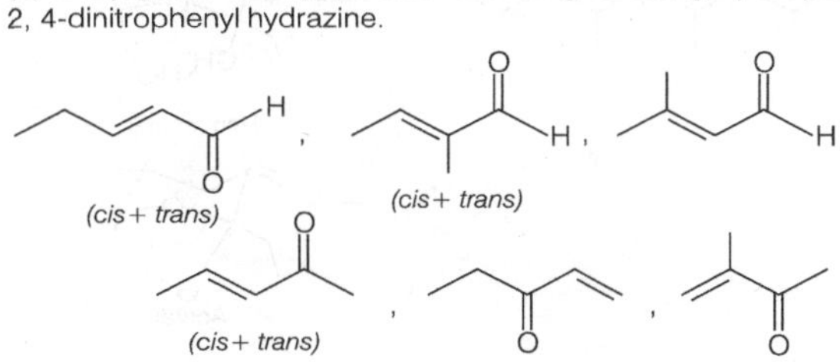

20. (4)

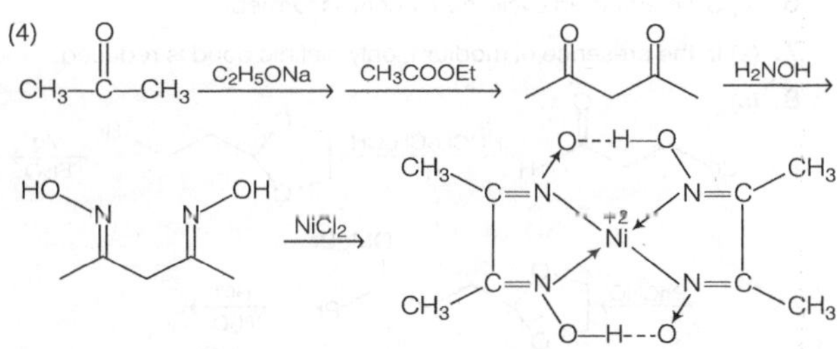

Has 4 monocyclic rings

21. (b)

(i) Benzaldehyde forms orange precipitate with 2, 4-dinitrophenyl hydrazine while all others form yellow precipitate.

(ii) Benzaldehyde, being conjugated, forms orange precipitate with 2, 4-dinitrophenyl hydrazine.

(iii) Aldehydes (aliphatic) and α-hydroxy ketones are oxidised by Fehling's solution giving red precipitate of Cu_2O.

(iv) $NaHSO_3$ forms white precipitate of bisulphite with all carbonyls.

Hence, (i) → (q,r,s), (ii) → (p), (iii) → (r,s), (iv) → (p,q,r,s)

Revisal Problems (JEE Main)

1. (b) Phenyl group is a locant on the parent chain containing aldehyde as principal functional group.

2. (b)

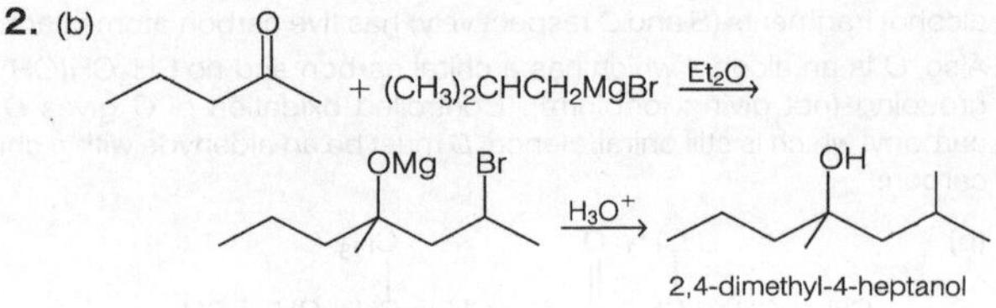

3. (a)

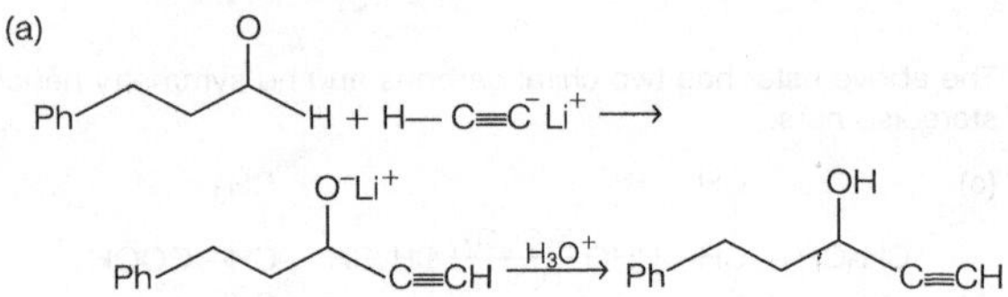

4. (c) Wittig reaction

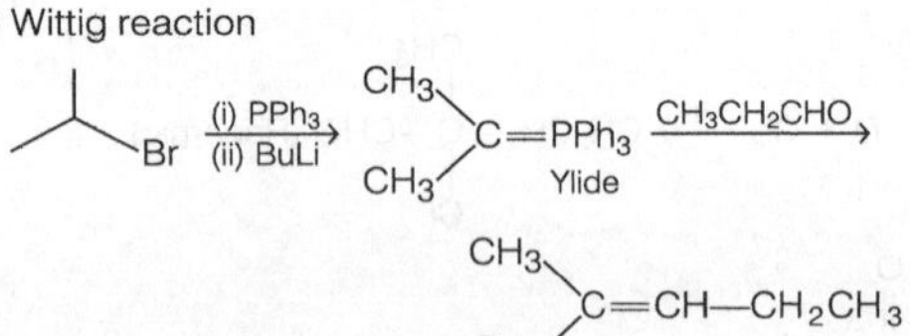

5. (a)

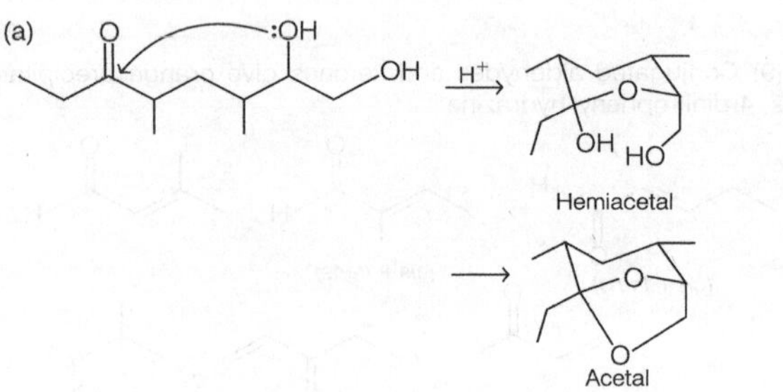

6. (c) Six membered cyclic hemiacetal is formed.

7. (a) In the presence of rhodium, only olefinic bond is reduced.

8. (a)

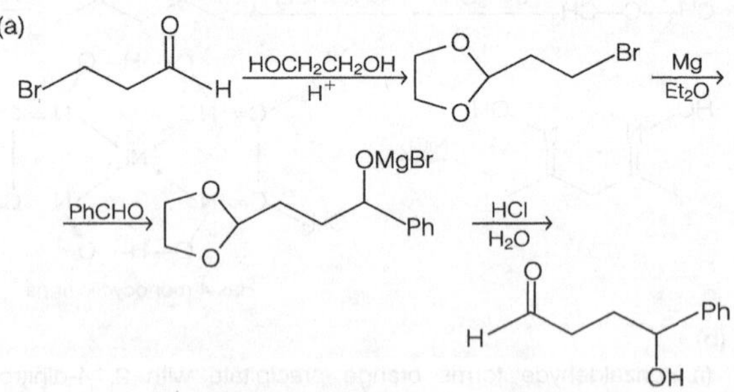

9. (c)

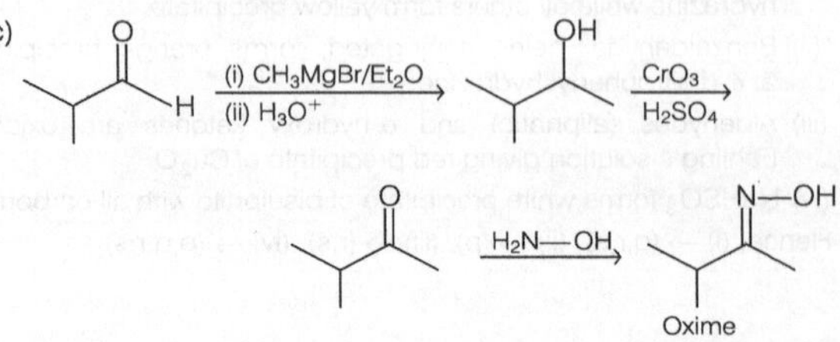

10. (c)

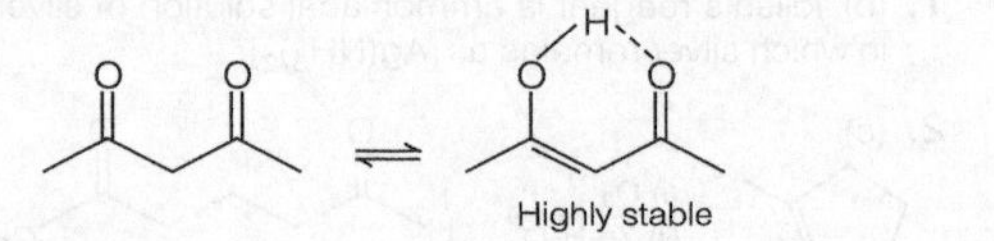

11. (b)

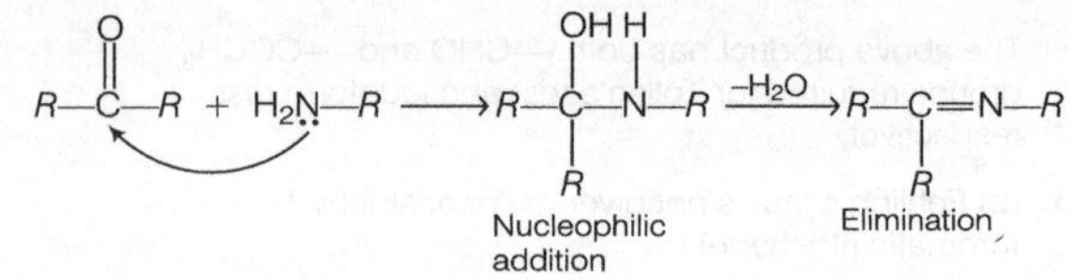

12. (b) Acetal formation occur in the presence of acid catalyst not in the presence of basic catalyst.

13. (b) α-halogenation occur *via* enol intermediate of aldehydes and ketones.

14. (d) All the three Grignard's synthesis can be used to prepare the given compound.

15. (c) $C_6H_5—\overset{O}{\overset{\|}{C}}—C_6H_5$ is a symmetrical ketone, gives a single oxime with hydroxylamine.

16. (d) It is more substituted and conjugated with carbonyl groups on both side.

17. (d)

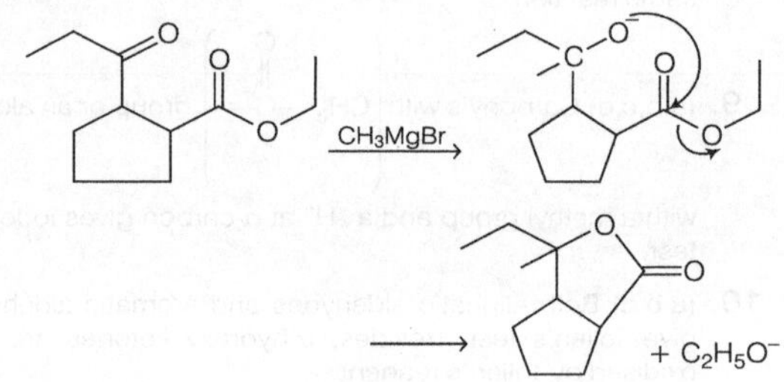

18. (c)

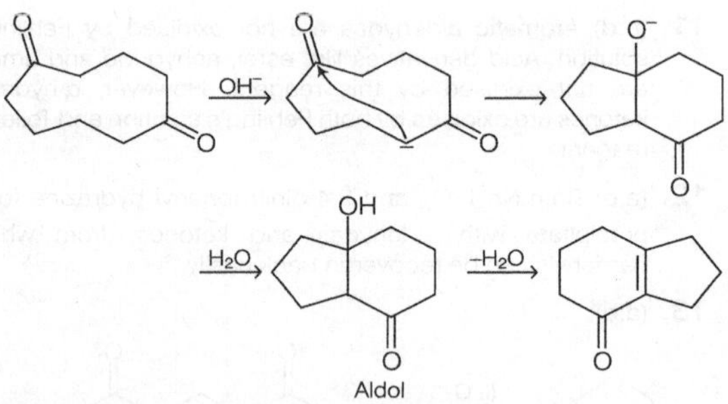

19. (d) $NaHCO_3$ gives effervescence with CH_3CH_2COOH.

$$CH_3CH_2COOH + NaHCO_3 \longrightarrow CH_3CH_2COONa + CO_2 + H_2O$$

20. (a) $\underset{1° \text{ alcohol}}{R—CH_2OH} \xrightarrow{[O]} R—CHO \xrightarrow{[O]} R—COOH$

21. (d) It is an example of intramolecular aldol condensation followed by dehydration of aldol.

22. (c)

$$R—COOH \xrightarrow{SOCl_2} R—\overset{O}{\overset{\|}{C}}—Cl \xrightarrow[-80°C]{Li[\{(CH_3)_3CO\}_3AlH]} R—\overset{O}{\overset{\|}{C}}—H$$

23. (b) 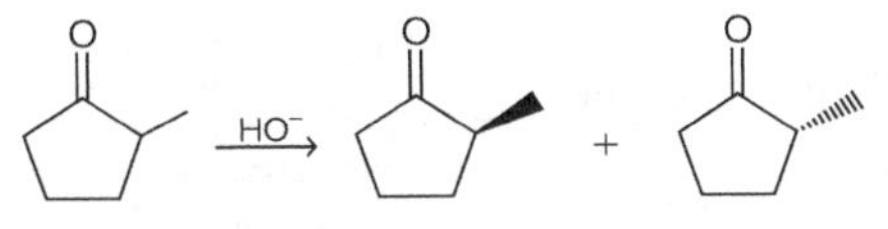

(I)

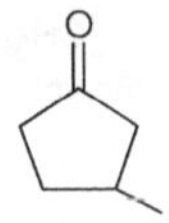

Not racemises on treatment with base because its α-carbon is not chiral.

(II)

24. (c)

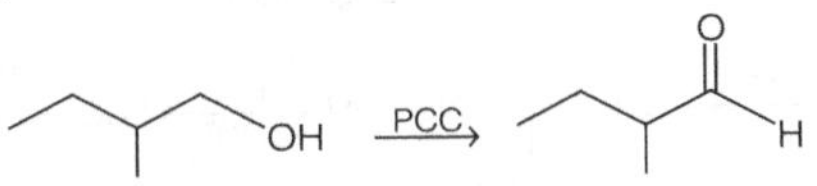

Optically active | Optically active (α-carbon chiral)

$\xrightarrow{OH^-}$ Racemic mixture

25. (c) It is an α-hydroxy aldehyde, which is never formed in aldol condensation.

26. (c) 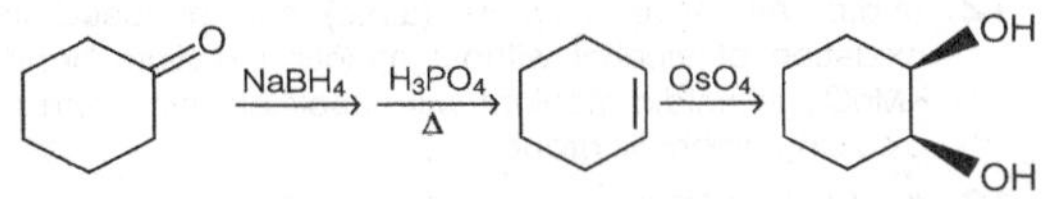

27. (d) Electron withdrawing group increases reactivity while electron donating group decreases reactivity in aldol condensation reaction.

28. (b) 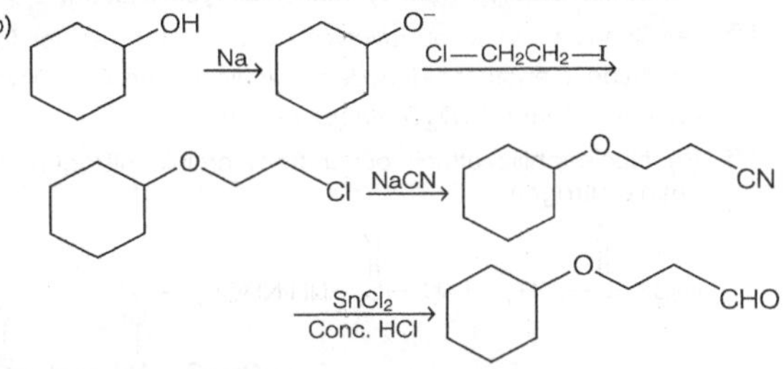

29. (c) In cross-Cannizaro reaction, formaldehyde is always oxidised, hence benzaldehyde is reduced in the present reaction.

30. (c) It is an example of intramolecular Cannizaro reaction.

Revisal Problems (JEE Advanced)

1. (b) Clemmensen reduction is most suitable here. If Wolff-Kishner reduction is used, OH^- may simultaneously bring about S_N2 and E2 reaction with halide group.

2. (a) α, β-carbon-carbon bond is formed in aldol condensation reaction.

3. (d) Hydrate formation involves nucleophilic addition at carbonyl carbon in the slow rate determining step. Therefore, electron withdrawing group increases reactivity while electron donating group decreases reactivity of carbonyls.

4. (c)

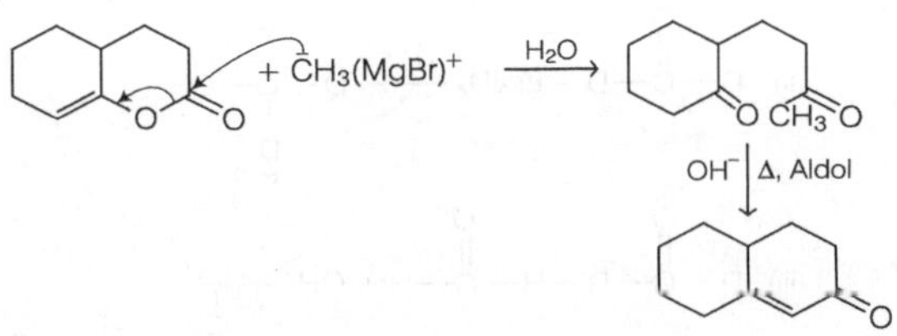

5. (b) 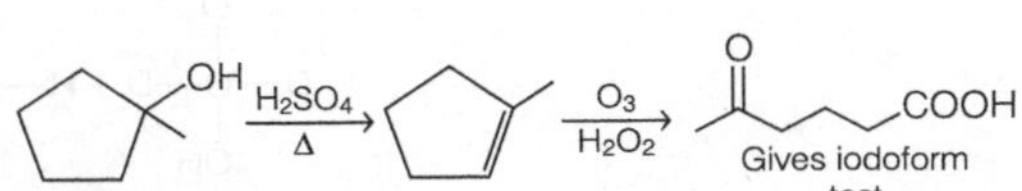

3°, not oxidised by $K_2Cr_2O_7/H^{\oplus}$

6. (d) Product is in conjugated system, more stable than non-conjugated reactant.

7. (d) It has eight α-H which can undergo exchange with deuterium.

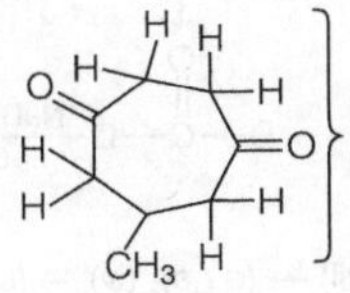

Eight exchangeable α-hydrogens

8. (b) 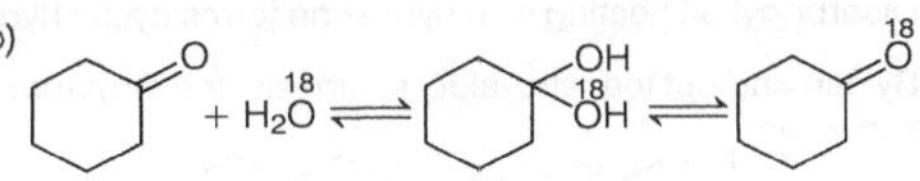

9. (a,d) 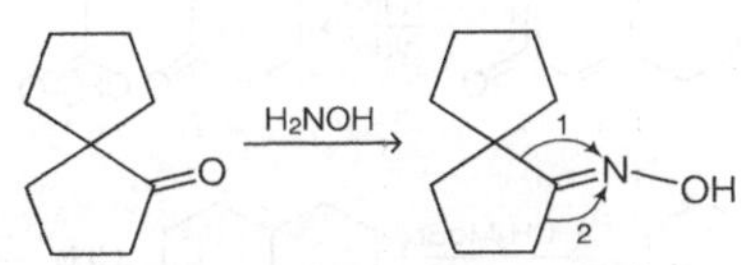

The oxime above undergoes Beckmann rearrangement on treatment with concentrated H_2SO_4. Rearrangement by route (1) gives (a) while rearrangement by route (2) gives (d).

10. (a, c) *A* is a 2° alcohol with $CH_3CH(OH)$— group.

11. (b,d) 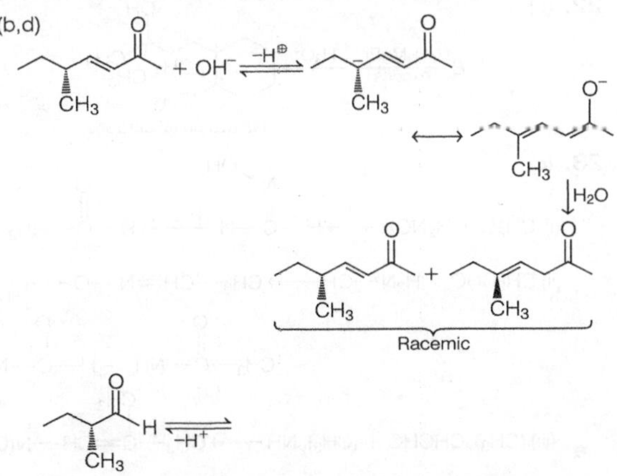

12. (a,b,c) All these reagents (a,b,c) can be used for selective oxidation of alcohol without affecting olefinic bond. However, $KMnO_4$ in acidic medium will attack olefinic bond too besides oxidising alcoholic group.

13. (b,c) In both option (b) and option (c) cases, the product enol is more stable than reactant keto, hence K_c is greater than 1.

14. (a,b,d) All these will produce diastereomeric salts with $NaHSO_3$, hence can be separated by fractional crystallisation.

15. (a) Cannizaro reaction involves hydride (H^-) transfer from one aldehyde to another aldehyde molecule, hence, C—D bond cannot be formed if D_2O/NaOD is used.

16. (d) Nucleophilic attack occur from aminic nitrogen not from amidic nitrogen.

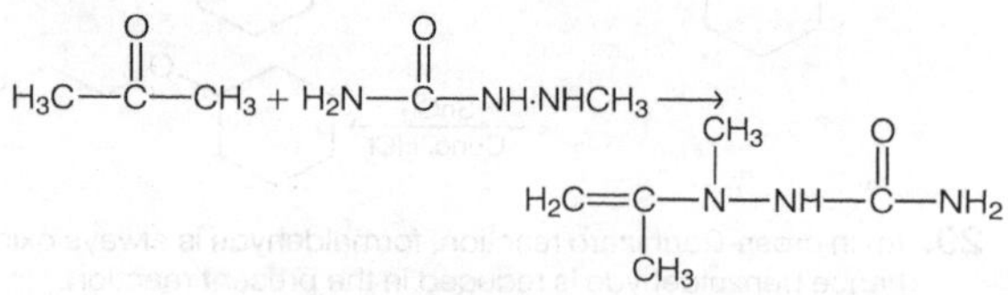

17. (d) In aldol condensation reaction followed by dehydration of aldol gives α, β-unsaturated carbonyl compound not the β, γ-unsaturated carbonyl as major product.

18. (b) In first step, the more reactive aldehyde group is acetylated and then ketonic group is reduced with $LiAlH_4$. In the final step, hemiacetal formation occur at aldehyde group.

19. (c) Dicarbonyl on heating with hydrazine forms cyclic hydrazone.

20. (c) By carrying out the retro-aldol synthesis, the structure of *B* can be known as

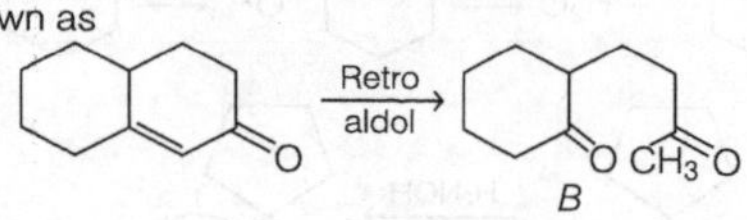

Hence,

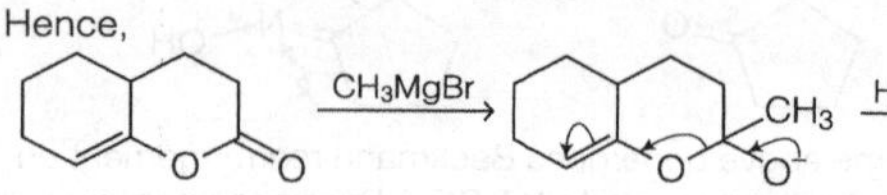

21. (b) Dicarbonyl *B* shown above on heating with N_2H_4 gives a cyclic hydrazone.

22. (c)

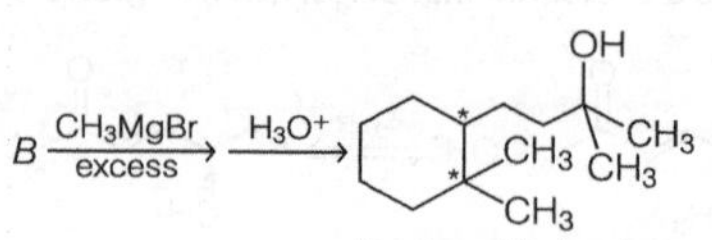

Has two chiral carbons

23. (c)

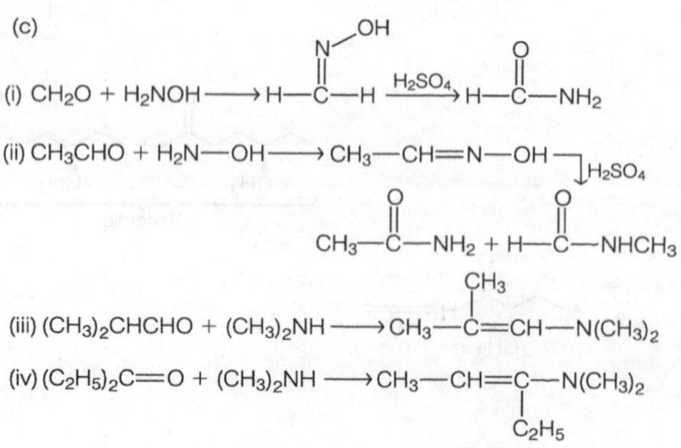

Hence, (i) → (p,A), (ii) → (p,B), (iii) → (q,C), (iv) → (q,D)

24. (b)

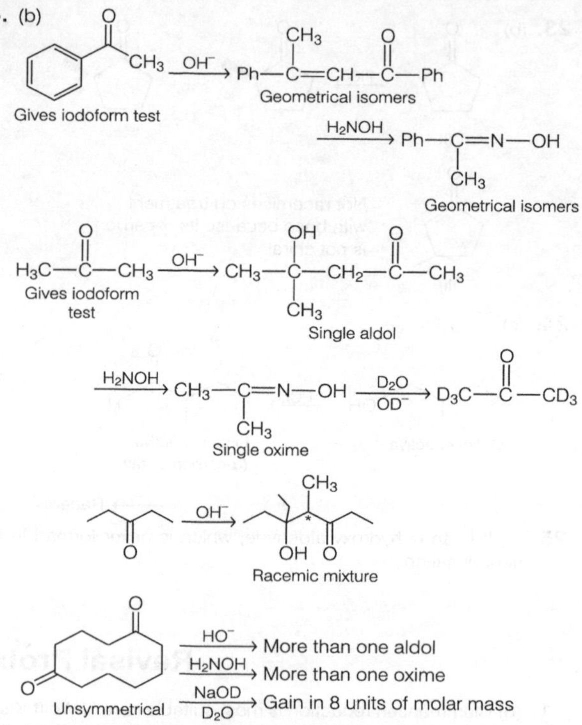

Hence, (i) → (q,r,t), (ii) → (p,q,s), (iii) → (t), (iv) → (r,s,t)

25. (d)

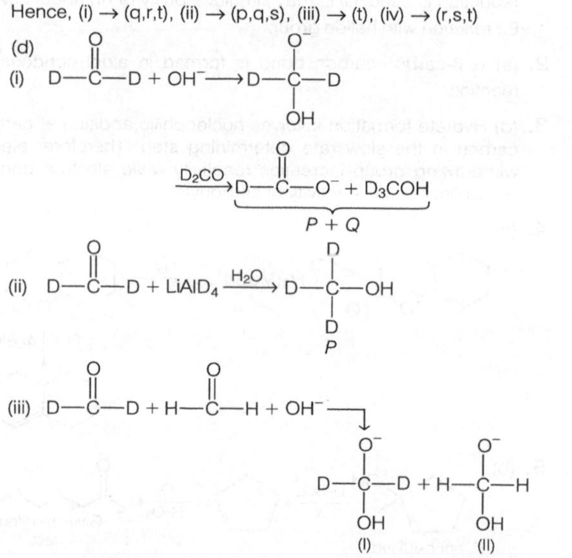

On hydride transfer with CH_2OI gives DCOOH and DCH_2OH while (II) on hydride transfer with CD_2O gives D_2CHOH.

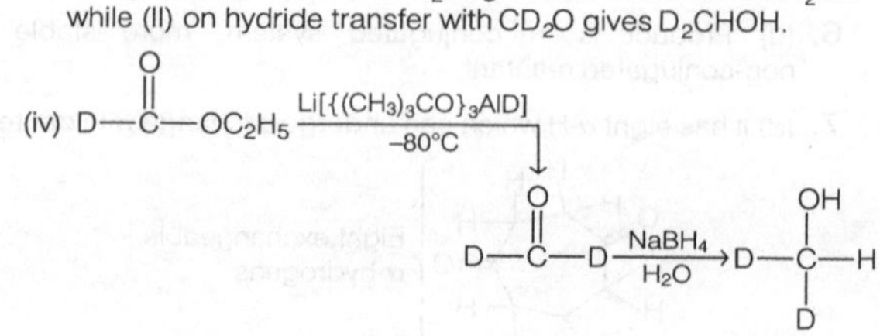

Hence, (i) → (p,q), (ii) → (p), (iii) → (q,r,s), (iv) → (s)

26. (4)

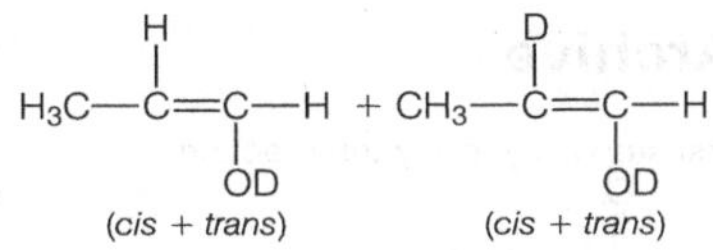

27. (6)

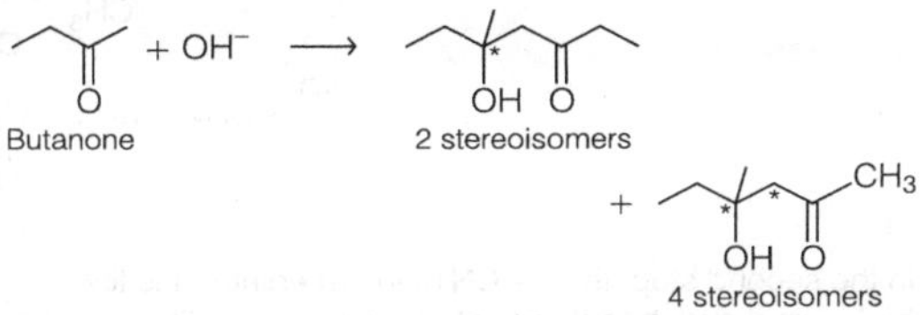

28. (4)

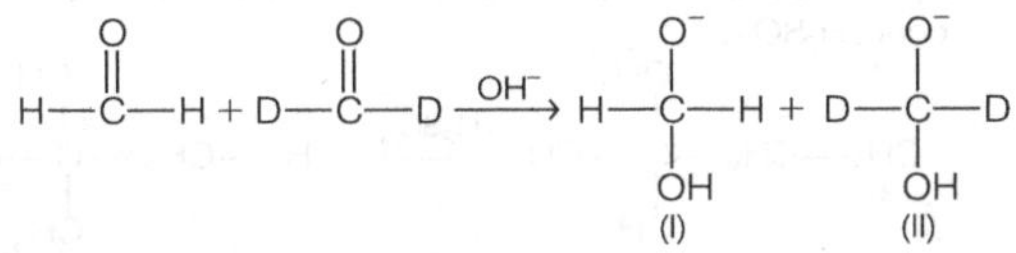

After H^- transfer, (I) gives CH_3OH and CD_2HOH and while (II) gives CD_3OH and CDH_2OH.

29. (4)

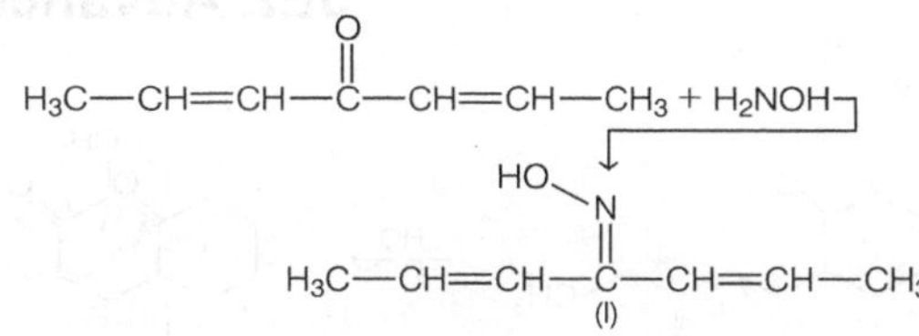

Four stereoisomers exist for I, *cis-cis*, *trans-trans* and *cis-trans* with OH *syn* to *cis* and OH *anti* to *cis*.

30. (4)

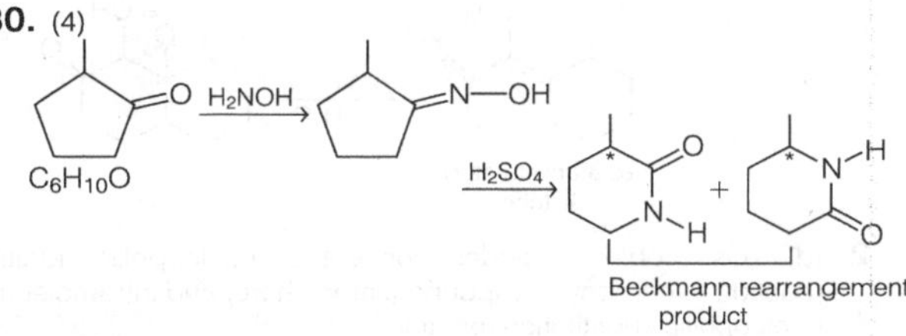

Both products have one chiral carbon each, hence two pairs of enantiomers are formed.

JEE Main & AIEEE Archive

1. (a)

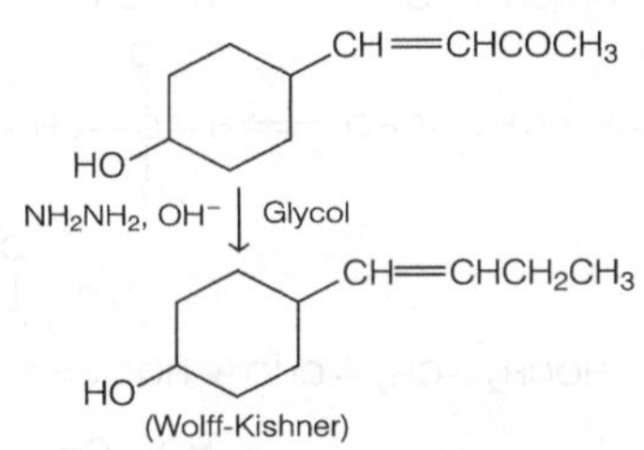

(b)

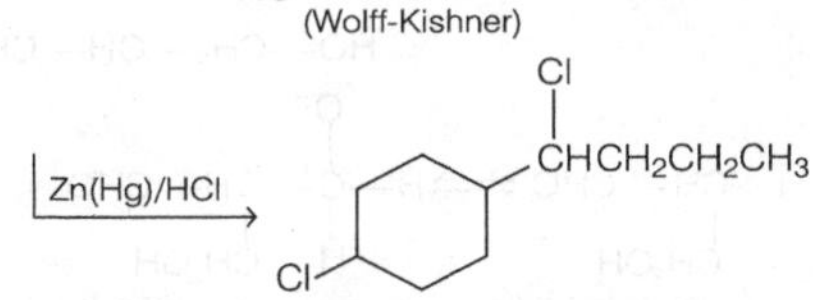

HCl attacks 2° alcohol as well as (C$=$C) bond.

(c)

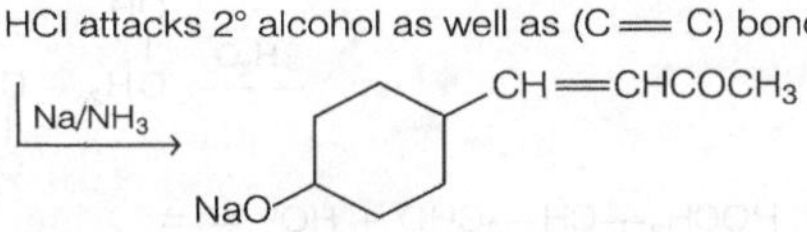

(d) $\xrightarrow{NaBH_4}$ Low yield

2. (a) Cannizaro reaction is given by aldehydes (RCHO) lacking H at α-carbon or lacking α-carbon (as in HCHO). With NaOH, there is formation of acid salt ($R\text{COO}^-$) by oxidation and alcohol (RCH_2OH) by reduction.

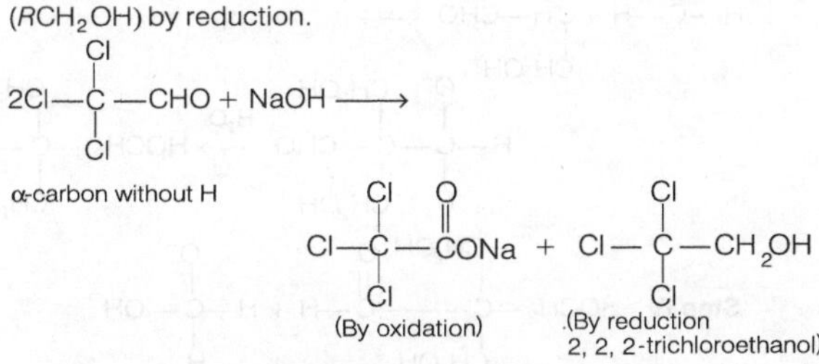

3. (a or c) All aldehydes including reducing sugar (as glucose, fructose) give silver mirror test (with Tollen's reagent).

$$RCHO + 2[Ag(NH_3)_2]^+ + 3OH^- \xrightarrow{\Delta} RCOO^- + 2Ag\downarrow\ (\text{silver mirror}) + 4NH_3 + 2H_2O$$

(R=H, CH_3)

Thus, acetaldehyde and formaldehyde give silver mirror test.

4. (d) $CH_3\overset{O}{\overset{\|}{C}}-Cl \xrightarrow{KOH\,(aq)} CH_3-\overset{O}{\overset{\|}{C}}-OH + KCl$ [SN reaction]

$CH_3-CH_2-Cl \xrightarrow{KOH\,(aq)} CH_3-CH_2-OH$ [S_N2 reaction]

$Cl-CH_2-CH_2-Cl \xrightarrow{KOH\,(aq)} \underset{OH}{CH_2}-\underset{OH}{CH_2}$ [S_N2 reaction]

$CH_3-\underset{Cl}{CH}-Cl \xrightarrow{KOH\,(aq)} CH_3-\underset{OH}{CH}-OH$ (Unstable) $\xrightarrow{-H_2O} CH_3-\overset{O}{\overset{\|}{C}}-H$ (Acetaldehyde)

5. (c) Addition of HCN is a nucleophilic reaction. Greater the electron deficiency of C atom of carbonyl group, higher the rate of reaction. Hence,

$$Ph-\overset{O}{\overset{\|}{C}}-Ph < Ph-\overset{O}{\overset{\|}{C}}-CH_3 < CH_3-\overset{O}{\overset{\|}{C}}-CH_3 < H-\overset{O}{\overset{\|}{C}}-H$$

6. (b)

$$CH_3-\underset{OH}{\overset{2^\circ}{CH}}-CH=CH-CH_3 \longrightarrow CH_3-\underset{O}{\underset{\|}{C}}-CH=CH-CH_3$$

Only suitable reagent is chromic anhydride in glacial acetic acid. Other will also affect (C$=$C) bond.

7. (d) C$=$C bond is cleaved and oxidised to —COOH,—CHO group is also oxidised to —COOH.

$$(CH_3)_2C=CHCH_2CHO \longrightarrow (CH_3)_2C=O + HOOC-CH_2-COOH$$

JEE Advanced & IIT JEE Archive

1. (a)

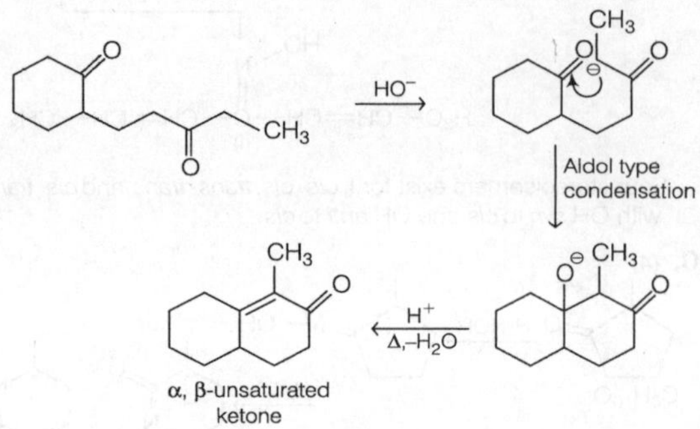

2. (d) This problem includes concept of nucleophilic addition reaction to carbonyl compound (ketone here) and intramolecular nucleophilic substitution reaction.

Complete reaction sequence is as shown below

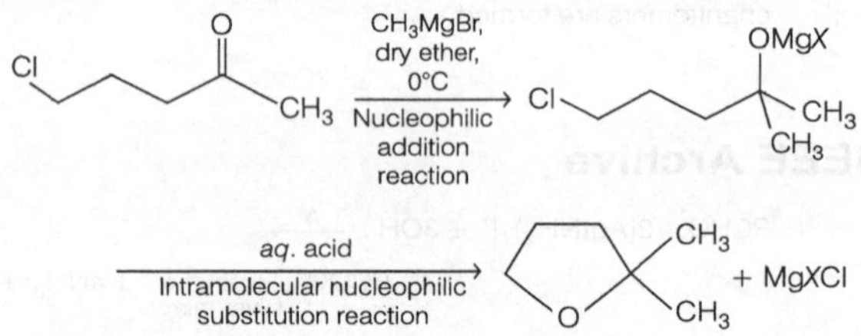

3. (5) Molecular weight of the ketone is 100.

So, molecular formula $= C_6H_{12}O$

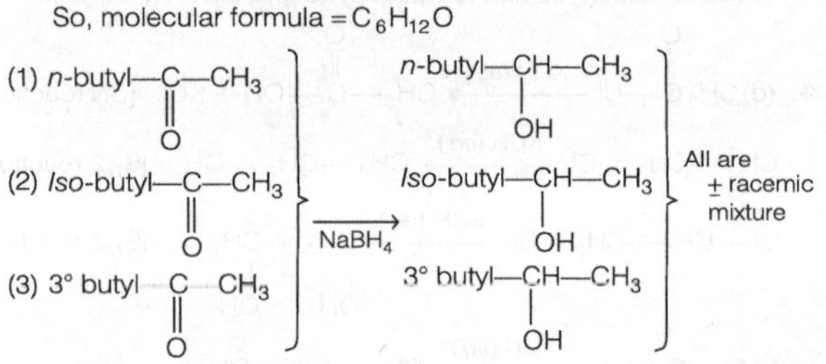

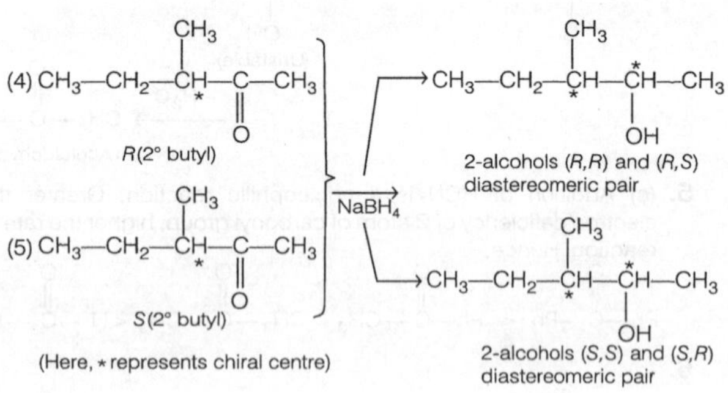

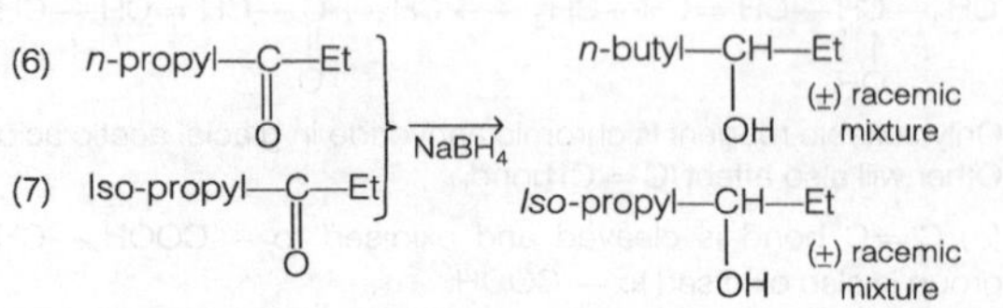

While in case of (4) and (5), they do not produce enantiomer due to the presence of stereogenic centre on ketone.

4. (a) The first step is cyanohydrin reaction.

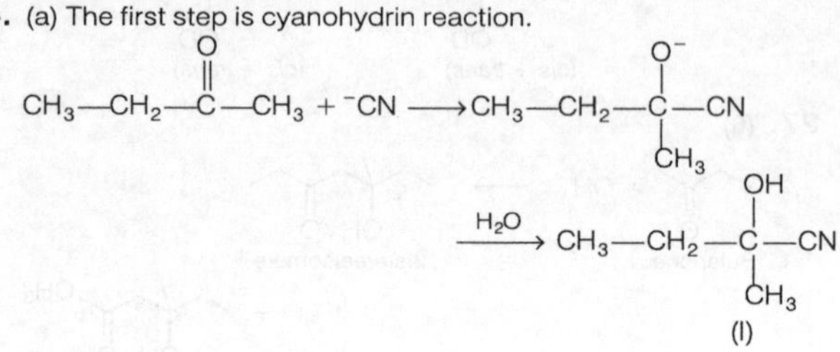

In the second step, the — CN of intermediate (I) is first hydrolysed and then dehydrated on heating in the presence of conc. H_2SO_4.

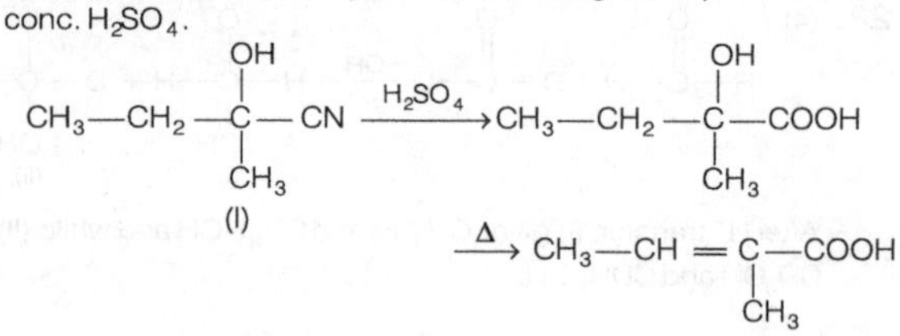

5. (c) The given reaction is an example of repeated aldol condensation followed by Cannizaro reaction.

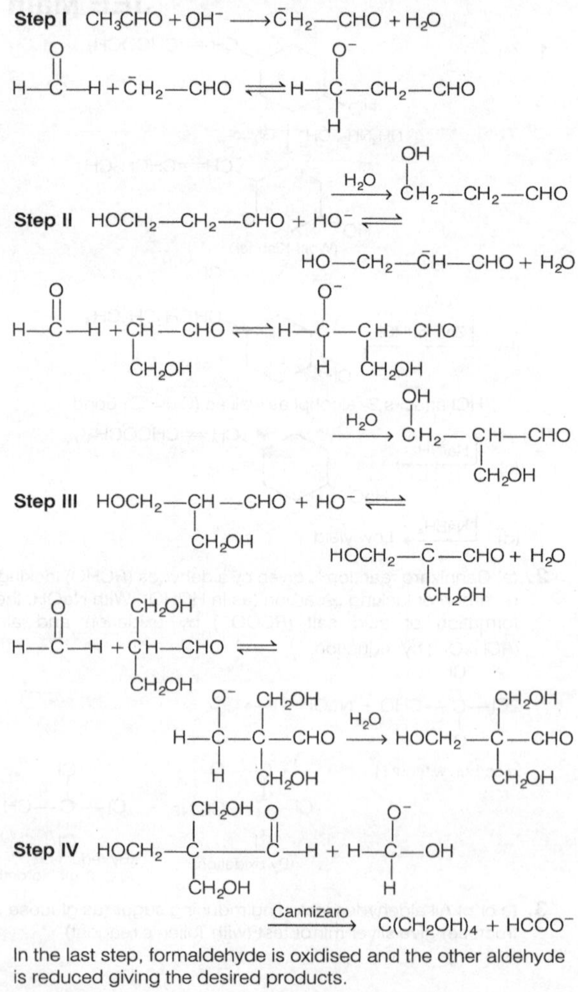

In the last step, formaldehyde is oxidised and the other aldehyde is reduced giving the desired products.

6. (1)

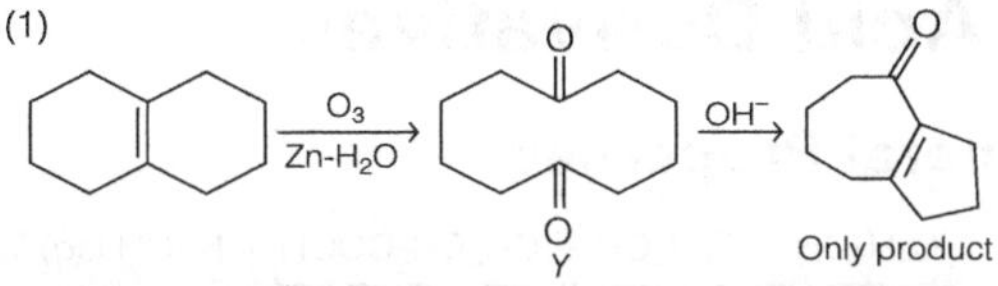

Only product

Explanations for Q. Nos. (7-9)

The given product is an ester, obtained by condensation of a hydroxy acid obtained through hydrolysis of a cyanohydrin.

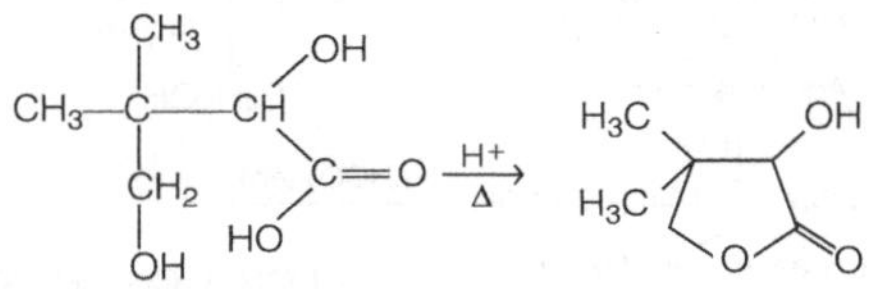

Acid above is obtained by acid hydrolysis of cyanohydrin *S* as

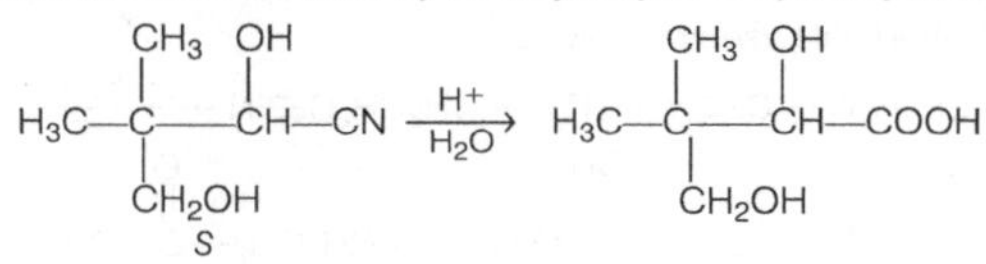

S is obtained by nucleophilic addition of HCN on *R*, hence *R* is

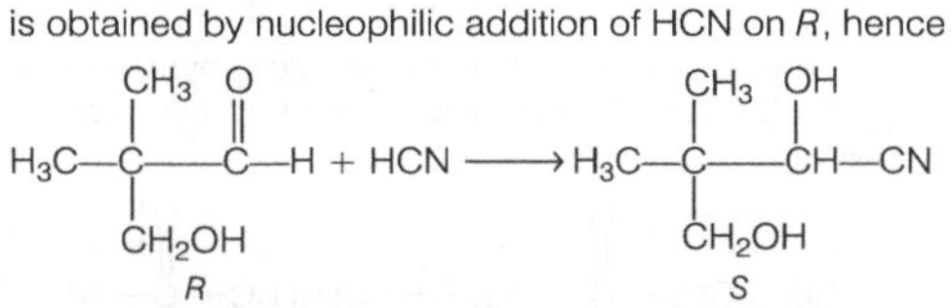

R is obtained by treatment of *P* and *Q* with aqueous K_2CO_3 through aldol condensation reaction as

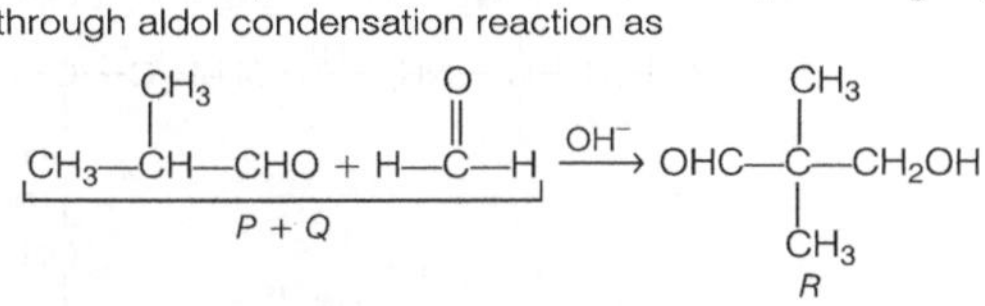

7. (b) **8.** (a) **9.** (d)

Explanations for Q. Nos. (10-12)

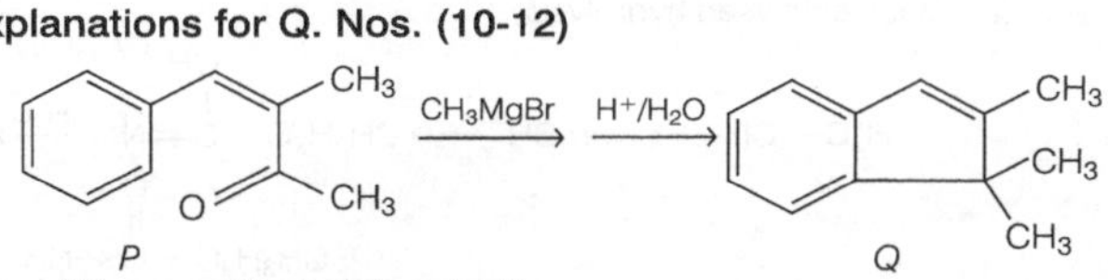

(Gives positive iodoform test)

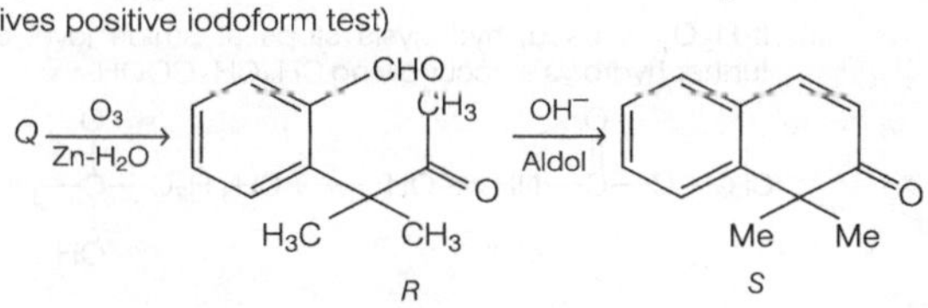

10. (b) **11.** (b) **12.** (a)

Explanations for Q. Nos. (13-15)

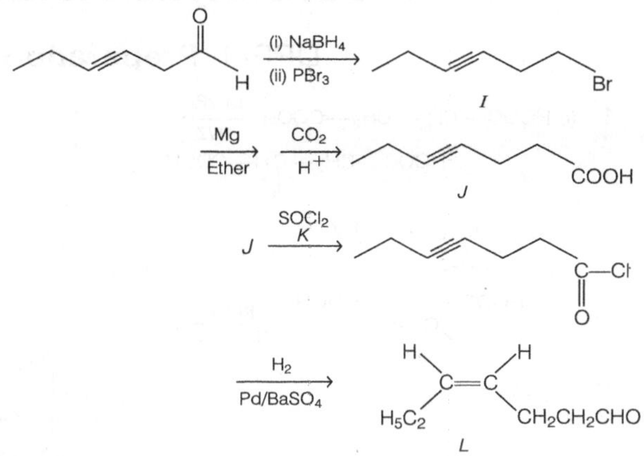

13. (d)

14. (a)

15. (c)

16. (b)

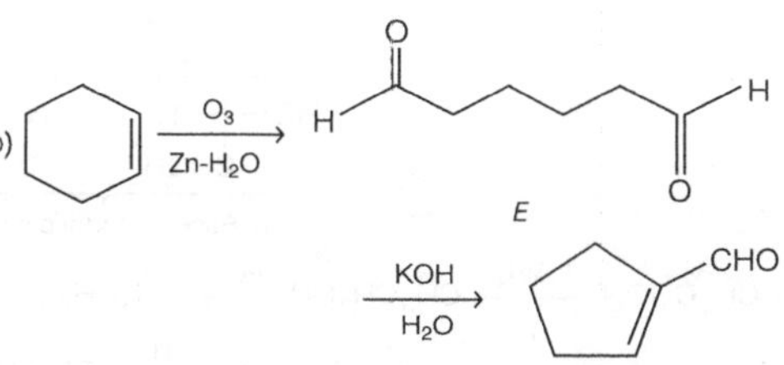

Intramolecular aldol condensation reaction

F

17. (b) $H_2C=O + H_2N-OH \longrightarrow$

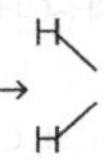

(Single)

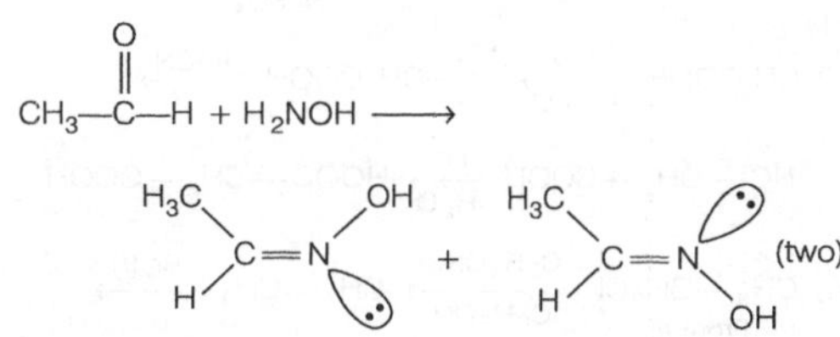

(two)

18. (c)

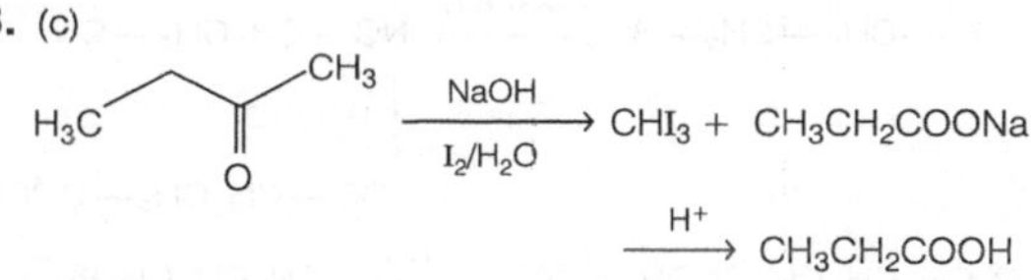

4. Carboxylic Acids and Acid Derivatives

DPP-1 Preparation and General Properties

1. (i) $HOOC—CH_2—CH_2—COOH \xrightarrow[HVZ]{Cl_2/P_4} HOOC—CH_2—\underset{Cl}{CH}—COOH$

$\xrightarrow[C_2H_5OH]{C_2H_5ONa\ (excess)} \xrightarrow{H_3O^+} HOOC—CH{=}CH—COOH$ (*cis* + *trans*)

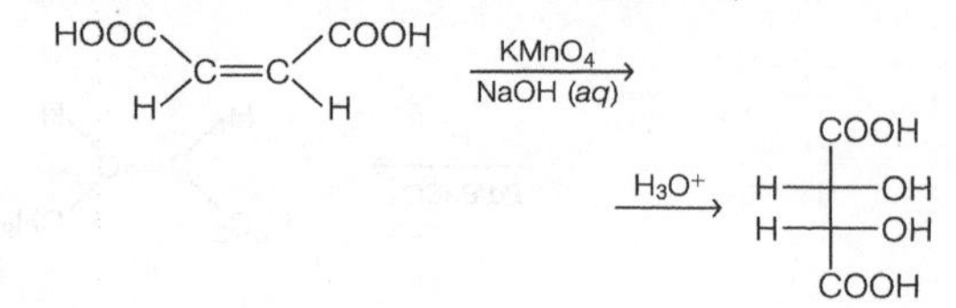

Meso tartaric acid

(ii)

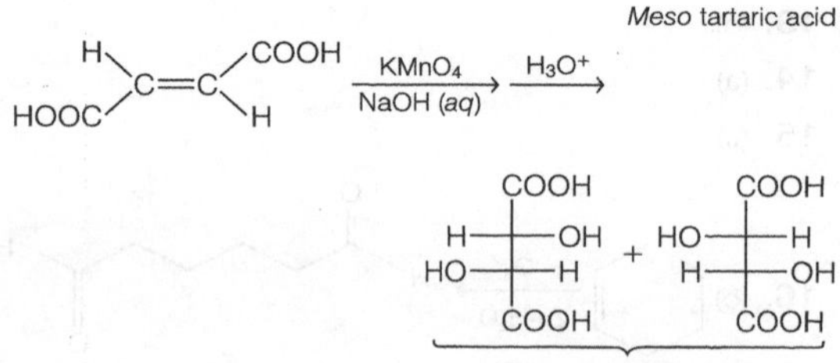

Racemic tartaric acids

2. (i) $CH_3COOH \xrightarrow{LiAlH_4} CH_3CH_2OH \xrightarrow{PCl_5} CH_3CH_2Cl \xrightarrow{NaCN}$

$CH_3CH_2—CN \xrightarrow[H_2O]{H^+} CH_3CH_2COOH$

(ii) $CH_3CH_2Cl \xrightarrow[ether]{Mg} CH_3CH_2MgCl \xrightarrow[H^+/H_2O]{HCHO}$

From (i)

$CH_3CH_2CH_2CH_2OH \xrightarrow[H_2SO_4]{CrO_3} CH_3CH_2CH_2COOH$

(iii) $CH_3COOH \xrightarrow[P\ (HVZ)]{Cl_2} Cl—CH_2COOH \xrightarrow{NaCN}$

$NC—CH_2—COOH \xrightarrow[H_2O]{H^+} HOOC—CH_2—COOH$

(iv) $CH_3—CH_2Cl \xrightarrow[C_2H_5OH]{C_2H_5ONa} CH_2{=}CH_2 \xrightarrow{Br_2(l)}$

From (i)

$Br—CH_2—CH_2—Br \xrightarrow{KCN(aq)} NC—CH_2CH_2—CN \xrightarrow{H^+/H_2O} HOOC—CH_2CH_2—COOH$

3. (i) $CH_3CH_2CH_2COOH \xrightarrow{Ag_2O} \xrightarrow[CCl_4/\Delta]{Br_2} CH_3CH_2CH_2Br$

$\xrightarrow[H_2O]{NaOH} CH_3CH_2CH_2OH \xrightarrow[H_2SO_4]{CrO_3} CH_3CH_2COOH$

(ii)

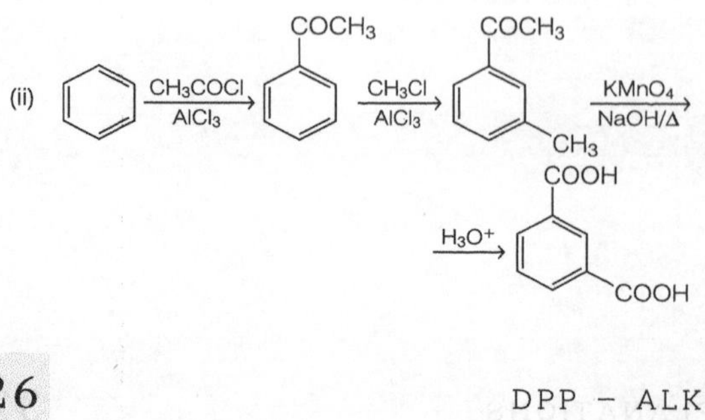

4. $C_4H_9OH + C_6H_5OH + CH_3CH_2COOH + NaOH\ (aq)$

Dissolve the mixture in diethyl ether

→ $C_6H_5ONa\ (aq)$, $CH_3CH_2COONa\ (aq)$ — Aqueous layer

→ C_4H_9OH in ether layer $\xrightarrow{\text{Fractional distillation}} C_4H_9OH$

Aqueous layer $\xrightarrow{HCl} C_6H_5OH + CH_3CH_2COOH$ (Extract in diethyl ether) $\xrightarrow{NaHCO_3\ (aq)} C_6H_5OH$ (In ether) $+ CH_3CH_2COONa\ (aq)$

5. (i) Acid catalysed hydrolysis

$CH_3H_2C—C{\equiv}N + H^+ \longrightarrow CH_3CH_2C{\equiv}\overset{+}{N}—H \xrightarrow[-H^+]{H_2O}$

$CH_3—CH_2—\underset{}{\overset{OH}{C}}{=}NH \rightleftharpoons CH_3CH_2—\overset{O}{\overset{\|}{C}}—NH_2$ (I)

If highly concentrated H_2SO_4 is used, hydrolysis stops I at level I (amide). If dilute acid is used, (I) hydrolysis further to acid.

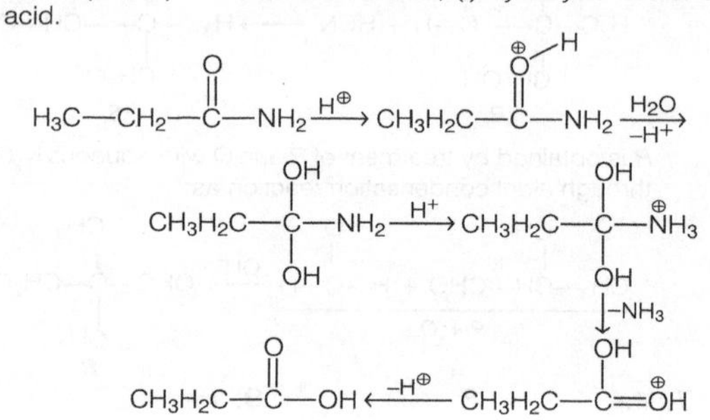

(ii) Base catalysed hydrolysis

$H_3C—CH_2—CN + OH^- \longrightarrow CH_3H_2C—\overset{OH}{C}{=}\ddot{N}:^- \xrightarrow{H_2O} CH_3H_2C—\overset{O}{\overset{\|}{C}}—NH_2$

If H_2O_2 is used, hydrolysis stops at amide level otherwise further hydrolysis occur giving CH_3CH_2COOH.

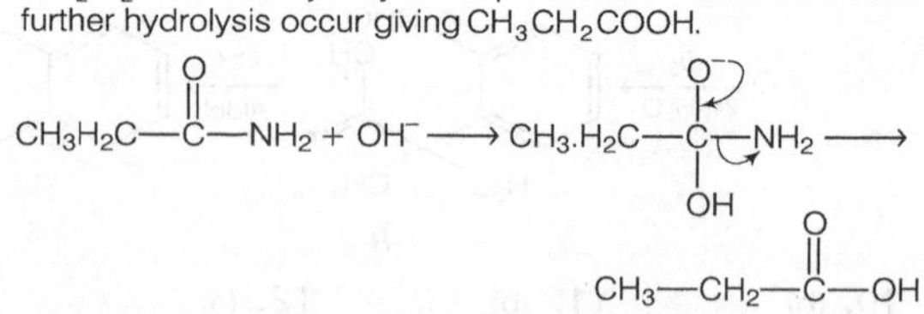

6. (c) Tollen's reagent is a mild oxidising agent, oxidises aldehyde to carboxylic acid but does not oxidise alcohols. All other oxidising agents mentioned here oxidises both aldehyde and alcohol groups.

7. (b) $H_5C_2—\underset{CH_3}{CH}—Br \xrightarrow[ether]{Mg} C_2H_5—\underset{CH_3}{CH}MgBr \xrightarrow[H^+]{CO_2} C_2H_5—\underset{CH_3}{CH}—COOH$

Racemic

Racemic Grignard reagent is formed which gives racemic product.

8. (d) 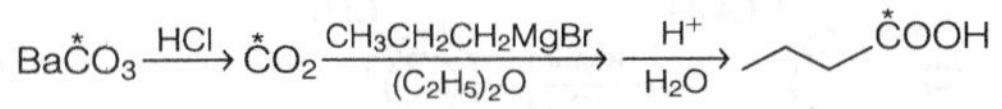

9. (c) 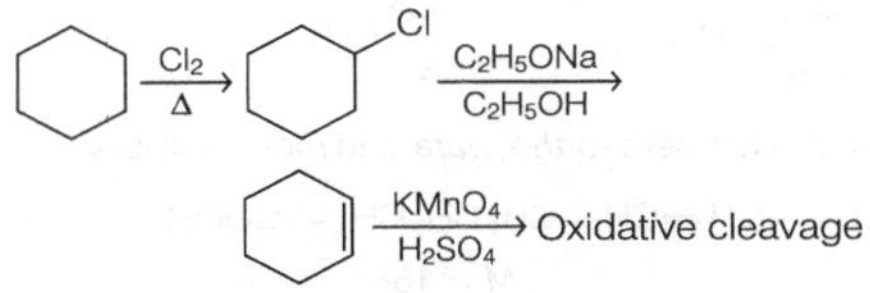

10. (b) 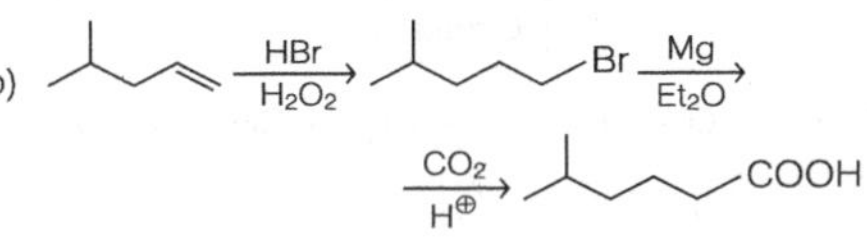

11. (a) 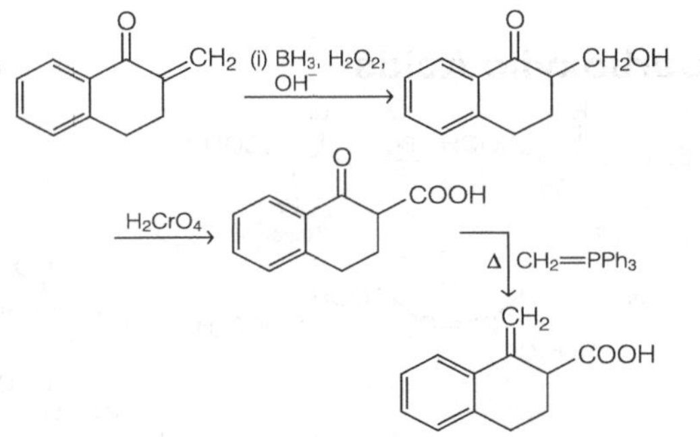

12. (c) 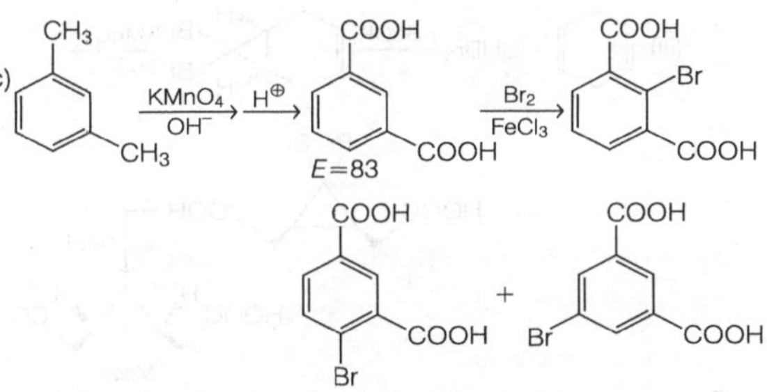

13. (c) 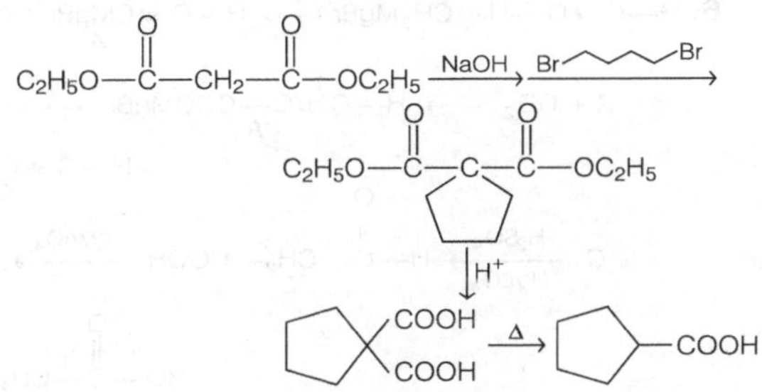

14. (a,c,d)

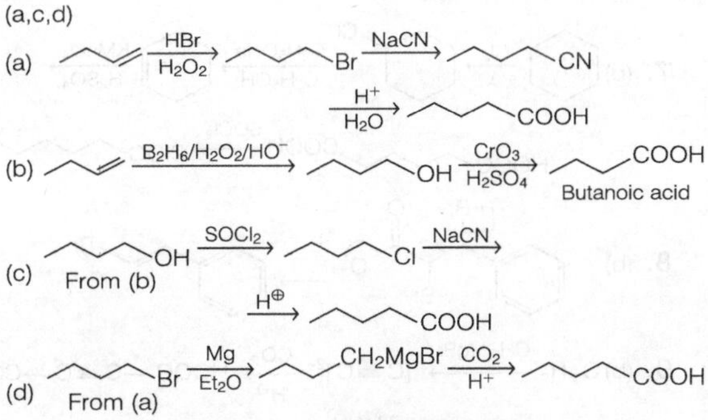

(b) is giving butanoic acid, all other gives pentanoic acid.

15. (c,d) Both NaOH and $NaHCO_3$ form salt with butanoic acid but not with butanol.

16. (a,b,c) All these have alkyl or partially oxidised alkyl groups present at *ortho* positions of benzene, oxidised to phthalic acid with alkaline $KMnO_4$.

17. (a,b,c)

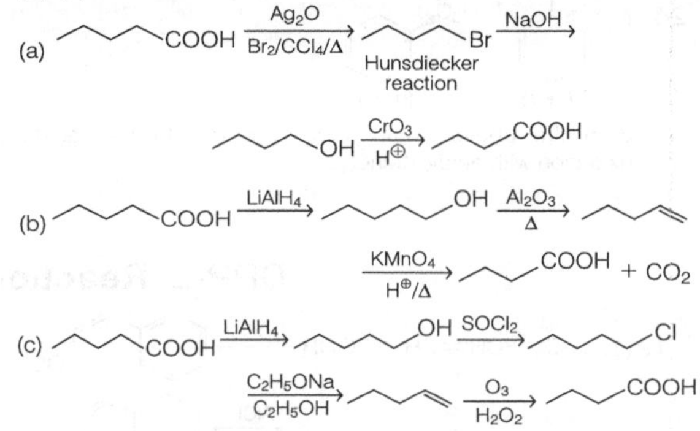

18. (b,c) 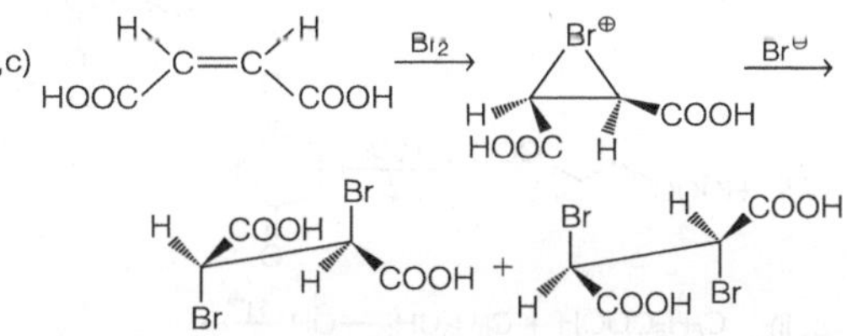

19. (b,c,d) Carboxylic or stronger acids give effervescence with $NaHCO_3$ but weaker acid like phenol does not. However if a strong electron withdrawing group is present at *para* position of phenol (nitro in the present case) it is neutralised by $NaHCO_3$.

Explanations for Q. Nos. (20–22)

Since, *A* on vigorous oxidation gives benzoic acid, it indicates that there is only one substituent on phenyl ring. Also *A* on reaction with HBr giving *B* (substitution of —OH by-Br) without change of configuration indicates the —OH group is not directly present on chiral carbon.

20. (c) From the above discussion, structure of *A* satisfying all the criteria is *c*.

21. (b) 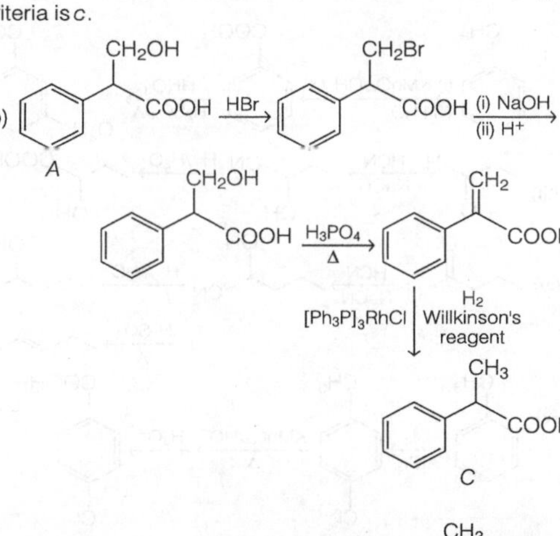

22. (b) 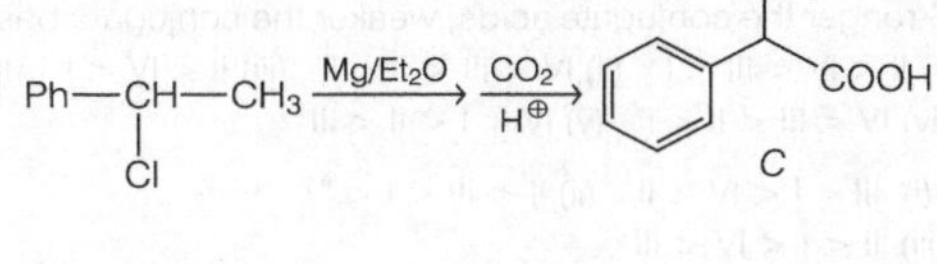

23. (2)

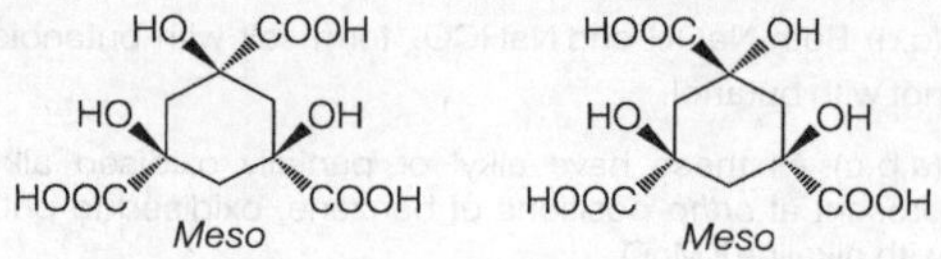

24. (4)

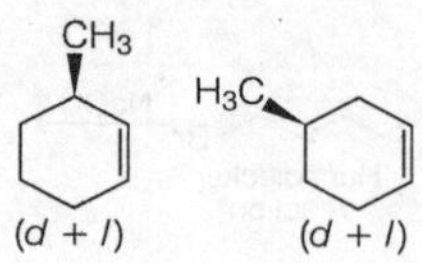

Both the above compounds give dicarboxylic acids upon oxidation with acidic $KMnO_4$.

25. (3)

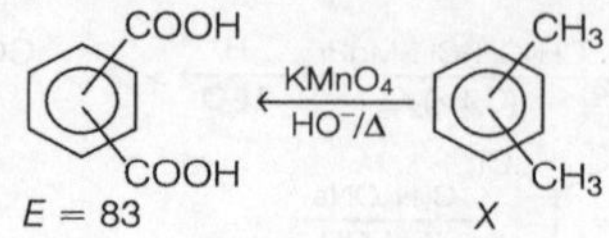

X has three isomers -*ortho*, *para* and *meta* isomers.

26. (3) $C_3H_7-CH=CH-CH_2CH_2CH_2-COOH$

$M = 156$

$\Big\downarrow$ Oxidation

$C_3H_7COOH + HOOC-CH_2CH_2CH_2-COOH$

DPP-2 Reactions of Carboxylic Acids

1. (i)

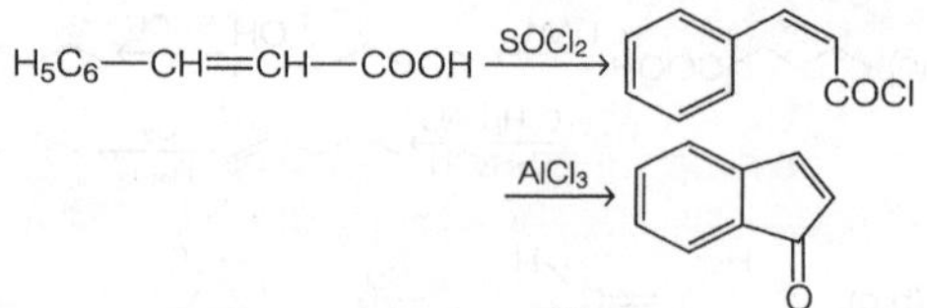

(ii)

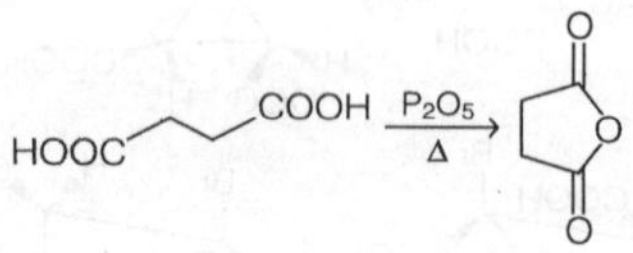

(iii)

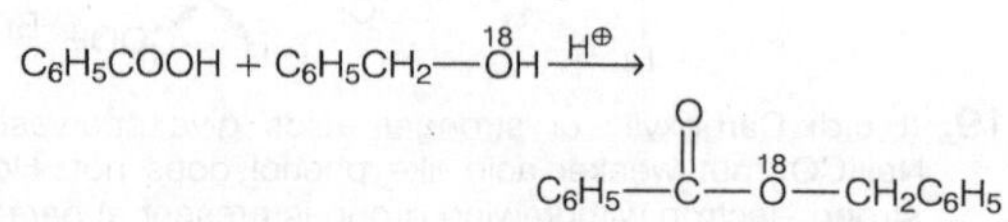

(iv)

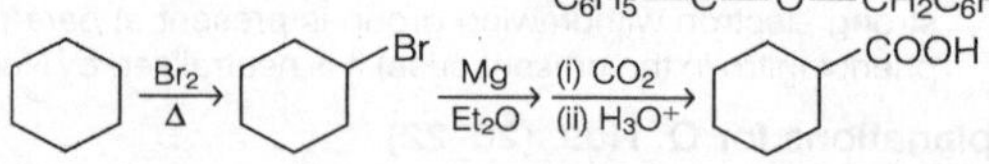

2. (i)

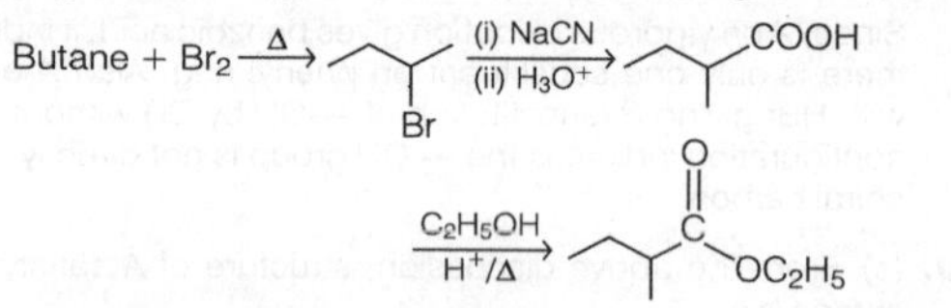

(ii)

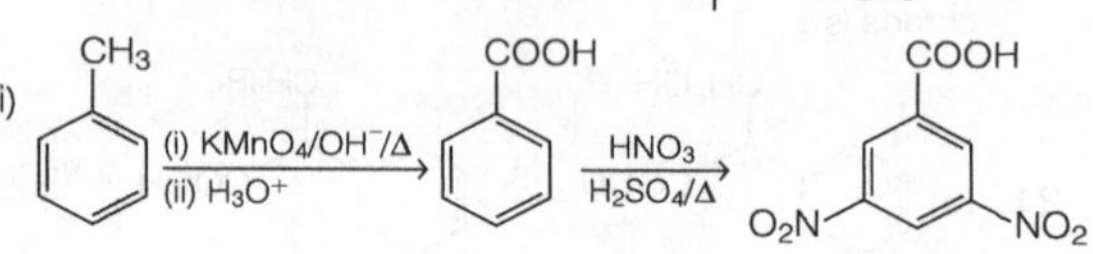

(iii)

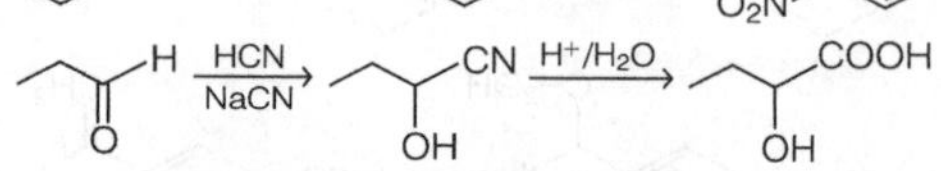

(iv)

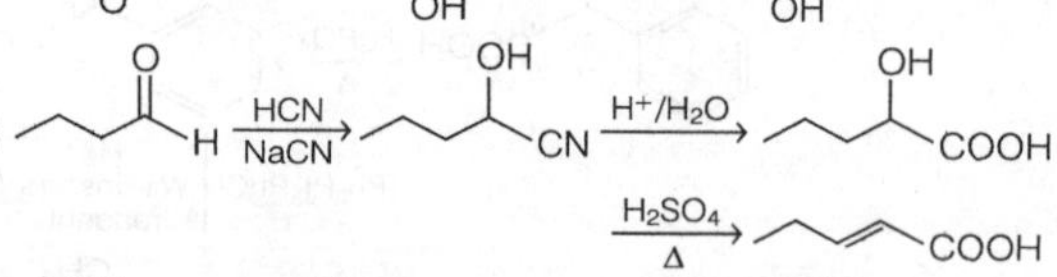

(v)

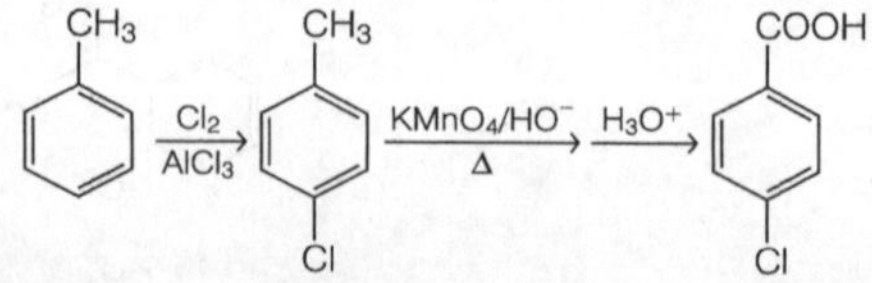

3. Stronger the conjugate acids, weaker the conjugate base.

(i) II < IV < III < I (ii) IV < III < II < I (iii) II < IV < I < III

(iv) IV < III < II < I (v) IV < I < II < III

4. (i) III < I < IV < II (ii) II < III < I < IV

(iii) II < I < IV < III

5. (i)

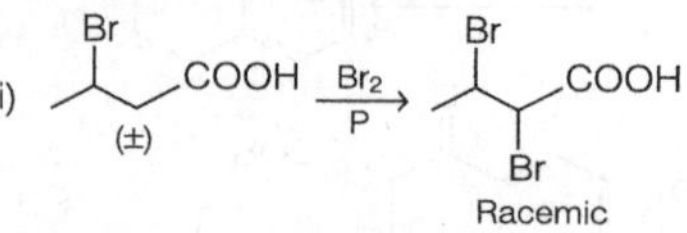

(ii)

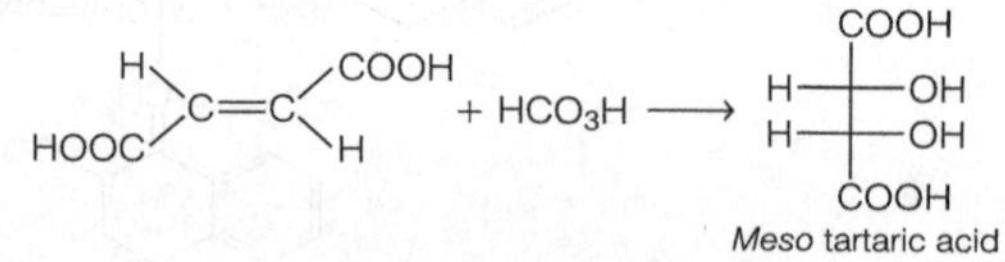

(iii)

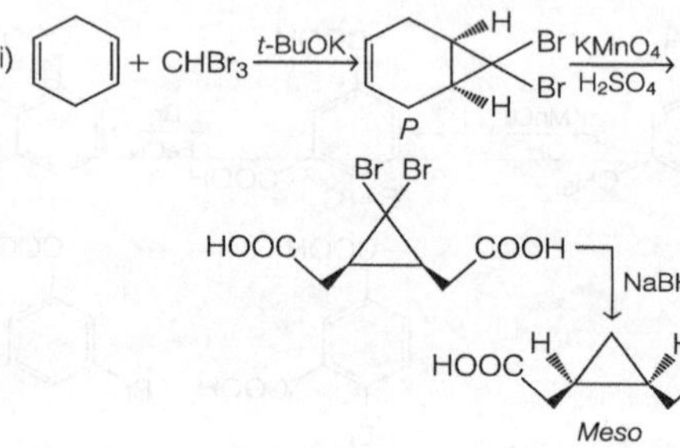

6. $H-C\equiv C-H + CH_3MgBr \longrightarrow \underset{A}{H-C\equiv CMgBr} + CH_4$

$A + CO_2 \longrightarrow \underset{B}{H-C\equiv C-COOMgBr} \xrightarrow{H_3O^{\oplus}} \underset{C}{H-C\equiv C-COOH}$

$C \xrightarrow[HgSO_4]{H_2SO_4} \underset{D}{H-\overset{O}{\overset{\|}{C}}-CH_2-COOH} \xrightarrow{KMnO_4} HO-\overset{O}{\overset{\|}{C}}-CH_2-COOH$

7. (b)

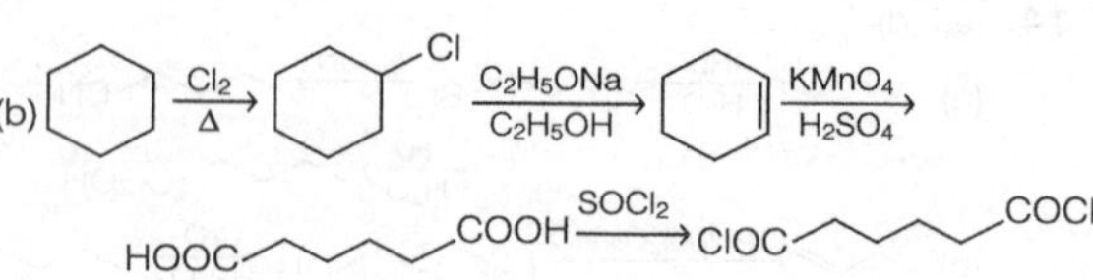

8. (b)

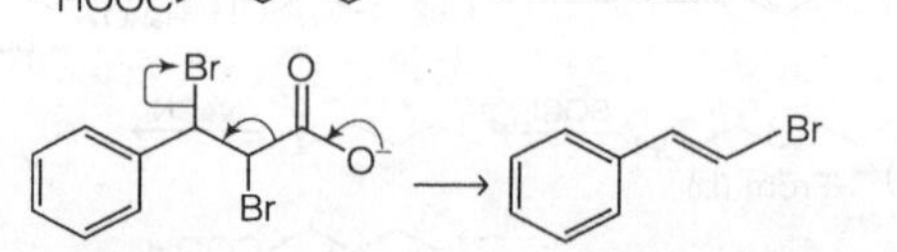

9. (b) $C_2H_2 \xrightarrow{CH_3MgBr} [C\equiv C]^{2-} \xrightarrow[H^{\oplus}]{CO_2} HOOC-C\equiv C-COOH$

$\xrightarrow[(ii)\ H_3O^{\oplus}]{(i)\ LiAlH_4} HO-CH_2-C\equiv C-CH_2-OH$

10. (c) The most stable conjugate base of ascorbic acid is

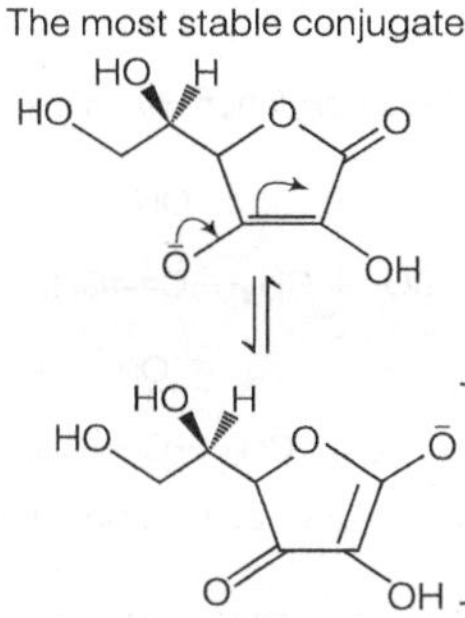

Resonance increases stability, hence acidic strength of conjugate acid.

11. (a) Ag_2O/NH_3 oxidises —CHO to —COOH without affecting triple bond or hydroxy group. Lindlar's catalyst in the subsequent step reduces selectively triple bond to *cis* configuration which is suitable for cyclisation during esterification.

12. (c)

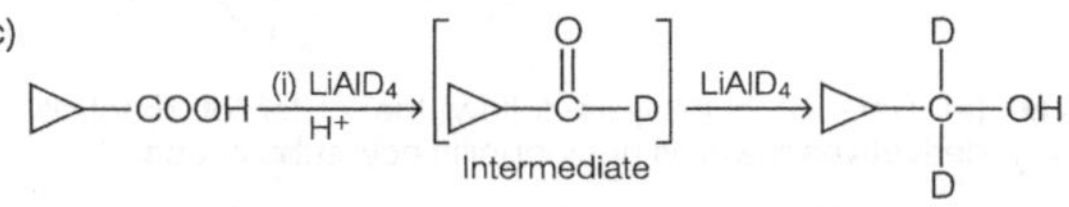

Reduction proceeds by deutride (D^-) transfer mechanism.

13. (b)

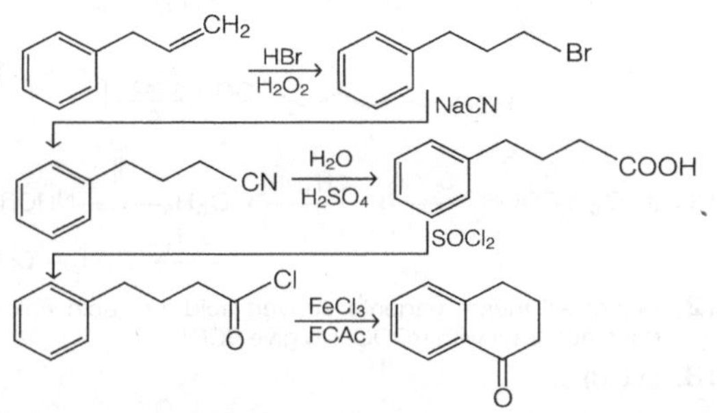

14. (c) Hydrazoic acid (NH_3) converts carboxylic acid group —COOH into amino group —NH_2 with the evolution of CO_2 and N_2.

15. (c,d)

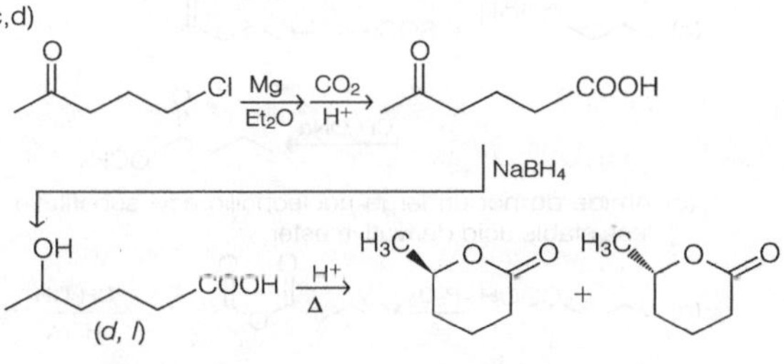

16. (c)

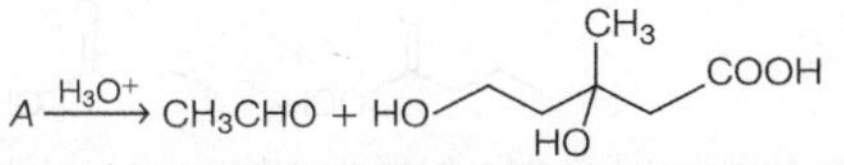

Acetal hydrolysis into acetaldehyde and corresponding diol.

17. (b,d) Oxidative cleavage occur at α-carbon of phenyl ring.

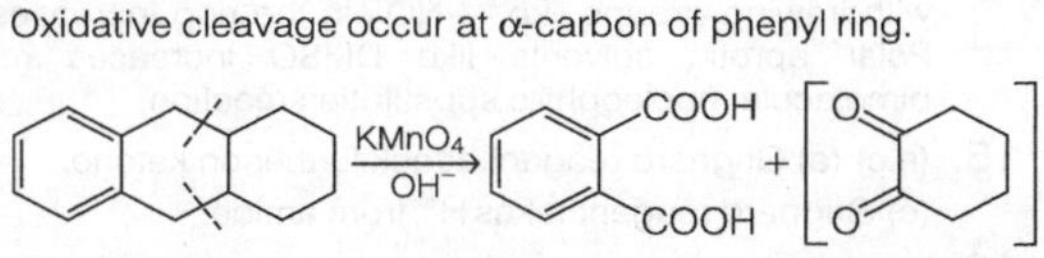

18. (a,b,d) Besides esterification, the tertiary alcohol also forms some carbocation (highly stable) which undergo Friedel-Craft alkylation with acid (giving "a") and with another alcohol (giving "d").

19. (a,b) During esterification (Fischer), nucleophilic attack occur from sp^3 oxygen of alcohol, hence configuration of α-carbon of alcohol is retained.

Explanations for Q. Nos. (20-22)

Since, *B* is a decarboxylation product of *A* and *B* on vigorous oxidation gives benzoic acid (neutralisation equivalent 122) indicates that *B* has only one alkyl group on phenyl ring. Also, *A* on vigorous oxidation gives a diacid (benzene dicarboxylic acid) whose neutralisation equivalent is 83. This indicates that *A* has an acid group and an alkyl group.

20. (b) From the above discussion and neutralisation equivalent (148) of *A*, its structure is derived to be ethyl benzoic acid.

21. (c)

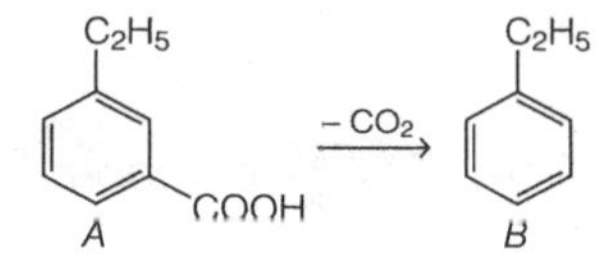

22. (d)

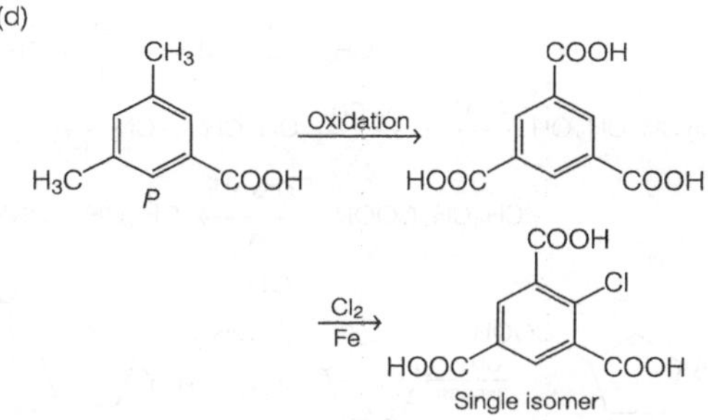

23. (4) $HOOC—CH_2—CH_2—COOH + 2HN_3 \longrightarrow H_2N—CH_2—CH_2—NH_2 + 2N_2 + 2CO_2$

24. (4) The formula of acid is $C_6H_{6-n}(COOH)_n$

Hence, neutralisation equivalent $= \dfrac{78 + 44n}{n} = 63.5$

$n = 4$

25. (6) $HOOC—\underset{\displaystyle CH_3}{\underset{|}{CH}}—CH_2—COOH + Br_2 \xrightarrow{P}$

$HOOC—\underset{\underset{(d + l)}{CH_3}}{\underset{|}{\overset{\overset{Br}{|}}{C}}}—CH_2—COOH + HOOC—\underset{\underset{(4\ stereoisomers)}{CH_3}}{\underset{|}{\overset{\overset{H}{|}}{C}}}—\overset{\overset{Br}{|}}{CH}—COOH$

26. (5)

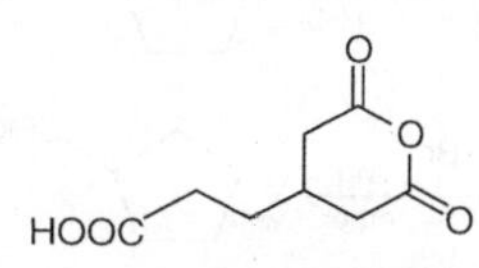

DPP-3 **Introduction to Acid Derivatives**

1. Acid chloride is more reactive because the carboxyl-carbon is sp^2 hybridised, provide less steric hindrance to nucleophilic attack. Also the carboxyl-carbon of acid chloride has greater partial positive charge (δ^+) than α-carbon of alkyl halides.

2. In nucleophilic acyl substitution, a more reactive acid derivative is converted spontaneously into more stable acid derivative. The order of stability of various acid derivatives is

Amide > Ester > Anhydride > Acid chloride

3. (i)

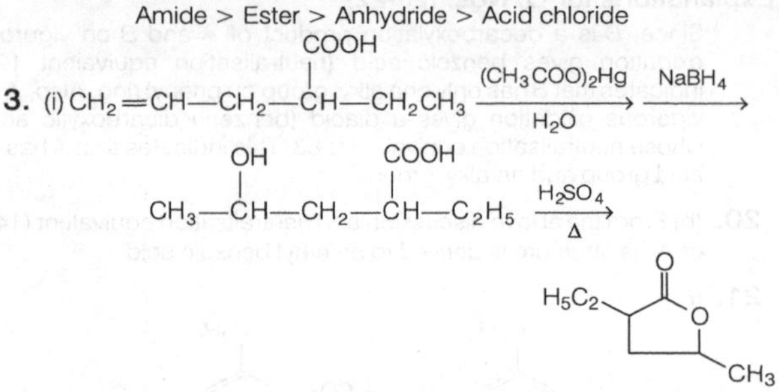

(ii) $CH_3—CH{=}CH_2 \xrightarrow[H^+/H_2O]{KMnO_4} CH_3COOH \xrightarrow[\Delta]{NH_3}$

$CH_3—\overset{O}{\overset{\|}{C}}—NH_2 \xrightarrow[\Delta]{P_2O_5} CH_3—CN$

(iii) $CH_3CH_2OH \xrightarrow{SOCl_2} \xrightarrow{NaCN} CH_3CH_2—CN \xrightarrow[H_2O]{H^+}$

$CH_3CH_2COOH \xrightarrow[\Delta]{C_2H_5NH_2} CH_3CH_2CONHC_2H_5$

4. (i)

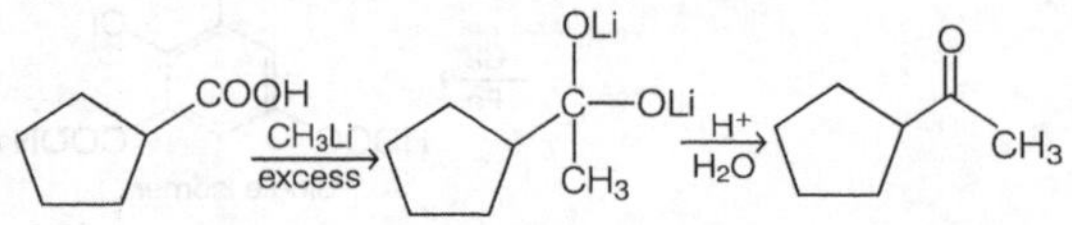

(ii)

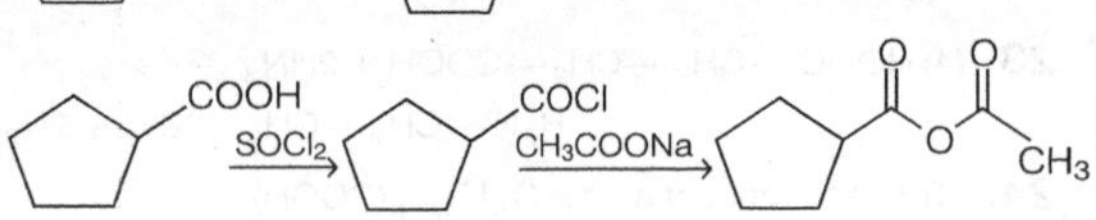

(iii)

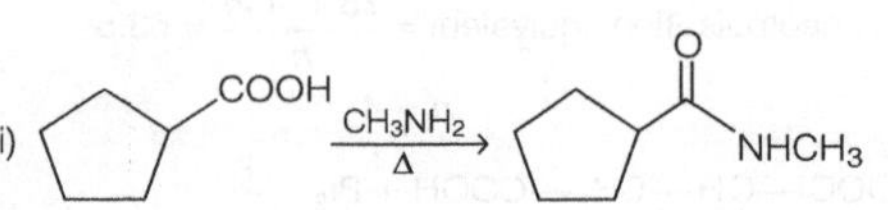

(iv)

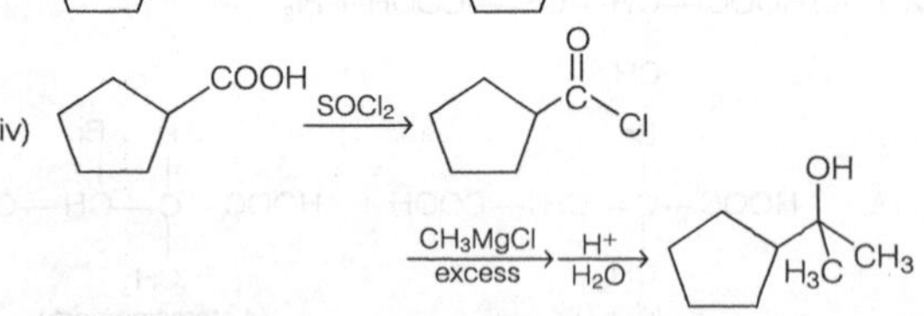

5.

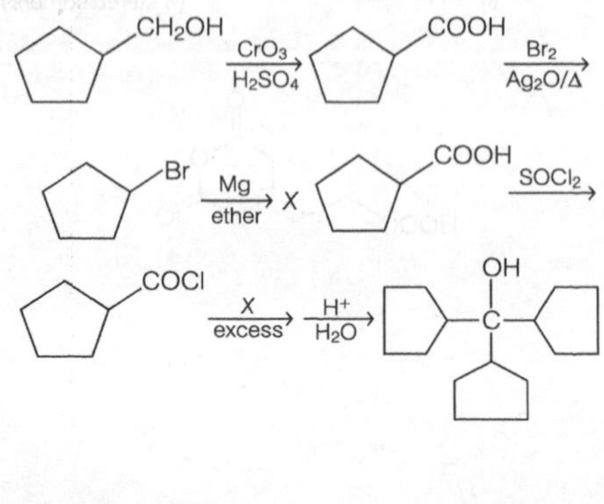

6. (c) β-keto undergo easier decarboxylation *via* six membered cyclic intermediate.

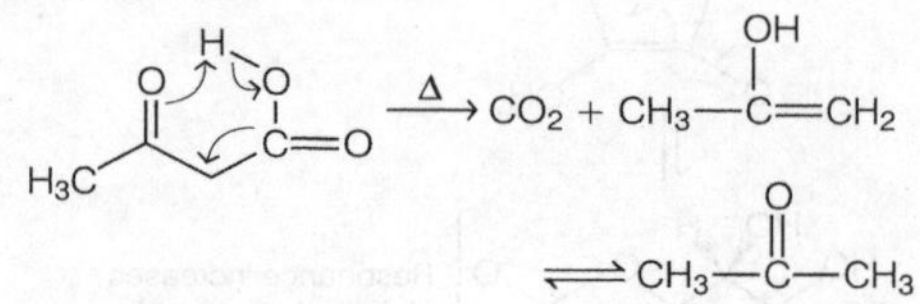

7. (d) Here the more stable acid derivative ester is being converted into less stable anhydride.

8. (a) The two —COOH groups are *syn*, can form cyclic anhydride.

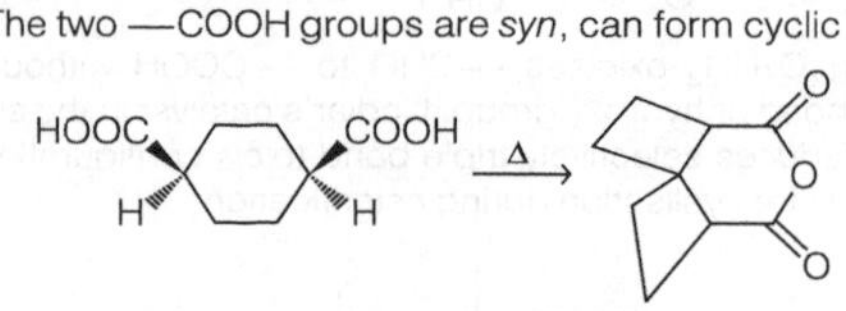

9. (a) Rate of hydrolysis follow the order of reactivity of acid derivatives towards nucleophilic acyl substitution.

10. (b)

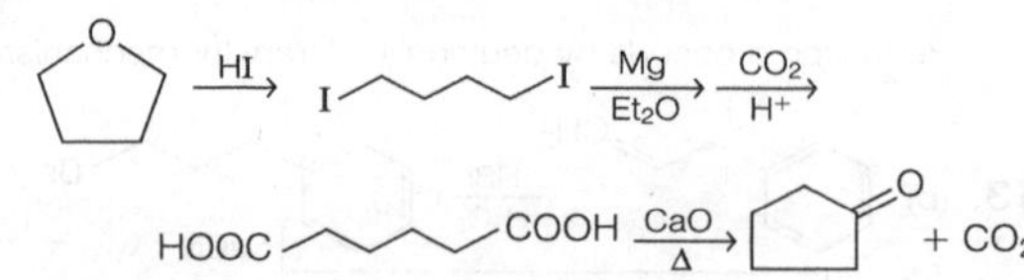

11. (b) $C_6H_5COOH \xrightarrow{SOCl_2} \xrightarrow{CH_3NH_2} C_6H_5—\overset{O}{\overset{\|}{C}}—NHCH_3$

$\xrightarrow{LiAlH_4} C_6H_5—CH_2NHCH_3$

12. (a,b,c) All these reagents convert acid into acid chloride. HCl does not react with RCOOH to give RCOCl.

13. (a,b,d)

(a)

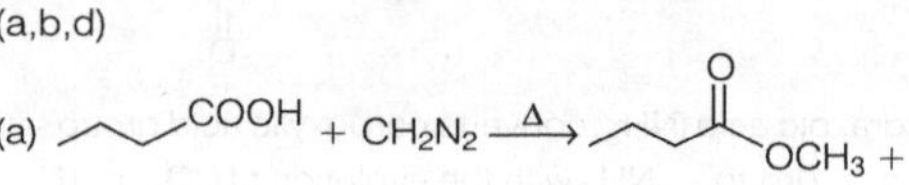

(b)

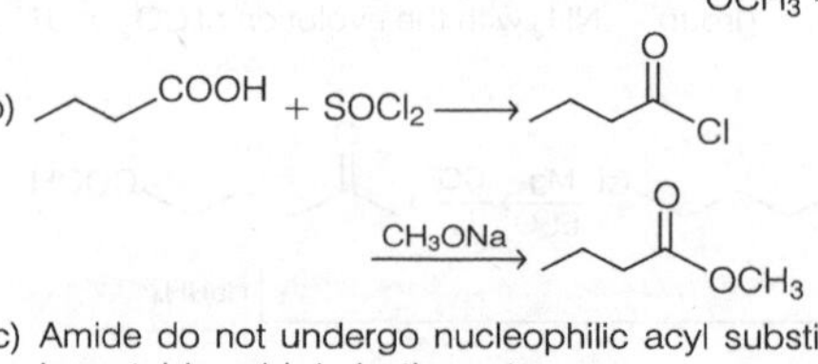

(c) Amide do not undergo nucleophilic acyl substitution to give less stable acid derivative ester.

(d)

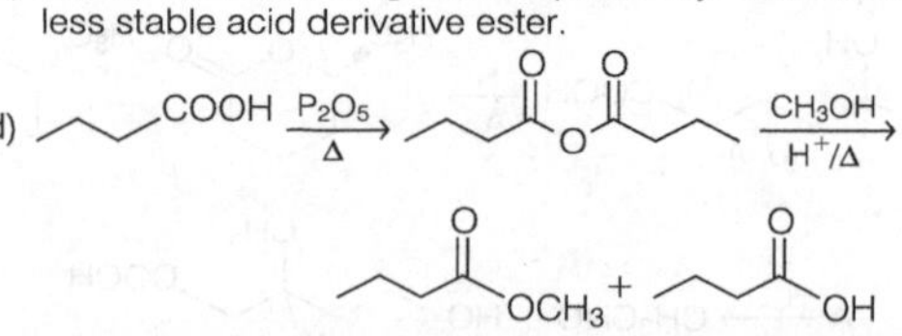

14. (a,c,d) Nucleophilic acyl substitution shown above is bimolecular, second order reaction. Therefore, presence of electron withdrawing groups like $—NO_2$ in the ring increases reactivity. Polar aprotic solvents like DMSO increases reactivity of bimolecular nucleophilic substitution reaction.

15. (a,c) (a) Grignard reagent attack further on ketone.
(c) Grignard reagent takes $H^{\oplus}$ from amide.

16. (b) Stronger bases are stronger nucleophiles (usually), hence poor leaving groups.

17. (d) Electron withdrawing group on phenyl ring increases electrophilicity of carboxyl group and *vice versa*. Hence, III is most reactive and II is most stabl

18. (b) $C_6H_5-\overset{O}{\overset{\|}{C}}-OH \xrightarrow{SOCl_2} C_6H_5-\overset{O}{\overset{\|}{C}}-Cl$

$$\xrightarrow[\text{excess}]{CH_3MgBr} \xrightarrow[H_2O]{H^+} C_6H_5-\underset{CH_3}{\overset{OH}{C}}-CH_3 \xrightarrow[\Delta]{H_2SO_4} C_6H_5-\overset{CH_3}{C}=CH_2$$

19. (a) In nucleophilic acyl substitution, a more reactive acid derivative is converted into a more stable acid derivative.

20. (a) $R-\overset{O}{\overset{\|}{C}}-OH \xrightarrow{H^\oplus} R-\overset{^+O-H}{\overset{\|}{C}}-OH$

$$\xrightarrow[-H^+]{R'-\overset{*}{O}H} R-\underset{OH}{\overset{OH}{C}}-\overset{*}{O}-R'$$

The above intermediate dehydrate to give ester in which oxygen (starred) of alcohol is retained in ester.

21. (b) Both are independently correct but formation of racemic mixture of ester is due to attack of acid on planar carbocation ($CH_3CH_2-\overset{\oplus}{C}H-CH_3$).

22. (b) Both are independently correct but formation of racemic mixture of esters is due to the formation of racemic mixture of alcohols (P) by hydride ion attack on planar carbonyl carbon of butanone.

23. (4) In the given reaction : $-COOH \longrightarrow -COOCH_3$

A net gain in molar mass of 14 unit is observed per —COOH present. Let there be "n" —COOH group (basicity) be present.

Hence, molar mass of acid = $65.5\,n$

Molar mass of ester = $65.5n + 14n = 79.5n = 318$

Hence, $n = \dfrac{318}{79.5} = 4$

24. (2) If n is less than 2, decarboxylation is preferred over cyclisation.

25. (4)

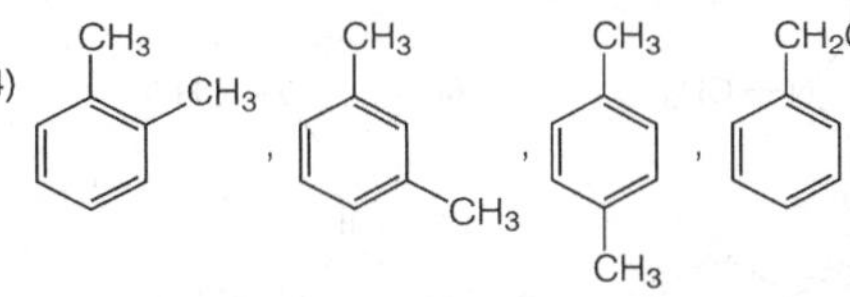

26. (10)

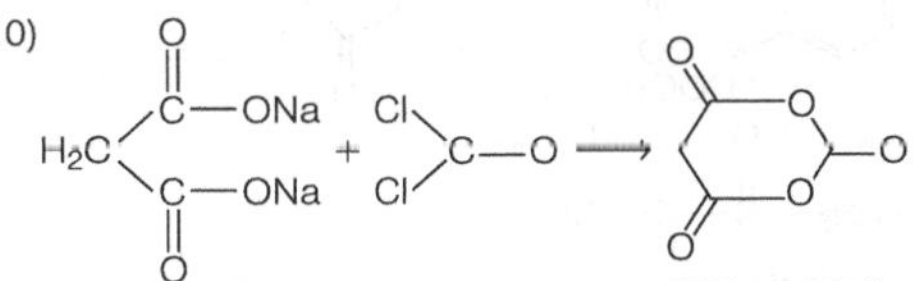

(10 lone pairs)

DPP-4 Acid Chlorides and Anhydrides

1. (i) $CH_3CH_2OH \xrightarrow{SOCl_2} CH_3CH_2Cl \xrightarrow[\text{ether}]{Mg} \xrightarrow[H_2O]{^*CO_2}$

$$CH_3CH_2-\overset{*}{C}OOH \xrightarrow{SOCl_2} CH_3-CH_2-\underset{*}{\overset{O}{\overset{\|}{C}}}-Cl \quad [^*C = C^{18}]$$

(ii) $CH_3CH_2CH_2COOH \xrightarrow[H_2O]{LiAlH_4} \xrightarrow{SOCl_2} CH_3CH_2CH_2CH_2Cl$

$$\xrightarrow[C_2H_5OH]{C_2H_5OK} CH_3CH_2-CH=CH_2 \xrightarrow[H^+/H_2O]{KMnO_4} CH_3CH_2COOH \xrightarrow{SOCl_2} CH_3CH_2COCl$$

(iii) $CH_3-COOH \xrightarrow[P]{Cl_2} Cl-CH_2-COOH \xrightarrow{NaCN}$

$$NC-CH_2-COOH \xrightarrow[H_2O]{H^+} HOOC-CH_2-COOH \xrightarrow{SOCl_2}$$

$$\xrightarrow[\text{xylene}]{Pd/BaSO_4/H_2} HOC-CH_2-CHO$$

(iv) $CH_3CH_2COOH \xrightarrow[H_2O]{LiAlH_4} CH_3CH_2CH_2-OH \xrightarrow{PCl_5} \xrightarrow{Li}$

$$CH_3CH_2CH_2Li \xrightarrow{CH_3CH_2COOH} CH_3CH_2-\underset{OLi}{\overset{OLi}{C}}-CH_2CH_2CH_3$$

$$\xrightarrow[H_2O]{H^+} CH_3CH_2-\overset{O}{\overset{\|}{C}}-CH_2CH_2CH_3$$

2. (i) $C_6H_5COOH \xrightarrow[\Delta]{P_2O_5} C_6H_5-\overset{O}{\overset{\|}{C}}-O-\overset{O}{\overset{\|}{C}}-C_6H_5$

(ii)

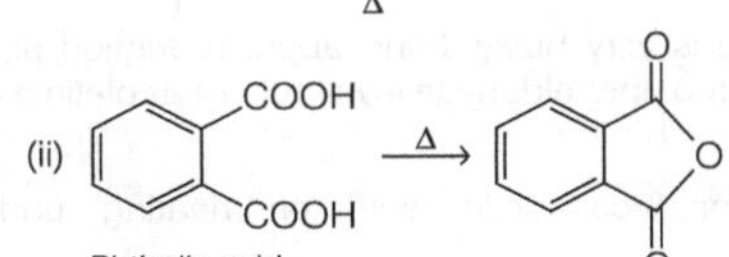

Phthalic acid

(iii)

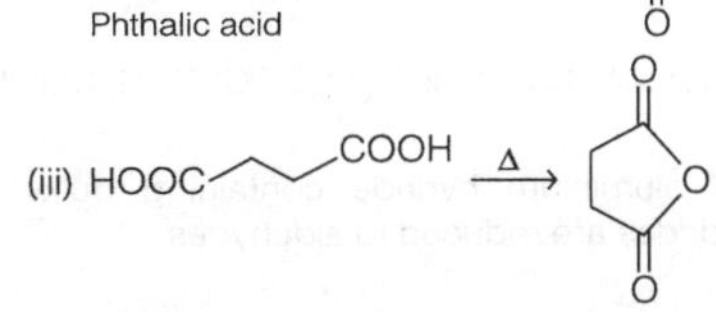

Succinic acid

(iv) $CH_3COOH \xrightarrow{SOCl_2} CH_3COCl \xrightarrow{CH_3CH_2COONa} CH_3-\overset{O}{\overset{\|}{C}}-O-\overset{O}{\overset{\|}{C}}-CH_2CH_3$

3. (i)

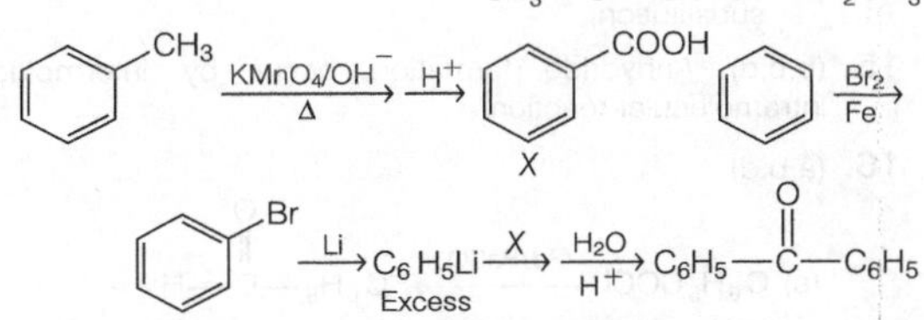

(ii)

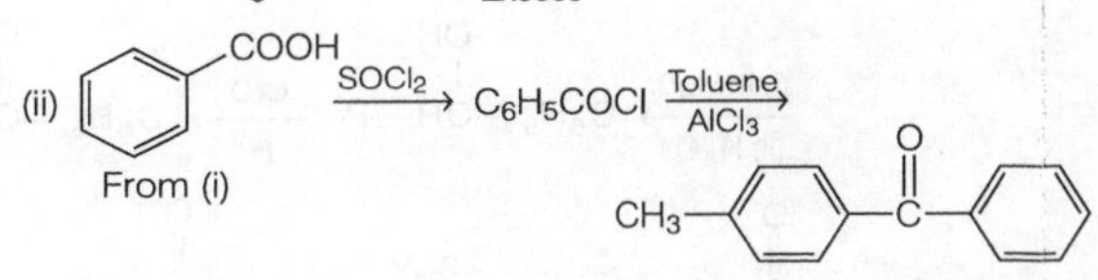

4. (i)

(ii)

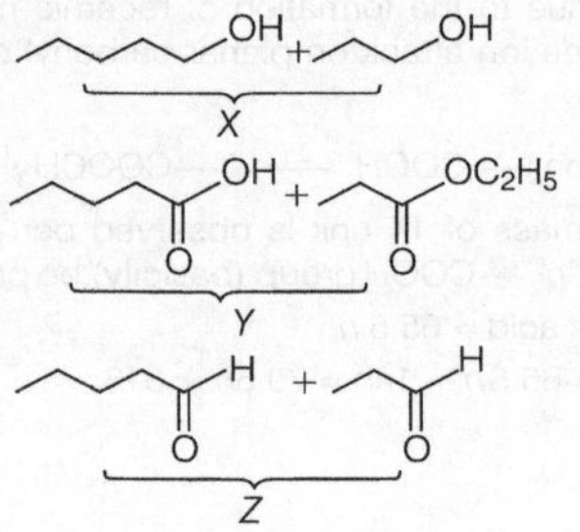

5.

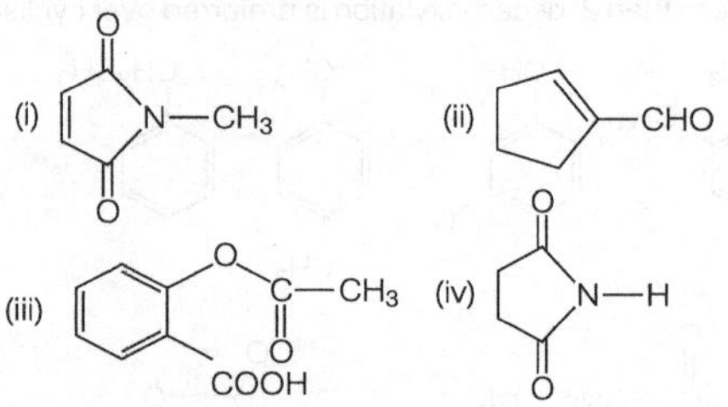

(v) $C_6H_5—\overset{O}{\overset{\|}{C}}—O—\overset{O}{\overset{\|}{C}}—CH_3$

6. (b) CF_3COO^- is better leaving group than CH_3COO^-.

7. (a) Nucleophilic attack occur at less hindered carboxyl group.

8. (c) More stable acid derivative amide is formed by nucleophilic attack of amino group at less hindered carboxyl group.

9. (c) R_2CuLi is a poor source of nucleophile, bring about S_NAc reaction at more reactive acid chloride but not at stable ester group.

10. (b) Given hydride is very bulky, bring about reduction at acid chloride and that too upto aldehyde level but not at olefinic triple bond.

11. (a) β-carbonyl or β-carboxylic acid on heating undergo decarboxylation.

12. (d) CH_3COCl is better substrate than C_6H_5COCl. C_6H_5COONa is better nucleophile.

13. (c) With complex aluminium hydride containing bulky alkyl groups, acid anhydrides are reduced to aldehydes.

14. (a,d)

(a) Acid amide is more stable acid derivative than ester, cannot be converted into ester by nucleophilic acyl substitution.

(d) Acid anhydride is more stable than acid chloride, hence cannot be converted into later by nucleophilic acyl substitution.

15. (b,c,d) Anhydride formation occur by intermolecular or intramolecular reaction.

16. (a,b,d)

(a) $C_6H_5COCl \xrightarrow{Pd/BaSO_4} C_6H_5—\overset{O}{\overset{\|}{C}}—H$

$\xrightarrow[\text{(ii) } H_3O^+]{\text{(i) PrMgBr}} C_6H_5—\underset{}{\overset{OH}{\overset{|}{C}H}}—Pr \xrightarrow[H^+]{CrO_3} C_6H_5—\overset{O}{\overset{\|}{C}}—Pr$

(b) $C_6H_5—\overset{O}{\overset{\|}{C}}—Cl + Pr_2CuLi \xrightarrow{S_NAc} C_6H_5—\overset{O}{\overset{\|}{C}}—Pr$

(c) It gives tertiary alcohol.

(d) $C_6H_5—\overset{O}{\overset{\|}{C}}—Cl \xrightarrow[-80°C]{Li[(t\text{-BuO})_3AlH]} C_6H_5—\overset{O}{\overset{\|}{C}}—H$

$\xrightarrow[\text{(ii) } H_3O^+]{\text{(i) PrMgBr}} C_6H_5—\overset{OH}{\overset{|}{C}H}—Pr \xrightarrow[H^+]{CrO_3} C_6H_5—\overset{O}{\overset{\|}{C}}—Pr$

17. (a,b,c)

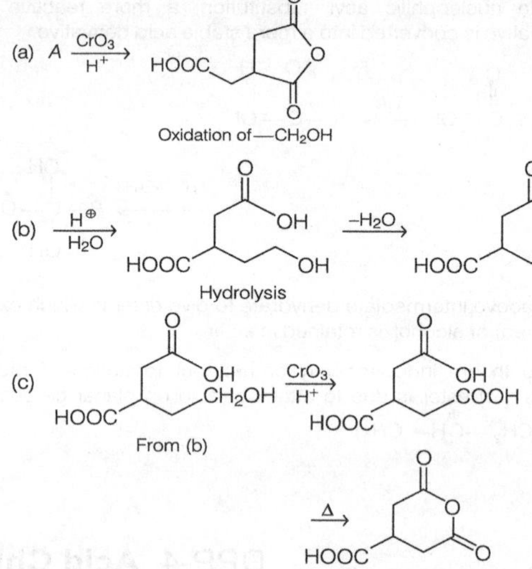

18. (a,b,c,d)

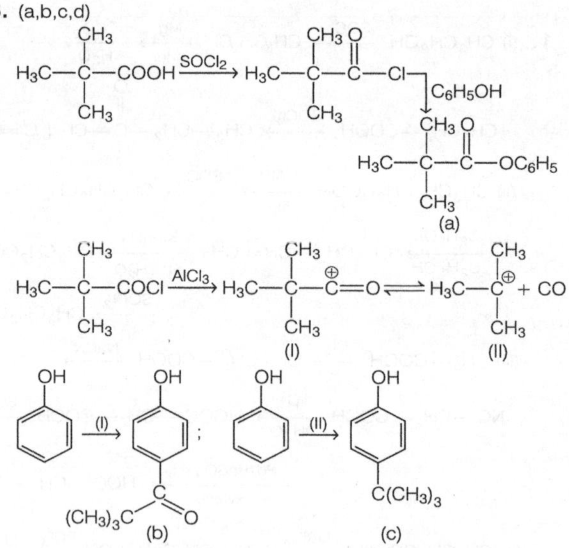

$C_6H_5—\ddot{O}—H + \overset{\oplus}{C}(CH_3)_3 \xrightarrow{-H^{\oplus}} C_6H_5—O—C(CH_3)_3$ (d)

19. (b,c)

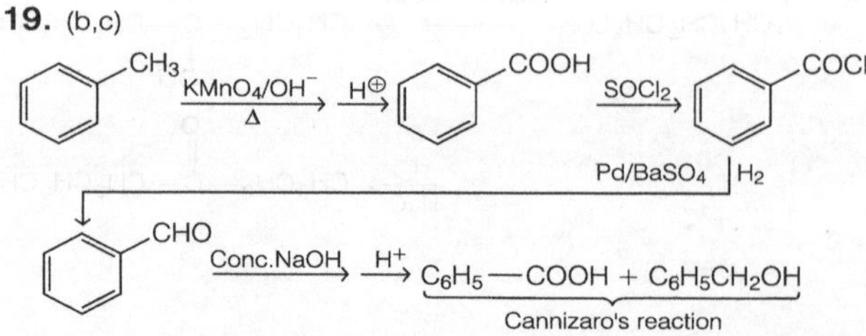

Explanations for Q. Nos. (20–22)

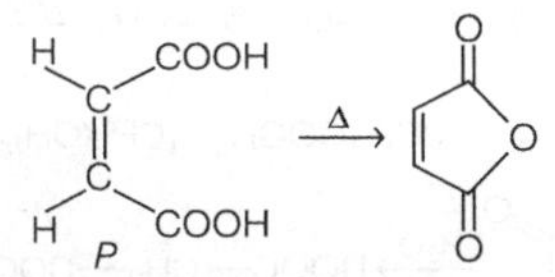

Therefore, Q is *trans* isomer (fumaric acid) of P.

20. (b) P gives optically inactive *meso* diol S and Q gives inactive (racemic) pair T and U.

21. (a)

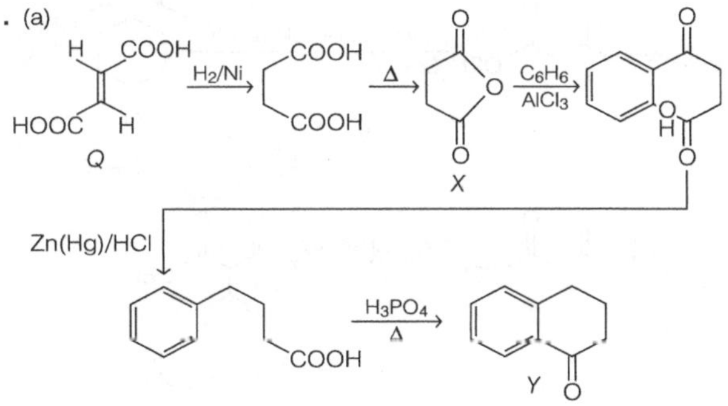

22. (c) Anti-dibromides are formed. Hence, *cis* isomer would give racemic mixture while *trans* isomer would give *meso*.

23. (3)

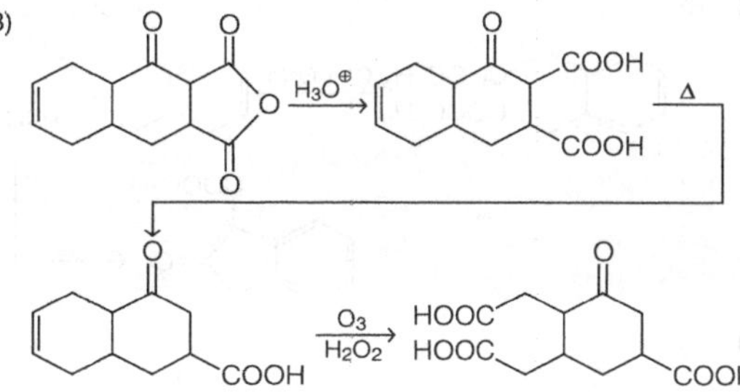

24. (2)

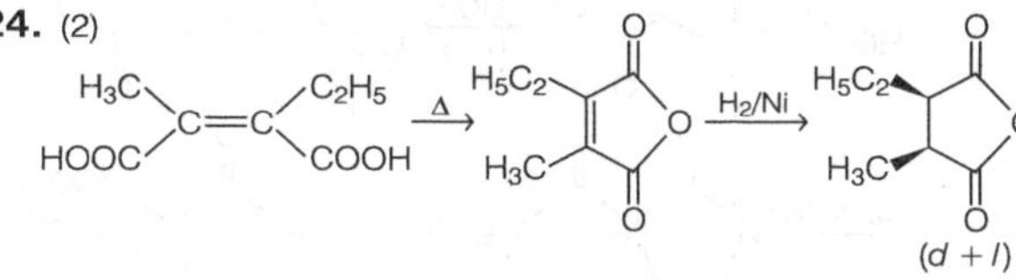

Syn hydrogenation occur giving —C_2H_5 and —CH_3 on the same side of plane only.

25. (3) Only *cis* isomers can give cyclic anhydride.

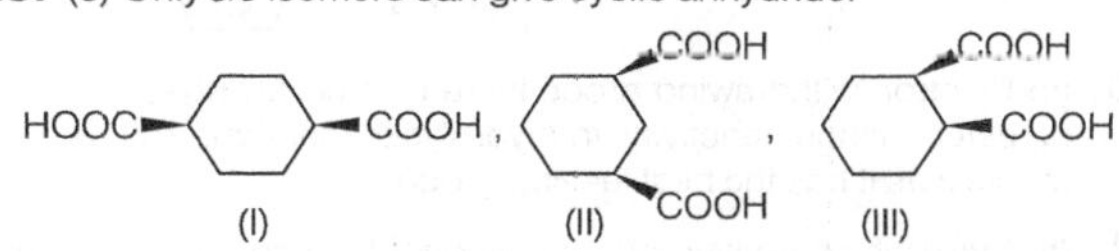

DPP-5 Esters

1. In basic solution, $RCOOH$ would be neutralised to form a nucleophile ($RCOO^-$) stronger than R—OH which otherwise acts as a nucleophile in Fischer's esterification.

However, hydrolysis is catalysed by both acid and bases as :

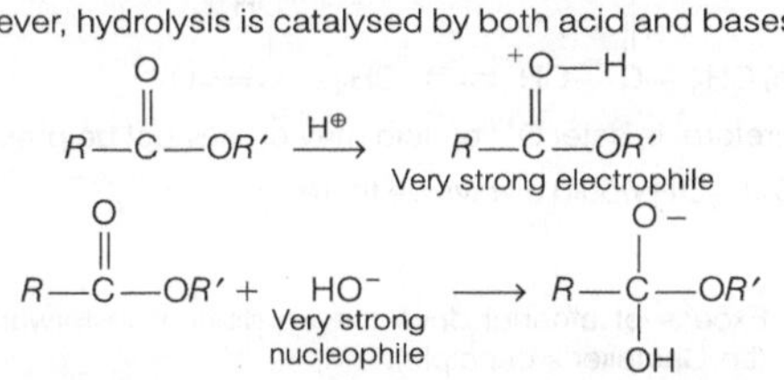

2. In acidic medium, the second —COOH is also esterified while the same does not happens in basic medium.

3. (a)

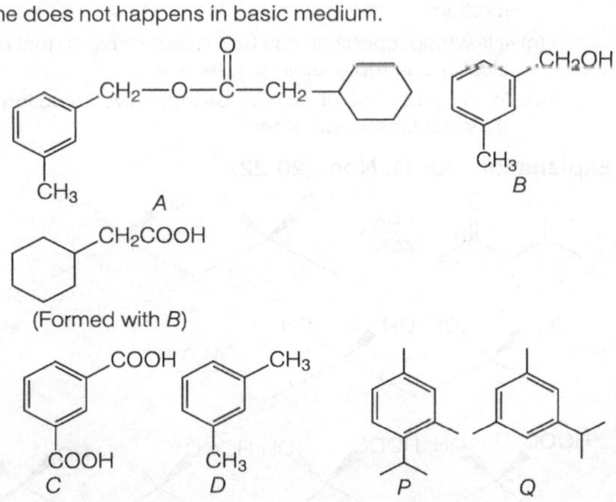

4. A : $HO—CH_2—CH_2—CH_2—OH$

B : $CH_3—CH(OH)—CH_2OH$ C : $CH_3OCH_2CH_2OH$

D : $CH_3O—CH(OH)—CH_3$ E : $CH_3O—CH_2—OCH_3$

5.

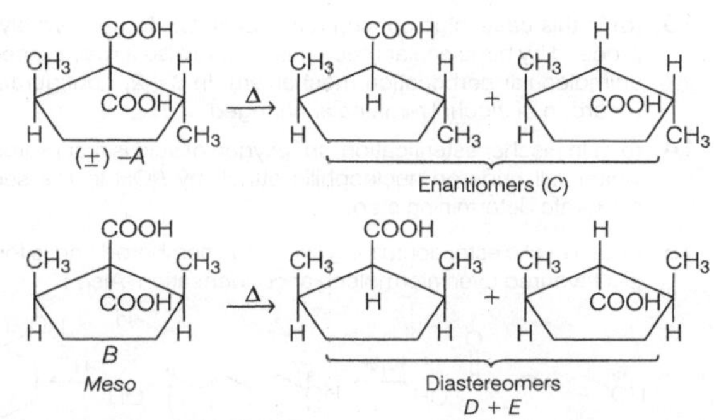

6. (a)

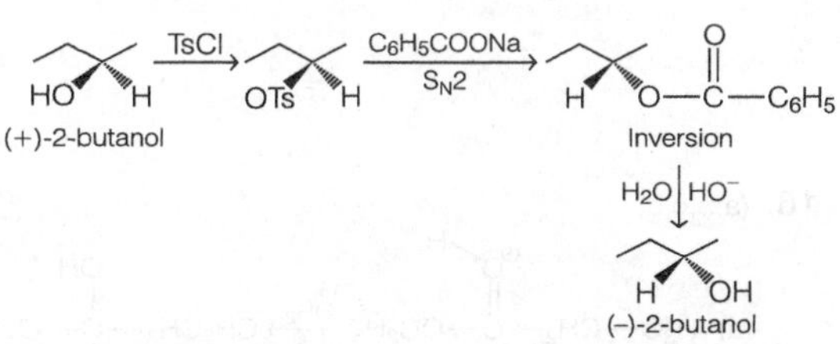

In ester hydrolysis, alkyl-oxygen bond is not cleaved, hence retention of configuration.

7. (b)

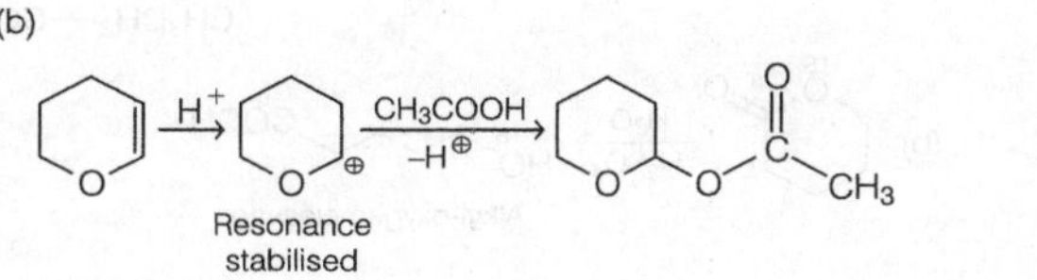

8. (b)

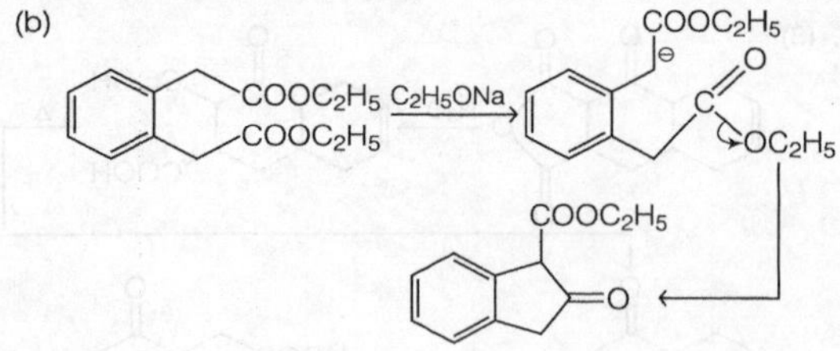

9. (d)

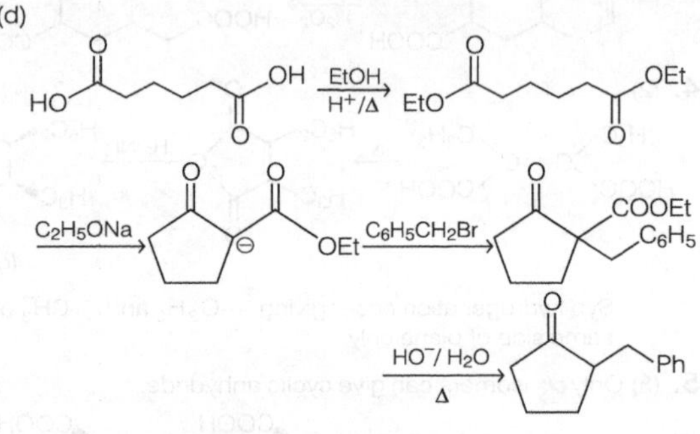

10. (c) Electron withdrawing group increases electrophilic character of ester, hence reactivity in hydrolysis reaction. (IV) is most reactive as it has the best leaving group.

11. (a) Ester which contain primary alcohol fraction can most easily be synthesised by Fischer's synthesis. Secondary and tertiary alcohols have chances of acid catalysed dehydration.

12. (d) Esters that produces a highly stable carbocation by alkyl-oxygen cleavage ($Ph_2CH^{\oplus}$ in present case) are more likely to undergo unimolecular acid catalysed hydrolysis.

13. (d) In this case, high concentration of NaOH forces hydrolysis to proceed by bimolecular mechanism otherwise it may proceed by unimolecular carbocation mechanism. In $B_{AC}2$, configuration at α-carbon of alcohol remains unchanged.

14. (a,c) In Fischer esterification, sp^2 oxygen of acid is first protonated which will undergo nucleophilic attack by *R*OH in the second, slow, rate determining step.

15. (a,b) If cyclic ester containing five or six membered ring is formed, it is favoured over intermolecular condensation. Also :

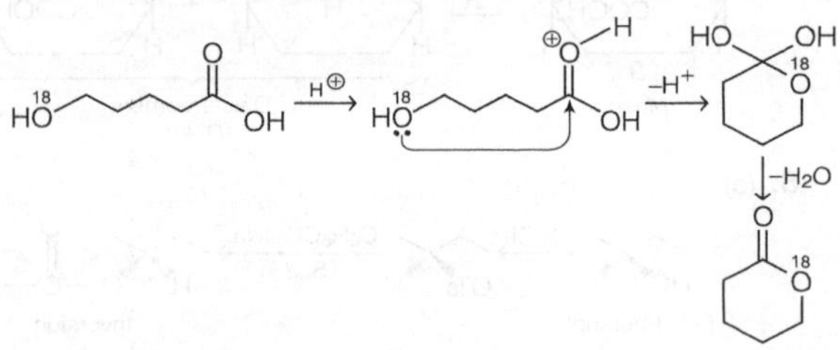

16. (a,b,c,d)

(a)

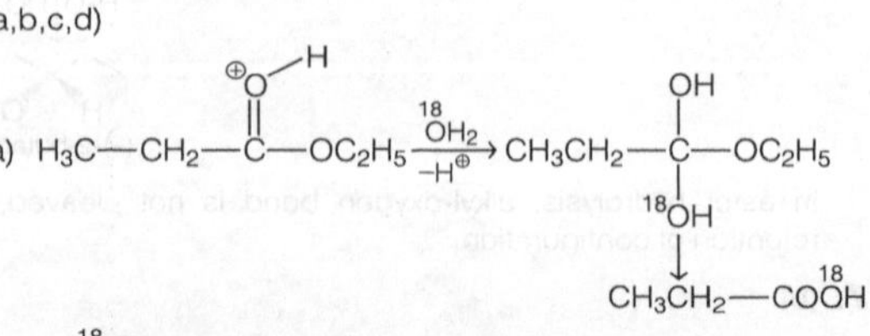

(b)

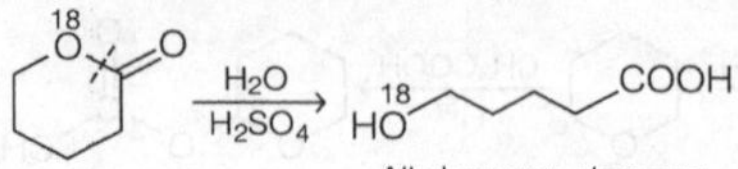

Alkyl-oxygen cleavage

(c)

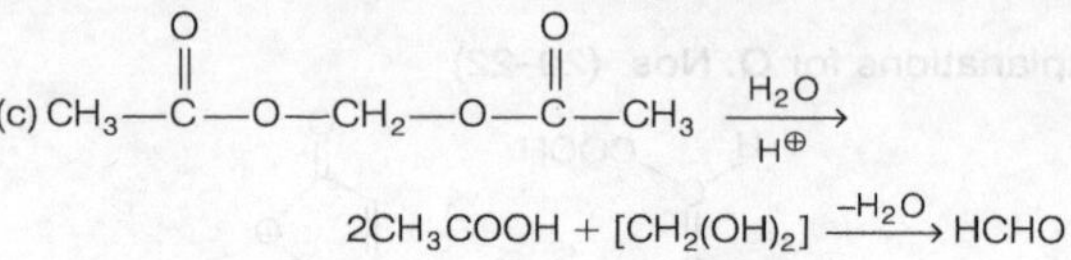

(d)

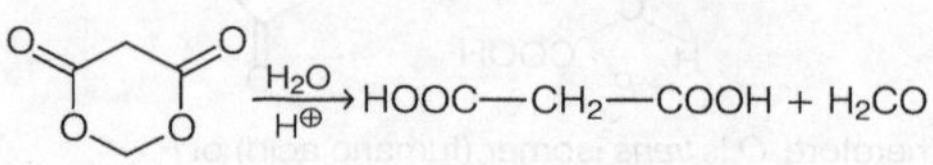

17. (b,c,d)

(b)

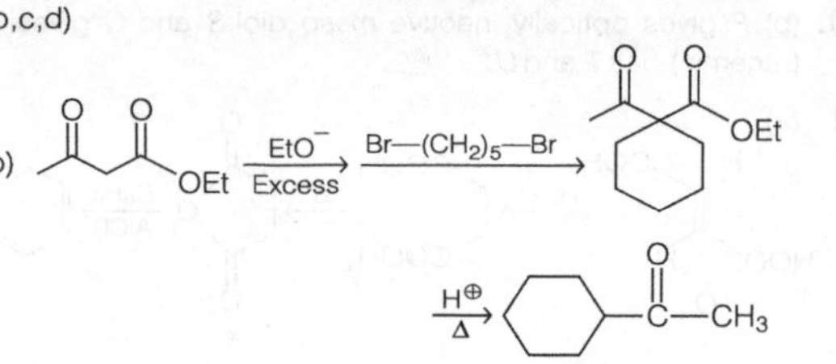

(c)

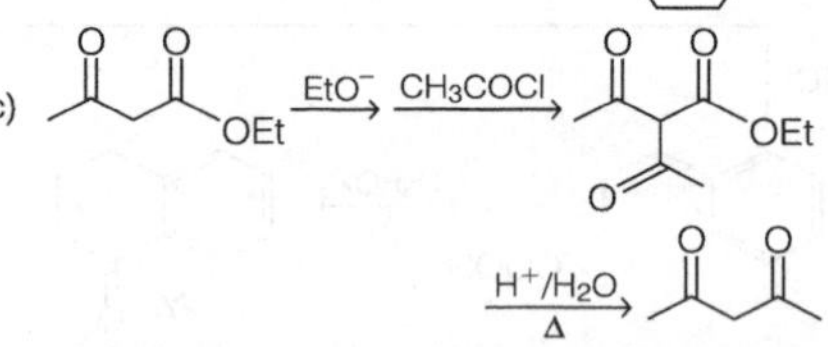

(d)

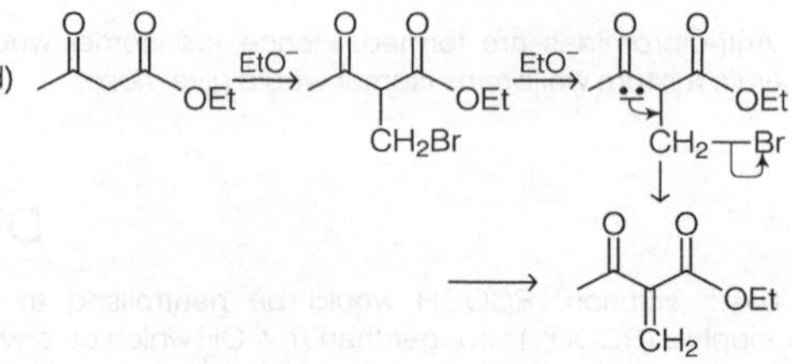

18. (b,c) $CH_3-\overset{O^{18}}{\overset{\|}{C}}-OH \rightleftharpoons CH_3-\overset{^{18}OH}{\overset{|}{C}}=O$

Therefore, in ester O^{18} of acid may or may not be present but O^{18} of $C_2H_5\overset{18}{O}H$ would always be there.

19. (a,b,d)

(a) Excess of ethanol drive the equilibrium in forward direction (Le-Chatelier's principle).

(b) Transesterification proceed well in both acidic and basic medium.

(c) It is wrong, condition can be made suitable so that even larger alcohol can replace the smaller one.

(d) In the presence of acid or base catalyst, reaction is always bimolecular second order.

Explanations for Q. Nos. (20-22)

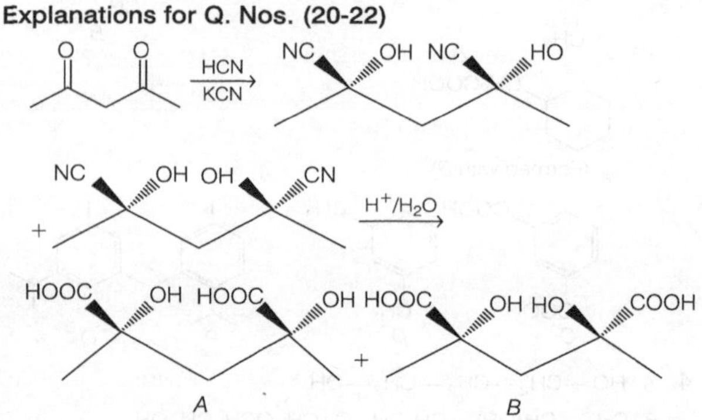

20. (b)

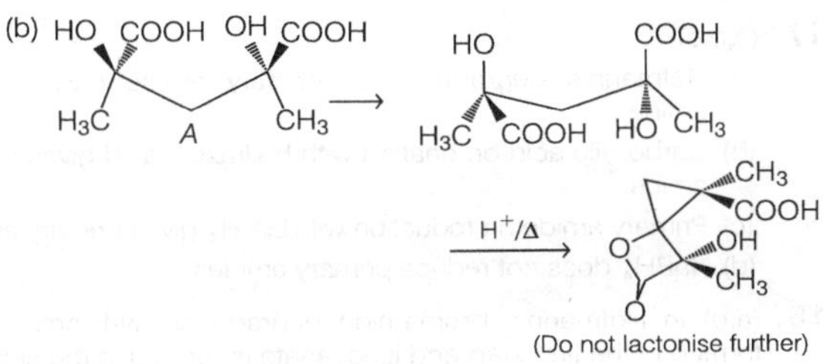

Second step of lactonisation is possible with *B*.

21. (c) Discussed in the previous question.

22. (c) Product shown in question-20. The —COOH and —OH are on anti sides, do not lactonise further.

23. (a) *R*COOH is neutralised to *R*COONa which itself is a nucleophile and it does not undergo nucleophilic attack by poor nucleophile *R*OH.

24. (c)

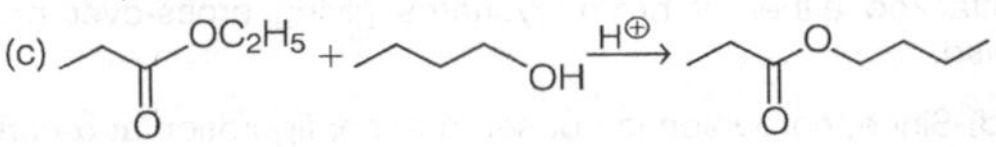

$+ CH_3CH_2OH$

Ethanol has lower boiling point than butanol, evaporates immediately after it is formed which enables the above reaction to proceed in forward direction.

However, suitable condition can be made to reverse the above reaction.

25. (a) At low hydroxide condition this hydrolysis proceeds as

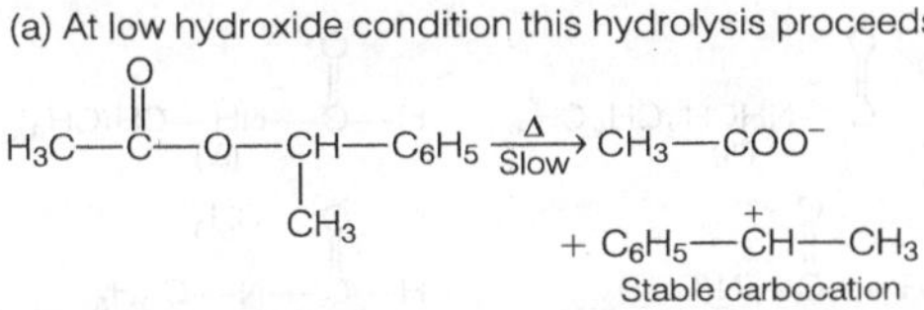

The planar carbocation then undergo nucleophilic attack by OH^- giving racemic mixture.

26. (a) Electron withdrawing nitro group increases electrophilic character of —COOH, hence increases reactivity in Fischer's esterification.

27. (6)

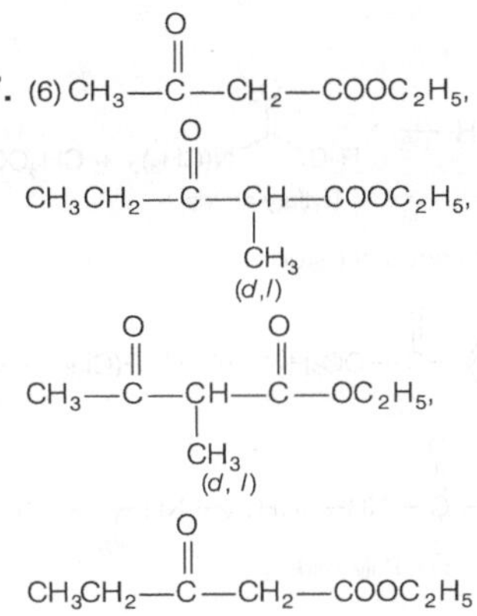

28. (7) Reactions (I), (III), (IV), (V), (VI), (VIII) and (IX) give ester.

29. (4)

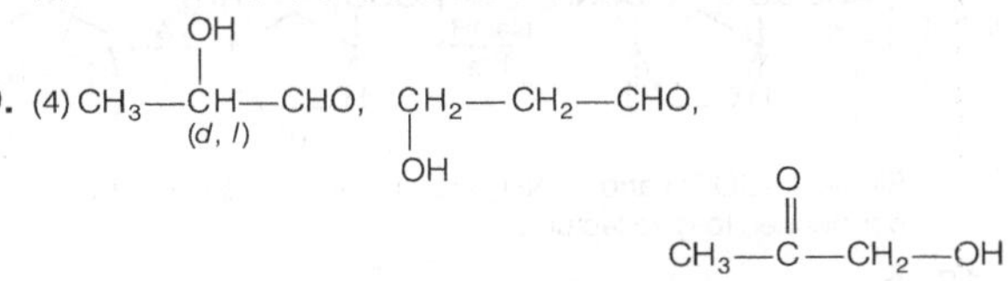

α-hydroxy ketones also reduces Tollen's reagent.

30. (7)

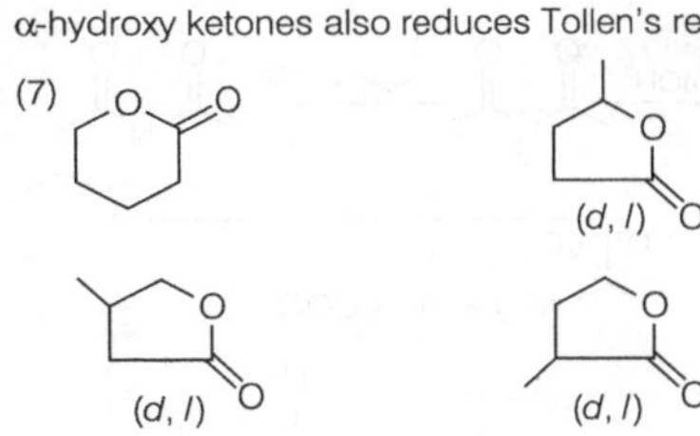

DPP-6 Acid Amides

1. $R—CN \xrightarrow{\text{Conc. } H_2SO_4} R—\overset{O}{\overset{\|}{C}}—NH_2$

If hydrolysis of nitrile is carried out in the presence of highly concentrated acidic medium, amide is formed. In dilute acidic medium, amide is further hydrolysed to acid.

$$R—CN \xrightarrow[H_2O_2]{NaOH} R—\overset{O}{\overset{\|}{C}}—NH_2$$

If alkaline hydrolysis is carried out in the presence of H_2O_2, partial hydrolysis occur giving amide as major product.

2. (i) < (iv) < (ii) < (iii).

Electron withdrawing group decreases reactivity while electron donating group increases reactivity in Hofmann's degradation.

3.

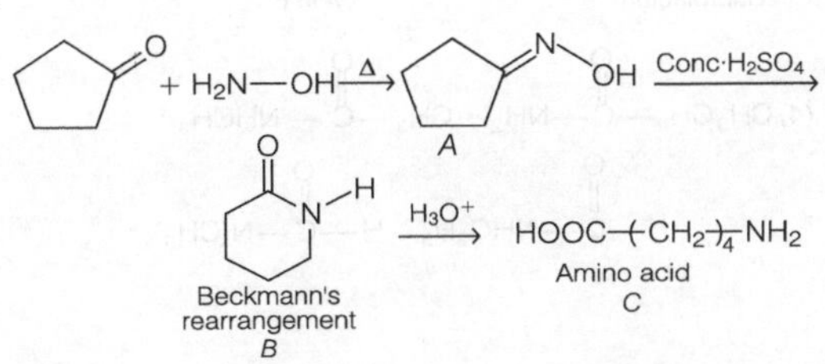

4. (i) $CH_3—\overset{O}{\overset{\|}{C}}—NH_2 \xrightarrow[H_2O/\Delta]{H^+} CH_3COOH \xrightarrow{SOCl_2} CH_3COCl$

(ii) $\underset{\text{From (a)}}{CH_3COOH} \xrightarrow[\Delta]{P_2O_5} (CH_3CO)_2O$

(iii) $\underset{\text{From (a)}}{CH_3COOH} \xrightarrow{LiAlH_4} CH_3CH_2OH \xrightarrow[\text{Conc. } H_2SO_4/\Delta]{CH_3COOH} CH_3COOC_2H_5$

5. The lone pair of nitrogen is part of conjugated system;

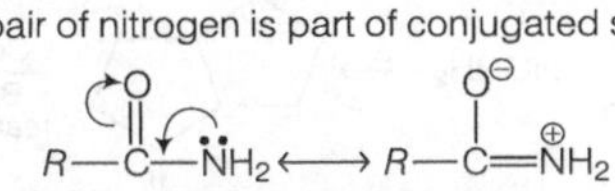

Due to above resonance delocalisation, lone pair of nitrogen is not available for donation to Lewis acid.

6. (b) Primary amide is dehydrated on heating with P_2O_5 to give nitrile.

7. (d)

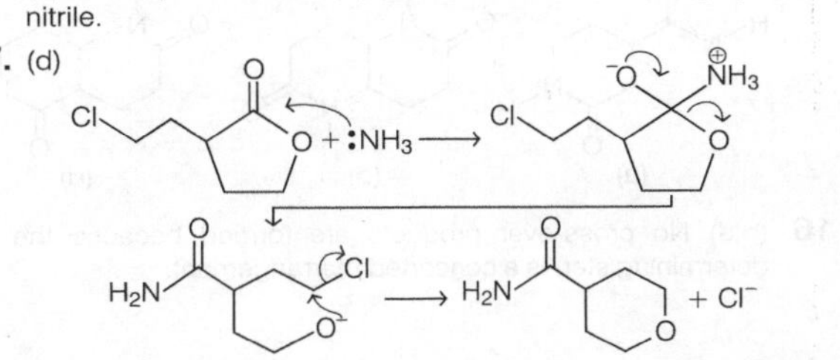

8. (b) Electron withdrawing group increases the electrophilicity of amide, hence increases reactivity in base (HO^-) catalysed hydrolysis.

9. (b)

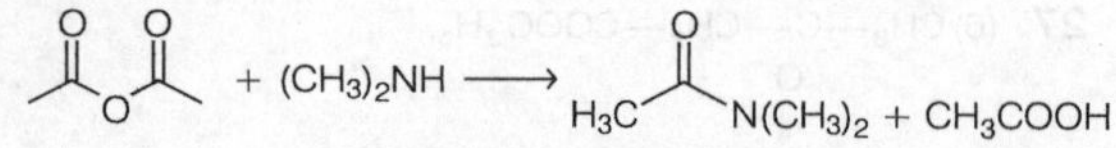

Tertiary amide

10. (c) Reaction occur at both ends of ester.

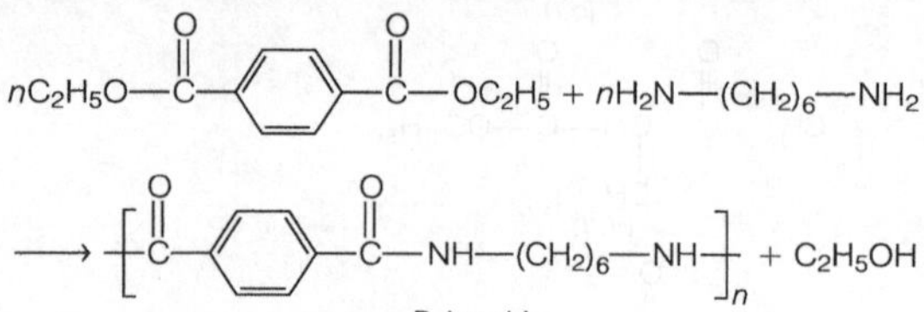

Polyamide

11. (b)

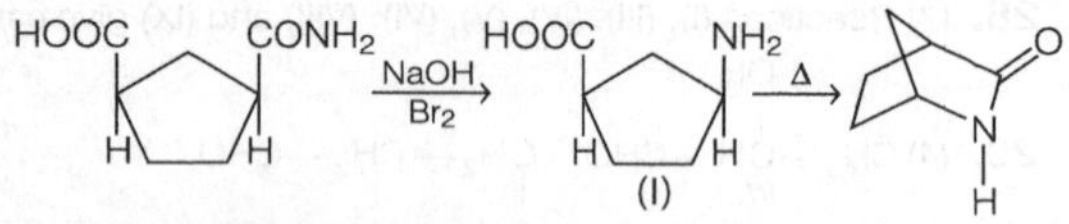

Since, —COOH and —NH_2 are on same side in (I), they condenses to give lactum.

12. (a)

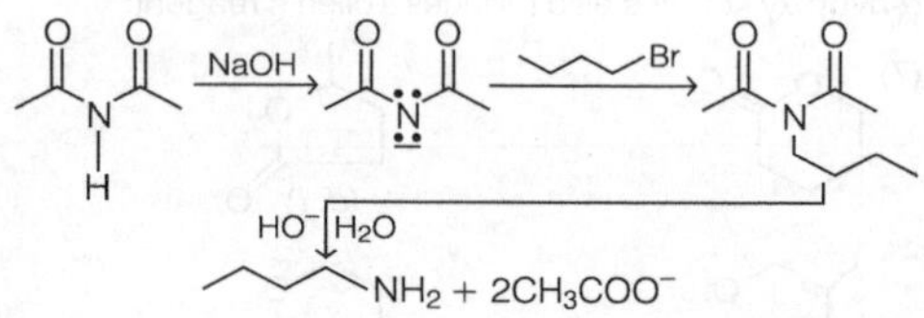

13. (b)

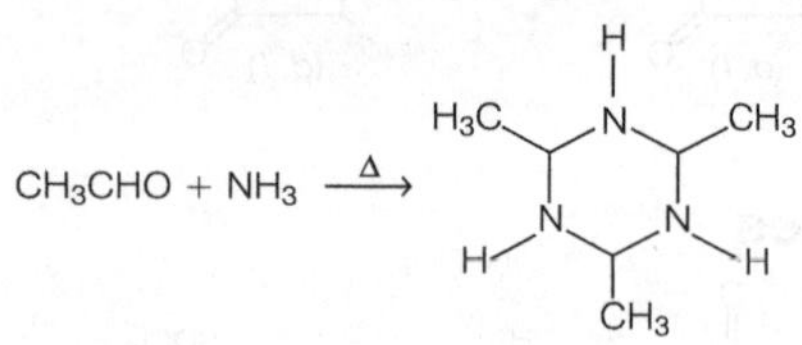

14. (a,b,c,d) All the given compounds react with amine to form stable acid derivative amide.

15. (a,b,d)

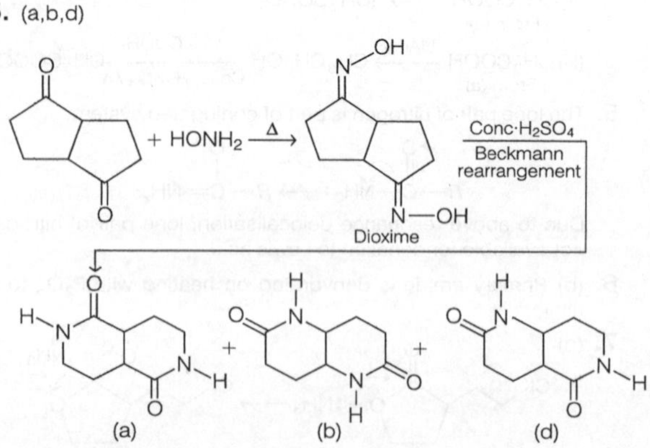

16. (b,d) No cross-over products are formed because the rate determining step is a concerted rearrangement.

17. (a,b,c)

(a) Hofmann's degradation of primary amide gives primary amine.

(b) Carboxylic acid on heating with hydrazoic acid gives primary amine.

(c) Primary amide on reduction with $LiAlH_4$ gives primary amine.

(d) $NaBH_4$ does not reduce primary amides.

18. (a,d) In Hofmann's bromamide degradation, N-bromamide is formed in the first step and isocyanate is formed in the slow rate determining step. Nitrine and hydroxamic acids are not formed.

19. (a,b) Amides can be hydrolysed in both acidic medium. During hydrolysis of amides, alkylnitrogen bond is not dissociated, hence configuration at α-carbon of amine is retained.

20. (b) Isocyanate is the most important intermediate formed in the slow rate determining step.

21. (c) Had the reaction occurred *via* step (V), carbanion could have attacked either of bromocyanates giving cross-over products also.

22. (d) Since, carbanion is not set free, configuration at α-carbon of amide is retained in amine.

23. (6) Secondary and tertiary amides do not undergo Hofmann's degradation.

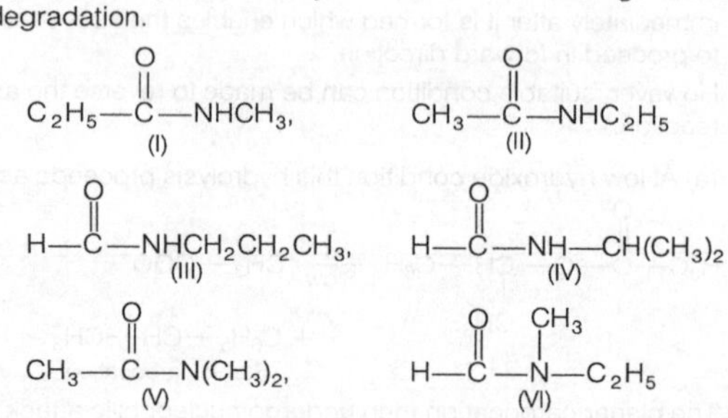

24. (5) Primary amides react with NaOH, Br_2 to give primary amines.

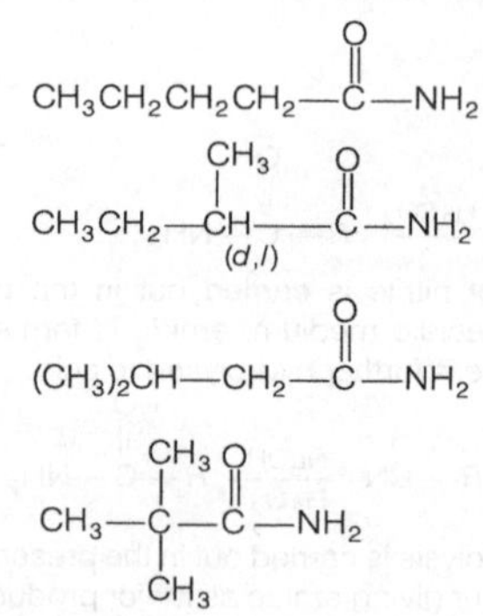

25. (6)

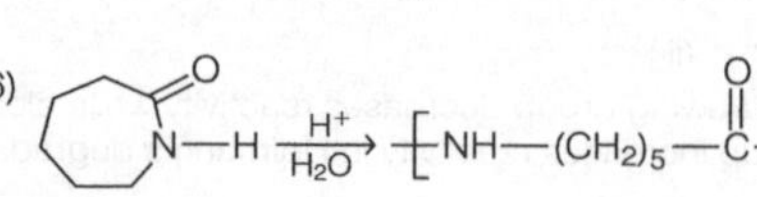

Caprolactum　　Nylon-6

26. (4)

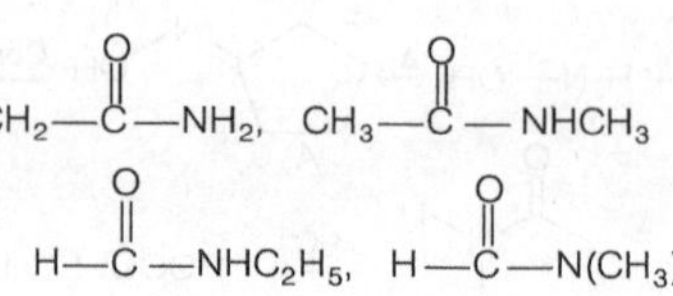

DPP-7 Preparation of Amines

1. (i)

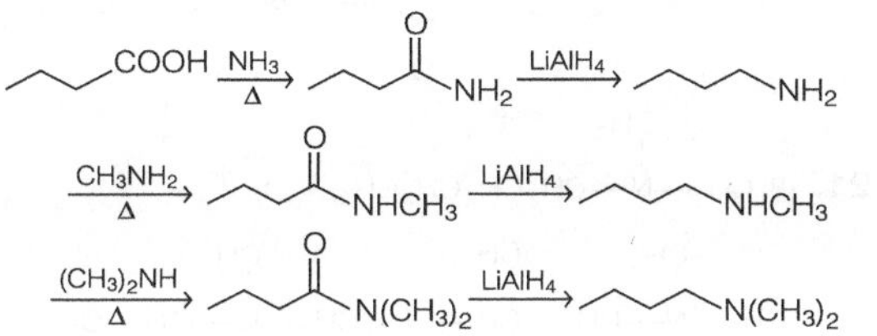

(ii)

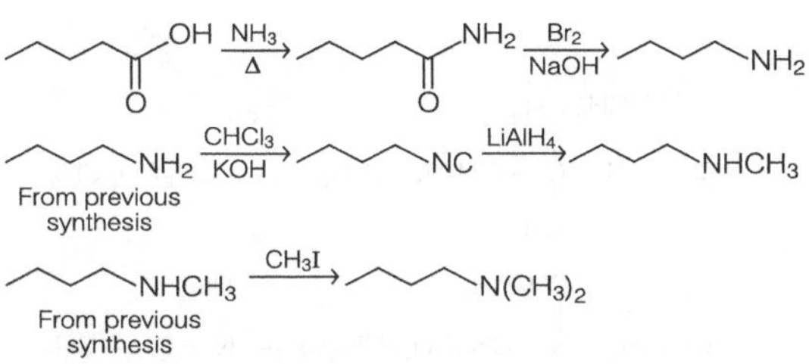

2.

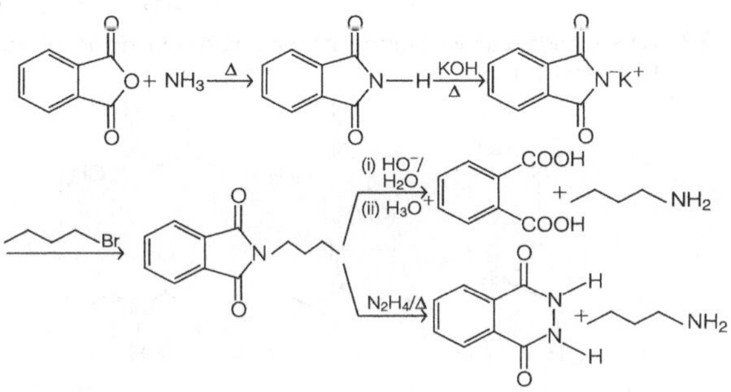

Hydrazine treatment in the last step is better method than hydrolysis of N-alkylimide.

3. (i)

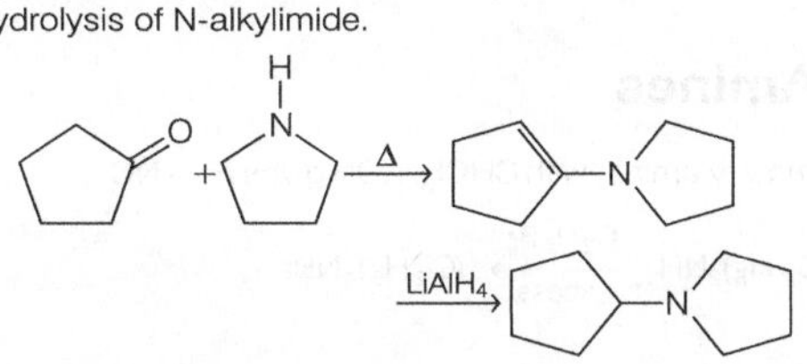

(ii)

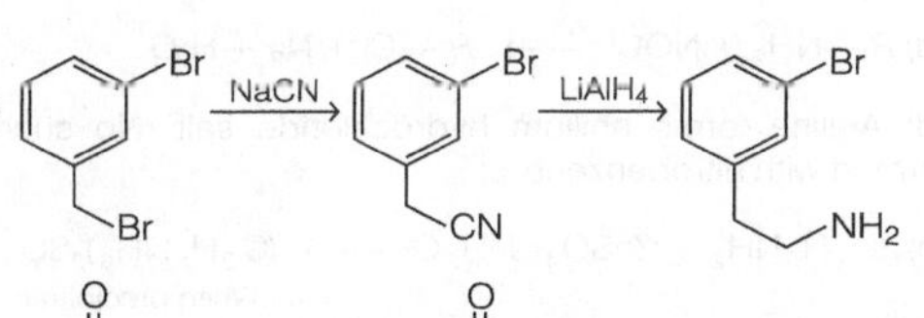

(iii)

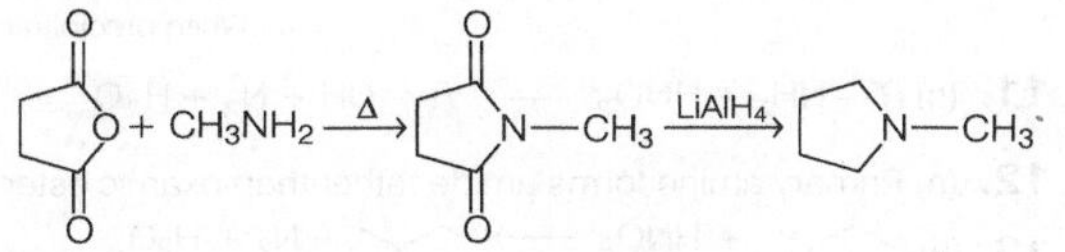

4. $(CH_3)_3C{-}\underset{A}{N(H)}{-}C_2H_5 \xrightarrow{CH_3I} CH_3{-}C(CH_3)_2{-}N^+(CH_3)_2{-}CH_2CH_3\ OH^-$ (B)

$B \xrightarrow{\Delta} \underset{C}{CH_3{-}C(CH_3){=}CH_2} + \underset{D}{C_2H_5{-}N(CH_3)_2}$

B on heating undergo β-elimination to give more substituted alkene (C).

$C \xrightarrow[H_2O]{Hg(OAc)_2} \xrightarrow{NaBH_4} \underset{E}{CH_3{-}C(CH_3)_2{-}OH}$

E is a tertiary alcohol, not oxidised by chromic acid solution.

5.

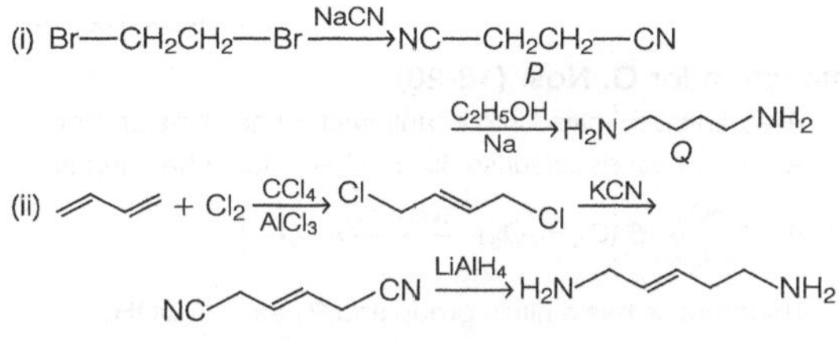

6. (d) In Hofmann's bromamide reaction, a primary amide reacts to give a primary amine.

7. (b) Alkyl cyanide gives primary amine on reduction.

$$R{-}CN \xrightarrow{LiAlH_4} R{-}CH_2NH_2$$

8. (c) It is Gabriel phthalimide synthesis of primary amine.

9. (b) $R{-}\overset{O}{\overset{\|}{C}}{-}NH_2 \xrightarrow{LiAlH_4} R{-}CH_2NH_2$

Same number of carbons as in amide

10. (a) $CH_3{-}CH{=}CH_2 \xrightarrow[CH_3CN]{H^+/H_2O} CH_3{-}CH{=}N{-}CH(CH_3)_2$

$$\xrightarrow{LiAlH_4} CH_3{-}CH_2{-}NH{-}CH(CH_3)_2$$

11. (c) Aniline gives a heavy precipitate of tribromo aniline with Br_2-H_2O. No such reaction and precipitate formation occur with $R{-}NH_2$.

12. (c)

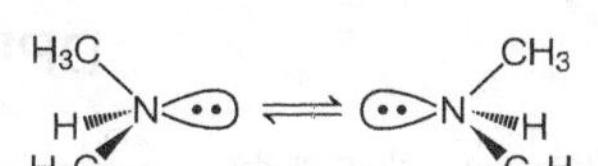

Fluxional inversion

Due to rapid interconversion of enantiomeric amines above, a pure enantiomer cannot be separated. Also they remain present in equal amount at equilibrium, it is a racemic mixture.

13. (c) Secondary amine, with cold, nitrous acid, produces yellow, oily liquid nitroso amine.

14. (a,b,c) (a) $R{-}CN \xrightarrow{LiAlH_4} R{-}CH_2NH_2$

(b) $R{-}CONH_2 \xrightarrow{LiAlH_4} R{-}CH_2NH_2$ } Primary amine

(c) $R{-}NO_2 \xrightarrow{LiAlH_4} R{-}NH_2$

(d) $R{-}X + NH_3 \longrightarrow R{-}NH_2 + R_2NH + R_3N + R_4NX$

15. (b,d)

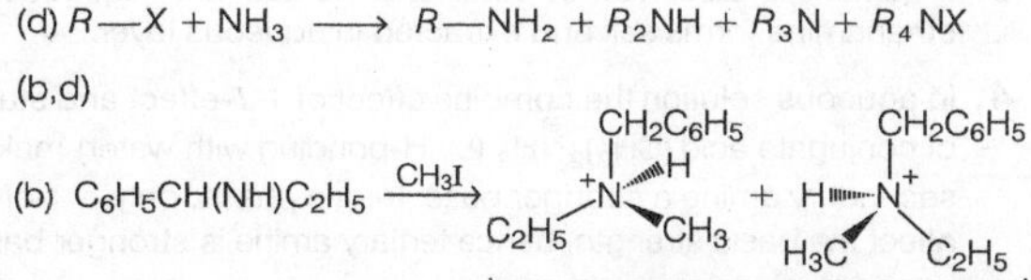

Racemic mixture

(d) $C_6H_5{-}CH_2{-}NH{-}C_6H_5 \xrightarrow{CH_3I} C_6H_5CH_2{-}\overset{\oplus}{N}(CH_3)(C_6H_5){-}H$

(Chiral produced as racemic mixture)

16. (b,c) (b) It is not a primary amine.

(c) For synthesis of aniline S_N2 reaction will have to occur at C_6H_5X which is very difficult.

17. (b,d) Secondary amine, with tosyl chloride forms sulphonamide that do not contain acidic hydrogen on nitrogen, do not dissolve in KOH solution.

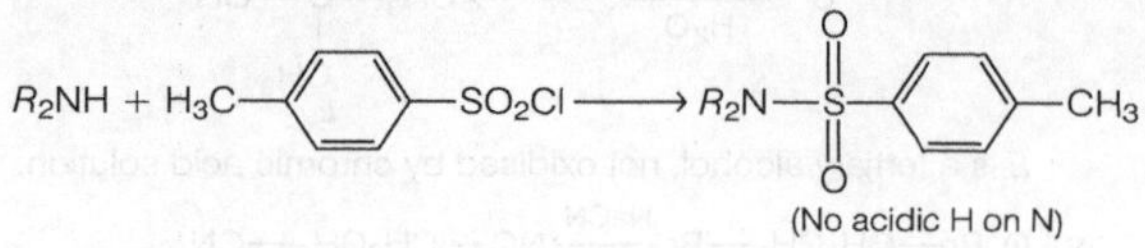

(No acidic H on N)

Explanation for Q. Nos. (18-20)

A— can be resolved into enantiomer = has chiral carbon.

A— does not decolourise Br_2 - H_2O solution = has ring structure.

$$A \xrightarrow[H_2O]{H_2SO_4} B\ (C_9H_{14}O_3) \xrightarrow{NaHCO_3} CO_2 \uparrow$$

Therefore, *A* has a nitrile group and *B* has — COOH.

$$B \xrightarrow{NaBH_4,\ H_2SO_4} C\ (C_9H_{14}O_3) \text{ sweet smelling liquid = ester.}$$

18. (c) *A* satisfying the above criteria is 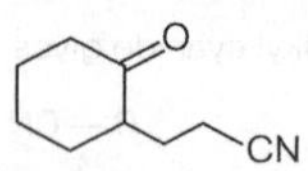

19. (c)

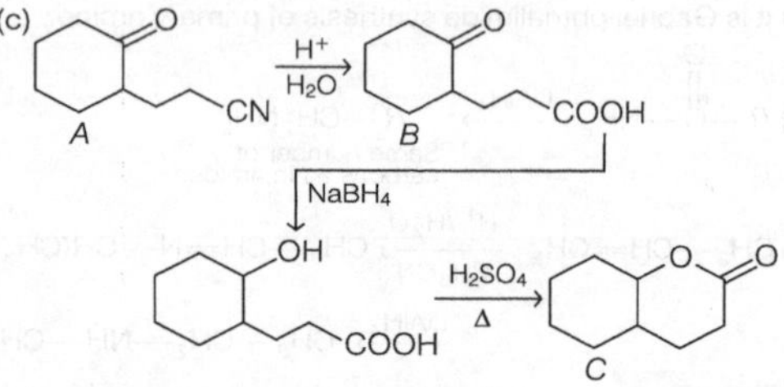

20. (d)

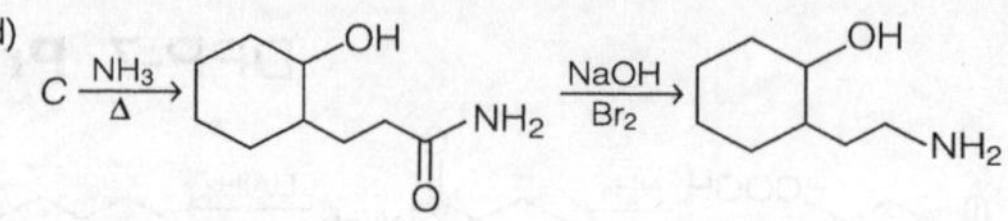

21. (8)

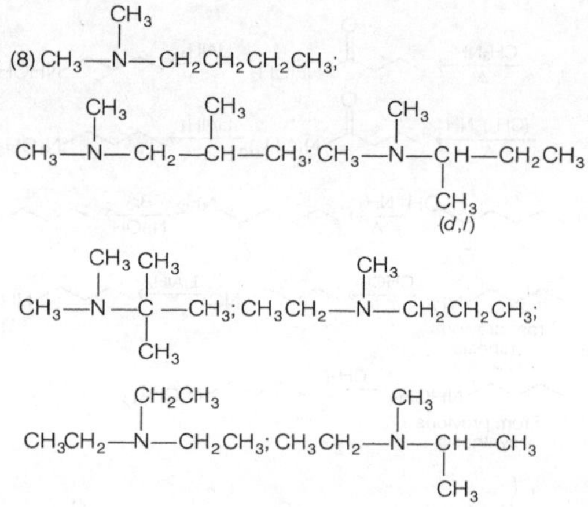

22. (7) Secondary amines form sulphonamides that do not dissolve in KOH solution.

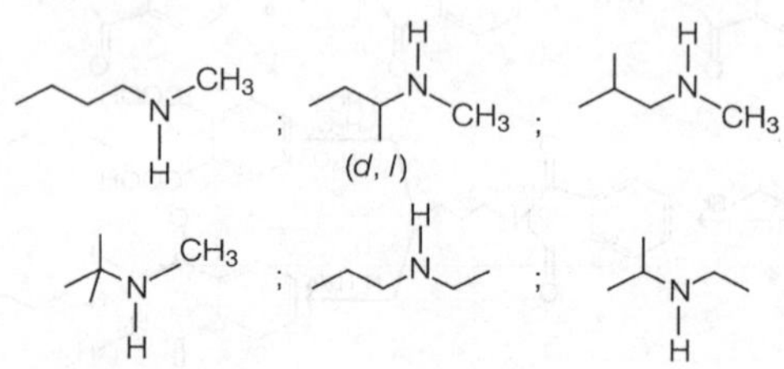

DPP-8 Reactions of Amines

1. 1° and 2° amine $\xrightarrow{TsCl}$ Precipitate (Sulphonamide) $\xrightarrow{KOH}$ Soluble (1°). or Insoluble (2°)

Tertiary amines do not form precipitate with tosyl chloride.

2. (i) $CH_3CH_2NH_2 \xrightarrow[KOH]{CHCl_3} CH_3CH_2NC$ (Carbyl amine reaction) $\xrightarrow{LiAlH_4} CH_3CH_2NHCH_3$

(ii) $CH_3CH_2CH_2NH_2 \xrightarrow[0°C]{HNO_2} CH_3CH_2CH_2OH \xrightarrow[H_2SO_4]{CrO_3} CH_3CH_2COOH$

3. If mixture is dissolved in ether and treated with aqueous HCl, ethanamine forms salt and extracted in aqueous layer.

4. In aqueous solution the combine effect of + *I*-effect and stability of conjugate acid $(CH_3)_2\overset{\oplus}{N}H_2$ (by H-bonding with water) makes secondary amine a stronger base. In gas phase, only + *I*-effect affect the basic strength, hence tertiary amine is stronger base.

5. $CH_3CH_2NH_2 + CS_2 \longrightarrow CH_3—CH_2—NH—\overset{SH}{\overset{|}{C}}=S \xrightarrow{HgCl_2} \underset{\text{Mustard oil}}{CH_3CH_2—NCS} + HgS + 2HCl$

No mustard oil test is given by secondary amines.

6. (b) Primary amine, with $CHCl_3$, KOH gives *R*—NC.

7. (c) $(C_2H_5)_2NH \xrightarrow[\text{Excess}]{C_2H_5Br} (C_2H_5)_4NBr$

8. (d) $R—NH_2 + NOCl \longrightarrow R—Cl + N_2 + H_2O$

9. (d) Aniline forms anilium hydrochloride salt. No such salt is formed with nitrobenzene.

10. (c) $2C_2H_5NH_2 + ZnSO_4 + H_2O \longrightarrow \underset{\text{White precipitate}}{(C_2H_5\overset{+}{N}H_3)_2SO_4} + ZnO$

11. (c) $R—NH_2 + HNO_2 \longrightarrow R—OH + N_2 + H_2O$

12. (a) Primary amine forms amide rather than oxamic ester.

13. (a)

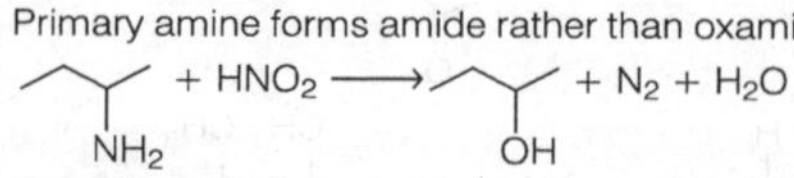

Optically active

14. (a,b,c) Primary and secondary amines form precipitate with tosyl chloride but tertiary do not form precipitate with tesyl chloride.

15. (a,c) Primary amine gives effervescence of N_2 with HNO_3 while secondary amines form yellow, oily liquid of nitroso amine. Tertiary amine do not react with HNO_2.

16. (a,b,c) Option (a) and option (b) are primary amine, form isocyanide in carbyl amine reaction.

$$\text{(c) } CHCl_3 \xrightarrow{NaOH} HCOOH \xrightarrow{(CH_3)_2NH} H-\overset{\overset{O}{\|}}{C}-N(CH_3)_2$$

17. (a,b) With CS_2, primary amine gives a sulphamic acid which reacts further with $HgCl_2$ giving R—NCS.

Secondary amine gives $R_2N-\underset{\underset{S}{\|}}{C}-SH$ which do not react with $HgCl_2$.

18. (b) In aqueous solution, the order of basic strength of amine is :

$$2° > 1° > 3°$$

Also, aliphatic amines are more basic than aromatic amines.

19. (b,c,d) Reaction proceed *via* carbocation intermediates.

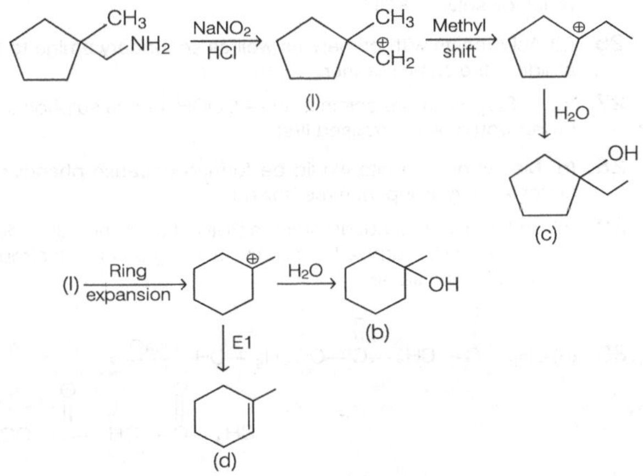

Explanations for Q. Nos. (20–22)

20. (c) In first step, dichloro carbene acts as an electrophile which is formed in α-elimination.

$$CHCl_3 + OH^- \longrightarrow {}^{\ominus}CCl_3 + H_2O$$
$${}^{\ominus}CCl_3 \longrightarrow Cl^- + :CCl_2$$

21. (c) Electron donating group increases the reactivity while electron withdrawing group decreases the reactivity.

22. (b)

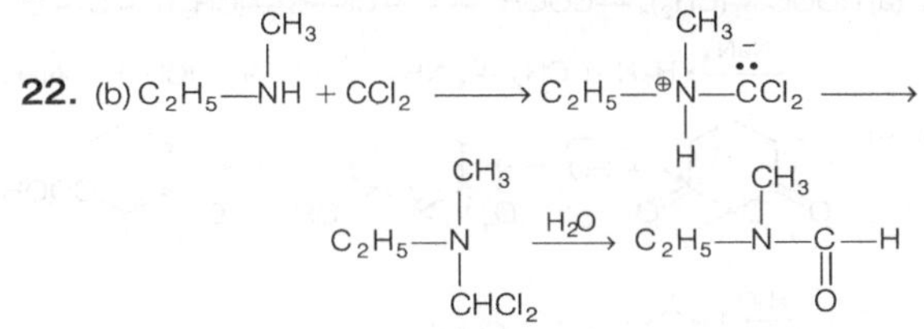

23. (4) $HOOC(-CH_2-)_4COOH + 2HN_3 \longrightarrow H_2N(-CH_2-)_4NH_2 + 2N_2 + 2CO_2$

24. (6) $H_2N(-CH_2-)_6NH_2 + 2CHCl_3 + 6KOH \longrightarrow NC(-CH_2-)_6NC + 6KCl + 6H_2O$

25. (3)

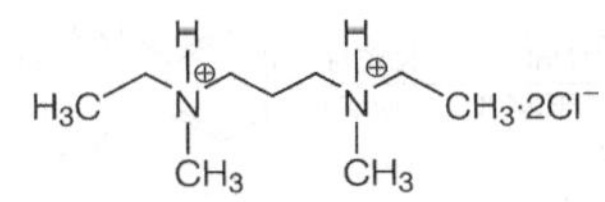

Symmetrical salt with two chiral centers

26. (5) It is mustard oil reaction which is given by primary amines only.

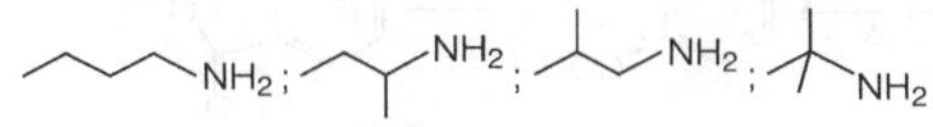

(d, l)

Revisal Problems (JEE Main)

1. (d) Fischer esterification is acid catalysed not base catalysed.

2. (b) Benzyl alcohol is not neutralised by NaOH, whereas benzoic acid forms salt C_6H_5COONa and goes into aqueous layer.

3. (d) Na_2CO_3 solution will neutralise unreacted $PhCH_2COOH$ and extract it into aqueous layer leaving ester is ether layer.

4. (c)

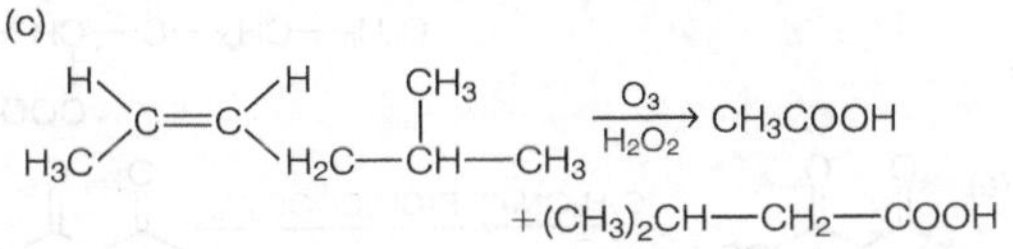

5. (c) Both I and IV give the same

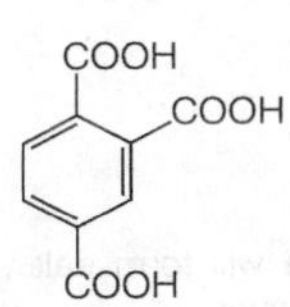

6. (c)

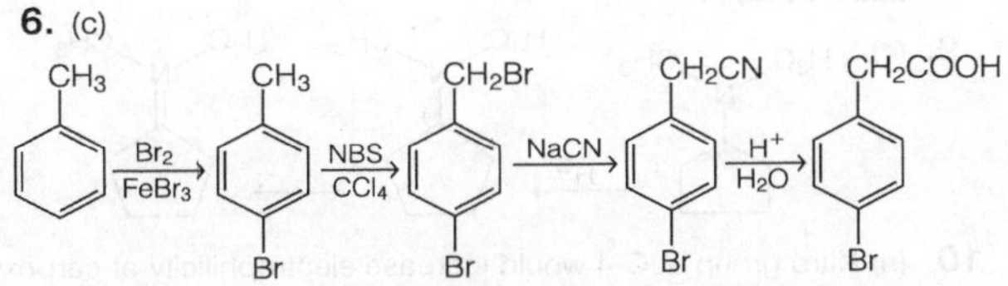

7. (a) Both III and IV bring about controlled reduction of acid chloride into aldehyde.

8. (c)

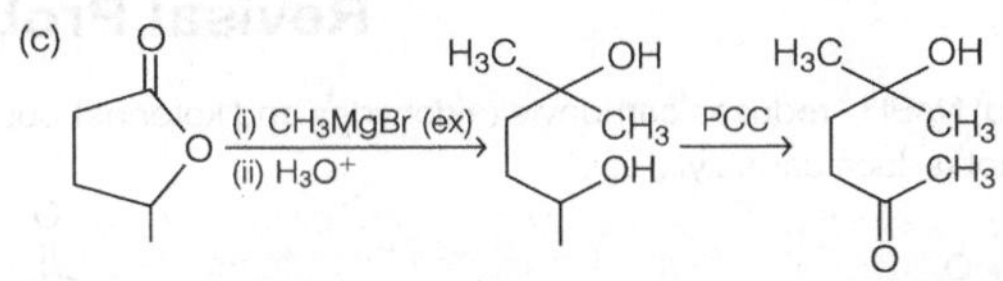

9. (d)

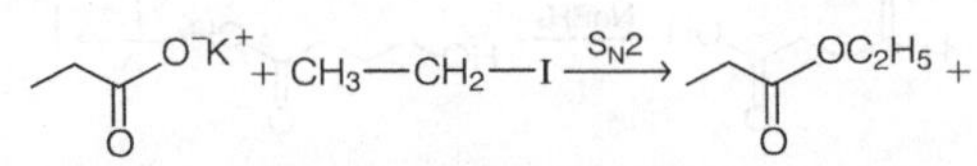

10. (b) The order of reactivity of acid derivatives towards nucleophilic acyl substitution is

Acid chloride > Anhydride > Ester > Nitrile

11. (d)

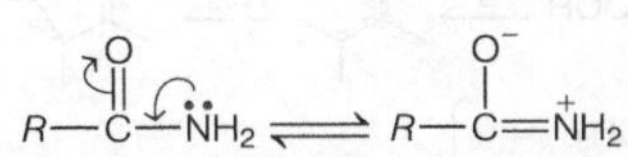

Pi bonding decreases the bond length of N—C bond.

12. (a) Amide is capable of intermolecular H-bonding but ester does not.

13. (c)

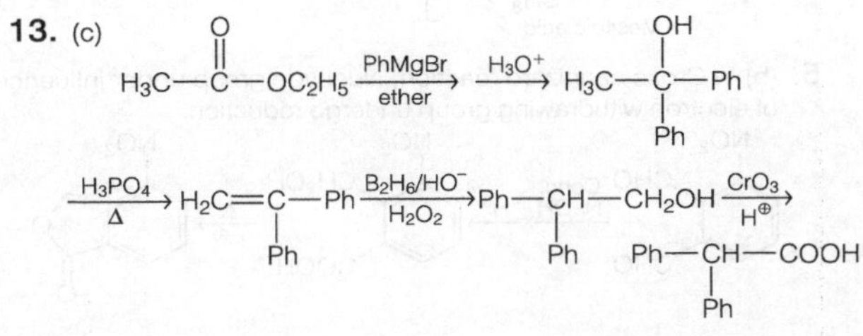

14. (b)

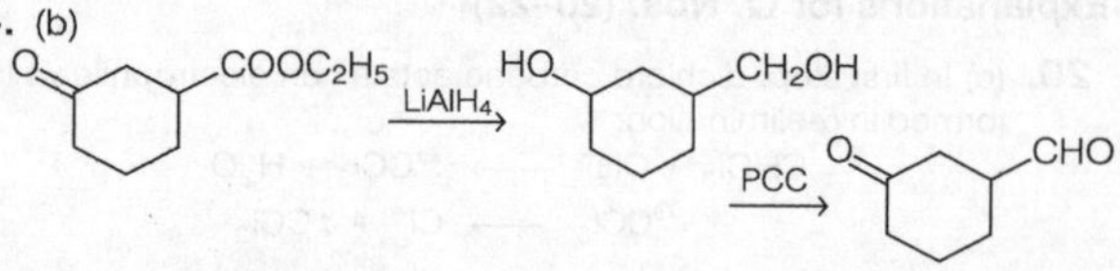

15. (a) $HOOC—(CH_2)_4—COOH \xrightarrow{SOCl_2} Cl—\overset{O}{\overset{\|}{C}}—(CH_2)_4—\overset{O}{\overset{\|}{C}}—Cl$

$\xrightarrow{NaN_3} H_2N—(CH_2)_4—NH_2 \xrightarrow{HCl} H_3\overset{+}{N}—(CH_2)_4—\overset{+}{N}H_3$

16. (d)

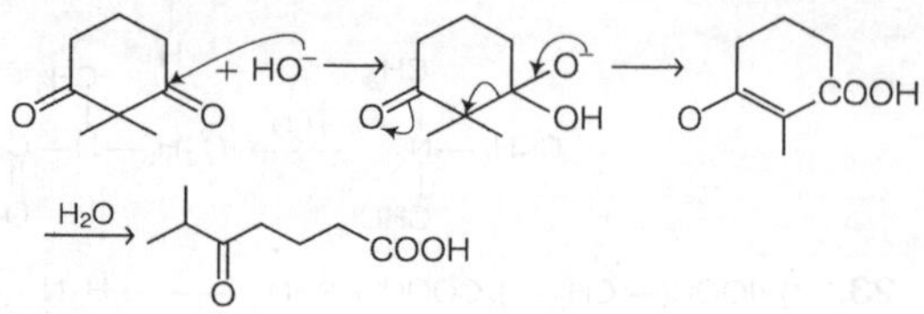

17. (a)

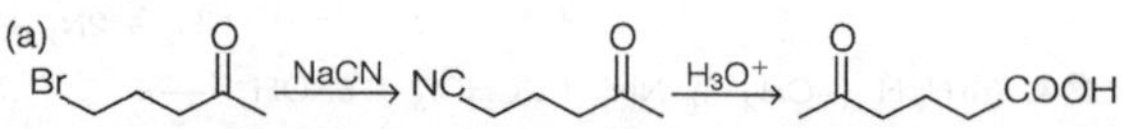

18. (d)

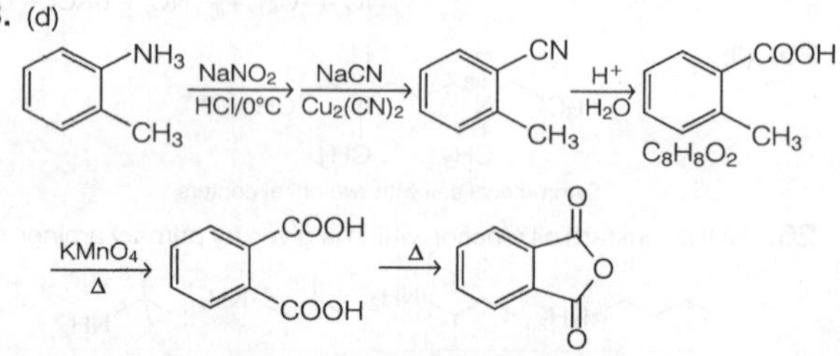

19. (d) It forms cyclic anhydride on heating.

20. (b) More stable acid derivative ester, not anhydride is formed.

21. (d) Planar α-carbon attack at the α-carbon of bromide to give racemic product.

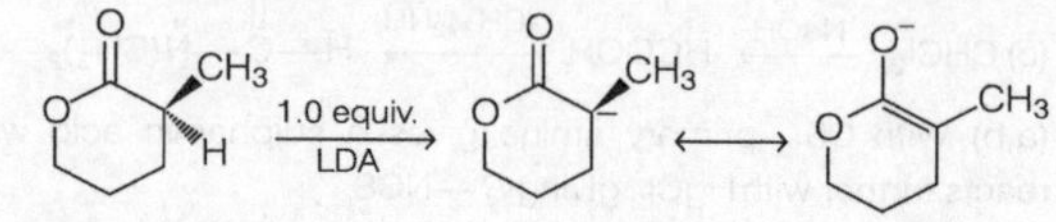

22. (a) The more stable amide is preferred over ester.

23. (c) If acid part is same, leaving group ability of alkoxide determines the reactivity.

24. (b) The phenolic —OH group is most easily acetylated to give ester.

25. (c) Primary amine reacts with tosyl chloride to give sulphonamide which dissolve in KOH.

26. (d) Acid reacts with primary as well as secondary amine to form amide in the same manner.

27. (d) $—SO_3H$ is more acidic than —COOH, hence sulphonic acid group would be neutralised first.

28. (d) Methyl propanoate would be formed because phenoxide is better leaving group than methoxide.

29. (b) Both are correct but greater reactivity of acid chloride is due to sp^2-hybridised carboxyl carbon and greater electrophilic character of the same.

30. (a) $CH_3—\overset{O}{\overset{\|}{C}}—CH_2—\overset{O}{\overset{\|}{C}}—OC_2H_5 + OH^- \xrightarrow{NaOH}$

$CH_3—\overset{O}{\overset{\|}{C}}—\overset{-}{C}H—\overset{O}{\overset{\|}{C}}—OC_2H_5 \quad Na^+$

Formation of above salt is responsible for solubility.

Revisal Problems (JEE Advanced)

1. (d) $NaBH_4$ reduces carbonyls (aldehydes and ketones) but does not reduce carboxylic acid.

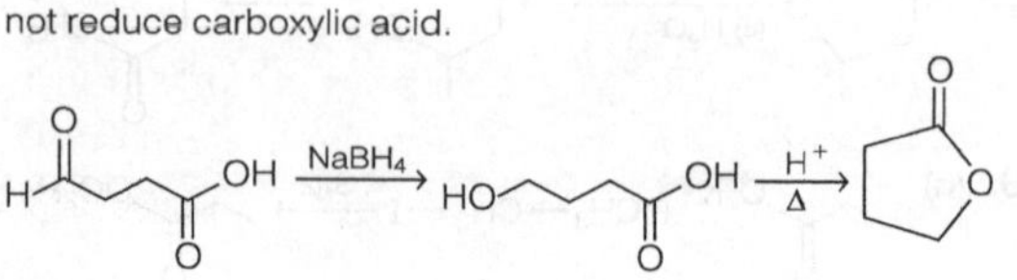

2. (d) S_N2 reaction usually fail to occur at tertiary α-carbon of alkyl halides.

3. (c)

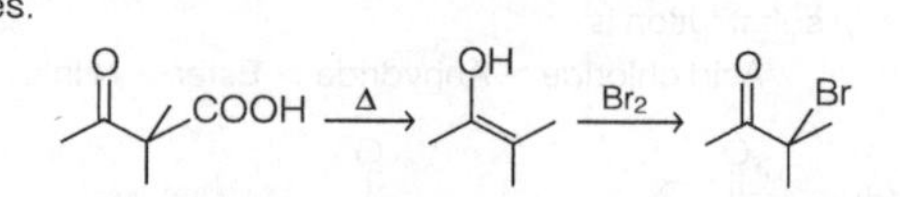

4. (a)

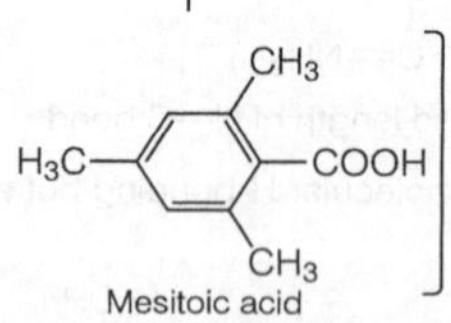

Mesitoic acid

The two *ortho* methyl group makes — COOH highly sterically hindered for attack of *R*OH.

5. (b) In Cross cannizaro reaction, aldehyde group under influence of electron withdrawing group undergo reduction.

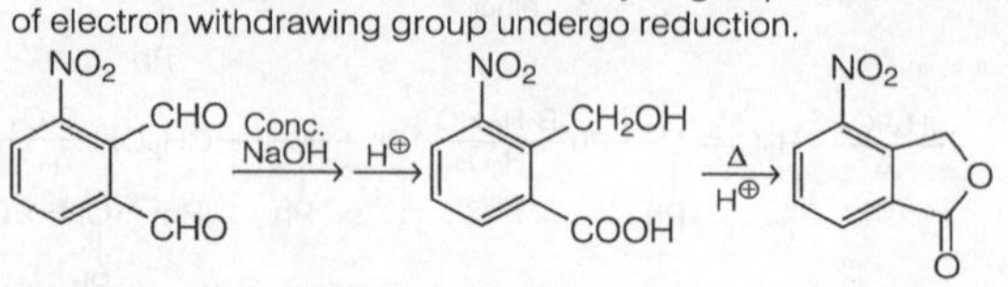

6. (d)

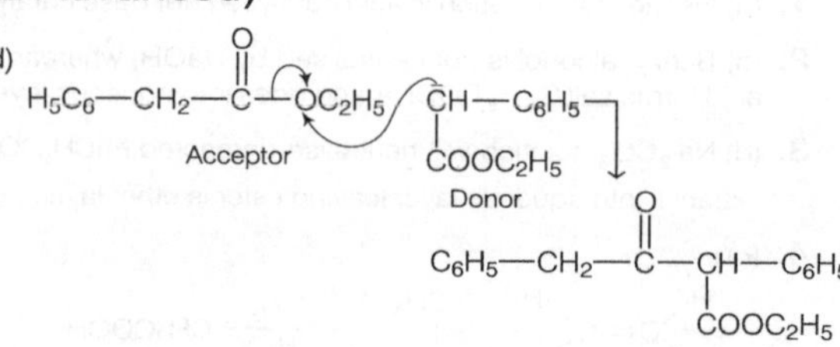

7. (a)

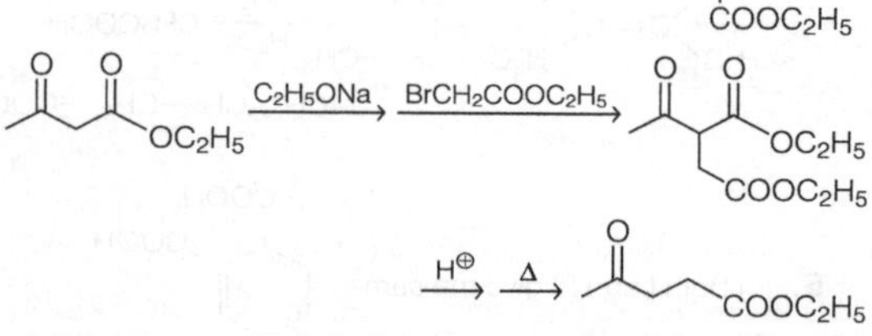

8. (b) Amine will form salt with HCl and it will be extracted into aqueous layer.

9. (c)

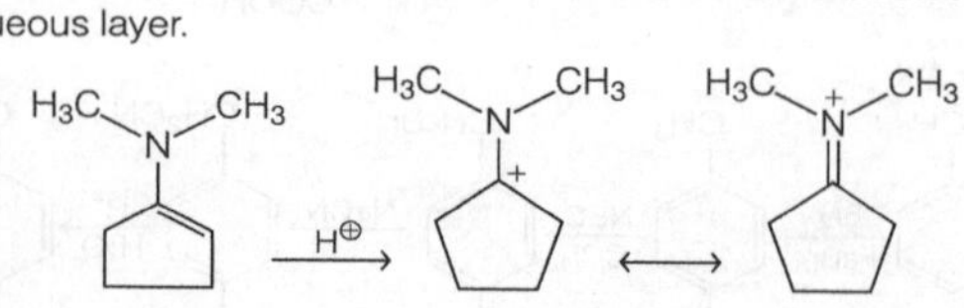

10. (a) Nitro group at C-4 would increase electrophilicity at carboxyl carbon at C-1.

11. (c,d) (c) $CH_3CH_2COOH \xrightarrow[P]{Br_2} CH_3CH(Br)—COOH \xrightarrow[\Delta]{OH^-}$

$CH_2=CH—COOH \xrightarrow[Zn\text{-}H_2O]{O_3} H—C(=O)—H + H—C(=O)—COOH$
Neutral (I)

(I) $\xrightarrow[\Delta]{H_2N—OH} H_2C=N—OH$
Single oxime

(d) $(CH_3)_2CH—CH_2COOH \xrightarrow[P]{Br_2} \xrightarrow[\Delta]{OH^-} CH_3—C(CH_3)=CH—COOH$

$\xrightarrow[Zn\text{-}H_2O]{O_3} CH_3—C(=O)—CH_3$
Neutral(II)

(II) $\xrightarrow[\Delta]{H_2NOH} (H_3C)_2C=N—OH$
Single oxime

Carbonyls formed from others (a, b) form more than one oximes.

12. (b,d) (b) $CH_3—C(=O)—NH_2 \xrightarrow{OH^-} CH_3COO^- + NH_3$
Neutral

(d) $CH_3—C(=O)—NH_2 \xrightarrow[\Delta]{P_2O_5} CH_3—CN$

13. (a,b,c,d) (a) $CH_3COCl + CH_3—CH(OH)CH_3 \longrightarrow CH_3—C(=O)—OCH(CH_3)_2$

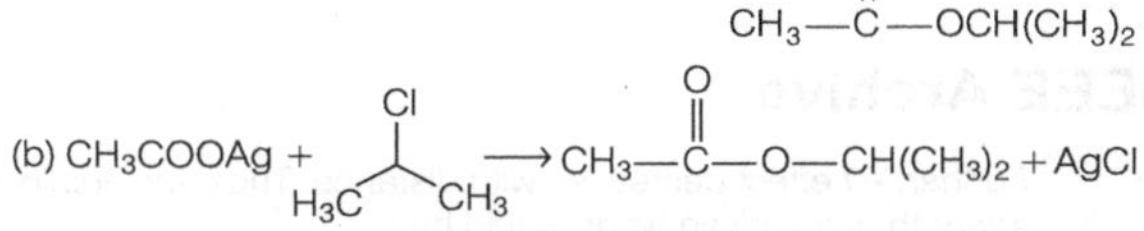

(c) $CH_3COOH + CH_3CH(OH)CH_3 \xrightarrow[H^{\oplus}]{\Delta} CH_3COOCH(CH_3)_2$

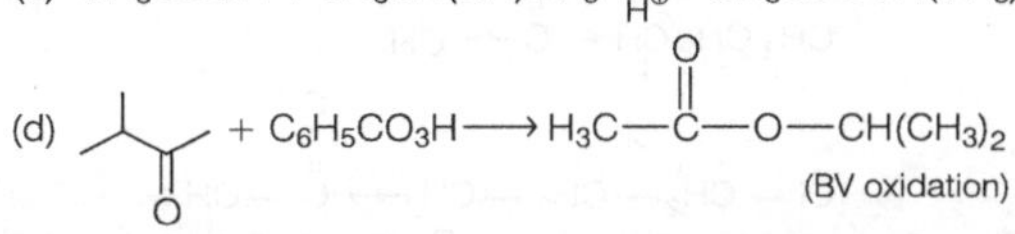

14. (a,b,c) Exchange of oxygen between acid and water takes place.

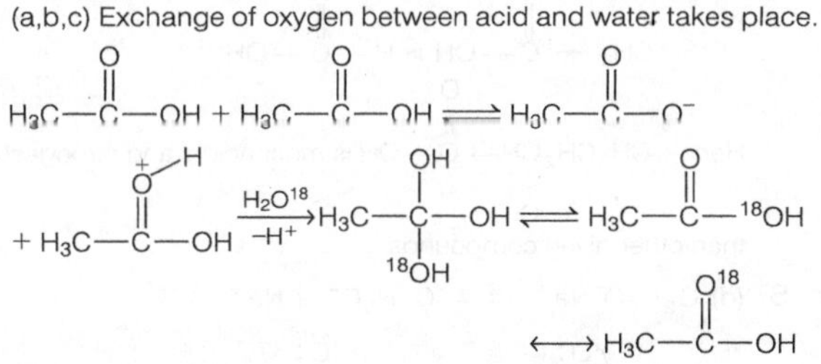

15. (a,b,c)
(a) Amides are more stable acid derivative than ester.

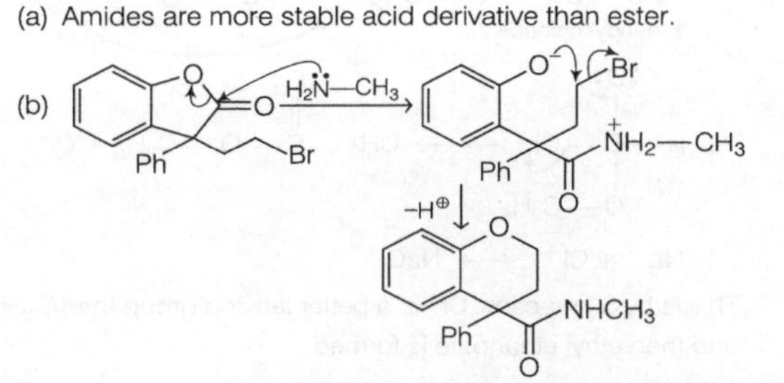

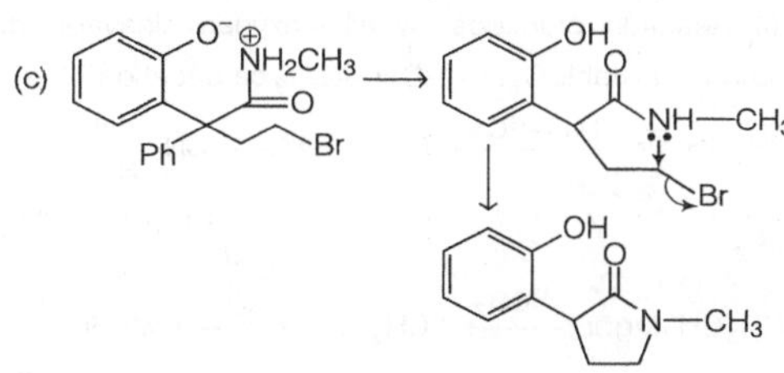

16. (b,d) In Hofmann's bromamide reaction, no cross-over products are formed.

17. (b) Electron releasing methoxy group from *para* position increases the reactivity in Hofmann's reaction.

18. (a) In nucleophilic acyl substitution (S_NAc), a more reactive acid derivative is always converted into more stable acid derivative.

Explanations for Q. Nos. (19-21)

A is an ester as it does not gives effervescence with $NaHCO_3$.

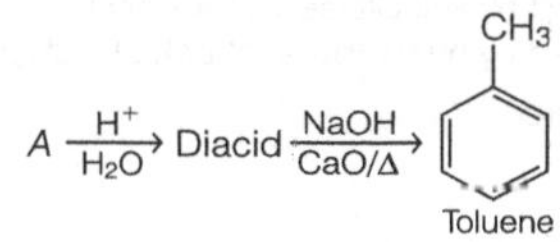

19. (c) Discussed in solution of Q. No. (19).

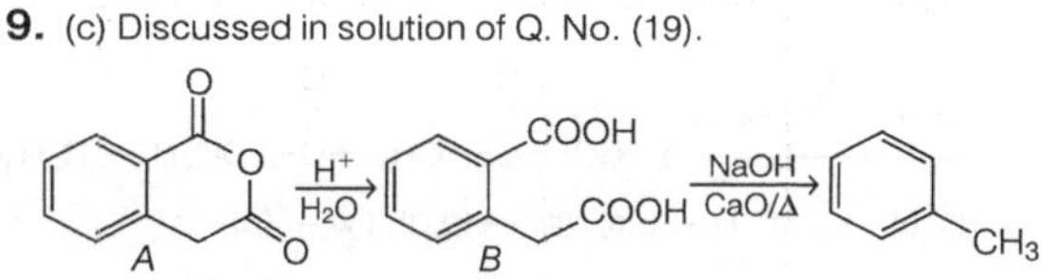

20. (d) Discussed in solution of Q.No. (19).

21. (c) Aliphatic ester is more electrophilic than aromatic ester.

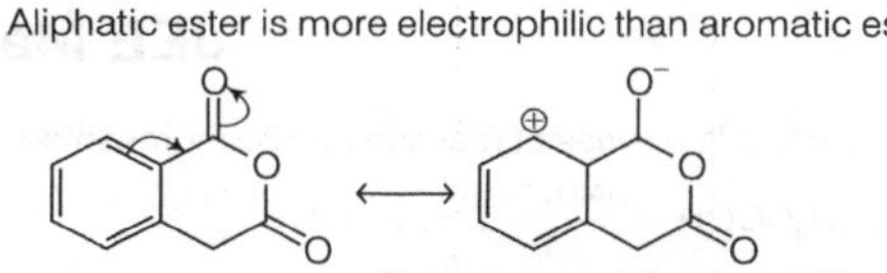

Hence,

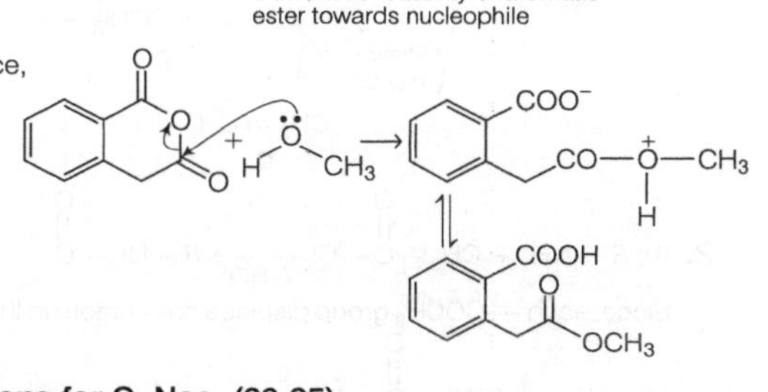

Explanations for Q. Nos. (22-25)

A has a —COOH group as well as it has a chiral centre. Also *D* (an alcohol) giving iodoform test, it must have a —CH(OH)CH_3 group.

22. (c)

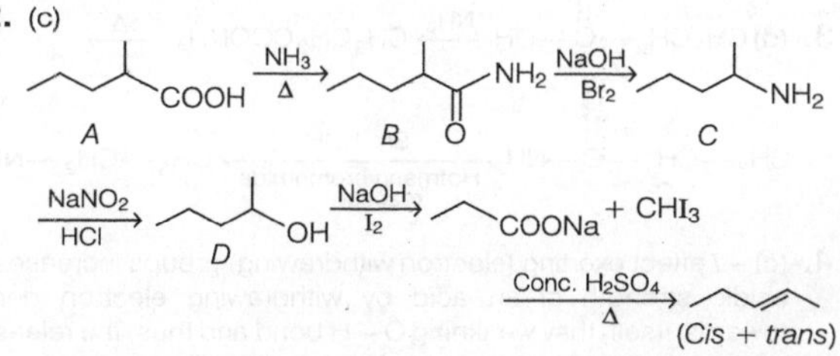

23. (c) *B* is chiral compound, shown in solution of Q. No. (22).

24. (d) Both *C* and *D* are shown above in solution of Q. No. (22).

25. (c) (i) Reaction proceeds by alkyl-oxygen cleavage due to formation of a stable $C_6H_5-\overset{\oplus}{C}H-CH_3$ carbocation.

(ii) $CH_3CH_2Br \xrightarrow{Na-^{13}CN} CH_3-CH_2-{}^{13}CN \xrightarrow[H_2O]{H^{\oplus}} CH_3CH_2-{}^{13}COOH.$

(iii) $CH_3CH_2MgBr \xrightarrow{^{13}CO_2} CH_3CH_2-{}^{13}\overset{O}{\overset{\|}{C}}-OMgBr \xrightarrow[H_2O]{H^+} CH_3CH_2-\underset{13}{\overset{O}{\overset{\|}{C}}}-OH$

(iv) Either of —COOH may undergo decarboxylation.

Thus, (i) → (r), (ii) → (s), (iii) → (s), (iv) → (q,r,s)

26. (b)

(i) No cross-over product is formed.

(ii) Nucleophilic attack may occur from either nitrogen of hydrazoic acid (N_3H).

(iii) P is formed by self Claisen condensation of ethylacetate while Q is formed by nucleophilic attack of $^-CH_2COOC_2H_5$ on ethyl propanoate.

(iv)
$$H_3C-\overset{O}{\overset{\|}{C}}-CH_3 + C_2H_5O^- \longrightarrow C_2H_5OH + CH_3-\overset{O}{\overset{\|}{C}}-\bar{C}H_2$$
$$\xrightarrow{C_2H_5O-\overset{O}{\overset{\|}{C}}-OC_2H_5} H_3C-\overset{O}{\overset{\|}{C}}-CH_2-\overset{O}{\overset{\|}{C}}-OC_2H_5 + C_2H_5O^-$$

Thus, (i) → (s,t), (ii) → (r,s), (iii) → (p,q), (iv) → (p)

27. (4)

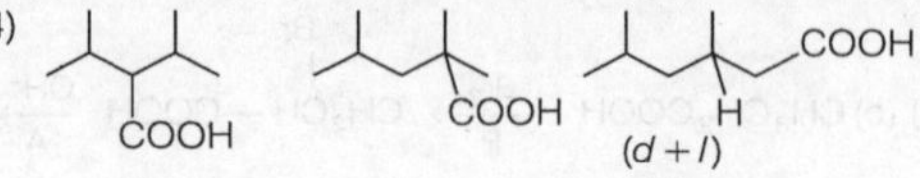

(I) (II) (III)

All of the above shown acids satisfy the given condition.

28. (9)

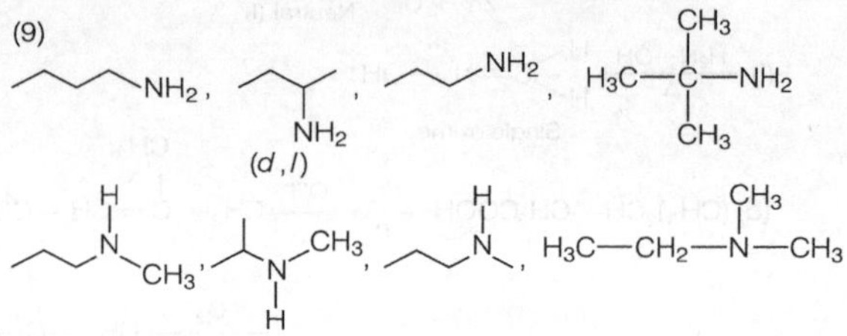

29. (6)

$$CH_3-\overset{O}{\overset{\|}{C}}-CH_2-\overset{O}{\overset{\|}{C}}-OC_2H_5,\ CH_3H_2C-\overset{O}{\overset{\|}{C}}-\underset{CH_3}{\underset{|}{CH}}-\overset{O}{\overset{\|}{C}}-OC_2H_5$$
(d + l)

$$H_3C-\overset{O}{\overset{\|}{C}}-\underset{CH_3}{\underset{|}{CH}}-\overset{O}{\overset{\|}{C}}-OC_2H_5,\ CH_3H_2C-\overset{O}{\overset{\|}{C}}-CH_2-\overset{O}{\overset{\|}{C}}-OC_2H_5$$
(d + l)

30. (4) Only primary and secondary amines (i, ii, iv and viii) give visible change with nitrous acid.

JEE Main & AIEEE Archive

1. (c) The complete series of reaction can be represented as

$$\underset{}{CH_3COOH} \xrightarrow{LiAlH_4} \underset{A}{CH_3CH_2OH} \xrightarrow{PCl_5} CH_3CH_2Cl + POCl_3 + HCl$$
$$CH_3CH_2Cl \xrightarrow[\text{(elimination reaction)}]{alc.KOH} \underset{C}{CH_2{=}CH_2}$$

2. (b) $R-NH_2 + CH_3-\overset{O}{\overset{\|}{C}}-Cl \xrightarrow[(-HCl)]{} R-NH-\overset{O}{\overset{\|}{C}}-CH_3$

Since, each $-COCH_3$ group displace one H-atom in the reaction of one mole of $CH_3-\overset{O}{\overset{\|}{C}}-Cl$ with one $-NH_2$ group, the molecular mass increases with 42 unit. Since, the mass increases by $(390 - 180) = 210$, hence the number of $-NH_2$ group is $\frac{210}{42} = 5$.

3. (d) $\underset{A}{CH_3CH_2-\overset{O}{\overset{\|}{C}}-OH} \xrightarrow{NH_3} \underset{B}{CH_3CH_2COONH_4} \xrightarrow{\Delta}$

$$\underset{C}{CH_3-CH_2-\overset{O}{\overset{\|}{C}}-NH_2} \xrightarrow[\text{Hofmann bromamide reaction}]{Br_2,\ KOH} CH_3-CH_2-NH_2$$

4. (c) $-I$ effect exerting (electron withdrawing) groups increases the acidic strength of an acid by withdrawing electron density towards itself, they weakning O—H bond and thus, the release of H^+ ion by an acid becomes easier.

Whereas $+I$ effect exerting (e^- releasing) groups decreases the acidic strength by donating electron density to O-atom.

Further, $-I$ effect decreases with distance. Thus, the acidic strength of the given acids would be :

$$CH_3\,CH_2\,\underset{Cl}{\underset{\downarrow}{CH}}\leftarrow\overset{O}{\overset{\|}{C}}\leftarrow OH$$
$$Cl-CH_2-CH_2-CH_2\twoheadrightarrow\overset{O}{\overset{\|}{C}}-OH >$$
$$CH_3\twoheadrightarrow\overset{O}{\overset{\|}{C}}-OH > H-\overset{O}{\overset{\|}{C}}-OH$$

Hence, $CH_3CH_2\underset{Cl}{\underset{|}{CH}}-\overset{O}{\overset{\|}{C}}-OH$ is most acidic and strongest acid than other given compounds.

5. (d) $C_2H_5O^-Na^+ \rightleftharpoons C_2H_5O^- + Na^+$

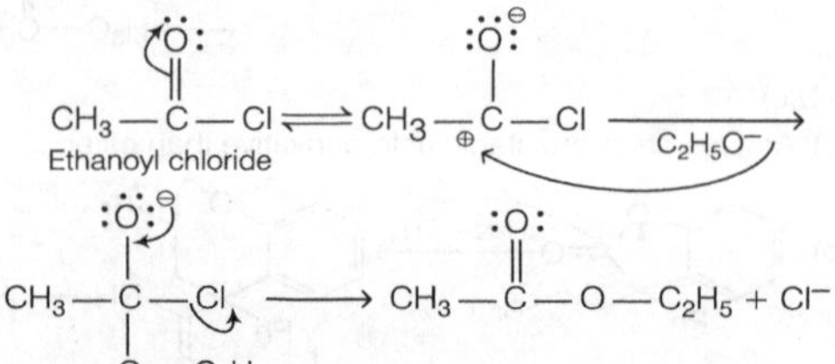

$Na^+ + Cl^- \longrightarrow NaCl$

This is by S_N reaction. Cl^- is a better leaving group than $C_2H_5O^-$ and then ethyl ethanoate is formed.

6. (d) A liquid + $C_2H_5OH \xrightarrow{\text{Conc. } H_2SO_4}$ Compound (Fruity smell)

Fruity smell is the characteristic property of ester, thus the above reaction leads to the formation of ester.

$$CH_3COOH + C_2H_5OH \xrightarrow{\text{Conc. } H_2SO_4} \underset{\text{Ethyl acetate (Fruity smell)}}{CH_3\overset{O}{\overset{\|}{C}}-OC_2H_5} + H_2O$$

This reaction is called esterification.

7. (b) $C_nH_{2n}O_2$ is general formula for open chain carboxylic acids and esters.

e.g. $n = 3$ $C_3H_6O_2$

Acid $CH_3CH_2-\overset{O}{\overset{\|}{C}}-O-H$

Ester $CH_3-\overset{O}{\overset{\|}{C}}-O-CH_3$

8. (b) $LiAlH_4$ reduces —COOH to —CH_2OH without affecting C ═ C bond.

9. (a) $C_2H_5NC + H_2O \xrightarrow{H^+} \underset{\text{Formic acid}}{HCOOH} + C_2H_5NH_2$

$C_2H_5NH_2 + H^+ \longrightarrow \underset{\text{Salt}}{C_2H_5NH_3^+}$

10. (c) $CH_3CH_2COOH \xrightarrow[\text{red P (HVZ reaction)}]{Cl_2} CH_3\underset{Cl}{\underset{|}{C}}HCOOH \xrightarrow[\text{(elimination)}]{\text{Alcoholic KOH}} \underset{\text{Acrylic acid}}{CH_2{=}CHCOOH}$

JEE Advanced & IIT JEE Archive

1. (a) This problem can be solved by using the stability of radical obtained after fragmentation of peroxyester.

Allylic radical are more stable than alkyl radical, so when there is a possibility of formation of allyl radical, it will undergo fragmentation through formation of allyl radical. i.e. fragmentation produces stable radical.

On the basis of stability of radical, fragmentation can be done as :

	Column I	Column II	Explanation
i.	p.	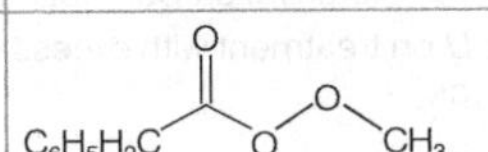	$C_6H_5-\dot{C}H_2 + CO_2 + CH_3\dot{O}$
ii.	r.	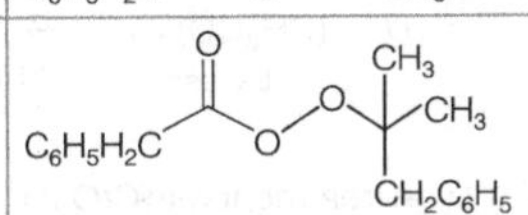	$C_6H_5-\dot{C}H_2 + CO_2 + Ph-CH_2-C(CH_3)_2-\dot{O}$ → $Ph-\dot{C}H_2 + CH_3-CO-CH_3$
iii.	s.	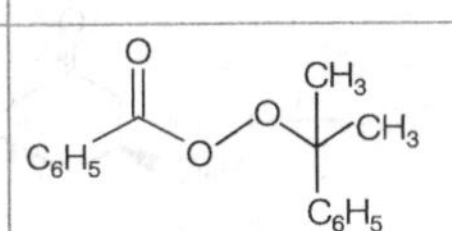	$C_6H_5-\dot{C}O_2 + CH_3-C(CH_3)(C_6H_5)-\dot{O} \xrightarrow{-CO_2} \dot{P}h + CH_3-CO-Ph + \dot{C}H_3$
iv.	q.	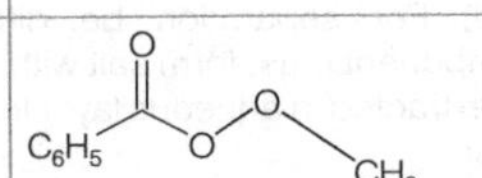	$C_6H_5-\dot{C}O_2 + \dot{C}H_3O$ → $C_6\dot{H}_5 + CO_2$

Thus, (i) → (p), (ii) → (r), (iii) → (s), (iv) → (q)

2. (2) **Plan** Reactant is cyclic anhydride and changes to dicarboxylic acid on hydrolysis. Also there is decarboxylation on heating if there is keto group w.r.t —COOH group. Ozonolysis cleaves (C ═ C) bond and H_2O_2 oxidises —CHO to —COOH group.

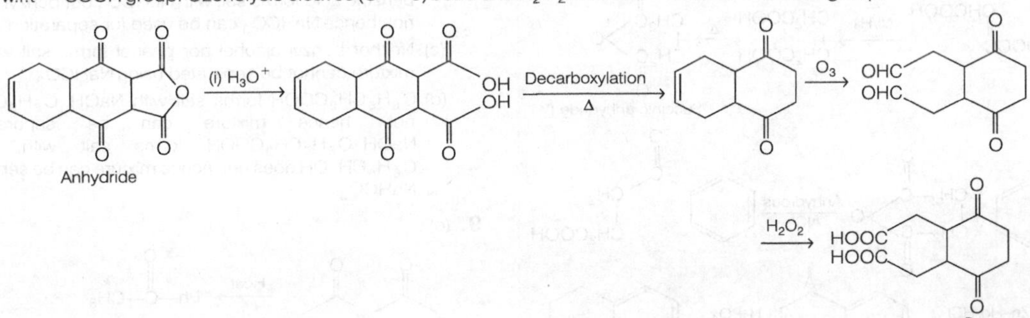

Thus, number of —COOH groups in $P = 2$.

3. (d) $NaHCO_3 \rightleftharpoons Na^+ + HCO_3^-$

HCO_3^- is decomposed by acid releasing CO_2

$$HCO_3^- + H^+ \longrightarrow H_2O + CO_2$$

If acid is stronger than HCO_3^- then CO_2 is released.

Phenol is less acidic and thus, does not liberate CO_2 with $NaHCO_3$.

4. (a) **Plan** Alkenes decolourise Br_2 water *cis*-isomer $\xrightarrow{\text{Dil. KMnO}_4}$ *Meso* isomers by *syn* addition

trans-isomer $\xrightarrow{\text{Dil. KMnO}_4}$ *d* (+) and *l* (−) isomers by *syn* addition thus, racemic mixture.

Formation of anhydride from dicarboxylic acid indicates *cis*-isomer. *P* and *Q* are isomers of dicarboxylic acids.

$$P, Q \xrightarrow{Br_2 \text{ water}} \text{Decolourised}$$

P and *Q* have (C = C) bond

$$P \xrightarrow{\Delta} \text{Anhydride}$$

Thus, *P* is *cis*-isomer.

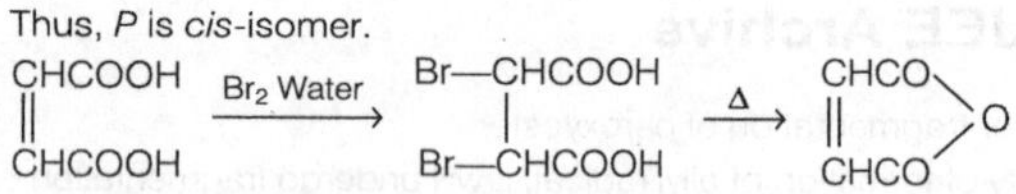

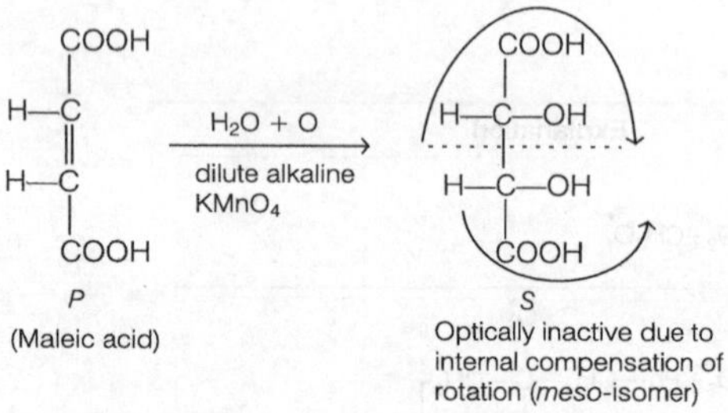

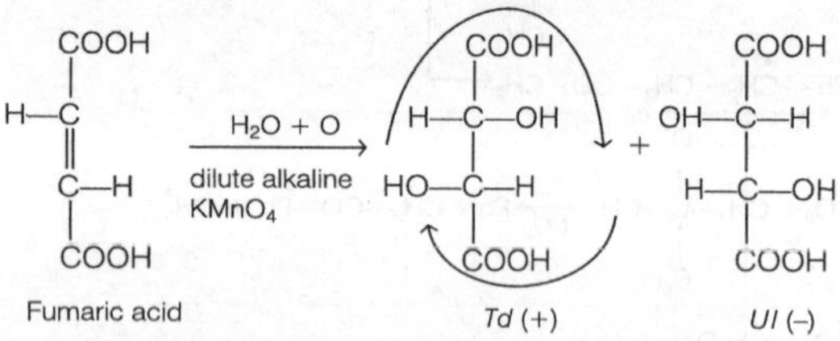

T and *U* (in 1 : 1 molar ratio) form optically inactive (racemic mixture) due to external compensation.

5. (a) **Plan** Ni/H_2 reduces (C = C) bond.

Benzene undergoes Friedel-Crafts reaction Zn-Hg/HCl reduces carbonyl group (Clemmensen reduction)

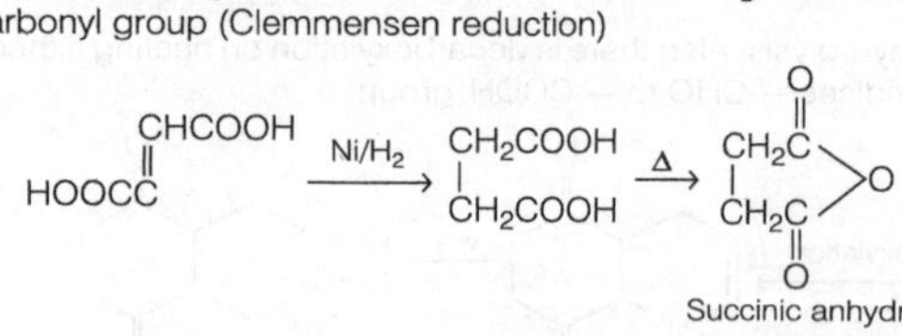

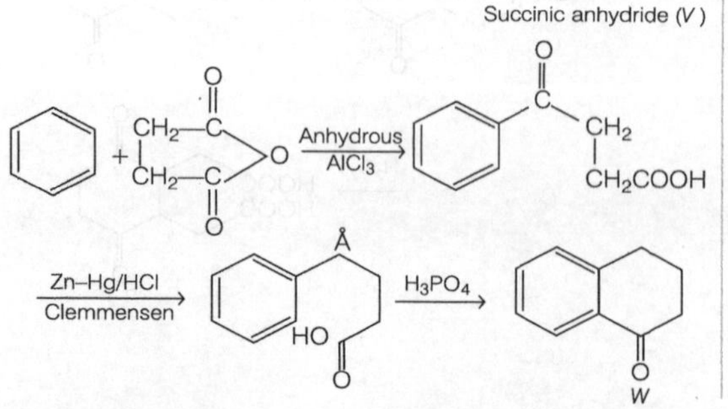

6. (b) It is a β-keto acid which undergo decarboxylation in very mild condition i.e. on simple heating. This occur through a six membered cyclic transition state as :

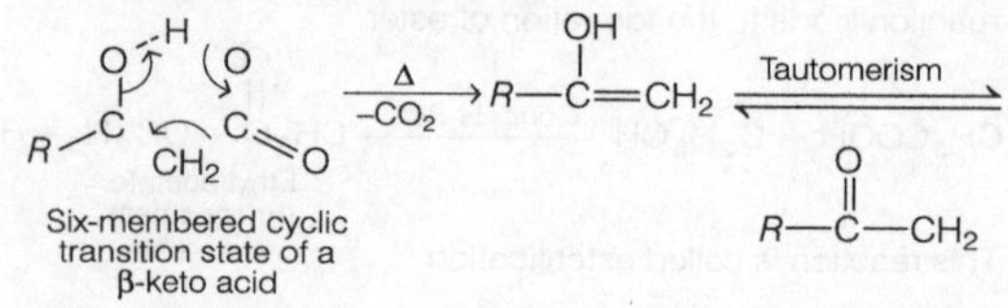

Note (i) Ordinary carboxylic acid require soda lime catalyst for decarboxylation.
(ii) Final step of decarboxylation in the above shown mechanism involve tautomerism, therefore, for decarboxylation of β-keto acid by above mechanism, the acid must contain an α-H].

7. (b) (a) Undergoes an ester hydrolysis in hot aqueous alkali as

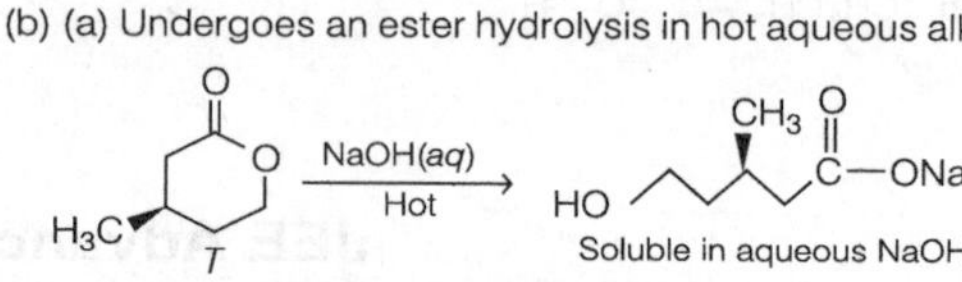

(b) $LiAlH_4$ reduces ester to alcohol as

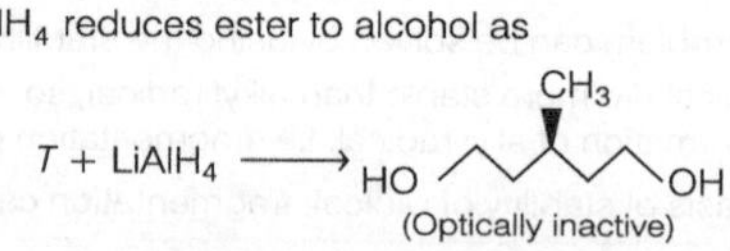

"*U*" No chiral carbon optically inactive.

(c) *U* on treatment with excess of acetic anhydride forms a diester as

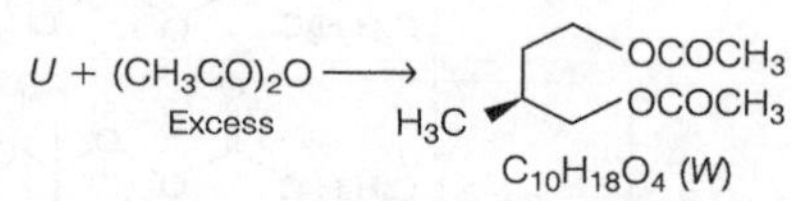

(d) *U* on treatment with CrO_3H^+ undergo oxidation to diacid which gives effervescence with $NaHCO_3$.

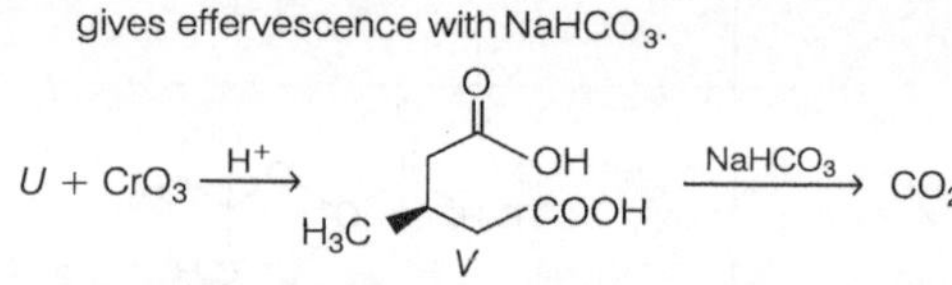

8. (b,d) For separation be differential extraction one of the component must form salt with the given base so that the salt will be extracted in aqueous layer leaving other component in organic layer.

(a) Both phenol and benzoic acid form salt with NaOH, hence this mixture cannot be separated.

(b) Benzoic acid forms salt with NaOH while benzyl alcohol does not, hence the mixture can be separated using NaOH. Also benzoic acid forms salt with $NaHCO_3$ but benzyl alcohol does not, hence $NaHCO_3$ can be used for separation.

(c) Neither benzyl alcohol nor phenol forms salt with $NaHCO_3$, mixture cannot be separated using $NaHCO_3$.

(d) $C_6H_5CH_2COOH$ forms salt with NaOH, $C_6H_5CH_2OH$ does not, hence mixture can be separated using $NaOH \cdot C_6H_5CH_2COOH$ forms salt with $NaHCO_3 \cdot C_6H_5CH_2OH$ does not, hence mixture can be separated using $NaHCO_3$.

9. (c)

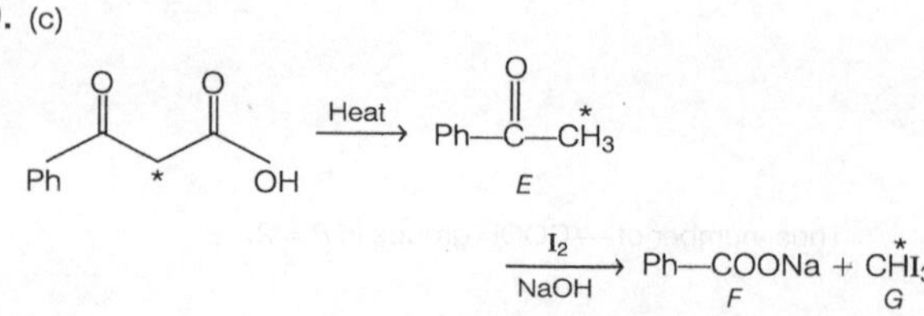

Last 3 Years' Questions
JEE Main & Advanced
(2016-2018) Volume 8

JEE Main

2016

1. 2-chloro-2-methylpentane on reaction with sodium methoxide in methanol yields

I. $C_2H_5CH_2C(CH_3)_2-OCH_3$ II. $C_2H_5CH_2C(CH_3)=CH_2$ III. $C_2H_5CH=C(CH_3)-CH_3$

[Alkyl Halide]

(a) Both I and III (b) Only III
(c) Both I and II (d) All of these

2. In the Hofmann-bromamide degradation reaction, the number of moles of NaOH and Br_2 used per mole of amine produced are **[Carboxylic Acid and Acid Derivatives]**

(a) four moles of NaOH and two moles of Br_2 (b) two moles of NaOH and two moles of Br_2
(c) four moles of NaOH and one mole of Br_2 (d) one mole of NaOH and one mole of Br_2

2017

3. Which of the following, upon treatment with *tert*-BuONa followed by addition of bromine water, fails to decolourise the colour of bromine? **[Aliphatic Alcohols, Ethers and Epoxides]**

(a) 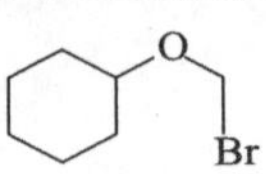

(b) 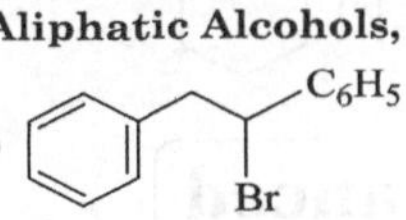

(c) 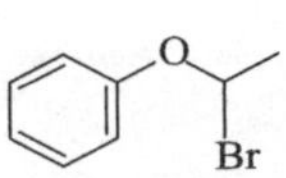

(d) 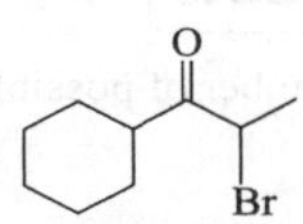

4. The correct sequence of reagents for the following conversion will be **[Aliphatic Aldehydes and Ketones]**

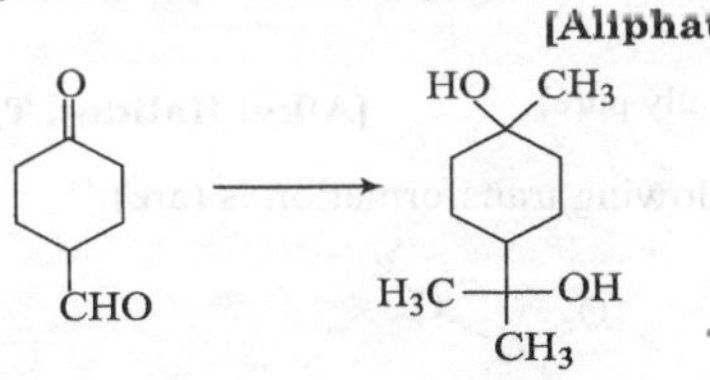

(a) $[Ag(NH_3)_2]^+ OH^-$, H^+ / CH_3OH, CH_3MgBr (b) CH_3MgBr, H^+ / CH_3OH, $[Ag(NH_3)_2]^+ OH^-$
(c) CH_3MgBr, $[Ag(NH_3)_2]^+ OH^-$, H^+ / CH_3OH (d) $[Ag(NH_3)_2]^+ OH^-$, CH_3MgBr, H^+ / CH_3OH

5. The major product obtained in the following reaction is

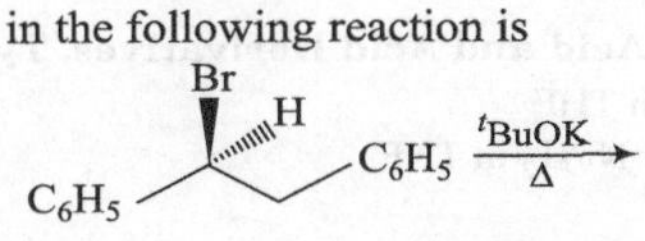

[Alkyl Halides]

(a) (±) $C_6H_5CH(O^tBu)CH_2C_6H_5$ (b) $C_6H_5CH=CHC_6H_5$
(c) (+) $C_6H_5CH(O^tBu)CH_2C_6H_5$ (d) (−)$C_6H_5CH(O^tBu)CH_2C_6H_5$

6. The major product obtained in the following reaction is

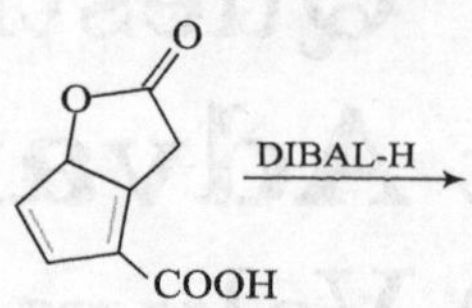

[Carboxylic Acid Derivatives]

(a) 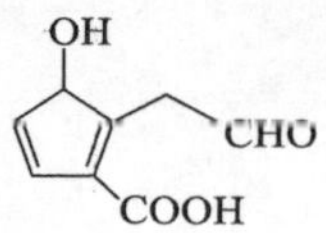 (b) 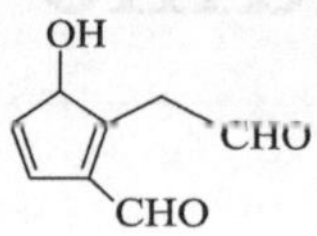 (c) 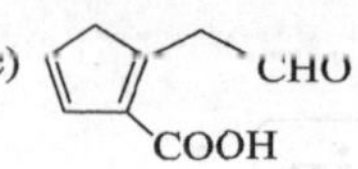 (d) 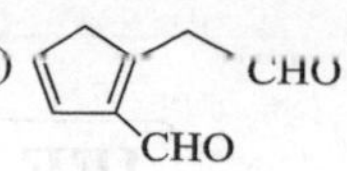

2018

7. The major product formed in the following reaction is

[Aliphatic Alcohols, Ethers and Epoxides]

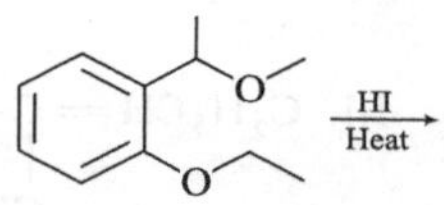

(a) 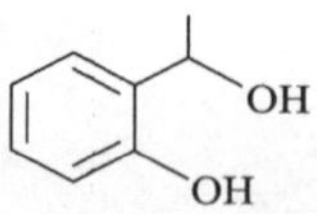 (b) 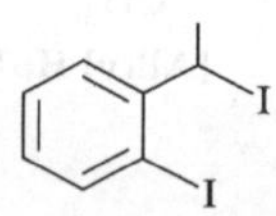 (c) 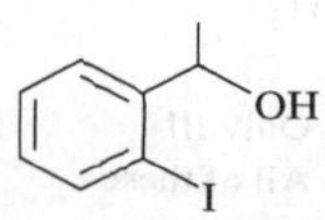 (d) 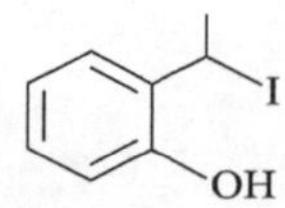

8. The major product of the following reaction is

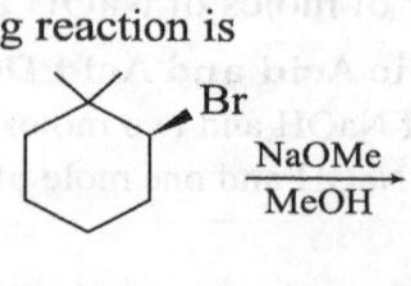

[Alkyl Halides]

(a) 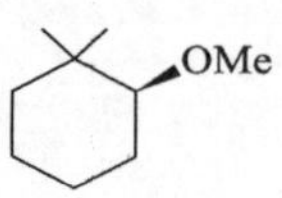 (b) 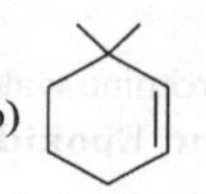 (c) 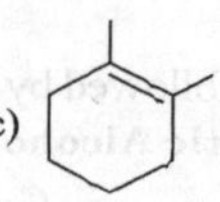 (d) 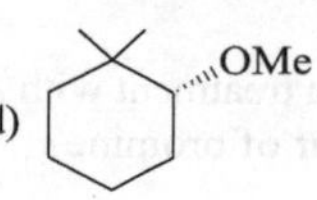

JEE Advanced

1. In the following monobromination reaction, the number of possible chiral product(s) is (are)...

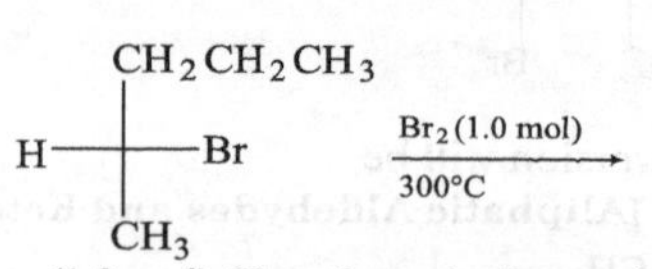

(1.0 mol) (Enantiomerically pure) **[Alkyl Halides, Type 3]**

2. Reagent(s) which can be used to bring about the following transformation is (are)

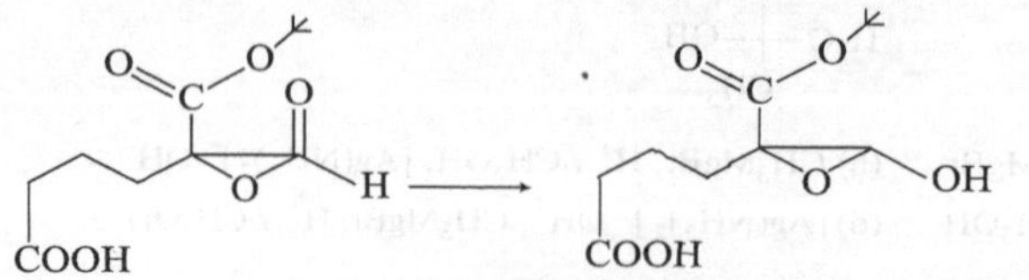

[Carboxylic Acid and Acid Derivatives, Type 1]

(a) $LiAlH_4$ in $(C_2H_5)_2O$ (b) BH_3 in THF

(c) $NaBH_4$ in C_2H_5OH (d) Raney Ni/H_2 in THF

Note ***Type-1*** *Only one correct option,* ***Type-2*** *More then one correct option,* ***Type-3*** *Single digit Answer,* ***Type-4*** *Numerical value based*

2018

3. In the following reaction sequence, the correct structure(s) of X is (are)

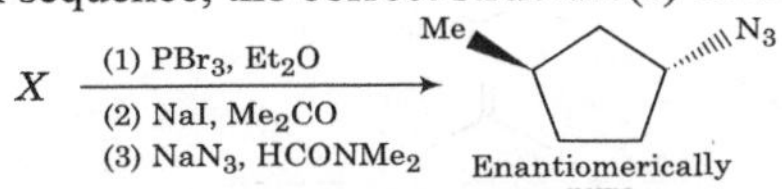

Enantiomerically pure

[Aliphatic Alcohols, Ethers and Epoxide, Type 2]

(a) 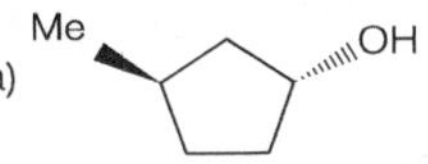 (b) 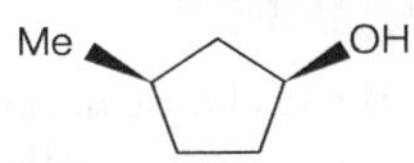

(c) 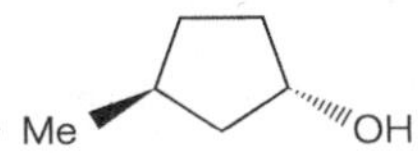 (d) 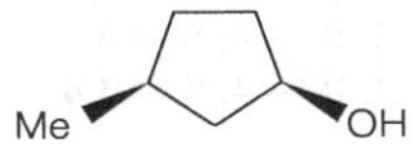

4. The desired product X can be prepared by reacting the major product of the reactions in List-I with one or more appropriate reagents in List-II.

(given, order of migratory aptitude : aryl > alkyl > hydrogen)

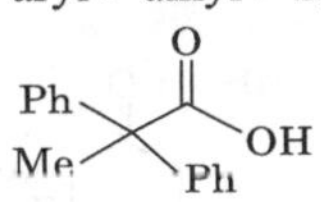

X

	List–I		List–II
P.	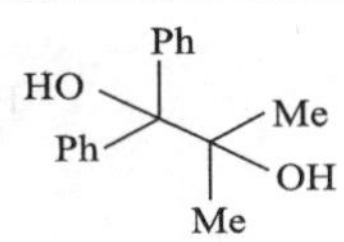 + H_2SO_4	1.	I_2, NaOH
Q.	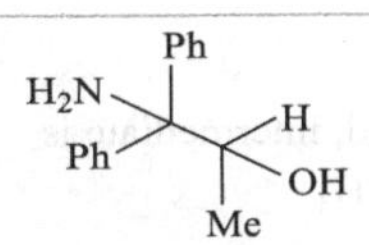 + HNO_2	2.	$[Ag(NH_3)_2]$ OH
R.	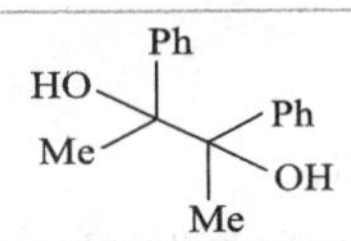 + H_2SO_4	3.	Fehling solution
S.	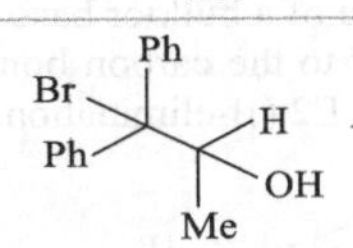 + $AgNO_3$	4.	HCHO, NaOH
		5.	NaOBr

The correct option is **[Aliphatic Alcohols Ethers and Epoxides, Type 2]**

(a) P → 1; Q → 2, 3; R → 1, 4; S → 2,4
(b) P → 1, 5; Q → 3,4; R → 4, 5; S → 3
(c) P → 1, 5; Q → 3,4; R → 5; S → 2,4
(d) P → 1, 5; Q → 2, 3; R → 1,5; S → 2,3

5. List-I contains reactions and List-II contains major products.

	List-I		List-II
P.	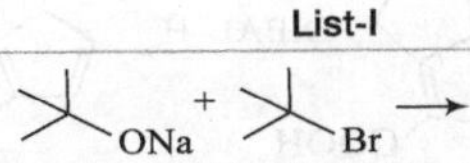	1.	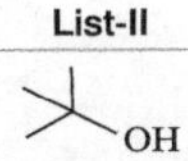
Q.	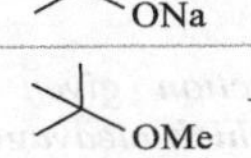 + HBr ⟶	2.	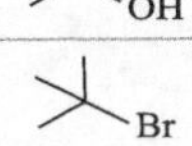

List-I		List-II
R.	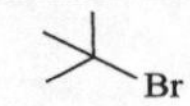 + NaOMe ⟶	3. 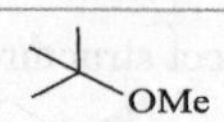
S.	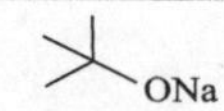 + MeBr ⟶	4. 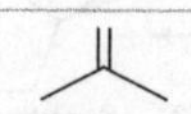
		5. 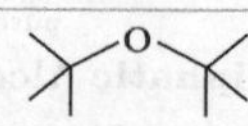

Match each reaction in List-I with one or more products in List-II and choose the correct option. **[Alkyl Halide]**

(a) P → 1, 5; Q → 2; R → 3; S → 4
(b) P → 1, 4; Q → 2; R → 4; S → 3
(c) P → 1, 4; Q → 1,2; R → 3,4; S → 4
(d) P → 4, 5; Q → 4; R → 4; S → 3,4

Answer *with* Explanations

JEE Main

1. *(d)* **Key concept** Strong nucleophile

($\bar{O}$Me) in polar solvent (MeOH) gives elimination products over substitution products but all products are possible in different yields.

$$CH_3-\underset{CH_3}{\overset{Cl}{C}}-CH_2CH_2CH_3 \xrightarrow[MeOH]{\overset{-\ +}{MeONa}} CH_3-\underset{CH_3}{\overset{OCH_3}{C}}-CH_2CH_2CH_3 \text{ (Less yield)} +$$

$$\underbrace{CH_3-\underset{CH_3}{C}=CH-CH_2-CH_3 + CH_2=\underset{CH_3}{C}-CH_2-CH_2-CH_3}_{\text{(More yield)}}$$

2. *(c)* Hofmann-bromamide degradation reaction is given as:

$$RCONH_2 + 4NaOH + Br_2 \longrightarrow \underset{(1^\circ \text{ amine})}{RNH_2} + Na_2CO_3 + 2NaBr + 2H_2O$$

Hence, four moles of NaOH and one mole of Br_2 are used.

3. *(a)* To show decolourisation, compound must be unsaturated.

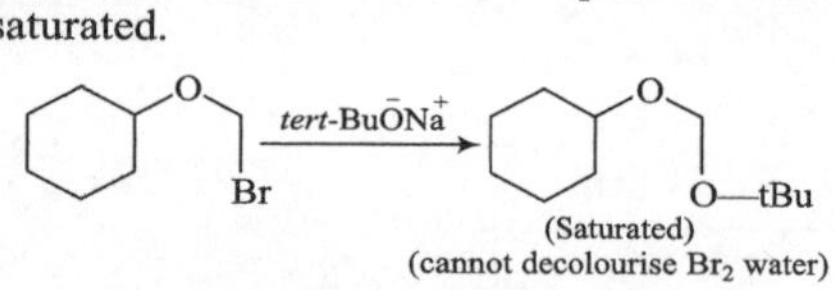

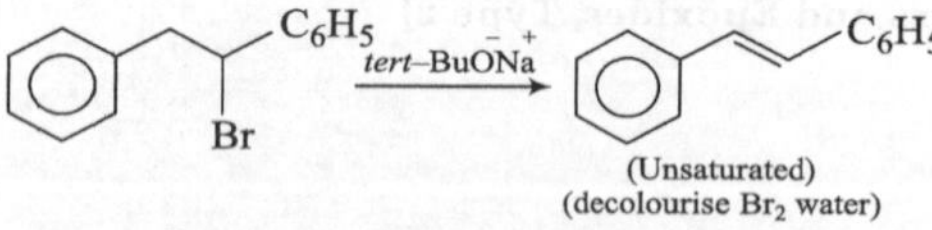

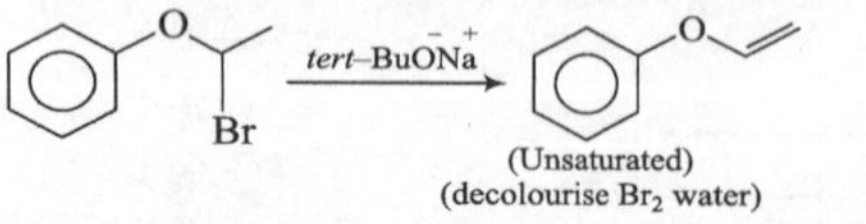

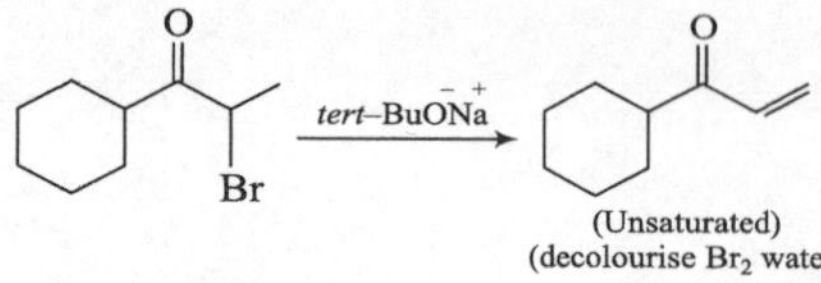

4. *(a)*

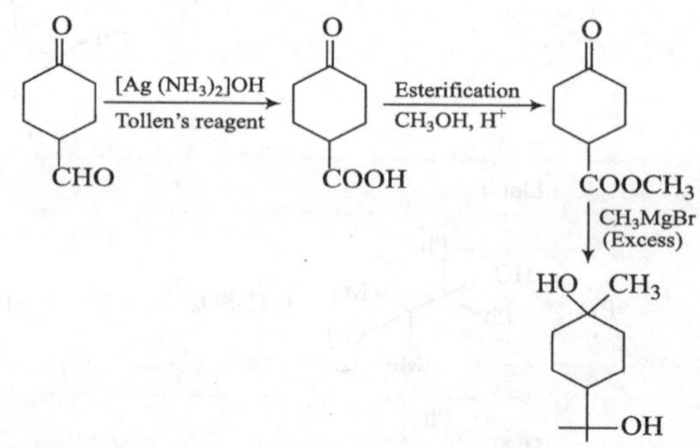

Before final product is formed, intermediate is

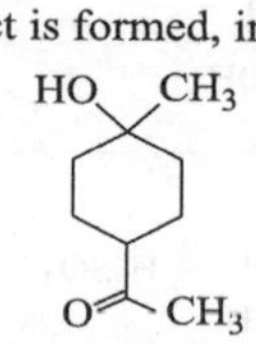

5. *(b)* An alkyl halide in presence of a bulkier base removes a proton from a carbon adjacent to the carbon bonded to the halogen. This reaction is called E2 (β-elimination reaction).

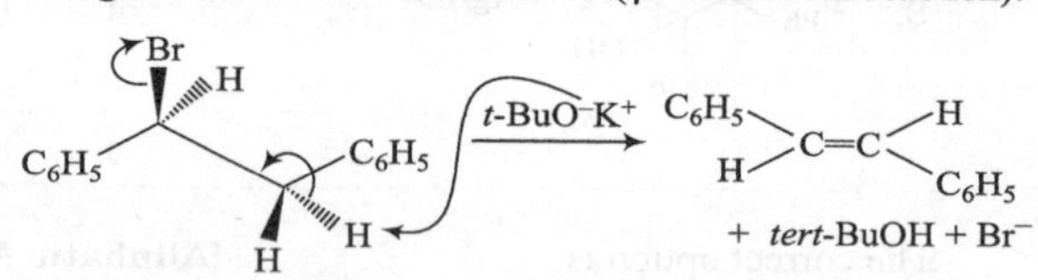

6. *(a)* DIBAL-H (Di-isobutyl aluminium hydride) is a reducing agent with formula. This is generally used for the preparation of aldehydes. Using DIBAL —H, Lactones are reduced directly to aldehydes.

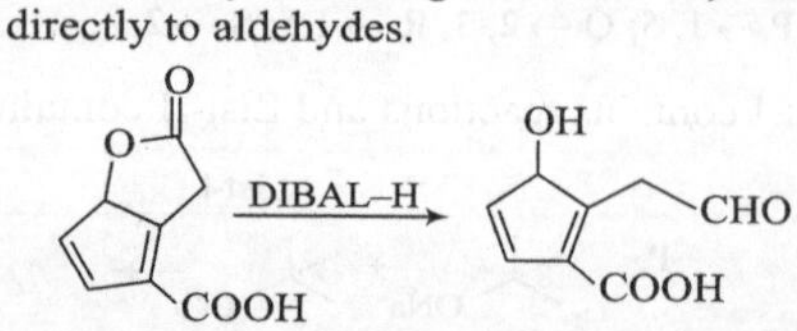

7. *(d)* **Key idea** *The reaction given is a nucleophilic substitution reaction in which cleavage at C—O bond is visible. The product formation can be visualised with the help of following analysis.*

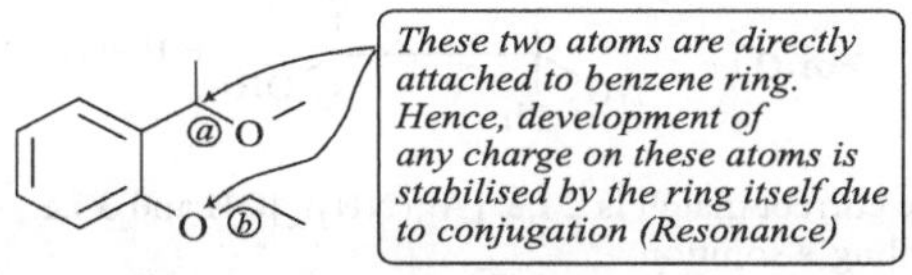

If any one properly visualise the fact written with figure above, than a conclusion can be made that C—O bonds marked (a) and (b) in the figure will undergo heterolysis during the reaction.

The reaction can be represented as

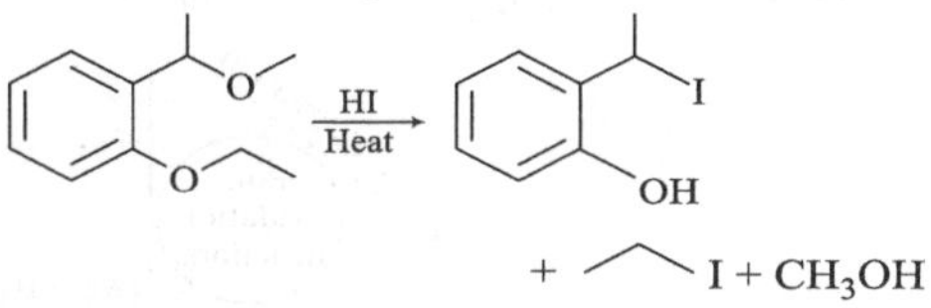

Mechanism

***Step* I** The reaction begins with the attack of H^+ of HI on oxygen to form oxonium ion as

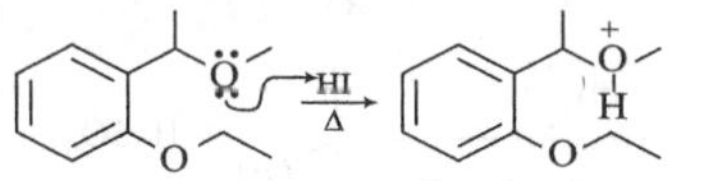

Oxonium ion

Step II This oxonium ion undergoes lysis and addition of I^- to form two products as

***Step* III** Similar pathway is followed at the other oxygen atom, which can be visualised as

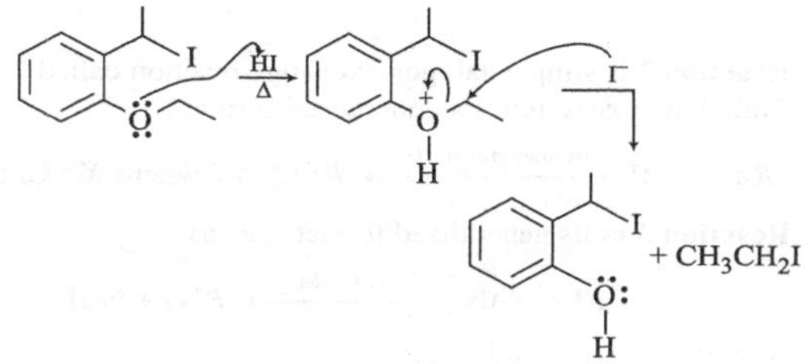

Note *Mechanism of a reaction is always a logical sequencing of events which may occur simultaneously as well.*

8. *(b)* Complete reaction can be represented as

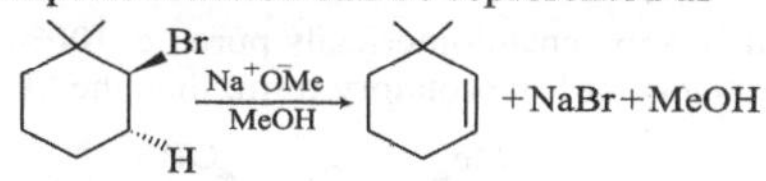

Thus, the given reaction is dehydrohalogenation which is a β-elimination proceeding through E_2 mechanism.

Mechanism The reaction proceeds through the formation of following transition state with simultaneous removal of Br and H atoms.

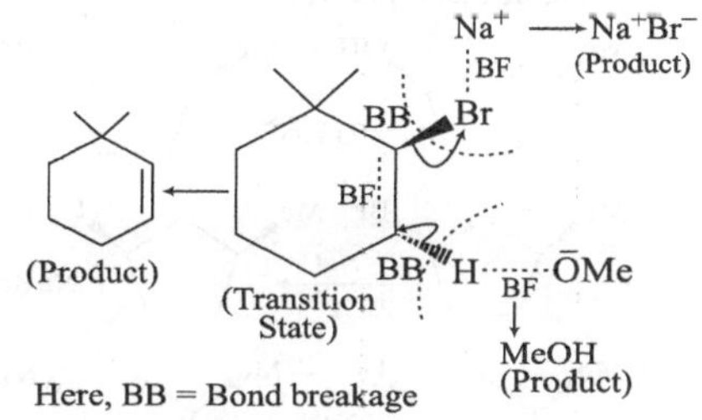

Here, BB = Bond breakage
BF = Bond formation

JEE Advanced

1. *(5)* Given compound undergoes free-radical bromination under given conditions, replacing H by Br.

C^* is chiral carbon.

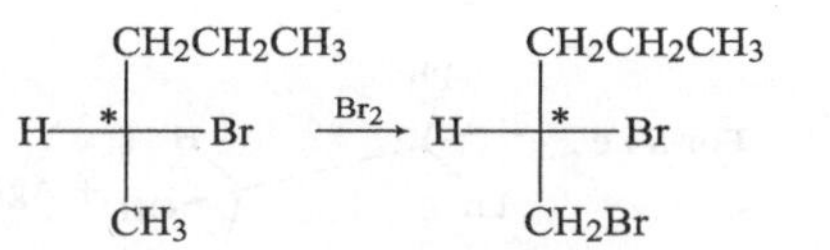

II. Chiral

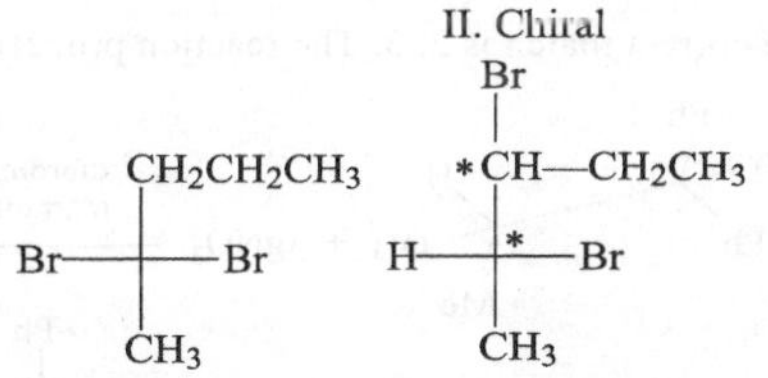

II. Achiral II. Chiral

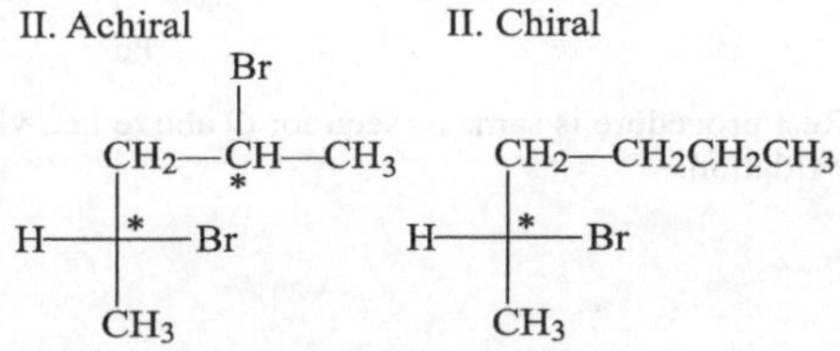

VI. Chiral V. Chiral

2. *(c)* Only — CHO group is to be reduced to —CH_2OH.

It can be done using $NaBH_4$ in C_2H_5OH.

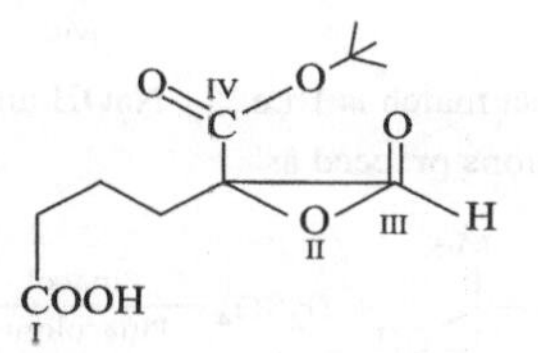

(a) $LiAlH_4 / (C_2H_5)_2O$ reduces I, II and III into —CH_2OH, and IV into diol.

(b) BH_3 / THF show same properties as (a).

(c) $NaBH_4 / C_2H_5OH$ reduces III into — CH_2OH.

(d) Raney nickel, same as (a) and (b), thus (c) is correct reagent.

3. *(b)* **Key idea** *All the reactions involved in the problem are Nucleophilic substitution i.e., S_N2 which biomolecular results in inversion of configuration at the chiral carbon.*

Reaction 1 in its generalised format is seen as

$$ROH \xrightarrow[\text{In diethyl ether } (Et_2O)]{PBr_3} RBr$$

Reaction 2 is simple halogen exchange reaction called **Finkelstein reaction**. Its generalised format is

$$RX + \text{NaI} \xrightarrow{\text{In acetone}(\text{Me}_2\text{CO})} RI + \text{Na}X \text{ where } X = \text{Cl or Br}$$

Reaction 3 in its generalised format seen as

$$RI + \text{NaN}_3 \xrightarrow{\text{HCONMe}_2} RN_3 + \text{NaI}$$

Now if the given product is

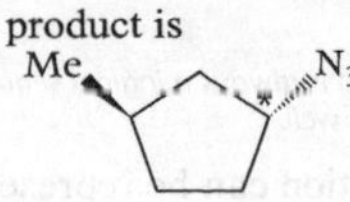

and which is too enantiomerically pure i.e. 100% either dextrorotatory or leavorotatory form, then the 'X' must be

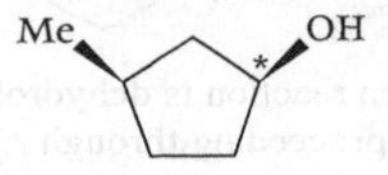

Note *The configuration at carbon * atom in 'X' becomes inverted due to S_N2 mechanism which is visible in the product as well.*

Thus, the probable reactions will be

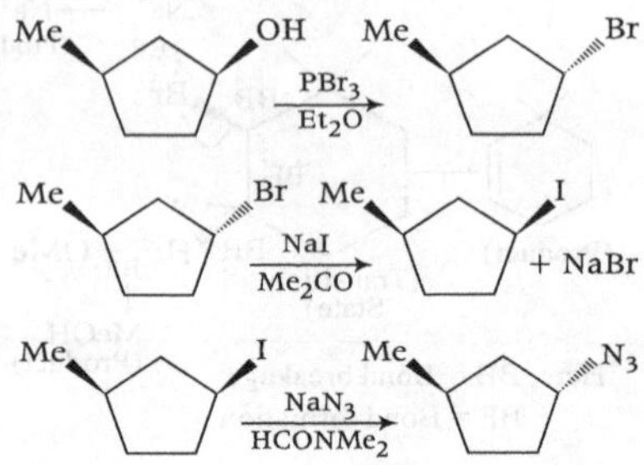

4. *(d)*

For ***P*** i.e.,

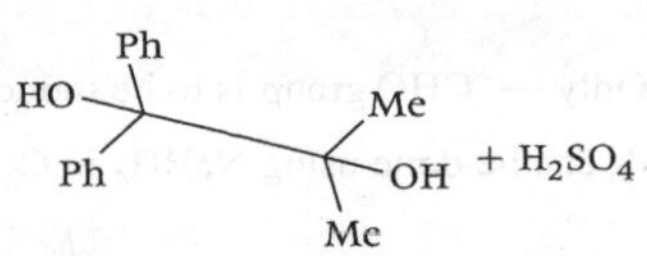

The correct match is 1 i.e., I_2, NaOH and 5 i.e., NaOBr

The reactions proceed as

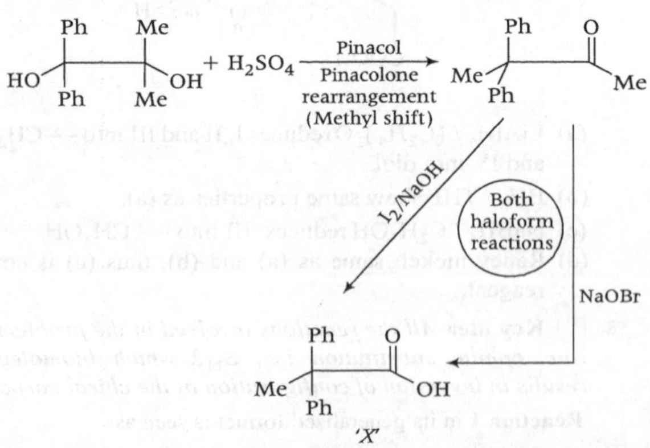

For ***Q*** i.e.,

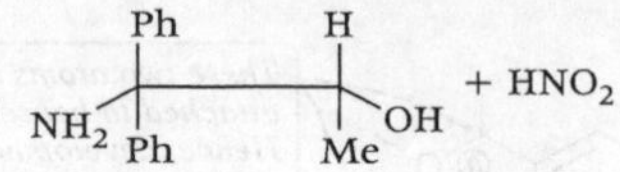

The correct match is 2 i.e. $[Ag(NH_3)_2]OH$ and 3 i.e., Fehling's solution.

The reactions proceed as

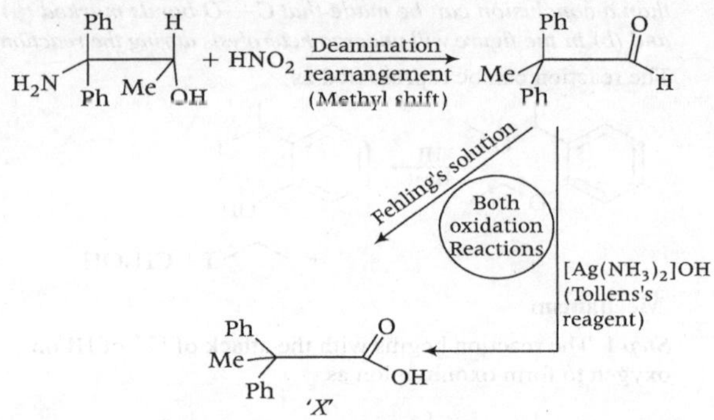

For ***R*** i.e.,

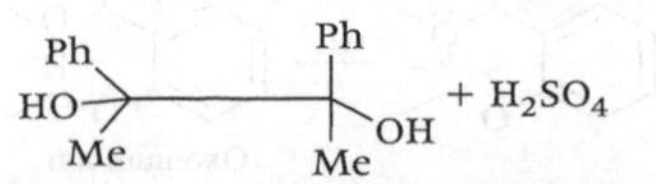

The correct match is 1, 5 again. The reaction proceed as

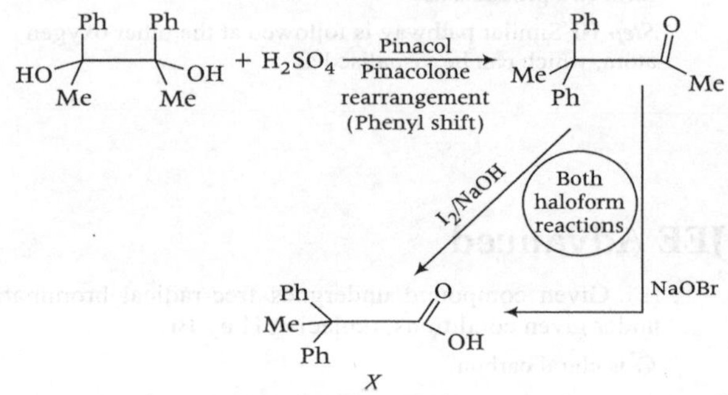

For ***S*** i.e.,

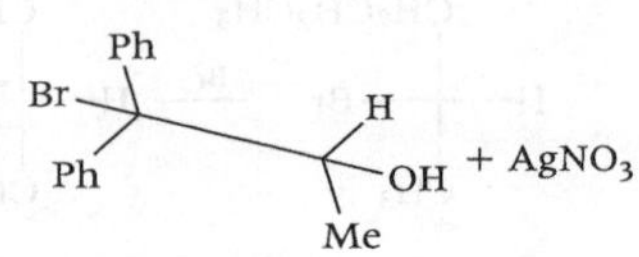

The correct match is 2, 3. The reaction proceed as

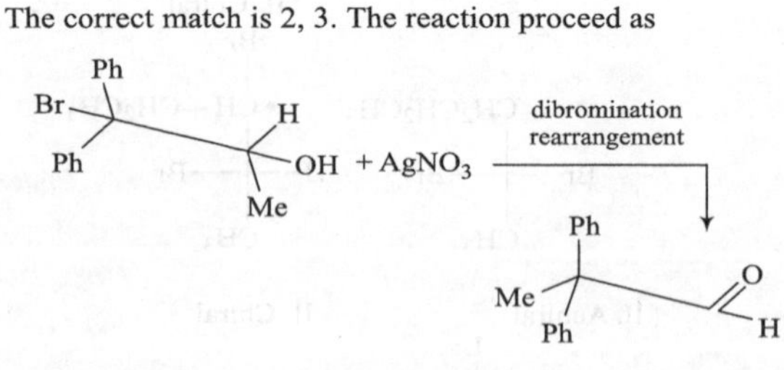

Rest procedure is same as seen for Q above i.e., via oxidation.

5. *(b)*

For *P*, i.e. 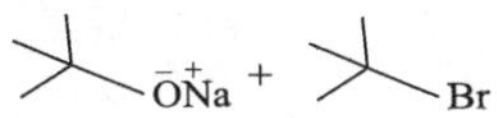

For this reaction 1 and 4 are probable products.

Product 1 i.e., 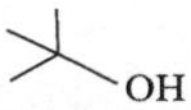 is formed due to substitution while product 4 i.e., 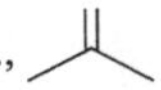 is formed due to elimination. A tertiary carbocation i.e, 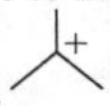 formed during the reaction. Remember for 3° carbocation ions elimination product predominates.

For *Q*, i.e. 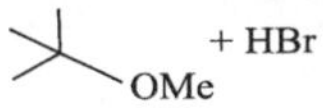

Correctly matched product for this reaction is 2 i.e., 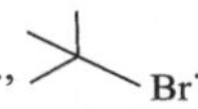.
The reaction proceeds as

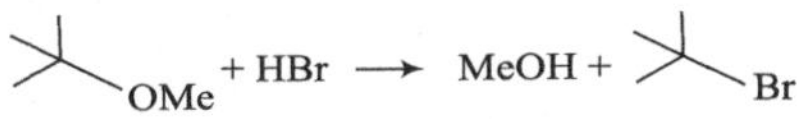

For *R* i.e., 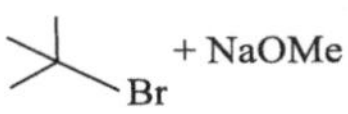

Correctly matched product is 4 i.e., 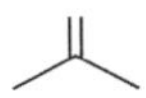. It is a normal elimination reaction and proceeds as

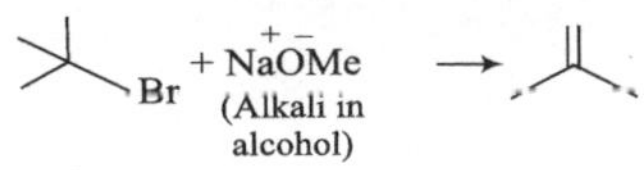

3° alkyl halide preferes elimination.

For *S* i.e., 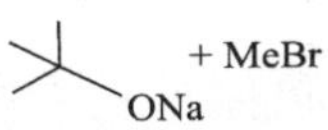

The correct match is 3 i.e., 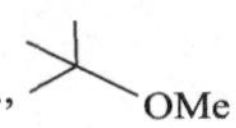. The reaction proceeds as

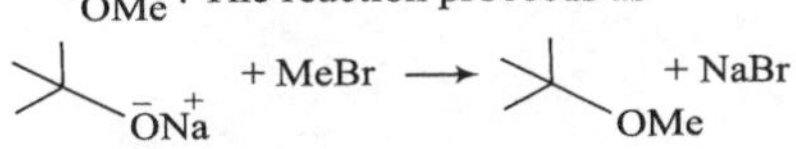